Mobile
Intelligent
Autonomous
Systems

Mobile
Intelligent
Autonomous
Systems

EDITED BY
Jitendra R. Raol • Ajith K. Gopal

CRC Press
Taylor & Francis Group
Boca Raton London New York

CRC Press is an imprint of the
Taylor & Francis Group, an **informa** business

CRC Press
Taylor & Francis Group
6000 Broken Sound Parkway NW, Suite 300
Boca Raton, FL 33487-2742

First issued in paperback 2017

© 2013 by Taylor & Francis Group, LLC
CRC Press is an imprint of Taylor & Francis Group, an Informa business

No claim to original U.S. Government works

Version Date: 20120612

ISBN 13: 978-1-138-07245-9 (pbk)
ISBN 13: 978-1-4398-6300-8 (hbk)

Library of Congress Cataloging-in-Publication Data

Mobile intelligent autonomous systems / editors, Jitendra R. Raol, Ajith K. Gopal.
    p. cm.
    Summary: "Written for systems, mechanical, aero, electrical, civil, industrial, and robotics engineers, this book covers robotics from a theoretical and systems point of view, with an emphasis on the sensor modeling and data analysis aspects. With the novel infusion of NN-FL-GA paradigms for MIAS, this reference blends modeling, sensors, control, estimation, optimization, signal processing, and heuristic methods in MIAS/robotics, and includes examples and applications throughout. The organization of the book is based on fundamental concepts, with sections covering fundamental concepts, methods and approaches, block/flow diagrams, and numerical examples. A MATLAB-based approach is used through selected case studies in the text"-- Provided by publisher.
    Includes bibliographical references and index.
    ISBN 978-1-4398-6300-8 (hardback)
    1. Intelligent control systems. 2. Motor vehicles--Automatic control. 3. Mobile robots. I. Raol, J. R. (Jitendra R.), 1947- II. Gopal, Ajith K.

TJ217.5.M63 2012
629.8'932--dc23
                                    2012021068

Visit the Taylor & Francis Web site at
http://www.taylorandfrancis.com

and the CRC Press Web site at
http://www.crcpress.com

*This book is dedicated
in loving memory to
Professor Dr. Naresh Kumar Sinha
(McMaster University, Hamilton, Ontario, Canada)
for his life-long contributions to Control Systems—Theory & Practice*

Contents

PART I Conceptual Foundations for MIAS

PART II MIAS and Robotics

PART III Allied Technologies for MIAS/Robotics

Preface

Mobile Intelligent Autonomous System (MIAS) is a fast-emerging and developing research area. Although it can be regarded as a general R&D area, in the literature it is mainly directed towards robotics, field or other type of robots. However, for the purpose of the present volume the MIAS should not be and is not only considered to mean the field of robotics. The MIAS here is meant to comprise theory and practice of several closely related technologies that have some elements of mobility, intelligence and/or autonomy operating and envisaged not only for robots, but also for other mobile vehicles: (i) micro-mini air vehicles (MAVs) and (ii) unmanned aerial vehicles (UAVs). Several important subareas within MIAS research and development are: (a) perception and reasoning, (b) mobility, autonomy and navigation, (c) haptics and teleoperation, (d) image fusion/computer vision, (e) mathematical modelling of robots and their manipulators, (f) hardware/software architectures for planning and behaviour learning leading to robotic sophisticated architectures, (g) vehicle–robot path and motion planning and control and (h) human–machine interfaces for interaction between humans and robots and other vehicles. The application of artificial neural networks (ANNs), fuzzy logic/systems (FLS), probabilistic and approximate reasoning (PAR), static and dynamic Bayesian networks (SDBN) and genetic algorithms (GA) to many of the above-mentioned problems is gaining impetus. Multi-sensor data fusion (MSDF) also plays a very crucial role at many levels of the data fusion process: (i) kinematic fusion (position, bearing only tracking), (ii) image fusion and tracking (for scene recognition), (iii) information fusion (for building world models) and (iv) decision fusion (for tracking and control actions). MIAS area as a technology is very useful for automation of complex tasks, surveillance in hazardous and hostile environment, human assistance in very difficult manual works, medical and field robotics, hospital reception systems, auto-diagnostic systems and many other related civil and military systems including mining robotics. Many other significant research areas for MIAS consist of sensor and actuator modelling, sensor failure detection, management and reconfiguration, object-scene understanding, knowledge acquisition and representation, and learning and decision making. Examples of dynamic systems generally considered within the MIAS are: (a) autonomous systems [unmanned ground vehicles (UGVs), unmanned aerial vehicles (UAVs), micro and mini air vehicles (MAVs), autonomous underwater vehicles (UWVs)], (b) mobile and fixed but autonomous robotic systems, (c) dexterous manipulator robots, (d) mining robots, (e) surveillance systems and (f) networked and multi-robot systems.

This volume on MIAS deals with many of the above aspects and applications (not necessarily all) across 35 contributed chapters. Many chapters deal with the topics that span new research and development of very recent past—last 3–4 years. There are some good books on various aspects of robotics; however, the treatment of such aspects as outlined above is somewhat limited or highly specialized (sometimes it is too general). The treatment in this volume is comprehensive and covers several related disciplines which would help in understating the robotics and MIAS from system-theoretic as well as practical point-of-view. On the whole the editors feel that there is a fair representation of various aspects and current issues related to the MIAS in the sense of the term clarified in the beginning. The main idea is that several of the disciplines considered and discussed in this book would be needed to build a good autonomous and/or intelligent mobile system; however, the mechanical and structural aspects of the vehicles are not addressed in this book. Where possible, the authors of some of the chapters illustrate certain concepts and theories with examples via numerical simulations coded in MATLAB®. At the outset it is clearly stated here that neither the editors (of this book) nor the authors (of various chapters in the volume) claim that the mathematical expressions, equations and/or formulae presented in the book are derived from first principles. For such derivations the readers may refer the respective cited references in the concerned chapters. Also it is not claimed that all the chapters discuss all the aspects of mobility, autonomy and intelligence. None the same the volume as a whole discusses various conceptual foundations, MIAS and robotics systems, and allied technologies to MIAS and robotics. The end users of this integrated technology of MIAS/robotics (theory and practice) will be systems-control-educational institutions, several R&D laboratories, mechanics-aerospace and other industries, transportation and automation industry, medical- and mining-robotics development institutions and related industries.

MATLAB® is the trademark of The MathWorks, Inc. For product information, please contact:

The MathWorks, Inc.
3 Apple Hill Drive
Natick, MA 01760-2098 USA
Tel: 508-647-7000
Fax: 508-647-7001
E-mail: info@mathworks.com
Web: www.mathworks.com

Acknowledgements

Several researchers and engineers all over the world have been making substantial contributions over the last four decades to this exciting and intriguing field which has caught the imagination of persons from all walks of life. It is emerging as an enabling technology to reckon with, especially for robotics, aerospace engineering and technological applications, and some industrial spin-offs. Certain interactions with the MIAS group of the CSIR (Council of Scientific and Industrial Research, in South Africa) and with a few colleagues of departments of electronics and communications engineering (E&CE), and instrumentation technology (IT) of M. S. Ramaiah Institute (MSRIT) have been very useful. The editor (J. R. Raol) is grateful to Dr. R. M. Jha (senior scientist, National Aerospace Laboratories (NAL), Bangalore) for his very useful initial tips a few years ago for publishing a book with international publishers, and continual guidance. He is also very grateful to Dr. Sethu S. Selvi (Head, E&CE), and Professor P. P. Venkat Ramaiah (Head, IT) for their moral support. He is especially very grateful to Ms. A. N. Myna (assistant professor, Information Science and Engineering, MSRIT) for her continual moral support during the writing of this book. The editors are very grateful to all the authors for their time, effort and patience and for ably contributing various chapters to this volume. We are also very grateful to CRC Press and especially to Mr. Jonathan Plant, Ms. Jennifer Ahringer, and Ms. Amber Donley for their full support during this book project. We are, as ever, very grateful to our spouses and children for their endurance, care, affection, patience, love and more—their additional spiritual presence in our lives and all that collectively inspires and gives us considerable strengths to go ahead in the midst of myriads of uncertainties and obstacles.

Jitendra R. Raol and Ajith K. Gopal

Editors

Jitendra R. Raol received B.E. and M.E. degrees in electrical engineering from M. S. University of Baroda, Vadodara, in 1971 and 1973, respectively, and a PhD (in electrical and computer engineering) from McMaster University, Hamilton, Canada, in 1986, where he was also a researcher and a teaching assistant. Dr. Raol taught for 2 years at the M. S. University of Baroda before joining the National Aeronautical Laboratory (NAL) in 1975. At NAL he was involved in the activities on human pilot modelling in fix- and motion-based research flight simulators. He re-joined NAL in 1986 and retired on 31 July 2007 as scientist-G (and head, Flight Mechanics and Control Division (FMCD) at NAL). He has visited Syria, Germany, the United Kingdom, Canada, China, the United States of America and South Africa on deputation/fellowships to work on research problems on system identification, neural networks, parameter estimation, multi-sensor data fusion and robotics, to present several technical papers at several international conferences, and deliver guest lectures at these places. He became Fellow of the IEE (United Kingdom) and a senior member of the IEEE (United States). He is a life-fellow of the Aeronautical Society of India and a life member of the System Society of India. In 1976, he won the K. F. Antia Memorial Prize of the Institution of Engineers (India) for his research paper on non-linear filtering. He was awarded a certificate of merit by the Institution of Engineers (India) for his paper on parameter estimation of unstable systems. He has received one best poster paper award for a paper on sensor data fusion and a gold medal and a certificate for a paper related to target tracking (from the Institute of Electronics and Telecommunications Engineers, India). He is also one of the recipients of the CSIR (Council of Scientific and Industrial Research, India) prestigious technology shield for 2003 for the contributions to the development of Integrated Flight Mechanics and Control Technology for Aerospace Vehicles in the country. The shield was associated with a plaque, a certificate and the prize equivalent of $67,000 for the project work.

Dr. Raol has published 110 research papers and several reports. He has guest-edited two special issues of Sadhana (an engineering journal published by the Indian Academy of Sciences, Bangalore) on advances in modelling, system identification and parameter estimation and on multi-source, multi-sensor information fusion. He has also guest-edited two special issues of the *Defense Science Journal* on MIAS and on aerospace avionics and allied technologies. He has guided six doctoral and eight master scholars and presently he is guiding six faculty members for their doctoral programmes. He has co-authored an IEE (UK) Control Series book, *Modeling and Parameter Estimation of Dynamic Systems* (2004), and a CRC Press (USA) book, *Flight Mechanics Modeling and Analysis* (2008). He has also authored a CRC Press book, *Multi-sensor Data Fusion with MATLAB* (2009). He has served as a member/chairman of several advisory, technical project review and doctoral examination committees. He is reviewer of several national and international journals. He is on the board of directors of a private R&D company as well as a private college in the state. His main research interests have been data fusion, system identification, state/parameter estimation, flight mechanics–flight data analysis, H-infinity filtering, ANNs, fuzzy systems, genetic algorithms and robotics. He has also authored *Poetry of Life* (Trafford Publishing, USA, 2009) and *Sandy Bonds* (Pothi.com, India, 2010) as the collection of his 140 poems.

Ajith K. Gopal qualified with a B.Sc. Eng. (Mechanical) from the University of Natal in South Africa in 1997. He subsequently completed the M.Sc. Eng. in 2000 in structural analysis of composite material and in 2003 obtained a PhD in engineering, from the University of Natal, for his work in energy absorption of smart materials. Dr. Gopal has 12 years of working experience, of which 8 years has been in the research and development domain. He is responsible for establishing the MIAS (Mobile, Intelligent, Autonomous Systems) research group at the CSIR in South Africa, in 2007, where he served as the research leader for 2007 and 2008.

Dr. Gopal has published four papers relating to composite and smart materials in the *Journal of Composite Materials* published by Elsevier and at *International Conferences on Composite Science and Technology*. He has also published a paper on path planning in the *Defense Science Journal*, a book chapter on data fusion in robotics in the CRC Press book *Multisensor Data Fusion with MATALB* (by J. R. Raol, 2010) and guest-edited a special issue of the *Defense Science Journal* on MIAS. He is currently an engineering/project manager at Land Systems South Africa where he is responsible for new technology strategy and development for the company.

Contributors

C. M. Ananda
ALD-Aerospace Electronics and Systems
    Division
CSIR-National Aerospace Laboratories
Bangalore, India

N. Ananthkrishnan
Independent Consultant
Mumbai, India

R. Ayyagari
Department of Instrumentation and Control
    Engineering
National Institute of Technology
Tiruchirappalli, India

Antoine Bagula
Department of Computer Science
University of Cape Town
Cape Town, South Africa

Victor M. Becerra
School of Systems Engineering
University of Reading
Reading, United Kingdom

Niteen Bhange
Coral Digital Technologies (P) Ltd.
Bangalore, India

Seetharama M. Bhat
Department of Aerospace Engineering
Indian Institute of Science
Bangalore, India

Dimitar Chakarov
Department of Mechanics and Multibody
    Systems
Institute of Mechanics
Bulgarian Academy of Sciences
Sofia, Bulgaria

Charu Chawla
Department of Aerospace Engineering
Indian Institute of Science
Bangalore, India

Jong Ho Choi
Industry University Collaboration Foundation
Chungnam National University
Daejeon, Korea

Julia H. Downes
School of Systems Engineering
University of Reading
Reading, United Kingdom

Ajith K. Gopal
Land Systems South Africa
Johannesburg, South Africa

Girija Gopalratnam
Flight Mechanics and Controls Division
CSIR-National Aerospace Laboratories
Bangalore, India

Nitin K. Gupta
IDeA Research and Development (P) Ltd.
Pune, India

Rong Haijun
Xi'an Jiaotong University
People's Republic of China

Mark W. Hammond
School of Systems Engineering
University of Reading
Reading, United Kingdom

R. M. Jha
Aerospace Electronics and Systems Division
CSIR-National Aerospace Laboratories
Bangalore, India

M. R. Kaimal
Department of Computer Science and
    Engineering
Amrita Viswha Vidyapeetham University
Kollam, India

C. Kamali
Flight Mechanics and Control Division
National Aerospace Laboratories
Bangalore, India

Kostadin Kostadinov
Department of Mechanics and Multibody
    Systems
Institute of Mechanics
Bulgarian Academy of Sciences
Sofia, Bulgaria

Marelize Kriel
CSIR Defence, Peace, Safety & Security
Pretoria, South Africa

Louise Leenen
Information Systems Research Group
Council of Scientific and Industrial
    Research
Pretoria, South Africa

Venkatesh K. Madyastha
Flight Mechanics and Controls Division
CSIR-National Aerospace Laboratories
Bangalore, India

Simon Marshall
School of Chemistry, Food Biosciences and
    Pharmacy
University of Reading
Reading, United Kingdom

M. Meenakshi
Department of Information Technology
Dr. Ambedkar Institute of Technology
Bangalore, India

A. N. Myna
Department of Information Science and
    Engineering
M. S. Ramaiah Institute of Technology
Bangalore, India

V. P. S. Naidu
Flight Mechanics and Control Division
CSIR-National Aerospace Laboratories
Bangalore, India

Slawomir J. Nasuto
School of Systems Engineering
University of Reading
Reading, United Kingdom

Isaac O. Osunmakinde
School of Computing
College of Science, Engineering and
    Technology
University of South Africa (UNISA)
Johannesburg, South Africa

Radhakant Padhi
Department of Aerospace Engineering
Indian Institute of Science
Bangalore, India

Ik Soo Park
Industry University Collaboration Foundation
Chungnam National University
Daejeon, Korea

Abhay A. Pashilkar
Flight Mechanics and Control Division
CSIR-National Aerospace Laboratories
Bangalore, India

Ambalal V. Patel
Integrated Flight Control Systems
    Directorate
Aeronautical Development Agency
Bangalore, India

A. Ramachandran
Department of Instrumentation Technology
M. S. Ramaiah Institute of Technology
Bangalore, India

Jitendra R. Raol
Department of Electronics and
    Communications Engineering
M.S. Ramaiah Institute of Technology
Bangalore, India

Vishal C. Ravindra
Flight Mechanics and Controls Division
CSIR-National Aerospace Laboratories
Bangalore, India

Herman le Roux
CSIR Defence Peace Security and Safety
Council of Scientific and Industrial
    Research
Pretoria, South Africa

Shobha R. Savanur
Department of Electrical Engineering
P. G. Halakatti College of Engineering and
    Technology
Bijapur, India

Motlatsi Seotsanyana
Mobile Intelligent Autonomous Systems
    Group
Council for Scientific and Industrial Research
Pretoria, South Africa

S. Seshadhri
Industrial Software Systems
ABB Corporate Research
Bangalore, India

Hema Singh
Aerospace Electronics and Systems Division
CSIR-National Aerospace Laboratories
Bangalore, India

G. K. Singh
Flight Mechanics and Control Division
CSIR-National Aerospace Laboratories
Bangalore, India

N. Sundararajan
School of Electrical and Electronic Engineering
Nanyang Technological University
Singapore

Alexander Terlunen
Department of Information Technology
University of Pretoria
Pretoria, South Africa

Bhekisipho Twala
Department of Electrical and Electronic
    Engineering Science
University of Johannesburg
Johannesburg, South Africa

R. Vrinthavani
Center for Airborne Systems
Defence Research and Development
    Organisation
Hyderabad, India

Rahee Walambe
Coral Digital Technologies (P) Ltd.
Bangalore, India
and
Department of Engineering
Lancaster University
Lancaster, United Kingdom

Kevin Warwick
Department of Cybernetics
University of Reading
Reading, United Kingdom

Benjamin J. Whalley
School of Chemistry, Food Biosciences and
    Pharmacy
University of Reading
Reading, United Kingdom

Dimitris Xydas
School of Systems Engineering
University of Reading
Reading, United Kingdom

Chika O. Yinka-Banjo
Department of Computer Science
University of Cape Town
Rondebosch, Cape Town, South Africa

Hyun Gull Yoon
Industry University Collaboration
    Foundation
Chungnam National University
Daejeon, Korea

Introduction

Jitendra R. Raol and Ajith K. Gopal

Over the last two decades, there has been considerable advancement of various technologies applicable to mobile vehicles, in general, and robots, in particular. Since it would be a formidable task to keep track of this technological progress, only a small and humble attempt has been made to present in this volume on MIAS, descriptions of some of the most important aspects of robotics and some recent progress not only in the field of robotics but also in the field of other types of mobile vehicles.

At the outset, a robot as a vehicle (either static or dynamic depending upon the intended purpose of its use) is defined variously by various people and/or groups [1] as: (i) 'an automatically controlled, reprogrammable, multi-purpose, manipulator programmable in three or more axes, which may be either fixed in place or mobile for use in industrial automation applications' (by the International Organization for Standardization [ISO 8373] and International Federation of Robotics [IFR]) and (ii) 'a reprogrammable multi-functional manipulator designed to move materials, parts, tools, or specialized devices through variable programmed motions for the performance of a variety of tasks' (by Robotics Institute of America [RIA]). However, Joseph Engelberger, the inventor of the first industrial robot in history says, 'I can't define a robot, but I know one when I see one'. An intelligent robot can be considered as machine or system that has an ability to extract information that it requires from its own environment and then use this information to plan its paths and trajectory so that it can move in its surroundings avoiding the obstacles in a purposeful manner. It can also be visualized as a system that autonomously senses it own environment and then acts in that surrounding. The so-called smart sensors also have similar abilities. Autonomous robot/vehicle will have an ability to make its own decisions and then take necessary action to execute a given task. This autonomy gives a robot the ability to sense its own situation and then act on it so that any task is executed with minimal or no intervention from any human operator. The full autonomy is in an autonomous robot/vehicles and a partial autonomy is in, say, teleoperated robots.

From the above definitions of a robot, the important aspects that emerge are handling or dealing with: (i) manipulator, (ii) several axes system, (iii) industrial automation, (iv) programmed motions, or un-programmed but behavioural motions and (v) autonomy. The manipulator implies that a robot is used as an extension of a human arm to carry out difficult manual or hazardous tasks. This extended arm would be stronger and longer (or even shorter) than the normal human arm, flexibility not being ruled out. The arm would move relative to the body of the robot where it is attached. The manipulator should be controllable and this will add to the number of degrees of freedom (DoF) beyond the basic coordinates of the robot movements (6DoF). Thus, the DoF in case of a robot might be more than six. An industrial robot is supposed to help in mundane industrial tasks of movement of objects, lifting of tiny or big mechanical parts, and handling assembly line works. It is most likely that many or most robots would be controlled or guided and navigated by some programmed commands (for initiating and continuing the motions), these being done remotely (by teleoperation) or from the onboard computers of the robots. Many of these activities are also common for other mobile vehicles, like mini-micro-air vehicles (MAVs), and to some extent to the unmanned aerial vehicles (UAVs) and/or underwater autonomous vehicles (UWAVs). Many unmanned ground vehicles (UGVs) also would have similar or some advanced features of programmed actions and resultant motions. So, from an immediate perspective and in a nutshell a robot is a mechanical device that performs various tasks. It could be controlled by a human being or is an autonomous mobile machine. However, since a robot is called upon to perform variety of tasks it would be equipped with sensors (cameras, acoustic devices, IR sensors), will have onboard computers on/in its body, and will have wires and cables (electrical or otherwise) connecting various subsystems. Hence, a robot can be visualized as a composite system that would mostly be a mechatronic system rather than only a mechanical system. Thus, a robot has basically a mechanical system, an electrical power supply (battery cells-) system, (vision, acoustic, EM-) sensors, actuators, onboard computers (or micro-processor-controller systems), data processing (algorithms) systems and control systems to command, guide and carry out movements of the mobile vehicle. Hence, such a system is an electro-mechanical (or even electro-hydraulic/pneumatic) system. The mechanical system of a general robot would comprise of the chassis, motors and wheels, the latter for the wheeled robots. The wheeled robot might have a drive of the type of a car, skid steer, differential drive, synchronous or pivot drive. Most robotic systems are powered by a set of batteries. The robots actuators convert electrical energy to mechanical energy/work which might be either rotational/angular (obtained by using electrical motors) or linear motion (obtained by using electromagnetic devices). These motors could be of various types: (a) AC (alternating current), (b) DC (direct current), (c) stepper motors and (d) servo motors; the latter are DC motors with some feedback capability and error compensation. In most robotic systems DC, stepper or servo motors are used.

We can visualize from the foregoing that the field of robotics, in a broader sense, encompasses the following disciplines: (i) sense, (ii) think and (iii) act. These three processes are equally applicable to other autonomous mobile vehicles. For example, a micro-air vehicle can be considered as a micro-air robot carrying out a specified task of monitoring an area for security reasons. The sensing involves measuring devices, and processing of the measured data; these data could be in the form of images and/or kinematic (position, etc.). These data are filtered using Kalman filters or even some advanced filtering techniques: derivative-free Kalman filter, particle filter or information filter. For real-time/online data processing one needs to resort to some simpler, computationally efficient and numerically stable filtering approaches. Some sensors would have in-built data processing units/facilities, the smart sensors. Examples of sensors are LIDAR (light detection and ranging), GPS (global positioning system), IMU (inertial measurement unit), Sonar (sound navigation and ranging), IR (Infra Red) camera and so on. These sensors give information of the obstacles in the path of the robot and/or the landmark locations, in addition to the robot's own locations. Some combination of these measured data would be utilized to obtain more accurate information about the sensed object, obstacles or environment. This calls for the study and use of multi-sensor data fusion (MSDF) and associated methods [2]. The images of the tracked objects and landmarks are required to be processed in image/vision processing algorithms (and image fusion algorithms) and used for further decision making. The tracking of the images might also be required. The MSDF can be accomplished at the sensor level/data level or at the data processing level in filtering algorithms. To handle complex and sophisticated tasks and movements we need to have smart sensors, specialized sensors with low-power consumptions and even low-cost sensors. The smart sensors perform certain data pre-processing tasks that are anyway routinely required before carrying out further data fusion operations and processing. A robot exists in its own perceptual space, despite the fact that its own 'perception' capability might be limited compared to that of the human. Thus, the robot's state space is both: its internal state and its external state. The robot's state of perception can be enhanced using some elements of artificial intelligence (AI).

In terms of robot path planning (the task that can be divided amongst the 'sense' part and the 'think' part, because often the path planning is very intimately based on the sensed environment) and data filtering we need to move from classical and conventional algorithms to the ones based on soft computing and ones that combine features of the classical (and hence proven) and the new algorithms to derive benefits from both methodologies. The robot needs to 'think' to decide on what actions to take by itself for its further motion to carry out a specified task. This capability is divided into five sub-category tasks: (a) path-motion/trajectory planning, (b) autonomy (either 'thinking' is pre-wired or being done online in real time while executing the task), (c) kinematics/dynamics, (d) perception ('intelligent sensing') and (e) localization (could include mapping also). For a robot to move from its initial location to its goal point it needs to know what path to follow and how to follow the chosen path. Much of the assigned tasks need to be carried out by the robot automatically. The knowledge of the robot's kinematics as well as the robot's motion model (mathematical modelling) is very important for successful robotic actions. These mathematical models are also needed to carry out simulation of robot's behaviour and its intended operations. This simulation helps designer in optimizing the robotic sub-systems (sensor and control management) to achieve a desired goal with certain specifications/accuracy and so on. That an autonomous mobile system needs to be equipped with 'perception' is very important for accurate path/motion planning, simultaneous localization and mapping, and accurate and logical decision making. The major difference in a human system (human being) and a robot is the lack of perception capability in a robot/mobile vehicle. Perception is built up by sensing and interpreting the environment, understanding the importance of the task to be carried out, learning from previous experience, and adapting to the new situations, hitherto not encountered by a robot. For any man-made system this is a formidable task indeed. The final task is that of action. This involves the driving mechanisms, motion control and even biometric (human type) motion management. Generally, the concepts adopted are based on wheeled, tracked and bipedal robots, but we need to evolve the robots that increasingly use the concepts of biometric movements like the smooth reptilian movements. The robot uses the actuators (also called effecters) to carry out its actions. Due to the differences in the types of actuations the robots are categorized as either mobile, manipulator or communication robots.

The main difference between the human and a robot is that the human brain interprets the sensed data ('sense'), derives from these data the required information for decision making ('think') and directs ('act') the human organs to execute certain tasks automatically. The idea of AI is then to build computer systems/software/algorithms/processes to mimic the human brain's capability of thinking and action. Thus, AI has a great scope in the field of robotics and other mobile vehicles. The robot applications are: (i) industrial robots, (ii) medical/surgical robots, (iii) mining/trucks robots, (iv) humanoid robots, (v) farming tractor, (vi) MARS Rover, (vii) games/sports robots and (ix) unmanned aerial/ground vehicles. The unmanned aerial vehicles/robots are used in military, aerospace applications, mining, health care and even in teaching/training. The so-called foraging/ rummaging robots are useful for (a) collecting samples from chemically hazardous places, (b) extraction of landmine, (c) exploring unknown regions, (d) collecting samples of rocks from other planets and (e) assembling parts on a manufacturing line. There are other uses of robotic systems: (i) an automatic mobile sweeper and (ii) an automatic car for children. The research in robotics and closely related fields (UAVs, MAVs, UGVs) encompasses several disciplines and technologies and their integration: (i) mechanical design, manipulators and actuation, (ii) mathematical modelling of robotic system (and other vehicles), including its environment, (iii) sensors, instrumentation and acquisition systems for measurements, (iv) pre-processing of measured data, filtering and data fusion, (v) image acquisition/fusion, processing and tracking, (vi) path and motion planning and algorithms to achieve optimal paths, (vii) algorithms for simultaneous localization and mapping, (viii) implementation of these and other algorithms onboard computers (micro-processors, embedded systems), (ix) implementation of sensor/actuator fault detection/management and reconfiguration schemes, (x) use of soft computing techniques for learning, adapting and decision making, (xi) guidance and navigation algorithms, (xii) multi-robot coordination mechanisms, and associated algorithms, (xiii) real time/online system identification and parameter estimation [3,4] for mathematical modelling of these mobile vehicles, (xiv) control algorithms and robotic systems' performance evaluation methods and (xv) hardware/software robotic architectural aspects for optimal configuration of the robotic system.

This volume covers many of the areas and disciplines mentioned above across 35 chapters contributed by experts from various fields and countries. Although each chapter can be read and studied independently, they are grouped into three logical parts to facilitate nearly smooth information flow throughout the book: Part I: Conceptual Foundations for MIAS, Part II: MIAS and Robotics and Part III: Allied Technologies for MIAS/Robotics. Many of the chapters attempt to present some new results in the form of novel techniques, algorithms and/or experimental/simulations/empirical data-analysis results, and other chapters either describe, in brief, the basic methods, technologies and/or review some literature in the relevant fields. Thus, the structure of this volume is a mixed one: it can be used as a recipe book, or a reference volume with some new results (like ones in a research/technical journal). It is not meant to be a text book for any course or syllabus however it can be used as a supporting volume and can be consulted for research projects at the post graduate as well as doctoral levels. At the outset it must be mentioned that neither the editors (of this volume) nor the authors (of various chapters in the volume) claim that the mathematical expressions, equations and/or formulae are derived from the first principles. For such derivations the readers can refer the respective cited references of the concerned chapters. Also, it is not claimed that all the chapters discuss all the aspects of mobility, autonomy and intelligent, nonetheless the volume as a whole discusses several basic concepts, MIAS and robotics systems, and allied technologies that would aid MIAS and robotics. The main idea is that the disciplines considered and discussed in this volume would be needed to build a good autonomous and/or intelligent mobile system, only that the mechanical and structural aspects of the vehicles are not addressed in this book. It must also be mentioned that the authors of various chapters seem to have taken enough care to present various theories, formulae/equations, simulations/real-data results and experimental results, it is important that the readers and users of this material, inferences, and results take enough care and precautions before applications of these to their own problems, case studies and systems. Any such endeavours would be at their own risk. Next, only a brief description of each chapter is given here for two reasons: (i) the importance of each work and related technology is clearly highlighted by the authors in the introductions of their chapters and (ii) the table of contents gives a good picture of what is being presented in the concerned chapters. The synergism connection diagram for the MIAS/Robotic System's theory and practice as presented in various chapters (whose numbers are indicated within parentheses) in this book is depicted in Figure I.1. Many of these technologies are the constituents to build a MIAS system, except that the mechanical structures and related aspects are not dealt with in this book.

In Chapter 1, some basic concepts of artificial neural networks, fuzzy logic and genetic algorithms are discussed. The scope of applications of these soft computing techniques as well as of AI to the robotic and other related systems is discussed. Also, certain combinations of ANNs, FL and GAs are described that are/can be used for robotic applications.

Chapter 2 by Ramachandran discusses mathematical modelling aspects for robot motion. Especially kinematics, dynamics, robot walking and probabilistic robot modelling based on available odometry measurements are considered.

In Chapter 3, Gopal discusses the need of data fusion in robotics. The author highlights the data fusion approaches and gives partial coding of these methods. The presentation is concise, but the importance of the data fusion applications to robots/MIAS should not be underestimated. Further aspects and application of multi-sensor data fusion are developed in Chapters 4, 5 and 30. In Chapter 4, Myna discusses image registration and fusion aspects. Some aspects of satellite image registration and fusion are presented with real imagery data. A good literature review of image registration is also provided. The concepts fuzzy logic type I and type II are highlighted for use in image fusion. Chapter 5 by Naidu presents discrete cosine transform-based method for image fusion. Also the performance evaluation results are presented using the image data from the open literature. In Chapter 6, Kaimal and the coauthor present a novel and improved motion segmentation process in the spectral framework. Specifically, they discuss motion detection, motion estimation clustering using spectral framework. The maximum likelihood motion framework is used. Extensive performance results of the approach are also presented.

In Chapter 7, the authors present some new work on formation control in multi-agent systems over packet dropping link. Since, data losses do occur in communications channels, be it the data transmission networks, multi-robot coordination data-networks, multi-target communications channels and/or wireless sensor networks, it is important to address this problem here.

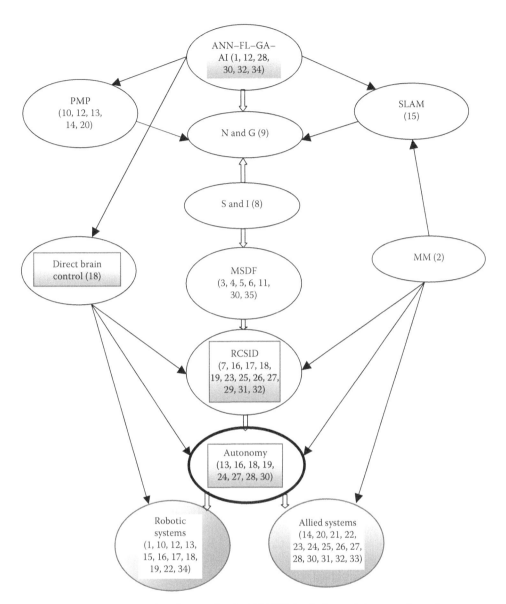

FIGURE I.1 Synergism connection diagram for the MIAS/Robotic System' theory and practice as presented in various chapters (indicated within parentheses) in this book. These technologies are constituents to build a MIAS system. (ANN–FL–GA–AI, artificial neural networks–fuzzy logic–genetic algorithm–artificial intelligence; PMP, path and motion planning; SLAM, simultaneous localization and mapping; N and G, navigation and guidance; S and I, sensors and instrumentation; MSDF, multi-sensor data fusion; MM, mathematical modelling; RCSID, reconfiguration control, system identification.)

In Chapter 8, the highlights of sensors and instrumentation systems that are used in many robotic systems are given. Many of these sensors are also used in other systems like UAVs/MAVs. The main emphasis has been on sensor types and sensor characteristics. Aspects of signal conditioning, data communications, MEMS and smart sensors are also briefly discussed.

Chapter 9 discusses the important problem of navigation and guidance which is a crucial and fundamental technology requirement for any mobile vehicle including robots for determining their pose and mapping their own environment. In the next few chapters, specific problems of vehicle path planning and simultaneous localization and mapping (SLAM) are described. In Chapter 10, a very important problem of vehicle path and motion planning is dealt with. Several classical as well as the so-called heuristic methods based on soft computing technologies are briefly reviewed and several methods are described. The program listing of D* algorithm is given in Appendix B, so that this D* algorithm can be compared with the one that is presented in Chapter 14. In Chapter 11, Twala presents some new solutions to the problem of missing and out-of-sequence measurements data. Specifically the approaches using Kalman filtering, and multiple imputation are discussed and some simulation experimental results are presented. In Chapter 12, Singh and Ajith present some new results of using genetic algorithm for robot path planning in the presence of dynamic obstacles. They present simulation results wherein the algorithm is able to avoid the obstacles in such an environment for a mobile robot. Chapter 13 by Motlasti presents a review kind of work on temporal logic motion planning and also suggests another approach for the same. Several logics and semantics are briefly presented. The author suggests a new methodology and presents a case study. In Chapter 14, Leenen, Terlunen and Roux present a new path planning solution based on the constraint programming approach. They present a new modified D* algorithm and some examples with the partial codes of the modified D* algorithm.

Chapter 15 discusses another equally important problem of simultaneous localization and mapping (SLAM) for mobile vehicles and robots. Several aspects of the SLAM are reviewed and a particular novel approach based on H-infinity filtering for joint state and parameter estimation in SLAM is presented. Some simulation results are also presented. The results of the H-infinity filter-based SLAM are found to be very encouraging and interesting.

In Chapter 16, several aspects of robotic architecture are briefly discussed. Specifically planning-based, behaviour-based and hybrid architectures are highlighted. These architecture philosophies are also very useful for other mobile vehicles. Alternative hybrid architecture with possibility of incorporating fuzzy logic and sensor data fusion is presented. In Chapter 17, Osunmakinde presents aspects of multi-coordination for robot. The challenges from single to multi-robot coordination and related software technologies are addressed.

In Chapter 18, Warwick and his team present a novel approach for control of a mobile robot—direct brain control. They experiment with a biological brain and establish that a robot can be practically controlled by such a biological brain (brain tissues from a rat). Their simulation and experimental studies are very convincing and encouraging, and paves a way to new and challenging studies which can be taken up for other vehicles like MAVs. In Chapter 19, Walambe presents novel work on development of safe and effective autonomous decision making in intelligent robotic system. She employs RCS–RMA (Real-Time Control System–Reference Model Architecture) as an architectural framework and presents extensive discussions and test results of such a system.

Padhi and Chawla present in Chapter 20, a novel approach to partially integrated guidance and control of unmanned aerial vehicles for reactive obstacle avoidance. The reactive nonlinear guidance algorithm is validated with full non-linear six-DoF model of UAV for single and two obstacles.

Kostadinov and Chakarov, in Chapter 21, consider impedance-controlled mechatronic system based on the redundancy of actuators. Specifically they consider a method for realization of certain approach for impedance control by redundancy actuation including some examples, and study serial–parallel manipulators with redundancy actuation. This actuation includes an approach specifying the desired stiffness in the operation space by means of a magnitude variation of the antagonistic driving forces. In Chapter 22, Professor Kostadin Kostadinov presents an enabling robot technology including web application for automation of the synthesis of closed structures for micro- and nano-applications. This application utilizes the advantages of tense piezo-actuators and closed robot kinematical structures.

In Chapter 23, Raol and his co-authors present a novel approach to system identification and parameter estimation for inherently unstable systems operating in a closed-loop control and apply this technology to the conditions of fault analysis and management of an aircraft. They also utilize the stabilized recursive estimation method for estimation of stability margins for a closed-loop aircraft-control system. In Chapter 24, the authors present the novel adaptive algorithms for acceleration in the context of smart antennas for mobile autonomous systems. In MIAS/robotics an obstacle negotiation can be achieved with the help of adaptive algorithms implemented in the sensors embedded within the robot. Also this could be the case with MAVs/UAVs. In adaptive arrays (algorithms) the pattern optimization is obtained by real-time active weighting of the received signal, and reconfigured according to the conditions under which it is operating. These optimum weights are obtained by employing an efficient adaptive algorithm, the extensive results of which are presented in this chapter.

In Chapter 25, the authors present an integrated modelling, simulation and controller design for an autonomous quad-rotor MAV that was designed and developed following a systematic approach under severe time/cost constraints. On the basis of the existing components and components readily available in the market, all sub-systems were modelled and integrated into a single-system model. This was used to test the performance of this MAV, its stability, response to external disturbances and control inputs. The quad-rotor was assembled from the available components. This modelling and simulation exercise helped cut down development time and cost by avoiding intermediate design changes.

The authors in Chapter 26 present an approach for determination of impact and launch points of a mobile vehicle using Kalman filter and smoother. Especially the Kalman filter and forward prediction are used to predict the impact point, whereas the Kalman filter and RTS fixed interval smoother and backward integration are used to predict the launch point. Algorithms are validated using simulated data of a target moving with constant velocity.

In Chapter 27, the authors present a novel stability augmentation system for MAV with an aim towards autonomous flight. They present a generic design methodology of robust fixed order H_2 controller and onboard computer for the MAV named Sarika-1. A Digital Signal Processor (DSP)-based onboard computer FIC (Flight Instrumentation Controller) was designed to operate under automatic or manual mode. This controller was ported on to the flight computer and was validated through the real-time hardware-in-loop-simulation (HILS). The responses obtained from the HILS compared well with those obtained from the off-line simulation thereby validating the design approach. The authors in Chapter 28 present a novel approach to neuro-fuzzy fault-tolerant aircraft auto-land controllers. The auto-landing problem consists of a high-performance fighter aircraft following a flight path which consists of flight segments: (i) a wing-level flight, (ii) a coordinated turn, (iii) descent on glide slope and (iv) the flare manoeuvre and touchdown on the runway. Hence, it was considered appropriate to study reconfigurable or intelligent systems that could detect failure and further utilize the available aerodynamic redundancy effectively in completing the mission safely.

In Chapter 29, Ananda considers the important problem of fault-tolerant system in the context of reconfiguration of control system for a civil aviation transport aircraft. Specifically reconfiguration issues, redundancy management and reconfiguration algorithm are discussed. The methods discussed in this chapter are equally applicable for fault tolerance and reconfiguration issues arising in robotic systems.

In Chapter 30, the authors study three different philosophies for automatic target recognition (ATR): classical, Bayesian and Neural Networks. Specifically, the authors review these approaches briefly and present some futuristic possibilities. They discuss several aspects of image processing, image sensors, target detection, target feature extraction, target tracking and sensor data fusion from the point of view of these philosophies of ATR.

The authors in Chapter 31 present work on real-time implementation of a novel fault detection and accommodation algorithm for an air-breathing combustion system. They use extended Kalman filter and related design process for analytical redundancy and fault detection and management. The novel approach is validated using simulation and real-time experiments, and these results show that the present design of the FDA algorithm is suitable to run on an embedded hardware, thereby showing potential application to other MIAS/Robotic systems. In Chapter 32, Patel and the co-author present analysis and simulation results of fuzzy logic–based sensor and control surface fault detection and reconfiguration for an aircraft. Starting with the application of EKF they develop fuzzy logic–based approach and employ it for reconfiguration of control law. They present extensive simulation results to establish the efficacy of their novel approach.

In Chapter 33, the authors present an approach for target tracking in 3D using 2D radar, and uses Doppler measurements for height as well as velocity activity estimation. Very accurate results were obtained by the presented method. This approach might be very useful for determining heights of various autonomous vehicles especially when they are flying in multi-coordinated scenarios.

In Chapter 34, Osunmakinde and his co-authors investigate the use of Bayesian Network and k-NN Models to develop behaviours for autonomous robots. They present (i) Bayesian Network and k-nearest-neighbour models, (ii) the modelling for behavioural and collision avoidance for robots and (iii) experimental evaluations of the approaches on number of comparative evaluations in static and dynamic environments. The average performance evaluation results of the models are compared based on the configuration of four sensors and presented in this chapter. In Chapter 35, Twala presents a new approach to the problem of out-of-sequence measurements data especially the copulas-based method and some simulation experimental results are presented. The work of this chapter can be considered as a complementary work of Chapter 11. In Appendix A, we give a brief description of a few statistical and numerical concepts and methods that would be useful in appreciating the analysis material presented in various chapters of the book. Finally, in Appendix B, a brief description of software and algorithms that might be useful in simulation/path planning for mobile robots is given. Also a brief description of aerospace vehicle simulation package is given.

REFERENCES

1. Martinez, W. Lecture notes on robotics, CECS-105, Dept. of Comp. Engg. and Comp. Sci., California State University, Long Beach, Presented at *Robotics 101* Webinar, April 2009. http://decibel.ni.com/content/ docs/DOC-4655. LA, USA. January 2011.
2. Raol, J. R. *Multisensor Data Fusion with MATLAB.* CRC Press, FL, USA, 2009.
3. Raol, J. R., Girija, G. and Singh, J. *Modelling and Parameter Estimation for Dynamic Systems*, IEE/IET Control Series Vol. 65, IEE/IET, London, UK, 2004.
4. Raol, J. R. and Singh, J. *Flight Mechanics Modeling and Analysis.* CRC Press, FL, USA, 2008.

Part I

Conceptual Foundations for MIAS

1 Neuro-Fuzzy–GA–AI Paradigms

Jitendra R. Raol

CONTENTS

1.1 INTRODUCTION

In the last five decades, research in the areas of artificial neural networks (ANNs), fuzzy logic (FL) and genetic algorithms (GAs) has seen many strides and when put together these independent looking technologies can be and are considered as a major part of the so-called soft computing paradigms. In most books on soft computing, many times only neuro-fuzzy-GA and its various aspects are only considered, support vector machines (SVM) sometimes being included. Also, in the area of artificial intelligence (AI) (mostly) ANNs are dealt along with other AI-related (languages, knowledge-based systems) aspects. In this chapter we discuss ANNs, FL (Type I and Type II), GAs [1–5] as a triad and study briefly their basic concepts, theories, and practical application possibilities for mobile intelligent autonomous systems (MIAS) and especially robotics. Also, we briefly touch upon the prevalent definitions of the AI and give a brief discussion on AIAS (an AI-based agent system). Where feasible we link the three soft computing approaches and see how these combinations have been exploited for solving certain problems in robotics.

At the outset we say that ANNs are used for learning the environment of the robot, the robot's own behaviour, and modelling purposes—the point is that an intelligent robot should build the model of its own environment and link this information/model with its own mathematical model (i.e., the model of a robot's behaviour) in conjunction with the measured data to navigate (Chapter 9) from one point to the next and so on to finally reach its ultimate goal.

The FL and associated description of the fuzzy membership functions, fuzzy implication functions, and fuzzy inference engine/system (FIES) [5] (see Figure 4.11) are used for representation and handling of vagueness properly, adequately (because of use of also Type II FL) and in fact formally in consistent mathematical form so that the uncertainties in the robot's environment and sensors' measurements can be formally modelled. The idea in using FL is to use heuristic knowledge and experiences of the domain experts, who have learnt the art and science of robotics and its engineering by trial-and-error methods, lots of experience gathered over a long period of time, and/or by other formal methods in robot's control management. This control management implicitly incorporates the robot path planning/motion planning (Chapter 10) and SLAM (simultaneous localization and planning, Chapter 15). No doubt, the G&N (robot guidance and navigation, Chapter 9) play a very crucial role in path/motion planning and SLAM, basically the PMP and SLAM feed into G&N as a whole for facilitating the motion of a mobile vehicle.

On the other hand, the GAs [6], based on the Nature's evolution mechanism are used for searching global minima (optimal) in the optimization problems related to robot's path/motion planning, and control problems. The GAs are simple and yet powerful methods for obtaining optimal solutions to many science and engineering problems. Robotics is a fertile field for systems engineering research and development and GAs can be utilized very effectively for optimizations of many problems that are encountered in the robotics' systems engineering.

In essence, while solving many real-world problems, for example, in MIAS and robotics, by neural networks (NNs), we encounter a variety of learning algorithms and a vast selection of possible ANN configurations and architectures [1]. There are a number of possible schemes for automatically optimizing the choice of individual ANNs and/or combining various architectures to obtain an optimal combination, such a combination might be required for classification of data (objects), scenes in an environment, and pattern recognition. There are two approaches to combining ANNs: (i) modular and (ii) ensemble-based approaches [1]. In the case of the former, a given learning/modelling problem is decomposed into a number of subtasks that are treated by specialist modules. These subnetworks carrying out the subtasks are local in the sense that the (neuronal) weights in one expert are decoupled from the weights in other subnetworks. In the second ensemble-based approach a set of networks is trained on what is essentially the same task, and then the outputs of the networks are combined—this can be called as ANN-based fusion of the resultant data/outcomes/decisions. This later configuration obtains a more reliable combined output than would be obtained by selecting the best network. One important observation is that one should not combine the networks of highly differing accuracies since this is not a very wise approach. A suitable approach to handle this aspect is the use of other two methods based on the so-called soft-computing techniques such as FL, GA and perhaps a hybrid of them. There have been a lot of attempts in the literature to integrate NN-Fuzzy, Fuzzy-GA, GA-NN and NN-Fuzzy-GA for deriving benefits from the merits of these individual soft-computing technologies. We discuss these approaches briefly in this chapter.

1.2 ARTIFICIAL NEURAL NETWORKS

Over the last five decades, the research and development in the area of ANNs have advanced at a rapid space. Modelled on the biological neural networks (BNNs of human brain and its nervous system) the ANNs are and can be devised, and even can be seen, as electrical/electronic (adaptive) circuits and systems which capture (like BNNs) some simple and basic ability to learn adaptively from the given test data, wherein the decision process is based on certain non-linear operations, although the linear operations are not uncommon. An illustration and comparison of biological neuron and artificial neuron are given in Figure 1.1 and Table 1.1 [2,4]. These ANNs actually are then the numerical methods or algorithms that can also be programmed on the conventional computers including PCs, and laptops, and many simulation tasks can be easily carried out easily. The ANN can be considered to provide some (orthogonal) basis functions for mathematical modelling activity from some empirical/experimental data. In themselves the ANNs are parallel computing structures like BNNs and hence can be easily programmed on parallel computing machines. One can directly use several neuronal architectures in some optimal manner and build computing machines using basic electronic circuits and switches, multiplies and adders (like analog computers of the past century) to carry out computational tasks for solving problems in science and engineering. For example, one can build ANN-based computing hardware to invert a matrix, and solve a least-squares estimation problem. In this way, one can build complex neuronal/architectural systems and hence a computer to carry out specific optimization and control computational tasks [2]. Even for certain specific tasks which are done repeatedly, the integrated circuits/chips (ICs) can be made or the working and evaluated code can be embedded on some firmware (embedded architecture, FPGA, the field programmable gate arrays).

The non-linearities in ANNs are useful in: (a) improving the convergence speed (of the algorithm), (b) providing a more general non-linear mapping between input–output signals and (c) reducing the effect of outliers in the measurements by curtailing their undesirable effects, especially using saturation-type non-linearity. Very successful ANNs have been the so-called feed-forward neural network (FFNN), and its variants. The FFNN has found successful applications in the signal analysis, pattern recognition, non-linear curve fitting/mapping, flight data analysis, aircraft mathematical modelling, adaptive control, system identification, parameter estimation, sensor

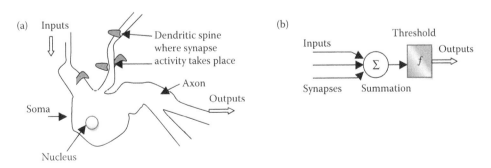

FIGURE 1.1 The artificial neuron imitates biological neuron in certain ways. (a) Biological neuron and (b) artificial neuronal model. (Adapted from *Multisensor Data Fusion with MATLAB.* J.R. Raol, CRC Press, FL, USA, 2009.)

TABLE 1.1

Comparison of Neural Systems

Biological Neuron Network/System (BNNS)		Artificial Neural Network/System (ANNS)
Neurons	$\rightarrow$	Nodes/units
Firing rate	$\rightarrow$	Activation levels
Signals received by dendrites and passed on to neuron receptive surfaces	$\rightarrow$	Data enter through input layer
Inputs are fed to the neurons through specialized contacts called synapse	$\rightarrow$	Weights provide the connection between the nodes in the input and output layers
Axons, dendrites, synapses	$\rightarrow$	Connections
All logical functions of neurons are accomplished in soma	$\rightarrow$	Non-linear activation function operates upon the summation of the product of weights and inputs
Output signal is delivered by the axon nerve fibre	$\rightarrow$	The output layer produces the network's predicted response
Excitatory/Inhibitory inputs	$\rightarrow$	Excitatory/Inhibitory inputs

data fusion, classification of objects, features, robot path and motion planning and SLAM. The other promising and very useful structure/architecture is the one called recurrent neural network (RNN). These RNN structures are also used in modelling, parameter estimation and control of dynamic systems [2,4]. Interestingly, the RRNs are very much amenable to parallelization and many filtering, signal processing, optimization and control methods and algorithms can be parallelized using this property of RNNs [2]. This will lead to building of truly parallel computing machines and will pave an assured way to autonomous systems by way of improving efficiency and cutting down the computational times to carry out complex and sophisticated tasks of control configurations, learning, adaptability and mobility. However, it is unfortunate that despite the existence of literature on the development of schemes of parallelization of the optimization and control algorithms using RNNs for more than two decades [2], not much work is reported in the open literature on application and building of neuronal parallel computers. The ANNs have, if not in all respects, some similarities to the biological neuron (neuronal) system. Biological neuron system (BNS) has unbelievably massive parallelism and consists of very simple processing elements but in very huge numbers (millions/billions). The FFNN and RNN are thus information processing system of a large number of simple processing elements, known as artificial neurons, and doing almost similar tasks (Figure 1.2). These neuronal elements are interconnected by links, which are represented by the so-called neuron (neuronal) weights (can be called coefficients also), and they cooperate to perform parallel distributed computations in order to carry out a desired task, the main difference lying in the numerical values of their weights. ANNs have only some resemblance to real NNs, although closer the similarity, the more complex would be the ANN as well more accurate in its representation of the biological computing power and ability in learning and adaptation. Thus, the ANNs are massively parallel adaptive circuits or filters and hence in true sense are the

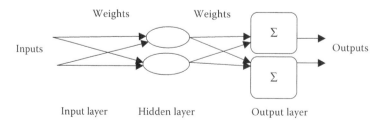

FIGURE 1.2   Feed-forward neural network structure with one hidden layer.

parallel computing machines by themselves. ANNs are used for input–output subspace (I/O) modelling because the basic NN functions can adequately approximate the system behaviour in some overall sense.

Thus, the basic elements of NNs are [1,2]: (i) basic structures and processes in connectionist models, (ii) representation using units and weights (which can be called even as coefficients) and (iii) supervised and/or unsupervised learning methods, in this case first supervised learning can be used, and then unsupervised learning depending upon the situation and the availability of the required data. The basic constituents of an ANN are nodes, these nodes are linked by excitatory and inhibitory connections (Table 1.1), and when the nodes are activated, say by stimulus, the activation spreads along the connections. The nodes in cognitive models are not (and need not be) equivalent to single neurons in the brain. They may act as approximation to a group of neurons. This group might become active under the same circumstances. Important aspects in the brain and BNN are [2]: (a) the firing rate of individual neurons and/or (b) the number of active neurons in a cooperative group of neurons—these both cases being indicative of degree of activation of the neural representation. This degree of activation represents one or more of the following: (i) stimulus intensity, (ii) certainty of the representation, (iii) strength of the response and (iv) tendency of the response. This is represented by the activation function which is generally non-linear.

1.3 FEED-FORWARD NEURAL NETWORKS

The feed-forward neural networks (FFNNs) have non-cyclic, layered and forward moving extended topology and hence can be considered to have structure free (in the conventional sense of polynomial model) non-linear mapping between input–output signals of a system (Figure 1.2). The chosen and specified network is first trained using the training set data and then it is used for prediction using a different input set, which belongs to the same class of the data. This second data set/segment is the validation set, the process is similar to the one used as cross-validation in system identification and parameter estimation literature. This cross-validation test is called real validation test for the successful performance of the FFNN. Of course there are many measures and metrics that can be used to ascertain the accuracy and satisfaction of the performance of the training of the NNs and its prediction capability. The weights of the network are estimated using the so-called back propagation (BP)/gradient-based optimization—this in the simplest term is the well-known steepest descent/gradient method. Because of the layered disposition of weights of the FFNN, the estimation of the weights requires propagation of the error of the output layer in backward direction and hence the name BP; however, it is not BP time-wise. The estimation/learning algorithms can be described using the matrix/vector notation for the sake of clarity and ease of implementation in PC MATLAB®, and in this way, it is very easy to understand and interpret the training algorithms. Even without the NN toolbox of MATLAB, the simulation and other related studies for learning algorithms and design of NNs can be easily and very efficiently carried out using the available and newly formulated (by the user/designer) dot-em (.m) files in MATLAB.

The FFNNs can be looked upon as non-linear black-box (modelling) structures, the parameters (weights/coefficients) of which can be determined, estimated by conventional optimization methods, including the extended Kalman filter (EKF). One can use some newer optimization methods like GA and its variants for training the NW (network). The FFNNs are found to be very suitable for system identification/parameter estimation, time-series modelling and prediction, pattern recognition/classification, sensor failure detection, reconfiguration of control laws and estimation of aerodynamic coefficients. In case of parameter estimation of dynamic systems using FFNN, actually the FFNN is used for predicting the time histories (after initial training) of aerodynamic coefficients, for example, and then a regression method is used to estimate the aerodynamic parameters (the aerodynamic stability and control derivatives) from the predicted aerodynamic coefficients' time histories [4]. A similar process can be adopted for the estimation of unknown parameters of the postulated mathematical model of a robot's environment.

Thus, ANN can be used as a mapping function 'f' or an operator between input (I)/output (O) data sets, that is, f that maps I into O; $f : I \to O$; or $y = f(x)$. Because the classification is a mapping from the feature/object space to some output classes (e.g., for the classification problem), that is, declaration of belong to some class, we can formalize the ANN [1], especially two-layered FFNN (trained with the generalized delta rule), as a feature classifier. Consider a two-layered NW classifier with (i) T neurons (number of features) in the input layer, (ii) H neurons in the hidden layer and (iii) c neurons (the number of classes) in the output layer. The number of hidden neurons, H is an appropriately selected number. This FFNN is fully connected between adjacent layers, and its operation can be thought of as a non-linear decision-making process, that is, given an unknown input $X = (x1, x2, \ldots, xT)$ and the class set $\Omega = \{\omega 1, \omega 2, \ldots, \omega c\}$, each output neuron is expected to produce yi of belonging to this class.

1.3.1 LEARNING ALGORITHM

For ANNs the learning algorithms are used to update the weights between the neuron units or nodes. There are mainly two types of learning approaches: (a) supervised learning and (b) unsupervised learning. In supervised learning NW's responses to the inputs are compared with the feedback from a teacher, the known data (images, etc.). This is expected to reduce the error in matching of the NW output to the desired output. In an unsupervised learning the weights are altered based on some internal criterion. In this case no feedback from a teacher or NW's performance is entertained. We assume that the FFNN has these variables: (i) u_0 as the input to the input layer of the NW, (ii) n_i as the number of input neurons, equal to the number of inputs u_0, (iii) n_i as the number of neurons of the first hidden layer of the NW, (iv) n_o as the number of output neurons equal to the number of outputs z, (v) W_1 ($n_h \times n_i$) coefficient/weight matrix between input and first hidden layer of the NW, (vi) W_{10} ($n_h \times 1$) bias weight/coefficient vector, (vii) W_2 ($n_o \times n_h$) weight/coefficient matrix between the first hidden layer and the output layer, (viii) W_{20} ($n_o \times 1$) bias coefficient/weight vector and (ix) μ as the learning rate or step size [4].

1.3.1.1 The BP Training Algorithm

This BP algorithm is based on steepest descent optimization, the simplest of the gradient methods. The forward pass signal processing and propagation are done using the following sets of equations. We know u_0 and the initial estimates of the weights [2–4]:

$$y_1 = W_1 u_0 + W_{10} \tag{1.1}$$

$$u_1 = f(y_1) \tag{1.2}$$

In Equations 1.1 and 1.2, y_1 is a vector of intermediate values and u_1 is the input to the first hidden layer. The sigmoid non-linear activation operator for a function is given by

$$f(y_i) = \frac{1 - e^{-\lambda y_i}}{1 + e^{-\lambda y_i}} \tag{1.3}$$

The signal between the first hidden layer and the output layer is expressed as follows:

$$y_2 = W_2 u_1 + W_{20} \tag{1.4}$$

$$u_2 = f(y_2) \tag{1.5}$$

u_2 is the signal at the output layer. A system of differential equations is given as

$$\frac{dW}{dt} = -\mu(t)\frac{\partial E(W)}{\partial W} \tag{1.6}$$

Equation 1.6 is the simplest steepest gradient (descent) approach for obtaining the optimal weights of the FFNN. Define the output error as $e = z - u_2$, and a suitable quadratic cost function based on it, then the expression for the weight gradient is given as follows:

$$\frac{\partial E}{\partial W_2} = -f'(y_2)(z - u_2)u_1^T \tag{1.7}$$

In Equation 1.7, u_1 is the gradient of y_2 with respect to W_2, which are the weights of the output layer. The derivative (f') of the node non-linear activation function f is given as

$$f'(y_i) = \frac{2\lambda_t e^{-\lambda y_i}}{(1 + e^{-\lambda y_i})^2} \tag{1.8}$$

Equation 1.7 is obtained from the quadratic function $E = \frac{1}{2}(z - u_2)(z - u_2)^T$ and using Equations 1.4 and 1.5. The modified error of the output layer is given as follows:

$$e_{2b} = f'(y_2)(z - u_2) \tag{1.9}$$

Finally, the recursive weight update/learning rule for the output layer is given by

$$W_2(i+1) = W_2(i) + \mu e_{2b}u_1^T + \Omega[W_2(i) - W_2(i-1)] \tag{1.10}$$

The Ω is the momentum factor to smooth any large weight changes that might occur and to accelerate the convergence of the algorithm to the steady state. The bracketed term is viewed as the approximate computation of the rate of change of the weights between any two instants of time, and hence to work as an anticipation factor for damping the large weight chances (this is like the rate feedback operation in the control system to increase the damping in the system [2]). Hence this artefact would stabilize the large weight excursions. The BP of the error and the weight update rule for W_1, the input layer are given as [2–4]

$$e_{1b} = f'(y_1)W_i^T e_{2b} \tag{1.11}$$

$$W_1(i+1) = W_1(i) + \mu\ e_{1b}u_0^T + \Omega[W_1(i) - W_1(i-1)] \tag{1.12}$$

The entire weight learning/ANN training process is recursive. It should be observed that the values of μ and Ω in Equations 1.10 and 1.12 need not necessarily be the same.

1.3.1.2 Recursive Least-Squares-BP Algorithms

We at present hear of one version of the recursive least squares—back-propagation (RLSBP) weight training algorithm [2]. It is based on the least-squares (LS) principle and uses forgetting factors and is considered as a special case of the conventional Kalman filter. The linear KF filter concept is directly used. The output signal is computed simply as

$$u_2 = y_2 \tag{1.13}$$

The computation of the Kalman gains is done as follows:
For layer 1, the updates for the gain K_1 and the covariance matrix P_1 are given as [4]

$$K_1 = P_1 u_0 \, (f_1 + u_0 P_1 u_0)^{-1} \tag{1.14}$$

$$P_1 = (P_1 - K_1 u_0 P_1)/f_1 \tag{1.15}$$

For layer 2, the updates for the gain K_2 and the covariance matrix P_2 are given as

$$K_2 = P_2 u_1 (f_2 + u_1 P_2 u_1)^{-1} \tag{1.16}$$

$$P_2 = (P_2 - K_2 u_1 P_2)/f_2$$

As the output layer is linear, the error for the output layer is given by

$$e_{2b} = e_2 = (z - y_2) \tag{1.17}$$

The BP of the output error to the inner layer gives

$$e_{1b} = f(y_1)W_2^T e_{2b} \tag{1.18}$$

Finally, the weight update rules are given as

$$W_2(i+1) = W_2(i) + e_{2b}K_2^T \tag{1.19}$$

$$W_1(i+1) = W_1(i) + \mu \, e_{1b}K_1^T \tag{1.20}$$

1.4   RECURRENT NEURAL NETWORKS

The other very familiar ANN structure is that of the recurrent neural networks (RNNs), based on the Hopfield neural network (HNN). These are the ANNs with the feedback from output variable to the input variable (Figure 1.3). Hence, RNNs are very suitable to model dynamic systems, since dynamic systems have an inherent feedback mechanism, since they have memory and have energy accumulating/dissipating nature. Also, the RNNs are used for the estimation of parameters of dynamic systems [2]. There are several variants of RNNs useful for this purpose, also the RNNs can be used for training and hence, several trajectory matching algorithms are available for this purpose [2].

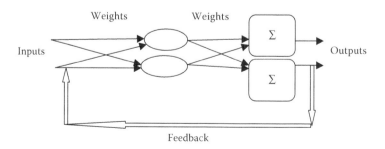

FIGURE 1.3   Recurrent neural network (RNN).

1.4.1 Some Variants of RNNs

Four variants of RNNs from the point of view of explicit parameter estimation are studied here. These are the variants of the basic Hopfield Neural Network structure of which at least three variants are related to each other by affine or linear transformation of their states [2] and are classified by the way in which the sigmoid non-linearity operates: (i) either on neuronal states (RNN-S), (ii) or weighted neuronal states (RNN-WS) or (iii) the residuals of the network output signal or forcing input (RNN-FI).

1.4.1.1 Basic Hopfield Neural Network

This is the RNN-S that has a number of mutually interconnected information processing units, the neurons and the outputs of the network are a non-linear function of the states of the network. The dynamics of the RNN-S are given as [2]

$$\dot{x}_i(t) = -x_i(t)R^{-1} + \sum_{j=1}^{n} w_{ij}\beta_j(t) + b_i, \quad j = 1, \ldots, n \tag{1.21}$$

Here, x is the internal state of the neurons, β are the output states $\beta_j(t) = f(x_j(t))$, w_{ij} are the weights, b is the bias input to the neurons and f is the sigmoid non-linearity. R is the neuronal impedance and 'n' is the dimension of the neuron state. Equation 1.21 is also written as

$$\dot{x}(t) = -x(t)R^{-1} + W\{f(x(t))\} + b \tag{1.22}$$

Equation 1.22 is a representation of classical neuro-dynamics and it obtains a simple system retaining essential features: (a) neuron as a transducer of input to output, (b) a smooth sigmoid response up to a maximum level of output and (c) a feedback nature of connections. This mathematical model has two aspects: dynamics meaning memory and non-linearity.

1.4.1.2 RNN-FI

The non-linearity operates on the forcing input: FI, that is, the weighted states and the input to the networks are taken as the modified input $= f(Wx + b)$. The dynamics are given by

$$\dot{x}_i(t) = -x_i(t)R^{-1} + f\left(\sum_{j=1}^{n} w_{ij}x_j(t) + b_i \right) \tag{1.23}$$

With $f(.) = f(FI)$. The RNN-FI is related to the RNN-S by an affine transformation.

1.4.1.3 RNN-WS

In this case the non-linearity operates on the weighted states. The dynamics of the NW are described as

$$\dot{x}_i(t) = -x_i(t)R^{-1} + f(s_i) + b_i \tag{1.24}$$

With $s_i = \sum_{j=1}^{n} w_{ij}x_j$. This NW structure is related to RNN-S by a linear transformation.

1.4.1.4 RNN-E

In this NW the non-linearity directly operates on the equation error. The function f or its derivative f' does not enter into the neuron dynamics. However, it affects the residual by way of quantizing these residual errors and reducing the effect of measurement outliers. The dynamics are given by

$$\dot{x}_i(t) = -x_i(t)R^{-1} + \sum_{j=1}^{n} w_{ij}x_j(t) + b_i \tag{1.25}$$

We say that the internal state x_i is β_i, the parameters of the general dynamic system.

1.5 PARAMETER ESTIMATION WITH HOPFIELD NEURAL NETWORK–RNN

The dynamic system is given by

$$\dot{x} = Ax + Bu; \quad x(0) = x_0 \tag{1.26}$$

For parameter estimation using HNN (RNN-S) the $\beta = \{A, B\}$ is the parameter vector to be estimated with n as the number of parameters. A suitable functional is associated with it the HNN which then iterates to a stable parameter estimation solution. In this NW the neurons change their states x_i according to Equation 1.21. The dynamics are affected by the non-linear function f, that is, $\beta_i = f(x_i)$. The parameter estimation equation that needs to be solved by some numerical integration method is

$$\dot{\beta}_i = -\frac{1}{\left(f^{-1}\right)'(\beta_i)} \frac{\partial E}{\partial \beta_i} = \frac{1}{\left(f^{-1}\right)'(\beta_i)}\left[\sum_{j=1}^{n} w_{ij}\beta_j + b_i\right] \tag{1.27}$$

The expressions for the weight matrix W and the bias vector b are given as follows [2]:

$$W = -\begin{bmatrix} \sum x_1^2 & \sum x_2 x_1 & 0 & 0 & \sum u x_1 & 0 \\ \sum x_1 x_2 & \sum x_2^2 & 0 & 0 & \sum u x_2 & 0 \\ 0 & 0 & \sum x_1^2 & \sum x_2 x_1 & 0 & \sum u x_1 \\ 0 & 0 & \sum x_1 x_2 & \sum x_2^2 & 0 & \sum u x_2 \\ \sum x_1 u & \sum x_2 u & 0 & 0 & \sum u^2 & 0 \\ 0 & 0 & \sum x_1 u & \sum x_2 u & 0 & \sum u^2 \end{bmatrix} \tag{1.28}$$

$$b = -\begin{bmatrix} \sum \dot{x}_1 x_1 \\ \sum \dot{x}_1 x_2 \\ \sum \dot{x}_2 x_1 \\ \sum \dot{x}_2 x_2 \\ \sum \dot{x}_1 u \\ \sum \dot{x}_2 u \end{bmatrix} \tag{1.29}$$

We can see that it is easy to compute the matrix elements in Equations 1.28 and 1.29, since it is assumed that the measurements of states, state derivatives and input are available. It is also assumed that these measurements are noise free, otherwise estimated parameters will be biased (Appendix A). The final algorithm for the parameter estimation of the dynamic system is: (i) compute W matrix, and the bias vector b from the measurements of x, $\dot{x}$ and u that are available (equation error formulation) for a certain time interval T, (ii) randomly assign the initial values of β_i and solve the differential equation 1.27 or 1.30. Since, $\beta_i = f(x_i)$ and the sigmoid non-linearity is a known function f, by differentiating and simplifying, one gets

$$\frac{d\beta_i}{dt} = \frac{\lambda(\rho^2 - \beta_i^2)}{2\rho}\left[\sum_{j=1}^{n} w_{ij}\beta_j + b_i\right] \tag{1.30}$$

$$f(x_i) = \rho\left(\frac{1 - e^{-\lambda x_i}}{1 + e^{-\lambda x_i}}\right) \tag{1.31}$$

The integration of Equation 1.30 yields the solution to parameter estimation problem for the structure of RNN-S. Equation 1.30 can be discretized and used. The scheme is non-recursive, since the required computations of elements of W and b are performed by considering all the available data. The discrete form is given as

$$\beta_i(k+1) = \beta_i(k) + \frac{\lambda(\rho^2 - \beta_i^2(k))}{2\rho}\left[\sum_{j=1}^{n} w_{ij}\beta_j(k) + b_j\right] \tag{1.32}$$

In summary, the ANNs have the following features: (a) they mimic some or certain behaviour of human brain, and have massively parallel architecture like that of BNNs; (b) they can be represented by adaptive (analog) circuits with input channel, weights (parameters/coefficients), one or at most two hidden layers and output channel with some non-linearities, while linearity is no exception; (c) these weights can be tuned to obtain optimal performance of the NN—network—be it in modelling of a dynamic system or non-linear curve fitting/mapping; (d) it requires to use good training algorithms to determine the weights; (e) the NWs can have feedback-type arrangement within the neuronal structure, and hence very useful for dynamic systems; (f) the trained NWs can be used for predicting the behaviour of a dynamic system; (g) the ANNs can be easily coded and validated using standard software procedures, or be realized using MATLAB tool boxes; (h) optimally structured NW architecture can be hard-wired/firm-wired and embedded into a chip (like FPGA, field programmable gate arrays) for practical applications—this will be the generalization of the crst while analog circuits-cum-computers and then the NW-based system can be truly termed as a new-generation powerful/parallel computer and can become a formidable tool for most of the modelling and control problems in robotics, especially in conjunction with FL and GAs. Thus, based on the above properties and features of ANNs (both FFNN and RNN), there are many ways in which the ANNs can be used for control augmentation for robots and MIAS: (i) conventional control can be aided by RBFNN controller (radial basis feed-forward NN with variable Gaussian functions) for online learning; (ii) ANN can attempt to compensate for uncertainty without explicitly identifying changes in the robot model; (iii) the ANN non-linearity can be truly made adaptive, by changing its slope and gain, and used in the desired dynamics block of the robot's/vehicle's controller and (iv) sensor and actuator failure detection, identification, isolation and management, including reconfiguration of MIAS controllers. Some of the many benefits that can be derived from using ANNs for modelling and control would be: (a) the robot/MIAS controller would become more robust and insensitive to the plant/system parameter variations and (b) the online learning ability of the ANNs would be very advantageous in handling certain unexpected behaviour.

1.6 FL AND FUZZY SYSTEM

FL was introduced by Lotfi Zadeh. The idea is to process data with allowance of partial/varying set membership instead of classical crisp membership. In classical set theory, the decisions are binary, 0 or 1, yes or no, on or off. The intermediate situations and decision are not allowed. The set membership is defined by a membership function. Once the function is defined it becomes crisp any way. But, it does introduce a range of values rather than one value, say yes or no, zero or one. Thus, the FL deals with noisy, imprecise, vague and /or ambiguous data or knowledge using a set membership function that has intermediate values, unlike classical set membership function (Figure 1.4). The FL provides for higher reliability in handling this imprecise information in the data. We do not require precise numerical inputs to the FL-based control systems. FL induces an empirical model and is based on the simple rule-based approach: IF x THEN y. Of course, compound rules are possible and often used. Here, no numbers but fuzzy set memberships, and fuzzy variables are used. For example, IF temp is too cool AND time is elapsed THEN cool fast. FL is used for imprecise but descriptive situations. FL-based theory can mimic human logic, for example, control strategy, the way a human expert would react to a changing situation and to control it. FL-based designs are or can be made robust, require little tuning, and can model non-linear processes and systems. The FL is used when these systems are generally very difficult to mathematically model, or when the model is very imprecise [1,3,5].

FL in fact can model any continuous function or system (its dynamics), and the quality of the approximation (if required) would depend on the quality of If–Then (fuzzy) rules. These rules can be formed by the domain experts, and/or an ANNs can be used to learn the rules from the problem-specific data. Alternatively one can use ANFIS (adaptive neuro-fuzzy inference system) for designing and specifying suitable fuzzy membership functions. As such the fuzzy engineering is a function approximation with fuzzy systems (which use FL and fuzzy operators). For example, if the FL is used in a conventional washing machine design, the m/c would save the energy, washing detergent and wear and tear on the clothes. This fuzzy approximation rests on mathematics of function approximation and statistical learning theory (SLT). The fuzzy system is also a natural and intuitive way to turn speech and measured actions into functions that approximate the hard tasks for which we need to build the fuzzy-logic-based systems, thereby making the difficult tasks as somewhat softer, cost-effective and time-saving tasks.

The basic unit of fuzzy FA is the 'If…Then' rule: If the 'robot is encountering the obstacle', Then 'turn right or left'. Fuzzy system is a set of If…Then rules that maps input sets like 'encountering' to output sets like 'move to right or left'. In additive fuzzy system (AFS), each input partially fires all rules in parallel and the system acts as an associative processor as it computes the output $F(x)$. The system then combines the partially fired rules and then puts fuzzy sets in a sum and converts this sum to a scalar or vector output. Thus, a match-and-sum fuzzy approximation can be viewed as a generalized AI expert system or as a (neural like) FAM (fuzzy associative memory)

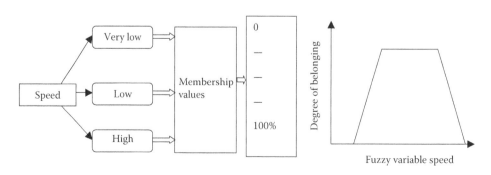

FIGURE 1.4 FL concept for speed variable.

structure that would act like an AI-based agent system (AIAS). The AFS are proven universal approximators for rules that use fuzzy sets of any shape and are computationally quite simple. A fuzzy variable is one whose values can be considered labels of fuzzy sets: pressure or tempera-ture $\rightarrow$ fuzzy variable $\rightarrow$ linguistic values such as very low, low, medium, normal, high, very high and so on $\rightarrow$ membership values (on the universe of discourse—Pascal or $°C$). The dependence of a linguistic variable on another such variable is described by means of a fuzzy conditional state-ment: R: If S1 (is true), then S2 (is true) Or S1 $\rightarrow$ S2. More specifically one can have: (a) If the load is small Then torque is very high, (b) If the error is negative-large, Then output is –ve large and (c) composite conditional statement as: R1: If S1 Then (If S2 Then S3) is equivalent to: R1: If S1 Then R2 AND R2: If S2 Then S3.

The knowledge necessary to control any plant or dynamic system is usually expressed as a set of linguistic rules of the form: If (cause) ... Then (effect). These are the rules with which new human operators are trained to control an actual process or an industrial plant and these rules constitute the knowledge base of the system. All the rules necessary to control a plant might not be elicited or in fact or known in advance. It is therefore necessary to use some technique capable of inferring the control action from these available rules:

A. Generalized Modus Ponens (GMP) $\rightarrow$
   Premise 1: x is A′;
   Premise 2: If x is A Then y is B;
   then consequence $\rightarrow$ y is B′.
   This is the forward data driven inference chain used in all fuzzy controllers, that is, given the cause try to infer what is the effect?

B. Generalized Modus Tollens (GMT) $\rightarrow$
   Premise 1: y is B′;
   Premise 2: If x is A Then y is B;
   then consequence $\rightarrow$ x is A′.
   This is directly related to backwards goal-driven inference mechanism. Infer the cause that leads to a particular effect.

A simple procedure to apply FL-based concepts to a control system is: (i) consider inputs, out-puts, system control and failure modes, if any; (ii) determine input/output relationships; (iii) use minimum number of input variables (error, i.e., control difference); (iv) the change-of-error (i.e., error derivative that can be obtained by a finite difference); (v) set up a system as a number of IF–THEN rules; (vi) create membership functions, giving meaning to input/output terms; (vii) create pre-/post-processing terms; (viii) test system to evaluate its performance, and if required tune the laws (rule base) or the membership functions and (ix) finally evaluate the design again and then release the control design when satisfactory results have been obtained. Much of the above is cap-tured by FIES [3,5].

The membership function is a graphical representation of magnitude of participation of each input (i.e., the degree of belongingness of the independent variable) to the qualifying variable, see Figure 1.4. It means that the membership function associates weighting or weight age with each input. It assigns the degree of belongingness to that variable input. It defines overlaps between inputs and determines output responses. These membership functions are used for fuzzification, that is, translating input values to fuzzy set memberships. Subsequently, the fuzzy inference engine/process/system (FIES) is applied that converts the intermediate outputs by considering the fuzzy implication functions and rule of aggregation/conjunctions/disjunctions. After this the final step is to defuzzify the variables, that is, to translate fuzzy output to crisp system output, that will give some numerical number. The shape of membership function is usually triangle, or trapezoi-dal, of course, a variety of shapes are possible, like Bell shape, saturation with slope and so on.

The height of the membership function shape is normalized to 1. Thus, the membership function value would vary from 0 to 1, taking any value between these two extremes, which then define the crisp set. Thus, the crisp set is a special case of the Fuzzy membership function. The width of the base of membership function could vary. Likewise, it should be possible to consider several results of the crisp set as special cases of the results from the FL/fuzzy set theory. The sum of all the membership values for a given case study and for a certain input value is always 1.0. The different value range for different input variables (for error and error rate) is feasible. The membership values, the antecedents are evaluated for the produced conclusion, consequent. The inferences are based on continuous membership values, from 0 to 1, this is the major deviation from Boolean logic, which uses true or false, 1 or 0. The logical sum is built using membership values in combination with rule-base matrix, fuzzy associative memory (FAM); the minimum of all antecedent values AND-part for output is used. The latter is from the If…Then fuzzy rule. The output combination, that is, the effects of these rules, is done by finding out the firing strength of each rule, and then by combining logical outputs for each rule before defuzzification process is completed. The process of combining rule outputs uses the MAX-MIN, MAX-DOT (MAX-PRODUCT), AVERAGING or ROOT-SUM-SQUARE paradigms [1,2,5,7]. These output combination rules are as follows:

1. MAX-MIN process: the magnitudes of all rules are tested, then highest one is selected, then horizontal coordinate of fuzzy centroid is used as output; this method does not combine individual results.
2. MAX-DOT process: scale each member function to fit under its peak value, take the horizontal coordinate of the fuzzy centroid of the composite area as the output. This shrinks all member functions to their peaks equal the magnitude of the respective function; this method produces a smooth, continuous output combining all active rules.
3. AVERAGING process: each function is clipped at the average value and the fuzzy centroid of the composite area is computed; this method does not give increased weight if multiple rules generate the same output member.
4. ROOT-SUM-SQUARE process: combination of several approaches—scale functions to respective magnitudes with root of sum of squares, compute fuzzy centroid of composite area; this method gives a good weighted influence to all firing rules.

The process of defuzzification is actually the defuzzification of combined results of inference process, that is, the output of the FIES. The result will be a crisp-numeric output. Mostly this is accomplished by the calculation of fuzzy centroid, that is, the centroid method, the process is: (a) to multiply the weighted strength of each output member function by the respective member function center points, (b) to add these values and (c) to divide area by the sum of the weighted member function strength. This crisp value is further used for getting error and then fuzzified at the input-side and the process continues. As we had mentioned earlier some tuning of the designed fuzzy system might be required in order to obtain the specified or desired performance. This is normally accomplished by changing rule antecedents, changing rule inference/conclusions (further broadening or narrowing the meanings and interpretations, depending upon the case study and application in the hand), changing/eschewing the centers of the input/output membership functions, adding additional membership functions, and/or by adding more rules, or even pruning the rules, and by avoiding the contradictions in the rules.

1.6.1   Fuzzy Inference Engine/System

In a fuzzy inference system (FIE/S) the defining parameters of membership function are tuned (adjusted) using either a BP algorithm alone, or in combination with a least-squares-type method.

This ANFIS procedure allows the fuzzy system learn the nature of fuzzy membership functions from the empirical input/output data of a system under consideration/design or analysis. It means that the membership function is adaptively tuned and determined by using ANN and the I/O data of the given system. In the ANFIS system the fuzzy membership structure and parameter adjustment are carried out as follows: (a) computation of these parameters (or their adjustment) is facilitated by a gradient vector, which provides a measure of how well the fuzzy inference system is modelling the empirical input/output data for a given set of parameters and (b) once the gradient vector is obtained, any of several optimization routines could be applied in order to adjust the parameters so as to reduce some error (measure of error usually defined by the sum of the squared difference between actual and desired outputs).

Consider a fuzzy system having rule base [3]: (1) If u_1 is A_1 and u_2 is B_1, then $y_1 = c_{11}u_1 + c_{12}u_2 + c_{10}$ and (2) If u_1 is A_2 and u_2 is B_2, then $y_2 = c_{21}u_1 + c_{22}u_2 + c_{20}$, where u_1, u_2 are crisp/non-fuzzy inputs, y is desired output, then let the membership functions of fuzzy sets A_i, B_i, $i = 1, 2$ be μ_{A_i}, μ_{B_i}.

Two inputs (u_1, u_2), and four fuzzy sets (A1, A2, B1, B2) (and 'pi' is the product operator) combine the 'AND' Process on A, B; C_{ij}-output membership function parameters: a, b, c and C_{ij}. That is to say that an equivalent pre-defined 'model' sort of is obtained by this ANFIS and it provides a 'predicted' output signal for given input signals. The steps involved in ANFIS are:

1. Each neuron i in layer 1 is adaptive with a parametric activation function. Its output is the grade of membership function to which the given input satisfies the membership function, that is, μ_{A_i}, μ_{B_i}. An example of a membership function is a generalized bell function: $\mu(u) = (1/1 + (|u - c/a|^{2b}))$, where $\{a, b, c\}$ are known as premise parameters.
2. Every node in layer 2 is a fixed node, whose output (w_i) is the product of all incoming signals $w_i = \mu_{A_i}(u_1)\mu_{B_i}(u_2)$, $i = 1, 2$.
3. Output of layer 3 for each node is the ratio of the ith rule's firing strength relative to sum of all rule's firing strengths $w_i = \mu_{A_i}(u_1)\mu_{B_i}(u_2)$, $i = 1, 2$.
4. Every node in layer 4 is an adaptive node with a node output: $\bar{w}_i y_i = \bar{w}_i(c_{i1}u_1 + c_{i2}u_2 = c_{i0})$, $i = 1, 2$, where $\{c_{i1}, c_{i2}, c_{i0}\}$ are known as consequent parameters.
5. Every node in layer 5 is a fixed node which sums all incoming signals: $y_p = \bar{w}_1 y_1 + \bar{w}_2 y_2$, where y_p is the predicted output.

ANFIS Learning Algorithm: When the premise parameters are fixed, the overall output is a linear combination of the consequent parameters. In symbols, the output y_p can be written as

$$y_p = \bar{w}_1 y_1 + \bar{w}_2 y_2$$
$$= \bar{w}_1(c_{11}u_1 + c_{12}u_2 + c_{10}) + \bar{w}_2(c_{21}u_1 + c_{22}u_2 + c_{20})$$
$$= (\bar{w}_1 u_1)c_{11} + (\bar{w}_1 u_2)c_{12} + \bar{w}_1 c_{10} + (\bar{w}_2 u_1)c_{21} + (\bar{w}_2 u_2)c_{22} + \bar{w}_2 c_{20}$$

which is linear in the consequent parameters c_{ij} ($i = 1, 2; j = 0, 1, 2$).

A hybrid algorithm adjusts the consequent parameters c_{ij} in forward pass and premise parameters $\{a_i, b_i, c_i\}$ in a backward pass. In forward pass, the network inputs propagate forward until layer 4, where the consequent parameters are identified by the least-squares method. In backward pass, the error signals propagate backwards and the premise parameters are updated by gradient descent method. The MATLAB Functions Used for FIS Generation and Training are given in the following steps:

Step 1. Generation of initial fuzzy inference system (block1): INITFIS = genfis1(TRNDATA), where, 'TRNDATA' is a matrix with $N + 1$ columns where the first N column contain data for each FIS input, and the last column contains the output data. INITFIS is single-output fuzzy inference system.

Step 2. Training of fuzzy inference system (block2): [FIS,ERROR,STEPSIZE,CHKFIS, CHKERROR] = anfis (TRNDATA, INITFIS, TRNOPT, DISPOPT, CHKDATA), where, vector 'TRNOPT' is used to specify training options, vector 'DISPOPT' is to specify display options during training, 'CHKDATA' (same data format as of training data) is to prevent over fitting of the training data set, and 'CHKFIS' is the final tuned fuzzy inference system. The Type 1 FL/S cannot handle certain uncertainties: (i) the meanings of the words that are used in the antecedents and consequents of rules can be uncertain (words mean different things to different people), (ii) consequents may have a histogram of values associated with them, especially when knowledge is extracted from a group of experts who do not all agree, (iii) measurements that activate a Type-1 FL/s may be noisy and therefore uncertain and (iv) the data that are used to tune the parameters of a Type-1 FLs may also be noisy. All these uncertainties translate into the uncertainties about fuzzy set membership functions. But in Type 1 FL the membership functions are themselves crisp and hence it cannot handle the uncertainties described above.

To summarize, the FL and FS have the features: (a) they are based on multivalued logic as against the bi-valued crisp logic; (b) they do not have any fixed architecture like NN; (c) they are based on certain rules, fuzzy If … Then rules that need to be *a priori* specified, decided; (d) FL is a machine learning/intelligent paradigm in which the desired behaviour is specified in rules by way of incorporating an expert's, design engineer's experience; (e) FLS deals with approximate reasoning in uncertain, in fact vague situations where truth is a matter of degree (of representation) and (f) FLS is based on the computational mechanism, algorithm, with which decisions can be inferred despite the incomplete knowledge, this being the process of inference engine, FIES. Thus, there are many ways FL/systems can be used to aid/augment the robot/MIAS control systems: (a) FL will approximately replicate some of the ways a human operator or human controller might respond to the machine, that is, robot's behaviour or movement, that is not behaving as expected due to a damage or failure; (b) the fuzzy If … Then rules can be used to create non-linear responses, if required from the controller or the dynamic system to bind the large errors; (c) to incorporate and implement the complex non-linear strategies based on the system design engineer's experience and intelligence within the control law by heuristics; (d) in case of rapidly changing dynamic situation that the vehicle might experience, adaptive fuzzy gain scheduling (AGS) strategy using the fuzzy relationships between the scheduling variables and controller parameters via FAM (fuzzy associated/association memory) can be adopted and (e) FL-based adaptive tuning of Kalman filter or any filter that is adopted for adaptive estimation and control for robots/MIAS.

Specifically, FL/Fuzzy Systems-based concepts can be applied to a number of control problems: robot motor control, vehicle driving, balancing of robot arms and walking in the context of mobile robots and MIAS.

1.7   GENETIC ALGORITHMS

GAs are new paradigms for solving optimization problems is science and engineering. They often provide global optimal solutions and are good at taking large, potentially large search spaces and navigating them, looking for optimal combinations of things, solutions [1,6,7]. Often such solutions cannot be found in our lifetime by using conventional approaches. One of the reasons is the fact that in the GA approach we search one solution in search space simultaneously. Another reason is that we use certain genetic operators that are useful in providing variation of the possibilities and we do not get restricted in our search space. Finally, we use the fitness function to determine the appropriateness of the intermediate results and searches so that we carry forward only the best strategies and solutions to complete our search process. The GAs are based on the principle of evolution and natural selection of the biological species and provide robust and lasting solutions to difficult problems

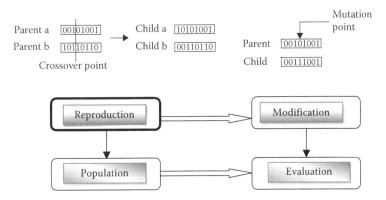

FIGURE 1.5 The GA operators and procedure: Crossover and mutation. (Notice in the XOVER first three digits are exchanged in this example; one can exchange the other digits instead.)

in a simple manner. The point is that inherent mechanism to obtain the optimal/global solution is very simple and yet very powerful. This is the basic and encouraging feature of the GAs. Hence, the GA is a directed search algorithm based on the mechanics of biological evolution. GA is useful in optimization and control of processes to understand the adaptive behaviour of the natural and man-made systems. By using GA we can design and build artificial systems, both HW and SW that would retain robustness of the naturally evolved systems. The GAs provide efficient, effective means for optimization and machine learning applications in robotics and many mobile intelligent systems. Interestingly the GAs are finding increasing applications in business management also.

The process of the GA is: (a) randomly initialize a population of chromosomes, (b) evaluate the fitness of each chromosome, (c) construct the phenotype (say, simulated robot), (d) corresponding to the encoded genotype (chromosome), (e) evaluate the phenotype (say, measure the simulated robot's walking ability in order to determine its fitness), (f) remove chromosomes with low fitness values, (g) generate new chromosomes, using certain selection schemes and genetic operators. So, in GA there are three most important functions: (i) fitness function, (ii) selection scheme and (iii) genetic operators like crossover and mutation. Figure 1.5 shows the cross over and mutation points and the process. The GA process is apparent from Figures 1.5 and 1.6. The components of a GA are: (i) an encoding technique via a gene, a chromosome; (ii) initialization procedure, that is, create samples or population; (iii) evaluation function, that is, fitness function; (iv) selection of parents, that is, reproduction; (v) use genetic operators, like mutation and, recombination; and parameter settings. The population could be (a) bit strings, (b) real numbers, (c) some permutations of elements, (d) list of some rules, (e) in genetic programming (the program elements) and (f) any data structure that can be used in the GA procedure defined in Figure 1.5. The reproduction process gives rise to the children where for the parents are selected at random with selection chances biased in relation to the chromosome evaluations. Then the chromosome modification is carried out by stochastically triggering them and using suitable genetic operators like crossover and mutation. The crossover is called recombination. The local modification of mutation is shown in Figure 1.6. The idea is to make a small movement in the local or global search space, and restore

FIGURE 1.6 Local modification in mutation in GA.

information, if it were lost in the population. The crossover as shown in Figure 1.5 is a very crucial operation in GA: (i) it accelerates search greatly early in evolution of a population and (ii) it yields effective combination of schemata (subsolutions on different chromosomes). In the process of evaluation the evaluator decodes a chromosome and assigns it a fitness measure/value. In the generational GA the entire populations are replaced with each iteration, and in the steady-state GA a few members are replaced in each generation. There are many important aspects in GA to be taken care of: (i) choosing basic implementation issues of representation, population size, mutation rate, selection, deletion policies, crossover and mutation operators; (ii) termination criteria; (iii) performance and scalability and (iv) the solution is only good as the evaluation function decides.

1.7.1 TYPICAL GA PROCESS

The long stretches of DNA that carry the genetic information of an individual (or a system) to build an organism are called chromosomes. In GA, the chromosomes represent encoding of information in a string of finite length and each chromosome consists of a string of bits (binary digit; 0 or 1), or it could be a symbol from a set of more than two elements; these strings could be real-valued numbers also. These chromosomes consist of genes and each gene represents a unit of information and it takes different values that are called alleles at different locations, which in turn are called loci. These strings which are composed of features or detectors, assume values like 0 or 1, which are located at different positions in the string. The total system is called the genotype or structure. The phenotype results when interaction of genotype with environment takes place. Thus, the GAs operate on population of possible sample solutions with chromosomes. The population members are known as individuals (meaning here a sample) and each sample individual is assigned a fitness value based on some objective function or cost function. Better solutions have higher fitness values and weaker solutions would have lower fitness values. In the initialization/reproduction phase, by randomly selecting information from the sample-search space and encoding it, a population of possible initial solutions is created. In the reproduction stage the individual strings are copied as per their fitness values. The strings with a greater fitness value are given higher probability of contributing one or more offspring to the next generation. In a crossover operation a site/location is selected randomly along the length of the chromosomes, and each chromosome is split into two pieces at the crossover site/location. The new samples are formed by joining the top piece of one chromosome with the tailpiece of the other. This crossover operation can be carried out in many different ways: it can be performed in the same string also, by swapping strings that are split at some location.

In Mutation operation, a single-bit in a string is changed at a random location, that is, '0' is changed to '1' or '1' is changed to '0'. In a string of real-valued numbers a small change is affected in that number. The main idea is to break monotony and add a bit of novelty, that is, obtain/provide new information; the mutation operation would help gain information not available to the rest of the population and thereby lends diversity to the population. Each iteration in the GA optimization process is called a generation, and in each generation pairs are chosen for crossover operation, the fitness is determined, and mutation is carried out during crossover operation (during or after has a subtle distinction). Then a new population evolves that is carried forward. The individuals may be fitter or weaker than some other population members; so the members are ranked as per their fitness values. In each generation, the weaker members are allowed to wither out/die out/discarded and the ones with good fitness values are continued to take part in further genetic operations. The net effective result is the evolution of the population toward the global optimum. In many practical optimization problems, the goal is to find optimal parameters to increase the production and/or to reduce the expenditure/loss; that is to get maximum profit by reorganizing the system and its parameters that affect the cost function. Since, this in effect, reflects on the cost, it is represented by the cost function. A carefully devised and convergent computational algorithm, like GA would eventually find an optimum solution to the problem. The parameters of the system that decide the cost are termed as decision variables. The search space is a Euclidean space in which parameters take different values and each point in the space is a probable solution.

1.7.1.1 Stopping Strategies for GAs

If the population size is fixed, then more generations might be needed for the convergence of a GA to the optimal and global solution. One way is to track the fitness value for no further improvement, that is, if the fitness value does not change much then one can stop the GA operations. As the algorithmic steps progress, a situation would occur wherein we need large number of generations to bring about a small improvement in the fitness value. One can define predetermined number of generation/ iterations to solve the problem. Also, insignificant change in the norm of estimated parameters/state can be tracked for a few consecutive iterations before stopping the search. It must be possible to do an effective search if one exploits some important similarities in the coding used in GAs. Another way is to evaluate gradient of the cost function and use the conventional approach for assessing the quality of the estimates for their convergence to true values. It is possible to use GA with gradient-based approach for evaluating the estimation accuracy as is done for other conventional estimation methods. Again, here as is true with all the other parameter estimation methods, the matching of time histories of the measured data and model responses is a necessary, but not sufficient, condition. Also an increase in the number of samples would generally increase the success rate.

1.7.1.2 GAs without Coding of Parameters

GAs become more complex because of (long-strings) coding the chromosomes, especially for more complex problems. We can instead use the real numbers and still use GAs on these numbers for solving optimization problems. Major change is in the crossover and mutation operations: averaging the two samples, that is, the two sets of parameter values can perform the crossover operation; and one can try varieties of averaging operations. After the crossover, the best individual is mutated, and in this operation a small noise is added. Assume that two individuals have β_1 and β_2 as numerical values of the parameters; then after crossover, we obtain the new individual as $(\beta1 + \beta_2)/2$, using standard averaging operation. For mutation we can have $\beta_3 = \beta_1 + d * v$, where d is a constant and v is a number chosen randomly between -1 and 1. Thus, all the GA operations can be performed by using real numbers like 4.8904, and so on, without coding the samples in the strings of '0' and '1'. This feature is extremely well suited for many engineering applications: filtering, parameter estimation, control, optimization and signal processing.

1.7.1.3 Parallelization of GAs

As such, the GAs are powerful and yet very simple strategies for optimization problems and can be used (a) for multimodal, multidimensional and multiobjective optimization problems and (b) in business and related fields. However, despite the fact that the computations required in GA operations are as such very simple, they become complex as the number of iterations grows and when the problem size also grows. This puts heavy demand on the computational power. The GA procedures can be parallelized and the power of the parallel computers can be used. Since GAs can work on population samples simultaneously, their natural parallelism can be exploited to implement GA processes on parallel computers.

1.7.1.4 Parameter Estimation Using GA

Most of the parameter estimation methods are based on the minimization of some quadratic cost function resulting in utilization of the gradient of the cost function for obtaining the estimated parameters. However, the application of GA to parameter estimation problem does not need utilization of the gradient of the cost function. Let the system equation be as follows [2]:

$$z = H\beta + v; \quad \hat{z} = H\hat{\beta} \tag{1.33}$$

The quadratic cost functional is given as

$$E = \frac{1}{2}\sum (z - \hat{z})^T (z - \hat{z}) = \frac{1}{2}\sum (z - H\hat{\beta})^T (z - H\hat{\beta}) \tag{1.34}$$

Instead of using a gradient-based algorithm one can use GA for estimation of the parameters of the model in Equation 1.33 by considering Equation 1.34 as a fitness function (it could be directly the numerical value of 'E' or it inverse '$1/E$'). The fitness value/function that can be used is given by

$$\left[\frac{1}{2}\sum_{k=1}^{N} (z(k) - \hat{z}(k))^T \hat{R}^{-1} (z(k) - \hat{z}(k)) + \frac{N}{2}\ln(|\hat{R}|)\right]^{-1} \tag{1.35}$$

Here, $\hat{R} = 1/N\Sigma_{k=1}^{N}(z(k) - \hat{z}(k))(z(k) - \hat{z}(k))^T$ is the measurement noise covariance matrix.

There are several good aspects/benefits of using GA: (i) concept is easy to understand; (ii) modularity, separate from application; (iii) supports multiobjective optimization; (iv) good for noisy environments; (v) always an answer/solution, it gets better with time/iterations; (vi) inherently parallel (like ANNs are), and hence easily distributed, can be easily programmed on parallel computers; (vii) many ways to speed up and improve a GA-based application as more knowledge about problem is obtained; (viii) easy to exploit previous obtained or alternate solutions; and (ix) flexible building blocks/modularity for hybrid applications.

1.7.2 Genetic Programming

In GP the idea is to evolve a program instead of bit-string for which Lisp program structure is best suited. The GP operators can do simple replacements of subtrees and all the generated programs can be treated as legal without any syntax errors. The GP process is: (1) randomly generate a combinatorial set of computer programs; (2) perform these steps iteratively until a termination criterion is satisfied: (a) execute each program and assign a fitness value to each individual, (b) create a new population with the steps: (i) reproduction—copy the selected program unchanged to the new population, (ii) crossover—create a new program by recombining two selected programs at a random crossover point and (iii) mutation—create a new program by randomly changing a selected program and (3) the best sets of individuals are deemed the optimal solution upon termination. The GP can be used for certain robot applications.

1.8 CLASSIFICATION AND IDEA OF COMBINING ANNs

The idea of combining ANNs is to develop n independently trained NNs which will have certain relevant features. These ANNs then can be used for classification of a given input pattern by using combination methods to decide the collective classification [1]. Two general methods approaches to combining multiple networks are based on: (i) a fusion technique and (ii) a voting technique. For the method based on the fusion process, the classification of an input X is based on empirical measurements, representing the probabilities that X comes from each of the c classes under the condition X. For the combined scheme, each network k estimates by itself approximations of those true values. One approach to combine the results on the same X by all n networks is to use the average value. This is taken as a new estimation of combined network and is considered an averaged Bayes classifier. To improve the estimation, the combiner is given the ability to bias the outputs based on *a priori* knowledge about the reliability of the NWs. The second method is based on voting technique. This considers the result of each network as an expert-valued judgment. A few voting procedures are adapted from group-decision-making theory of (i) unanimity, (ii) majority and plurality.

1.8.1 FL-BASED METHOD TO COMBINE ANNs' OUTPUTS/RESULT OF CLASSIFICATION

Yet another method utilizes the fuzzy integral (FI) for combining NNs, in fact the outputs of the ANNs. The FI is a non-linear functional that is defined with respect to a FM (fuzzy measure), that is, $g\lambda$-fuzzy measure [1,5,7]. It is a good measure to combine the results of multiple sources of information. Here, 'g' is not necessarily additive, the property of monotonic behaviour is substituted for the additive property of the conventional measure. From the definition of a fuzzy measure, the so-called $g\lambda$-fuzzy measures were introduced that satisfy an additional property, which specifies that the measure of the union of two disjoint subsets can be directly computed from the component measures. Using the notion of fuzzy measures, the concept of the fuzzy integral was developed. This FI is a non-linear functional that is defined with respect to a fuzzy measure, that is, $g\lambda$-fuzzy measure. Specify X as a finite set, and let $h: X \rightarrow [0,1]$ be a fuzzy subset of X. Then FI over X of the function h with respect to a fuzzy measure g can be defined [1,3]. The $h(y)$ is the measure of the degree to which the concept h is satisfied by y. The min of $h(y)$, then measures the degree to which the concept h is satisfied by all the elements in 'E'. The value $g(E)$ is then a measure of the degree to which the subset of objects 'E' satisfies the concept measured by 'g'. The value obtained from comparing these two quantities in terms of the 'min' operator indicates the degree to which 'E' satisfies both the criteria of the measure 'g' and 'min' of $h(y)$ over 'E'. Finally, the 'max' operation is used to take the biggest of these terms. The FI can be interpreted as finding the maximal grade of agreement between the objective evidence and expectation. We have $\Omega = \{\omega 1, \omega 2, \ldots, \omega c\}$ as the set of classes of interest. Even each of ωi may be a set of classes on its own. We have $Y = \{y1, y2, \ldots, yn\}$ as a set of ANNs, and A is the object chosen for recognition. Also, we have $hk:Y \rightarrow [0,1]$ the partial evaluation of the object A for class ωk, that is, $hk(yi)$ is an indication of how certain one can be in the classification of object A to be in class ωk using the network yi. The '1' indicates the absolute certainty that the object A is really in class ωk and 0 implies the absolute certainty that the object is not in the class ωk.

1.8.2 GA-BASED METHOD TO COMBINE ANNs' OUTPUTS/RESULTS

In general, the GA operators are used to create new individuals from an initial sample set for a given optimization problem [1,7]. Unlike this method, we may want to optimize the weights of the ensemble networks (i.e., several ANNs) with GA operations. In that case a string must encode $n \times c$ real-valued parameters thereby obtaining an optimal combination of the coefficients for combining NNs; here each coefficient is encoded by 8 bits and scaled between 0 and 1. Then GA is used to manipulate the most promising strings in its search for improved solutions. GA operates through a simple cycle of stages: (a) creation of a population of real-valued strings, (b) evaluation of each string with recognition rate on training data, (c) selection of good strings and (d) genetic/mutation manipulation to create the new population of strings. This latter operation is carried out as: (i) one-point crossover with probability 0.6 and (ii) standard mutation with probability 0.01, wherein these numbers could vary depending upon the problem at hand. This cycle can stop as the recognition rate gets better no more. In effect, the GA approach takes pieces of weighting coefficients to combine NNs as such strings. These ANNs could have been the results of training for object classification or for a multirobot coordination scenario.

1.8.3 GA: FL HYBRID METHOD

In the literature, the FL and GA have been proposed for achieving some aspect of the intelligent system [7]. However, these soft-computing technologies have certain important differences, which have prompted the AI researchers to try combining them to produce more powerful systems, the idea being the utilization of strengths of the two methodologies. Each method has its own pros and cons. To produce more powerful system, some integration and synthesis approaches are needed. ANNs can be used as a baseline system, because these are well recognized as powerful I/O mapping operators, but human operators cannot easily incorporate their knowledge about the problem into the ANNs.

Hence, FL would be useful by way of incorporating this knowledge in the fuzzy If…Then rule base (ITRB). Thus, the FL gives a possibility to utilize top-down knowledge from any designer of the system. Thus, Human operators can enhance the ANNs by incorporating their heuristic knowledge also with fuzzy membership functions; wherein these functions are modified through learning process ably supported by ANN-learning algorithm for fine tuning of the defining parameters of the fuzzy membership functions. Once can use here ANFIS. After the learning is accomplished, the human operators may be able to understand the acquired rules much better. On the other hand, GA is a powerful tool for structure optimization of FL and ANNs which provide/support the evaluation functions for the GA. Thus, the complementary nature of FL and GA leads us to believe that a further refined genetic-fuzzy-neural (GFN) system will definitely improve the state-of-the-art pattern recognition/classification and other used of control and optimization for MIAS applications.

1.9 SOME POSSIBLE APPLICATIONS OF NEURO-FUZZY–GA TO ROBOTICS

In this section we discuss possible combinations of neuro-fuzzy–GA that might be useful in MIAS/ robotics applications.

1.9.1 ANN-Based Camera: Robot Coordination

This system is that of fixed cameras and a robot arm wherein the visual system should identify the target as well as determine the visual position of the end effector (robotic) [8]. The target position is Xtarget and the visual position of the hand is Xhand and these are inputs to the NN controller $N(\cdot)$. This NN controller generates a joint position θ for the robot: $\theta = N(\text{Xtarget,Xhand})$. The neurally generated θ is compared with the optimal $\theta 0$ generated by a fictitious perfect controller $R(\cdot)$: $\theta 0 = R(\text{Xtarget,Xhand})$. The NN learning makes an output 'close enough' to $\theta 0$. The situations that arise are: (i) generating appropriate learning samples which should be in accordance with $\theta 0 = R(\text{Xtarget,Xhand})$. This is non-trivial task. This is because $R(\cdot)$ is often an unknown function. Here, a form of self-supervised or unsupervised learning strategy might be required; and (ii) constructing the NN mapping from the available learning samples, here the input space is of a high dimensionality, and the samples are randomly distributed. Eventually a good and efficient learning algorithm might be required.

1.9.2 ANN for Robot Path Planning

ANN can be applied to path planning and intelligent control of an autonomous robot which is required to move safely in a partially structured environment (the environment may involve any number of obstacles of arbitrary shape/size). Some of these obstacles might be moving. In ref. [9], an approach to solve this path/motion-planning problem for a mobile robot control using ANN is studied. A collision-free path for moving robot among obstacles is constructed based on two ANNs: (i) one NW is used to determine the free space using ultrasound range finder data and (ii) the second NW finds a safe direction on the path in the workspace while avoiding the nearest obstacles. The approach is based on using the sensor data from the environment, and making the first NW to learn these situations to obtain a free segment of space for safe path as output. This NW is the principal component analysis network (PCANW), which combines unsupervised and supervised learning in the same topology. The second NW is a multilayer perceptron (MLP) typically trained with static BP learning algorithm. The aim of this NW is to determine the robot azimuth for the next move from the output of first network and from the specified goal coordinates. This NW contains three layers: input, hidden and output. For this NW we give the known free space segment as the output of the first NN and the goal segments in which the coordinates of the robot goal position should be situated. Then from the output layer of this NW we obtain the information about robot motion direction (azimuth) in the next step which in turn is given to the control unit.

1.9.3 Integrated GA-FL Synthesis

One GA-FL architecture for intelligent control and obstacle avoidance has these basic steps [1,7]: (i) the input/output spaces (I/O/S) of the robot system to be controlled is divided into fuzzy regions; (ii) the I/O regions are encoded into bit strings; (iii) GA is used as an optimization-based learning procedure to generate a set of fuzzy rules; (iv) these generated fuzzy rules are used for determining performance, and a fitness value based on fitness function is assigned; (v) a negative penalty is assigned to the fitness function if a robot enters into a wall or an avoidance region; (vi) if the stopping criterion is not fulfilled, then go to step (iii) and (vii) determine a mapping from the I/S to the O/S based on the combined fuzzy rule base using a defuzzification method. Important features of this GA-FL system are that [1,7]: (a) it is a self-learning adaptive method like ANN; (b) it is able to learn the fuzzy control rules without any prior knowledge, it needs to have only the knowledge of the performance of the system; (c) it can be used for building FLCs (Fuzzy Logic Controller) for obstacle avoidance and (d) it is flexibility in choosing the If ... Then rules for the FLC.

1.10 ARTIFICIAL INTELLIGENCE

In the field of AI the attempt is to make computers more intelligent and to better understand human intelligence via four major approaches [10]: (i) study the thought process of humans to understand how humans think, (ii) observe and study the actions of the humans to understand how humans carry out various actions, (iii) try to build a mathematical or behavioural model of human to capture major aspects of one's behavioural pattern and then use this model to build AI-based systems and (iv) study further behaviour of humans from rationality point of view, that is, to capture one's rational behaviour that is beyond the normal or casual behaviour. Hence, the AI is broadly divided into four classes [10]: (i) the systems, both hardware and software (like computers, and algorithms/programs) that think like humans, (ii) the systems (HW and SW) that act like humans act, (iii) the systems/SW that think rationally like humans think and (iv) the systems that act rationally like humans.

1.10.1 Thinking Like Humans

In this class the efforts are made to make computers think like the way the humans think while solving their own problems. That is we are aiming to have machines with minds (of their own). This can be taken in full and literal sense. We would also like the machines/computers to perform activities like that humans associate and perform: (i) decision making, (ii) problem solving and (iii) learning new things and so on. The cognitive science combines the computer models and certain experimental techniques from the discipline of psychology, and cognitive psychology, to construct precise and testable theories of the working of the human mind: here it is important being able to solve problems like a human being would do.

1.10.2 Acting Like a Human

Here, we want that an AI-based computer passes a certain test. This computer/SW would/should have certain capabilities in order to be acting like a human: (i) natural language processing, (ii) knowledge representation, (iii) automated reasoning, (iv) machine learning, (v) computer vision and (vi) robotics to manipulate objects and so on. The prototype is Turing test.

1.10.3 Rationally Think or Rational Thinking

Here, we are concerned with asking questions like: what are the laws of thought and how should we think? There are certain laws of thought that are supposed to govern the operation of the mind, and this study initiated the field of logic. The logistic tradition, within the context of AI, helps build on certain programs to create intelligent systems; however, it is faced with the problems of presentation of problem description using a formal notation, and computability.

1.10.4   RATIONAL ACT

This is based on the rational agent approach—an agent is something that acts and carries out certain simple tasks or even complex ones. The computer agents should have many more attributes like: (1) operating under autonomous control, (2) perceiving their environments, (3) persisting over a prolonged time duration, (4) adapting to a change and (5) capability to take on another's goals. Thus, a rational agent acts to achieve the best outcome, or the best expected outcome, under the assumption that its impressions of the world and its convictions are correct. Rational thinking is a prerequisite for rational acting; of course it is not a necessary condition.

1.10.5   PERSPECTIVE ON AI

The areas of application of AI are: (a) systems that can understand and generate speech (nearly as humans do), (b) systems that can understand images, (c) robotics and (d) the systems that work as assistants, and agents. The methods of study and applications of AI are based on interdisciplinary relationships to and among mathematics, philosophy, psychology (computer-) linguistics, biology and engineering sciences and utilize the following approaches: (a) problem solving and searching, knowledge representation and processing, (b) action planning, (c) handling uncertain knowledge, (d) machine learning and (e) NNs. For AI in essence we are trying to develop and make algorithms, software, machines or/and systems that in some way, and perhaps in many more ways think, act and behave like reasonable humans. And it has been so far agreed upon that it is feasible to do this even partially by using ANNs, FL and GAs and including other variants of these new paradigms, or various combination of these three streams. Thus, neuro-fuzzy-GA can be intelligently employed to construct AI programs and machines which can be then called as intelligent machines. In their functionality actions they should be good at the tasks that the humans routinely perform. It will take yet decades and even centuries before we can build such machines that will nearly exactly think, rationally think, act and rationally act like humans. That will be a real miracle. However, we can build the machines (thinking!), or AI machine in the sense of probably approximately correct algorithm, probably approximately exact performance, and/or probably approximately perfect behaviour. This gives us an idea for a slightly different definition of AI which is doable and practical rather than trying to achieve a nearly impossible goal.

The new or novel perceptive for the AI should be: Make an attempt to build algorithms, software, machines and/or systems that perceive, think, reason out, make decision and act rationally in the sense of achieving 'probably approximately correct algorithm/performance/perfect behaviour' in the areas of problem solving and learning as said earlier above. Then gradually approach toward the perfection by ever refining these artefacts. This definition of AI is practically achievable and should be universally acceptable. Also, this will give definite measure of evaluating the performance of the AI systems and one can calibrate the nearness of it to the human's thinking and acting. The point of this definition is that it gives a means of measure in the sense of achieving the higher and higher probability in the sense of/for lesser and lesser approximation, higher probability for more and more correct algorithm and so on. We can see from Figure 1.7 [1] that it contains all the abstract level of AI components. ANN provides a basic structure for the AI systems. The heuristic and high-level knowledge on the plant or system which is to be supported by and designated as an AI-based system can be incorporated via FL. Then the GA can be used for optimization of any component/algorithm. Thus, these three major constituents of the AI play a major role in designing and even operation of the system. One complex and sophisticated system/SW would be automatic target recognition (ATR) which not only can be powered by all the components of AI, but also by all the levels of data

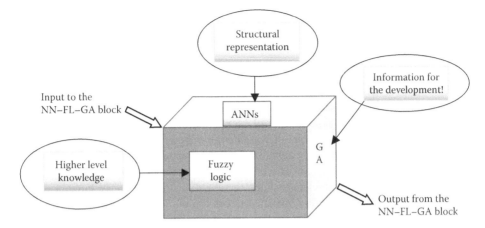

FIGURE 1.7 Building of an intelligent system ('cube', AI) using soft computing paradigms: ANN–FL–GA with their respective roles.

fusion. Thus, such an ATR system would be truly autonomous with AI for adaptive capabilities and decision making. There is a tremendous scope for applications of AI to the field of mobile intelligence autonomous systems and robotics in particular. In building such a system one can use all the constituent components of the AI: ANNs, FL and GAs as shown in Figure 1.7.

1.10.6 INTELLIGENT AGENTS

In this computer and information age we everyday encounter intelligent agents but often we are not aware of them and we utilize these agents routinely. These intelligent agents are: (i) computational algorithm, (ii) some formulae to compute a simple result from complicated data sets, (iii) Internet browsing and e-business, (iv) PCs themselves as systems of extended computations, (v) extended memories of PCs and microprocessors and (vi) automated office to assist us in carrying out certain routine tasks. Many of these agents might not be operating on the principle of AI. But they carry out some basic tasks and assist the user to carry out his/her task in some efficient manner and with some expediency so that in an overall sense the task is carried out very smartly and in a very less laborious way. Thus, AI-based agents are supposed to be highly powerful, advanced and very efficient. An agent is an intermediate assistant (assistance) that helps the used of the agent system to carry out simple or complex tasks that the user herself may not be able to do or does not know the rules of carrying out that task, for example, an insurance agent who will explain some policy rules and then get us the insurance policy for which the user will routinely pay the premium at some period times. The user is not much aware of the rules and even not worried or concerned much about the procedure of getting the policy. Thus, the agent has to be smarter than its user and must be able to carry out the assigned task that the user might not be able to do ever, and still enjoys the benefits of the agent system's efficiency. The AI-based agent system (AIAS) could be software, an algorithm or a piece of hardware system with embedded AI functional units, or even everything of these. These components interact with each other and perform an assigned job in an efficient and timely manner for the user of the agent system. Thus, the AIAS has to be flexible highly adaptable, and should have learning ability without which it would not be or remain adaptable. It should also have some built-up fault-tolerant mechanism so the service of the AIAS is consistent, persistent and reliable. The AIAS is an information agent also in the sense that it does need updated information and also it can give new information when sought, for example, browsing the internet websites. A really smart agent (AIAS) should have the properties: (i) it should be able to cooperate with the user, and with the sub agents, (ii) it should have the ability to learn and adapt to the new environment and (iii) it should have at least partial autonomy. The AIAS could be static or mobile, it might come up to action when called for, or it could be all in one and based on the input conditions it would respond appropriately. An AIAS can replace a secretary in the manager's office, can carry out many routine tasks, can be a forefront medical assistant to the main doctor or a specialist, can be receptionist in a director's office and so on. The AIAS could have any of all of the above-mentioned components of the AI definitions. Interestingly, defining an AI-based agent has the same problems as the problems faced by various definitions of AI itself. But, we can safely define an intelligent agent as an AIAS that has at least one component of the AI (definition) working in the system or is based on at least one of these aspects of AI. Thus, AIAS should be able to either act like human, think like human, rationally think like human or rationally act like human. Depending upon its level of such a competence the AIAS can be judged as smart, talented or intelligent. However, without being so much and too much concerned about its definition, we should be able to define and built an AIAS that serves the intended purpose and is able to carry out the functioned in a safe and efficient manner, thereby reducing burden the humans themselves. For example, even a classical autopilot greatly reduces the burden of a pilot during cruise flight, though here the autopilot is not intelligent at all. It performs its functionality based on what is already programmed or hard wired in the controller of the autopilot. However, if this autopilot has some aspects of learning (via ANNs), some aspects of logical decision making (via FL/ANFIS) or some evolutionary-algorithmic capacity then the new autopilot can be called basic AIAS system. One can think of human's capability and build an AIAS by incorporating more and more human like attributes into the functionality of the AIAS so that one can have the required ability and responses from thus designed and built AIAS. A true AIAS should have many capabilities of a typical smart human: (a) learning from environment, (b) learning from the feedback from the subagents, or components of the AIAS, (c) flexibility, (d) modularity, (e) adaptability, (f) responsiveness, (g) information-sharing ability, (h) fault tolerance, (i) two-way communications, (j) partial or full autonomy, (k) mobility, (l) accurate and reliable, (m) distinguishing between blurred commands and tasks directions and (n) good overall exuberance that gives security and dependability.

1.10.7 Intelligent MIAS/Robotic Systems

In a nutshell, an intelligent MIAS/robotic would use a vision system and higher-level smart sensors and have an ability to learn from the environment and avoid static/dynamic obstacles in/on its path to its goal. These systems should have 3D/4D outer level functionality, and would need more power and mobility to cope with these additional requirements. These systems should have adequate mobility, adequate computational memory, and processing capability and speed to carry out complex algorithmic tasks, and must have accurate path planning ability. The IRS (intelligent robotic system)/MIAS should have (i) ability to learn, (ii) creativity based on behaviour-based learning architecture, (iii) ability of reasoning based on expert-controller systems supervising its actions, (iv) ability to use language for interacting and communicating with the humans, (v) higher level of vision systems, 3D and infrared cameras, sonar and EM systems and (vi) FL-based knowledge incorporation for use of human experts' experience for firm and good decision making. As compared to FL type I, the Type 2 FL/S can handle rule uncertainties and measurement uncertainties [11]. In probability/statistical theory the variance provides a measure of dispersion about the mean, and similarly—the Type 2 FL/S provides similar measure of dispersion. This new dimension can be thought as related to a linguistic confidence interval, and the Type 2 has more degrees of freedom and is expected to give better performance than Type 1 FL/S. The Type 2 FL is used for [11] (a) classification of coded video streams, (b) co-channel interference elimination from non-linear time-varying communication channels, (c) control of mobile robots and decision making, (d) equalization of non-linear fading channels, (e) extracting knowledge from questionnaire surveys, (f) forecasting of time-series and function approximation, (g) learning linguistic membership grades, (h) pre-processing radiographic images, (i) relational databases, (j) solving fuzzy relation equations and (k) transport scheduling. The Type 2 FL/s seem to be applicable when [11] (i) the data-generating system is known to be time-varying but the mathematical description of the time-variability is unknown (e.g., as in mobile communications), (ii) measurement noise is non-stationary and the mathematical description of the non-stationarity is unknown (e.g., as in a time-varying SNR), (iii) features in a pattern recognition application have statistical attributes that are non-stationary and the mathematical descriptions of the non-stationarities are unknown, (iv) knowledge is mined from a group of experts using questionnaires that involve uncertain words, (v) linguistic terms are used that have a non-measurable domain. Thus, it is believed that a truly AI-based MIAS/Robotic system can emerge with synergism of advanced NNs, FL Type II systems, advanced evolutionary algorithms and AI-based agent systems.

REFERENCES

1. Cho, S.-B. Fusion of neural networks with fuzzy logic and genetic algorithm, *Integrated Computer-Aided Engineering* 9, 363–372, IOS Press, 2002.
2. Cichocki, A. and Unbehanen, R. *Neural Networks for Optimisation and Signal Processing*, John Wiley and Sons, New York, 1993.
3. Raol, J.R. *Multi-Sensor Data Fusion with MATLAB*, CRC Press, FL, USA, 2009.
4. Raol, J.R., Girija, G., and Singh, J. *Modelling and Parameter Estimation of Dynamic Systems*. IEE/IET Control Series Book Vol. 65, IEE/IET, London, UK, 2004.
5. Sugeno, M. *Fuzzy Measures and Fuzzy Integrals: A Survey, Fuzzy Automata Dec. Proc.*, Amsterdam, North-Holland, pp. 89–102, 1977.
6. Cernic, S., Jezierski, E., Britos, P., Rossi, B., and García Martínez, R. *Genetic Algorithms Applied to Robot Navigation Controller Optimization*, Buenos Aires Institute of Technology, Madero 399. (1106) Buenos Aires, Argentina, e-mail: rgm@itba.edu.ar.
7. Masoud, M. and Stonier, R.J. *Fuzzy Logic and Genetic Algorithms for Intelligent Control and Obstacle Avoidance*. Complexity International, 2, 1995. http://www.complexity.org.au/ ci/vol02/ mm94n2/mm94n2.html, Jan, 2010.
8. Anon. http://www.learnartificialneuralnetworks.com/robotcontrol.html, NeuroAI, Artificial Neural Networks, Digital Signal Processing, Algorithms and Applications, January 2011.
9. Janglová, D. Neural networks in mobile robot motion, *International Journal of Advanced Robotic Systems*, 1(1), 15–22, ISSN 1729-8806, 2004.
10. Russell, S. and Norvig, P. *Introduction to AI: A Modern Approach*, www.cs.berkeley.edu/ ~russell/intro.html, accessed July 2011.
11. Mendel, J.M. Why we need Type-2 fuzzy logic system. www.informit.com/articles/article.aspx?p = 21312-32k. May 2001.

2 Mathematical Models of Robot Motion

A. Ramachandran

CONTENTS

2.1 INTRODUCTION

To evaluate any design of a robot by simulation, the kinematic and dynamic mathematical models of the robot including its physical characteristics are required. Several related mathematical tools and algorithms are also required for: (i) spatial localization, (ii) rotation matrix, (iii) homogeneous transformation matrix and its composition, (iv) robot dynamics algorithms, (v) robot system identification, (vi) robot simulation, (vii) optimization algorithms and (viii) robot/arms/links control algorithms [1–16]. In this chapter, basic kinematics and dynamic models for robot/rigid body are studied. In general, the mathematical kinematic model is obtained to implement the simulation of, say biped's, robot kinematic, and the kinematics model is obtained by homogeneous transformation matrix applying the Denavit Hartenverg (DH) method [1]. Forward and inverse kinematic transformations also play an important role in robot kinematics. Next, we discuss terms that are used frequently in the literature on robot modelling. The kinematics is the study of motion of an object without considering the effect of forces acting on it. The term degrees of freedom (DoF) implies the number of independent position variables needed in order to completely specify motions of a robot. Robot can have an extended DoF because of the extra robot arms, as extensions. A robotic manipulator is a collection of links that are interconnected by flexible joints. At the end of the robot there is a tool or end effector. The robot workspace is defined as the volume of space which can be easily reached by the end effector. The dexterous workspace is defined as the volume of the space where the end effector can be arbitrarily oriented and/or positioned. The reachable workspace is a volume of space which the robot can reach in at least one orientation. The kinematic problem is a mathematical description of the position and orientation of the links of robot, including its legs and arms with respect to time. The position and orientation of an end effector with respect to a reference coordinate system can be computed for given joint angle and link parameters. This is referred to as forward or direct kinematic problem. The computation of the link joint angles for a given position and orientation of the end effector and link parameters is referred to as an inverse kinematic problem.

2.2 ROBOT SPATIAL LOCATION

Robot manipulation implies its movement in certain space. A coordinate system is required to describe position and movement in this specified space. To adequately locate a robot in some coordinate space, it is necessary to have some mathematical tools that allow the space localization of its points, which is called spatial localization [1]. In any single plane, the position has 2-DoF and hence a robot's position will be defined by two independent components. In 3-DoF space it will be necessary to use three components (see Figure 2.1). The location of a link of the robot is defined with respect to a fixed reference coordinate system such that each link of a robot can rotate or translate (i.e., linear displacement) with respect to the reference coordinate system or body-attached coordinate system. The DH is a minimal representation for a line which is used by the robotic engineers to help them describe the positions of links and joints unambiguously [1,16]. Here, every link gets its own coordinate system and a few rules to consider in choosing the coordinate system are: (i) the z-axis is in the direction of the joint axis; (ii) the x-axis is parallel to the common normal: $x_n = z_n \times z_{n-1}$, if there is no unique common normal (parallel z axes), then d (defined below) is a free parameter and (iii) the y-axis follows from the x- and z-axis by choosing it to be a right-handed coordinate system. The inter-link transformations are uniquely described (once the coordinate frames are specified), by the parameters: (a) θ is an angle about previous z, from old x to new x; (b) d is an offset along previous z to the common normal; (c) r is the length of the common normal; assuming a revolute joint, this is the radius about previous z and (d) α is the angle about common normal, from the old z-axis to the new z-axis.

2.2.1 POSITION OF A ROBOT

The classical form used for specifying the position of a point is the Cartesian coordinate system. Also the polar coordinates for two dimensions and the cylindrical and spherical of three dimensions are often used depending on the description required. The reference system is defined by the perpendicular axes between them specified. In 2DoF the reference system corresponding OXY is defined by two coordinated vectors OX and OY perpendicular between them with a point 'O', the origin, of common intersection (Figure 2.2) [1]. For the three-dimensional space, the Cartesian system OXYZ is compound by three orthonormal vectors of coordinates OX, OY and OZ.

2.2.2 ORIENTATION OF A ROBOT AND ITS LINKS

For a solid body it is necessary to define the orientation of the body with respect to a reference frame. In the case of a robot it is necessary to indicate the orientations of its links. In the case of a robot that has to kick a ball it is necessary to know the orientation with which it must kick the ball. An orientation in the 3D space is defined by 3-DoF or three linear-independent components. To describe the orienta-

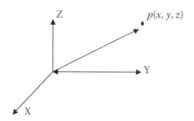

FIGURE 2.1   Description of a position in a fixed reference frame.

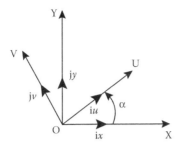

FIGURE 2.2 Orientation of OU-axis by angle with respect to OX-axis.

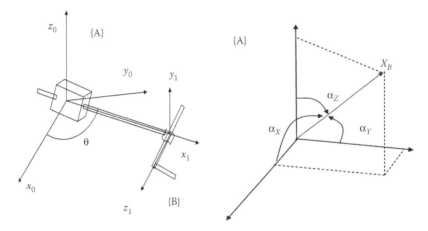

FIGURE 2.3 Robot links and rotation coordinates from {A} to {B}.

tion of an object with regard to a reference system, it is a practice to have a new system, and then to study the existing space relationship between the two systems. Figure 2.3 illustrates the robot links and its orientation description. A description of {B} with respect to {A} suffices to give orientation. Various defining expressions for this rotation are given in terms of the angles between the corresponding axes with a dot product between respective vectors as given below:

$$\cos(\alpha_X) = X_A \cdot X_B$$
$$\cos(\alpha_Z) = Z_A \cdot X_B \qquad (2.1)$$
$$\cos(\alpha_Y) = Y_A \cdot X_B$$

$${}^A X_B = \begin{bmatrix} X_A \cdot X_B \\ Y_A \cdot X_B \\ Z_A \cdot X_B \end{bmatrix}$$

2.2.3 ROTATION MATRIX

The rotation matrices provide a simple algebraic description of the rotations of a rigid body, and are used for computations in geometry and computer graphics. In 2D they are determined by the angle θ of rotation, and in 3D in addition an axis of rotation is involved. The angle and axis are implicitly represented by the entries of the rotation matrix. The rotation matrix is a very convenient operation for the description of orientation. Assume that there are two reference systems OXY and OUV with the same origin O. The OXY is the fixed reference system and the OUV is the mobile system. The unitary vectors of these axes system are ix, jy and iu, jv, respectively, thus giving the vector representation as shown in expression (2.1).

$$pxy = [px; py]' = px \ \mathrm{i}x + py \ \mathrm{j}y$$
$$puv = [pu; pv]' = pu \ \mathrm{i}u + pv \ \mathrm{j}v \tag{2.2}$$

The so-called rotation matrix is defined from the above relationships: the orientation of the OUV system with respect to the OXY system. For a 2D system the orientation is defined by an independent parameter. If the relative position of the OUV system is considered rotated by angle α over the OXY system then R matrix is represented as

$$R = \begin{bmatrix} \cos\alpha & -\sin\alpha \\ \sin\alpha & \cos\alpha \end{bmatrix} \tag{2.3}$$

If the coordinated axes of both systems coincide, the R matrix will correspond to the unitary matrix. In a 3D space, the OXYZ and OUVW systems are coincident in the origin, the OXYZ being the fixed reference system, and OUVW the mobile reference system. If the OU axis is coincident with OX, the rotation matrix for the 3D system is given as

$$R = \begin{bmatrix} 1 & 0 & 0 \\ 0 & \cos\alpha & -\sin\alpha \\ 0 & \sin\alpha & \cos\alpha \end{bmatrix} \tag{2.4}$$

Depending upon which axes are coincident the R matrix will change appropriately. If the OV and the OY are coincident, then the R matrix is given by

$$R = \begin{bmatrix} \cos\phi & 0 & \sin\phi \\ 0 & 1 & 0 \\ -\sin\phi & 0 & \cos\phi \end{bmatrix} \tag{2.5}$$

If the OW and the OZ are coincident then the R matrix is given by

$$R = \begin{bmatrix} \cos\theta & -\sin\theta & 0 \\ \sin\theta & \cos\theta & 0 \\ 0 & 0 & 1 \end{bmatrix} \tag{2.6}$$

These three matrix equations 2.4 through 2.6 are the basic rotation matrices for orientation transformation in the 3D axes coordinate system. The rotation matrices are square matrices, with real entries. They are characterized as orthogonal matrices with unit determinant: $R^{\mathrm{T}} = R^{-1}$, det $R = 1$.

2.2.4 Homogeneous Transformation

The rotation matrices represent the orientations of a solid in the space. For a combined representation of the position and orientation the homogeneous coordinate transformation matrix or homogeneous transformation matrix (HTM) is required. The sequence of homogeneous transformation is depicted in Figure 2.4 [12]. This HTM representation by homogeneous coordinates for the solid body localization in n-dimensional space is derived through coordinates of an $(n + 1)$-dimensional space. The n-dimensional space is represented by homogeneous coordinates in $(n + 1)$ dimensions, in such a way

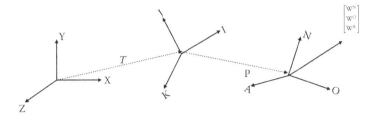

FIGURE 2.4 Homogeneous transformation as a series of rotation and translation.

that the $p(x, y, z)$ vector will be represented by $p(wx, wy, wz, w)$. Here, 'w' has an arbitrary value and it represents a scale factor. A vector $p = ai + bj + ck$, where i, j and k are the unitary vectors of OX, OY and OZ axes of OXYZ reference system is represented in homogeneous coordinates by some column vector. On the basis of the definition of the homogeneous coordinates the homogeneous transformation matrix is evolved. The HTM 'T' is a 4×4 dimensions matrix and represents the transformation of homogeneous coordinates—vector from one coordinate system to another coordinate system. This HTM 'T' is given by

$$T = \begin{bmatrix} \text{Rotation matrix } (3 \times 3) & \text{Position vector } (3 \times 1) \\ \text{Perspective transformation } (1 \times 3) & \text{Scaling } (1 \times 1) \end{bmatrix} \tag{2.7}$$

If the perspective is null and the scale value is unitary then the HTM T is given by

$$T = \begin{bmatrix} \text{Rotation matrix } (3 \times 3) & \text{Position vector } (3 \times 1) \\ 0 & 1 \end{bmatrix} \tag{2.8}$$

This then represents the orientation and position of an OUVW coordinate system rotated and translated with respect to the OXYZ reference system. Essentially the rotation and translation can be expressed in terms of homogeneous coordinates, that is, a single matrix–vector product produces rotation and this transformation. For a complete HTM description and various numerical examples the details in [1] will be very useful.

2.3 ROBOT'S KINEMATICS

In order to specify and adjust the robot's controller, the kinematic and dynamic mathematical models of the robot are needed. The kinematics is the study and analysis of the robot's movements with respect to a chosen reference coordinate system. It is an analytic specification and description of the spatial movement of the robot. This is a relationship between the position and the orientation of the robot's links and the joint coordinates. It is possible to determine the position and the orientation of the robot's end links based on its geometric relations. The inverse kinematics is the process of determining each joint coordinates from the specified position/orientation of the robot end links. Table 2.1 and Figure 2.5 show the functionality of the robot's kinematics. The kinematic model of a robot is represented by the HTM as discussed in Section 2.2.4. This HTM process is necessary for a robot with more than 2DoF. A robot of n-DoF is formed by n-links assembled by n-articulations in such way that each articulation-link constitutes one DoF. To each of these links a reference system is associated and the homogeneous transformations are used to represent the rotations/relative translations from the different links which compose the robot. It is thus possible to represent the translations and relative rotations between the different links.

TABLE 2.1
Robot Kinematic Functional Matrix

Kinematic	What Is the Situation	What Is Derived/Obtained
Direct	Each joint coordinate	Position and orientation of each joint at the end of the robot's link
Inverse	Position and orientation of each joint at the end of the robot's link	Each joint coordinate

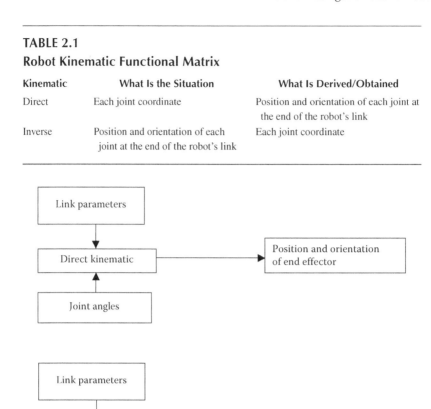

FIGURE 2.5   Direct and inverse kinematics.

2.4   DYNAMIC MODEL

Robot dynamics is the study of the relationship between the various forces acting on a robot mechanism and their accelerations. Robot dynamics can be studied under two broad classifications such as forward dynamics and inverse dynamics. In forward dynamics study, the accelerations are computed for given underlying forces. This is mainly used for simulation study and analysis of robot's design and/or controller's design for the robot. In the inverse dynamics the forces are computed for the given accelerations. This process is widely used for online control of a robot which includes motion control, path planning and control of forces. In robot dynamics the other important aspects are: (a) computation of the coefficients of the equation of motion; (b) identification of inertia parameter, that is, estimating the inertia parameters of a robot mechanism from measurements of its dynamic behaviour and (c) given the forces at some joints and the accelerations at others, determining the unknown forces and accelerations [2,6].

2.4.1   EQUATIONS OF MOTION

The description of the mechanism of the robot movement is given in terms of its component parts: bodies, joints and the parameters that characterize them. A dynamic model consists of (a) a kinematic model of the robot mechanism and (b) a set of inertia parameters. In fact, 10 inertia parameters are required to define the inertia of a single rigid body (mass, location of centre of mass and six rotational inertia parameters), that is, a dynamic model will normally contain 10 inertia parameters per body. When the bodies are connected together to form a mechanism, they tend to lose a few degrees of motion freedom. In that case some of their inertia parameters may have little or no effect on the dynamic behaviour of the system. We next discuss the equation of motion (EOM) for a typical robot. The EOM for a robot mechanism can be written as

$$\tau = H(q)\ddot{q} + c(q,\dot{q},f_{ext}) \tag{2.9}$$

In Equation 2.9, q, $\dot{q}$, $\ddot{q}$ and τ are vectors of joint position, velocity, acceleration and forces variables of n-dimensional coordinate vectors, respectively. These force variables are defined such that $\dot{q}^T\tau$ is the power delivered by force τ to the system. Thus, $\dot{q}$ and τ qualify as a set of generalized velocity and force variables, and f_{ext} denotes an external force acting on the robot because robot is travelling in an environment. This is very appropriate if the robot makes contact with its environment via its end effector. H is the joint-space inertia matrix, and it is a symmetric and a positive-definite matrix. c is the joint-space bias force, which is the joint-space force that must be applied to the system in order to produce zero acceleration. H and c are the coefficients of the equation of motion. The kinetic energy of the robot mechanism is given by

$$T = \frac{1}{2}\dot{q}^T H\dot{q} \tag{2.10}$$

The other two equations of motion are used for robot motion and path control system:

$$\tau = H(q)\ddot{q} + C(q,\ \dot{q})\dot{q} + \tau_g(q) + J(q)^T f_{ext} \tag{2.11}$$

In Equation 2.11, the bias force is broken into three parts. The term $C(q,\ \dot{q})\dot{q}$ is obtained taking into account the Coriolis and centrifugal forces. The gravity forces are taken care in the term $\tau_g(q)$. The effect of the external force is considered in $J^T f_{ext}$. The matrix J is the Jacobian of the end effector that satisfies the equation

$$v_{ee} = J\dot{q} \tag{2.12}$$

Here, v_{ee} is the spatial velocity of the end effector or the combined effect of the spatial velocities of the end effectors. The second equation of motion is the operational-space formulation and is given by the following equation:

$$\Lambda(x)\dot{v} + \mu(x,v) + \rho(x) = f \tag{2.13}$$

Here, x is a vector of position coordinates, v is a spatial velocity vector of the end effector and f is the spatial force acting on the end effector. Λ, μ, ρ are the coefficients of the equation of motion, where Λ is the operational-space inertia matrix, μ has the Coriolis and centrifugal force terms, ρ the gravity terms and f is the sum of an external force acting on the end effector and the projection onto the end effector of the actuator forces acting at the joints. Several algorithms are available for solving the dynamic problems of robot [4–6]: (i) the recursive Newton–Euler algorithm (RNEA), (ii) the articulated-body algorithm (ABA) and (iii) the composite-rigid-body algorithm (CRBA). These numerical algorithms are required because various relationships between the robot body and its links might be non-linear and complex.

2.5 ROBOT'S WALKING

The physical movement of a robot (or its arm, i.e., legs) from one place to another is termed as robot walking. A robot built with four wheels/legs is easy to control because it maintains static equilibrium throughout its motion. In this machine, control of speed and direction is greatly simplified. Dynamic walking is an approach that emphasizes the mechanics of the legs, attempting to understand them for the purpose of making control simpler and more economical [1]. For the robots that use static walking the control criterion is to maintain the projection of the centre of gravity (COG) on the ground, inside of the foot support area (FSA). In this approach only slow walking speeds can be achieved and only on the flat surfaces, since for walking on rough terrain this will be a severe limitation. For the dynamic walking robot the COG (or centre of mass, COM) can be outside of the support area. The ZMP (zero-momentum/moment point) criterion is generally used to generate biped control algorithms and is a very successful method for controlling robotic locomotion.

The walking problem can be divided into two aspects: (i) the balance control and (ii) the walking sequence control. In balance control, a feedback-force system at each robot's foot can be implemented to calculate the ZMP and then feed it into the incremental fuzzy PD controller to decrease the ZMP error [1]. The controller's aim is to adjust the lateral robot's position to maintain the ZMP point always inside of the support area. The walking sequence control of a biped robot is determined by controlling the hip and foot trajectories [1]. In order to achieve the stable dynamic walking, the change between simple supports walking phase and double supports walking phase should be smooth. One can use the cubic polynomials algorithms to control the sagittal motion to guarantee a smooth change between the walking phases [1]. The robot's stability at dynamic walking can then be achieved by applying the ZMP criterion in the incremental fuzzy PD controller to guarantee the balance control at walking [1]. The system that acts like two coupled pendula is a simple planar mechanism with two legs that can make walk a robot stably down a slight slope with no other input or control. The stance leg would act like an inverted pendulum, and the swing leg would act like a free pendulum attached to the stance leg at the hip. For a sufficient mass at the hip, the system will have a stable limit cycle. The limit cycle is nominal trajectory that repeats itself and will return to this trajectory even if perturbed slightly. An extension of the two-segment passive walker includes knees that would provide natural ground clearance without any need for additional mechanisms. A hexapod walking robot is a mechanical vehicle that walks on six legs, and since a robot can be statically stable on three or more legs, a hexapod robot has greater flexibility in its movement. The merit is that when it becomes partially disabled it will be able to walk. Since all the six legs of the robot are not needed for stability, the other legs can be used to reach new foot placements or manipulate a payload.

2.6 PROBABILISTIC ROBOT MOTION MODELLING

As we know, the robot motion is inherently uncertain due to: (a) its mathematical model might not be known accurately, (b) there are uncertainties in the measurements obtained from the sensor being used for sensing robot's position and control feedback and (c) there might be other inaccuracies in the signal processing and transmission channels and systems. In such a situation it would be prudent to utilize the probabilistic approach to robot motion modelling. Also, one can use the concepts of ANNs and FL (Chapter 1) for robot modelling and representation of the heuristic domain knowledge, respectively.

One approach is to utilize dynamic Bayesian network (DBN) for representing controls, robot dynamic states and robot sensation states. Figure 2.6 depicts the outline of this DBN [15] wherein the robot motion states are X, robot control inputs are U and the sensor outputs are Z. These states are interconnected such that input states/control commands drive the robot states and in turn the DBN generates probable positions that are measured by the sensing devices and the output states are Z. It is presumed that there are uncertainties in every stage: (i) control input

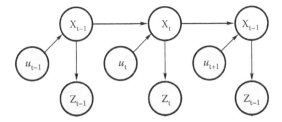

FIGURE 2.6   Dynamic Bayesian network for probabilistic motion modelling.

commands, (ii) state/process noise and (iii) measurements noise. These are modelled by probabilistic concepts and are incorporated into the DBN at appropriate points/nodes. This approach is called probabilistic robot motion modelling (PRMM). In order to implement the Bayesian filter (DBN) one needs the transition model $p(x \mid x', u)$ that specifies a posterior probability. This signifies that action u propels the mobile robot from its current state x' to the next state x. The main point here is to be able to specify the probability of transition probability, $p(x \mid x', u)$, from the current state to the next state and that is to be done by using and incorporating the mathematical model of the robot motion, that is, the EOMs of the robot. As we have seen in Sections 2.2 and 2.4, the configuration of a robot is described by six parameters; (i) 3D Cartesian coordinates and (ii) three Euler angles called pitch, roll and tilt. Then the state space of the robot system 3D with components as (x, y, θ) defines its motion variables. The robot motion models could be based on (a) odometry and (b) velocity measurements (i.e., dead reckoning). The models based on odometry can be used for the systems that have wheel encoders, and the models based on velocity can be applied without encoders. The idea is to compute the new positions/locations of the robot based on the measured velocities and the time elapsed.

2.6.1   Odometry-Based Motion Modelling

When the robot moves from one position to another its motion state undergoes a transition from state $\langle \bar{x}, \bar{y}, \bar{\theta} \rangle$ to state $\langle \bar{x}', \bar{y}', \bar{\theta}' \rangle$

The odometry data are captured in the input command as

$$u = \left\langle \delta_{\text{rot1}}, \delta_{\text{rot2}}, \delta_{\text{trans}} \right\rangle$$

These component input commands are given by the following equations:

$$\delta_{\text{trans}} = \sqrt{(x' - x)^2 + (\bar{y}' - \bar{y})^2}$$
$$\delta_{\text{rot1}} = a\tan 2(\bar{y}' - \bar{y}, \bar{x}' - \bar{x}) - \bar{\theta} \tag{2.14}$$
$$\delta_{\text{rot2}} = \theta' - \bar{\theta} - \delta_{\text{rot1}}$$

Here, x and y are the linear positions of the robot and θ is the robot tilt angle. However, the measurements are generally affected by measurement noise. The odometry model taking into account these noise effects are given as

$$\hat{\delta}_{\text{rot1}} = \delta_{\text{rot1}} + \varepsilon_{\alpha_1 |\delta_{\text{rot1}}| + \alpha_2 |\delta_{\text{trans}}|}$$
$$\hat{\delta}_{\text{trans}} = \delta_{\text{trans}} + \varepsilon_{\alpha_3 |\delta_{\text{trans}}| + \alpha_4 |\delta_{\text{rot1}} + \delta_{\text{rot2}}|}$$
$$\hat{\delta}_{\text{rot2}} = \delta_{\text{rot2}} + \varepsilon_{\alpha_1 |\delta_{\text{rot2}}| + \alpha_2 |\delta_{\text{trans}}|}$$

The errors in the rotations/transitions arising due to some misalignments in wheels and non-smooth surfaces on which the robot is moving are modelled as Gaussian noise processes. Then based on the assumption of normal probability distribution function (pdf) one can compute the posterior probability given the current state and input (from the odometry measurements). One can use the conventional approach of normal density function with mean and known covariance matrices for the noise processes. Alternatively, one can use the concept of the sampling-based density function. The former process is applicable if the pdf is symmetrical, and the latter is used if the pdf does not have the usual normal form of the distribution. This approach can be used for any type of arbitrary distribution function since it is based on the sampling of the distribution function.

The process is given by the following equations:

$$u = \langle \delta_{rot1}, \delta_{rot2}, \delta_{trans} \rangle, \quad x = \langle x, y, \theta \rangle$$

Given:
Compute the following new estimated observations components based on the samples of the distribution function:

$$\hat{\delta}_{rot1} = \delta_{rot1} + sample(\alpha_1 \mid \delta_{rot1} \mid + \alpha_2 \, \delta_{trans})$$
$$\hat{\delta}_{rot2} = \delta_{rot2} + sample(\alpha_1 \mid \delta_{rot2} \mid + \alpha_2 \, \delta_{trans})$$

Then compute the following new estimated/projected robot motion-state components based on the estimated/projected measurements components obtained above:

$$y' = y + \hat{\delta}_{trans} \sin(\theta + \hat{\delta}_{rot1})$$
$$x' = x + \hat{\delta}_{trans} \cos(\theta + \hat{\delta}_{rot1})$$
$$\theta' = \theta + \hat{\delta}_{rot1} + \hat{\delta}_{rot2}$$

In this way, a new robot pose is computed using the odometry measurements that incorporate the Gaussian probability models for noise processes and the current robot pose. It must be mentioned here that the EOMs of the robot motion are not explicitly involved, though these can be included if deemed so.

2.6.2 Dead Reckoning

In a similar way as for the odometry-based measurements one can use the velocity measurements and time elapsed to derive the new robot pose using the concept of probabilistic modelling [15]. Dead reckoning (or ded for deduced/DR) is the process of calculating a robot's current position by using a previously determined position. Then the robot advances its position based on known/estimated speeds over elapsed time. The modern inertial navigation systems also depend upon dead reckoning and are widely used. Thus, dead reckoning is the process of estimating the value of any variable quantity by using an earlier value and adding the incremental changes that have occurred in the meantime/elapsed time. A demerit is that since new values are computed from previous values, any errors and uncertainties are cumulative and these will grow with elapsed time. Hence, here the probabilistic robot motion modelling concept discussed in Section 2.6.1 can be favourably used. The difference between the odometry approach and the DR is that now the speed/velocity measurements of the robot are available and these are used in the PRMM process [15]. Again the merit of this approach is that as such there is no need to incorporate the real mathematical model of the robot in determining the pose of the vehicle. The PRMM approach is useful if the knowledge of the mathematical model of the robot is very poor or it is very difficult to obtain such a mathematical model. The point is that the pose of the robot can be simply determined and ascertained by using the real measurements from the sensors on board the robot using the PRMM as explained in Section 2.6.1. One needs to relate the sensor measurements to the robot coordinates that determine its position and orientations. These coordinates are the ones that need to be determined by using the sensor measurements in the PRMM approach with appropriate models of the noise processes and any known statistics.

REFERENCES

1. Navarro, D. Z. *A Biped Robot Design*. Doktorarbeit, Freie Universität Berlin, Fachbereich Mathematik und Informatik, December 2006. http://www.diss.fu-berlin.de/diss/receive/FUDISS_thesis_000000002504, Accessed August 2008.
2. Siciliano, B. and Khatib, O. (Eds.). *Springer Handbook of Robotics*. Berlin: Springer, 2008.
3. Khalil, W. and Dombre, E. *Modeling, Identification and Control of Robots*. New York: Taylor & Francis, 2002.
4. FU, K. S., Gonzalez, R. C. and Lee, C. S. G. *Robotics Control, Sensing, Vision and Intelligence*. New York: McGraw-Hill Book Company, 1987.
5. Featherstone, R. *Rigid Body Dynamics Algorithms*. Boston: Springer, 2007.
6. Featherstone, R. *Robot Dynamics Algorithms*. Boston: Kluwer Academic Publishers, 1987.
7. Featherstone, R. and Orin, D. E. Robot dynamics: Equations and algorithms. *IEEE Int. Conf. Robotics and Automation*, 1:826–834, 2000.
8. Sporns, O. Complexity. *Scholarpedia*, 2(10):1623, 2007.
9. Featherstone, R. *Robot Dynamics Algorithms*. Boston: Kluwer Academic Publishers, 1987. http://www.scholarpedia.org/article/Robot_dynamics, Scholarpedia open-access encyclopedia, 2008.
10. Meiss, J. Dynamical systems. *Scholarpedia*, 2(2):1629, 2007.
11. Izhikevich, E. M. Equilibrium. *Scholarpedia*, 2(10):2014, 2007.
12. Melamud, R. *An Introduction to Robot Kinematics*, generalrobotics.org/ppp/Kinematics_final.ppt, October 2011.
13. Spatial modeling—Some fundamentals for robot kinematics, http://doc.istanto.net/ppt/1/spatial-modeling–some-fundamentals-for-robot-kinematics.htm, October 2011.
14. Anon. *Kinematics, Advanced Graphics (and Animation)*, Spring 2002, ppts, Index of /~gfx/Courses/2002/Animation.spring.02/Lectures.www.cs.virginia.edu/~gfx/Courses/2002/Animation.spring.02/Lectures, Accessed October 2011.
15. Thrun, S., Burgard, W. and Fox D. *Probabilistic Robotics*, The MIT Press, Massachusetts, 2005, http://www.robots.stanford.edu/probabilistic-robotics/ppt/motion-models.ppt. Stanford University, Accessed October 2011.
16. Stengel, R. F. *Robotics and Intelligent Systems*. Princeton, NJ: Princeton University, 2009. http://www.princeton.edu/~stengel/MAE345Lecture1.pdf7

3 Data Fusion in Mobile Intelligent Autonomous Systems

Ajith K. Gopal

CONTENTS

3.1 INTRODUCTION

Intelligent field robotic systems are mobile robots that operate in an unconstrained environment that is constantly changing with regard to both internal and external states. Further, unlike their conventional manufacturing counterparts that operate in controlled environments, these internal and external states are not part of a finite defined set and hence the challenge in field robotic systems lies in the significantly more complex levels of perception, navigation, planning and learning required to achieve their mission.

Perception in the context of robotics addresses the challenges associated with understanding the environment in which the robots are operating. This understanding is achieved through the creation of physical and knowledge models that are truthful representations of the robotic system's operating environment. Many types of sensors can be used for this world modelling task, including sonar, laser scanners and radar; however, the primary focus of MIAS research with regard to perception is on the acquisition and interpretation of vision data. Navigation uses knowledge of the perceived external environment, as well as additional information of the robots own physical state from other onboard sensors, to address the challenges of localizing the system within the environment. Localization is the core construct required for accurate control of the platform actuators to maintain a given trajectory. Mapping of an unknown environment while simultaneously localizing within the same environment is also a part of the navigation challenge. This simultaneous mapping process is known as SLAM (Simultaneous Localization And Mapping) in modern literature.

The primary objective of the planning function is to integrate knowledge of the high-level goals with the system's current understanding of its environment and position within the environment to generate appropriate behaviours that will take the system closer to successfully achieving the goals. Thus, the end result of planning is the generation of behaviour options and decisions on which behaviour is to be executed. Learning closes the loop by incorporating knowledge of previous states, behaviours and results, to improve the system's reasoning capability and increase the probability of deciding on the optimal behaviour. Learning, reasoning or adaptive behaviour generation is what differentiates mobile field robotic systems from its industrial counterparts; learning is not required when the environment is constrained to the extent that pre-programming is sufficient for the robot to achieve its objectives.

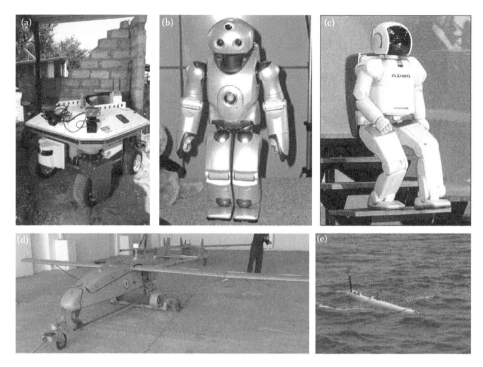

FIGURE 3.1   (a) CSIR (SA) driver-less vehicle research platform (www.csir.co.za/mias). (b) Sony's QRIO humanoid. (Data from www.en.wikipedia.org/wiki/Qrio.) (c) Honda's Asimo humanoid. (Photo taken by the author at the Auto Africa motor show in SA.) (d) Israel Aircraft Industries RQ-2 Pioneer UAV. (Data from www.en.wikipedia.org/wiki/RQ-2_Pioneer.) (e) Bluefin-12 AUV/Bluefin Robotics in the USA (www.en.wikipedia.org/wiki/AUV). (From A. Gopal, Chapter 12: Overview of data fusion in mobile intelligent autonomous systems, in *Multi-Sensor Data Fusion with MATLAB*, J. R. Raol (ed.), CRC Press, FL, USA, 2009.)

These four MIAS areas overlap to some degree and need to function synergistically for the robotic system to be capable of fulfilling its intended purpose. The common thread among these four functional areas is suitable knowledge representation. A unified framework definition for interfaces and data flow (system architecture) is also required to facilitate integration of all the MIAS sub-systems into a single demonstrator system. It is at this system-level integration that the benefits of multi-sensor data fusion are most applicable and clearly distinguishable. Mobile intelligent autonomous systems not only operate in an environment that changes, but also in an environment that changes unpredictably. As a result, they have to firstly detect if and when a change has occurred and then assess the impact of the change on their task/objective. Depending on the result of the assessment the system then has to take appropriate planning and actuation decisions to compensate for the change and still achieve its objective. Examples of MIAS research internationally is captured in Figure 3.1 and include the driver-less car and underground mine safety initiatives at the Council for Scientific and Industrial Research (CSIR) in South Africa, the Sony (QRIO) and Honda (Asimov) humanoid programmes, unmanned military ground and air vehicles and the Bluefin-12 autonomous underwater vehicle (AUV) from Blue Fin Robotics in the United States.

3.2   WHY DOES MIAS NEED DATA FUSION

Data fusion is required in MIAS because the environment in which the system is operating is unconstrained and not always fully observable and the sensors used to observe the environment do not provide complete information, and in many instances missing information. Further, the real world is spatially and temporally linked and data fusion provides a useful methodology for realizing this relationship. A common example of an environment that is not fully observable is when a mobile robot needs to navigate in a cluttered environment with many obstacles obscuring its sensors, or to a point that is beyond the range of its exteroceptive sensors. In this case better path planning strategies can be achieved if the system localization is fused with an *a priori* map (see Chapter 15) of the environment. Mobile robots rely on sensors to generate a description of the internal states and external environment. Internal state sensors are also referred to as proprioceptive sensors and sensors for measuring the external environment are referred to as exteroceptive sensors. Proprioceptive sensors are typically used to measure robot velocity, acceleration, attitude, current, voltage, temperature and so on, and are used for condition monitoring (fault detection), maintaining dynamic system stability and controlling force/contact interaction with the external environment while exteroceptive sensors are typically used to measure range, colour, shape, light intensity and so on, and are used to construct a world model to interact with the environment through navigation or direct manipulation. The requirement for sensor fusion arises because the exteroceptive sensors measure very specific information such as range, or colour, and thus only provide a partial view of the real world which needs to be fused together to provide the complete picture. Also, information that needs to be fused does not need to be limited to numerical format (as is typical of measurement sensors), but can be linguistic (as well as in other formats) which allows the system to build a more descriptive world model that can be used to determine if the end objective has been achieved. In addition, data fusion increases accuracy (in the sense of improving prediction accuracy) and robustness of information and reduces noise and uncertainty—Dead-reckoning errors for mobile robot localization is cumulative and fusion with exteroceptive sensor information is required to reset these errors and determine the robot location more accurately. The fusion of data also facilitate the detection and extraction of patterns in proprioceptive sensor data that are (apparently) not present in the un-fused data and are consequently very useful for system condition monitoring and target tracking. Lastly, robotic systems operating under a decentralized processing architecture require data fusion to incorporate data communicated from neighbouring nodes with local observations.

3.3   A BRIEF OVERVIEW OF DATA FUSION APPROACHES IN MIAS

In general, there are three types of data fusion based on when the processing of the sensor information is executed (see Figure 3.2). In addition, data fusion in mobile robotics occur at different levels; there is time-series data fusion of a single sensor, fusion of data from redundant sensors (same type of sensors), multiple sensor (different sensors) data fusion and fusion of sensor and goal information. Each level of fusion has its own associated challenges dependent on the type of data being fused, requirements of the application and reliability of outputs. The approaches to data fusion in any MIAS application include statistical analysis (such as the Kalman filter), heuristic methods (such as Bayesian networks), possibility models (such as fuzzy logic and Dempster–Shafer), mathematical models, learning algorithms based on neural networks and genetic algorithm-type evolutionary algorithms and hybrid systems [1]. A brief overview of some of the research being conducted internationally is presented here to give the reader a broad perspective of the field and is not intended to be a comprehensive mathematical description of all the available data fusion methods, many of which are covered in Chapters 4, 5 and 30 of the book.

While data fusion is a relatively new field of research, its theoretical basis is founded on 'old' paradigms. Bayes theorem, which is based on the theory of probability, enjoys the widest support, while the more modern methodology of fuzzy logic (based on the concept of membership functions) is gaining popularity, albeit much slower than expected. The Dempster–Shafer theory, which is based on beliefs and its level of support and plausibility, is another 'old' paradigm used in data fusion, but research using this basis is scattered with long periods of inactivity. The generalized form of Bayes theorem is given by

$$p(A|B) = p(B|A)p(A)/p(B) \tag{3.1}$$

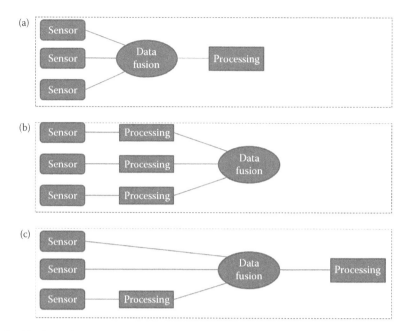

FIGURE 3.2 Configuration types of data fusion: (a) centralized data fusion, (b) distributed data fusion and (c) hybrid data fusion, which is a combination of centralized and distributed.

Since

$$p(A,B) = p(B,A)$$

where $p(A,B)$ is the probability that events A and B occur, and $p(A|B)$ is the probability of event A occurring given that event B has already occurred. When applying Bayes rule in data fusion the more common methods to estimate unknown parameters are the Kalman filter, maximum likelihood estimate and maximum a posterior estimate. Kalman filtering is by far the most popular. Bayesian estimators were very prominent in earlier works in sensor fusion for mobile robot applications, but research in this method seems to have stagnated for some years before again recently increasing in popularity. Work by Matthies and Elfes in 1988 (cited in ref. [2]) uses a Bayesian estimator to fuse sonar and stereovision range data for marking cells in an occupancy grid as occupied, empty or unknown. More recently, in ref. [3], the central problem of path planning is modelled in mobile robots with a partially observed Markov decision process, with continuous states and actions, applied to a visually guided robot. Because of the complexity of the model, a Bayesian optimization technique is used to approximate the cost function required for the control algorithm. The advantage of Bayesian optimization over other search approaches is that it finds multiple minima with as few cost evaluations as possible. Martinez-Cantin et al. detail a Bayesian optimization algorithm incorporating data fusion, for planning T-steps ahead and this is presented here, for ease of reference and comparison with the other fusion methodologies.

Algorithm 1: Bayesian optimization algorithm incorporating data fusion (after [3])

Require: An initial policy $\pi_0(\theta)$

  1. **For** j = 1:*MaxNumberOfPolicySearchIterations*
  2. **for** i = 1:*M*

3. Sample the prior states $x^{(i)}_0 \sim p(x_0)$
4. **For** $t = 1:T$
5. Use a motion controller regulated about the path $\pi(\theta)$ to determine the current action $u^{(i)}_t$
6. Sample the state $x^{(i)}_t \sim p(x_t|u^{(i)}_t, x^{(i)}_{t-1})$
7. Generate observations $y^{(i)}_t \sim p(y_t|u^{(i)}_t, x^{(i)}_t)$
8. Compute the belief state $p(x_t|y^{(i)}_{1:t}, u^{(i)}_{1:t})$ using a SLAM filter
9. Evaluate the approximate average mean square error cost function using simulated trajectories
10. Choose a new promising set of policy parameters θ using Bayesian optimization
    a. Update the expressions for the mean and variance functions of the Gaussian process $(C^\pi(\theta); \mu(\theta), \sigma^2(\theta))$ using the data $D_{1:N}$
    b. Choose $\theta_{N+1} = \arg \max_\theta EI(\theta)$
    c. Evaluate $C^\pi_{N+1} = C^\pi(\theta_{N+1})$ by running simulations
    d. Augment the data $D_{1:N+1} = \{D_{1:N}, (\theta_{N+1}, C^\pi_{N+1})\}$
    e. $N = N + 1$

The Dempster–Shafer theory on the other hand does not use probabilities, but analogous masses that are assigned to all the subsets of the system. Given sensor information A and B, the resultant mass of the combined observation, C, as given by Dempster's combination rule is

$$m(c) = \left[\sum_{A \cap B = C} m_A(A) m_B(B) \right] \bigg/ \left[1 - \sum_{A \cap B = \varnothing} m_A(A) m_B(B) \right] \tag{3.2}$$

where m_A is the measure of confidence in the evidence from sensor A and m_B is the measure of confidence in the evidence from sensor B. The Dempster–Shafer notions of support and plausibility are considered as loose lower and upper limits, respectively, to the uncertainty of an element being in a specific state [4]. Support and plausibility are defined from ref. [4] as

$$spt(A) = \sum_{B \leq A} m(B) \tag{3.3}$$

$$pls(A) = \sum_{A \cap B \neq \varnothing} m(B) \tag{3.4}$$

Wu et al. [2] introduce a weighting factor to each mass, m, to account for variations in sensor accuracy, and apply the methodology to fuse video and audio information in a tracking problem. No performance improvement was achieved over the conventional Dempster–Shafer (unweighted) methodology, which is in hindsight expected as the mass accounts for the overall confidence level of the sensor information. Bendjebbour et al. [5] use the Dempster–Shafer fusion model in the context of Hidden Markov Models to fuse a noisy radar image and an only partially observable optical image to enhance the information content of the scene. This is an important contribution to the Dempster–Shafer methodology as it allows contextual information to be taken into account in the fusion process, something which has been used in the Bayesian methodology decades before and no doubt contributed to the research bias in favour of the Bayesian approach.

The Kalman filter-based approach to data fusion is probably the most significant and widely studied methods to date. It is dependent on linear state-space models and Gaussian probability density functions; the extended Kalman filter is based on a Jacobian linearization and can be applied to non-linear models. However, errors introduced in the linearization process in real applications such as navigation, where the underlying processing are typically non-linear, can cause the state-vector fusion approach to suffer from inaccuracies. Hu and Gan [6] assert that in these cases the measurement fusion approach is preferred to state-vector fusion. Measurement fusion directly fuses sensor data to obtain a weighted or augmented output which can then be input into a Kalman filter (the data from multiple sensors are combined using weights to reduce the mean-square-error estimate and the dimension of the observation vector remains unchanged). In the state vector approach the data from multiple sensors are added to the observation vector and results in increased dimension (a separate Kalman filter is required for each sensor observation and the results of the individual estimates are then fused to give an improved estimate). According to Hu and Gan [6], the state vector fusion method is computationally more efficient but the measurement fusion method has more information and should be more accurate; the net result is that both methods are theoretically functionally equivalent, if the sensors have identical measurement matrices. An example of Kalman filter-based fusion is described in ref. [7] where the issue of cumulative errors in odometry for localization is addressed through the fusion of laser range data and image intensity data. An EKF is used to determine the location of vertical edges, corners and door and window frames that is then used to match against an *a priori* map to obtain a better estimation of the robot location. Grandjean and Vincent used the Kalman filter to fuse laser and stereo range data to model the environment using 3D planes [8]. Kim et al. [9] use the extended Kalman filter to fuse ultrasonic satellite data and inertial sensor data to improve the localization accuracy of a mobile robot. Their localization system consisted of 4 U-SAT transmitters, 1 U-SAT receiver, 2 wheel encoders and a gyroscope. While the results achieved by Kim et al. showed only a slight improvement in localization accuracy, their presentation of the Kalman filter algorithm used for data fusion is very useful and applicable in this context. A pseudo code version of their graphical representation is presented here, in generalized form.

Algorithm 2: Generic Kalman filter data fusion algorithm

Require: V_L, V_R, θ, U

1. **Set** P_k(index, x, y, θ) = U(index, x, y, θ)
2. **Repeat until end**
3. Calculate P_{k+1} using system kinematic model and inertial sensor measurements
4. Estimate P_k by applying the Kalman filter on P_{k+1}
5. **if** P_k(index) = U_{k+1}(index) **and**
6. **if** Dist($U_{k+1}(x, y) \cdot P_k(x, y)$) < 10 cm **then**
7. Current localization is $P_{k+1} = U_{k+1}$
8. **else** current localization is $P_{k+1} = P_k$
9. **end if**
10. **else** current localization is $P_{k+1} = P_k$
11. **end if**
12. **end**

where P(index, x, y, θ) is the posture of the robot determined from the inertial sensors and robot kinematics and U(index, x, y, θ) is the posture of the robot determined from the U-SAT sensors.

The more modern approach of fuzzy logic-based sensor fusion uses membership functions to fuzzify the individual sensor inputs, applies a scheme to combine the fuzzified values of the various sensors and finally defuzzifies the output through specific rules. Escamilla-Ambrosio and Mort [10] present a multi-sensor data fusion architecture that uses an additional fuzzy inference scheme to dynamically adjust the measurement noise covariance matrix. Further, they propose two alternative defuzzification methods; the centre of area and winner takes all methodologies. Using a confidence measurement, c, as the basis of fusion, the centre of the area schema is

$$\widehat{x_k^i} = \frac{\sum_{j=1}^{N} \widehat{x_k^{i(j)}} c_k^{i(j)}}{\sum_{j=1}^{N} c_k^{i(j)}} \tag{3.5}$$

where $\widehat{x_k^i}$ is the ith element in the fused state vector and j is the sensor number with N being the total number of sensors. The winner takes all schema and essentially chooses the state element with the highest confidence value for that time estimate:

$$\widehat{x_k^i} = \arg\ \max_j(c_k^{i(j)}) \tag{3.6}$$

A generalized algorithm based on the work of Escamilla-Ambrosio and Wort [10] is presented below.

Algorithm 3: Fuzzy logic-based data fusion algorithm

1. **repeat** until end
2. Timestep, $k = k + 1$
3. Estimate state vector x_k using Kalman filter
4. Input sensor measurement (instrument) noise covariance matrix, R
5. Calculate actual noise covariance, C
6. Dynamically adjust R using C through fuzzy inference system
7. Update x_k using adjusted R
8. Calculate confidence value, c, of x_k through fuzzy inference system using R, C and the theoretical covariance, S
9. **end**
10. **for** $i = 1{:}m$ (m elements in state vector)
11. $\widehat{x_k^i} = \arg \max_j(c_k^{i(j)})$
12. **end**

In another example, Ortiz-Arroyo and Christensen [11] uses fuzzy logic and data fusion mechanisms to retrieve information in an open-domain question-answering system. A Tellex Modified data fusion method is used where a re-ranking process is added to the methodology introduced by Tellex et al. [12]. The original methodology fuses the rank of passages retrieved from the various passage retrieval systems with the combined total number of passages retrieved. The result of the fusion process is a score that is a measure of the relevance of the retrieved information. The Tellex Modified methodology re-ranks the union of the top passages retrieved. In their investigation of fuzzy logic-based sensor fusion for tracking systems, Zhang et al. [13] use fuzzy clustering algorithms, initially introduced by Bezdek, to allow each data point partial membership to fuzzy sets.

Particle filters is another modern approach to sensor fusion and an example is the work of Germa et al. [14], where vision and RFID (radio frequency identification) data are fused to allow a mobile robot to track people in crowds. The essence of particle filters is the representation of the posterior distribution of the state vector by a set of weighted particles that are updated over time by new sensor measurements. The work of Germa et al. [14] is also particularly useful because it provides a generic particle filtering algorithm that is repeated here for convenience and comparison purposes.

Algorithm 4: Generic particle filtering data fusion algorithm

Require: $[\{x^{(i)}_{k-1}, w^{(i)}_{k-1}\}]^{N}_{i=1}, z_k$

13. **if** $k = 0$ **then**
14. Draw $\boldsymbol{x}^{(1)}_0, \ldots, \boldsymbol{x}^{(i)}_0, \ldots, \boldsymbol{x}^{(N)}_0$ i.i.d according to $p(\boldsymbol{x}_0)$, and set $w^{(i)}_0 = 1/_N$
15. **end if**
16. **if** $k \geq 1$ **then** $[-[\{\boldsymbol{x}^{(i)}_{k-1}, w^{(i)}_{k-1}\}]^{N}_{i=1}$ being a particle description of $p(\boldsymbol{x}_{k-1}|z_{1:k-1})-]$
17. **for** $i = 1, \ldots, 10$ **do**
18. 'Propagate' the particle $\boldsymbol{x}^{(i)}_{k-1}$ by independently sampling $\boldsymbol{x}^{(i)}_k \sim q(\boldsymbol{x}_k|\boldsymbol{x}^{(i)}_{k-1}, z_k)$
19. Update the weight $w^{(i)}_k$ associated to $\boldsymbol{x}^{(i)}_k$ according to $w^{(i)}_k \propto w^{(i)}_{k-1}((p(z_k|\boldsymbol{x}^{(i)}_k)p(\boldsymbol{x}^{(i)}_k|\boldsymbol{x}^{(i)}_{k-1})/ q(\boldsymbol{x}^{(i)}_k|\boldsymbol{x}^{(i)}_{k-1}, z_k))$
20. Prior to a normalization step so that $\sum_i w^{(i)}_k = 1$
21. **end for**
22. Compute the conditional mean of any function of $\boldsymbol{x}_k$, for example, the medium mean square estimate $Ep(\boldsymbol{x}_k|z_{1-k}[\boldsymbol{x}_k]$, from the approximation $\sum^{N}_{i=1} w^{(i)}_k \delta(\boldsymbol{x}_k-\boldsymbol{x}^{(i)}_k)$ of the posterior $p(\boldsymbol{x}_k|z_{1:k})$
23. At any time or depending on an 'efficiency' criterion, resample the description $[\{\boldsymbol{x}^{(i)}_k, w^{(i)}_k\}]^{N}_{i=1}$ of $p(\boldsymbol{x}_k|z_{1:k})$ into the equivalent evenly weighted particles set $[\{\boldsymbol{x}^{(s(i))}_k, 1/_N\}]^{N}_{i=1}$, by sampling in $\{1, \ldots, N\}$ the indexes $s^{(1)}, \ldots, s^{(N)}$ according to $P(s^{(i)} = j) = w^{(j)}_k$; set $\boldsymbol{x}^{(i)}_k$ and $w^{(i)}_k$ to $\boldsymbol{x}^{(s(i))}_k$ and $1/_N$
24. **end if**

where $\boldsymbol{x}_k$ is the state vector, w_k is the weight (or probability) and z_k is the sensor measurements.

Examples of mathematically based data fusion methods include exponential mixture density (EMD) models and vector field histograms. Julier et al. [15] investigate theoretical properties of exponential mixture density models to determine their applicability to the robust data fusion in distributed networks. They take advantage of previous usage of EMDs in expert data fusion and the adaptation of nonlinear filtering methods, such as particle filters, to suit the EMDs computational requirements and develop upper and lower bounds on the probabilities of the result. The result they claim is a consistent and conservative data fusion method for information with unknown correlations. Conner et al. [16] use the evidence grid methodology (also referred to as occupancy grids, certainty grids or histogram grids), in conjunction with vector field histograms (VFH) to fuse CCD camera data and laser rangefinder data for navigation. The data fusion was successfully demonstrated on their Navigator wheeled platform. The evidence grid methodology can also implement probabilistic measurements as demonstrated by Martin and Moravec from CMU (cited in ref. [16]).

Research in data fusion is also being conducted from the system architecture perspective. A data fusion process model developed by the Joint Directors of Laboratories (JDL) Data Fusion Working Group (created in 1986) defines relationships between data sources processing types in an attempt to define a common framework for data fusion implementation. Six levels of processing were defined: (i) source pre-processing—manipulation of the input for interfacing with the other levels, (ii) object refinement—refinement of initial object identifications, (iii) situation refinement—determination of relationships between identified objects, (iv) threat refinement—inference of the future state of the system, (v) process refinement—optimization of the current data and level of information, (vi) data management—storage and retrieval methodologies of the data. More recently, Carvalho et al. [1] developed a general data fusion architecture based on the Unified Modeling Language (UML) that allows for dynamic changes in the framework to respond to changes in a dynamic environment. The approach introduces taxonomy for data fusion and facilitates various levels of fusion. Six levels of processing are defined: (i) pre-procession of individual sensor data, (ii) low-level data fusion—fusion of data before data analysis, (iii) data analysis—analyses fused data and outputs quantified variables for high-level fusion, (iv) high-level data fusion—fusion of data after some form of data analysis, (v) mixture-level data fusion and analysis—fusion of low-level data and high-level variables, (vi) variable interpretation—fuses variables from different sources into a single variable, or multiple views characterizing the sensed environment. Multiple instances of low-level, high-level and mixture-level fusion can exist. The output of the data fusion process is sent to a decision module that then decides the control function of the various actuators, or modifies the sensing requirement from the individual sensors if insufficient information is available to make a control decision.

3.4 CONCLUDING REMARKS

Although research into data fusion is advancing at a steady rate, current challenges to accurate data fusion for mobile, intelligent and autonomous systems remains predominantly the characterization of the uncertainty in sensor measurements and whether the quantitative gain from data fusion can justify the cost implications of additional sensors and processing time. In general, Dempster–Shafer and Bayes are very similar methodologies, though Dempster–Shafer is slightly expensive on the processing side; Dempster–Shafer, on the other hand, is not as heavily dependent on the prior state of information as Bayes. Fuzzy logic is much less process intensive, but is very dependent on the way the membership functions are defined; slight variations can lead to significant changes in results. The bottom line when it comes to deciding on a fusion methodology is to consider each application individually and in the context of its operating environment. When implementing data fusion in MIAS, the following guidelines may be useful: (i) fused data from many poor-quality sensors is worse than fused data from a few good-quality sensors, (ii) manage errors in initial processing to a minimum as the difficulty to correct these downstream increases exponentially, (iii) qualify all assumptions for the specific application—do not rely on general assumptions without substantiating evidence, (iv) allocate sufficient (more than you initially estimated) data for training learning algorithms, (v) fusion of incoming data is a dynamic process.

REFERENCES

1. H. S. Carvalho, W. B. Heinzelman, A. L. Murphy and C. J. N. Coelho, A general data fusion architecture, *Proceedings of the Sixth International Conference of Information Fusion*, 2, 1465–1472, 2003.
2. H. Wu, M. Siegel, R. Stiefelhagen and J. Yang, Sensor fusion using Dempster–Shafer theory, *Proceedings of the 19th IEEE Instrumentation and Measurement Technology Conference (IMTC/2002)*, 1, 7–12, 2002.
3. R. Martinez-Cantin, N. de Freitas, E. Brochu, J. Castellanos and A. Doucet, A Bayesian exploration–exploitation approach for optimal online sensing and planning with a visually guided mobile robot, *Autonomous Robots*, 27(2), 93–103, 2009.
4. S. Challa and D. Koks, Bayesian and Dempster–Shafer fusion, *Sadhana*, 29(Part 2), 145–176, 2004.
5. A. Bendjebbour, Y. Delignon, L. Fouque, V. Samson and W. Pieczynski, Multisensor image segmentation using Dempster–Shafer fusion in Markov fields context, *IEEE Transactions on Geoscience and Remote Sensing*, 39(8), 1789–1798, 2001.
6. H. Hu and J. Q. Gan, Sensors and data fusion algorithms in mobile robotics, Technical report: CSM-422, Department of Computer Science, University of Essex, UK, January 2005.
7. J. Neira, J. D. Tardos, J. Horn and G. Schmidt, Fusing range and intensity images for mobile robot localization, *IEEE Transactions on Robotics and Automation*, 15(1), 76–84, 1999.
8. P. Grandjean and A. R. Robert de Saint Vincent, 3-D modeling of indoor scenes by fusion of noisy range and stereo data, *Proceedings of the IEEE International Conference in Robotics and Automation*, pp. 681–687, Scottsdale, 1989.
9. J. Kim, Y. Kim and S. Kim, An accurate localization for mobile robot using extended Kalman filter and sensor fusion, *Proceedings of International Joint Conference on Neural Networks*, Hong Kong, pp. 2928–2933, 2008.
10. P. J. Escamilla-Ambrosio and N. Mort, Multisensor data fusion architecture based on adaptive Kalman filters and fuzzy logic performance assessment, *Proceedings of the Fifth International Conference on Information Fusion*, 2, 1542–1549, 2002.
11. D. Ortiz-Arroyo and H. U. Christensen, Exploring the application of fuzzy logic and data fusion mechanisms in QAS, *Lecture Notes in Computer Science*, 4578, 102–109, 2007.

12. S. Tellex, B. Katz, J. Lin, G. Marton and A. Fernandez, Quantitative evaluation of passage retrieval algorithms for question answering, *Proceedings of the 26th Annual International ACM SIGIR Conference on Research and Development in Information Retrieval*, Toronto, Canada, 2003.

13. D. Zhang, X. Hao and H. Zhao, Data fusion approach for tracking systems based on fuzzy logic, Technical Report, Communication and Information System Institute, School of Information Science & Engineering, North Eastern University, Shenyang, China, Web access: http://isif.org/fusion/proceedings/fusion01CD/fusion/searchengine/pdf/TuB24.pdf, Last accessed 28/04/2011.

14. T. Germa, F. Lerasle, N. Ouadah and V. Cadenat, Vision and RFID data fusion for tracking people in crowds by a mobile robot, *Computer Vision and Image Understanding*, 114(6), 641–651, 2010.

15. S. J. Julier, T. Bailey and J. K. Uhlmann, Using exponential mixture models for suboptimal distributed data fusion, *Nonlinear Statistical Signal Processing Workshop*, pp. 160–163, IEEE, UK, 2006.

16. D. C. Conner, P. R. Kedrowski and C. F. Reinholtz, Multiple camera, laser rangefinder, and encoder data fusion for navigation of a differentially steered 3-wheeled autonomous vehicle, *Proceedings of SPIE, the International Society for Optical Engineering*, Vol. 4195, pp. 76–83, Bellingham, 2001.

17. A. Gopal, Chapter 12: Overview of data fusion in mobile intelligent autonomous systems, in *Multi-Sensor Data Fusion with MATLAB*, J. R. Raol (ed.), CRC Press, FL, USA, 2009.

4 Image Registration and Fusion

A. N. Myna

CONTENTS

4.1 INTRODUCTION

Image registration and image fusion processes are very important operations in (i) remote sensing applications, (ii) robotics, (iii) target tracking/identification and (iv) medical imaging. Image registration is the process of superimposing two or more images of the same scene taken at different times from different viewpoints by the same or by different sensors. These images may have relative geometric transformations between them like translation, rotation, scale and so on. The aim of registration is to geometrically align the two images—the *source* and the *target* images. Registration algorithms try to align a target image over a source image so that pixels present in both images are in the same reference frame. This process is useful in the alignment of an acquired image over a template or a time series of images of the same scene. Two major applications of this process are the alignment of radiology images in medical imaging and alignment of satellite images for environmental study. The latter images of the environment and other surrounding vehicles can be used for guidance of autonomous vehicles including robots.

In image fusion the idea is to obtain a fused image from two or more individual images of the same object or scenario to enhance the overall information. Image fusion is the process of combining relevant information from two or more images into one composite image. The input images may be captured by different sensors or by the same sensor under different conditions. The resulting image will contain more accurate descriptive information than any of the input images. The images have to be registered before they are fused. Image fusion techniques fall into the category of being pixel-level based, feature based or decision based according to the level at which they are applied. One example is the fusion of magnetic resonance imaging (MRI) and computer-aided tomography (CT) images in medical imaging. As these images contain complementary information, the fused image gives an enhanced image with more details.

4.2   IMAGE REGISTRATION

Differences between two images generally occur due to: (i) geometric transformations between images are caused due to spatial mapping, such as translation, rotation and scaling from one image to the other. Changing the orientation or parameters of the imaging sensor can cause such differences; (ii) occlusion occurs when part of an image moves out of the image frame or new data enter into the image frame due to an alignment difference. Sometimes sensor errors produce identifiably invalid data in an image. Occlusions also occur when an obstruction comes between the imaging sensor and the object being imaged. For example, in satellite images, clouds frequently occlude the earth; (iii) noise occurs due to sampling error and background noise in the sensor, and also from unidentifiably invalid data introduced by sensor error and (iv) temporal change are actual differences between the objects or scenes being imaged at different timings. In satellite images, lighting, erosion, construction and deforestation are examples of differences due to change. Image registration is usually followed by generation of large panoramic images for viewing and analysis. Image mosaics are created by warping and blending together several registered overlapping images. Similar registration tasks include producing super-resolution images from multiple images of the same scene, detection of change, motion stabilization, topographic mapping and multi-sensor image fusion. If $I_S(x, y)$ and $I_T(u, v)$ are the source image and target images, respectively, then the relationship between these images is

$$I_T(u, v) = T_2\{I_S(T_1\{(x, y)\})\} \tag{4.1}$$

where T_1 is a two-dimensional (2-D) geometric transformation operator that relates the (u, v) coordinates in I_T to the (x, y) coordinates in I_S. T_2 is the intensity function. The process of image registration determines the geometric transformation and intensity function. Diagrammatically, it can be represented as shown in Figure 4.1, which shows an example of image registration using three transforms: rotation, x-axis translation, and y-axis translation. The estimation of the intensity function T_2 is useful when images are captured from different sensors or when illumination is changed by automatic gain exposure of a camera. As the aim of image registration is to detect the changes in a scene, successful registration detects differences due to alignment, occlusion and noise while preserving

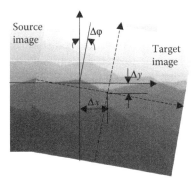

FIGURE 4.1   Image registration with transformations (rotation $\Delta\varphi$, Δx and Δy translation) between source and target images.

differences due to change. Registration algorithms must assume that change is small with respect to the content of the image. In addition, a sufficient amount of the object or scene must be visible in both images. At least 70% of the content of the source image must be present in the pattern to be registered against it. In practice, medical and satellite sensors can usually be oriented with enough precision for images to share 90% or more of their content. Image registration is widely used in the fields of remote sensing, medical imaging and computer vision. In general, depending on the way the images are obtained, the applications can be divided into four main groups. (a) Various viewpoints: Images of the same area are acquired from various viewpoints in order to gain a larger 2D view or 3D representation of the same area in remote sensing—mosaicing of images of the scene. (b) Various times: Images of the same area are acquired at different times, usually on regular basis, and possibly under different conditions in order to find and evaluate changes in the area that appear between the consecutive image acquisitions; examples of applications: In remote sensing—monitoring of global land usage, landscape planning. In computer vision—automatic change detection for security monitoring, motion tracking. In medical imaging—monitoring of the tumor evolution and treatment. (c) Various sensors: In order to gain more complex and detailed representation of the scene, images acquired from one or more sensors are integrated; examples of applications: In remote sensing—fusion of images from sensors with different characteristics offer better spatial resolution, spectral resolution or radar images independent of cloud cover and solar illumination. Medical imaging—registration and fusion of images obtained from various sensors like MRI, ultrasound or CT, positron emission tomography (PET), single photon emission computed tomography (SPECT) or magnetic resonance spectroscopy (MRS). The registration technique should take into account not only the assumed type of geometric deformation between the images but also radiometric deformations, noise corruption, required registration accuracy and application-dependent data characteristics.

4.2.1   IMAGE REGISTRATION STEPS

In general, image registration consists of the four major steps. (i) Detection of control points: Control points are significant points or structures detected automatically or manually by a domain expert. They are a set of selected pixels that contain important information. Since manual identification of control points is time consuming and tedious, several automatic techniques have been developed. (ii) Control point matching: In this step, the correspondence between the control points detected in the target and the source images is established. This involves finding the matching control points in both the images. Here for each control point in the source image, the matching control point is searched in the target image. Matching methods are based on cross-correlation, mutual information, contour matching and so on. (iii) Estimation of the transformation parameters: After the matching control points are detected, the next step is to estimate the parameters that caused the deformation in the target image, for example, horizontal translation, vertical translation, rotation angle, scaling factor and so on. (iv) Resampling and transformation: The final step is to geometrically transform the target image into the reference coordinate system of the source image. Using the estimated parameters, the target image is transformed to align over the source image. Based on the parameters detected in the previous step, the target image is aligned so that both the source and the target images are now in the common reference frame. The target image is transformed by means of the mapping functions. Image values in non-integer coordinates are computed by the appropriate inter-polation technique.

Each of these registration steps has its typical problems during implementation. The control points detected should be uniformly spread over the images and should be easily detectable. They must have enough common elements in both the images even when there are object occlusions or other unexpected changes. They must be stay at a fixed position during the whole experiment. The detection methods should have good localization accuracy. A good detection algorithm is one which can detect the common control points in all projections of the scene regardless of the particular image deformation. During control points matching step, corresponding control point features may be dissimilar due to the different imaging conditions or due to the different spectral sensitivity of the sensors. The choice of the similarity metric has to be stable, robust and efficient to consider these factors. Even aloof control points without corresponding counterparts in the other image should not affect its performance.

Transformation parameters must be estimated based on the *a priori* information about the acqui-sition process and expected image degradations. Otherwise, the technique should be general enough to handle all possible degradations that might appear. The accuracy of the methods used to detect control points and match them and the acceptable approximation error must also be considered. The decision about which control point pair must be removed to make accurate estimation has to be made. It is desirable not to remove the important control point pairs. This decision is very important and extremely difficult. The choice of resampling technique depends on the amount of accuracy required during interpolation and the computational complexity. Commonly used technique are nearest-neighbour or bilinear interpolations. Some applications require more precise methods.

4.2.1.1   Control Point Detection

The two sets of features in the source and the target images are represented by control points. There are two types of control points, intrinsic and extrinsic. Intrinsic control points are marked in the image irrespective of the data. They are often marked by the user or placed on the sensor only for registration purposes and are easily identified. Extrinsic control points are determined from the data, either manually or automatically. Manual control points are points such as identifiable land-marks or anatomical structures recognized by human intervention. They are selected by domain experts. When there is a large amount of data, this will be practically very difficult. Therefore, sev-eral algorithms for automatic registration have been proposed. Features selected may be significant regions (like forests, lakes, fields), lines representing region boundaries, coastlines, roads, rivers or points representing region corners, line intersections, points on curves with curvature, local extrema of the wavelet transform and so on. These points or features are later matched using their spatial relations or various descriptors of features. Control points must be uniquely found in both images and more tolerant of local distortions. As the estimation of proper transformation parameters depends on these features, a sufficient number of control points must be detected to perform the calculation. But too many features will make feature matching more difficult and time consuming. As the accuracy and the efficiency of point matching methods strongly depend on the total number of control points, they must be carefully selected. To detect the control points some algorithms directly use image pixel values, some operate in the frequency domain, while some algorithm use low-level features such as edges and corners, others use high-level features such as identified objects, or relations between features.

4.2.1.2   Control Point Matching

In this step, for each control point on the source image, the best matching control point on the target image is determined. The aim is to find the pair wise correspondence between the selected control points using appropriate descriptors. The methods available can be classified as follows:

- *Using spatial relations*: These methods exploit the spatial relations among the control points. The distance between the control points and about their spatial distribution is exploited. Graph matching algorithms are used to find the matching points after applying particular transformation to the target image and selecting those that fall within a given range next to the control points in the source image. Clusters based on the assumed transformations can also be constructed.
- *Using invariant descriptors*: In this method, the matching control points are estimated using the invariant to the expected image deformation. Control points from the source and target images with the most similar invariant descriptions are paired as the matching ones. The common description is the *image intensity function* itself, limited to the close neighbourhood of the feature. If the detected features are points, then the similarity measures that can be used are cross correlation, phase correlation, moment-based invariants, mutual information and so on. A small window of points around a control point in the source image is statistically compared with windows of the same size around each control point in the target image. The measure of match is based on the similarity measure selected. The centres of the matched windows are control points that can be used to solve for the transformation parameters between the two images.
- *Pyramids and wavelets*: To reduce the computational cost due to the large image size, the pyramidal approach is widely used. It starts with the source and target images on a coarse resolution and gradually improves the estimates of the correspondence of the mapping function parameters while going up to the finer resolutions. This coarse-to-fine hierarchical strategy employs the usual registration methods.

A multiresolution pyramid consists of a set of images representing an image in multiple resolutions. The original image sitting at the base of the pyramid is down-sampled by a constant scale factor in each dimension to form the next level. Level 0, at the base of the pyramid, is referred to as the finest level. Level $n-1$, at the tip of the pyramid, is known as the coarsest level. At every level, search space is considerably decreased and the computational time is saved. The registration with respect to coarse features is achieved first and then small corrections are made for finer details. To overcome the problem of identifying a false match at a coarse level, a backtracking or consistency check should be incorporated into the algorithms. Several pyramids that are available in literature are summing pyramid, median pyramid, averaging pyramid, cubic spline-based pyramid, multi resolution Laplacian pyramid, Gaussian pyramids and so on. Wavelet decomposition has inherent multiresolution character of the images, because of which it is recommended for the pyramidal approach. The image is successively filtered with two filters, a low-pass filter L and high-pass filter H, both working along the image rows and columns. These methods decompose the image recursively into four sets of coefficients (LL, HL, LH, HH). At every level one or a combination of these coefficients can be used for matching.

4.2.1.3   Estimation of the Transformation Parameters

After establishing the feature correspondence between source image I_S and target image I_T, the mapping function is constructed. The mapping function should be based on the assumed geometric deformation of the target image, the method of image acquisition and the required accuracy of the registration process. The type of the mapping function and its estimated parameters are required for the transformation of the target image to overlay it over the source image. Mapping functions can be broadly divided into two global mapping functions and local mapping functions, area of the image they are intended to transform. If the transformation of the entire image is needed, then global models are appropriate. They use all the control points for estimating one set of the mapping functions parameters which is valid for the entire image. If the transformation is to be applied only to a small area of the image then local mapping functions are applied. They treat the image as a composition of small areas and the functions parameters depend on their location in the image. In this case, the parameters of the mapping function are computed for each area separately. Based on the accuracy of superimposing of the control points used for computation of the parameters, mapping functions can also be categorized into interpolating functions and approximating functions. Interpolating functions map the target image control points onto the source image control points exactly whereas approximating functions compute the parameters of the transformation so that the matched points superimpose as nearly as possible. If the number of matching control points is large, then the approximation method is found more suitable. For intrinsic or manual control points, there are usually fewer but more accurate matches. In this case, interpolation will be more applicable. The relationship between a point (X,Y) in one image and its corresponding point ($X^!$, $Y^!$) in the other image is expressed by a 2-D transformation:

$$\begin{bmatrix} X^! \\ Y^! \end{bmatrix} = s \begin{bmatrix} \cos\theta & -\sin\theta \\ \sin\theta & \cos\theta \end{bmatrix} \begin{bmatrix} X \\ Y \end{bmatrix} + \begin{bmatrix} \Delta X \\ \Delta Y \end{bmatrix} \tag{4.2}$$

where parameters (s, θ, ΔX, ΔY) represent scaling factor, rotation angle and translations along the horizontal and vertical directions, respectively. The transform parameters are estimated using a set of matched control points $\{(X_i, Y_i)\}$ and $\{(X_i^!, Y_i^!)\}$ through linearization of the above equations. In most of the remote sensing applications the images have a certain level of geometrical correction that enables the use of this kind of transformation.

4.2.1.4   Resampling and Transformation

The final step is to geometrically transform the target image into the reference coordinate system of the source image. Based on the parameters detected in the previous step, the target image is aligned so that both the source and the target images are now in the common reference frame. The target image is transformed and thus registered according to the mapping model constructed during the previous step. This transformation can be done in a forward or backward manner. In forward approach, each pixel from the target image is directly transformed using the estimated mapping functions. In the backward approach, the registered image data from the target image are determined using the pixels from the source image and the inverse of the estimated mapping function. The image interpolation takes place in the target image on the regular grid. The nearest-neighbour function, the bilinear and bicubic functions, quadratic splines, cubic B-splines, higher-order B-splines, Gaussians and truncated *sinc* functions belong to the most commonly used interpolants. Even though the bilinear interpolation is outperformed by higher-order methods in terms of accuracy and visual appearance of the transformed image, bilinear interpolation offers probably the best trade-off between accuracy and computational complexity and thus it is the most commonly used approach. Cubic interpolation is recommended when the geometric transformation involves a significant enlargement of the sensed image. The simplest scheme for grey-level interpolation is based on the nearest-neighbour approach called zero-order interpolation. But the nearest-neighbour interpolation yields undesirable artefacts such as stair-stepped effect around diagonal lines and the curves. Bilinear interpolation produces the output images that are smoother and without the stair-stepped effect. It is a reasonable compromise between smoothness and computational cost.

4.2.2 Satellite Image Registration

The study of time series of remotely sensed (satellite) images is an important task in many remote sensing applications where the objective is to study different environmental phenomena. For such applications, a co-registration of the satellite images acquired at different times is important. This co-registration is often performed using a combination of manual and automatic registration techniques. However, for a multi-temporal problem where the number of images becomes large, manual correction of images is often not feasible. Hence, a fully automatic procedure would be desirable. The aim of satellite image registration is to align the satellite images obtained at different points of time so that changes such as movement of clouds, growth of vegetation and so on can be detected. In remote sensing applications, users generally use manual registration, which is not feasible in cases where there is a large amount of data. Thus, there is a need for automated techniques that require little or no operator supervision. As the amount of imaging data generated by various earth observing satellites grows rapidly, it becomes essential to develop reliable automatic algorithms for both on-the-ground and on-board processing of these data. However, before images generated by different sensors and/or at different times could be used for such high-level tasks as change detection or data fusion, these images have to be accurately registered. It is an important operation needed in remote sensing that basically involves the identification of many control points. As the manual identification of control points may be time consuming and tedious, automated techniques have been developed. The increased volume of satellite images has reinforced the need for automatic image registration methods. Since the performance of a methodology is dependent on specific application, sensor characteristics and the nature of composition of the imaged area, it is unlikely that a single registration scheme will work satisfactorily for all different applications. In this work, a system for automatic registration of satellite images has been developed.

4.2.2.1 Satellite Images

The very-high-resolution radiometer (VHRR) and the data relay transponder (DRT) together provide: (i) round-the-clock, regular, half-hourly synoptic images of weather systems including severe weather, cyclones, sea-surface temperature and cloud surface temperatures, water bodies, snow and so on, over the entire territory and adjoining land and sea areas; (ii) collection and relay of meteorological, hydrological and oceanographic data from unattended remote platforms; (iii) timely warning of impending disasters due to cyclones, floods, storms and so on; and (iv) dissemination of meteorological information including processed images of weather systems to the forecasting centres in a broadcast mode. The images of earth sent by the satellites are received at regular intervals of time. Sources of geometric distortion are earth rotation, panoramic effects, earth curvature, scan time skew, variation in platform altitude, velocity and aspect ratio. Shifts may appear in the image either due to disturbance in the satellite altitude or due to imaging payload. The image data are made up of grey values on the co-ordinate system defined by lines and frames. Certain control points are chosen in the image which is cloud free for most part of the year. These points are called ground control points (GCPs). A GCP is a point on the surface of the earth where both image co-ordinates and map co-ordinates can be identified. It is the geographical location where there is a sharp transition between water and land. Currently the transformation is computed with respect to source image by computing the geometrical transformation in these GCPs. Manually this is done by identifying GCPs in the given target image using an input device like mouse. Its X and Y co-ordinate values are obtained. These values are compared with X and Y co-ordinate values of reference image to find the rotation angle and also the horizontal and vertical shifts. This leads to the problem of not locating the GCPs accurately and may lead to erroneous calculations. This process can be automated. A portion of the sub-sampled/averaged image data downloaded regularly shall suffice as input to the developed system. The process of control point extraction and comparison is automated. The transformations causing the deformation can be computed using the matched control points.

4.2.3 LITERATURE REVIEW

Registration is one of the basic tasks in image processing. Given two images representing the same or even analogous objects, it consists of geometrically transforming one image (the target image) to the other (the source image) so that the pixels representing the same physical structure may be superimposed. It is a crucial step in all image analysis tasks in which the final information is gained from the combination of various data sources like in image fusion, change detection and multichannel image restoration. Image registration establishes spatial correspondence, that is, the process of registration will establish the correspondence as to which point on one image corresponds to a particular point on the other image. In remote sensing, registration is required for multispectral classification, environmental monitoring, change detection, image mosaicing, weather forecasting, creating super-resolution images, integrating information into geographical information system (GIS) and so on. During the last few decades, growing amount and diversity of images obtained have invoked the research on automatic image registration. Because of its importance in various application areas as well as its complicated nature, image registration has been the topic of much recent research. A vast amount of published literature is available on image registration. A relatively less number of investigations deal with satellite image registration in the presence of clouds. The following sections give a brief background and some of the important pertinent studies that were done recently. These are reviewed elaborately and discussed to identify the lacunae in the existing literature. A comprehensive survey of image registration methods is provided in ref. [1]. It gives a framework of all the registration techniques. A survey [2] provides an extensive review of the recent developments. As discussed earlier, registration methods consist of the following four steps, namely, control point detection, control point matching, transformation model estimation, resampling and transformation. A few relevant works in the above areas are discussed in the following sections.

4.2.3.1 Control Point Detection

Several techniques are available in the literature for performing this step. Control points can be selected manually or automatically using control point detection methods. Since manual identification of control points is time consuming and tedious, several automatic techniques have been developed. Some automatic control point detection techniques take the number of control points to be detected, as an input parameter [3]. Fonseca et al. [4] suggest the use of optical flow ideas for the extraction of features. The optical flow approach was originally motivated by estimation of relative motion between images as proposed in ref. [5]. The class of optical flow registration covers a very large number of methods. The line intersections as the control points are used in ref. [6]. But line intersections cannot be found in satellite images. Centroids of water regions, oil and gas pads are also used as control points as suggested in ref. [7]. Local curvature discontinuities detected using the Gabor wavelets are used as control points in the works in ref. [8]. Local extrema of wavelet transform are considered as control points in ref. [9]. A similar technique is used in ref. [10]. Corners are used as control points in ref. [11]. Corners are points of high curvature on the region boundaries. Much effort has been spent in developing precise, robust and fast method for corner detection. The corners are used as control points [12]. Corners are widely used as control points because of their invariance to imaging geometry and are well perceived by a human observer. Zitova [13] suggests the selection corner-like dominant points in blurred images. Recently Zitova et al. [14] proposed a parametric corner detector method which does not employ any derivatives and which was designed to handle blurred and noise data.

Line features are used in many algorithms. Line correspondence is usually expressed by pairs of line ends or middle points. Object contours are used in refs. [15,16]. Coastal lines are considered in ref. [17]. Roads are used as features by Li et al. [18]. In ref. [19], roads are selected as the features. Region-like features are generally high-contrast closed-boundary regions of an appropriate size. They are often represented by their centres of gravity, which are invariant with respect to rotation, scaling and skewing and are stable under random noise and grey-level variations. Holm [20] consider water reservoirs and lakes as regions of interest. Forests are considered in ref. [21] and urban areas are used in ref. [22]. Li [16] proposes to use the centres of closed boundaries as control points. For open contours, the salient segments are taken. For images obtained by the same sensor, control points can be selected using the Forstner interest operator and when images from different sensors are to be registered, polygons representing real-world objects can be used as proposed in ref. [23]. In ref. [24], the use of structural knowledge is investigated, where the knowledge about the relationship between the objects and their connection to the features apparent in the image data is represented efficiently by semantic nets. The Ground Control Points are acquired from maps [25]. They are acquired by field surveying using global positioning system (GPS). Tuo [26] uses local entropy vector as the feature. Wavelet decomposition of the images is used to detect the control points in order to save time and to select reasonable number of control points. Fonseca and Costa [9] detect the modulus maxima of LH and HL coefficients. Djamdji et al. [27] use just HH coefficients as control points. You and Kaveh [28] use maximum compact fuzzy sets of wavelet coefficients as features. The complete theory is given [29] on multiresolution signal decomposition and the wavelet representation. A comparative study of several registration methods was performed and it was concluded that wavelet-based modulus maxima approach is well suited for remotely sensed images [30]. It is particularly useful for registering images of the same sensor as proposed in ref. [3]. Control points are selected at the lowest level of the wavelet transform so that not too many control points are selected. Corvi and Nicchiotti [31] use the maxima and the minima of the residues of the discrete wavelet transforms of the images as control points. More information on wavelets and the wavelet transforms can be found in refs. [29,32]. Several wavelet pyramids are evaluated that may be used both for invariant feature extraction and for representing images at multiple spatial resolutions to accelerate registration [33]. They show that bandpass wavelets obtained from the steerable due to Simoncelli performs best in terms of accuracy and consistency.

4.2.3.2   Control Point Matching

The correspondence between the control points in the source and the target images is established using the matching methods. Here for each control point in the source image, the matching control point is searched in the target image. The matching methods are either area based or feature based.

4.2.3.2.1   Area-Based Methods

These methods merge the control point detection step with the matching part. These methods deal with the images without attempting to detect the control points. Windows of predefined size or even entire images are used for the correspondence as estimated in ref. [34]. The classical representative of the area-based methods is the Normalized Cross Correlation. Hanaizumi and Fujimara [35] compute cross correlation for each assumed geometric transformation of the target image. The computational load grows very fast with the increase of the transformation complexity. Drawbacks of the correlation-like methods are the flatness of the similarity measure maxima due to the self-similarity of the images and high computational complexity. But their hardware implementation is very easy. Therefore they are still often in use. To accelerate the computational speed, Fourier methods are preferred rather than the correlation-like methods. They exploit the Fourier representation of the images in the frequency domain. The phase correlation method was proposed for the registration of translated images. Scientists have investigated FFT-based approaches for image registration for many years. Kuglin and Hines [36] developed a method called phase correlation by using certain properties of the Fourier transform. In ref. [37], a way was discovered to use the Fourier transform to determine rotation as well as well as shift. Cideciyan et al. [38] determine the phase-correlation function for each discrete rotation value and choose the parameter set resulting in the highest phase correlation. Reddy and Chatterjee [39] improved on the algorithm of Castro and Morandi [37] by greatly reducing the number of transformations needed. They propose that if the images are represented using log-polar mapping, phase correlation could be used to match the images even if rotation and scaling were present. Hongjie Xie et al. [40] give the implementation details of phase correlation. A similar technique was used in ref. [30]. Owing to the global transform, this approach cannot be used to determine local geometric distortions. To convert the image from rectangular to log-polar co-ordinates, the steps of the angle and logarithm base can be calculated using the method as specified in ref. [41]. In order to attain high-accuracy, Holm [39] proposes that the polar plane must have the same number of rows as the rectangular plane. Affinely distorted images are registered by means of phase correlation and log-polar mapping as investigated in ref. [42]. Siavash Zokai and George Wolberg [43] demonstrate the superior performance of the log-polar transform in featureless image registration in the spatial domain, which yields the eight parameters of the perspective transformation that best aligns the two input images.

4.2.3.2.2 Feature-Based Methods

After the two sets of features in the source and target images are represented by control points, matching control points can be found using their spatial relations, using invariant descriptors, relaxation methods or combining cross correlation with wavelets.

4.2.3.2.2.1 Methods Using Spatial Relations
These methods use the spatial relations among the control points. The information about the distance between the control points and about their spatial distribution is exploited. Stockman et al. [6] compute, for every pair of control points from both the source and the target images, the parameters of the transformation which maps the points on each other and represented as a point in the space of transform parameters. The parameters of transformation that closely map the highest number of features tend to form a cluster, while mismatches fill the parameter space randomly. The cluster is detected and its centroid is assumed to represent the most probable vector of matching parameters.

4.2.3.2.2.2 Methods Using Invariant Descriptors
The correspondence of control points can also be estimated using their description invariant to the expected image deformation. Flusser [44] proposes the matching likelihood coefficients. The simplest feature description is the image intensity function itself, limited to the close neighbourhood of the feature. Abdelsayed et al. [45] estimate cross correlation on these neighbourhoods.

Many authors use closed-boundary regions as the features. Li and Manjunath [16] use a chain code representation of contours as the invariant description and a chain code correlation-like measure is used for finding the correspondence. A large group of methods use moment-based invariants for description of closed-boundary region features. Zitova [13] proposes the computation of a vector of invariants for each control point over its circular neighbourhood of the radius of 60 pixels. The point with the minimum distance between its invariant vector and the invariant vector of the control point counterpart is found and set as matching points. In refs. [19,24], the knowledge about the control points which is represented explicitly using semantic nets and rules is used. The best correspondence between the Geographical Information System (GIS) data and the image is found by an A* algorithm. Since registration methods are problem specific, Eikvil et al. [46] propose an adaptive registration approach that automatically selects the appropriate registration methods based on image characteristics by using a neural net. Dare et al. [23] match the polygons which are selected as features by storing the edges of the polygons as line segments. The algorithm then works through the line segments in a stepwise manner, projecting each line segment one by one, from the target image to the source image. A search area is set up around each pixel on the projected line segments and a search is conducted for a corresponding matching pixel in the source image. Matching is performed by minimizing a cost function based on the edge strength and edge direction. Tuo et al. [26] divide both the source and target images into blocks and local entropy vector is computed for each block, mean square error is used as the similarity measure. The centres of the matched sub-images are the matched control points. Fedorov et al. [3] take windows about each control point which have been rotated so that their central gradient points downward. In ref. [25], the correlation coefficient is calculated for each of the RGB component and by adding the three of them, the co-ordinates of the highest correlation is selected.

4.2.3.2.2.3 Pyramids and Wavelets   To reduce the computational cost due to the large image size, feature matching is performed by means of pyramidal approach. Rosenfield and Vanderbrug [47] investigated the use of both the source and target images at a coarser resolution and then, on locations with small error measure to match higher-resolution images. LeMoigne [48] uses wavelet decomposition in which the effect of Daubechies wavelet is investigated. Here cross correlation is used as a similarity measure. Fonseca and Costa [9] achieve feature point matching by maximizing the correlation coefficient over small window surrounding the points within the LL sub-bands of the wavelet transform. Ilya Zavorin and Jacqueline LeMoigne [33] use a search scheme that is based on a gradient descent approach.

4.2.3.3   Transformation Model Estimation

Here the task to be solved consists of choosing the type of the mapping function and its estimation. Given a sufficient number of matching control points, the unknown scaling parameter 's', rotation angle 'θ' and translation parameters (Δx, Δy) can be retrieved using the least-squares regression method as used in ref. [26]. In ref. [49] an algorithm called the Marquardt–Levenberg algorithm to implement this method is given. Refs. [33,43] provide the modified versions of the Marquardt–Levenberg algorithm. The relaxation technique described in ref. [50] can be used to register images under translation. In this case, the point matching and the determination of the best spatial transformation are accomplished simultaneously. Each possible match of points defines a displacement which is given a rating according to how closely other pairs would match under this displacement. The procedure is then iterated, adjusting, in parallel, the weights of each pair of points based on their ratings. The clustering technique described in ref. [6] is similar in that the matching determines the spatial transformation between the two images. In this case, the transformation is a rotation, scaling and translation although it could be extended to other transformations. For each possible pair of matching points, the parameters of the transformation are determined which represent a point in the cluster space. By finding the best cluster of these points, using classical statistical methods, the transformation which most closely matches the largest number of points is found.

4.2.3.4   Image Resampling and Transformation

The mapping functions constructed are used to transform the target image and thus to register the images. Inverse of the estimated mapping function is applied to the target image on the same coordinate system as the source image. The image interpolation takes place so that neither holes nor overlaps occur in the output image. The interpolation itself is usually realized via convolution of the image with an interpolation kernel. The nearest-neighbour function, the bilinear and bicubic functions, quadratic functions as described in refs. [51,52] belong to the most commonly used interpolants. Bilinear interpolation offers the best trade-off between accuracy and computational complexity.

4.2.4   Satellite Image Registration Methodology

Registration algorithms attempt to align a target image over a source image so that pixels present in both images are in the same coordinate system. This process is useful in the alignment of an acquired image over a template, a time series of images of the same scene, or separate bands of a composite image. Two practical applications of this process are the alignment of radiology images in medical imaging and alignment of satellite images for environmental study. Figure 4.2 shows the context diagram of satellite image registration. The whole process is divided into three phases. First, the images are loaded. In this phase, the images are pre-processed to improve their visual quality and to adjust their resolution. The second phase is the control point extraction and matching, which

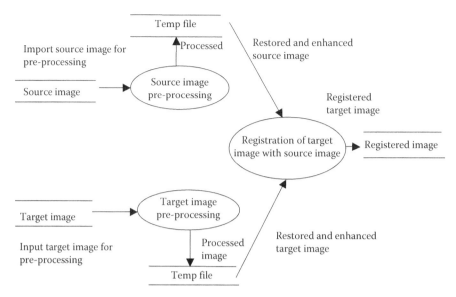

FIGURE 4.2 Context diagram of the image registration system.

is performed by using wavelet-based approach. The third phase consists of generating an output image, which is a simply registered image or a mosaic. This phase involves estimating the transformation parameters and mapping the target image on to the source image coordinates. The work has been implemented in MATLAB® [53]. The proposed approach requires that the source image and the target images are square images of resolution $2^j \times 2^j$ pixels where j is an integer. If they are not so, then parts of the images are taken such that they have 2^j resolution. For example, if the image size is 300×300 pixels then sub-image of size 256×256 pixels can be taken, for an image of size 600×600 pixels the sub-image size can be taken as 512×512 pixels. The overview of the registration process in the developed system is shown in Figure 4.3.

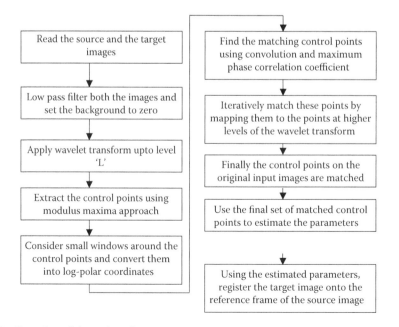

FIGURE 4.3 Overview of the registration process.

4.2.4.1 Pre-Processing

Pre-processing is the set of operations performed on the images at the lowest level of abstraction, both input and output are intensity images. These images are of the same kind as the original data captured by the sensor, with an intensity image usually represented by a matrix of image function values. Pre-processing does not increase image information content but it is very useful in a variety of situations since it helps to suppress information that is not relevant to the specific image processing or analysis task [54]. The aim of pre-processing is an improvement of the image data that suppresses undesired distortions or enhances some image features important for further processing. Pre-processing methods are classified into image enhancement and image restoration. The main objective of image enhancement is to process an image so that the result is more suitable than the original image for specific application. Image enhancement refers to accentuation or sharpening of image features such as edges, boundaries or contrast to make a graphic display more useful for display and analysis. Although the enhancement process does not increase the inherent information content in the data, it does increase the dynamic range of the chosen features so that they can be detected easily [55]. The aim of image enhancement is to improve the interpretability or perception of information in images for human viewers or to provide better input for other image processing techniques.

Image enhancement includes grey level and contrast manipulation, noise reduction, noise reduction, edge crisping and sharpening, filtering, interpolation and magnification and so on. The greatest challenge in image enhancement is quantifying the criterion for enhancement. Therefore, a large number of image enhancement techniques are empirical and require interactive procedures to obtain satisfactory results. They also depend on the applications. For example, a method that is quite useful for enhancing x-ray images may not necessarily be the approach for enhancing pictures of Mars transmitted by the space probe. Image enhancement is a very important task because of its usefulness in virtually all image processing applications. There is no general theory of image enhancement. When an image is processed for visual interpretation, the viewer is the ultimate judge of how well a particular method works. Visual evaluation of image quality is a highly subjective process, thus making the definition of a 'good image' an elusive standard. Even in situations when a clear-cut criterion of performance can be imposed on the problem, a certain amount of trial and error is usually required before a particular image enhancement approach is selected. Image enhancement approaches fall into two broad categories (a) spatial-filtering methods, and (b) frequency-domain methods [56]. The term spatial filtering refers to the image plane itself, and approaches in this category are based on direct manipulation of pixels in an image. This aspect of spatial filtering is also very useful in image fusion [57–60] using various techniques. Frequency domain processing techniques are based on modifying the Fourier transform of an image. The basic idea in image enhancement is to sharpen the image to enhance fine details. The following image enhancement techniques that have been implemented as a part of the developed system:

i. Contrast stretching: The idea behind contrast stretching is to increase the dynamic range of the grey levels in the image being processed so that the image grey level occupies the entire dynamic range available.

ii. Smoothing using ideal low-pass filter: The smoothing or blurring operation is used primarily to diminish the effect of spurious noise and false contours that may be present in the digital image.

iii. Clipping: In the developed system, after the source and the target images are low pass-filtered, the scattered portion of the clouds will either blend with the background or their intensity will diminish considerably. It is typical to follow an operation like this with clipping to eliminate objects based on their intensity. The pixels whose grey levels are less than a threshold value 'T' are set to zero so that these pixels are not chosen as control points.

iv. Image restoration: Degradation of images can have many causes: defects of optical lenses, non-linearity of the electro-optical sensor, graininess of the film material, relative motion between an object and camera, wrong focus, atmospheric turbulence in remote sensing or astronomy, scanning of photographs and so on. The objective of image restoration is to reconstruct the original image from its degraded version. In the developed system, the filters used are: (i) median filter and (ii) Wiener filter.

4.2.4.2 Control Point Extraction

Control points are significant points or structures in the images. They should be distinct, spread all over the image and efficiently detectable in both images. They are expected to be stable in time, to stay at fixed positions during the whole experiment [2]. Further, not too many of them must be selected since it is time consuming to match them. The proposed approach uses the wavelet-based modulus maxima approach to automatically extract the control points. In this approach, since the control points are extracted at the lowest level of the wavelet transform, time required is very less and not too many points are selected. The number of control points required can be controlled by the user by specifying only one parameter α. Figure 4.4 shows the steps involved in automatically detecting the control points, in the developed system. The wavelet transform is obtained using Mallat's pyramidal approach [29] for the pre-processed images. This scheme saves considerable amount of memory and computations, which is very much necessary in remote sensing applications where large images are used. The scaling and wavelet functions used are 'Haar' functions. The control points are detected from the modulus maxima of the wavelet transform, applied to the source and the target images. The wavelet transform is applied till level 'L'. The lowest level 'L' of the wavelet transform is chosen such that the image reduces to size 32×32 pixels. For example, if the images are of size 512×512 pixels, 'L' is chosen to be 4. At the lowest level L, let LH_L, HL_L and HH_L denote the images consisting of the vertical, horizontal and diagonal components of the image, respectively. The corresponding components of the source and the target images are used to detect the control points in both the source and the target images. The horizontal and vertical components at level L can be used to select a set of control points [9]. They can also be selected by using only the diagonal components [27]. The image consisting of the control points denoted as 'I_C' is obtained as follows. To identify the features that are present in both the images, the modulus maxima of the wavelet transform is used to detect sharp variation points, which correspond to edge points in the images. Let 'I_M' denote the modulus image, which is obtained by taking the square root of the sum

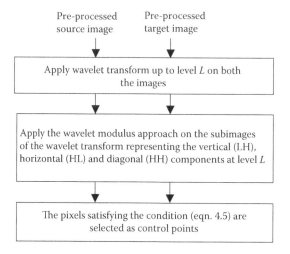

FIGURE 4.4 Steps involved in control point detection.

of squares of the vertical, horizontal and diagonal components of the wavelet transform at each pixel of the source/target image. Computation of I_M is expressed as

$$I_M = \sqrt{LH_L^2 + HL_L^2 + HH_L^2} \qquad (4.3)$$

Next, a thresholding procedure is applied on the wavelet transform modulus image I_M in order to eliminate non-significant feature points. A threshold value 'T' is calculated as follows:

$$T = \alpha(\sigma + \mu) \qquad (4.4)$$

where α is a constant provided by the user, σ is the standard deviation, and μ is the average grey level of the modulus image I_M obtained using Equation 4.3. The selection of α depends on the number of control points required. If a very large number of control points are extracted, the registration process takes a long time. But extracting less number of control points may lead to the problem of not finding appropriate matches in the source and the target images. Therefore, a moderate number of control points must be extracted. Typical values for α are 1.5, 2.0 or 2.5. Finally, all the pixels in the image I_M whose greylevel is greater than the threshold value T computed as in Equation 4.4 are selected as control points. The condition can be expressed as

$$I_C = I_M > T \qquad (4.5)$$

where I_C denotes the image containing the control points. The X and Y coordinates of each of the control point on the source image are stored in a two-dimensional array. Let 'SCP' denote the set of control points in the source image. It contains two columns. The first column is to store the 'X' coordinate and the second column is to store the 'Y' coordinate of the control point. The number of rows is equal to the number of control points selected. The X and Y coordinates of the control points selected in the target image are stored in another similar array. Let TCP denote this array. During the matching process, a mask of size $m \times m$ around each control point in the source image and a window of size $n \times n$ around each control point in the target image are considered, and then convolution is performed to check if they match. This is clearly explained in the next section. To aid the matching process, those control points on the source image which lie near the edges and for which mask cannot be selected within the image are deleted from the array, SCP. Similarly, those control points on the target image which lie near the edges and for which window cannot be selected within the range of the target image are deleted from the array, TCP. The remaining control points in the arrays are carried on to the next step.

4.2.4.3 Control Point Matching

In this implementation, matching of control points is performed in two phases. In the first phase, the matching control points are found by exhaustive search process. This phase is performed only at the lowest level of the wavelet transform. In this phase, the LL images of both the source and target images at the lowest resolution level, are divided into 'N' non-overlapping square blocks, $B_1, B_2, ..., B_N$ of size $b \times b$ pixels as shown in Figure 4.5. Control points in each block 'B_i' of the LL_L source image are compared with only the control points in the corresponding block 'B_i' of the LL_L target image. This initial comparison greatly saves the time taken for execution. The selection of 'b' depends on the transformation. This step assumes that each pixel in block 'B_i' of the source image falls into the corresponding block 'B_i' of the target image after the transformation. For example, if the images are divided into four blocks, it is assumed that the source image is not translated by more than half of its size and not rotated by more than 90°. Here each control point in each block of the source image is compared with each and every control point in the corresponding block of the target image. The

Compared with

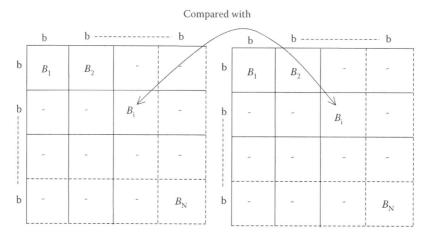

FIGURE 4.5   (a) B_N blocks of LL image at level L of the DWT of source image, (b) B_N blocks of LL image at level L of the DWT of target image.

coordinates of the matched control points are stored in a separate array. The similarity measure used is the phase correlation. In the second phase, comparison is performed iteratively at progressively higher resolution, which allows for faster implementation and higher registration precision. The candidate control points are only those that have matched in the previous iteration. The first iteration directly takes the control points from the previous phase. In each subsequent iteration since the size of the images increases by 2 in each dimension, the matched control points in the previous iteration must be mapped to the next higher level of the wavelet transform in order to compare them. The 'X' and 'Y' coordinates of the control points at level L are mapped to the previous level L–1 as follows:

$$X(L-1) = X(L)*2-1 \qquad (4.6)$$

$$Y(L-1) = Y(L)*2-1 \qquad (4.7)$$

This is the reason for considering square images of resolution 2^j. Only the control points for which the phase correlation ρ is greater than a preset threshold value are considered as matching control point pairs. At each level this threshold value is increased.

4.2.4.3.1 Matching Process
The actual control point matching in both the phases is achieved by maximizing the phase correlation coefficient over small windows surrounding the points within the LL subbands of the wavelet transform. First a mask of size $m \times m$ pixels is taken around each control point in the LL source image and a window of size $n \times n$ pixels is taken around each control point in the LL target image. m and n are assumed to be odd numbers. The window size must be greater than the mask size, that is, $n > m$. This is based on the assumption that the matching control point on the target image may be in the vicinity of the considered point rather than at the center of the window. The mask is moved along the window (convolution). As it is moved, the overlapping pixels in the mask and the window are converted into the log polar coordinate system and then their phase correlation is found. If the maximum phase correlation value exceeds a preset threshold value, then the corresponding coordinates of the pixel are taken as the matching control point pairs. Since rotation and scaling appear as translations in this new coordinate system, it helps in matching the mask and the window pixels even in the presence of rotation and scale. Zokai and Wolberg [43] suggest that, this method

performs better compared with other methods such as Levenberg–Marquardt algorithm or Fourier–Mellin transform. In an extension of the phase correlation technique, Castro and Morandi [37] have proposed a technique to register images, which are both translated and rotated with respect to each other. Rotational movement, by itself without translation, can be deduced in a similar manner as translation using phase correlation by representing the rotation as a translational displacement with polar coordinates. Rotation and scaling will appear as translations in the log-polar coordinates. Therefore, as the mask moves along the window, they are converted into log-polar coordinates and then the phase correlation is computed. If 'img' is the image to be converted into the image in log-polar coordinate system to be called as 'imglp' the following algorithm as proposed in ref. [40] can be used:

```
Let M = number of rows and N = number of columns
    Δθ = ∏/M; Base b = 10^(log(N/2)/N);
    for i = 1 to M
      θ = (i-1) * Δθ
      for j = 1 to N
            Radius r = b^j - 1
            x1 = r * cos (θ) + N/2
            y1 = r * sin (θ) + M/2
            t = fractional part of x1
            u = fractional part of y1
            x = integer part of x1
            y = integer part of y1
          imglp (i, j) = img (x,y) * (1-t) * (1-u) + img (x+1,y) * t * (1-u) +
                           img (x,y+1) * (1-t) * u + img (x+1,y+1) * t * u
      end for
    end for
```

During convolution, after converting the subparts of the images which are to be compared, into log-polar coordinates, the phase correlation ρ between two images img1 and img2 is calculated as follows:

$$\rho = F^{-1}\left(\frac{F(img1) * conj(F(img2))}{\| F(img1) * conj(F(img2)) \|} \right) \quad (4.8)$$

where F is the Fourier transform, and 'conj' is the complex conjugate. Later the maximum phase correlation value is found among the pixels in the window surrounding the control point. If the maximum value does not occur at the centre of the window, then the control point is moved to the location where the maximum value occurs. All the control points where the maximum correlation value is greater than a threshold correlation value are considered as matching points. To verify that the match is consistent, matching is performed in the reverse direction also. This reversed verification reduces the number of mismatched pairs in the matching process. Nevertheless, some false matches will inevitably occur. Therefore, a consistency checking procedure is performed.

4.2.4.3.2   Consistency Check
Since the matching process is performed iteratively, discarding inconsistent matches in the initial stages will speed up the process of matching. Since the proposed approach assumes only the presence of translation and rotation, the distance between two sets of control points on the source image denoted as P_{S1} (x_1, y_1) and P_{S2} (x_2, y_2), respectively, must be approximately equal to the distance between their corresponding matching control point pairs in the target image denoted by P_{T1} (x_1, y_1) and P_{T2} (x_2, y_2), respectively, that is, $P_{S1}P_{S2} = P_{S1}P_{S2}$. Using this fact, different bins are used to store the points based on the distances between the matching control point pairs with the other pairs in

the source and the target images. A matching pair, (P_S, P_T) is put into a bin 'i' only if distance between P_S and P_{Sj} is equal to the distance between P_T and P_{Tj}, where P_{Sj} and P_{Tj} represent each control point pair in bin i containing n such control point pairs. If the condition is not satisfied for any one of the point pair in the bin, then the pair, (P_S, P_T) is stored in a new bin. Only those points in the bins, containing sufficient number of control point pairs are carried to the next level. The matching process is repeated at each level of the wavelet transform; finally matching is performed on the original source and target images. Since the control points are fine tuned at each level, the set of points matched in the original images will be accurate. This is another advantage of using the wavelet decomposition. The final set of matching control point pairs are used to estimate the transformation parameters.

4.2.4.4 Estimation of Transformation Parameters

After the matching is performed at all the levels of the wavelet transform, the finally matched control points in the source image and the target image are stored in separate arrays. Given a sufficient number of points, the parameters of the transformation can be derived through approximation or interpolation. The proposed system uses the approximation method in which the least-squares regression analysis is used to estimate the parameters, that is, horizontal shift (Δx), vertical shift (Δy), rotation angle (θ), which can be computed as follows. It is assumed that the transformation between the target image and the source image is a rigid transform composed of a combination of rotation, translation and scale. If (X,Y) is the control point in the source image, and (X^I, Y^I) is the corresponding control point in the target image, the transform can be written as

$$\begin{bmatrix} X^I \\ Y^I \end{bmatrix} = s \begin{bmatrix} \cos\theta & -\sin\theta \\ \sin\theta & \cos\theta \end{bmatrix} \begin{bmatrix} X \\ Y \end{bmatrix} + \begin{bmatrix} \Delta X \\ \Delta Y \end{bmatrix} \tag{4.9}$$

where s is the scale factor, θ is the anti-clockwise rotation angle, and $(\Delta X, \Delta Y)$ are the translations. Derivation is as shown below. Let $\{(X_i, Y_i): i = 1, 2, \dots N\}$ represent the control points on the source image and $\{(X_i^I, Y_i^I): i = 1, 2, \dots N\}$ represent the corresponding matched control points on the target image. Let N be the number of matched control points. From Figure 4.6, we have

$$r_1 = s \cos\theta \tag{4.10}$$

$$r_2 = s \sin\theta \tag{4.11}$$

$$s = \sqrt{r_1^2 + r_2^2} \tag{4.12}$$

$$\theta = \cos^{-1}(r_1/s) \tag{4.13}$$

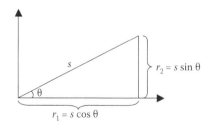

FIGURE 4.6   Representing a point (r_1, r_2) in terms of s and θ.

From Equation 4.9

$$X^| = s(X\cos\theta - Y\sin\theta) + \Delta Y \tag{4.14}$$

$$Y^| = s(X\sin\theta + Y\cos\theta) + \Delta Y \tag{4.15}$$

In general, substituting (r_1, r_2) for (s, θ) in Equations 4.14 and 4.15, we get

$$X_i^| = X_i r_1 - Y_i r_2 + \Delta X \tag{4.16}$$

$$Y_i^| = Y_i r_1 + X_i r_2 + \Delta Y \tag{4.17}$$

The mean square error is defined as follows:

$$\text{MSE} = \Sigma \left[\left(X_i^| - X_i \right)^2 + \left(Y_i^| - Y_i \right)^2 \right] \tag{4.18}$$

By minimizing the MSE and simplifying, we get

$$r_1 = \frac{\Sigma_{i=1}^{N} \left(\left(X_i X_i^| + Y_i Y_i^| \right) - N\bar{X} * \bar{X}^| - N\bar{Y} * \bar{Y}^| \right)}{\Sigma_{i=1}^{N} \left(X_i^2 + Y_i^2 \right) - N\bar{X}^2 - N\bar{Y}^2} \tag{4.19}$$

$$r_2 = \frac{\Sigma_{i=1}^{N} \left(\left(X_i Y_i^| + Y_i X_i^| \right) - N\bar{Y} * \bar{X}^| - N\bar{X} * \bar{Y}^| \right)}{\Sigma_{i=1}^{N} \left(X_i^2 + Y_i^2 \right) - N\bar{X}^2 - N\bar{Y}^2} \tag{4.20}$$

$$\Delta X = \bar{X}^| - \bar{X} * r_1 + \bar{Y} * r_2 \tag{4.21}$$

$$\Delta Y = \bar{Y}^| - \bar{Y} * r_1 + \bar{X} * r_2 \tag{4.22}$$

where $\bar{X}, \bar{Y}, \bar{X}^|, \bar{Y}^|$ are the average values of $\{X_i\}, \{Y_i\}, \{X_i^|\}, \{Y_i^|\}$, respectively. Scale s and angle θ according to Equations 4.12 and 4.13, respectively.

4.2.4.5 Image Resampling

Using the transformation parameters computed in the previous step, the target image must be transformed and thus be registered. The registered image data from the target image is determined using the inverse of the estimated mapping function. The image interpolation takes place in the target image on the regular grid. The interpolation itself is realized through convolution of the image with an interpolation kernel. The interpolant used here is the bilinear function. It offers the best trade-off between accuracy and computational complexity and thus it is the most commonly used approach. Bilinear interpolation approach of image resampling uses the grey levels of the four nearest neighbours. This approach is straightforward. Because the grey level of each of the four integral nearest neighbours of a non-integral pair of coordinates $(x^|, y^|)$ is known, the grey-level value at these

Source initial

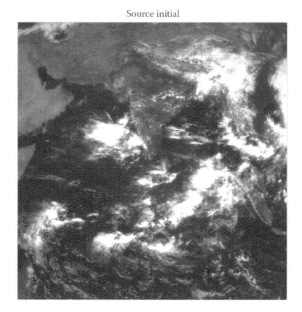

FIGURE 4.7   Source image (a portion of the large image taken by a Geostationary satellite of Indian sub-continent).

coordinates, denoted as $v(x^1, y^1)$, can be interpolated from the values of its neighbours by using the relationship

$$v(x^1, y^1) = ax^1 + by^1 + cx^1y^1 + d \qquad (4.23)$$

Target initial

FIGURE 4.8   Target image (taken on a different day) (a portion of the large image taken by a Geostationary satellite of the Indian sub-continent).

Registered

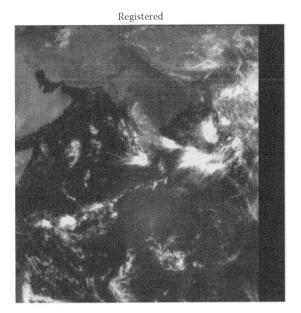

FIGURE 4.9 Registered target image.

where the four coefficients are determined from the four equations in four unknowns that can be written using the four known neighbours of (x^l, y^l). When these coefficients have been determined, $v(x^l, y^l)$ is computed and this value is assigned to the location in $f(x,y)$ that yields the spatial mapping into location (x^l, y^l). Figures 4.7 and 4.8 show the source and target satellite images obtained on two different days [61]. Figure 4.9 shows the registered target image with reference to the source image using the approach described earlier. Figure 4.10 shows the superimposed source and target images.

Superimposed

FIGURE 4.10 Superimposed images.

4.3 IMAGE FUSION

Image fusion is the process of combining relevant information from two or more images into one composite image. The input images may be captured by different sensors under different conditions. The resulting image will contain more accurate descriptive information than any of the input images. Image fusion of multisource images will have effects of denoising (reduction in noise), resolution improvement, definition improvement (image quality, detail highlighting and texture features) and compensation for loss or failure from one sensor. The most important issue concerning image fusion is to determine how to combine the different sensor images. Various applications of Image Fusion are:

- Concealed weapon detection
- Medical diagnosis
- Defect inspection
- Intelligent robots
- Remote sensing
- Military surveillance
- Image classification
- Robot vision
- Digital camera application
- Aerial and satellite imaging
- Multi-focus image fusion

4.3.1 MODES OF IMAGE FUSION

4.3.1.1 Single Sensor Image Fusion

It consists of fusion of images obtained from a single sensor such as a digital camera. Two common single sensor fusion systems using are given below.

4.3.1.2 Fusion of Multi-Exposure Images

It is very difficult to obtain a single image of a scene where all parts of the scene are well exposed. Some details are under-exposed while some are over-exposed. Different shutter speeds are required to capture the details of interesting areas of a scene. A set of such multi-exposure images of a scene obtained by a single camera can be fused into a single image so that all image areas appear well exposed.

4.3.1.3 Fusion of Multi-Focus Images

Sharp images contain better information than blurred images. Because of the differences in a scene's depth, it is not possible to capture an image where all parts of the scene are equally sharp. Only scene areas that are at the focus plane appear sharp, and other areas in front of or behind the focus plane appear blurred. Fusion of such multi-focus images produces a sharp image. A scene area at and near the focus plane appears sharp in an image and has a higher contrast. This information can be used for fusion.

4.3.1.4 Multisensor Image Fusion

In many cases, images from one sensor cannot give the complete picture. Different sensors exploit different regions of the image. Therefore, a multi-sensor image fusion system can take full information from the individual sensors and produce a better image that contains more information than the original ones.

4.3.2 IMAGE FUSIONS AT DIFFERENT LEVELS

There are three levels of image fusions, pixel-level image fusion, feature-level image fusion and decision-level image fusion according to Song et al. [57] based on the phase at which the image is fused.

4.3.2.1 Pixel-Level Image Fusion

This is the lowest level of processing; Pixel-level fusion method focuses on fusing the data from all sensors. The sets of pixels in the source images are merged pixel to pixel. This merging of pixels takes place according to a defined decision rule to form the corresponding pixel in the fused image. Fusion at this level requires accurate spatial registration of the images from different sensors prior to applying the fusion operator. Pixel-based fusion schemes work by combining the pixel values of the two or more images to be fused in a linear or non-linear way. According to Malviya and Bhirud [58] they range from simple averaging of the pixel values of registered images to more complex multiresolution pyramid and wavelet methods.

4.3.2.2 Feature-Level Image Fusion

Here the relevant features are first extracted using segmentation procedures from the data. They are then fused based on features that match some selection criteria. The features can be differentiated by characteristics such as size, shape, contrast and texture. As the fusion is based on identified features in the sources the resulting probability of detecting useful features in the fused image increases.

4.3.2.2.1 Decision-Level Image Fusion

Decision image fusion, also called symbol image fusion, is a high-level information fusion. At this level, decisions/detections are based on the outputs from the individual sensors. The outputs are fused together and used to reinforce common interpretation or resolve any differences. Decision-level fusion combines the results from multiple algorithms to yield a final fused decision. This method requires a high abstract and lower homogeneity of data source. In actual applications the features of the images at different levels chosen and combined for optimal fusion effect.

4.3.3 IMAGE FUSING METHODS

Some of the well-known image fusion methods are high-pass filtering technique, weighted average method, Laplacian pyramid method, IHS transform-based image fusion, PCA-based image fusion, wavelet transform image fusion, pair-wise spatial frequency matching and fuzzy logic-based methods. The research activities are mainly in the area of developing fusion algorithms that improve the information content of the composite imagery and for making the system robust to the variations in the scene, such as dust or smoke, and environmental conditions like day or night [57]. Several types of pyramid decomposition or multi-scale transform are developed for image fusion, such as Laplacian pyramid, ratio-of-low-pass pyramid, morphological pyramid, gradient pyramid and multi-scale wavelet decomposition [59]. Park et al. [60] use Daubechies wavelet basis to enhance image sharpening and preserve spectral information. Wavelet-based image registration can be performed using the work proposed by Myna et al. [61]. The basic strategy of fusion using pyramidal approach is to first fuse the information at the same level to obtain the fused information at a higher level. The fusion is then carried out at corresponding level. Multilevel image fusion is essentially an information processing process of integrating and level-by-level abstracting the multi-source information from a low level to a high level. Fusion of CT and MRI has been achieved by many researchers. CT and nMRI are complementary on reflecting human body information. Teng et al. [62] fuse CT and MRI images and show that there is a need to fuse the effective information in order to provide more useful information for clinical diagnosis.

4.3.4 EVALUATION OF IMAGE FUSION PERFORMANCE

Evaluation techniques are broadly classified as subjective evaluation of image fusion performance, objective evaluation criteria of image fusion performance. According to Song et al. [57], subjective evaluation is based on the performance of the fused images as usually observed and evaluated by humans. Their perception for images depends on not only the content of the images but also the mental state of the observer. It may be affected by the environment condition, visual performance and knowledge level. These observers may be naïve without any training or may be observers experienced in image technologies. Therefore, subjective evaluation may be complicated but it provides information about the visual quality of the images as perceived by humans. Objective evaluation criterion uses objective performance indicators to evaluate the image fusion effect. For comparing the fusion methods, evaluation indicators such as root mean square error, cross entropy, mutual information and combination entropy are used. If the purpose is to increase the spatial resolution, then indicators such as image average value, standard deviation and spatial resolution are used. If the purpose is to improve the resolution, then evaluation indicators such as standard deviation, average gradient, spatial frequency and contrast variation. To evaluate whether the information content of a fused image is increased or not, indicators such as entropy, cross entropy, mutual information, combination entropy and standard deviation. If the purpose is to fuse the spectral characteristic of the image, then the indicators are deviation index, correlation coefficient and spectral distortion.

4.3.5 Image Fusion Using Fuzzy Logic

Recently fuzzy approaches have been used to account for uncertainty and to incorporate heuristic knowledge. Fuzzy logic and fuzzy sets give the formal tools to reason about incomplete, imprecise, uncertain information. Elements of a fuzzy set have membership degrees to that set (see Figure 1.4). Fuzzy set theory defines fuzzy operators on fuzzy sets. Fuzzy logic uses If–Then rules to apply the appropriate fuzzy operator. Fuzzy image processing using Type-1 Fuzzy Logic (T1FL) has three main stages: image fuzzification, inference process and image defuzzification. Fuzzification process is concerned with finding a fuzzy representation of non-fuzzy input values. The membership function is the essence of fuzzy sets. A membership function is used to associate a degree of membership of each of the elements of the domain to the corresponding fuzzy set. The membership function for fuzzy sets can be of any shape or type as determined by experts in the domain over which the sets are defined. Inference process is to map the fuzzified inputs to the rule base, and to produce a fuzzified output for each rule. A degree of membership to the output sets are determined based on the degree of membership in the input sets and the relationship between the input sets. Defuzzification process is to convert the output of the fuzzy rules into a scalar, or non-fuzzy value [63]. Fuzzy logic can be applied at various levels of the image fusion process. A multiresolution image fusion scheme has been proposed by Liu et al. [64] based on fuzzy region feature and the fusion process has been implemented in fuzzy space. The source images are segmented into important regions, subimportant regions and background regions. K-means clustering algorithm can be used for segmentation according to the pixel grey-level distribution. The importance of image region is relative. One cannot very definitely determine whether a region is important or not. The region feature importance is a fuzzy concept. Therefore, it is necessary to fuzzy the importance attribute of regions. The fuzzification is performed as follows. Suppose the highest grey level and lowest grey level is L_{max} and L_{min} respectively, default is 255 and 0. Then defining the function of membership i region belongs to j is

$$\mu_{i,j} = \exp\left[\frac{-(ME_i - E_j)^2}{(L_{max} - L_{min})/2}\right] \tag{4.24}$$

where $E_1 = L_{min}$, $E_2 = (L_{max} - L_{min})/2$, $E_3 = L_{max}$; and ME_i is the mean of pixel grey level within region i; $\mu_{i,1}$, $\mu_{i,2}$, $\mu_{i,3}$ are the values of membership in important region, subimportant region and background region. E_1, E_2 and E_3 are the three respective attributes of the image important region, subimportant region and background region. $ME_i = E_1$ indicates that the region i is background, the fusion result F_1 is the corresponding region of image B; $ME_i = E_2$ indicates that the region i is subimportant, the fusion result F_2 is obtained by single pixel-based fusion algorithm; $ME_i = E_3$ indicates that the region i is important, the fusion result F_3 is the corresponding region of image A. Finally, depending on the feature of every region, the membership of each pixels is defined as $\mu_{i,1}$, $\mu_{i,2}$ and $\mu_{i,3}$. The final fusion result is achieved by the defuzzification process using the membership

$$F = \sum_{i=1}^{3} \mu_{i,j} F_j / / \sum_{i=1}^{3} \mu_i \qquad (4.25)$$

where F is the multiresolution representation of the fused image. The final fused image is obtained by performing inverse discrete wavelet frame transform. Meitzler et al. [65] applied both Mamdani and ANIFS methods for pixel image fusion. Images are converted into column form. Using Mamdani Fuzzy Logic an FIS file is made taking two input images. The number and type of membership functions for both the images are decided by tuning functions. Input images in antecedent are resolved to a degree of membership from 0 to 255. Rules are made for two input images to resolve the two antecedents to a single number from 0 to 255. On every pixel fuzzification is applied using the rules developed previously which gives a fuzzy set represented by a membership function and results in output image in column format. Later the column form is converted to matrix and the fused image can be displayed. The ANFIS technique uses training data which is a matrix with three columns and entries in each column are from 0 to 255 in steps of 1. A check data which is a matrix of pixels of two input images in column format is formed. For training, FIS structure which is generated by *gesfis1* command is needed with training data, number of membership functions and type of membership functions as input. To start training, *anfis* command is used which takes generated FIS structure and training data as input and returns trained data. On every pixel, fuzzification is applied using the generated FIS structure with check data and trained data as inputs which returns output image in column format. Later the column form is converted to matrix form and the fused image is displayed. Yan Na et al. [63] use fuzzy logic approach to fuse CT and MRI images. For designing fuzzy inference rules, the contents of two source images are analysed with histograms. According to the imaging mechanism of CT image and MRI image, CT image provides bone information while MRI image provides soft tissue information of human body. So the fusion target is to fuse bone information from CT image and soft tissue information from MRI image. For better design fuzzy inference rules, the content of two images is analysed with histogram. Nine sections are used in the calculation of histograms. Nine fuzzy sets mf1, mf2, mf3, mf4, mf5, mf6, mf7, mf8 and mf9 are used in fuzzification process. The input membership function is Gaussian function and the output membership function is trapezoidal function. Following are 17 inference rules designed by using fusion target and histograms of CT image and MRI image [63]:

    1. If (ct is mf1) and (mri is mf1) Then (output1 is mf1)
    2. If (ct is mf1) and (mri is mf2) then (output1 is mf2)
    3. If (ct is mf1) and (mri is mf3) then (output1 is mf4)
    4. If (ct is mf1) and (mri is mf4) then (output1 is mf5)
    5. If (ct is mf1) and (mri is mf5) then (output1 is mf6)
    6. If (ct is mf1) and (mri is mf6) then (output1 is mf6)
    7. If (ct is mf1) and (mri is mf7) then (output1 is mf7)
    8. If (ct is mf1) and (mri is mf8) then (output1 is mf8)
    9. If (ct is mf1) and (mri is mf9) then (output1 is mf9)
   10. If (ct is mf2) then (output1 is mf2)
   11. If (ct is mf3) then (output1 is mf3)

12. If (ct is mf4) then (output1 is mf4)
13. If (ct is mf5) then (output1 is mf5)
14. If (ct is mf6) then (output1 is mf6)
15. If (ct is mf7) then (output1 is mf7)
16. If (ct is mf8) then (output1 is mf8)
17. If (ct is mf9) then (output1 is mf9)

Pixel-level fusion based on fuzzy and neuro-fuzzy algorithms have been implemented by Singh et al. [66] to fuse a variety of images. Another pixel-level algorithm based on fuzzy logic has been proposed in ref. [67]. Here the pixel values are fuzzified, membership modification is performed and then defuzzified. The application of the full fuzzy transform on image fusion has been tested in ref. [68]. Firooz Sadjadi [69] compares the effectiveness of several image fusion algorithms. They have also defined a set of measures of effectiveness. The benefit of using the fuzzy logic approach in image sensor fusion is that simple rules can be exploited to achieve the fusion of several bands [65]. A content analysis-based medical images fusion with fuzzy inference is presented in [63].

4.3.6 TYPE-2 FUZZY LOGIC IN IMAGE FUSION

The fuzzy logic technique, especially Type 1, for image fusion is now being widely used. Type-2 Fuzzy sets are being increasingly used for modelling uncertainty and imprecision in a better way than Type-1 [70]. Type-1 fuzzy sets are not able to directly model some uncertainties because their membership functions are totally crisp. On the other hand, Type-2 fuzzy sets are able to model such uncertainties because their membership functions are themselves fuzzy. Membership functions of Type-1 fuzzy sets are two-dimensional, whereas membership functions of Type-2 fuzzy sets are three-dimensional. It is the new third-dimension of Type-2 fuzzy sets that provides additional degrees of freedom that make it possible to directly model uncertainties [71]. The following methodology can be applied for multiresolution image fusion using Type-2 Fuzzy Logic (T2FL) as depicted in Figure 4.11: (i) apply wavelet transform (WT) on the input image series, (ii) then determine the important regions at the lowest level of the WT using interval T2FL approach, (iii) using interval T2FL FIS (fuzzy inference system/engine) fuse the images based on the important regions determined from step 2. Use the type reduction algorithm to get FL Type I and then the defuzzification process is applied, (iv) repeat these steps at every level of the wavelet transform, and then apply inverse wavelet transform to get the final fused image. Unfortunately, there is not much open literature on the application of Type II fuzzy logic to image fusion problem. Further work on image segmentation and image fusion is dealt with in Chapters 5 and 6.

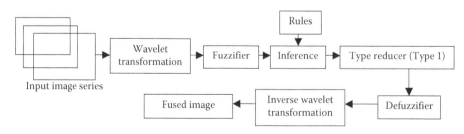

FIGURE 4.11 Application of Type-2 fuzzy logic to image fusion, fuzzifier uses the Type II fuzzy membership function.

REFERENCES

1. L.G. Brown, A survey of image registration techniques, *ACM Computer Surveys*, 24, 1992, 325–376.
2. B. Zitova and J. Flusser, Image registration methods: A survey, *Image Vision and Computing*, 21(11), 2003, 977–1000.
3. D. Fedorov, L.M.G. Fonseca, C. Kenney, and B.S. Manjunath, Automatic registration and mosaicing system for remotely sensed imagery, in *9th International Symposium on Remote Sensing, International Society for Optical Engineering*, Sept. 2002, Crete, Greece.
4. L.M.G. Fonseca, G. Hewer, C. Kenney and B.S. Manjunath, Registration and fusion of multispectral images using a new control point assessment method derived from optical flow ideas, *Proceedings of SPIE*, 3717, 104–111, April 1999, Orlando, FL.
5. S.S. Beuchemin, and J.L. Barron, The computation of optical flow, *ACM Computing Surveys*, 27, 1995, 433–467.
6. G.C. Stockman, S. Kopstein, and S. Beneth, Matching images to models for registration and object detection via clustering, *IEEE Transactions on Pattern Analysis and Machine Intelligence*, 4, 1982, 229–241.
7. J. Ton, and A.K. Jain, Registering Landsat images by point matching, *IEEE Transactions on Geoscience and Remote Sensing*, 27, 1989, 642–651.
8. B.S. Manjunath, C. Shekhar, and R. Chellappa, A new approach to image feature detection with applications, *Pattern Recognition*, 29, 1996, 627–640.
9. L.M.G. Fonseca and H.M. Costa, Automatic registration of satellite images, *Proceedings of the Brazilian Symposium on Computer Graphic and Image Processing*, Brazil, 1997, 219–226.
10. J.W. Hsieh, H.Y.M. Liao, K.C. Fan, and M.T. Ko, A fast algorithm for image registration without predetermining correspondence, *Proceedings of the International Conference on Pattern Recognition ICPR'96*, Vienna, Austria, 1996, pp. 765–769.
11. Y.C. Hsieh, D.M. McKeown, and F.P. Perlant, Performance evaluation of scene registration and stereo matching for cartographic feature extraction, *IEEE Transactions on Pattern Analysis and Machine Intelligence*, 14, 1992, 214–237.
12. C.Y. Wang, H. Sun, S. Yadas, and A. Rosenfeld, Some experiments in relaxation image matching using corner features, *Pattern Recognition*, 16, 1983, 167–182.
13. B. Zitova, Image registration of blurred satellite images, http://staff.utia.cas.cz/zitova/registration.htm
14. B. Zitova, J. Kautsky, G. Peters, and J. Flusser, Robust detection of significant points in multiframe images, *Pattern Recognition Letters*, 20, 1999, 199–206.
15. X. Dai, and S. Khorram, Development of a feature-based approach to automated image registration for multitemporal and multisensor remotely sensed imagery, *International Geoscience and Remote Sensing Symposium IGARSS'97*, Singapore, 1997, pp. 243–245.
16. H. Li and B.S. Manjunath, A contour-based approach to multisensor image registration, *IEEE Transactions on Image Processing*, 4(3), 1995, 320–334
17. H. Maitre and Y. Wu, Improving dynamic programming to solve image registration, *Pattern Recognition*, 20, 1987, 443–462.
18. S.Z. Li, J. Kittler, and M. Petrou, Matching and recognition of road networks from aerial images, *Proceedings of the Second European Conference on Computer Vision ECCV'92*, St Margherita, Italy, 1992, pp. 857–861.
19. S. Growe and R. Tonjes, A knowledge based approach to automatic image registration, *IEEE Intl. Conference on Image Processing (ICIP '97)*, Santa Barbara, CA, USA, Vol. 3, pp. 228–231, Oct. 1997.
20. M. Holm, Towards automatic rectification of satellite images using feature based matching, *Proceedings of the International Geoscience and Remote Sensing Symposium IGARSS'91*, Espoo, Finland, 1991, pp. 2439–2442.
21. M. Sester, H. Hild, and D. Fritsch, Definition of ground control features for image registration using GIS data, *Proceedings of the Symposium on Object Recognition and Scene Classification from Multispectral and Multisensor Pixels*, CD-ROM, Columbus, OH, 1998.
22. M. Roux, Automatic registration of SPOT images and digitized maps, *Proceedings of the IEEE International Conference on Image Processing ICIP'96*, Lausanne, Switzerland, 1996, pp. 625–628.
23. P. M. Dare, R. Ruskone, and I.J. Dowman, Algorithm development for the automatic registration of satellite images, *Image Registration Workshop*, NASA Goddard Space Flight Center, pp. 88–88, Nov. 1997.
24. H. Koch, K. Pakzad, R. Tonjes, Knowledge based interpretation of aerial images and maps using a digital landscape model as partial interpretation, in: *Semantic Modeling for the Acquisition of Topographics Information from Images and Maps*, Birkhäuser Verlag, Basel, pp. 319, 1997.

25. T. Kadota and M. Takagi, Acquisition method of ground control points for high resolution satellite imagery, *Lecture Papers of the 25th Asian Conference on Remote Sensing*, 2002. http: //www. infra.kochi-tech.ac.jp/takagi/Papers/acr02_tka.pdf, accessed July 2011.
26. M. Tuo, L. Zhang, Y. Liu, Multisensor aerial image registration using direct histogram specification, *IEEE Intl. Conference on Networking, Sensing and Control*, Taipei, Taiwan, pp. 807–812, March 2004.
27. J.P. Djamdji, A. Bajaouri, and R. Maniere, Geometrical registration of images: The multiresolution approach, *Photogrammetric Engineering and Remote Sensing*, 59(5), 1993, 645–653.
28. Y. You and M. Kaveh, A regularization approach to joint blur identification and image restoration, *IEEE Transactions on Image Processing*, 5, 1996, 416–428.
29. S.G. Mallat, A Theory for multiresolution signal decomposition: The wavelet representation, *IEEE Trans. on Pattern Analysis and Machine Intelligence*, 2(10), 1989, 674–693.
30. V. Rao, K.M.M. Rao, A.S. Manjunath, and R.V.N. Srinivas, Optimization of automatic image registration algorithms and characterization, geoimagery bridging continents, *XXth ISPRS Congress*, 12–23 July 2004, pp. 698–703, Istanbul, Turkey.
31. M. Corvi and G. Nicchiotti, Multiresolution image registration, *in Proceedings* 1995 *IEEE Conference on Image Processing*, Vol. 3, pp. 224–227, 23–26 October 1995, Washington DC, USA.
32. C. Chui, *An Introduction to Wavelets*, C. Chui (Editor), Academic Press, New York, 1992.
33. I. Zavorin and J. Le Moigne, Use of multiresolution wavelet feature pyramids for automatic registration of multisensor imagery, *IEEE Transactions on Image Processing* 14(6), 2005, 770–782.
34. R.J. Althof, M.G.J. Wind, and J.T. Dobbins, A rapid and automatic image registration algorithm with subpixel accuracy, *IEEE Transactions on Medical Imaging* 16, 1997, 308–316.
35. H. Hanaizumi and S. Fujimura, An automated method for registration of satellite remote sensing images, *Proceedings of the International Geoscience and Remote Sensing Symposium IGARSS'93*, Tokyo, Japan, 1993, pp. 1348–1350.
36. C.D. Kuglin and D.C. Hines, The phase correlation image alignment, *Proceedings of the IEEE International Conference on Cybernetics and Society*, New York, 1975, pp. 163–165.
37. E.D. Castro and C. Morandi, Registration of translated and rotated images using finite fourier transform, *IEEE Transactions on Pattern Analysis and Machine Intelligence* 9, 1987, 700–703.
38. A.V. Cideciyan, S.G. Jacobson, C.M. Kemp, R.W. Knighton, and J.H. Nagel, Registration of high-resolution images of the retina, *SPIE Medical Imaging VI: Image Processing* 1652, 1992, 310–322.
39. B.S. Reddy and B.N. Chatterji, An FFT-based technique for translation, rotation and scale-invariant image registration, *IEEE Transactions on Image Processing* 5, 1996, 1266–1271.
40. H. Xie, N. Hicks, G.R. Keller, H. Huang, V. Kreinovich, An IDL/ENVI implementation of the FFT-based algorithm for automatic image registration, *Intl. Journal of Computers and Geosciences*, 29, 2003, 1045–1055
41. D. Young, Straight lines and circles in the log-polar image, *Proceedings of the 11th British Machine Vision Conference*, Bristol, UK, 2000, pp. 426–435.
42. G. Wolberg and S. Zokai, Robust image registration using log-polar transform, Proceedings of the IEEE International Conference on Image Processing, Canada, Sept. 2000.
43. S. Zokai and G. Wolberg, Image registration using log-polar mappings for recovery of large-scale similarity and projective transformations, *IEEE Transactions on Image Processing*, 14(10), Oct. 2005, 1422–1434.
44. J. Flusser, Object matching by means of matching likelihood coefficients, *Pattern Recognition Letters* 16, 1995, 893–900.
45. S. Abdelsayed, D. Ionescu, and D. Goodenough, Matching and registration method for remote sensing images, *Proceedings of the International Geoscience and Remote Sensing Symposium IGARSS'95*, Florence, Italy, 1995, pp. 1029–1031.
46. L. Eikvil, P.O. Husoy, and A. Ciarlo, Adaptive image registration, http://earth.esa.int/rtd/Events/ESA_EUSC_2005, Oct. 2005.
47. A. Rosenfeld and G.J. Vanderbrug, Coarse–fine template matching, *IEEE Transactions on Systems, Man and Cybernetics*, 7, 1977, 104–107.
48. J. Le Moigne, Parallel registratin of multi-sensor remotely sensed imagery using wavelet coefficients, *Proceedings of the SPIE: Wavelet Applications*, Orlando, FL, 2242, 1994, pp. 432–443.
49. D.W. Marquardt, An algorithm for least-squares estimation of non-linear parameters, *Journal of the Society for Industrial and Applied Mathematics*, 11(2), 1963, 431–441.
50. S. Ranade and A. Rosenfeld, Point pattern matching by relaxation, *Pattern Recognition*, 12(2), 1980, 269–275.

51. N.A. Dodgson, Quadratic interpolation for image resampling, *IEEE Transactions on Image Processing* 6, 1997, 1322–1326.
52. K. Toraichi, S. Yang, and R. Mori, Two-dimensional spline interpolation for image reconstruction, *Pattern Recognition* 21, 1988, 275–284.
53. R.C. Gonzalez, R.E. Woods, and S.L. Eddins, *Digital Image Processing Using MATLAB*, Pearson Education, Inc.
54. M. Sonka, V. Hlavac, and R. Boyle, *Image Processing, Analysis and Machine Vision*, Cengage Learning, India, 2008.
55. A. K. Jain, *Fundamentals of Digital Image Processing*, Prentice-Hall, Inc., Englewood, Cliffs, NJ, October, 2000.
56. R.C. Gonzalez and R.E. Woods, *Digital Image Processing*, 3rd Edition, Pearson Education, Inc., Upper Saddle River, NJ, Indian Edition by Dorling Kindersley India Pvt. Ltd., 2009.
57. B. Song and Y. Fu, *The Study on the Image Fusion for Multisource Image*, 2nd International Asia Conference on Informatics in Control, Automation and Robotics, Wuhan, pp. 138–141, 6–7 March 2010.
58. A. Malviya and S.G. Bhirud, Image fusion of digital images, *International Journal of Recent Trends in Engineering*, 2(3), 2009, 146–148.
59. E. Fernandez, *Image Fusion*, Project Report, University of Bath, June 2002.
60. J.-H. Park, K.-O. Kim, and Y.-K. Yang, Image fusion using multiresolution analysis, *IEEE International Symposium on Geoscience and Remote Sensing*, Sydney, Australia, pp. 864–866, Vol. 2, 2001.
61. A.N. Myna, M.G. Venkateshmurthy, and C.G. Patil, Automatic registration of satellite images using wavelets and log-polar mapping, *First International Conference on Signal and Image Processing*, Hubli, Vol. 1, pp. 446–451, 7–9 Dec. 2006.
62. J. Teng, S. Wang, J. Zhang, X. Wang, Fusion algorithm of medical images based on fuzzy logic, *Seventh International Conference on Fuzzy Systems and Knowledge Discovery*, Yantai, Shandog, pp. 546–550, 10–12 Aug. 2010.
63. Y. Na, H. Lu, and Y. Zhang, Content analysis based medical images fusin with fuzzy inference, *Fifth International Conference on Fuzzy Systems and Knowledge Discovery*, Shandong, pp. 37–41, 18–20 Oct. 2008.
64. G. Liu, Z.-L. Jing, and Shao-Yuan, Multiresolution image fusion scheme based on fuzzy region feature, *Journal of Zhejiang University Science A*, 7(2), 2006, 117–122.
65. T.J. Meitzler, D. Bednarz, E.J. Sohn, K. Lane, and D. Bryk, Fuzzy logic based image fusion, *Aerosense 2002*, Orlando, FL, April 2–5, 2002.
66. H. Singh, J. Raj, G. Kaur, and T. Meitzler, Image fusion using fuzzy logic and applications, *IEEE International Conference on Fuzzy Systems*, Budapest, Hungary, pp. 337–340, 25–29 July 2004.
67. R. Maruthi and K. Sankarasubramanian, Pixel level multifocus image fusion based on fuzzy logic approach, *Asian Journal of Information Technology*, 7(4), 2008, 168–171.
68. M. Dankova and R. Valasek, Full fuzzy transform and the problem of image fusion, *Journal of Electrical Engineering*, 57(7), 2006, 82–84.
69. F. Sadjadi, Comparative image fusion analysis, *IEEE Computer Society Conference on Computer Vision and Pattern Recognition*, San Diego, 25 June 2005.
70. J.R. Castro, O. Castillo, L.G. Martinez, Interval type-2 fuzzy logic toolbox, *Engineering Letters*, 15(1), 2007, EL_15_1_14.
71. J.M. Mendel, I. Robert, and B. John, Type-2 fuzzy sets made simple, *IEEE Transactions on Fuzzy Systems*, 10(2), 2002, 117–127.

5 Multi-Sensor Image Fusion Using Discrete Cosine Transform

V. P. S. Naidu

CONTENTS

5.1 INTRODUCTION

Multi-sensor image fusion (MSIF) is a procedure that combines two or more registered source images to produce a single image of the observed scene. The resultant fused image is supposed to have/would have more information (in some overall sense) than any of the source images. Recently, MSIF has emerged as an innovative and promising research area in digital image processing discipline. The application areas of MSIF are (i) battlefield monitoring, (ii) remote sensing, (iii) machine vision, (iv) robotic vision, (v) airfield/airport surveillance, (vi) enhanced vision system and (vii) medical imaging. The MSIF merges the information contents from several base images (or images from different sensor modalities) taken from the same scene in order to construct and accomplish a single image that contains most of the useful information from the original registered source images [1]. The information (or uncertainty) that is not required in particular is minimized. Therefore, in general, the fused image is of higher quality or contains more information compared to the original source images. The MSIF can be performed at three different levels: pixel level, feature level and/or decision level [2,3]. In this chapter, pixel-level-based MSIF using multi-resolution discrete cosine transform (MDCT) is presented along with results. The basic aspects of image fusion, registration, feature extraction and some methods for image fusion have been considered in Chapter 4.

A simple pixel-level MSIF is to take the mean (average or mathematical expectation in the probabilistic sense) of the grey-level registered source images pixel-by-pixel (PBP). This algorithm is computationally very simple but it could produce some undesired effects and reduced feature contrast in the resultant fused image. To overcome these problems, multi-resolution techniques such as (a) wavelets [1,4,5], (b) multi-scale transforms—image pyramids [3,6], (c) signal processing—spatial frequency [7] and (d) statistical signal processing [8,9] have been studied and used by many researchers. The method of wavelet transforms (WT) provides a good localization in both spatial and frequency domains and hence are used in many signal processing applications.

The discrete wavelet transform (DWT) provides directional information in decomposition levels and contains unique information at different resolutions [4,5]. Using these characteristics of WT, image fusion algorithms were presented earlier [1–9]. Ref. [10] presents a novel image fusion methodology using fuzzy set theory. Similar characteristics, as that of DWT, can also be accomplished using MDCT. Hence, in this chapter, a novel MDCT-based image fusion algorithm is developed to fuse the registered source images. In the image registration process, the information in the source images is adequately aligned and registered prior to merging of the images (means the source/base images are properly registered). We assume here in this chapter that the source images are registered.

5.2 BASICS OF DISCRETE COSINE TRANSFORM

In image processing discipline, discrete cosine transform (DCT) is an important and widely used transform for researchers and technologists. In this transform domain, large DCT coefficients are concentrated in the low-frequency region and therefore it is known to have good energy compaction properties. The DCT $X(k)$ of a 1D signal $x(n)$ of length N can be represented as [11–13]

$$X(k) = \alpha(k) \sum_{n=0}^{N-1} x(n) \cos\left(\frac{\pi(2n+1)k}{2N}\right), \quad 0 \le k \le N-1 \tag{5.1}$$

where

$$\alpha(k) = \begin{cases} \sqrt{\dfrac{1}{N}} & k = 0 \\[2mm] \sqrt{\dfrac{2}{N}} & k \ne 0 \end{cases} \tag{5.2}$$

n = sample index
k = frequency index (normalized)

At $k = 0$, Equation 5.1 becomes $X(0) = \sqrt{1/N}\sum_{n=0}^{N-1} x(n)$. It is the mean of the image signal $x(n)$ and is known as DC (direct coefficient), and other coefficients ($X(k)$, $k \ne 0$) are known as AC (alternative coefficients). The inverse discrete cosine transform (IDCT) is defined as

$$x(n) = \sum_{k=0}^{N-1} \alpha(k) X(k) \cos\left(\frac{\pi(2n+1)k}{2N}\right), \quad 0 \le n \le N-1 \tag{5.3}$$

Generally, Equation 5.1 is called the analysis formula or forward transform and Equation 5.3 is called the synthesis formula or inverse transform. In Equations 5.1 and 5.3, the (orthogonal) basis sequence $\cos(\pi (2n + 1)k/2N)$ is real and displays discrete time (co-)sinusoids. Similarly, the 2D discrete cosine transform $X(k_1, k_2)$ of an image signal $x(n_1, n_2)$ of size $N_1 \times N_2$ is defined as [11]

$$X(k_1, k_2) = \alpha(k_1)\alpha(k_2) \sum_{n_1=0}^{N_1-1} \sum_{n_2=0}^{N_2-1} x(n_1, n_2) \cos\left(\frac{\pi(2n_1+1)k_1}{2N_1}\right)$$
$$\times \cos\left(\frac{\pi(2n_2+1)k_2}{2N_2}\right), \quad \begin{array}{l} 0 \le k_1 \le N_1 - 1 \\ 0 \le k_2 \le N_2 - 1 \end{array} \tag{5.4}$$

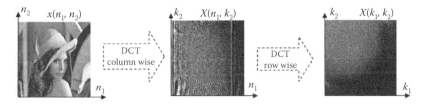

FIGURE 5.1 Two-dimensional DCT computation using separability property.

and the 2D-IDCT is defined as

$$
\begin{aligned}
x(n_1, n_2) = \sum_{k_1=0}^{N_1-1} \sum_{k_2=0}^{N_2-1} \alpha(k_1)\alpha(k_2) X(k_1, k_2) \cos\left(\frac{\pi(2n_1+1)k_1}{2N_1} \right) \\
\times \cos\left(\frac{\pi(2n_2+1)k_2}{2N_2} \right), \quad
\begin{array}{l} 0 \le n_1 \le N_1 - 1 \\ 0 \le n_2 \le N_2 - 1 \end{array}
\end{aligned}
\tag{5.5}
$$

Here, $\alpha(k_1)$ and $\alpha(k_2)$ are similarly defined as in Equation 5.2. Both 2D-DCT and 2D-IDCT are separable transformations and the advantage of this property is that 2D-DCT or 2D-IDCT can be computed in two steps by successive 1D-DCT or 1D-IDCT operations on columns and then on rows of an image signal x of size (n_1, n_2), as illustrated in Figure 5.1.

5.3 MULTI-RESOLUTION DISCRETE COSINE TRANSFORM

The MDCT is very much similar to the WT. In WT, the signal is filtered separately by low-pass and high-pass finite impulse response (FIR) filters and the output of each filter is decimated by a factor of two to obtain the first level of decomposition. The decimated low-pass-filtered output is then filtered separately by low-pass and high-pass filters followed by decimation with a factor of two to obtain the second level of decomposition. The successive levels of decomposition can be achieved by repeating this procedure. The inspiration and idea behind the MDCT process are to substitute the FIR filters by DCT [14,15]. The schematic of MDCT (one level of decomposition) is shown in Figure 5.2. The image to be decomposed is transformed into frequency domain by applying DCT column wise: (i) take the IDCT on first 50% of points (0 to 0.5π) to obtain the low-pass image L; (ii) similarly, take

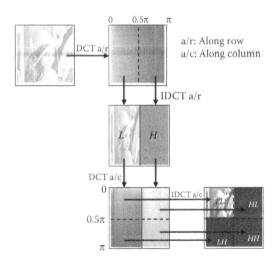

FIGURE 5.2 Information flow diagram of multi-resolution image decomposition using DCT.

the IDCT on second 50% of points (0.5π to π) to obtain the high-pass image H; (iii) the low-pass image L is transformed into frequency domain by applying DCT in row wise; (iv) take the IDCT on first 50% of points (in row wise) to obtain the low-pass image LL; (v) similarly take IDCT on the remaining 50% to obtain the high-pass image LH; (vi) the high-pass image H is transformed into frequency domain by applying DCT in row wise; (vii) take the IDCT on first 50% of points (in row wise) to obtain the low-pass image HL and (viii) similarly take IDCT on the remaining 50% to obtain the high-pass image HH. The LL contains the average image information corresponding to the low-frequency band of multi-scale decomposition. It could be considered as smoothed and sub-sampled version of the source image. It represents the approximation of source image. The LH, HL and HH are detailed sub-images which contain directional (horizontal, vertical and diagonal) information of the source image due to spatial orientation. Multi-resolution could be achieved by recursively applying the same algorithm to low-pass coefficients (LL) from the previous decomposition. The MATLAB® code of MDCT used for image decomposition (one level) is given below:

```
function[I] = mrdct(im)
% multi-Resolution Discrete Cosine Transform
% VPS Naidu, MSDF Lab, NAL
% input: im (input image to be decomposed)
% output: I (Decomposed image)

[m,n] = size(im);
mh = m/2; nh = n/2;
for i = 1:m
  hdct(i,:) = dct(im(i,:));
end
for i = 1:m
  hL(i,:) = idct(hdct(i,1:nh));
  hH(i,:) = idct(hdct(i,nh+1:n));
end
for i = 1:nh
  vLdct(:,i) = dct(hL(:,i));
  vHdct(:,i) = dct(hH(:,i));
end
for i = 1:nh
  I.LL(:,i) = idct(vLdct(1:mh,i));
  I.LH(:,i) = idct(vLdct(mh+1:m,i));
  I.HL(:,i) = idct(vHdct(1:mh,i));
  I.HH(:,i) = idct(vHdct(mh+1:m,i));
end
%END
```

The image can be reconstructed by reversing the previously described procedure in Figure 5.2. Figure 5.3a shows the ground truth image (lena.png) used in multi-resolution analysis (freely available in the open literature). The first and second levels of decomposition of Figure 5.3a are shown in Figure 5.3b. The reconstructed image from the second level of decomposition is shown in Figure 5.3c (left side) and the error image (difference between the true image and the reconstructed image) is shown in Figure 5.3c (right side). One can observe from Figure 5.3 that the reconstructed image (almost) exactly matches the ground truth image. It means that there is no information loss with this analysis. The MATLAB code of IMDCT used for image reconstruction is given below:

```
function[im] = imrdct(I)
% Inverse Multi-Resolution Discrete Cosine Transform
% VPS Naidu, MSDF Lab, NAL
% input: I (decomposed image)
```

FIGURE 5.3 (a) Ground truth image: lena.png. (b) Multi-resolution image decomposition (left side: first level of decomposition; and right side: second level of decomposition). (c) Reconstructed image from second level of decomposition and the error image.

```
% output: im (reconstructed image)

[m,n] = size(I.LL);
m2 = m*2;
n2 = n*2;
for i = 1:n
  ivLdct(:,i) = [dct(I.LL(:,i));dct(I.LH(:,i))];
  ivHdct(:,i) = [dct(I.HL(:,i));dct(I.HH(:,i))];
end
for i = 1:n
  ihL(:,i) = idct(ivLdct(:,i));
  ihH(:,i) = idct(ivHdct(:,i));
end
for i = 1:m2
  hdct(i,:) = [dct(ihL(i,:)) dct(ihH(i,:))];
end
for i = 1:m2
  im(i,:) = idct(hdct(i,:));
end
%END
```

5.4   MULTI-SENSOR IMAGE FUSION

The information flow diagram for the MDCT-based pixel-level image fusion scheme is shown in Figure 5.4. The registered source images I_1 and I_2 are decomposed into D ($d = 1,2, \ldots, D$) levels using MDCT. The resultant decomposed images from I_1 are $I_1 \rightarrow \left\{ {}^{1}LL_D, \left\{ {}^{1}LH_d, {}^{1}HH_d, {}^{1}HL_d \right\}_{d=1,2,\ldots,D} \right\}$ and from I_2 are $I_2 \rightarrow \left\{ {}^{2}LL_D, \left\{ {}^{2}LH_d, {}^{2}HH_d, {}^{2}HL_d \right\}_{d=1,2,\ldots,D} \right\}$. At each decomposition level ($d = 1, 2, \ldots, D$),

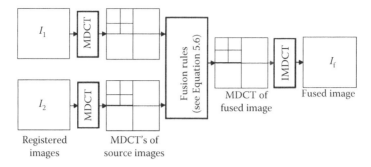

FIGURE 5.4 Schematic diagram for the MDCT-based pixel-level image fusion scheme.

the fusion rule will select the larger absolute value of the two MDCT detailed coefficients, since the detailed coefficients corresponds to sharper brightness changes in the images such as edges and object boundaries, and so on. These coefficients fluctuate around zero. At the coarsest level ($d = D$), the fusion rule is to take average of the MDCT approximation coefficients since the approximation coefficients at the coarser level are the smoothed and sub-sampled version of the original image. The complete set of fusion rules are

$$^{f}LH_d = \begin{cases} ^{1}LH_d & |^{1}LH_d| \geq |^{2}LH_d| \\ ^{2}LH_d & |^{1}LH_d| < |^{2}LH_d| \end{cases} \tag{5.6a}$$

$$^{f}HH_d = \begin{cases} ^{1}HH_d & |^{1}HH_d| \geq |^{2}HH_d| \\ ^{2}HH_d & |^{1}HH_d| < |^{2}HH_d| \end{cases} \tag{5.6b}$$

$$^{f}HL_d = \begin{cases} ^{1}HL_d & |^{1}HL_d| \geq |^{2}HL_d| \\ ^{2}HL_d & |^{1}HL_d| < |^{2}HL_d| \end{cases} \tag{5.6c}$$

$$^{f}LL_D = 0.5(^{1}LL_D + {}^{2}LL_D) \tag{5.6d}$$

The fused image I_f can be obtained using IMDCT.

$$I_f \leftarrow \left\{ ^{f}LL_D, \left\{ ^{f}LH_d, {}^{f}HH_d, {}^{f}HL_d \right\}_{d=1,2,\dots,D} \right\} \tag{5.7}$$

The MATLAB code used for image fusion is given below:

```
function[imf] = mrdctimfus(im1,im2)
% Image fusion using MDCT
% VPS Naidu, MSDF Lab, NAL
% input: im1 & im2 (images to be fused)
% output: imf (fused image)

% multi-resolution image decomposition
X1 = mrdct(im1);
X2 = mrdct(im2);
```

```
% Fusion
X.LL = 0.5*(X1.LL + X2.LL);
D = bdm(X1.LH,X2.LH);
X.LH = D.*X1.LH + (~D).*X2.LH;
D = bdm(X1.HL,X2.HL);
X.HL = D.*X1.HL + (~D).*X2.HL;
D = bdm(X1.HH,X2.HH);
X.HH = D.*X1.HH + (~D).*X2.HH;
% fused image
imf = imrdct(X);
%END
```

5.5 FUSION QUALITY EVALUATION METRICS

When the reference image is available, the performance of image fusion algorithms can be evaluated using the following metrics:

1. Mean absolute error (MAE) [15,16]: This metric is the mean absolute error of the corresponding pixels in reference and fused images. It is also known as spectral discrepancy and it can be used to measure the spectral quality of the fused image. This value will increase when the dissimilarity between the fused and ground truth images increases:

$$MAE = \frac{1}{MN} \sum_{x=1}^{M} \sum_{y=1}^{N} \left| I_r(x,y) - I_f(x,y) \right| \tag{5.8}$$

Here, I_r is the reference image, (x, y) is the pixel index and M and N are the sizes of the image.

2. Peak signal-to-noise ratio (PSNR) [17]: It is given as

$$PSNR = 20 \log_{10} \left(\frac{L^2}{\frac{1}{MN} \sum_{x=1}^{M} \sum_{y=1}^{N} \left[I_r(x,y) - I_f(x,y) \right]^2} \right) \tag{5.9}$$

Here, L is the number of grey levels in the image. Its value will be high when the fused and reference images are similar. Higher value implies better fusion.

When the reference image is not available, the following metrics could be used to evaluate the performance of the fused algorithms.

1. Standard deviation (SD) is given as

$$SD = \sqrt{\frac{1}{MN} \sum_{x=1}^{M} \sum_{y=1}^{N} \left(I_f(x,y) - \bar{I}_f \right)^2} \tag{5.10}$$

Here, $\bar{I}_f = (1/MN)\sum_{x=1}^{M}\sum_{y=1}^{N} I_f(x,y)$. It is known that standard deviation is composed of the signal and noise parts; hence, this metric would be more efficient in the absence of noise. It measures the contrast in the fused image. An image with high contrast would have a high standard deviation.

2. Spatial frequency (SF) [14,18]: Spatial frequency criterion is

$$SF = \sqrt{RF^2 + CF^2} \qquad (5.11)$$

Here, row frequency of the image is given as

$$RF = \sqrt{\frac{1}{MN} \sum_{x=1}^{M} \sum_{y=2}^{N} [I_f(x,y) - I_f(x,y-1)]^2}$$

and the column frequency of the image is given as

$$CF = \sqrt{\frac{1}{MN} \sum_{y=1}^{N} \sum_{x=2}^{M} [I_f(x,y) - I_f(x-1,y)]^2}$$

This frequency in spatial domain indicates the overall activity level in the fused image. And (x, y) is the pixel index. The fused image with high SF would be preferred.

The MATLAB code for evaluating the fusion quality metrics is given below:

```
function [MAE,PSNR,SD,SF] = pereval(imt,imf)
% fusion quality evaluation metrics
% imt: true image
% imf: fused image
[M,N] = size(imt);

% mean absolute error (MAE)
MAE = sum(sqrt((imt(:)-imf(:)).^2))/(M*N);

% Peak signal to noise Ratio (PSNR)
L = 256;
RMSE = sqrt(sum((imt(:)-imf(:)).^2)/(M*N));
PSNR = 10*log10(L^2/RMSE);

% standard deviation SD
If = mean(imf(:));
Id = (imf(:)-If(:)).^2;
SD = sqrt(sum(Id)/(M*N));

% spatial frequency criteria SF
RF = 0; CF = 0;
for m = 1:M
  for n = 2:N
    RF = RF + (imf(m,n)-imf(m,n-1))^2;
  end
end
RF = sqrt(RF/(M*N));

for n = 1:N
  for m = 2:M
    CF = CF + (imf(m,n)-imf(m-1,n))^2;
  end
```

```
end
CF = sqrt(CF/(M*N));

SF = sqrt(RF^2 + CF^2);
%END
```

5.6 COLOUR IMAGE FUSION

Generally, the colour image is the RGB colour model which consists of red, green and blue components. However, this model is not suitable for colour image fusion since there is correlation between the image channels. The RGB colour image is transformed to other colour model such as YCbCr where the correlations between the channels are very minimal. In the YCbCr colour model, the intensity information component is represented by Y, and Cb and Cr represent the colour information. The transformation (T) of the RGB colour model to YCbCr is

$$\begin{bmatrix} Y \\ Cb \\ Cr \end{bmatrix} = T \begin{bmatrix} R \\ G \\ B \end{bmatrix} + b \tag{5.12}$$

Here,

$$T = \begin{bmatrix} 65.481 & 128.553 & 24.966 \\ -37.797 & -74.203 & 112 \\ 112 & -93.786 & -18.214 \end{bmatrix}$$

$$b = \begin{bmatrix} 16 \\ 128 \\ 128 \end{bmatrix}$$

The MATLAB code to convert the RGB colour model to the YCbCr colour model is given below:

```
function[y] = RGB2YCbCr(T,b,r)
% conversion from RGB toYCbCr color space
% input: RGB color image
% output: y - YCbCr color space image
[M,N,O] - size(r);
for i = 1:M
  for j = 1:N
    a = [r(i,j,1);r(i,j,2);r(i,j,3)];
    y(i,j,:) = T*a+b;
  end
end
%END
```

Similarly, the transformation of the YCbCr colour model to the RGB colour model is given as

$$\begin{bmatrix} R \\ G \\ B \end{bmatrix} = T^{-1} \left(\begin{bmatrix} Y \\ Cb \\ Cr \end{bmatrix} - b \right) \tag{5.13}$$

The MATLAB code to transform the YCbCr colour model to the RGB colour model is given as

```
function[r] = YCbCr2RGB(T,b,y)
% conversion from YCbCr to RGB color space
% input: y - YCbCr color space image
% output: RGB color image
[M,N,O] = size(y);
for i = 1:M
  for j = 1:N
    a = [y(i,j,1);y(i,j,2);y(i,j,3)];
    r(i,j,:) = inv(T)*(a-b);
  end
end
%END
```

The fusion process is very much similar to the previous process described in Section 5.3. The images to be fused are transformed from RGB to YCbCr. The fusion process is done only on intensity or luminance components. It is assumed that the images to be fused have similar saturation and hue (S–H); hence, the chrominance components can be averaged. This assumption reduces the computational complexity. The fusion process is given in the following steps and is shown in Figure 5.5: (i) Transform the registered RGB colour images (I_1 and I_2) to be fused into the YCbCr colour models. (ii) The intensity or luminance images from YCbCr colour models, that is, Y_1 and Y_2, are used in the fusion process as described in Section 5.3. The resultant is the fused intensity component Y_f. (iii) Average the chrominance components from the images to get fused chrominance components $Cb_f = 0.5(Cb_1 + Cb_2)$ and $Cr_f = 0.5(Cr_1 + Cr_2)$. (iv) Transform the fused luminance and chrominance components to obtain the fused image I_f in the RGB colour model.

The MATLAB code for the colour image fusion is given below:

```
function[imf] = cif(im1,im2)
% Color image fusion
% input: im1&im2 - color images to be fused
% output: imf - fused color image
T = [65.481 128.553 24.966; -37.797 -74.203 112; 112 -93.786 -18.214];
b = [16;128;128];

YY1 = RGB2YCbCr(T,b,im1);
YY2 = RGB2YCbCr(T,b,im2);
Y1 = YY1(:,:,1); Cb1 = YY1(:,:,2); Cr1 = YY1(:,:,3);
Y2 = YY2(:,:,1); Cb2 = YY2(:,:,2); Cr2 = YY2(:,:,3);

Yf = mrdctimfus(Y1,Y2);
Cbf = 0.5*(Cb1 + Cb2);
```

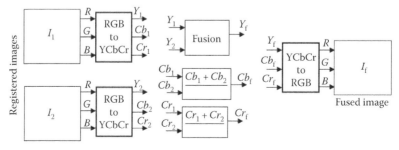

FIGURE 5.5   Information flow diagram of colour image fusion algorithm.

```
Crf = 0.5*(Cr1 + Cr2);
y(:,:,1) = Yf; y(:,:,2) = Cbf; y(:,:,3) = Crf;
imf = YCbCr2RGB(T,b,y);
%END
```

The MATLAB programme to evaluate the fusion quality is given as

```
function[MAE,PSNR,SD,SF] = CIFpereval(imt,imf)
% color image fusion quality evaluation metrics
% inputs: imt - true image & imf - fused image
% outputs: MAE - mean absolute error, PSNR - peak signal to noise ratio
% SD - standard deviation & SF - spatial frequency
[M,N,K] = size(imt);

% mean absolute error (MAE)
MAE = sum(sqrt((imt(:)-imf(:)).^2))/(M*N*K);

% Peak signal to noise Ratio (PSNR)
L = 256;
RMSE = sqrt(sum((imt(:)-imf(:)).^2)/(M*N*K));
PSNR = 10*log10(L^2/RMSE);

% standard deviation SD
If = mean(imf(:));
Id = (imf(:)-If(:)).^2;
SD = sqrt(sum(Id)/(M*N*K));

% spatial frequency criteria SF
for j = 1:K
  RF = 0; CF = 0;
  for m = 1:M
    for n = 2:N
      RF = RF + (imf(m,n,j)-imf(m,n-1,j))^2;
    end
  end
  RF = sqrt(RF/(M*N));

  for n = 1:N
    for m = 2:M
      CF = CF + (imf(m,n,j)-imf(m-1,n,j))^2;
    end
  end
  CF = sqrt(CF/(M*N));

  SF(j) = sqrt(RF^2 + CF^2);
end
SF = mean(SF);
%END
```

5.7   RESULTS AND DISCUSSIONS

Figure 5.6a is the considered reference image I_r to evaluate the performance of the proposed MDCT fusion algorithm. The out-of-focus input images I_1 and I_2 are taken to evaluate the fusion algorithm and are shown in Figure 5.6b and c. The first column in Figures 5.7 through 5.10 show fused images and the second column shows the error images. The error (difference) image is computed by taking the corresponding pixel difference of reference image and fused image, that is, $I_e(x, y) = I_r(x, y) - I_f(x, y)$.

FIGURE 5.6   (a) Reference image I_r. (b) Source images (left side: first source image I_1; and right side: second source image I_2).

FIGURE 5.7   Fused and error image with one level ($D = 1$) of decomposition using MDCT.

FIGURE 5.8   Fused and error image with one level ($D = 1$) of decomposition using wavelets.

The fused and error images by one level of decomposition using MDCT and wavelet fusion algorithms [15] are shown in Figures 5.7 and 5.8, respectively. Similarly, the fused and error images by five levels of decomposition using MDCT and wavelet fusion algorithms are shown in Figures 5.9 and 5.10, respectively. It is observed that the fused images of both MDCT and wavelet are almost similar for these images. The reason could be because of taking the complementary pairs. The performance metrics for evaluating the image fusion algorithms are given in Table 5.1. The values of the metrics shown

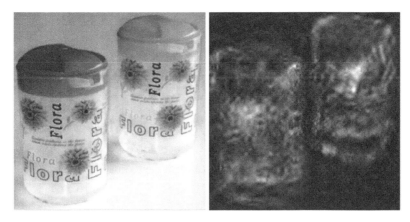

FIGURE 5.9 Fused and error image with two levels ($D = 5$) of decomposition using MDCT.

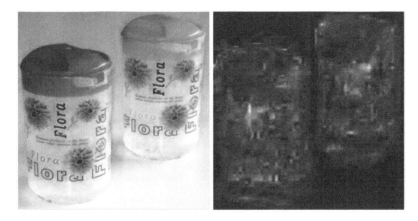

FIGURE 5.10 Fused and error image with two levels ($D = 5$) of decomposition using wavelets.

TABLE 5.1
Fusion Quality Evaluation Metrics

Decomposition Levels	Algorithm	MAE	PSNR	SD	SF
$D = 1$	MDCT	7.1248	37.6342	53.9064	16.2644
$D = 1$	Wavelets	6.9549	37.6967	54.0076	16.6209
$D = 2$	MDCT	6.4423	38.1421	54.6416	19.2544
$D = 2$	Wavelets	6.2495	38.2648	54.8007	19.4432
$D = 5$	MDCT	5.7912	**39.3969**	56.1548	**20.2316**
$D = 5$	Wavelets	**5.6431**	39.3240	**56.2166**	20.2296

in bold font are better compared to others for the same column. The performance of MDCT is almost similar to that of wavelets. From the results, it is observed that image fusion with higher level of decomposition performs superior fusion quality. Figure 5.11a is considered a reference image to evaluate the colour image fusion algorithm. The images to be fused are shown in Figure 5.11b. The fused and error colour images with five levels of MDCT are shown in Figure 5.12. The colour image fusion quality evaluation metrics are given in Table 5.2. As expected, higher levels of decomposition in the fusion process provides better fusion results.

FIGURE 5.11   (a) Colour ground truth image. (b) Colour images to be fused.

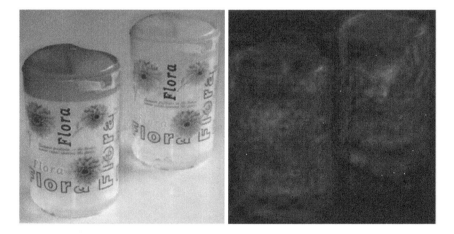

FIGURE 5.12   Fused and error colour images with five levels MDCT.

TABLE 5.2
Fusion Quality Evaluation Metrics:
Colour Images

Decomposition Levels	MAE	PSNR	SD	SF
$D = 1$	7.2556	37.4812	65.7277	16.4103
$D = 2$	6.6018	37.9464	66.2346	19.3407
$D = 5$	**5.9675**	**39.0010**	**67.1970**	**20.2991**

5.8   CONCLUDING REMARKS

Pixel-level image fusion by MDCT algorithm has been implemented and evaluated in this chapter. The performance of this algorithm is compared with a well-known image fusion technique of wavelets. It is inferred that image fusion by MDCT performs almost similar to or slightly better than that by the wavelets. However, MDCT is computationally simple and it could be well suited for real-time applications. Image fusion by higher level of decomposition provides better fusion results. The proposed algorithm has been extended to fuse the colour images and the results seem promising.

REFERENCES

1. G. Pajares and J. Manuel de la Cruz, A wavelet-based image fusion tutorial, *Pattern Recognition*, 37, 1855–1872, 2007.
2. P.K. Varsheny, Multisensor data fusion, *Electronics and Communication Engineering Journal*, 9(12), 245–253, 1997.
3. P.J. Burt and R.J. Lolczynski, Enhanced image capture through fusion, *Proceedings of the 4th International Conference on Computer Vision*, Berlin, Germany, pp. 173–182, 1993.
4. S.G. Mallet, A theory for multiresolution signal decomposition: The wavelet representation, *IEEE Transactions on Pattern Analysis and Machine Intelligence*, 11(7), 674–693, 1989.
5. H. Wang, J. Peng and W. Wu, Fusion algorithm for multisensor image based on discrete multiwavelet transform, *IEEE Proceedings—Vision Image and Signal Processing*, 149(5), 283–289, 2002.
6. F. Jahard, D.A. Fish, A.A. Rio and C.P. Thompson, Far/near infrared adapted pyramid-based fusion for automotive night vision, *IEEE Proceedings of the 6th International Conference on Image Processing and Its Applications (IPA97)*, pp. 886–890, 1997.
7. S. Li, J.T. Kwok and Y. Wang, Combination of images with diverse focuses using the spatial frequency, *Information Fusion*, 2(3), 167–176, 2001.
8. R.S. Blum, Robust image fusion using a statistical signal processing approach, *Image Fusion*, 6, 119–128, 2005.
9. J. Yang and R.S. Blum, A statistical signal processing approach to image fusion for concealed weapon detection, *IEEE International Conference on Image Processing*, Rochester, New York, pp. 513–516, 2002.
10. A. Nejatali and L.R. Ciric, Novel image fusion methodology using fuzzy set theory, *Optical Engineering*, 37(2), 485–491, 1998.
11. N. Ahmed, T. Natarajan and K.R. Rao, Discrete cosine transform, *IEEE Transactions on Computers*, 32, 90–93, 1974.
12. Wolfram Mathematica Documentation Center. http://reference.wolfram.com/legacy/applications/digitalimage/FunctionIndex/DiscreteCosineTransform.html, accessed on April 2012.
13. G. Strang, The discrete cosine transform, *SIAM Review*, 41, 135–147, 1999.
14. V.P.S. Naidu and J.R. Raol, Fusion of out of focus images using principal component analysis and spatial frequency, *Journal of Aerospace Sciences and Technologies*, 60(3), 216–225, 2008.
15. V.P.S. Naidu and J.R. Raol, Pixel-level image fusion using wavelets and principal component analysis—A comparative analysis, *Defense Science Journal*, 58(3), 338–352, 2008.
16. V.P.S. Naidu, G. Girija and J.R. Raol, Evaluation of data association and fusion algorithms for tracking in the presence of measurement loss, *AIAA Conference on Navigation, Guidance and Control*, Austin, USA, 11–14 August 2003.
17. G.R. Arce, *Nonlinear Signal Processing—A Statistical Approach*, Wiley-Interscience Inc., Publication, Hoboken, New Jersey, USA, 2005.
18. A.M. Eskicioglu and P.S. Fisher, Image quality measures and their performance, *IEEE Transactions on Communications*, 43(12), 2959–2965, 1995.

6 Motion Segmentation Using Spectral Framework

M. R. Kaimal and R. Vrinthavani

CONTENTS

6.1 INTRODUCTION

In the field of robotics and micro-air vehicles, the importance of image detection, capturing, its analysis (image processing) and fusion need not be over-emphasized. These vehicles use image data to detect and avoid static as well as moving obstacles and locate other targets and/or landmarks. Often, it is required to detect, identify and locate an object of interest in a sequence of video strings. Hence, various aspects related to image synchronization, image registration, background separation, feature extraction and image segmentation are very important processes related to image analysis and image fusion. In computer vision which is profusely utilized in robotics as well as in other autonomous vehicles, the motion (spatio-temporal) segmentation refers to segmenting a sequence of images based on the coherent motion of a single object or multiple objects in a dynamic scene. Most methods of motion segmentation aim at decomposing a video into moving objects and the background. This decomposition is an essential step for applications like traffic monitoring, robotics, video surveillance, automated inspection and many other similar systems. The literature on motion segmentation and the various approaches applied to it is very large. An article by Zappella et al. [1] gives an excellent review on the subject. The approaches can be generally classified into four groups: image difference, statistical approaches, optical flow, and factorization. Wavelet-based methods are also popular. This division is not absolute and some of the algorithms could be placed in more than one category. Changes in the sequence of images are computed either by considering the pixel-wise motion (which are known as Dense-based approaches) or by considering some salient features, like edge points, corner points, colour, textures and so on (these are known as feature-based methods). Also, a single algorithm does not address all the situations involved in motion segmentation problem. One of the elegant approaches to segmentation of points in feature space is based on mixture models (e.g., Gaussian mixture model (GMM)). These models represent the points in terms of a set of parameters of the various processes or features of the scene under consideration and the number of components in the mixture. Hence, estimating the various parameters of the models turns out to be the major task in solving the segmentation problem which is generally computationally expensive. Another idea that has been widely used in segmentation is based on the eigenvector decomposition of 'affinity matrix'. The property of 'pair-wise affinity' between points in a feature space defines the affinity matrix. In this chapter, instead of using a direct affinity matrix, a probability model for the affinity is defined and then iterative optimization scheme is applied to the associated likelihood function to achieve the eigenvector decomposition. So, this approach provides a more general and natural mechanism for analysing the motion space and explains how better segmentation can be achieved in identifying moving objects in video sequences.

This chapter presents a new motion segmentation method using iterative graph spectral framework. This is achieved through defining an appropriate probability model for the affinity matrix, associated likelihood function and an iterative scheme for obtaining the eigenvector decomposition. The method consists of two steps. In the first step, motion regions are detected and motion vectors are computed for these detected regions. In the second step, a similarity matrix is computed from the motion vectors and motion segmentation is done by iteratively optimizing a maximum likelihood function. The method has been tested using real-world motion sequences and is found to perform better than existing methods and gives only very low error rate and it also detects the slow-moving objects.

6.2   MAJOR MOTION SEGMENTATION APPROACHES

Image difference is one of the simplest and widely used techniques for detecting changes. Pixel-by-pixel intensity difference of two frames is computed first and then thresholding is applied to get a group of pixels having a similar property. The idea is to compute a rough sketch of the changing areas, and for each such area extract spatial or temporal information in order to track the region. This technique is really sensitive to noise and the method yields no useful information when the frame rate is not high enough. Most of the existing motion segmentation methods using image difference do not accurately detect slow-moving objects from the video sequences. One popular method used for motion segmentation is background subtraction. The purpose of a background subtraction algorithm is to distinguish moving objects (or foreground) from static, or slow-moving parts of a scene (called background). An important drawback of this technique is that it uses the same threshold for every pixel. This way, a moving object is likely to disappear in the background when entering a darker (shaded) area in the scene. Many background subtraction techniques have been proposed with as many models and segmentation strategies [2]. Statistical concepts are widely used in the motion segmentation field. In these methods, each pixel is modelled using suitable probability density functions, generally Gaussian. GMM are also popular where the distribution of values observed at each pixel over time is represented by a weighted mixture of Gaussians [2]. In statistical approaches, motion segmentation is treated as a classification problem where each pixel has to be classified as background or foreground. Statistical approaches are further divided depending on the framework used. Common frameworks are maximum *a posteriori* probability (MAP), particle filter (PF) and expectation maximization (EM). MAP is based on Baye's rule. The main aim of PF is to track the evolution of a variable over time. The basis of the method is to construct a sample-based representation of the probability density function. PF is an iterative algorithm wherein iteration is composed of a prediction and an update. After each action the particles are modified according to a model (prediction) and then each particle weight is re-evaluated according to the information extracted from an observation (update). The particles with small weights are eliminated at each iteration. Another commonly used method for segmentation is the EM algorithm, which is again an iterative mechanism. The EM algorithm computes the maximum likelihood (ML) estimate in the presence of missing or hidden data. ML is used in estimating the model parameter(s) that most likely represent the observed data. Each iteration of EM is composed by the E step and the M step. In the E step, the missing data are estimated using the conditional expectation, while in the

M step, the likelihood function is maximized. Convergence is assured since the algorithm is guaranteed to increase the likelihood at each iteration. When EM algorithm [3–4] is used each object is represented by a separate Gaussian distribution. The EM algorithm [4] has proved to be cumbersome to use in practice, due to the problems of estimating the parameters of the motion mixture model and of controlling its structure.

Ever since the factorization technique was introduced by Tomasi and Kanade [5] to recover structure and motion using features tracked through a sequence of images, the methods have become very popular. The idea is to factorize the trajectory matrix W (the matrix containing the position of the P features tracked throughout F frames) into two matrices: motion M and structure S. If the origin of the world coordinate system is moved at the centroid of all the feature points, and in the absence of noise, the trajectory matrix is at most rank 3. Exploiting this constraint, W is decomposed and truncated using singular value decomposition (SVD). A closely related approach for segmentation and grouping [6–8] is graph spectral [9] methods. These methods all share the feature of using the eigenvectors of a weighted adjacency matrix to locate salient groupings of objects. At the level of image segmentation, several authors have used algorithms based on the eigenmodes of an affinity matrix to iteratively segment image data. For instance, Sarkar and Boyer [7] proposed a method which uses the leading eigenvector of the affinity matrix, and this locates clusters that maximize the average association. This method is applied to locating line segment groupings. Perona and Freeman [8] gave a similar method which uses thresholding of the first largest eigenvector of the affinity matrix for segmenting. The method of Shi and Malik [6], on the other hand, uses thresholding of the second generalized eigenvector of the affinity matrix which gives the normalized cut that balances the affinity between clusters (through the cut) and the affinity within the clusters (of the associations). Clusters are located by performing a recursive bisection using the eigenvector associated with the second smallest eigenvalue of the Laplacian matrix (the degree matrix minus the adjacency matrix), that is, the Fiedler vector. Weiss [10] has presented a unified view of four popular algorithms that analyse segmentation using eigenvectors of affinity matrices.

Kelly and Hancock [11] have developed an iterative spectral framework for pair-wise clustering. They have used the maximum likelihood method [12] to detect moving objects by performing pair-wise clustering on a set of motion vectors. There are two problems with this approach. First, in order to reduce the motion vector noise, it uses a multi-resolution block matching method to estimate the motion field. Hence, computational cost increases. Another drawback is that it is unable to detect slow-moving objects in the video sequence. These problems are addressed in this work. The method explained in this article also uses the iterative spectral framework [11]. In order to reduce noise without increasing computational complexity, the motion regions are detected and motion vectors are computed only for these regions using a block matching algorithm (BMA). Instead of using only one previous frame to detect motion pixels, the proposed method uses a set of m frames to detect the motion region. Hence, the slow-moving objects are also detected. The proposed segmentation method has two steps. In the first step, motion estimation is done by finding the motion region and applying the BMA to obtain the motion vector. In the second step, iterative spectral framework [11] is used to cluster the motion regions.

6.3 COMPUTATION OF MOTION VECTOR

The computation of the motion vectors is done using single resolution BMA using spatial/temporal correlation [13]. This BMA is based on predictive search that reduces computational complexity and provides a reliable performance. The method measures the similarity of motion blocks using spatial correlation. It uses a predictive search to efficiently compute block correspondences in different frames. This BMA assumes that the translational motion from frame to frame is constant. The current frame is divided into non-overlapping blocks which are then matched with a block in the destination frame by shifting the current block over a predefined neighbourhood of pixels in the destination frame. At each shift, mean-squared distances between the grey values of the two blocks are computed. The distance is calculated by

$$D(A, B) = \frac{1}{n} \sum_{i=1}^{n} (A(i) - B(i))^2 \tag{6.1}$$

Here, D is the distance, A is the current block, B is the block reference frame and n is the total number of pixels in the block. The shift which gives the smallest distance is considered the best match. In order to reduce the computational burden, the motion vector of the current block is predicted from that of the neighbour blocks in the temporal or spatial direction. Because the computational complexity is much lower than the optical flow equation and the pel-recursive methods, block matching has been widely adopted as a standard for video coding and hence it provides a good starting point.

6.4   MAXIMUM LIKELIHOOD FRAMEWORK

These 2D motion vectors (for the extracted motion blocks) are characterized using a matrix of pairwise similarity weights. Let $\hat{n}_\alpha$ and $\hat{n}_\beta$ be the unit motion vectors for the pixel blocks indexed α and β. The elements of this weight matrix $W_{\alpha,\beta}$ are given by

$$W_{\alpha,\beta} = \begin{cases} \frac{1}{2}(1 + \hat{n}_\alpha \cdot \hat{n}_\beta) & \text{if } \alpha \neq \beta \\ 0 & \text{otherwise} \end{cases} \tag{6.2}$$

The aspect of grouping motion blocks into coherent moving objects is treated as that of finding pair-wise clusters [8]. The aim is to locate the updated set of similarity weights which partition the image into regions of uniform motion.

Let V denote the index set of the detected motion blocks in the image and suppose that Ω is the set of pair-wise clusters, that is, distinct moving objects, to which these blocks are to be assigned. The initial set of clusters is defined by the eigenmodes of the link-weight matrix $W^{(0)}$. Sarkar and Boyer [7] have shown how the positive eigenvectors of the matrix of link-weights can be used to assign objects to perceptual clusters. Using the Rayleigh–Ritz theorem, they observe that the scalar quantity $\underline{x}^T W^{(0)} \underline{x}$ is maximized when $\underline{x}$ is the leading eigenvector of $W^{(0)}$. Moreover, each of the subdominant eigenvectors corresponds to a disjoint pair-wise cluster. They confine their attention to the same-sign positive eigenvectors (i.e., those whose corresponding eigenvalues are real and positive, and whose components are either all positive or all negative in sign). If a component of a positive same-sign eigenvector is non-zero, then the corresponding object belongs to the associated cluster of motion blocks. The eigenvalues λ_t of $W^{(0)}$ are the solutions of the equation $|W^{(0)} - \lambda I| = 0$, where I is the identity matrix. The corresponding eigenvectors $\underline{x}_{\lambda 1}, \underline{x}_{\lambda 2}, \dots$ are found by solving the equation $W^{(0)} \underline{x}_\omega = \lambda_\omega \underline{x}_\omega$. Let the set of positive same-sign eigenvectors be represented by $\Omega = \{\omega \mid \lambda_\omega > 0 \wedge [(\underline{x}_\omega^*(i) > 0 \forall i) \vee (\underline{x}_\omega^*(i) < 0 \forall i)]\}$, where $\underline{x}_\omega^*(i)$ is the ith component of the eigenvector indexed ω. Kelly and Hancock [11] commence a simple model of the cluster formation process based on a series of independent Bernoulli trials. The linkage of each pair of nodes within a cluster is treated as a separate Bernoulli trial. The link-weight for the pair of nodes is treated as the success probability of the trial. The similarity weight $W_{\alpha,\beta}$ is taken as the parameter of the Bernoulli distribution. The probability that the block association are correct is $W_{\alpha,\beta}$ while the probability that it is in error is $1 - W_{\alpha,\beta}$. Here, they introduce a cluster membership indicator $s_{\alpha\omega}$ which represents the degree of affinity of the object indexed α to the cluster with index ω. The random variable associated with the trial is taken as the product of these cluster indicators for the pair of nodes, that is, $s_{\alpha\omega} s_{\beta\omega}$; this indicates whether the two nodes belong to the same cluster. This is unity if both blocks belong to the same object or cluster and is zero otherwise. Using the property, Bernoulli distribution is given as

$$P(s_{\alpha\omega}, s_{\beta\omega} \mid W_{\alpha,\beta}) = W_{\alpha,\beta}^{s_{\alpha\omega} s_{\beta\omega}} (1 - W_{\alpha,\beta})^{(1 - s_{\alpha\omega} s_{\beta\omega})} \qquad (6.3)$$

This distribution takes on its largest values when either the motion vector similarity weight $W_{\alpha,\beta}$ is unity and $s_{\alpha\omega} = s_{\beta\omega} = 1$ or $W_{\alpha,\beta} = 0$ and $s_{\alpha\omega} = s_{\beta\omega} = 0$. Using this model, a joint likelihood function is developed for the link-weights and the cluster membership indicators. This likelihood function can be used to make both a maximum likelihood re-estimate of the link-weight matrix and a MAP estimate of the cluster membership indicators. In the case of re-estimating the link-weight matrix, the cluster indicators are treated as data. Applying this model to log-likelihood function for the observed set of motion vector similarity weight, we get

$$L = \sum_{\omega \in \Omega} \sum_{(\alpha,\beta) \in \Phi} \left\{ s_{\alpha\omega} s_{\beta\omega} \ln(W_{\alpha,\beta}) + (1 - s_{\alpha\omega} s_{\beta\omega}) \ln(1 - W_{\alpha,\beta}) \right\} \qquad (6.4)$$

The above log-likelihood function can be optimized using an EM-like process. To maximize the log-likelihood function with respect to the link-weights and the cluster membership indicators we take the derivatives of the expected log-likelihood function with respect to the elements of the link-weight matrix and cluster membership variables. In the E step, the cluster membership probabilities are updated according to the formula

$$s_{\alpha\omega}^{(n+1)} = \frac{\prod_{\beta \in V} \left\{ \dfrac{W_{\alpha,\beta}^{(n)}}{1 - W_{\alpha,\beta}^{(n)}} \right\}^{S_{\beta\omega}^{(n)}}}{\sum_{\omega \in \Omega} \prod_{\beta \in V} \left\{ \dfrac{W_{\alpha,\beta}^{(n)}}{1 - W_{\alpha,\beta}^{(n)}} \right\}} \qquad (6.5)$$

Once the revised cluster membership variables are available, the M step of the algorithm is applied to update the similarity weight matrix. The updated weights are given by

$$W_{\alpha,\beta}^{(n+1)} = \sum_{\omega \in \Omega} s_{\alpha\omega}^{(n)} s_{\beta\omega}^{(n)} \qquad (6.6)$$

and these steps are interleaved and iterated to convergence.

6.5 PROPOSED METHOD FOR MOTION SEGMENTATION

The demerit of the block matching scheme is that while the high-resolution field of motion vectors obtained with small block sizes capture fine detail, it is susceptible to noise. At low resolution (for large block sizes), the field of motion vector is less noisy but the fine structure is lost. Also, it will not detect the moving objects with slow motion. In order to remove this drawback, a new algorithm for detecting the moving object using maximum likelihood framework is proposed. The proposed algorithm has two steps: first, motion detection and motion estimation, and second, clustering. Instead of performing BMA to the entire frame, first, the moving objects are detected by taking the difference of maximum intensity values and minimum intensity values for each pixel of a set of frames. Then, find the motion vectors of these regions.

6.5.1 MOTION DETECTION

A set of consecutive frames from the static camera video is considered. The range of pixel values that a particular location (x, y) can take will vary significantly if that pixel belongs to a moving object. If it belongs to the background then the range of the pixel value at (x, y) in the consecutive set of frames will be small. So, this technique can be used to detect the motion in a set of frames. Let a frame t in a video sequence be represented as $f(t)$. A pixel at location (x, y) in the frame t be represented as $f(x, y, t)$. Let MAXP(x, y) represents the maximum intensity value of a pixel at location (x, y) in a set of frames $S = \{f(t + i)\} : i = 0, \ldots, \mathrm{d}t\}$, where t is current frame and $\mathrm{d}t$ is the number of frames in the set S. Similarly, MINP(x, y) represents the minimum intensity value at location (x, y) in S. Let the difference between them be denoted as

$$\mathrm{d}(x, y) = \mathrm{MAXP}(x, y) - \mathrm{MINP}(x, y) \tag{6.7}$$

If the object at location (x, y) is not moving, then $\mathrm{d}(x, y)$ will have a very small value. If the object is moving, then $\mathrm{d}(x, y)$ will be greater. Figure 6.1a, b shows the original frame 1 and 5 of the traffic sequence. MAXP and MINP for the set of frames 1–5 is shown in Figure 6.1c, d. The difference between the MAXP and MINP is shown in Figure 6.1e, f, showing the image after thresholding. The problem of direct application of the threshold to the difference image is that some noise will be present in the resulting image. If the object is moving very slowly, then it is difficult to identify whether the detected pixel is a noise or an object. If we apply morphological operation to the thresholded image, the slow-moving object will also disappear. In order to solve this problem we can take gradient of difference image along the horizontal and vertical direction. Take the Euclidean distance of the gradient images. Apply threshold to the gradient image. The threshold selected is standard deviation of the gradient image. Modified morphological dilation and erosion operations are then performed to the binary image. For dilation, four connected neighbouring pixels are set to 1 if the current pixel is 1, where 1 indicates that the pixel belongs to object mask. For erosion, four connected neighbouring pixels are set to 0, where 0 indicates that the pixel belongs to the background. The modified dilation and erosion operations are defined below:

FIGURE 6.1 Comparison of image frames: (a) original frame 1; (b) frame 5; (c) MAXP image for the set of frames 1–5; (d) MINP image; (e) difference image; and (f) image after applying the threshold $T = 10$ to the difference image. (From Vrinthavani R. and M. R. Kaimal. In *Sp. Issue, Mobile Intelligent Autonomous Systems*, Eds. J. R. Raol and A. Gopal, *Defence Science Journal*, 60(1), January 2010. With permission.)

FIGURE 6.2   (a) Gradient image of the difference image of Figure 6.1e. (b) Binary image after applying the modified morphological operations. (c) Corresponding motion-detected image. (From Vrinthavani R. and M. R. Kaimal. In *Sp. Issue, Mobile Intelligent Autonomous Systems*, Eds. J. R. Raol and A. Gopal, *Defence Science Journal*, 60(1), January 2010. With permission.)

For dilation, if $g(x, y) = 1$, then

$$\begin{aligned} \text{Pixel}(x, y) &= \text{Pixel}(x+1, y) = \text{Pixel}(x, y+1) \\ &= \text{Pixel}(x-1, y) = \text{Pixel}(x, y-1) = 1 \end{aligned} \tag{6.8}$$

For erosion, if $g(x, y) = 0$, then

$$\begin{aligned} \text{Pixel}(x, y) &= \text{Pixel}(x+1, y) = \text{Pixel}(x, y+1) \\ &= \text{Pixel}(x-1, y) = \text{Pixel}(x, y-1) = 0 \end{aligned} \tag{6.9}$$

The morphological operations are performed in the order erosion followed by dilation. When erosion is performed isolated, noisy pixels will be removed, and then when dilation is performed it will highlight the object pixels very well. Figure 6.2a shows the gradient image of the difference image shown in Figure 6.1e. Figure 6.2b shows the motion-detected binary image after applying the morphological operation, and Figure 6.2c shows corresponding motion-detected image. The parameter affecting the motion detection is the number of frames in the set. If the number of frames used in the set is less, then small motions will not be detected. If the set contains a large number of frames then noise will also be detected as motion. Figure 6.3c shows motion detection with five frames in the set. Figure 6.4c shows the motion-detected images with 10 frames in the set. The difference between Figures 6.4a and b, which is shown in Figure 6.4c has more noise. As the number of frames are large small noises in the frame may appear as moving.

FIGURE 6.3   Results of motion detection with five frames in the set S: (a) first frame of the set, frame 26; (b) last frame of the set, frame 30; (c) motion-detected image. (From Vrinthavani R. and M. R. Kaimal. In *Sp. Issue, Mobile Intelligent Autonomous Systems*, Eds. J. R. Raol and A. Gopal, *Defence Science Journal*, 60(1), January 2010. With permission.)

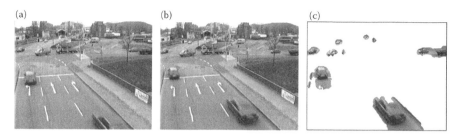

FIGURE 6.4 Results of motion detection with 10 frames in the set S: (a) first frame of the set, frame 20; (b) last frame of the set, frame 30; (c) motion-detected image. (From Vrinthavani R. and M. R. Kaimal. In *Sp. Issue, Mobile Intelligent Autonomous Systems*, Eds. J. R. Raol and A. Gopal, *Defense Science Journal*, 60(1), January, 2010. With permission.)

6.5.2 MOTION ESTIMATION

Once the motion pixels are detected, the motion is to be estimated for these pixels. For the motion estimation the BMA discussed in Section 6.3 is used. The difference is that instead of applying the algorithm to the whole frame it is applied only to the motion regions, thereby reducing the computational complexity to a great extent. Figure 6.5b shows the motion map obtained by passing the original frames to BMA. Figure 6.5c shows the motion map obtained by passing the motion-detected image to BMA. In Figures 6.5a and b, the block size is 4×4 and the threshold used is 5.

6.5.3 CLUSTERING USING SPECTRAL FRAMEWORK

The spectral framework of Ref. [11] is used in this step. The 2D velocity vectors for the extracted motion blocks are then characterized using a matrix of pair-wise similarity weights W. The link-weight matrix is then calculated using Equation 6.2. The clustering is done by maximizing the log-likelihood function described in Section 6.4 with respect to weight matrix and the cluster membership indicator. This is done using an EM-like process. In the E step, cluster membership probabilities are updated and in the M step, similarity weight matrix is updated. The same-sign eigenvectors are extracted from the current link-weight matrix W. These are then used to compute the cluster membership matrix S using the equation

$$\hat{s}_{i\omega} = \frac{\left|x_{\omega}^*(i)\right|}{\displaystyle\sum_{i \in V}\left|x_{\omega}^*(i)\right|} \tag{6.10}$$

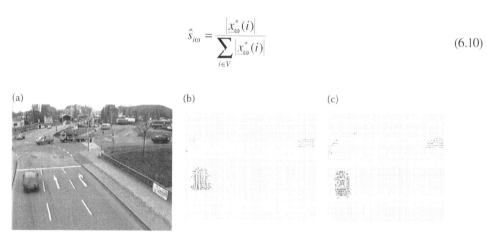

FIGURE 6.5 (a) Frame 10 of the traffic sequence, (b) motion map with original frames, (c) motion map with motion-detected image. (From Vrinthavani R. and M. R. Kaimal. In *Sp. Issue, Mobile Intelligent Autonomous Systems*, Eds. J. R. Raol and A. Gopal, *Defense Science Journal*, 60(1), January 2010. With permission.)

The number of same-sign eigenvectors determines the number of clusters for the current iteration. Using the cluster membership matrix S, link-weight matrix W is updated. This is done as follows. For each cluster, compute the link-weight matrix $\hat{W}_\omega = s_\omega s_\omega^T$. Perform an eigen decomposition on each cluster link-weight matrix to extract the non-zero eigenvalue $\hat{W}_\omega$ and the corresponding eigenvector φ_ω^*. Since the matrix $\hat{W}_\omega$ is rank one since it is defined as the product of two vectors, the computation of the first eigenvector can be regarded for computational purposes as a normalization of the vectors s_ω. In practice, the link-weight matrix may be noisy and hence the cluster structure may be subject to error. In an attempt to overcome this problem, the updated link-weight matrix must be refined with a view to improving its block structure. The aim here is to suppress structure which is not associated with the principal modes of the matrix. This is done by applying the following equation:

$$W^* = \sum \frac{\lambda_\omega^*}{|\Omega|} \varphi_\omega^* (\varphi_\omega^*)^T \tag{6.11}$$

The link-weight matrix is then used to update cluster membership matrix. An updated matrix of cluster membership variable $\hat{S}$ is computed by applying the following equation to the revised link-weight matrix W^*:

$$\hat{s}_{i\omega} = \frac{\prod_{j \in V} \left\{ W_{ij}/(1 - W_{ij}) \right\}^{s_{j\omega}}}{\sum_{i \in V} \prod_{j \in V} \left\{ W_{ij}/(1 - W_{ij}) \right\}^{s_{j\omega}}} \tag{6.12}$$

The updated cluster membership matrix $\hat{S}$ is used to compute the updated link-weight matrix $\hat{W} = (1/|\Omega|) S S^T$. Once the updated link-weight matrix is in hand, it is then again used to compute same-sign eigenvectors and the whole process is repeated till convergence.

6.6 DISCUSSION OF THE RESULTS

The presented new method/algorithm is tested using video sequences with known ground truth. Figure 6.6a shows the final link-weight matrix obtained for the 40th frame with the original method.

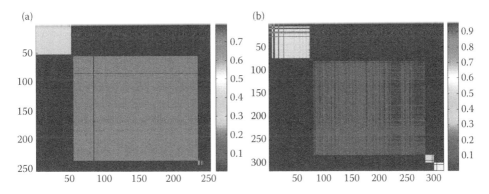

FIGURE 6.6 (a) Final link-weight matrix obtained after applying the original method; (b) final link-weight matrix obtained after applying the improved motion segmentation algorithm. (From Vrinthavani R. and M. R. Kaimal. In *Sp. Issue, Mobile Intelligent Autonomous Systems*, Eds. J. R. Raol and A. Gopal, *Defense Science Journal*, 60(1), January 2010. With permission.)

Here, four block structures are visible. Each block structure represents one cluster. Figure 6.6b shows the final link-weight matrix obtained for the 40th frame with the improved method. Here, six block structures are visible. Figure 6.7 shows the result of the motion segmentation algorithm with the traffic sequence. Figure 6.7a shows the original frames 10, 20 and 40. Figure 6.7b shows the ground truth of these frames. Figure 6.7c shows the motion map obtained with the original method. Figure 6.7d shows the result of the motion segmentation algorithm using the original method. Here, for the 10th frame, three motion objects are detected; for the 20th frame, five motion objects are detected and for the 40th frame, four motion objects are detected. Figure 6.7e shows the output of the motion detection step of the improved algorithm. Figure 6.7f shows the motion map obtained with improved algorithm. Motion map is obtained using the motion-detected images. Figure 6.7g shows the result of the motion segmentation algorithm using the improved method. Here, for the 10th frame, five motion objects are detected, for the 20th frame, six motion objects are detected and for the 40th frame, six motion objects are detected. Here, the number of frames used in the motion detection step is five. The block size used is 4×4. The algorithm converged in an average of three iterations. Figure 6.8 shows the result of the motion segmentation algorithm with the taxi sequence. Figure 6.8a shows the original frames 10, 20 and 30. Figure 6.8b shows the ground truth of these frames. For all the three frames, the number of moving objects is four. The four objects are the left car, the middle car, the right vehicle and the pedestrian. Figure 6.8c shows the motion map obtained with the original method. Figure 6.8d shows the result of the motion segmentation algorithm using the original method. Here, for the 10th frame, four motion objects are detected, for the 20th frame, three motion objects are detected and for the 30th frame, three motion objects are detected. Figure 6.8e shows the motion map obtained with improved algorithm. The motion map is obtained using the motion-detected images. Figure 6.8f shows the result of the motion segmentation algorithm using the improved method. Here, for all the three frames, four motion objects are detected. The complexity of the proposed algorithm is reduced because of the motion detection step. This step will detect the motion and only the detected motion pixels are further passed to the next step. In this way, the noise will be eliminated as well as the number of pixels passed to the next step is also decreased compared to the original method. The motion detection step has the complexity of $O(n)$, where n is the number of pixels in a frame. Table 6.1 shows the quantitative analysis of the result. The table lists the number of objects detected, the percentage of correctly classified pixel in the total number of object pixels in the ground truth (true positive rate), percentage of pixels wrongly detected as object pixels to the total number of pixels in the background in the ground truth (false negative rate) and the total number of pixels classified correctly in the whole image (percentage of correct classification) comparison between the two algorithms. The percentage of correct classification is almost same using both methods but the true positive rate is much higher in the improved method compared to the original method. Also, the number of moving objects detected is greater in the improved method compared to the original method.

6.7 CONCLUDING REMARKS

In this chapter, an improved spectral framework for the affinity matrix defined using a probability model for associations and the appropriate likelihood function is iteratively optimized to achieve better motion segmentation. The proposed method/algorithm first detects the motion using a set of frames and uses that information for computing the motion vector. The merit of this step is that it can detect moving objects with very slow motion and also reduce the time complexity in computing the motion vector. Using the motion vectors we then define a matrix of pair-wise similarity weights. Subsequently, a probabilistic association model is defined for cluster membership of the motion blocks. With the assumption that similarity weights follow a Bernoulli distribution, a log-likelihood function is used to update the similarity matrix. The eigenvectors of the matrix are also extracted from the link-weight matrix. The performance of this method seems to be better than existing methods.

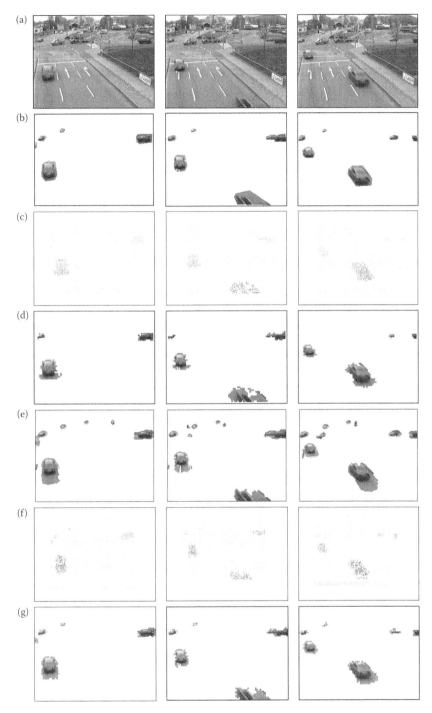

FIGURE 6.7 (a) Original frames (10, 20 and 40) of the traffic sequence. (b) Ground truth of the corresponding frames. (c) Corresponding motion map with the original method. (d) Result of the motion segmentation algorithm with the original method. (e) Motion-detected images using the improved algorithm. (f) Motion map detected using the improved algorithm. (g) Result of the improved motion segmentation algorithm. (From Vrinthavani R. and M. R. Kaimal. In *Sp. Issue, Mobile Intelligent Autonomous Systems*, Eds. J. R. Raol and A. Gopal, *Defense Science Journal*, 60(1), January 2010. With permission.)

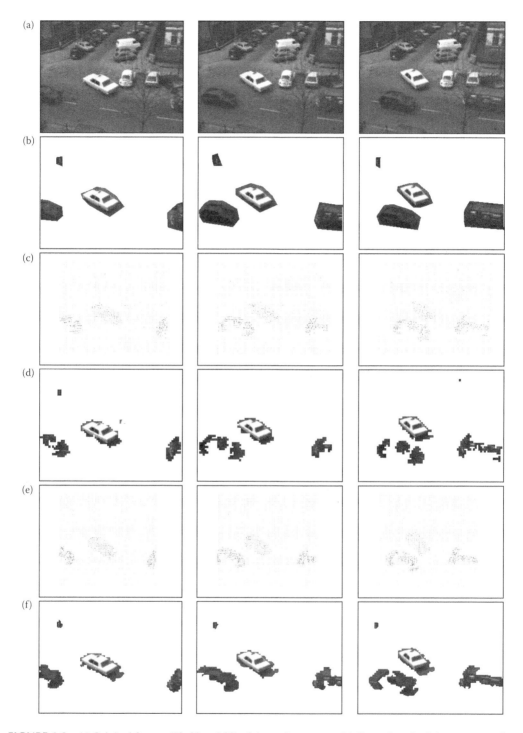

FIGURE 6.8   (a) Original frames (10, 20 and 30) of the taxi sequence. (b) Ground truth of the corresponding frames. (c) Corresponding motion map with the original method. (d) Result of the motion segmentation algorithm with the original method. (e) Motion map detected using the improved algorithm. (f) Result of the improved motion segmentation algorithm. (From Vrinthavani R. and M. R. Kaimal. In *Sp. Issue, Mobile Intelligent Autonomous Systems*, Eds. J. R. Raol and A. Gopal, *Defense Science Journal*, 60(1), January 2010. With permission.)

TABLE 6.1
Comparison of Performance Numbers of the Original and the New Methods

	Traffic Sequence			Taxi Sequence		
Frame number	10	20	40	10	20	30
Number of moving objects in the ground truth	5	6	6	4	4	4
Original method						
Number of objects detected	3	5	4	4	3	3
True positive rate	78.25	78.60	80.30	67.25	50.14	50.90
False negative rate	0.60	1.66	1.27	1.30	1.20	1.60
% of correct classification	98.40	97.08	97.70	95.64	91.87	91.30
Improved method						
Number of objects detected	5	6	6	4	4	4
True positive rate	90.09	81.49	89.40	78.59	63.40	64.88
False negative rate	0.95	1.09	1.90	1.18	1.20	2.30
% of correct classification	98.64	97.80	97.60	96.88	93.70	92.80

Source: From Vrinthavani R. and M. R. Kaimal. In *Sp. Issue, Mobile Intelligent Autonomous Systems*, Eds. J. R. Raol and A. Gopal, *Defense Science Journal*, 60(1), 37–49, January 2010. With permission.

REFERENCES

1. L. Zappella, X. Llado and J. Selvi. New trends in motion segmentation. In *Pattern Recognition*, In TECH, Ed. Intechweb.org, 2009.
2. C. Stauffer and W. E. L. Grimson, Adaptive background mixture models for real time tracking. In *Proceedings. 1999 IEEE Conference on CVPR*, Ft. Collins, CO, USA, Vol. 2, pp. 2246–2252, 1999.
3. A. D. Jepson, W. J. MacLean and R. C. Frecker. Recovery of ego-motion and segmentation of independent object motion using the EM algorithm. In *Proceedings of the British Machine Vision Conference*, University of York, York, UK, pp. 175–184, 1994.
4. E. H. Adelson and Y. Weiss. A unified mixture framework for motion segmentation: Incorporating spatial coherence. In *Proc. IEEE Computer Vision and Pattern Recognition*, San Francisco, USA, pp. 321–326, 1996.
5. C. Thomasi and T. Kanade, Shape and motion from image streams under orthography: A factorization method. *International Journal of Computer Vision*, 9(2), 137–154, 1992.
6. J. Shi and J. Malik. Normalized cuts and image segmentations. In *Proc. IEEE CVPR*, San Juan, Puerto Rico, pp. 731–737, 1997.
7. S. Sarkar and K. L. Boyer. Quantitative measures of change based on feature organization: Eigenvalues and eigenvectors. *Computer Vision and Image Understanding*, 71(1), 110–136, 1998.
8. P. Perona and W. T. Freeman. Factorization approach to grouping. In *Proc. ECCV*, Freiburg, Germany, pp. 655–670, 1998.
9. Fan R. K. Chung. *Spectral Graph Theory*. American Mathematical Society, Providence, Rhode Island, 1997.
10. Y. Weiss. Segmentation using eigenvectors: A unifying view. In *Proc. IEEE International Conference on Computer Vision*, Kerkyra, Corfu, Greece, pp. 975–982, 1999.
11. A. Robles-Kelly and E. R. Hancock. A probabilistic spectral framework for grouping and segmentation. *Pattern Recognition*, 37(7), 1387–1405, 2004.
12. A. Robles-Kelly and E. R. Hancock. Maximum likelihood motion segmentation using eigen decomposition. In *Proc. IEEE Image Analysis and Processing*, Ravenna, Italy, pp. 63–68, 2001.
13. J. S. Shyn, C. H. Hsieh, P. C. Lu and E. H. Lu. Motion estimation algorithm using inter-block correlation. *IEE Electronics Letters*, 26(5), 276–277, 1990.
14. R. Vrinthavani and M. R. Kaimal, An improved motion segmentation algorithm using spectral framework. In *Sp. Issue, Mobile Intelligent Autonomous Systems*, Eds. J. R. Raol and A. Gopal, *Defence Science Journal*, 60(1), 37–49, January 2010.

7 Formation Control in Multi-Agent Systems over Packet Dropping Links

S. Seshadhri and R. Ayyagari

CONTENTS

7.1 INTRODUCTION

Proliferation of communication channels into control loops has enabled plethora of applications that abstractly engaged the attention of control engineers in the past. Moreover, the presence of a communication channel has necessitated a paradigm shift in the way control loops have been analysed in the past. The usual assumptions of synchronous availability of sensor information and actuation are no longer valid in the presence of communication channels. This is primarily due to packet loss and delay associated with communication channels. Traditional control loops assume continuous availability of feedback data. Data losses that occur in the communication channels make this assumption invalid. Furthermore, the tools employed for classical control need to be reformulated; the proposed new tools for control loops need to be integrated with communication channels. Control loops integrated with communication channels for information exchange are called networked control systems (NCSs). A detailed review of NCSs can be found in Refs. [1–15] (and the references therein). Researchers have investigated NCSs in the recent past because these NCSs offer some advantages in certain applications [16–25]. NCSs have many advantages such as flexibility, modularity, ease of implementation, reduced wiring, reduction in cost, modularity to name a few. It is now possible to embed control and computing capabilities all along a distributed scenario using pervasive communication channels. Furthermore, the scale of the control loops has also increased significantly.

Motivated by these developments, control scientists have investigated coordinated control among various computing nodes in the network. This has brought into existence the concept of consensus, coordination and formation control and these are widely investigated problems in the multi-agent domain. The aforesaid problems are generally classified as coordinated control problems in the multi-agent domain. The main advantage of coordinated control is its operational efficiency, and the enhanced functionality that can be obtained from the distributed control loop. Teams of agents acting in coordination can produce greater operating efficiency and functionality than individual agents performing solo missions. As an example, consider the case of intelligent vehicle highway system (IVHS), maintaining traffic by treating vehicles as agents that coordinate to maintain a pre-specified formation improves the throughput and reduced congestion. Coordination among vehicle (agents) also helps reduce accidents or in other words improve the safety and throughput of the traffic. One may conclude from this discussion that coordination among teams of agents can be used to improve efficiency and can be used to realize goals that were not realizable earlier with individual agents [13–20]. Formation and coordination control are the two problems being widely investigated by control scientists. Consensus control involves a team of agents agreeing on a common metric such as distance, velocity, and so on. The agreement translates into agents meeting or being closer to each other. Rendezvous points are used to indicate the agreement among the agents. Formation control has been widely investigated both in the control and in the multi-agent systems paradigm. Accurate maintenance of a geometric configuration among multiple agents moving as a team promises less expensive, more capable systems that can accomplish objectives which might have been impossible with a single agent. The concept of formation control has been studied extensively in the literature with application to the coordination among robots [21–30], UAVs [31], AUVs [32], satellites [33], aircraft [34] and spacecraft [35]. There are multiple advantages to using formations of agents. These include cost, feasibility, flexibility, accuracy, robustness and energy efficiency. As an example, one may consider the surveillance problem using aircrafts, wherein coordination reduces the time required for completing the operation [36].

Various strategies and approaches have been proposed for formation control. These approaches can be roughly categorized into three broad categories: (i) leader-following, (ii) behavioural and (iii) virtual leader or virtual structure approaches [36]. In the leader-following approach, one of the agents is designated as the leader while others are being designated as followers. The basic idea is that the followers track the position and orientation of the leader with some offset. This approach is also called separation-bearing control (SBC) in autonomous robotic consensus [37,38]. There are many variations of this theme including designating multiple leaders, formation of a chain and other tree topologies. A detailed review of the other two methods, namely, behavioural and virtual structure, is available in [36]. Our investigation is closely related to the leader-based strategy or the SBC in consensus control of robots. One limitation in the proposed approaches is the assumption that all agents in the team are informed, that is, the agents know the orientation and position of all the other agents in the team [36]. This requires coordination information to be available to all the agents in the team at any given instant.

One major challenge in the implementation in multi-agent systems paradigm is the presence of communication channels that are used for achieving coordination among various agents in the team. As communication channels are usually associated with packet loss, maintaining a formation can become increasingly difficult in the presence of packet loss. This is mainly due to the loss of coordination information among various agents in the network. Furthermore, when wireless communication channels are used for implementing formation control loops it has been found in practice that wireless sensor networks are more pronounced to packet loss than their wired counterparts. Typically, in such applications wireless or radio communication is preferred for implementation issues. It is easy to verify that the presence of communication channel makes the assumption that all agents being informed are invalid. As one might expect that, in order for the formation control algorithm to work, it should be robust to link failures. Furthermore, packet-dropouts can also result in catastrophic outcomes. As an example consider the IVHS [39], wherein data-loss may result in collision of vehicles. Thus, robustness to packet loss is an important attribute required in any formation control algorithm. In our investigation, we first show that the formation control problem is intractable in the presence of packet losses. Later, we propose an estimation-based formation algorithm and finally, we investigate the data to be transmitted in the event of a packet-dropout to reduce the estimation error covariance. Teams of autonomous agents working in coordination achieve greater efficiency and operation capability than agents performing solo-missions. Multi-agent systems have been investigated widely in the recent past owing to their wide applications and advantages. The distributed consensus in multi-vehicle cooperative control has been treated in Ref. [40]. Formation control is one among the problems being investigated in both control and multi-agent systems paradigms. In formation control teams of agents moving together are required to maintain a pre-defined geometric configuration. Formation control problems have application in vehicle control, unmanned air-craft vehicles, consensus and formation control of robots, in industrial robots to name a few. In order to maintain formation, agents in a team need to exchange information like relative displacement, velocity and so on. These variables that are exchanged among agents in a team for maintaining formation are called coordination variables, and are used to achieve coordination among agents. Hence, there arises a need to transmit these coordination variables among all the agents in the network. One may visualize that any loss in coordination variable can jeopardize the formation. Communication channels are used for information exchange among agents and are the enabling factor of formation control algorithms. A major challenge in implementation of formation control problems stems from the packet loss that occurs in these shared communication channel. In the presence of packet loss the coordination information among agents is lost. Moreover, there is a move to use wireless channels in formation control applications. It has been found in practice that packet losses are more pronounced in wireless channels, than their wired counterparts. In our analysis, we first show that packet loss may result in loss of rigidity. In turn, this causes the entire formation to fail. Later, we present an estimation-based formation control algorithm that is robust to packet loss among agents. The proposed estimation algorithm employs a minimal spanning tree algorithm to compute the estimate of the node variables (coordination variables). Consequently, this reduces the communication overhead required for information exchange. Later, using simulation, we verify the data that are to be transmitted for optimal estimation of these variables in the event of a packet loss. Finally, the effectiveness of the proposed algorithm is illustrated using suitable simulation example.

7.2 PROBLEM FORMULATION

Consider a formation $\Im$ of agents that are connected using communication links as in Figure 7.1a. The given formation can be conveniently represented as a graph, $G_p = \{N, E, W\}$. Here N is the set of nodes, E is the edge set and W the weights and is the intra-separation distance between the agents as in Figure 7.1b.

Definition 1

An agent is said to undergo a rigid motion along a trajectory, only when the Euclidean distance between the agents in the team remains constant all along the trajectory of the agent.

Definition 2

The graph in Figure 7.1b is said to be rigid, if for all position assignments of the nodes, each and every move of the agent preserves the distance between the positions of any pair of vertices in a graph. This condition may be expressed as

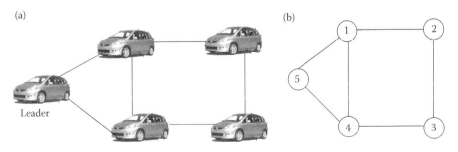

FIGURE 7.1   (a) Representation of an agent-based system; (b) graph of the team of agent.

$$\left\| x_i - x_j \right\| = C_{ij} \quad \forall \{i, j\} \in E \tag{7.1}$$

Here, C_{ij} is the predetermined distance between i and j in the team or the intra-spacing between the agents and x_i and x_j are the relative positions of the agents i and j in the team. The intra-spacing between the agents is similar to the separation in SBC mentioned in Section 7.1. Stated otherwise, rigid motion is the only kind of motion the team can undergo. Thus, it is possible to 'maintain formation' by keeping the intra-distance spacing constant. This requirement can be given as

$$\nabla = \left\| x_i(t) - x_j(t) \right\| = \left\| x_i(\tau) - x_j(\tau) \right\| = C_{ij} \quad \forall [t, \tau] \in \Re^+, \forall \{i, j\} \in E \tag{7.2}$$

Intuitively rigidity gives the information regarding the minimum number of edges that are needed for maintaining the formation. Rigidity of graphs has been studied for a long time now. One approach to ascertain the rigidity of a 2-D planar graph is proposed in Refs. [41,42]. We now investigate the conditions for rigidity for the formation control framework. From Equations 7.1 and 7.2, we have

$$\frac{1}{2} \left\| x_i - x_j \right\|^2 = C_{ij} \quad \forall \{i, j\} \in E \quad t \geq 0 \tag{7.3}$$

Assuming smooth trajectory, differentiation of Equation 7.3 gives

$$\frac{d}{dt} \left(\frac{1}{2} \left\| x_i - x_j \right\|^2 \right) = (x_i - x_j)^T (\dot{x}_i - \dot{x}_j) \quad \forall \{i, j\} \in E \quad t \geq 0 \tag{7.4}$$

A slight manipulation of Equation 7.4 leads to

$$R(q)\dot{q} = 0 \tag{7.5}$$

Here, $q = [x_1, x_2, \ldots, x_n] \in \Re^{dn}$, n is the cardinality of the vertex set and d is the dimension of the vector. The rigidity matrix given by $R(q)$ is of the order $\Re^{m \times nd}$ with m being the number of edges of the given graph. Given that q_0 is a feasible formation, the graph is generically rigid if and only if [41,42]:

$$\begin{aligned} Rank(R(q_0)) &= 3n - 6 \quad \text{if } d = 3 \\ &= 2n - 3 \quad \text{if } d = 2 \end{aligned} \tag{7.6}$$

One may visualize from Equations 7.1 and 7.2 the need for transmitting the relative displacements at any given instant to all the other agents in team. Invariably, communication channels are used for transmitting coordination information like relative displacement, velocity and so on. Data loss in communication channels causes the coordination information to be lost. This makes formation intractable as the rigidity of the formation is lost and this can be ascertained from Equation 7.6. Now, consider the set of points given by

$$d_0 = [d_{01}, d_{02}, \ldots, d_{0n}] \quad d_{0i} \in \Re^{dn} \tag{7.7}$$

Assuming d_0 to meet the rigidity constraints in Equation 7.6, let us now define the relative error as

$$r_i(t) = x_i(t) - d_{0i} \tag{7.8}$$

A possible strategy to maintain the formation is to run consensus on as per Equation 7.8.

Assuming the links to be healthy, we have

$$\dot{r}_i(t) = -\sum_{j \in N_i(t)} (r_i - r_j) \tag{7.9}$$

Equation 7.9 illustrates the consensus of Equation 7.8. It can be verified that

$$\dot{r}_i(t) = \frac{d}{dt}(x_i - d_{0i}) = \dot{x}_i \tag{7.10}$$

With

$$(r_i - r_j) = (x_i - d_{0i}) - (x_j - d_{0j}) \tag{7.11}$$

The formation control equation can thus be written as

$$\dot{x}_i = \sum_{j \in N_i(t)} (x_i - x_j) - (d_{0i} - d_{0j}) \tag{7.12}$$

The main drawback of Equation 7.12 is that it requires more communication as all the agents should know the position of the other agents in the team. This requires all the links to be healthy. In the presence of packet loss the assumption that all agents are informed is invalid. One may conclude from the preceding discussion that the formation control problem becomes intractable in the presence of packet losses. Thus there is a need to devise an algorithm that is more robust to link failures. In our analysis, we propose an estimation-based formation control algorithm among a team of multiple agents connected over a lossy link. We also investigate the data to be transmitted in the event of a packet loss for optimal estimation.

7.3   ESTIMATION-BASED FORMATION CONTROL ALGORITHM

The first step in the algorithm is to construct a minimum spanning tree (MST) by considering the healthy links in the team. The main requirement to maintain formation from Equation 7.12 is that the graph G_p should be connected and can be inferred by creating a MST from the healthy links. In our analysis, we employ a greedy algorithm to construct the tree. It is generally seen that as the distance between the agents increases, the packet loss, as well as energy and delays in the channel, also increases. The MST is constructed after leaving out the links with packet dropouts at each time epoch. The algorithm is shown in Table 7.1. The MST with packet-loss between node 1 and node 2 is shown in Figure 7.2, and the MST constructed using the algorithm with packet loss between nodes 5 and 3 is shown in Figure 7.3. The next step in the algorithm is to generate the estimate of the various node variables w.r.t. the leader node. Now consider the illustrated graph in Figure 7.1b again. With reference to one-hop neighbours, the relative displacements available to node 2 may be given as

$$Z_{21} = x_2 - x_1 + \varepsilon_{21}$$
$$Z_{24} = x_2 - x_4 + \varepsilon_{24}$$

Or more generically as

$$Z_{ij} = x_i - x_j + \varepsilon_{ij} \tag{7.13}$$

TABLE 7.1

Minimum Spanning Tree (MST) Algorithm with Packet Losses

Function MST(n,e,w)

Input:

n: Number of agents in the team

e: Number of links

w: Weight associated with the links

T: Tokens from the nodes

G: Graph of the team in terms of edge-set

Initialize Q (arbitrary graph), $N(q) = 0$ is the number of healthy links

For $i = 1$:e

If $T(i) == 1$

$N(q) = N(q) + 1$

Q: Add edge to Q

elseif $T(i) == 0$

Q: remove the edge from (Q)

end

Return $Q, N(q)$

Input: $a(v)$ leader node edge with minimum weight in Q

$W = []$;

$Q, N(q)$

Initialize the tree Tr

While (Tr $< N(q) - 1$)

$a(v)$ be the set containing u

$b(v)$ be the set containing v

if $a(u) != b(v)$

then

add edge $(v,u) \leftarrow$ Tr

Merge $a(u)$ and $b(v)$ into one cluster

Return Tr

Equation 7.13 can be written as

$$Z = Hx_i + \varepsilon \tag{7.14}$$

where H is the incidence matrix and it gives the relative displacement of agent i w.r.t. its one-hop neighbours. The least-square estimate of the node variable or the coordination variable can be computed using

$$\hat{x} = (H^T H)^{-1} H^T z \tag{7.15}$$

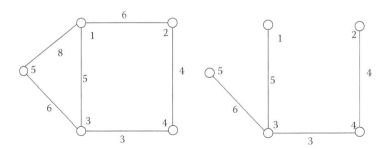

FIGURE 7.2 MST with packet loss in link between nodes 1 and 2.

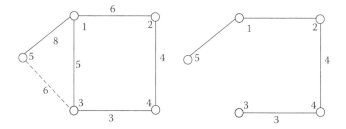

FIGURE 7.3   MST with packet loss in link between nodes 5 and 3.

Let P be the covariance matrix, then the best linear unbiased estimate (BLUE) is given as

$$\hat{x} = \underbrace{(H^T P^{-1} H)^{-1}}_{L_p} H^T P^{-1} z \qquad (7.16)$$

where L_p is the graph Laplacian. It is easy to verify that the position of an arbitrary agent at any given time is a linear combination of its own position moves and the position moves of its one-hop neighbours. Thus, by considering the reference or leader node, Equation 7.16 can be modified as Equation 7.18.

$$Z = H_r x_r(k) + H_b x_b(k) + \varepsilon$$

Or

$$Z - H_r x_r(k) = H_b x_b(k) + \varepsilon \qquad (7.17)$$

The best linear unbiased estimate (BLUE) of the one-hop neighbour nodes can then be estimated as

$$\hat{x}_b^{\,*} = (H_b^T P^{-1} H_b)^{-1} H_b^T P^{-1} (Z - H_r^T x_r) \qquad (7.18)$$

where H_b represents the partitioned matrix containing agents other than the reference node, and H_r is the partitioned incidence matrix considering the reference node. It is seen that Equation 7.18 is similar to the one obtained in Ref. [43]. The next step in the algorithm is to construct the estimate of the node variables with the missing links using Equation 7.18. Once the estimate is available to reference or leader node formation can be maintained using Equation 7.12. It is easy to visualize from Equation 7.9 that the position of the agent in the formation depends on its own displacement and that of its one-hop neighbours.

7.4   DATA TO BE TRANSMITTED IN THE EVENT OF A LOST PACKET FOR OPTIMAL ESTIMATION

Whereas the graphs and rigidity aspects have been dealt with in Refs. [44–46], the two strategies have been proposed in the past for dealing with packet loss in NCSs [47,48]. The switching topologies and time delays in networks of agents are treated in Ref. [49]. The strategies for dealing with the packet loss are: (i) transmitting zero and (ii) transmitting previous value of the control input in the event of a lost packet. These strategies are called 'to zero' and 'to hold', respectively, in Ref. [48]. It has been reported in Refs. [47,48] that none of the strategies above can be claimed to be superior to the other. This necessitates simulation or experimentation to select one of the above strategies.

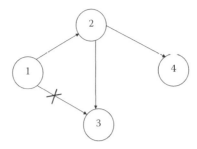

FIGURE 7.4 Team of robotic agents with packet loss in link between agents 1 and 3.

Our simulation studies indicate that transmitting a linear combination of the present measurement alongside the estimate of the sensor measurement in the past instant outperforms the 'to hold' and 'to zero' strategy investigated in Ref. [48]. The above result has also been proved theoretically in Refs. [2,40]. In Ref. [2], the above result has been extended to the consensus problem among robotic agents, wherein the network of robotic agents shown in Figure 7.4 is considered. Now, assume that there is a packet loss in link between agents 1 and 3. The simulation of the positions of the robotic agents is shown in Figure 7.5. The estimation-based formation control algorithm is shown in Table 7.2.

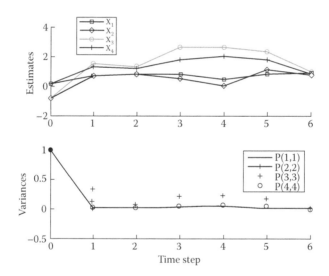

FIGURE 7.5 Position of agents with packet loss using the proposed methodology.

TABLE 7.2

Algorithm for Formation Control over Packet Dropping Links

Input:

MST (n,e,w), G-the graph of the formation

Compute $Q1$: $G - \mathrm{Tr}$

$Q2 = G - Q1$

Determine Z_{b}

Compute $\hat{x}$ using Equation 7.19 for $Q2$

Compute Z_{r}

7.5   RESULTS AND DISCUSSIONS

Consider the formation alongside the initial conditions shown in Figure 7.6. Given that there is a link failure between agents 1 and 2, we compare the performance of the proposed algorithm with that of the other two strategies widely employed in literature—(i) to transmit zero or (ii) to transmit past value of the control input [44,45]. The position of agents 3 and 5, and the estimation error covariance after 2 position moves after 50 iterations by transmitting zero are shown in Figure 7.7. The position of agents 2 and 5 by transmitting the measurement available in the previous instant and the estimation error covariance is shown in Figure 7.8. It is seen that the method of transmitting either zero or the past measurement is not suitable for maintaining the formation. The positions of agents 3 and 5 after two position moves by using the estimation scheme in Equation 7.18 is shown in Figure 7.9. It is seen that the proposed estimation scheme is able to maintain the formation in the presence of packet losses. The positions of agents 3 and 5 with packet dropouts in one-link over 50 iterations are shown in Figure 7.10. It is seen that the scheme performs well and can maintain formation even in the presence of one-link loss over the entire estimation period. The

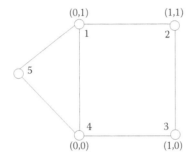

FIGURE 7.6   Team of agents with initial formation.

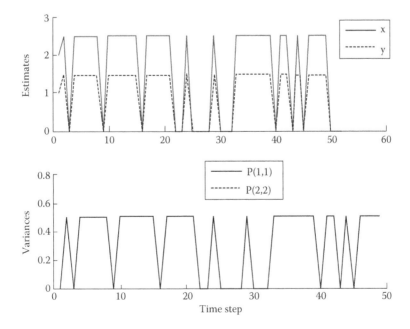

FIGURE 7.7   Position of agents 3 and 5 after two position moves by transmitting zero.

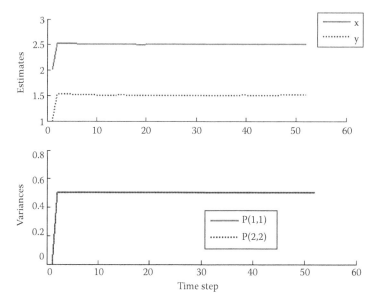

FIGURE 7.8   Position of agents 3 and 5 after 2 position moves by transmitting the measurement available at the previous instant.

position of agents 3 and 5 after 2 position moves by transmitting a linear combination of the past measurement and present estimate is shown in Figure 7.11. It may be observed that the estimation error covariance is reduced by using the proposed information scheme. Further discussions on various applications of NCSs are (i) distributed process control [2,10], (ii) wireless sensor networks [8], (iii) consensus and cooperation control [40], (iv) intelligent vehicle highway systems (IVHS) [43] and (v) tele-robotics [22,50].

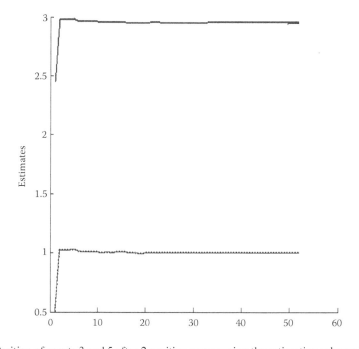

FIGURE 7.9   Position of agents 3 and 5 after 2 position moves using the estimation scheme in Ref. [19].

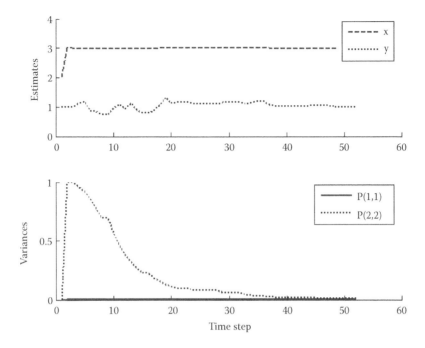

FIGURE 7.10   Estimated position of agents 3 and 5 after 2 position moves with two-packet dropping link for 50 iterations using the estimation scheme in Ref. [19].

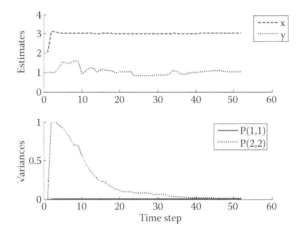

FIGURE 7.11   Position of agents 3 and 5 with two packet dropping links for 50 iterations using the proposed information scheme.

7.6   CONCLUDING REMARKS

An estimation-based formation control algorithm for agents connected over packet dropping links has been proposed in this investigation. It was also shown that the proposed algorithm is robust to link failures. Simulation results show that transmitting a linear combination of past measurement and present estimate reduces the estimation error covariance. Investigations into the maximum packet loss rate that can guarantee proposed formation control algorithm to maintain formation and dynamic formations wherein nodes get attached and tethered are future extensions of this investigation.

REFERENCES

1. J. Baillieul and P. J. Antsaklis, Control and communication challenges in networked real-time systems, *Proceedings of IEEE*, 95(1), 2007, 9–28.

2. S. Seshadhri, *Control and Estimation Methodologies for Networked Control Systems Subjected to Communication Constraints*, PhD dissertation, Department of Instrumentation and Control Engineering, National Institute of Technology-Tiruchirappalli, India, Dec. 2010.

3. J. P. Hespanaha, P. Naghshtabrizi, and Y. Xu, A survey of recent results in networked control systems, *Proceedings of IEEE*, 95(1), 2007, 138–162.

4. F. Lian, *Analysis, Design, Modeling and Control of Networked Control Systems*, PhD dissertation, Department of Mechanical Engineering, University of Michigan, 2001.

5. J. J. C. van Schendel, *Networked Control Systems: Simulation and Analysis*, Traineeship Report, Technical University of Eindhoven, 2008.

6. M. Pohjola, *PID Controller Design in Networked Control Systems*, Master's thesis, Department of Automation and Systems Technology, Helsinki University of Technology, Jan. 2006.

7. W. Zhang, *Stability Analysis of Networked Control Systems*, PhD dissertation, Department of Electrical and Computer Science, Case Western Reserve University, August 2001.

8. M. Bjorkbom, *Wireless Control System Simulation and Network Adaptive Control*, PhD dissertation, School of Science and Technology, Department of Automation and Systems Technology, Altoo University, Oct. 2010.

9. J. Nilsson, *Real-Time Control Systems with Delays*, PhD dissertation, Department of Automatic Control, Lund Institute of Technology, 1998.

10. F-Y. Wang and D. Liu (Eds.), *Networked Control Systems: Theory and Applications*, Springer-Verlag, London, 2008.

11. D. Hristu-Varsakelis and W. S. Levine, *Handbook of Networked and Embedded Control Systems*, Birkhauser, Boston, 2005.

12. S. Seshadhri and R. Ayyagari, Hybrid controllers for systems with random communication delays, In *Proceedings of International Conference on Advances in Recent Technologies in Communications and Computing (ARTCom'09)*, Kottayam, India, pp. 954–958, 2009.

13. M. Chow and M. Tipsuvan, Network based control systems: A tutorial, In *Proceedings of the 27th Annual Conference of the IEEE Industrial Electronics Society (IECON'01)*, Denver, USA, pp. 1593–1602, 2001.

14. A. Ray and Y. H. Halevi, Intergrated communication and control systems, *Journal of Dynamic Systems, Measurements and Control*, 110, 367–373, 1988.

15. G. C. Walsh, H. Ye, and L. Bushneil, Stability analysis of networked control systems, *Proc. American Control Conference*, San Diego, CA, pp. 2876–2880, 1999.

16. Z. Wei, M. S. Branicky, and S. M. Philips, Stability of networked control systems: Explicit analysis of delay, *IEEE Control System Magazine*, 21(1), 84–99, 2001.

17. Y. Shi and Y. Bo, Output feedback stabilization of networked control system with random delays modeled by Markov chains, *IEEE Trans. Automatic Control*, 54(7), 1668–1674, 2009.

18. S. Xi-Ming, L. Guo-Ping, D. Rees, and W. Wang, Stability of systems with controller failure and time varying delay, *IEEE Trans. Automatic Control*, 53(10), 2391–2396, 2008.

19. G. A. Kamnik, R. Schechter-Glick, and V. Sadov, Using sensor morphology for multi-robot formation, *IEEE Transactions on Robotics*, 24(2), 271–282, 2008.

20. W. Ren, Collective motion from consensus with Cartesian coordinate coupling, *IEEE Transactions on Automatic Control*, 54(6), 1330–1335, 2009.

21. T. Samad, J. S. Bay, and D. Godbole, Network centric systems for military operations in urban terrain: The role of UAVs, *Proceedings of the IEEE*, 95(1), 92–107, 2007.

22. D. Sorid and S. K. Moore, The virtual surgeon, *IEEE Spectrum*, 37(7), 26–31, 2000.

23. M. Chiang, S. H. Low, A. R. Calderbank, and J. C. Doyle, Layering as optimization decomposition: A mathematical theory of network architectures, *Proceedings of IEEE*, 95(1), 255–312, 2007.

24. R. Madhan, N. B. Mehta, A. F. Molisch, and J. Zhang, Energy-efficient routing in wireless networks, *IEEE Transactions on Automatic Control*, 54(3), 512–527, 2009.

25. S. Seshadhri and R. Ayyagari, Consensus among robotic agents over packet dropping links, In *Proceedings of IEEE 3rd International Conference on Bio-Medical Engineering and Informatics (BMEI-2010)*, Yantai, China, pp. 2636–2640, 2010.

26. T. Balch and R. Arkin, Behavior-based formation control for multi-robot teams, *IEEE Transactions on Robotics and Automation*, 14(6), pp. 926–939, 1998.

27. F. Fahimi, Sliding mode formation control for under-actuated surface vessels, *IEEE Transactions on Robotics*, 23(3), 617–622, 2007.
28. J. A. Fax and R. M. Murray, Information flow and cooperative control of vehicle formation, *IEEE transactions on Automatic Control*, 49(9), pp. 1465–1474, 2004.
29. R. Fierro, A. K. Das, V. Kumar, and J. P. Ostrowski, Hybrid control of formations of robots, In *Proceedings of the IEEE International Conference on Robotics and Automation*, Seoul, Korea, May 2001, pp. 157–162.
30. S. Seshadhri and R. Ayyagari, Platooning over packet dropping links, *International Journal of Vehicle Autonomous Systems*, 9(1/2), 46–62, 2011.
31. P. Ogren, M. Ergerstedt, and X. Hu, A control Lyaponov function approach to multi-agent coordination, *IEEE Transactions on Robotics and Automation*, 18(5), 847–851, 2002.
32. P. Ogren, E. Fiorelli, and N. E. Leonard, Formations with a mission: Stable coordination of vehicle group maneuvers, In *Proceedings of the 15th International Symposium on Mathematical Theory of Networks and Systems*, Notre Dame, IN, 2002.
33. T. Sugar and V. Kumar, Decentralized control of cooperating mobile manipulators, In *Proceedings of the IEEE International Conference on Robotics and Automation*, Leuven, Belgium, May 1998, pp. 2916–2921.
34. P. K. C. Wang, Navigation strategies for multiple autonomous mobile robots moving in formation, *Journal of Robotic Systems*, 8(2), 177–195, 1991.
35. F. Giulietti, L. Pollini, and M. Innoceti, Autonomous formation flight, *IEEE Control Systems Magazine*, 20(6), 34–44, 2000.
36. D. J. Stiwell, and B. E. Bishop, Platoons for underwater vehicles, *IEEE Control Systems Magazine*, 20(6), 45–52, 2000.
37. J. R. Carpenter, Decentralized control of satellite formations, *International Journal of Robust and Nonlinear Control*, 12, 141–161, 2002.
38. M. R. Anderson and A. C. Robins, Formation flight as cooperative game, In *Proceedings of AIAA Guidance, Navigation, and Control Conference*, Boston, MA, August, 1998, pp. AIAA-98–4124.
39. F. Y. Hadaegh, W-.M. Lu, and P. K. C. Wang, Adaptive control of formation flying spacecraft for interferometry, *In the Proceedings of the IFAC Symposium on Large Scale Systems: Theory and Applications*, Patras, Greece, 1998, pp. 97–102.
40. W. Ren and R. W. Beard, *Distributed Consensus in Multi-Vehicle Cooperative Control: Theory and Applications*, Verlag Springer, London, 2008.
41. S. Carpin and L. Parker, Cooperative leader following in distributed multi-robot formations, In *Proceedings of the IEEE International Conference on Robotics Automation (ICRA 2002)*, Washington, DC, 2002, pp. 2994–3001.
42. J. P. Desai, A graph theoretic approach for mobile robot team formations, *Journal of Robotic Systems*, 19(11), 511–525, 2002.
43. P. Varaiya, Smart cars on smart roads, *IEEE Transactions on Automatic Control*, 38(2), 195–207, 1993.
44. G. Laman, On graphs and rigidity of plane skeletal structures, *Journal of Engineering Mathematics*, 4, 331–340, 1970.
45. T. Eren, P. N. Belhumeur, B. D. O. Anderson and S. Moorse, A framework for maintaining formations based rigidity, *Proceedings of the IFAC Congress*, Barcelona, Spain, 2002.
46. P. Barooah, and J. P. Hespanaha, Estimation on graphs form relative measurements, *IEEE Control Systems Magazine*, 2007, pp. 57–74.
47. L. Schenato, Optimal estimation in networked control systems subjected to random delay and packet dropouts, *IEEE Transactions on Automatic Control*, 53(5), 1317–1331, 2008.
48. L. Schenato, To zero or hold control inputs with lossy links?, *IEEE Transactions on Automatic Control*, 54(5), 1093–1099, 2009.
49. R. Olfati-Saber and R. M. Murray, Consensus problems in networks of agents with switching topologies and time-delays, *IEEE Transactions on Automatic Control*, 49(9), 1520–1533, 2004.
50. R. Oboe and P. Fiorini, A design and control environment for internet based tele-robotics, *International Journal of Robotic Research*, 17(4), 443–449, 1998.

Part II

MIAS and Robotics

8 Robot's Sensors and Instrumentation

A. Ramachandran

CONTENTS

8.1   INTRODUCTION

In this chapter, we provide an overview of sensors and instrumentation from the point of view of their usage in robotics. An engineer/scientist/researcher working in the field of robotics should know about sensors and their characteristics, since the sensors are the 'eyes' for any mobile intelligent autonomous vehicle. Some basic knowledge on analogue signal processing is also required to select proper signal conditioners for a given sensor. The sensors, microelectronic and mechanical systems (MEMS) and even smart sensors are the key elements and component in robotics for gathering information about the environment as well as about robots themselves [1–6].

The important elements of a simple autonomous robotic system are shown in Figure 8.1. Sensors play a major role in gathering information about the robot as well as the environment where the robot moves because these very sensors provide perception that is needed by the robot to carry out its mission. The information gathered by the sensors and communicated to the robot helps the robot to ascertain its world's model and this in turn is used to navigate the robot. Sensors and instrumentation systems are an interface between the control system and the real world for a mobile vehicle. Once a robot gets an adequate world model, its planner sets the robot to an assigned task by giving directions to follow an optimal path (from PMP, Chapter 10). The final commands to robot come from the decision maker to initiate the motion of the robot. Thus, sensors, instruments and actuators are very important ingredients in any mobile autonomous systems. Hence, a basic instrumentation system comprises some sensor, signal processing/conditioning device/system/circuit and an output device that interfaces with some other systems of the robotic system that will need this output from the instrumentation system. Certain sensor/instrumentation systems are integrated systems so that it will not be possible to separate these aspects of sensing/signal conditioning and output device. This output device may be a simple amplifier gain device or some registration/synchronization device. Also, the actuators and sensors have their own dynamic characteristics that should be known in advance. Their mathematical models might be required to incorporate into the dynamic simulations for flight vehicles or robots. In particular, for mobile intelligent systems we have control effectors that might be electrical, pneumatic or hydraulic. The output sensors for the same systems might be sensors for measurements of linear positions, rates, accelerations and angular rates and positions. The navigation devices/sensors are radio, inertial measurement unit (IMU) and global positioning system (GPS). We commonly need the measurements of the following quantities, although all of them might not be required for the robotic

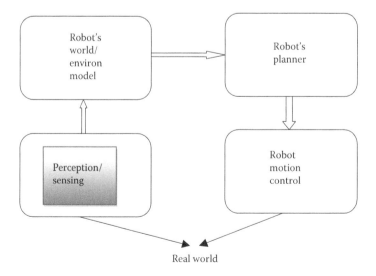

FIGURE 8.1   Autonomous robotic system.

system: temperature, speed, force, stress and strain, mass, size, volume, pressure, flow rate, position, velocity and accelerations (both linear and angular).

8.2 SENSOR CLASSIFICATION

Sensors are classified based on either what to measure or how to measure. Sensors are classified as internal sensor (proprioceptive sensors) and external sensor (exteroceptive sensors) based on what parameters are being measured [1]. Internal sensors are used to measure the parameters related to the robot itself such as motor speed, wheel load, heading of the robot, battery status and so on. The external sensors are used to gather information about the environment where the robot is moving and traversing, for example, distance of the obstacle from the robot, ambient light intensity (if the robot is moving in dark areas or in the night), surface type (on which the robot is moving, that is, say it might be a rough terrain) and so on. Sensors are also classified as passive sensors and active sensors based on how the parameters are measured. Passive sensors do not require any external energy (usually DC supply) to measure the physical parameters. Some examples of the passive sensors are thermocouple, piezoelectric sensors and so on. Active sensors require some external energy (usually DC supply) to measure the physical parameters. Some examples of the active sensors are resistive sensors, sonars (sound navigation and ranging), strain gauges and so on.

Transducer is a device that converts one form of energy (mechanical or chemical) into another form of energy, usually in electrical voltage or current. A sensor is a device that senses the change in physical parameters to be measured. An actuator is a device that converts an electrical signal into some mechanical action. A transmitter is a device that converts one form of energy (mainly the electrical form) into another form of energy (mainly the electrical form) via the electromagnetic energy and it is able to transmit the converted energy over a long distance. In practice, there is not much difference between a transducer, sensor and transmitter, especially now smart sensors are available that do much of the common work within the structure and the architecture of the smart sensor configuration itself. A typical structure of transducer/transmitter and measurement system is depicted in Figure 8.2. Certain sensors that are useful in robotics might be based on on/off operations of switches that detect the presence or absence of obstacles in the vicinity of a robot. For example, the proximity of an obstacle to the robot can be detected by a proximity switch. Next, some important sensor devices are briefly discussed. In general, various sensors can be classified as (a) acoustic, (b) chemical, (c) electrical/electronic, (d) magnetic, (e) electromagnetic (EM radiation), (f) mechanical, (g) thermal and (h) optical. Many types of sensors/sensor mechanisms used/useful in robotics and mobile intelligent autonomous system (MIAS) systems are illustrated in Figure 8.3.

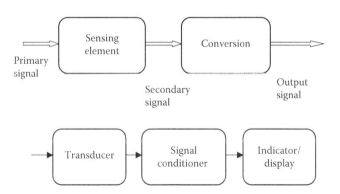

FIGURE 8.2 Process and stages of transducer and measurement system.

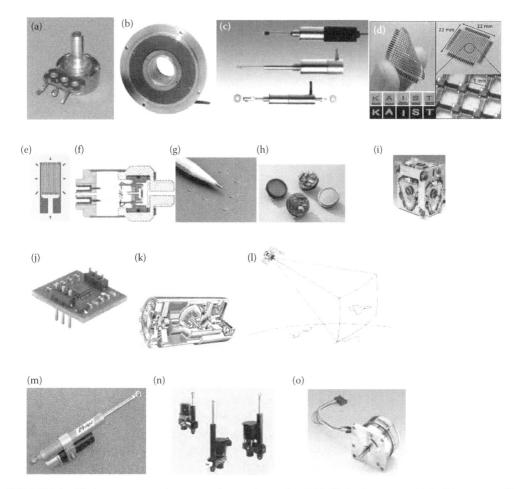

FIGURE 8.3 Various sensors and actuators that can be used in MIAS/robotic systems. (a) Position sensor, (b) angular encoder, (c) LVDT, (d) tactile sensors, (e) strain gauge, (f) pressure sensor, (g) thermistor, (h) ultrasonic range finder, (i) optical angular range sensor, (j) accelerometer, (k) rate gyroscope, (l) GPS segment, (m) a hydraulic actuator, (n) a pneumatic actuator and (o) electric motor. (Adapted from Stengel, R. F. *Robotics and Intelligent Systems,* Princeton University, Princeton, New Jersey, USA, 2009. http://www.princeton.edu/~stengel/MAE345Lecture1.pdf.)

8.2.1 Capacitive and Inductive Devices

A parallel-plate capacitor is used as a sensor to detect displacement, force, acceleration and pressure. The inductive devices are LVDT or RVDT (linear/reluctance variable differential transformer). The advantages of LVDT are that large displacements can be measured and the device is highly linear. The disadvantages are that they have limited frequency response (due to the AC excitation frequency and mass of the moving part) and they require complex electronic signal conditioners. The main applications of the inductive devices are measurement of displacement, acceleration, liquid level and so on.

8.2.2 Piezoelectric Transducer

In this sensor, a crystalline material is used that produces an electrical charge on its surface when the sensor material is mechanically pressed and hence stressed [2]. This electric charge is converted into a voltage. When this device is placed inside a pressure transducer, the application of force (and hence pressure) is transformed into an electrical signal. The major applications of this sensor are measurement of acceleration, pressure, force and so on. This sensor is not a suitable static measurement. Charge amplifiers are used as a signal conditioner for such transducers.

8.2.3 FORCE-SENSING RESISTORS

Force-sensing resistor (FSR) responds to the physical pressure applied on its surface. When no force is applied, the FSR depicts maximum resistance, and when force is applied, its resistance becomes less. The FSR is basically made of two layers: (i) the resistive (semi-conductive polymer) top layer and (ii) the 'inter-digitating' conductive electrodes on the bottom layer upon which the top layer rests. When no force is applied, the digitating area makes very little contact with each other and hence resistance is high. But when the force is applied, the top and bottom layers get squashed together, causing an increase in conductance and hence a decrease in resistance (conductance is inversely proportional to resistance).

8.2.4 BEND/FLEX SENSORS

In this type of sensor, a change in resistance is noted when the sensor is bent or flexed. Hence, the amount of bending/flexing causes the top resistive layer and bottom 'digitating' layer to change the resistance value of the sensor. This type of sensor is then useful in sensing the robot's arm movement which will be used for gripping control and the tightness of the grip per se.

8.2.5 SONAR SENSORS

This is basically a sonar sensor in a class of acoustic sensor. This uses ultrasonic sound waves around 40 kHz (above human hearing range). It transmits the sound waves and waits for the echo. If no echo is received then there is no object nearby. If echo is detected, the time from when the signal is sent and when the echo/reflected signal is received determines the distance of an object. This time multiplied by $1/2v$ gives the distance of the object from the robot, wherein v is the speed of sound wave. Thus, sonar sensor is very crucial in a robotic system to detect the obstacles in the path of a mobile vehicle. The sonar can also be used in conjunction with other vision sensors to improve the accuracy and extended range of measurements.

8.2.6 INFRARED SENSORS

The infrared (IR) radiation exists in the electromagnetic spectrum and is not visible to the naked eye (it has a longer wavelength than visible light; see Figure 8.4). The IR sensors are a type of thermal sensor. IR sensors commonly house a pyroelectric crystal formed as thin-plate capacitors which produce a surface electric charge when the device is exposed to IR radiation. Pyroelectric materials generate temporary change in energy when heated or cooled. The electric charge causes a voltage drop in the capacitor. The charge itself is induced as pyroelectric crystals which inherently have spontaneous electrical polarization—IR changes the polarization which in turn causes induction of charges. When the amount of the radiation falling on the crystal changes, the amount of charge that can be measured accordingly gets affected. There usually is a preamplifier circuit built in as the high-impedance source of the pyroelectric set-up that produces very little current voltage follower. These IR sensors are very useful for night vision for mobile robots. The IR sensors can also be used in conjunction with vision sensors as well as sonars in the multi-sensor data fusion architecture to enhance the predictive accuracy of the robot's overall vision system and can be very useful for avoiding multiple obstacles moving in an uncertain fashion in the robot's environment.

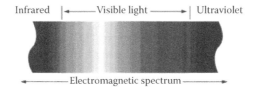

Infrared | ←——Visible light ——→ | Ultraviolet

←———————— Electromagnetic spectrum ————————→

FIGURE 8.4   Infrared, visible light and ultraviolet rays spectrum.

8.2.7 Photoresistors/Photoconductor

Photoresistor sensors are sensitive to light intensity change. When no light falls on this sensor, it shows maximum resistance, usually called dark resistance. When light falls on this sensor, a decrease in the sensor's resistance value is observed. The measured change in resistance gives the measure of the light intensity.

8.2.8 Wheel Sensors

These sensors are used to measure position or speed of the wheels or steering of a mobile robot. The optical encoders are used to measure the velocity/position of the wheel. This concept is depicted in Figure 8.5.

8.2.9 Speed Transducers

The speed transducers/sensors are used to measure output and speed of the rotating part of a mobile system/robot. There are several types of speed sensors: (i) optical, (ii) magnetic pickups and (iii) tachometers.

8.2.10 Vision Sensor

Robotic vision is a complex sensing and perception process. It involves extracting, characterizing and interpreting information from several images in order to detect, identify and describe objects/obstacles in a robot's environment. A vision sensor (e.g., a digital camera) converts the visual information (light rays emanating from an illuminated object) into an image and then to electrical signals. This electrical signal is further processed by electronics system and gives the digital image which is easily read by the computer. At present charge coupled device (CCD) image sensors are widely used in robotic applications because of their small size, light weight, robustness and better electrical performance. Camera should be mounted above the working area of the robot which eliminates reduced resolution, parallax error and robot hand obstructing the field of view. The digital image produced by a vision sensor is further processed to get meaningful visual effects of an object. This is done by using digital image processing techniques which involve a series of steps:

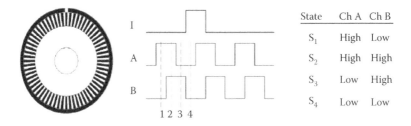

FIGURE 8.5   The encoder concept for measuring the position and speed of a wheel of a robot.

(i) image registration, (ii) synchronization, (iii) preprocessing to remove the background noise, (iv) image segmentation, (v) image description, (vi) image recognition and (vii) interpretation to decipher the object under investigation. For further details on image processing, segmentation and image fusion, see Chapters 4 through 6.

8.3 OTHER SENSORS AND THEIR FEATURES AND USAGES

8.3.1 TOUCH AND TACTILE SENSORS

When physical contact is made with an object, the sensor senses this. An example of touch sensor is a simple micro-switch that either turns on or off as contact is made. A force sensor is also used as a touch sensor which not only detects the presence of a contact but also senses the amount of pressure/force applied on it. A collection of touch sensors which provides information about the contact and about the size of an object in contact is called a tactile sensor. All displacement sensors like micro-switches, LVDTs, pressure sensors and magnetic sensors can be used as touch sensors.

8.3.1.1 Proximity Sensors

A proximity sensor senses the presence or absence of an object close to the robot. It could be a non-contact-type sensor. This sensor information is useful for navigational task of the robot. Different types of non-contact proximity sensors used in robot are magnetic, eddy current and Hall effect, optical, ultrasonic, inductive and capacitive.

8.3.1.2 Sniff Sensors

This is similar to a gas sensor. This sensor is sensitive to particular gases and provides information about gas detected while the robot is on its path. This information provided by this sensor is useful for safety as well as search and detection purposes.

8.3.1.3 Voice Recognition Devices

When a voice recognition system recognizes a word, it sends a signal to the robot controller. The controller directs the robot to the desired location and orientation on its path. These devices are more useful for disabled persons and for medical robot. Voice recognition system involves 'what is said' and taking an action on the perceived information. It works on the frequency content of the spoken words. Voice synthesizer is realized in two ways: (i) to create a word combined with both phonemes and vowels and (ii) to record the words either in digital or in analogue form. These words are accessed by the system as and when they are required.

8.3.1.4 Range Finders

Laser-based range finders (LADAR) measure the distance of an object from the robot by different methods, namely, direct time delay, indirect amplitude modulation and triangulation. Triangulation is the most accurate and gives the highest-resolution results for short-distance measurements.

8.4 SALIENT SENSOR CHARACTERISTICS

Some important parameters to be kept in mind while selecting the sensors for use in robotics are (i) accuracy (closeness of the reading to the true value), (ii) repeatability/precision (repeatability of the same quantity of an error), (iii) resolution (ability to distinguish between two closely spaced objects), (iv) linearity of the sensor (the relationship between the measured quantity and the output of the transducer), (v) range of the sensor (say 400 km range of a radar's capacity to detect an object) and (vi) range in spectrum of the sensor's dynamic responses. These aspects are further stressed next.

8.4.1 ACCURACY

It is related to the measuring quality of the sensor. It is defined as how much the physical parameter read/detected by any sensor is close to the true value of the physical parameter. However, here the true value may not be known, but it is based on some primary standard. Sensor accuracy is determined by static calibration. Often it is provided by the manufacturer in the charts that accompany the sensor device when it is procured.

8.4.2 PRECISION

It is the capability of the sensor that should give out the same value (i.e., the same reading) when the sensor is repetitively used for measuring the same quantity under the same prescribed conditions (environmental, operator). Here, it does not matter even if the accuracy of the sensor is less. That is to say even if the measured value is wrong, on repeated measurement the sensor should give the same wrong value—this is the precision.

8.4.3 REPEATABILITY

It is the closeness of agreement between successive results obtained with the same method under the same conditions and in a short time interval.

8.4.4 RESOLUTION

It is the minimum step size within the range of measurement of a sensor. Anything with the distance between the two neighbouring objects cannot be accurately measured or distinguished.

8.4.5 SENSITIVITY

It is defined as the change in output response divided by the change in input response.

8.4.6 LINEARITY

Normally there should be a linear relationship between output value of the sensor and input variable to be measured. It is always not feasible that this relationship is linear, but when the markings on the scale are provided, these should be such that the user is not worried whether this relationship is linear or non-linear. The non-linearity can/should be masked by the display.

8.4.7 RANGE

It is the difference between the minimum and maximum value of sensor input variable to be measured. The range can also be considered as accurate measurements that a sensor (e.g., radar) can provide for detecting an object quite far away from the location of the sensor.

8.4.8 FREQUENCY RESPONSE

The frequency response is the range in which the system's ability to resonate (respond) to the input remains relatively high. The broader the range of frequency response, the better the ability of the system to respond to inputs of varied frequency range. Similarly, it is important to consider the frequency response of a sensor and determine whether the sensor's response is fast enough under all operating conditions, in particular for military, underwater and aerospace applications. However, the broader the frequency response, that is, if the bandwidth (BW) of the sensor is higher, there are sure chances that the sensor outputs might get contaminated with random noise, since the noise frequencies are usually broader in their spectra. So, the BW has a trade-off with the effect of the noise.

8.4.9 RELIABILITY

It is the ability of the sensor to work properly over a period of time without any failure/deviation from its specifications. The sensor's measurements should be reliable for being able to be trustworthy, since many decisions would depend on the sensors' measurements. The reliability also plays a crucial role in the sensor data fusion, since the data fusion from the unreliable sensors is of no value addition to the overall precision accuracy of the measurements.

8.4.10 INTERFACING

This is related to what type of interface is provided by the manufacturer to read the sensor data for further processing. This interface may be digital or analogue signal type. If the output is analogue type, it needs additional interface/electronics to interface with microcontroller and/or stand-alone data acquisition and control units. If it provides a digital interface, it is possible to interface directly with the microcontroller and/or stand-alone data acquisition and control units.

8.4.11 SIZE, WEIGHT AND VOLUME OF THE SENSOR

Size is a critical consideration for joint displacement sensors. When robots are used as dynamic machines, the weight of the sensor is also important. Volume or space is also critical in micro-robots and mobile robots used for surveillance. Cost is important especially when the number of sensors involved is large in any end application.

8.4.12 ENVIRONMENTAL CONDITION

The sensor should be capable of working and providing reasonably very accurate results even in a harsh environment (e.g., higher operating temperatures, higher vibration levels, occasional shocks and high altitude where temperatures are negative).

8.5 ACTUATORS AND EFFECTORS

The three major types of actuators used for the MIAS/robotics are (i) electrical, (ii) pneumatic and (iii) hydraulic. The main electric/electrical effectors are DC motors and motors. The merits of these are [3] as follows: (i) fast and very accurate, (ii) one can possibly apply sophisticated control techniques to motion, (iii) relatively inexpensive and (iv) the new rare earth motors have reduced weight, high torques and fast response times. The demerits of these electric actuators are as follows: (i) due to high speed, some gear transmission is always needed to obtain lower speed and lower torques, (ii) precision is limited by the gear backlash, (iii) possibility of electric arcing, (iv) possibility of over-heating and (v) for locking into positions, brakes are required.

The hydraulic actuators have the following merits [3]: (i) they have large lifting capacity, (ii) moderate speeds, (iii) power-to-weight ratio is high, (iv) good servo control can be obtained, (v) can operate in stalled conditions without any damage, (vi) they have fast response and (vii) they have smooth operation at low speeds. The demerits are as follows: (a) these systems are expensive, (b) maintenance is often required, (c) high-speed cycling is limited, (d) miniaturization is difficult and (e) power source needed.

The pneumatic actuators have the following merits: (i) they are relatively less expensive, (ii) they have high speeds, (iii) there is no pollution due to non-usage of oil/fluid, (iv) there no need of the return line for any fluid to flow back, (v) very suitable for modular robot designs and (vi) they can stall without damage. The merits of these actuators are as follows: (a) air is compressible, and hence can limit the control and accuracy, (b) the air can leak and (c) air-drying and filtering might be required.

8.6   SIGNAL CONDITIONING

The sensors' output signals are required to be pre-processed and even before that these signals are required to be conditioned, amplified and filtered. The signal conditioner also performs signal conversion and level shifting operations after amplification. The output of signal conditioner is directly interfaced to a microcontroller through ADC (analogue to digital converter) or to stand-alone data acquisition and control unit. Some signal conditioner provides excitation voltage for sensors. Most of the signal conditioner units use instrumentation amplifier for amplification because of its better characteristics like high CMRR (common mode rejection ratio), high input impedance, low output impedance and higher bandwidth and gain. If a sensor outputs current as a measured value, a current amplifier or current-to-voltage convertor is used. For restive-type sensor, a bridge circuit amplifier is used. For piezoelectric-type sensor, a voltage amplifier or charge amplifier is used as a signal conditioner. The length of cable connecting sensor and amplifier between sensor and amplifier affects the accuracy of the measurement when voltage amplifier is used as signal conditioner. To overcome this problem, a charge amplifier is widely used as a signal conditioner for piezoelectric sensors. To eliminate aliasing to error (due to under-sampling) some signal conditioners have built in anti-aliasing filter that is usually a single pole/double low-pass filter. The Nyquist theorem states that a signal should be sampled at a rate of at least twice the highest frequency component of the expected measured signal. But, if the signal has high-frequency components at or near sampling rate, it is possible to read a high-frequency signal as a much lower frequency one due to aliasing effect. The aliasing can be avoided if the redundant high-frequency signal (in general, it is a high-frequency noise) is filtered before sampling/recording at the analogue stage itself, otherwise the signal has to be sampled at a very high sampling rate, and then it is thinned for practical use. Many signals have false high-frequency components, usually caused by noise on the system and due to working environment, which must be identified and eliminated, otherwise the measured data will be inaccurate. Thus, either increase the sampling rate of the system or use an anti-aliasing filter at the output of the signal conditioner. But the oversampling method is more expensive as it needs high-performance ADC, more memory and a high-bandwidth bus. The other best option is the use of hardware analogue filter before the signal is fed to ADC. This hardware analogue filter is realized with operational amplifiers. In summary, all measurement chain analysis begins with the sensor. In order to design an effective signal measurement chain circuit, the designer must have knowledge about the sensor characteristics and requirements.

8.7   SIGNAL/DATA COMMUNICATIONS AND SENSOR NETWORK

Often, the information like physical parameters, and control commands are transmitted from one device to another which is located far away and it has independent power circuitry. When information is transmitted, some distortion and noise affect the transmitted signal and sometimes crucial information may be lost. Hence, the signal-to-noise ratio decides maximum data transfer rate. The information may be transmitted via twisted pair copper wires, coaxial cables, fibre optic or by wireless transfer. There are several industry standard communication interfaces available such as RS-232 (EIA-232), RS-485 (EIA-485), Modbus, HART, deviceNet, Profibus, Foundation Fieldbus, Industrial Ethernet and TCP/IP. A sensor network is composed of a large number of sensor nodes, which are densely deployed either inside the object under study or very close to it. The number of nodes in a sensor network can be of several orders of magnitude higher than the nodes in an ad hoc network. The features of a sensor node are limited in power, computation and memory capability. They are prone to failure and lead to frequent change in network topology. While designing a sensor network the factors to be kept in mind are (i) fault tolerance, (ii) scalability, (iii) hardware constrains, (iv) sensor network topology, (v) environment, (vi) transmission media and (vii) power consumption. The sensor networks could

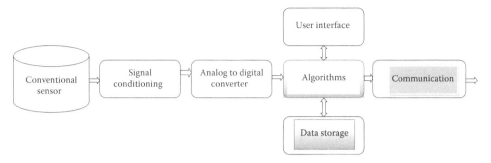

FIGURE 8.6 Mini smart sensor system—MSSS (in conventional robotic instrumentation system there might be more such independent systems; in MSSS it would be an integrated system).

be wired or wireless networks (WSNWs). These WSNWs could be arranged in centralized or decentralized configurations depending upon the need for a given robotic/vehicle system/scenario. The present trend is that the use of wireless sensor networks and sensor data fusion plays very a important role in target/object/obstacle tracking, image tracking/image fusion and decision making. In a decentralized WSNW, every node would be a signal-processing node and there is no central node. WSNWs are very useful for health monitoring of the robotic/mobile vehicle system.

8.8 ROBOTIC INSTRUMENTATION

The major components that any robotic instrumentation system would have are (i) its body-mounted sensors, (ii) data acquisition unit, (iii) signal conditioner and (iv) interface electronics between the real world and the controller. The instrumentation system acquires data from various sensors and processes it for further use. The processed data may be stored or transmitted to a remote central location. The processed data will be used to navigate the robot with the help of a controller, in a path/motion planning mode or for simultaneous localization and mapping functionality. A robotic instrumentation system with one sensor and its processing units is shown in Figure 8.6.

8.9 MICROELECTRONIC AND MECHANICAL SENSORS/SYSTEMS

The MEMS is the study of small electromechanical devices, sensor and systems. These devices range in their size from a few microns to a few millimetres. The field has several alternative names: MEMS, micromechanics, Micro System Technology (MST), Micro Machines and/or simply Micro. Now, there are NEMS due to the surge of nanotechnology, wherein the devices/sensors range in their sizes from a nanometre to a micron. As such the field encompasses all aspects in science, engineering and technology that are involved with things on smaller dimensions, the miniature scale, in a similar way the IC (integrated circuits) evolved a few decades ago. Electronics and other related processes behave quite differently in/at the micro/nano domains/scales. The forces related to volume, like weight and inertia, tend to substantially decrease in significance whereas the forces related to surface area, such as friction and electrostatics tend to become large due to the decrease in surface area, in turn due to the reduction in the size of the device. The forces like surface tension that depend upon an edge become enormous in proportion. Some examples of small systems in the nature are (i) an ant carrying many times its weight and (ii) a water bug walking on the surface of a pond.

8.10 SMART SENSORS

The smart sensors are the devices that communicate digitally and have certain signal/data processing ability wired into them in some way. These could be a single-chip sensing mechanism with some associated processing capabilities. A smart sensor can also be defined as a device that has some intelligence or logic built into its mechanism [5]. One can also think of a smart sensor as a device that is not only a sensing mechanism but is a processing device and can calibrate itself. It can also be made to carry out other processes/procedures: equalization, recording, scaling of the data and computation of a few statistics of the incoming signals. In fact a smart sensor should be appropriately called a smart device or a mini system, rather than only a smart sensor. Then, this smart device/mini smart sensor system (SD/MSSS) would be called upon to react to the readings that the MSSS has taken by altering the process it is monitoring. It would do this via actuators or other controls and then would communicate that knowledge to a sensor network. It would very appropriately perform its multiple roles in a wireless sensor network (WSN). The MSSS should know what it is and how it works. This is done by storing information on the sensor and adding calibration data. If the manufacturer or the user can perform an initial calibration and store the data on board the sensor, the MSSS can use this information to correct for certain sensor errors. In this way, the MSSS is using the information about itself to improve its own performance thereby acquiring some smartness.

Next, small microprocessors can be put onto/into the MSSS. Then, calibration can be carried out by some algorithms implemented in the microprocessors. These microprocessor-based MSSS can perform many simple and small tasks. This embedded small microprocessor can convert the secondary variable (say, thermistor's resistance) to the primary variable (i.e., temperature) via an algorithm. In a similar manner, the raw signals/data can then be converted to the data that are in engineering units so that these are directly useable in further pre-processing algorithms; these again might also be housed in a small microprocessor in the MSSS. Thus, the smart sensors, as such it would be wise to call them MSSS, have one or more of the following features [6]: (i) they have or work as electronic data sheets, (ii) they can do self-identification, that is, try to know themselves, what they are and what they are doing, (iii) they can do self-calibration and compensation if required and (iv) they carry out the task of data/results communications to other sensor or control nodes. A typical smart sensor system (MSSS), often and usually termed as a smart sensor, is shown in Figure 8.6 [6].

REFERENCES

1. Raol, J. R. *Multi-Sensor Data Fusion with MATLAB*, CRC Press, Boca Raton, Florida, USA, 2009.
2. Dunn, D. J. Sensors and primary transducers: Tutorial 2. http://www.freestudy.co.uk/instrumentation/tutorial2.pdf, October 2011.
3. Stengel, R. F. *Robotics and Intelligent Systems*, Princeton University, Princeton, New Jersey, USA, 2009. http://www.princeton.edu/~stengel/MAE345Lecture1.pdf.
4. Dr. Bill Trimmer MEMS tutorial, http://mmadou.eng.uci.edu/Edu_Services/MEMSEdu.htm, accessed October 2011.
5. Mark Clarkson, http://archives.sensorsmag.com/articles/0597/smartsen/index.htm, accessed October 2011.
6. Wiczer, J. Smart Sensors, Sensor Synergy, Inc. Sensors Expo 2002, San Jose, CA, May 2002.

9 Robot Navigation and Guidance

Jitendra R. Raol

CONTENTS

9.1 INTRODUCTION

The word navigation originates from the Latin word 'navis', meaning ship. Thus, navigation is sea-faring [1]. It means the act of passing on water in ships or some other vehicles on water. It has three connotations: the guidance of ships or even airplanes from a place to place, the ship traffic and the work of a sailor. In present time with the advancement of mobile technology, and invention and usage of various kinds of vehicles, both the ground-based and air-space-based vehicles, the meaning of navigation is expanded as finding and following a suitable route, including determining one's own location during the journey. According to ancient (and perhaps, very profound and nearly perfect) language Sanskrit, the word navigation originates from the Sanskrit word 'Navgati' (meaning science of sailing); the 'Nav' ('Naav') means sailor or ship and 'gati' means pace or speed in Sanskrit [2]. In the present context, we can include the mobile robots among other vehicles and apply the concepts of navigation and guidance to the study of path planning (Chapter 10) and simultaneous localization and mapping (SLAM, Chapter 15) for robots. Hence, in this chapter only very rudimentary aspects of guidance and navigation (G&N) are presented. The sea-faring presupposes availability of some measurements because there are no landmarks on the open ocean. However, the direction is the easier measurement to obtain, since in the night the North Star (Polaris) shows the north direction [1], whereas in the daytime, the sun can be followed. During a cloudy day the polarization of sky light is used to locate the sun. The use of a magnetic compass would make finding North easier under many conditions. Latitude is also easy to obtain. Longitude, however, poses a problem, because it presupposes the use of an accurate time standard, chronometer [1].

Thus, since the prehistoric times, we have been trying to figure out/find out a reliable way to tell ourselves, that is, guide us where we are and how we got to our destination, and further how to get back to the origin (or get back to our home/or get-home concept in aerospace control systems), that is, to navigate ourselves from where we, in the first place, started or set out onto the journey [3]. This profound knowledge has a great impact on our survival and our economic power. People in earlier times marked trails (part of guidance) when they got out (part of navigation) hunting for food. Later on, people began making maps (guidance paths/landmarks). Much later the people developed the use of the latitude (the location on the Earth measured to the north or to the south from the equator), and the longitude (the location on the Earth measured to east or to west of a designated prime meridian, interestingly 'time' element comes in the picture). This they did as a way of locating places on the earth. Currently, the prime meridian used worldwide runs through the GOE (Greenwich Observatory in England) [3]. The mariners of the earlier times learned to plan their course by following the stars (for guidance), and then they could venture out (navigate) into the open seas, until that time they followed only the coast lines (for guidance) so that they did not lose their paths (navigation). However, the stars are only visible at night and that too in clear (weather) nights, and sometimes, the lighthouses provided a light to guide these mariners at night and warn them of any hazards. Subsequently, the magnetic compass and the sextant were invented. The needle of a magnetic compass always points to the magnetic North Pole, so it gives us our 'heading', or the direction we are going. Whereas the sextant uses adjustable mirrors to measure the exact angle of the stars, the moon and the sun above the horizon, and using these angles and an 'almanac' (records) of the positions of the sun, the moon and the stars, one can determine one's latitude in clear weather. However, determination of longitudes was still a serious issue. In mid-eighteenth century a cabinet maker (John Harrison) developed a chronometer, and for the next two centuries, sextants and chronometers were used in combination to provide latitudes and longitudes [3]. So, we also observe from the foregoing discussions that the problem of guidance and navigation is an integrated or very closely related one. In recent times the meaning of navigation is to find and to follow a suitable route or path, including determination of one's own location during the journey. Navigation is related to geodesy, in geodesy the positions of points are usually treated as constants or gradually changing. The differences between navigation and geodetic positioning are in navigation [1]: (i) the location data are needed immediately or at least after certain delay and (ii) the position data are variable, that is, time dependent. The navigation nowadays covers airplanes, missiles and spacecraft as well as vehicles that move on the land, and even the pedestrians often navigate with the aid of modern technological devices, for example, a GPS (global positioning system) receiver, and so on—this is because of the two modern technologies: GPS and inertial navigation. Also, the advanced data processing technologies have been developed, especially the Kalman filter which determines/estimates the states (positions, velocities, accelerations and headings, bearings of the moving vehicles, including robots) using the measured data of the range to the vehicle from the observer's station (where the measurement device, like, a radar is located), azimuth and elevation. This needs to be done in real time/online in order that the estimated states are communicated to the ground control and monitoring stations to take further action to guide the vehicle on its path, and carry out any mid-course corrections if needed. From the foregoing it can be seen that the navigation and guidance go hand-in-hand. Nowadays, variety of measurement devices, radars and vision- and acoustic-based sensors are available for observing the moving vehicles, targets and landmarks.

Several decades ago, many radio-based navigation systems were developed and used widely, and a few of these ground-based radio-navigation systems are still in the vogue today. One drawback of using ground-based radio waves is that we have only two choices [3]: (i) a system that is very accurate but does not cover a wide area (spatial limitation) and (ii) a system that covers a wide area but is not very accurate. High-frequency radio waves can provide accurate position location but can only be picked up in a small, localized area, whereas the lower frequency radio waves (like FM radio) can cover a larger area but would not give very accurate position estimates.

Hence, scientists and engineers thought to provide accurate coverage for the entire world by placing high-frequency radio transmitters in space, high above the Earth. Such a transmitter would broadcast a high-frequency radio wave with a special (message-)coded signal that could cover a large area and still reach Earth. This led to the development of the GPS system. Now several such systems are operational worldwide. The GPS brings together the advances (of hundreds of years) in navigation by providing precisely located kind-of-lighthouses, the GPS satellites, in the space that are all synchronized to a common time standard. The GPS system provides us our location anywhere on or above the Earth to within about 10 m (of accuracy). Even greater accuracy, usually within less than 1 m, can be obtained with differential corrections calculated by a special GPS receiver at a known fixed location, this scheme is called DGPS (differential GPS).

A fixed industrial robot is a mechanical structure, wherein one end is firmly fixed to the floor and the other end (the effector) is free to move under some programme control [4]. A few sensors are attached to the moving parts of the robot so that the position of the end-effector can be calculated since the lengths of the links are known (relative to a fixed frame of reference with its origin at the base). However, a mobile robot is in a moving frame of reference, this frame being originated somewhere on to the robot's body itself, and its position must be determined relative to a fixed frame of reference somewhere in the surroundings of the robot. This difference between a fixed and a mobile robot dictates that, in order to control the mobile robot reliably, one of the following conditions must be satisfied [4]: (a) the motion of the test-robot must be greatly constrained, by fixing the paths of the robot; (b) a fixed frame of reference must be provided for the robot, so that it continually refers to this fixed frame using on-board sensors, and therefore it can know its precise position in its own environs. Hence, the robot can detect and correct any deviations in its path (in comparison to the commanded path stored onboard the robot); and (c) a fixed map of the surroundings should be put in the onboard computer of the robot. The robot would then periodically produce a current map of the surroundings (mapping), during its motion, using the data measured from the onboard sensors. The current map is then compared to the fixed map (stored in an onboard processor) to determine the current robot position (localization). Using this technique (essentially SLAM, Chapter 15), the robot would dynamically recognize and avoid static or moving obstacles, hence determining its own path to the commanded destination. This combined problem is called simultaneous localization and mapping, SLAM, which is linked to navigation and guidance by recognizing that the guidance is provided by the stored map and the navigation is provided by the measured data from the onboard sensors, where for some estimator like Kalman filter is used. Thus, we can see that the ability to navigate is vital to the successful application of mobile robots, and this navigation requires [5]: (i) an environmental representation, that is, a good model of the surroundings, (ii) position estimation (localization) and (iii) robot-path planning. In fact, the robot is required to (a) possess good knowledge of its environment, model of the environment or maps, encapsulated in an internal representation or map, (b) know its current position on the map and (c) be able to plan a route or a path from one point on the map to another. So the essential component of a successful robot-navigation system is an appropriate environmental representation. This is an internal model of the world stored (for guidance) by the mobile robot. So far as the position estimation is concerned, the degree of localization required for the current task is very important. When travelling longer distances the robot may not need an extremely accurate estimate of its position. For short-distance (passing through a door) tasks, a more accurate estimate of position of/for the robot is required. Finally, the path planning determines how the robot would move from one point to another, usually in the shortest manner possible, in shortest time, avoiding static and/or dynamic obstacles (Chapters 10 and 12).

A typical robot guidance technique generally consists of following the underground-buried cables or painted lines on the surface onto which a robot is supposed to traverse a path. These techniques are classical and very reliable and fairly easy to implement, however, they greatly

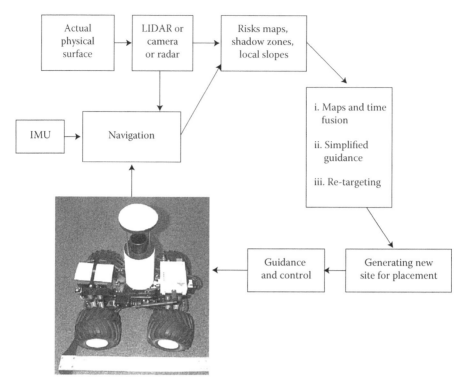

FIGURE 9.1 Navigation, guidance and control functionality block diagram for a mobile robot based on terrain maps (mapping of the actual surface).

constrain the motion of the robot. It must be mentioned here that in some literature the concepts and/ or methods of navigation and guidance are discussed interchangeably, that is, the discussion on one topic can be found in the discussion on the other topic and vice versa, even though both are distinct in some sense. Hence, the field of guidance and navigation should be understood to be taken in the integrated way as one complementary to the other, one aiding to the other. In robotics the problem of SLAM (Chapter 15) can be thought of combining the aspects of both navigation and guidance. The map (or sub-map) provides the guidance for the robot to execute its motion, the robot moves from a point A to B, this means it navigates between two points. When it thus navigates, its new positions and the new measurements can be utilized (combined with previous estimates) to refine the coordinates of the map, thereby improving the guidance (provided by the map's coordinates). This guidance is further utilized by the robot for improved navigation, and the cycle continues until the goal is reached. This is in fact the main aspect of G&N, and hence it is studied mostly together. So from the context it should be clear as to whether one is talking about guidance or navigation, or both together in an integrated way. Since, the concepts of path/motion planning and SLAM are discussed in Chapters 10 and 15, we just briefly discuss the concepts and simple methods/features of navigation and guidance in this chapter (interestingly PMP/SLAM and G&N feed to each other). Figure 9.1 depicts the navigation, guidance and control functionality for a mobile vehicle including a robot.

9.2 ROBOT NAVIGATION

The main difference between navigation requirements for robot and human is the definite and sizeable difference in perceptual capabilities [4]. Humans can detect, classify and identify environmental features under widely varying ambient conditions that to quite independent of relative orientation and distance. However, the robots while being able to detect obstacles before they collide to them, might have limited perceptual and decisional capabilities. These abilities of perception and decision making need to be built into robot's hardware/software architectures in the form of (i) computational algorithms, (ii) fuzzy logic-based inference system/engine (FIES), (iii) rule base, (iv) knowledge representation and (v) training by ANNs. These architectures could be (a) robot's guidance/navigational one, (b) robot path and motion planning and/or (c) robot control, the latter being mainly the real-time control (RCS), for autonomous or remote-controlled robots. The decisions to utilize the new emerging technologies, like those of soft computing, must be based on a critical analysis of technical risk and cost involved in augmenting the robot with these additional features and abilities. In many situations it is almost mandatory to provide such capabilities to a mobile robot. The robot navigation involves [4]: (i) global navigation—this is the ability to determine robot's own position in absolute or map-referenced coordinates, and to move it to a desired destination; (ii) local navigation—the ability to determine robot's own position relative to objects, stationary or moving, in the vicinity, and not collide with them and (iii) personal navigation—this involves being aware of the positioning of the various parts that make up robot itself, in relation to each other and handling objects, since the robot would have extended arms. The local navigation feature supports the global navigation in the mapped surroundings. Because knowing the robot's own position relative to a known mapped feature helps determine its absolute position.

9.2.1 Inertial Navigation

For inertial navigation very suitable sensing device is a gyroscope (Chapter 8). When a gyroscope is used to aid the navigation task, the axis of a gyroscope is set parallel to the direction of motion of the robot [4]. In the situation that the robot deviates from the (guided) path, the acceleration will be produced perpendicular to the direction of the robot's motion. This acceleration is detected by the gyroscope, and this acceleration is integrated twice to give the position deviation from the path, which is then corrected by control action (by the robot's computer). The problem is that path deviation at constant velocity cannot be corrected, since the acceleration would be zero. Also, the axis of a gyroscope would tend to drift with time giving rise to errors, and these errors would accentuate with double integration.

9.2.2 Position Determination via Fixed Beacons

The beacons are fixed at appropriate locations in the environment and the precise locations of which are made known to the robot. As the robot moves, it uses some onboard device/sensors to measure its distance and direction from any one beacon. Thus, the robot is able to calculate its own precise position in the environment.

9.2.3 Optical or Ultrasonic Imaging

In this method an absolute map of the surroundings is created and stored onboard (in the computer of) the robot [4]. The robot periodically generates a current-map of its environment while it is moving, using an onboard video camera and/or ultrasonic transducers. Various objects on/in the absolute map are recognized in the current map. Then by using the cross-correlation method the estimates of the robot position are obtained. Several of these estimates are averaged to give the current position of the robot. The demerits of this system are [4]: (i) it involves time for scene analysis computations, (ii) optical imaging suffers from the requirement of critical ambient lighting (can be overcome by using infrared cameras), (iii) the ultrasonic imaging suffers from fuzzy images due to stray multiple reflections of the transmitted pulse and (iv) the ultrasonic imaging relies on the knowledge of the speed of sound. This speed depends on the ambient temperature and humidity, and because of the presence of the temperature gradients in the environs, fuzziness of the images would increase.

9.2.4 OPTICAL STEREOVISION

Here, the idea is to view the same object in the surroundings using two onboard cameras from two different incidental angles. The angular disposition of each camera is measured, and since the inter-camera distance is known, the distance to the object can be estimated. This is the case of fusion of two images to obtain the depth information of the object, that is, to get an estimate of the third dimension of the object. Once the object is recognized in the absolute map, the position of the robot can be estimated. By repeating this with several objects one can obtain better estimates. The demerits of this method are [4]: (a) instructing both the cameras to view the same point on an object and (b) processing the vast amounts of data fast enough for real-time operation; however, for the latter some fast real-time estimations methods can/should be used.

9.3 ROBOT GUIDANCE

Present robotics research is concerned with the development of methods that will allow improved implementations for the robots [4]: (a) which can understand a problem, resolve it and utilize an appropriate solution, starting with incomplete information—this route can be the planning-based robotic guidance architecture and (b) which use information derived from the environment to modify their behaviour, dynamically, in order to attain the assigned goals—this route can be called the behaviour-based robotic guidance architecture. However, many use an appropriate mixed robotic architecture called hybrid architecture (Chapter 16).

9.3.1 WIRE-GUIDED ROBOT

In this approach the buried cables are arranged in some closed loops. Each closed loop carries a signal with a different frequency. Small magnetic disks are fixed to the ground at junctions and before and after sharp bends. The alternating current (AC) flowing in the buried wires generates a magnetic field, it being stronger near the wire. This magnetic field is picked up by two coils and the difference is amplified and adapted for further use for guiding of the mobile robot. This arrangement allows detection of potential danger points by the robot and for appropriate speed reduction while crossing these junctions. The system also includes some communication nodes/points along the paths where the robot can send its status report to the main computer. The computer co-ordinates the robot's planned routes to avoid any collisions. This approach has certain demerits [4]: (i) the paths cannot be easily changed because the cables are buried in channels (may be about 1 cm deep) and (ii) the cost of laying cables into the ground is relatively high.

9.3.2 PAINTED-LINE-GUIDED ROBOT

The vehicle follows the lines on the floor. These lines are painted using visible or invisible fluorescent dye. This dye fluoresces when ultraviolet light (UVL) is projected onto it and hence the lines. The merit of this approach over wire-guidance method is that the paths can be fixed rapidly and are easy to change, by repainting these paths. The demerits are [4]: (i) the path networks should be kept simple, since junctions might be quite complicated to manage, (ii) the dye-paint is subjected to wear and tear, and hence erosion and the lines require to be painted often and (iii) the lines can be obscured by other objects that will disable the robot guidance.

9.3.3 DEAD RECKONING

This is a method of estimating the position of a mobile vehicle without astronomical measurements, and is the process of estimating the robot current position based upon a previously estimated position-fix. This position is advanced (i.e., an incremental value is added to the present location) based on the known or estimated speeds over elapsed time. Although the classical methods of dead reckoning are no longer in vogue for the navigation, some of the modern inertial navigation systems/ methods that also depend on dead reckoning concept are very widely used. In this method the precise rotation of each robot drive wheel is measured periodically by using optical shaft encoders, then the robot calculates its expected/next position from it known starting point of motion. However, the drive wheel of the robot may slip causing an error in dead reckoning. When this slippage occurs at the drive wheel, the encoder on that wheel would register a wheel rotation, and yet that wheel has not driven the robot relative to the ground because of the slippage of the wheel. These and other errors could accumulate.

9.3.4 TACTILE DETECTION

This is carried out through interaction between the robot and its environment such that the geometry of the environment could be recognized. This means physical (tactile) contact and it requires that [4] (i) contact should not harm the robot/environment, (ii) the contact be identified and (iii) a strategy for scanning the environment be developed. For human–robot interaction a tactile sensor system monitors and limits the forces of interaction when humans and robot come into contact with each other. This system is implemented as a pressure-sensitive skin on a robot, such that when it detects and localizes the contact, the robot responds by stopping immediately, and if a collision occurs the integrated damping elements will absorb the forces of the robot's exigency.

9.3.4.1 Isolated Binary Contact Sensor

This approach uses a two-position switch (or its equivalent), informing of present state—contact or no contact. If the switch is placed at the strategic points on a moving arm, the obstacles can be encountered when the arm is in motion and then appropriate decisions can be made.

9.3.4.2 Analogue Sensors

In the 'Hill and Sword gripper', the gripper has sensors/buttons on it, which when pressed activate a screen which obscures, as a function of stress which it undergoes, a ray issued from an LED. This ray is picked up by a phototransistor and this information is then used to give an indication of clamping force and its form.

9.3.4.3 A Sensors' Matrix

It consists of a matrix of elementary digital and analogue sensors that are commonly used to produce shape information. The important aspects for this matrix approach are [4]: (a) the sensors should be well located, (b) the information obtained should be less noisy and (c) interpretation can be computed in real time.

9.3.5 DETECTION OF PROXIMITY

Since a tactile detection system can create some problems due to their dependence on physical contact, proximity detection or remote sensing can be used at the cost of losing some position accuracy in comparison with that of tactile detection. The proximity sensors can be used when: (i) the object is able to naturally transmit a signal, for example, radioactive transmission, (ii) the object is mounted with its own transmitter and (iii) a signal is transmitted to the object to be detected and then it is received back after the reflection from the object. This signal can be of natural origin, for example, reflection of ambient light, or can be artificial. In the first two methods at (i) and (ii) the detector is a passive receiver. This is also true for detection of the signal (of natural origin) as in (iii). However, if the signal is artificial, it indicates that there is an artificial transmitter and a receiver, and when these two are placed on the same sensor, an active sensor is created. The common active sensors in use are: (a) ultrasound, (b) radio frequency-waves emitters and (c) luminous radiation sensors.

9.3.5.1   Infrared Proximity Detection

In case of the sensor positioned directly facing the surface, the light received by the (detecting-) photodiode produces a signal that would be a function of the distance between the sensor and the surface. However, the difficulties would be [4]: (i) the whiteness of the surface has to be known, for the reflecting surface and (ii) the sensor-axis should be normal to the surface. Often the proximity sensors are used for the detection of the presence of an object (rather than measuring distance). The proximity detection sensors do not provide a total solution to the problem of position awareness (guidance) and navigation on their own. Yet, they play a key role in navigation of free-ranging robots as these sensors can be used effectively for obstacle avoidance.

9.4   SUMMARY OF THE NAVIGATION TASK

Thus, from the previous sections as well as from the discussions in Chapters 10 and 15, we can ascertain that the mobile robot navigation involves the following aspects: (i) localization and map-building (SLAM), (ii) local and global navigation depending on the area to be covered by the vehicle and (iii) command (and control) decision-making ability to engineer the path. Combining these one can evolve some integrated approaches to the mobile robot navigation and guidance. The goal for the mobile robot, and to a great extent, for any mobile autonomous ground or air vehicle, is then (a) to reach a chosen destination, (b) to follow a planned trajectory; that is, based on planning-based architecture or behaviour-based architecture and (c) to explore and map an assigned area/surrounding/environ. There are several subtasks: (i) to identify the present or current pose/location, (ii) to avoid any collisions which might be static or dynamic and (iii) to determine a path to follow. As we have seen in Chapter 15, the localization and map building are related to (i) identifying the robot's position in its environment and (ii) constructing an internal mathematical/perceptual or a qualitative model of any unknown aspect or special feature in the environment. In the local navigation phase, the mobile robots would rely on the most recent measurement data and try to avoid any collision. In the global navigation phase the objective is broader than that of the local navigation task, and the mobile robot is expected to cover long-range paths, thereby needing inputs from more measurements from variety of sensors and other information. The local navigation is/can be accomplished by using one these methods: (i) rule base already stored onboard the robot computer, herein one can use fuzzy logic based If … Then rules for taking some intermediate decisions, (ii) artificial potential field created around the obstacles and (iii) any other virtual force field to help avoid the obstacles, and/or guide the robot onto the path. The local navigation is generally measurement data based, reactive (reacting to some input action) and usually fast, whereas the global navigation is map-based, deliberate and relatively slow because if the map is not accurate the robot has to create an accurate map. The objective of the global navigation is to find an optimal path in some forced manner or using some probabilistic methods. Finally, the command and control decision making would utilize either layered control system/action and/or distributed topology and architecture for mobile navigation of a robotic system. In the integrated approach to the mobile robot navigation the problem of motion is considered as a single combined task. The main idea is to build a model of the robot's environment. Then find a path from the initial point to the target point on the navigation landscape.

9.5 MATHEMATICS FOR NAVIGATION AND GUIDANCE

For inertial navigation the acceleration of the vehicle is given as [1]:

$$\frac{d^2 X(t)}{dt^2} = \begin{bmatrix} \dfrac{d^2 x(t)}{dt^2} \\[2mm] \dfrac{d^2 y(t)}{dt^2} \\[2mm] \dfrac{d^2 z(t)}{dt^2} \end{bmatrix} \tag{9.1}$$

Here, x, y, z are vehicle's three-dimensional coordinate (x, y, z coordinates in a three-axis system), and the accelerations are measured continually.

The attitude angles (roll angle, pitch angle and heading/yaw angle/attitude) of the vehicle are given as [6]

$$\dot{\phi} = p + q \tan \theta \sin \phi + r \tan \theta \cos \phi$$
$$\dot{\theta} = q \cos \phi - r \sin \phi \tag{9.2}$$
$$\dot{\psi} = r \cos \phi \sec \theta + q \sin \phi \sec \theta$$

Here, p is the roll rate, q the pitch rate and r is the yaw rate of the vehicle in the body axis system of the vehicle. The transformation between the global (Xg) and the vehicle's coordinates system is given as

$$X(t) = \text{attitude angle matrix} * Xg(t) \tag{9.3}$$

The attitudes are defined by the three unknowns and are functions of time and vary with the movement of the vehicle. Before the vehicle's journey begins, some accurate knowledge of the attitudes is determined and used and during the motion of the vehicle, the attitude changes due to the movement of the vehicle are measured with the help of gyroscopes, and then are integrated to obtain the attitude of the vehicle. The first integration of Equation 9.1 gives the vehicle's velocity and the second integration give the vehicle's position in the three-axis system of coordinates. One can realize that the accuracy of position $x(t)$ gets progressively poorer with time, because the acceleration measurements are imprecise and the errors accumulate due to integration. It would be prudent here to use a combination of some filtering technique in combination with the double integration of the acceleration equation (see Figure 9.2).

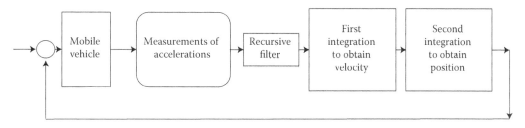

FIGURE 9.2 Recursive filter (could be Kalman filter) to reduce the integration error in the robot guidance scheme.

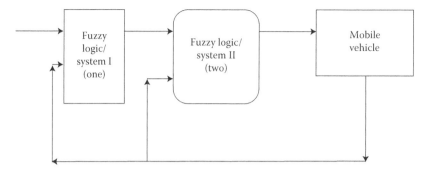

FIGURE 9.3 Mobile vehicle's fuzzy logic-based two-level path tracking control.

9.6 MOBILE ROBOT TRACKING USING FUZZY LOGIC

For a mobile robot, path tracking is a very important aspect and this involves sensing and control actions [7]. The tracking problem involves the creation of a steering command (and a velocity command) that is sent to the low-level controller that in turn initiates motion/action. The lateral motion of the vehicle is commanded by the steering command input and the longitudinal motion is commanded by the velocity input. Often these motions of the vehicle are coupled. The fuzzy logic can be applied in autonomous path tracking and can have the features: (a) modelling of the inaccuracies in the measurements, (b) sensor data fusion can be used from the context of fuzzy logic, (c) incomplete environment can be captured in the fuzzy membership functions and fuzzy If . . . Then rules and fuzzy Type I as well as Type II logic, (d) the required non-linear control laws can be derived from the some experts (say human-driver/operator) of the domain knowledge and incorporated into fuzzy If . . . Then rules and (e) the input–output data can be recorded from the operator's driving experiments and these data can be used in ANFIS system to learn the driving/control strategies for the mobile vehicle path tracking. A two-level fuzzy logic-based path tracking is illustrated in Figure 9.3 [7]. The fuzzy logic-based system is used to update the parameters of the pure-pursuit and generalized predictive path following/tracking in real time. The scheme can be implemented in a sequential manner. The input commands to this scheme are: (i) errors in lateral and heading directions, (ii) vehicle curvature and (iii) vehicle velocity. The required curvature and velocity are the controller outputs.

REFERENCES

1. Vermeer, M. Method of navigation. Maa-6.3285, 12. Huhtikuuta, Course Notes, 2010 (martin.vermeer@ tkk.fi, July 2011).
2. Sthapati, G. Hindu Mayan Connection, Vastu Vedic Research Foundation. http://indiansrgr8.blogspot.in/ 2011/05/hindu-mayan-connection.html, July 2011.
3. Anon. What is navigation? The Aerospace Corporation, http://www.aero.org/education/primers/gps/ navigation. July 2011.
4. Oliver, H. Where am I going and how do I get there—An overview of local and personal robot navigation, 20 May 1997. http://www.doc.ic.ac.uk/~nd/surprise_97/journal/vol1/oh. April 2011.
5. Winters, N. A holistic approach to mobile robot navigation using omni-directional vision, Ph.D. thesis, University of Dublin, Trinity College, Dublin, October 2001.
6. Raol, J.R. and Singh, J. *Flight Mechanics Modeling and Analysis*. CRC Press, Taylor & Francis, FL, 2009.
7. Driankov, D. and Saffiotti, A. (Eds.) *Fuzzy Logic Techniques for Autonomous Vehicle Navigation*. Physica-Verlag, A Springer-Verlag Company, New York, NY, 2001.

10 Robot Path and Motion Planning

Jitendra R. Raol

CONTENTS

10.1 INTRODUCTION

A mobile robot or any such mobile vehicle is required to move from an initial point, say A, to an intermediate point, and then to the final goal point, say B; this is the domain of path planning problem (PP). If there are obstacles on the path, the mobile robot/vehicle should be able to avoid these without touching these obstacles. These obstacles/objects in the path space could be static or dynamically moving. In most cases the obstacles' locations might be known in advance, but this is not always the case. Often these obstacles' movement is uncertain, we do not know when and how the obstacles will start moving, and the robot (i.e., its onboard computational algorithm) should have an ability to account for this uncertainty. Normally, this would be done by using some filtering algorithm in-built in the path planning algorithm, or by using a fuzzy logic to account and model this uncertainty (Chapter 1). The robot starts from point A and then moves ahead (the aspect of motion-trajectory planning, MP) avoiding any obstacles coming in its path and then reaching its goal point B. It is most probable that there could be more than one path for a robot to follow, these paths being equally feasible paths, but only one would be an optimal path [1–3]. In that case the robot should be programmed such that it follows this optimal (or at least suboptimal) path for its travel to the goal position. Thus, one can visualize two main optimizing (cost-)functions: (a) obstacle function to help the robot to avoid it (to avoid an obstacle) and (b) the goal function that would continually attract the robot to the goal. This is something like push–pull strategy: pushing/repulsing away from an obstacle, and pulling/attracting towards the goal. This aspect is very well captured in the potential field method of robot path planning, wherein there is an attracting (to the goal) potential function and a repelling (from the obstacle) potential function. Thus, the robot path planning problem entails (a) determining the free path routes where there are no obstacles and (b) the optimization of the path, that is to say it selects one good path between the initial and the goal points, for the robot to follow. This goodness of the path has to be specified *a priori* in terms of minimum cost with respect to length or spending less time or minimum effort in following the path; this being captured in a suitable cost function.

Thus, a mobile robot needs to navigate from an initial point to its final goal or destination point. These two points are assumed connected by a path, which need not be straight (but should be hassle free, because in some cases these connections might not be straight lines), and naturally there could be infinite number of possible and feasible paths between these two end points. So, the immediate problem is to find a good path that the robot should follow to reach its goal and that to while ensuring that the robot's system spends minimum energy/effort on the part of the mobile robot. Thus, the navigation (Chapter 9) is the act of carrying out locomotion primitives to move between points at the same time avoiding static and/or dynamic obstacles, and that there could be some uncertainty about/around these obstacles (Chapter 12). These obstacles (in the vicinity of the mobile robot) could be of a variety of types with certain uncertainties about them (Chapter 12). Thus, in the process of path planning, we need to define (i) the state—a model of the environment, and also the model of the robot and (ii) the actions—a model of the robot's motion primitives, that is, what action the robot (or its onboard guiding/tracking/control algorithm) would or should take. An environment model is composed of (a) knowledge of the robot's location and

(b) knowledge of the existence of location of obstacles. There are several factors to be kept in mind in solving the path planning problem [1–4]: (i) number of dimensions (robot's dynamics and its degrees of freedom, DOF), (ii) number of obstacles and complexity of their (location/size) geometry, and knowledge of any uncertainty, in terms of their statistics (i.e., mean values and covariance matrices), (iii) complexity of robot state (robot's mathematical model) and (iv) error or uncertainty from sensors (measurement errors, because the robot, in fact its software, would use these measurements to guide itself). In robot's action models, the knowledge of an effect of an action on the environment must be known *a priori*, that is, without even the action taking place, that is, *a priori* information should be available. Then, other associated important factors to be kept in mind are [1–4]: (a) constraints on robot actions (because the robot has to carry out certain tasks within certain domain of the assigned task specific to its application, that is, medical robot, or mining robot, constraints for the both being of different dimensions, accuracy, and performance), (b) motion (kinematic) constraints (e.g., car-like robots), bounded velocity and acceleration, (c) dynamic effects at high speeds (e.g., vibration effect that will change the robot's intended operational domain, and perhaps, the stable operation of the robot) and (d) error or uncertainty in the robot's actions.

An autonomous mobile system or robot has to make its own decisions, perhaps and most probably embedded in the onboard computer, wherefrom a suitable decision to achieve its goal or next move could originate—this move will then direct an actuator to impart motion to the robot. A major component of such an autonomous system is its path planner, and following the assigned optimal path gets the robot from one location or its current state to another location, and eventually to the final goal point, while simultaneously avoiding all kind of the obstacles [1–4], this being an involved process. Path/motion planning (PMP) is also used by the transport engineers along with traffic-flow mathematical models to model the flow of traffic, and the games (playing/sports games) also use some form of path planning to move computer-controlled players around the game environment, requiring a fast path planner, since multiple paths need to be planned at the given time instances [1–4]. With respect to robots, the time and space complexity (in computational algorithms) of robotic path planning algorithms should be kept low, since in most cases embedded processors have limited computational power and the path planner should give the shortest collision free path [1–4] and in addition to these aspects there are other objectives: (i) to avoid danger to the moving robot, (ii) to reduce fuel consumption and/or (iii) to cover an entire area, used by domestic robots; these latter robots, that is, the service robots, clean floors or are security robots that periodically patrol a specified area.

In this chapter, path- and motion planning methods and algorithms for a mobile robot are briefly described. We hasten to add here that the problems of path planning and motion planning are very closely related and hence in the open literature these names are used interchangeably, without making any difference between two planning aspects. We also add here that the motion planning is a path planning problem in itself, with an additional computational aspect of 'defining' or 'coordinating' the specific movement or its behaviour in the 2D space, with very subtle distinction between path and motion planning problems. Various aspects about path planning such as the configuration space (CS), localization and mapping are discussed. The combined problem of robot localization and mapping (i.e., the map building, based on some knowledge of the landmarks along the path of the robot) is termed as SLAM (simultaneous localization and mapping) and is discussed in Chapter 15. The related problem of robot guidance and navigation is described in Chapter 9. A brief description of the classical approach to path/motion planning is given in Section 10.2, and in Section 10.3 some heuristic methods are briefly described. The PMP problem is then revisited in Section 10.4 wherein a detailed exposition is given in Sections 10.4 through 10.6. A simple distinction between local and global path planning approaches is presented in Table 10.1. It must be mentioned here that the partial algorithmic listings of various classical algorithms (A*, D*, Bug algorithm (BA), etc.) can be found in Ref. [1].

TABLE 10.1
Robot Path Planning Methods

Global Path Planning	Local Path Planning
It is a map-based process	It is a sensor-based approach
It is a deliberate system	It is a reactive system
Requires complete knowledge of the workspace area	It does not require complete knowledge
Obtains a feasible path that leads to the goal	Moves towards a target following a path and avoiding obstacles in the path
It has relatively slower response	It has a fast response

Source: Adapted from Buniyamin, N. et al. *International Journal of Mathematics and Computers in Simulations*, 5(1), 9–16, 2011.

10.2   CLASSICAL METHODS

The usual approach is to formulate an approach based on some algorithm, by defining the initial and goal points, and assigning the *x*, *y* position-locations for all the static obstacles, or in case of the moving obstacles, defining the precise model of the obstacle movements [1,5]. This approach/algorithm has to solve the PP problem, that is, give to the robot information about what is the next location ($x+$; $y+$) for it to go onto that location. Subsequently, this algorithm is implemented using some programming language. This classical intuitive approach has been very successful in many cases. However, this method can fail when the knowledge about the problem environment is poor or incomplete, that is, if the knowledge of the obstacles is sketchy. A computer algorithm based on the classical robot architecture/path planning situation has no inherent intelligence (unless some AI is built into its functionality), and hence it does what to do as told by a programmer, how much clever one might be. Also, the amount of computation required to provide the feasible and optimal solution may be very large, this being the case if the algorithmic iterations and/or the logical loops (if ... then else) are too many. This is true of multi-dimensional problems for robots with several DOF; here, DOF refers to the freedom of movement which a robot has, because the robot might have some extended arms. A translational robot which can move horizontally and vertically has only two DOFs, whereas a complex robot arm or with extended accessories might have six DOF. This DOF is the number of independent parameters required to specify the movement of a particular robot. This will also help in defining the robot configuration or CS for the robotic environment. Some classical methods for robot PMP are Roadmap, Cell decomposition, Potential Field and Mathematical Programming, a few of which are briefly described in the following sections.

10.2.1   ROADMAP APPROACH

In this roadmap (RM) approach, the free C-space (set of feasible motions/CS) is reduced to or mapped onto a network (NW) of 1D lines, the approach also being called the Retraction, Skeleton, or Highway approach [5]. In this RM case, the search for a solution is limited to the NW, and the PMP problem degenerates to a graph-searching problem. There are four popularly known RMs approaches [5]: (i) Visibility graph (VG), (ii) Voronoi diagram (VD), (iii) Silhouette (S) and (iv) Subgoal NW (SNW). The VG, as the collection of lines in the free C-space, connects a feature of an object to the feature of another object. These features are vertices of polygonal obstacles; there are the edges in the visibility graph, and the VG is constructed in 2D. In the VD, there is a collection of objects, and the VD is a partition of space into numerous cells. Each of these cells consists of the points generally closer to one particular object (than any other objects). In the approach called Silhouette, one projects an object (which is) in a higher dimensional space to a lower dimensional space and then traces out the boundary curves of the projection, similar to tracing out the silhouette of a person. In the SNW method an explicit representation of the configuration obstacles is not built, but a list of reachable configurations from the starting configuration is maintained. On reaching the goal configuration (i.e., when it is reachable) the PMP problem is solved. This reachability (of one configuration from another) is determined by a simple local MP algorithm. This local operator entails the moving of the robot in a straight line between the configurations.

10.2.2   CELL DECOMPOSITION

In this method-cum-algorithm the free C-space is decomposed into a set of simple cells. Then the adjacent relationships among the cells are computed. Subsequently, a collision-free path between the starting point and the goal configuration is found. This is carried out first by identifying the two cells containing the start point and the goal point and then connecting them with a sequence of intermediate connected cells.

10.2.3   POTENTIAL FIELDS

In this case a robot is treated as a point in a CS, that is, as a particle under the influence of some artificial potential field, like an electrical or gravitational (PF) field, and then a potential function is defined over free space as the sum of an attractive potential force that pulls the robot towards the goal point and a repulsive potential force that pushes the robot away from the obstacles [1,5,6]. This attractive potential (function/force) is defined as a function of the relative distance between the robot and the target point and it is assumed that the target is a fixed point in space. However, it is possible to incorporate the velocities of the robot and target into the potential field to expand the degrees of freedom so that the robot is capable of managing varieties of tracking tasks [6]. Thus, a potential function that depends on the relative distances and velocities is more versatile and would obtain better optimal paths. Likewise, it is also feasible to incorporate the relative positions and velocities between the robot and the obstacles for constructing the obstacle functions or the repulsive potential function.

10.2.4   MATHEMATICAL PROGRAMMING

In this mathematical programming (MP) approach to path planning, the requirement of obstacle avoidance is represented with a set of inequalities on the configuration parameters (CP) [5]. The PM planning is then formulated as a mathematical optimization problem. The solution of the PMP problem finds a curve between the start and goal points minimizing a certain scalar quantity.

10.3   HEURISTIC METHODS

In general, the classical methods are so called because they are well proven and tested to obtain reasonable results which might not be optimal but must be working solutions within the given constraint of the problem definition and of available knowledge about the robotic environment. Thus, the so-called classic or conventional approaches have a few limitations [5]: (i) high time complexity in high dimension computations, that is, these methods would take longer to solve the PMP problem of higher dimensions and (ii) these methods might get often trapped in some local minima. Hence, these algorithms become inefficient in practice. There are other methods/algorithms that improve the efficiency of the classic methods: (a) probabilistic roadmaps (PRM) and (b) rapidly exploring random trees (RRT). These algorithms have advantage of high-speed implementation taking less time for higher dimensional computations in algorithmic steps and iterations. As a result of the above-mentioned limitations some newer approaches have been developed over the last several years. There arc several methods/algorithms which are called heuristic algorithms compared to the classical PMP methods [5]: (1) a combination of the simulated annealing (SA) and potential field (PF), (2) artificial neural network (ANN), (3) genetic algorithms (GA), (4) particle swarm optimization (PSO), (5) ant colony (ACO), (6) stigmergy, (7) wavelets, (8) fuzzy logic (FL) and (9) tabu search (TS). It must be mentioned here that although the solutions to the PMP problems are not guaranteed by these algorithms, however if the solutions are found, then these are obtained in less time compared to those of the classical/deterministic methods.

10.3.1 ARTIFICIAL NEURAL NETWORKS

It is possible to use ANNs (Chapter 1) for real-time collision-free path/motion planning in a dynamic environment in which may be, the obstacles are moving in the surroundings of the robot. This approach can be applied to several types of robots [5]: (a) point mobile robots, (b) manipulator robots, (c) car-like robots and (d) multi-robot systems. In this approach the state space of the ANN is taken as the CS of the robot. The dynamic activity landscape of the ANN then represents the dynamically varying environment of the robot. Like in the potential field method, the target attracts the robot in entire state space. The obstacles keep locally pushing the robot always away from them to avoid any collisions. Thus, the real-time robot motion to the goal is planned through the dynamic activity landscape of the ANN without [5]: (i) explicitly searching over the free space and/or the collision-free paths, (ii) explicitly optimizing any chosen cost function, (iii) any prior knowledge of the environment, (iv) any learning process and (v) any local collision checking steps. Such a scheme would be computationally very efficient.

10.3.2 GENETIC ALGORITHMS

The application of GA (Chapter 1) to the robot PMP problem requires (a) the development of a suitable chromosome of the path, (b) a path guidance strategy, (c) a method for obstacle avoidance and (d) an appropriate cost or fitness function that would help achieving an optimal and smooth path. The environment is assumed to be static and known. Further discussion on the use of GA for path/motion planning in the presence of uncertain environment is given in Chapter 12.

10.3.3 SIMULATED ANNEALING

This approach, in fact, is used to pull out an optimization process that gets trapped in a local minimum while searching a global solution. A new solution (e.g., x-new coordinate for the candidate of next position) is randomly chosen at each step of the iteration. This is chosen from a set of neighbours of the current solution x, because the algorithm is trapped here [5]. One accepts the new solution unconditionally if the new position of the robot has lower potential energy, $U(x\text{-new}) \leq U(x)$ [or else with uphill move—probability of $\exp(-\Delta/T)$ with $\Delta = U(x\text{-new}') - U(x)$], where U is the cost function (i.e., potential function) and T the temperature. If the chosen new solution x-new is not accepted, the algorithm proceeds to the next step. The temperature T is decreased by cooling rate r. This is repeated until a small value near zero is reached or when the solution has escaped from trap of the local minimum.

10.3.4 PARTICLE SWAM OPTIMIZATION

The PSO method is primarily based on the study of graceful movements of swarms of flying birds [5]. It is based on the idea that the collective intelligence behaviour exhibited by certain biological populations can be exploited for solving other related problems in robotics. In the PSO method, a set of (stochastically generated) solutions (called initial swarms) propagates in the design space towards the optimal solution over a number of iterations (moves). This strategy utilizes a large amount of information about the design space. This information is assimilated and shared by all the members of the swarm population engaged in flying or foraging tasks. Thus, the PSO method is inspired by the ability of flocks of birds, schools of fish and herds of animals which continually move to adapt to their changing and newer environment. This swarm-optimized behaviour ('acquired' due to evolution over thousands of years) helps these groups of animals/insects/birds to find rich sources of food, and avoid predators. Inherently this is achieved by practicing a process of information sharing and thus developing an evolutionary advantage and an evolutionary stable strategy. It is possible here to augment the PSO strategy with fuzzy logic to incorporate the information-sharing process by way of fuzzy membership functions and/or fuzzy If … Then rules.

10.3.5   ANT COLONY

An ACO method for path planning takes the advantage of collective behaviour of ants which engage in a task of foraging from a nest to a source of food [5]. In this method the two individual groups of ants are placed at both the nest and food sources, respectively. A number of ants (called agents) are released from the nest, that is, from the start configuration, and then they begin to forage, that is, search the food and reach goal configuration. Each ant has a certain quantity of pheromone (a substance) which it drops along the path while it is traversing along the path. The other ants track down this pheromone (trails) that were previously dropped by the nest's ants. These pheromone trails could contain information like lower cost, and nearness of obstacle (defined by the obstacle function) while the ants travel in forward direction to the goal position. The results from some preliminary tests show that the ant colony-based robot path/motion planning is capable of reducing the intermediate configurations between the initial and goal configuration that too in an acceptable running time [5].

10.3.6   STIGMERGY

The Stigmergy concept describes the indirect communication that takes place among individuals in social insect societies [5]. The notion is that the regulation and coordination of the building activity do not depend on the workers themselves but are mainly achieved by the nest. This is to say that a stimulating configuration triggers a response of, say a termite worker, transforming the configuration into another configuration. This action may trigger, in turn, another and yet a different action performed by the same termite or any other worker in the colony. By a careful design process and integration of the robot sensing, actuation and control features, it would be possible to utilize Stigmergy concept in task-oriented robot path/motion planning. This powerful mechanism of coordination being attractive requires minimal capabilities of the individual robots. No direct communication is required for the robots. The idea here seems to be that the nest building configuration that is being evolved itself elicits a certain response in the group of workers such that these workers start modifying the current configuration and develop a new one that might be a better configuration. This process can be replicated for robot PMP situations. It is like saying that when a robot finds good going on its path then it is inspired to do well and perhaps much better and then it evolves a better solution and a path, it is like quickly learning not only from the environment but from its own successful action a next optimal or efficient move. It is suggested here that GA can be used in conjunction with the Stigmergy concept so that the fitness of the stimulation can be evaluated and the quality of configuration change or modification can be ascertained. This approach parallels the behaviour-based robotic architecture.

10.3.7   WAVELET THEORY

In this approach, the terrain is represented in a multi-resolution manner by using wavelets. Then a path is planned hierarchically through the sections that are relatively smooth and reasonably well approximated on coarser levels [5]. The preferred terrain sections are distinguished by using the hierarchical approximation errors in the chosen cost function. In order to compute this error the corresponding wavelet coefficients are used. Another non-scalar path-cost measure, based on the sorted terrain costs, can also be used in the global path search algorithms. This obtains the paths that avoid high-cost terrain areas. It is also possible to add some constraints for specific robot tasks to obtain an efficient motion planning on a rough terrain surface.

10.3.8   FUZZY LOGIC

The idea of using fuzzy logic for robot path planning can be envisioned for: (i) modelling the uncertainty about/around the moving obstacles, (ii) taking the uncertainty of the measurements from various sensors into account and (iii) using certain rule base in the fuzzy inference system (FIS). For example, if the sensor measurement says the obstacle is very near, then turn right or left (is a command decision to the robot), and so on. One can use Type I or Type II (Interval Type II) fuzzy logic. The fuzzification of the robot's environment can be done and suitable membership functions chosen. It is also possible to use ANN architecture for robot map learning and subsequent navigation in a dynamic environment, alternatively one can use adaptive neuro fuzzy inference system (ANFIS).

10.3.9   TABU SEARCH

This is a meta-heuristic approach and the corresponding algorithm does not suffer from the local minimum problem [5]. It counts the number of times that a specific move is repeated, and then the algorithm determines the weights of moves. These weights are listed in a table as a guide to the path planning. By using the TS (tabu search, a set of tabu, i.e., the forbidden) the moves are defined every iteration of the search to confine the robot's navigable locations, and guide it towards the goal destination. The range-sensor measurements are used and the value of the cost function is defined, such that the robot is attracted to certain obstacle vertices. It then moves along a path consisted of lines which connect the vertices of different obstacles. The planner also takes advantage of random moves when the solution gets trapped in dead-ends. Several different exercises and experiments have shown a good efficiency of this approach.

10.4   ROBOT PATH PLANNING

First, the robot's configuration is defined to be a point in its environment, which is given by its position and rotation (see Section 10.2.1 and Figure 10.1). The path planning problem is to determine the next (best/optimal) move. This move is from the current configuration (point) to the next point and further on. With this, the robot will eventually assume the goal configuration (the final/destination point) [1].

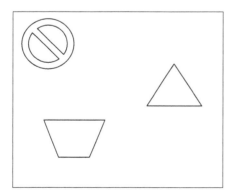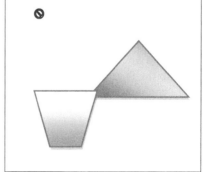

FIGURE 10.1   Robot's CS: A robot with circle around it (left subplot) and the robot with the obstacles expanded to allow the robot to be modelled as a point-size object (right subplot).

The choice of the path, as specified by the path planner, to be followed by the mobile robot should minimize a suitable metric, for example, some norm of the distance travelled, from the initial point to the goal point, or some form of energy (which will be spent by the robot) should be minimized. The information available in and gathered from (regarding changes in) the environment should be used, and based on this information the next best move should be determined. The foregoing definition does not require (any specification in advance of) a complete path from the starting configuration to the goal configuration—it only requires that the path planner is able to repeatedly give the next best move to robot to reach the goal point. Thus, the robot moves by/in discrete amounts, since it is impossible for a robot to make continuous adjustments to its drive system, because even if another instruction is issued immediately, some definite time will elapse between the two consecutive commands, hence, it can be said that the robot moves continually rather than continuously.

10.4.1 Configuration Space

We need to model the robot (by a suitable mathematical model, Chapter 2) and the environment in order to plan paths for a robot within that environment. We assume that an N-dimensional vector describes the state of the robot within the environment. This vector defines the robot's configuration (space) and should be complete enough to accurately describe the state of the robot within its environment [1]. At the same time it should be simple enough to keep its dimension relatively small so that the path planning is computationally feasible. A fixed-body robot can move in any direction within its 3D (three-dimensional space) and has six degrees of freedom (6DOF): its position in x-, y- and z-axis directions and associated rotations, pitch, yaw and roll. For a non-fix-bodied robot, for example, with an external manipulator arm, the robot's configuration dimension should be expanded to include this arm, if it has a significant impact on the operation or movement of the robot. Thus, finally the set of all such possible robot's configurations (its XYZ position and rotations, external arm dimensions and protruding objects' dimensions) is known as the CS. We can restrict the dimensional space to give us three degrees of freedom (3DOF): the x and y position of the robot and the direction that the robot is facing, that is, the yaw, heading. Thus, the given robot can be enclosed by a minimum radius and be considered circular or be designed to have a circular shape. With this definition of the CS, it is a continuous 2-D Euclidean space. For all practical purposes, the robot can be regarded as a point in space within the CS (Figure 10.1). Also it is possible to regard the robot's own space as any other regular shape, depending upon its dimensions and extensions. Conversely, the robot can be regarded as a 'point' in the CS and the obstacles' size can be increased appropriately to obtain an equivalent CS (the right-side diagram of Figure 10.1).

10.4.2 Graphs

The robot's environment is modelled using a directed graph. A directed graph, $G(V, E)$, is defined to be a set of vertices V and directed edges E [1]. An edge $e = (v, w)$ (here e is a component of the set E) is a directed link from the source vertex v (here v is a component of set V), to the destination vertex w, in the V space. For each vertex v, the set of vertices, the destination of some edge on that vertex, is given by $out(v) = \{w|e = (v,w)|E\}$, and the set of vertices, the source of some edge on that vertex, is given by $in(v) = \{w|e = (w,v)|E\}$ [1]. The start vertex represents the robot's initial physical position denoted by vS, and the goal vertex by vG. In a 2D environment, the graph can be given in a grid pattern with edges between adjacent grid cells. In case of the obstacle lying between the two adjacent vertices, the edge could be removed or its cost could be set to a prohibitively high value.

10.4.3 Breadth First Search

Breadth first search (BFS) algorithm assumes that all the edges (Section 10.4.2) have the same weight and it obtains optimal paths [1]. The algorithm processes vertices in the order of the number of edges between the given vertex and the start by maintaining two lists of vertices: (i) the list containing the vertices that must be processed in the current level and (ii) the list containing the vertices that would be processed in the next level. The BFS algorithm is initialized with the start vertex (marked as visited) and placed in the first list. The second list is left empty. While the first list is not empty, a vertex is removed from it and expanded, and when a vertex is expanded, it adds all its neighbours which have not been visited to the second list (and marks them as visited). Once all the vertices in the first list are expanded, the two lists, first and second, are swapped and the process is continued. When the goal is reached (it is added to the second list) the algorithm can stop as a path has been found. If both the lists are empty and the goal is not reached, this means that there is no path between the start and the goal. The lists can be implemented as linked lists. The insertion and removal in the lists can be carried out in continual time. The vertex tags and the back pointers are stored as part of the vertex data. These are given the required initial values when the vertex is initialized. Thus, getting and setting of vertex tags require some constant time. In an extreme case the graph might consist of two connected components (one of which contains only the goal vertex and the other every remaining vertex), then each vertex except the goal will be visited. In this way, all the edges will be examined. The algorithm is the simplest graph algorithm that will provide optimal paths; if all the edge-costs are the same, however, a more general algorithm is needed to allow for different edge costs.

10.4.4 DIJKSTRA'S ALGORITHM

This (D-)algorithm allows for different edge costs. It uses a cost function $g(v)$ which is an estimate cost from the starting vertex to the vertex 'v', and it is initialized as zero for the start vertex and infinity for all the other remaining vertices. The vertices are visited in the ascending order of their 'g' values, and when a vertex is visited (it is marked as visited and) the 'g' values of all its unvisited neighbours are updated. This is done if the 'g' value of the vertex plus the edge cost to the neighbour is less than the 'g' value of the neighbour. When a vertex is marked as visited, its 'g' value contains the exact cost from the starting vertex to that vertex. When and if the goal vertex is visited then a path is found. However, if there are no more vertices to visit then there is no path. It seems that due to various logical enunciations being required in both of the above algorithms it might be prudent to build a rule-based expert system SW or algorithm to improve the efficiency of these algorithms.

10.5 OTHER ALGORITHMS

In this section, we describe several other path planning approaches and algorithms.

10.5.1 BUG ALGORITHM

The BA is designed for use in/on a robot with sensors which have a limited range. The algorithm directs/moves the robot directly towards the goal (see Figure 10.2). The obstacles are considered as polygonal in shape and are either convex or concave. As shown in Figure 10.2, when an obstacle

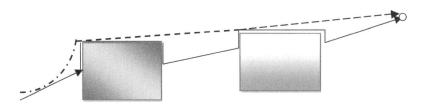

FIGURE 10.2 Robot path planning with the BAs (the straight lines are due to the BA and the dashed lines are due to the TBA).

blocks the robot's path, the robot must go/move around the edge of the obstacle until it reaches another point on its original path to the goal, then the robot continues towards the goal. In a situation that the robot travels fully around the obstacle encountered, without being able to move towards the goal, this means a path to the goal does not exist. The BA does not provide an optimal path. The Tangent Bug algorithm (TBA) as an extended version of the standard BA uses the sensor information within a given range of the robot. The TBA requires that a robot has the range sensors all around it. The algorithm operates in two modes [1]: (i) the motion to target mode and (ii) boundary following mode. In both the modes the TBA models the obstacles within its sensor range as thin walls, and hence the algorithm does not model the depth of an obstacle it encounters (it only recognizes that an obstacle exists). These thin walls are constructed by detecting discontinuities in the measurements from these sensors. If the range changes gradually and suddenly increases to the maximum range, the angle for which the sudden increase occurs signifies/indicates the vertex of an obstacle. Thus, with the information from these sensors, the algorithm constructs a local tangent graph from the obstacles. These local tangent graphs contain vertices on the edges of obstacles and edges from the robot's position to some of these vertices. During a motion to a target mode, the robot moves in a straight line towards the goal. It can also move towards a vertex of an obstacle (for which the path through that vertex is the shortest known path). The TBA gives a path that is much closer to the optimal path. These BAs are designed to work in environments with stationary obstacles. The Dynamic Bug algorithm (DBA) optimizes the TBA for dynamic environments.

10.5.2 Potential Functions

A potential function to create a potential field is used in which the goal is an attraction/attractive point and the obstacles are repulsion/repulsive points. The total potential function (TPF) is constructed by summing all the individual potential functions (for the goal and all the obstacles): $U(p)$ is TPF, p is any point in the environment, U0 is the potential function for the goal and Ui (for i in $[1, n]$) are the potential functions for the 'n' obstacles [1]. In this method the direction of the movement of the mobile robot is determined by using the gradient decent method on the TPF, and hence this method gives the direction for the robot to move ahead. The potential function (PF) U must be smooth; if not, the robot's path may oscillate around points that are not smooth. Because a continuous vector field is used to plan the robot's path, it can be used to control the robot's drive system directly; however with the PMP algorithms that give the discrete solutions, there would be a need to interpolate these results to provide a smooth path. A problem with this method is that there might be local minima, for which some known solutions can be used, for example, one can use SA approach to come out of the local minimum. Hence, often the potential function method is seen used in conjunction with other methods. One can use variety of mathematical representations for defining obstacle functions that are most suitable for the type of obstacles. Also, one can use different types of potential functions for a given environment to represent different types and kinds of obstacles or the potential function itself could be formulated such that it is a time variant function to model moving obstacles like, say humans in the environment of the mobile robot. Some decision mechanism can also be used to choose one function from the other function in a highly dynamic environment with the obstacles that have uncertainty around them.

10.5.3 A* Algorithm

This algorithm makes use of a heuristic function to determine the order in which to process the graph vertices [1] (Section 10.4.2). If Euclidean coordinates for each vertex are known, the heuristic function is defined as the Euclidean distance function (EDF). If the heuristic function is closer to the exact distance, the algorithm improves in terms of processing a fewer vertices. The algorithm produces optimal paths and determines shortest paths by processing the fewest possible graph vertices, of course for a given heuristic. If the shortest path distances are overestimated, then A* does not guarantee optimal paths, but one can find paths quickly. The A* algorithm uses similar data types as used in Dijkstra's algorithm. The A* algorithm uses (i) an estimated cost function 'g', (ii) a labelling function 't', (iii) a back pointer function 'b', (iv) a priority queue called the open list and (v) a heuristic function 'h' that gives the heuristic distance between a given vertex and the goal vertex. The priority queue is ordered by the function $f(v) = g(v) + h(v)$ for each vertex v of V. To realize the path planning, the algorithmic steps followed by/in the A* algorithm are given as follows [1]:

   i. Initialization: all vertices are labelled as NEW, while all other vertex functions are undefined.
   ii. The algorithm is initialized by labelling the starting vertex as OPEN.
   iii. The cost 'g' is assigned a zero value and it is incorporated into the priority queue.
   iv. The algorithm processes the priority queue until it is empty or the goal vertex is labelled as CLOSED.
   v. After a vertex is removed from the priority queue, the cost function of all neighbours that are labelled as NEW or OPEN are considered.
   vi. For any NEW neighbours, the cost is set to be the cost of the current vertex plus the edge cost to the neighbour.
   vii. The same calculation, as above, is used for OPEN neighbours, but this computed value is only used to update the cost of the neighbour if it improves on the current cost.
   viii. For the NEW neighbours, and neighbours with improved cost values, the back pointers are set to the current vertex.
   ix. These are labelled as OPEN and are reinserted into the priority queue.

With the heuristic function set to zero, this algorithm is equivalent to Dijkstra's algorithm. The A* algorithm will first expand the vertices on the straight line from the start to the goal. Then it will expand the vertices neighbouring this line. An important aspects is the direction in which the algorithm processes its vertices, and for this reason, the A* algorithm should also be investigated in the reverse order of processing from the goal vertex to the start vertex.

10.5.4   D* Algorithm

It is 'Dynamic A*', without using any heuristics, and can be considered as a dynamic version of Dijkstra algorithm [1]. In the Dijkstra's algorithm, as shown in the left subplot of Figure 10.3, grid cells are processed in radial directions, whereas in the A* algorithm (right subplot of Figure 10.3) the grid cells are processed in the direction of the goal if a good heuristic is used. The filled cells have been processed while the open cells are on the open list to be processed. The algorithm starts processing vertices from the goal vertex and not the start vertex. If a change in the environment occurs, the altered vertices are revisited and the change is propagated outwards (often only after processing a few vertices). Thus, the whole graph is not explored every time a new obstacle is discovered or an obstacle moves within the environment. The D* algorithm makes use of (i) a current

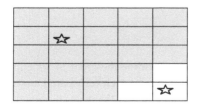

FIGURE 10.3   In Dijkstra's algorithm (left subplot) grid cells are processed in radial directions. In the A* algorithm (right subplot) the grid cells are processed in the direction of the goal if a good heuristic is used. The filled cells have been processed while the open cells are on the open list to be processed.

cost function 'g', (ii) a vertex tag function 't' and (iii) an open list. It defines a previous cost function 'p' that is used to maintain the order in which the vertices in the open list are processed after a change has occurred, the order being determined by using the key function $k(v) = \min[g(v), p(v)]$. The previous cost function 'p' can be integrated into the key function 'k', thereby making the cost function 'p' redundant.

10.5.5   FOCUSED D*

The D* algorithm does not use heuristics as the A* algorithm does. The Focused D* algorithm (FDA) [1]: (a) uses the same notation as in the D* algorithm (DA), (b) does not maintain the previous cost values 'p' and (c) uses the open list key value 'k' to store the same information. The 'h' refers to the current cost of a vertex and 'g' as the heuristic cost. The FDA is almost identical to the DA, however, the main difference is in the sorting of the open list, this is done by using $f(v) = k(v) + h(v) + fb$. Here, fb is the bias of the robot when the vertex was already inserted into the open list, it being initially zero. Then the bias is increased by the heuristic cost between the robot's previous and current positions whenever it moves. This avoids the heuristic values becoming invalid whenever the robot moves.

10.5.6   D* LITE

A recent algorithm Lifelong Planning A* (LPA*) algorithm makes use of both previous path calculations and heuristics, and was developed by adapting the A* algorithm to reuse information if the environment changed [1]. The implementation is simple and easy to understand, but it is only for a fixed start and goal position. And hence it cannot be used directly for robot path planning. The D* Lite (DL) algorithm was derived from LPA*. It works much like D* as a path planner; however, the actual algorithm is quite different. The D* Lite algorithm makes use of an open list (as such, it is an ordered list of vertices) that need to be processed. The open list only contains the goal vertex and vertices are processed from the list, in order. This is carried out until the termination conditions are met. The DL algorithm uses the common definitions for the current cost of a vertex 'g' and the heuristic cost of a vertex 'h' and it also defines a look-ahead value 'la' which is the next value for the current cost. The open list is sorted by the key vector (a,b) where $a = \min[g(v), la(vaa)] + h(v) + km$, $b = \min[g(v), la(v)]$ and km is similar to FD-robot's bias value. The first element of the key vector is used to order the vertices. The ties, if any, between the vertices are broken using the second element. When the robot moves, the km value is increased by the heuristic distance between the robot's previous and current positions, since the km value changes, once a vertex is processed its actual key value may be different from the one it was inserted into the open list with.

10.6   GRAPH REPRESENTATIONS

The placement of the graph vertices and the connections between these vertices is dictated by the graph's representation of the environment. Many other aspects, such as how the robot actually moves between graph vertices, are discussed.

10.6.1   GRID REPRESENTATION

An environment is approximated by a grid, each cell of which will be represented by a vertex in the graph [1]. The cell's neighbours will have appropriate edges within the graph. If a neighbourhood of each cell is too small, the algorithm will have a limited range of motion. If the neighbourhood of each cell is too large, the algorithm will take too long for processing all the neighbours for each cell. The three common neighbourhoods are: (i) the von Neumann neighbourhood that includes only those cells which are directly North, South, East and West of a given cell, (ii) the Moore neighbourhood in addition to (i) above also includes the diagonal neighbours and (iii) the extended Moore neighbourhood includes the Moore neighbourhood and their own Moore neighbours too. The centre of a chosen neighbourhood is the cell from which this neighbourhood is constructed. The radius r of a neighbourhood is appropriately defined so that the number of rows/columns in a neighbourhood is equal to $2r + 1$. The robot might move along the edge of an obstacle until it reaches the obstacle's corner. If the outer corner of the robot's neighbourhood is unoccupied, it will attempt to move through the obstacle. This is solved by forming a hierarchy where the neighbours closer to the centre of the neighbourhood determine if the outer neighbours are free.

10.6.2 VISIBILITY MAPS

In the visibility map/graph (VMG) the vertices are connected if (and only if) there is an unobstructed line of sight (LOS) between them [1]. The cost is given as the actual distance between the vertices. In a simple visibility map the vertices of the graph are placed on the corners of the obstacles as well as on the starting and goal positions, and it is referred to as a visibility graph. Here, the assumption is that the obstacles in the environment are polygons (or they can be contained within polygons). To obtain the VMG the visibility between every pair of vertices in the graph must be determined. Each vertex is taken and the visibility between that vertex and every other vertex is calculated; this is done using a rotational plane sweep algorithm thereby reducing the number of polygon edges to consider when testing the visibility between two vertices.

10.6.3 QUAD TREES

An approach to reduce the number of cells within a grid is to use a multi-resolution grid, such as a quad tree (QT) that represents the environment by splitting it into four cells [1]. Each of which is a node under the root node of the quad tree. Each cell of the grid is also split into further four cells by creating four child-nodes, repeated until a desired depth is reached. In case of four identical children, they are represented by the single parent cell. The QT allows for a high resolution around the edges of obstacles. Then for the rest of the graph low resolution is permitted. In the QT the partitions are not necessarily divided into equal parts. Also, it is not necessary for the partition lines to be perpendicular. These lines are positioned anywhere within the node, as long as they divide the node's space into four parts. Then an optimal division is found/can be found in which the number of nodes in the tree is minimized. A neighbourhood must also be defined for the quad tree cells. This neighbourhood does not have a fixed number of cells. The cost of travelling from one neighbour to another is taken as the Euclidean distance between the two cells. Finally, the robot moves between the cells of the quad tree in the same way as it does for a grid.

10.6.4 DIRECTIONAL REPRESENTATION

A third dimension is added to a regular grid as the direction in which the robot is facing. This 3D grid, a directional grid (DG), is constructed from a traditional grid [1]. This grid is split into 'n' layers, 'n' being the number of cells in the neighbourhood (excluding the centre cell of the neighbourhood). These layers are arranged such that the direction of a neighbour changes by $360°/n$ degrees from one layer to the next layer. After the arrangement, a cell within a layer is linked to the corresponding cells in the two layers adjacent to it, the chosen cell. The first and last layers are linked in the same manner so that the directional axis is wrapped around. In the completed DG, movement between the layers rotates the robot and movement (to be imparted) within a layer moves the robot in a single direction. The distance between the two cells (on a directional grid) is not the actual/real distance between the positions that are represented by these cells. The merit of using the DG is that it has an additional cost to rotating the robot which should provide paths with fewer turns.

The demerit of DG is that it increases the dimension of the grid and it would require the PPA (path planning algorithm) to process many more cells. The DG ensures that the different objectives can be amalgamated into the graph representation.

10.7 DYNAMIC ALGORITHMS

These algorithms should satisfy a number of requirements: (a) they must deliver optimal or near-optimal paths and (b) the algorithms must find these optimal paths with reasonable computational cost in time and efficiency, since the mobile robots usually have limited computing power.

10.7.1 DYNAMIC BUG ALGORITHM

The DBA adapts the TBA for a dynamic environment [1]. It also functions with the two modes: the motion to the goal mode and the boundary following mode. The DBA does not observe all the obstacles around itself, since, if an obstacle is blocking the robot's path to the goal, the DBA only looks as far as the left and right edges of the obstacle. When both the edges of the obstacle are found, the robot moves past the edge which gives the shortest path to the goal. As is done in the TBA, if the robot moves away from the goal, the boundary-following mode is initiated, and in this mode the DBA will note the shortest distance between any point on the obstacle and the goal. The algorithm first checks (before moving around the obstacle) if there is any point to move to which is closer to the goal than any point on the obstacle. If such a point exists, the robot move towards this point. Then the boundary following mode is terminated; otherwise, the robot will continue to move around the obstacle in the direction of the last move to goal movement. DBA does not maintain any information about the environment and hence it does not require any more computation when the obstacles move. It is a suboptimal algorithm.

10.7.2 A* ALGORITHM

The A* algorithm does not handle changes within a graph, and hence it is adapted to perform as a dynamic algorithm, by re-planning the entire path every time the robot wants to move by one more step [1]. The re-planning works well on robots that have powerful processors onboard and these processors repeatedly run a SPC (sense → plan → act) cycle of/in the robotic configuration (Chapter 16). This technique is also necessary for graph representations that depend on the robot's position, for example, visibility graphs. In another approach the A* algorithm is run again when a change in the environment is observed. These two methods allow the robot to follow an optimal path, but the computations are more expensive than usually needed. It follows, from the manner in which the A* algorithm plans its paths, that only when certain vertices in the graph are modified the algorithm is required to re-plan its path. The algorithm only utilizes the vertices that were open on its list at some stage to determine the shortest path. If a change occurs to an edge with the property that neither the source nor the destination of the edge were on the open list then the current optimal path does not change. The algorithm is then not required to be executed. If the changes increase the weight of edges, another optimization need be made. If the weight of an edge is raised the algorithm only needs to re-plan its path. This is done if the raised edge is on the remainder of the current optimal path to be traversed by the mobile. These new conditions as to when to re-run the algorithm can ensure a significant reduction in the computational cost of utilizing the A* algorithm in any dynamic environment.

10.7.3 SUBOPTIMAL A* ALGORITHM

In this suboptimal algorithm the restriction of path optimality is relaxed. A heuristic that overestimates the exact distance is specified. If the Manhattan heuristic is used on a Moore neighbourhood, it does not underestimate the exact distance [1]. Additionally one can ignore some information about the environment. Thus, a variant of the A* algorithm was developed in [1] and is referred to as the Blind A* algorithm (BAA). The BAA functions exactly like the A* algorithm but it only re-runs the A* algorithm when and if the current path within a certain range is blocked. The choice of when to re-plan the path for the algorithm is favourable, in a dynamic environment, since a far-away obstacle may no longer obstruct the robot. The range of the algorithm for re-planning its path is relatively small. However, it is sufficiently large enough for the robot to move around obstacles. The BAA ignores possible path obstructions until it is closer to them, and hence it is expected that the paths which BAA produces would be on average longer than the A* algorithm.

10.7.4   D* Algorithm

In the D* algorithm, if there is a change in the cost of a vertex, it is propagated along the back pointers. If a change raises a vertex, by causing its 'g' value to be greater than its 'k' value, then neighbouring vertices are also raised. This is done until a better path with less cost is found. When this path is found, the numerical 'g' values of the vertices are reduced again. This up-down cycle signifies that each vertex in this cycle will be processed twice. Further details on D* algorithm can be found in Ref. [1].

10.7.5   Robot Motion-Trajectory Planning

In fact, motion planning is aimed at providing the robots with the capabilities of automatically deciding and executing a sequence of motions in order that the robot performs a task without colliding with any of the other objects in the robot's environment. In the path planning, as we have seen in several previous section, the design of only geometric, that is, kinematic details of the locations and orientations of the robot are taken into account, whereas in the trajectory planning the design of the linear and angular velocities is taken into consideration. Autonomous agents like mobile robots that sense, plan and act in real and/or virtual worlds, need path planning and motion-trajectory planning algorithms. These algorithms and systems are utilized for representing, capturing, planning, controlling and rendering motions of several physical objects individually or collective (as discussed in Chapter 17) [3], and the applications are: (i) manufacturing, (ii) mobile robots, (iii) computational biology, (iv) computer-assisted surgery and (v) digital actors. Thus, the goal of motion planning is to compute motion strategies, for example: (a) geometric paths, (b) time-parameterized trajectories and (c) sequence of sensor-based motion commands. Also the aim is to achieve high-level goals, for example: (a) go to destination D without colliding with obstacles, (b) assemble a product P, (c) build map of environment E and (d) detect and find an object O.

Many times in motion planning the problem stated is that of the basic path planning problem: to compute a collision-free path for a rigid, articulated object (the robot), or a mobile robot among static (and dynamic) obstacles with the inputs: (a) geometry of robot and obstacles, (b) kinematics of robot (degrees of freedom) and (c) initial and goal configurations (robot's placements) and with the output as continuous sequence of collision-free robot movements connecting the initial and goal (robot) configurations. However, the basic path planning problem can be made more complex, more sophisticated and more general by extension and incorporating the following additional aspects: (i) environment of moving obstacles with uncertainties, (ii) multiple robots (multi-robot coordination and path planning), (iii) movable objects, (iv) deformable objects, (v) enhancing goal task to gather data by sensing, (vi) nonholonomic constraints, (vii) dynamic constraints, (viii) optimal planning as against the heuristic or suboptimal planning and (ix) recognizing uncertainty in control and sensing that will have effect on accuracy of trajectory traversed. Thus, the path planning problem is extended to encompass the motion planning with more sophistication. These extensions (at least a few of them) are/will be useful [3–6] (a) in the design of manufacturing and related processes, (b) robot programming and proper placement where really needed, like their stable positioning, (c) checking building code, (d) generation of instruction sheets, (e) model construction by the mobile robot, (f) graphic animation of digital actors, (g) computer-assisted surgical planning, (h) prediction of molecular motions, (i) Mars exploration rovers, (j) military vehicle movements, (k) assembly maintainability, (l) virtual environment and games, (m) computational biology and chemistry application and (n) RoboCup. The representation concepts common to path/motion planning problems are (as we have discussed in Sections 10.2 through 10.5): (i) state-space (CS), position, velocity with respect to time; (ii) composite configuration/state spaces (extended robot arms, protrusions); (iii) stability regions in configuration/state spaces and (iv) visibility regions in configuration/state spaces.

10.8   COMBINATION OF A* ALGORITHM AND FUZZY LOGIC

In this joint approach it is assumed that all the obstacles are static, and the robotic map and the locations of the obstacles are provided [7]. The A*FIS (fuzzy inference system) algorithm provides a collision-free path for the given robot. The A* provides a coarser level path planning and the FIS does the finer level planning. Individually the two algorithms have the following features: (a) the A* has merits in path optimality, and the deadlocks, and demerits in non-holonomic constraints, time complexity and input size; (b) whereas, the FIS has merits in non-holonomic constraints, time complexity and input size, and demerits in path optimality and deadlocks. In the joint A*FIS algorithm these latter demerits of the FIS are compensated by the use of A* algorithm. In this approach first the preliminary coarser path is found by the A* algorithm. Then initial FIS is generated that is further optimized by using GA. It means that the FIS parameters are optimized by a suitable GA. The trained FIS is used in the FIS path planner, wherein the intermediate points are used as goal and the result is added to the path. Thus, the finer adjustments are made by the trained FIS which itself has been optimized by using GA. In the joint algorithm this FIS method is repeated for all the points in the solution initially provided by the A* algorithm. Additionally, the GA maximizes the performance for small-sized benchmark maps.

10.9   MULTI-AGENT SYSTEMS PATH PLANNING

The multi-agent system path planning is a complex problem because of the large number of agent systems involved (Chapter 1). In multi-agent systems several agents within an environment cooperate for a part of time to carry out an assigned task. Hence, the common methods might not be very suitable for path planning in a multi-agent system which has special features and requirements [8]: (i) collision avoidance not only of obstacles but among the multiple agents/systems/robots, (ii) avoiding deadlock situation (no solution found, or the algorithm getting trapped in a local minimum), (iii) computations are likely to be very large, (iv) lot of information exchange might be required between several agent systems and (v) hence, the communication overheads are heavy. The possible approaches then for the multi-agent systems path planning are [8]: (a) centralized, (b) decoupled and (c) some combined methods. In the centralized approach all the robots are considered as if they are operating in one aggregated system. One tries to find a complete and optimal solution provided if such a solution exists by using the complete information. Here, the computational capacity required is very high and would increase exponentially with the number of the robots in the composite system. In the decoupled approach one generates paths for the individual robots, with this the interactions are taken care of. In this case the computational time is proportional to the number of the neighbouring robots, and the method is found to be robust [8]. In the combined method one uses the cumulative information to have a global path planning, and then the local information is used for local path planning. In the combined approach to path planning for multi-agents system, the main goal is to plan a complete path from the present location to the final location. The method used could be A* algorithm, Wavefront or probabilistic roadmaps (PRM). This combined approach needs graph representation that can be achieved by cell decomposition method, and for the local path planning in the context of the combined approach, the goal is to avoid obstacles, and utilize cooperation. For the local path planning one can use the potential field method or vector field method. In this case, no global information and no map representation are required. One can use the rule bases, that is, If 'a condition is satisfied', Then 'the left agent should move first' and so on. However, one needs to do the resource allocation due to the multi-agents systems. This might lead to suboptimal solutions to the path planning problem. Other important aspects for the combined approach are [8]: (i) a leader and the several followers concept is feasible, (ii) one can specify a hierarchy of leaders, or have a virtual leader, (iii) one can use the concept of virtual dampers and springs and (iv) one can assign dynamic information to the edges and vertices.

The foraging and rummaging robots in a cooperative environment and tasks are one of the examples of several multi-agent systems for which one of the above methods can be used for path planning. In this case aerial images are taken of the sites/objects and processed to determine their locations to aid robot path planning. Alternatively the clustering and evolutionary algorithms (GAs) can be used for determining an optimal path.

10.10   ILLUSTRATION OF PATH PLANNING PROBLEM WITH A* ALGORITHM

An illustration of the working of the A* algorithm written in MATLAB® [9] is presented here. First a 2D map array is defined. This array stores the coordinates of the map and the objects in each coordinates. Next, the map is initialized with appropriate input values. This part is then made interactive, so the user can assign target locations and the obstacles locations onto the displayed 2D graphical (Figure 10.4). This also allows a selection of the vehicles' initial and final position. Once the algorithm has been run, the optimal path is generated by starting of at the last node, that is, the target node, and then identifying its parent node until the start node is reached. This is supposed to be the optimal path for the given CS with static obstacles specified. A distance function is used to compute the distance between any two Cartesian coordinates. A function is used to take a node and return the expanded list of successors along with computed function values. The criterion used is that none of the successors are on the 'closed' list. A function to populate the 'open' list is also used. A function to return the Node with minimum 'fn' takes the list 'open' as its input and returns the index of the Node that has the least cost value. The MATLAB

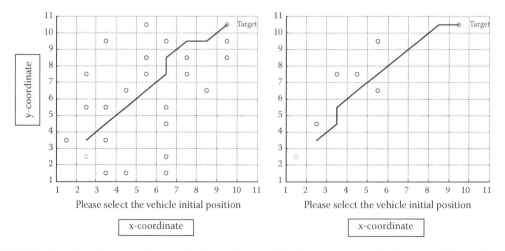

FIGURE 10.4   Path Planning Results of A* algorithm with 22 (left subplot) and 5 obstacles (right subplot).

sub-plots resulting after running this A* algorithm are shown for 22 and 5 obstacles in Figure 10.4, which is self-explanatory.

REFERENCES

1. Crous, C.B. Autonomous robot path planning. MSc thesis, University of Stellenbosch, South Africa, March 2009.
2. Simon, K. Evolutionary approaches to robot path planning. Doctoral thesis, Brunel University, UK, March 1999.
3. Sziebig, G. Interactive vision-based robot path planning. MSc thesis, Budapest University of Technology and Economics, May 2007.
4. Jean-Claude, L. Motion planning: A journey of robots, digital actors, molecules and other artifacts, Computer Science Department, Stanford University, Stanford, CA.
5. Masehian, E. and Sedighizadeh, D. Classic and heuristic approaches in robot motion planning—A chronological review. *World Academy of Science, Engineering and Technology*, 23, 101–106, 2007.
6. Ge, S.S. and Cui, Y.J. *Dynamic Motion Planning for Mobile Robots using Potential Field Method, Autonomous Robots*. Kluwer Academic Publishers, The Netherlands, 13, 207–222, 2002.
7. Kala, R., Shukla, A. and Tiwari, R. Fusion of probabilistic A* algorithm and fuzzy inference system for robotic path planning. *Artificial Intelligence Review*, Springer Publishers, 33, 4, 275–306, 2010.
8. Kaplan, K. Path planning for multi agent systems, 2005. http://www.robot.cmpe.boun.edu.tr/robsem/robsem/MASPP_KEMAL.ppt. July 2011.
9. Paul, V. Path planning A* algorithm in MATLAB (code, April 2005), http://www.yasni.com/vivian+paul+premakumar/check+people, July 2011.
10. Buniyamin, N., Sariff, N., Wan Ngah, W.A.J. and Mohamad, Z. Robot global path planning overview and a variation of ant colony system algorithm. *International Journal of Mathematics and Computers in Simulations*, 5(1), 9–16, 2011.

11 Out-of-Sequence Measurements
Missing Data Problem

Bhekisipho Twala

CONTENTS

11.1 INTRODUCTION

There has been an emergence of several approaches to sensing applications in recent years. Several applications require the maintenance of a high-fidelity estimate of the state of a dynamic system based on a sequence of noisy observations. Such applications demand the use of filtering mechanisms such as the Kalman filter (KF) [1,2] to fit the observation sequence to a given model of the system dynamics. Our primary motivation for this chapter comes from problems that arise in multiple targets tracking with distributed sensor networks. One central problem in multiple target tracking (MTT) is out-of-scope measurements (OOSM) [3–14] which is sometimes referred to as the problem of tracking with random sampling delays [15–17] and the problem of incorporating time-delayed measurements [18].

Much of the work on tracking and filtering is built on the assumption that measurements are available immediately to an agent. However, there might be situations in which measurements are subject to non-negligible delays, such that the lag between measurement and receipt is of sufficient magnitude to have an impact on estimation or prediction [19]. This can be caused by a number of specific effects: communication delays from the sensor to the tracker; different sensors observing the current state of the target at different times; delays in sending tracks to the data fusion node (often because the sensor is a rotating radar measurement with specific time stamps) and unsteady pre-processing times of the observed data, depending on the system load, can vary from one measurement to another. Delayed measurements can create difficulties, especially for discrete time filtering. The delayed observations are classified into two categories: (1) constant delays and (2) random delays. The constant delays involve measurements being delayed by the same constant lag. In this way, measurements are never observed out-of-sequence; they are simple and consistent. Such behaviour could be induced, for example, by a constant bandwidth restriction on a sensor network. In contrast, random delays provide a number of possibilities, including that measurements are delayed with a constant probability but fixed lag, or constant probability with a random lag. Such problems could arise as a result of intermittent bandwidth restrictions on a sensor network. All models of random delay have the potential to cause OOSM. In this chapter, we are more interested in the problem of random delays, hence, OOSM.

Yet another problem related to OOSM is that of incomplete data or missing values. In fact, skipping the 'correction' step and proceeding straight to the 'prediction' step when using a KF for estimation purposes is the equivalent of considering the delayed measurement as missing [20–22]. The presence of missing values is commonplace in large real-world databases. This has become one of the most important problems in academic research since most learning systems and statistical analyses in the early stages were not designed to handle missing data (incomplete vectors). There are several reasons why there are missing values in data. An item could be missing because it was unavailable or arises by 'default' in data recording activities. Missing values could also occur because of confusing questions in the data gathering or because of sensor malfunction. In some situations, the missingness could be caused by the relationships between the attribute variables themselves. That is, the information that is missing on a given attribute variable could be a result of its relation to values of other attribute variables in the dataset. An extreme case is that the missing value could be a result of its relation to an unobserved (missing value) in the dataset.

In this chapter, we explore the resilience of OOSM, on the one hand, and imputation, on the other, to the various forms of imperfections in sensor data in order to increase the awareness of the impact delayed measurements could have when building robotic prediction models. Our main focus is on the application of OOSM and multiple imputation (MI) to multi-target tracking prediction. The topic is significant because (i) the OOSM is an emerging technology that can aid in the handling of delayed measurements in single- or multi-target tracking prediction; (ii) the delayed observations problem is related to the incomplete data problem, hence, the use of imputation procedures (single or multiple imputation) could be utilised; (iii) the MI method has an advantage over single imputation strategies due to the fact that it overcomes the under-representation of uncertainty about which value to impute (i.e., single imputation methods underestimate the true variance of the values they are attempting to fill in or impute) and (iv) since there is a lack of adequate tools to deal with delayed measurements, machine learning techniques have been used to tackle such problems, including either single or multiple target tracking or navigation prediction.

Much of the work on tracking and filtering is built on the assumption that the measurements are immediately available to an agent. However, it is not difficult to conceive situations in which the measurements are subject to non-negligible delays, such as the lag between measurement and receipt is of sufficient magnitude to impact on estimation. In such situations, the classical assumption that observations are available immediately is easily violated [19]. One direct solution to the OOSM problem is simply to ignore and discard the OOSM in the tracking process more like the listwise deletion is a standard default approach for dealing with missing data in most statistical packages. This solution leads obviously to a loss of the information contained in the discarded OOSM. To avoid this drawback, several alternative methods have been proposed in the literature to deal with the OOSM problem, especially for random delays. It is also striking that most of the methods proposed to handle delayed measurements have in common that delayed measurements are always ultimately incorporated into the filtering process. In the case of time delay, one common approach is related to solving a partial differential equation and boundary condition equations which do not have an explicit solution in general [23–27]. In the case of discrete time systems (and especially for random delays), the problem has been investigated via a standard Kalman filtering [28] and by augmenting the system accordingly [28–30]. Matveev and Savkin [31] consider an iterative form of state augmentation for random delays with a random lag. Larsen et al. [32] address the OOSM problem by recalculating the filter through the delayed period. In the same context, Larsen et al. [32] further propose a measurement extrapolation approximation using past and present estimates of the KF and calculating an optimal gain for this extrapolated measurement. Thomopoulos and Zhang [17] examine the case of random delay under the name of the fixed sampling and random delay filter that is shown to be equivalent to constraining the lag to a value of 1. Alexander [33] and later Larsen et al. [32] suggest using the delayed measurements to calculate a correction term and adding this to the filter estimate. Zhang et al. [34] proposed algorithms that try to minimise the information storage in an OOSM situation. Challa et al. [9] formulated the OOSM problem in a Bayesian framework. The above methods are described in more detail in Section 11.3.

11.2 BACKGROUND PROBLEM INFORMATION

The presentation herein is based on the KF equations for a discrete linearised time-varying system with state vector x_k, input vector u_k and output vector y_k. KF is the optimal recursive data processing algorithm for a discrete linear system corrupted with noise in the states and measurements. What a KF requires is knowledge of the system and measurement dynamics, a statistical description of the system and measurement noises, uncertainty in the dynamic models and any available information about the initial conditions of the variable of interest. Based on this knowledge, it gives the optimal estimate of the state variables under observations [35,36]. Since its inception, KF has become a subject of extensive research and application, particularly in areas of autonomous, assisted navigation or target tracking. To give detailed analysis of KF is out of the scope of this work. For more detailed study of KF's probabilistic origin, the author recommends Maybeck [36]. The following lists the equations for notation purposes introduced by Geld [38], improved by Bar-Shalom and Li [39] and later by Julier and Uhlman [40].

11.2.1 System and Model Descriptions

The KF solves the problem of estimating the state $x \in \Re^n$ of a discrete time controlled process that is assumed to evolve over time t_{k-1} to t_k and governed by the linear stochastic difference equation

$$x(k) = \boldsymbol{F}(k, k-1)\boldsymbol{x}(k-1) + v(k, k-1) \tag{11.1}$$

where $x(k)$ is the state vector at time k, $\boldsymbol{F}(k, k-1)$ is the state transition matrix to time t_k from t_{k-1} and $v(k, k-1)$ represents the (cumulative effect of the) process noise for this interval. The order of the arguments in both $\boldsymbol{F}$ and v is according to the convention for the transition matrices. Typically, the process noise has a single argument, but here the two arguments will be needed for clarity. The time τ, at which the OOSM was made, is assumed to be such that

$$t_{k-l} < \tau < t_{k-l+1} \tag{11.2}$$

This will require the evaluation effect of the process noise over an arbitrary non-integer number of sampling intervals. Note that $l = 1$ corresponds to the case where the lag is a fraction of a sampling interval; for simplicity this is called the '1-step-lag' problem, even though the lag is really a fraction of a time step. The measurement $Z \in \Re^m$ and thus measurement or observational model is

$$z(k) = \boldsymbol{H}(k)\boldsymbol{x}(k) + w(k) \tag{11.3}$$

where $z(k)$ is the observation vector, $w(k)$ is the observation noise vector and $\boldsymbol{H}(k)$ is the observation matrix. The noise vector $v(k, k-1)$ and $w(k)$ are assumed to be independent (of each other), white, and with normal probability distributions

$$p(w) \sim N(0, Q) \tag{11.4}$$

$$p(v) \sim N(0, R) \tag{11.5}$$

The process noise covariance $Q(k)$ and measurement noise covariance $R(k)$ are mutually uncorrelated and they are given as

$$E[v(k, j)v(k, j)'] = Q(k, j) \quad E[w(k)w(k)'] = R(k) \tag{11.6}$$

Similar to Equation 11.1, one has

$$x(k) = F(k, \kappa)x(\kappa) + v(k, \kappa) \tag{11.7}$$

where κ is the discrete time notation for τ. The above can be written backward as

$$x(\kappa) = F(\kappa, k)[x(k) - v(k, \kappa)] \tag{11.8}$$

where $F(\kappa, k) = F(k, \kappa)^{-1}$ is the backward transition matrix.

11.2.2 FUSION PROCESS FOR TIME-DELAYED MEASUREMENTS

We denote a cumulative set of measurements as $Z^k \triangleq \{z(i)\}_{i=1}^k$, then the OOSM problem (up to time instance $t = t_k$, and excluding a measurement $z(\tau)$ with a time stamp $t_\tau < t_k$ as shown in Figure 11.1) reduces to the problem of computing the conditional mean estimate of the target state

$$\hat{x}(k \mid k) \triangleq E[x(k) \mid Z^k] \tag{11.9}$$

and its associated error covariance

$$P(k \mid k) \triangleq cov[x(k) \mid Z^k] \tag{11.10}$$

With the assumption that the initial state x_v is Gaussian, the conditional mean estimate $\hat{x}(k \mid k)$ of the target state, which is optimal in the minimum variance sense, can be computed recursively using

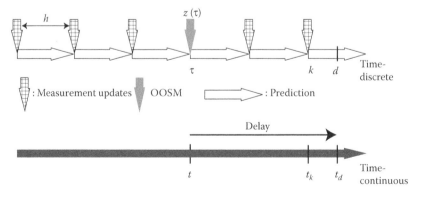

FIGURE 11.1 Timing diagram of out-of-sequence measurements.

the KF. Also, it is assumed that a measurement z is collected and used to update the track at the time interval h. The basic KF algorithm can then be extended to multi-sensor systems where the data are assumed to arrive at known times and in correct time sequence. Suppose that a given measurement corresponding from time τ (denoted with discrete time notation κ),

$$z(\kappa) \triangleq z(\tau) = H(\kappa) \, x(\kappa) + w(\kappa) \tag{11.11}$$

arrives with a certain delay after Equations 11.9 and 11.10 have been computed, as shown in Figure 11.1. Here, one faces the problem of updating the state estimate and its covariance with the delayed measurements, that is, to compute:

$$\hat{x}(k \mid \kappa) \triangleq E[x(k) \mid Z^\kappa] \tag{11.12}$$

and

$$P(k \mid \kappa) \triangleq cov[x(k) \mid Z^\kappa] \tag{11.13}$$

where

$$Z^\kappa \triangleq \{Z^k, z(\kappa)\} \tag{11.14}$$

Equation 11.13 provides a simple, intuitive interpretation of the weight in the time-delayed KF. The weight assigned to a measurement is a function of the degree to which the measurement is correlated with the current state of the system.* Therefore, the difficulty in implementing the time-delayed KF is in calculating $P(k \mid \kappa)$. Solutions to the delay measurement problem are presented in Section 11.3.

11.3 VARIOUS EXISTING METHODS

11.3.1 OOSM

There are a number of solutions proposed in the literature to solve this OOSM problem. Most existing solutions to the problem are based on retrodiction, where backward prediction of the current esti-mated state is used to incorporate the OOSMs at appropriate time instants. However, in recent years, some researchers have tackled the OOSMs problem without the need of backward prediction. See, for example, Ref. [41]. Here, we are more interested in the backward prediction solutions that are described below. In Ref. [17], the authors examine the case of random delays of the complete mea-surement vector which arrive out-of-sequence, where the lag is restricted to a value of 1. The mea-surements arrive at the fusion with random delays that can be due to queuing at the sensor buffer and to delays in the transmission time as well as in the propagation time. Optimal filters for the estimation of target tracks based on measurements of uncertain origin received by the fusion at random times and out-of-sequence are derived for the cases of random sampling, random delay, and both random sampling and random delay.

In Refs. [33] and [32], the authors suggest using the delayed measurement to calculate a correction term and adding this to the filter estimate, again considering the complete observation vector being

---

* The result is algebraically the same as derived by Larsen et al. (1998). However, the interpretation is very different. Larsen considered taking an observation and extrapolating its value forward to the current time step in the filter. Julier and Uhlmann (2005) considered calculating the correlation backward from the current time to the time when the obser-vation was made.

delayed. In the same context [32], a measurement extrapolation approach was proposed to ensure the optimality of the filter and at the same time address issues like changing measurement and state noise covariance matrices. Another approach [31] that addresses the random delayed measurements problem is called state augmentation. This approach was designed to handle a linear discrete time partially observed system perturbed by white noises. The reduced-order linear unbiased estimator was designed via iterative state augmentation. By so doing [31], the authors managed to solve the minimum variance state estimation problem and further show how their proposed approach is exponentially stable under natural assumptions. Ref. [42] follows up this approach by proposing a variable dimension filter which handles only essential past states not just past states that are up to some maximum delay like the approach in Ref. [31].

Ref. [40] considers the problem of applying a KF to estimate the state of a dynamic system using a sequence of observations that are not precisely time stamped. They argued that the problem has analogies with the identity ambiguity problem that arises in MTT applications. They further described a way in which multiple hypotheses and covariance union (CU) methods could be utilised for this kind of problem and compared them with the probabilistic data association filter (PDAF) method. Their results showed the PDAF yielding the most accurate results, however, at a higher computational cost. The strong dependence of PDAF on the accuracy of the likelihood model is another of its weaknesses. Although CU requires the evaluation of two KF updates, its advantage lies in its ability not to rely on specific assumptions as to the veracity of the likelihood model.

More recently, in Ref. [34], two algorithms with three cases of different information storage for the state estimation update with OOSM were proposed. Both algorithms are optimal in the linear minimum mean square error sense for the information available at the time of update. Their proposed algorithms (based on the linear minimum mean square error) try to minimise the information storage in an OOSM situation using different minimum storage of information concerning the occurrence time of single OOSMs. Further, they extend the single OOSM update algorithms to the case of arbitrarily multiple OOSMs. Ref. [9] formulated and solved the OOSM problem in the Bayesian framework. They established that the solution involved the joint probability density of current and past states or the state corresponding to the delayed measurement. For the case of multiple delays, the authors show that the solution involves a Bayesian recursion for the joint probability density of an augmented state vector. Based on this, the augmented state Kalman filter (AS-KF) and its variable dimensions extension (VDAS-FK) are proposed as the fundamental solution to this problem in the linear Gaussian case. AS-FK handles noise-target state cross-correlation implicitly and can be readily extended to handle clutter. The idea of VDAS-FK is that the augmented state only carries the current state and the past state for which there was a missing measurement. The filter will reduce to a normal KF if there is no OOSM. Further, a new augmented state probabilistic data association filter (AS-PDA) is proposed in Ref. [9]. This filter is meant to deal with data association issues arising from the presence of clutter in the OOSM problem. Simulation results are used to demonstrate the effectiveness of these algorithms. The results show the proposed solutions as computationally expensive when compared with existing methods but straightforward to implement and they also yield significant improvements in terms of performance. A more principled way to handle this problem is to extend Challa's Bayesian formalism [9,37] to include uncertainties in the time delays. This is analogous to a problem that arises in MTT [43]. MTT occurs when a tracking system receives an observation of one of several different targets, but the exact identity of the observed target is not known.

11.3.2 MULTIPLE IMPUTATION APPROACH

Imputation is the substitution or replacement of some value of a missing data point or missing component of a data point [44–46]. MI is one of the most attractive methods for general-purpose handling of missing data in multivariate analysis as a three-step process. First, sets of M plausible values ($M = 5$ in Figure 11.1) for missing instances are created using an appropriate model that

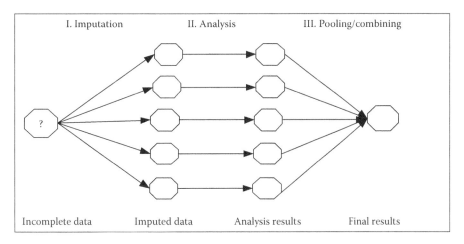

FIGURE 11.2   Multiple imputation process—three steps.

reflects the uncertainty due to the missing data. Each of these sets of plausible values is used to 'fill-in' the missing values and create M 'complete' datasets (imputation). Second, each of these M datasets can be analysed using complete data methods (analysis). Finally, the results from the M complete datasets are combined, which also allows the uncertainty regarding the imputation is taken into account (pooling or combining).

For example, replacing each missing value with a set of five plausible values or imputations (as it was the case in our illustration in Figure 11.2) would result in building five decision trees (DTs) [47], and the predictions of the five trees would be averaged into a single tree, that is, the average tree is obtained by MI. MI retains most of the advantages of single imputation and rectifies its major disadvantages as already discussed. There are various ways to generate imputations. Schafer [46] has written a set of general-purpose programs for MI of continuous multivariate data (NORM), multivariate categorical data (CAT), mixed categorical and continuous (MIX) and multivariate panel or clustered data (PNA). NORM includes an expectation maximisation (EM) algorithm for maximum likelihood estimation of means, variance and covariances. NORM also adds regression-prediction variability by using a Bayesian procedure known as data augmentation (DA) [48] to iterate between random imputations under a specified set of parameter values and random draws from the posterior distribution of the parameters (given the observed and imputed data) [49]. These two steps are iterated long enough for the results to be reliable for multiple imputed datasets [46]. The goal is to have the iterates converge to their stationary distribution and then to simulate an approximately independent draw of the missing values. The algorithm is based on the assumptions that the data come from a multivariate normal distribution and are missing at random (MAR). MAR essentially says that the cause of missing data may be dependent on the observed data but must be independent of the missing value that would have been observed.

Although not absolutely necessary, it is almost always a good idea to run the EM algorithm [50] before attempting to generate MIs. The parameter estimates from EM provide convenient starting values for DA. Moreover, the convergence behaviour of EM provides useful information on the likely convergence behaviour of DA. Therefore, EM estimates of the parameters are computed and then the number of iterations required is recorded, say t. Then, a single run of DA algorithm of length tM using the EM estimates as starting values is performed, where M is the number of imputations required. The convergence of the EM algorithm is linear and is determined by the fraction of missing information. Thus, when the fraction of missing information is large, convergence will be very slow due to the number of iterations required. However, for small missing value proportions, convergence is obtained much more rapidly with less strenuous convergence criteria. The EM and

DA processes are described below. The core idea of the EM algorithm is to introduce some unobserved variables Z, appropriate for the model under consideration, such that if Z were known, the optimal value of θ could be computed easily. Then, the complete conditional probability density (including the missing variables) can be written as

$$L(\theta \mid X, Z) = \sum\nolimits_{i=1}^{N} \sum\nolimits_{j=1}^{M} z_{ij} \log f(x_i \mid z_i; \theta) f(z_i; \theta) \qquad (11.15)$$

The usual approach is to regard Z as missing data and estimate it iteratively. The intuition behind the EM algorithm is that we would like to maximise the complete data likelihood but it cannot be utilised directly, so we maximise its expectation, denoted by $Q(\theta \mid \theta')$, instead. As shown by Dempster et al. [50], $L(\theta \mid X)$, the complete data likelihood, can be maximised by iterating the following steps:

1. Initialise parameters randomly. Set $t = 0$.
2. *E*-step: Determine $Q(\theta \mid \theta^{(t)}) = E[L(\theta \mid X, Z) \mid X, \theta^{(t)}]$.
3. *M*-step: Set $\theta^{(t+1)} = \arg \max_{\theta} \{Q(\theta \mid \theta^{(t)})\}$, where $\theta^{(t)}$ are the current parameter estimates in time step t.
4. Iterate steps 2 and 3 until convergence.

Assume that the dataset $X = \{x_1, \ldots, x_N\}$ is divided into an observed X_{obs} and missing X_{miss} components, respectively. To handle missing values we can rewrite the EM algorithm as follows:

1. Initialise parameters randomly. Set $t = 0$.
2. *E*-step: Determine $Q(\theta \mid \theta^{(t)}) = E[L(\theta \mid X_{obs}, X_{miss}, Z) \mid X_{obs}, \theta^{(t)})]$.
3. *M*-step: Set $\theta^{(t+1)} = \arg \max_{\theta} \{Q(\theta \mid \theta^{(t)})\}$, where $\theta^{(t)}$ are the current parameter estimates in time step t.
4. Iterate steps 2 and 3 until convergence.

The expectation (*E*) step computes the expected values for the sufficient statistics, given a model and values for model parameters θ, that is, the expected value of the complete data likelihood with respect to the missing data, given the observed data and the current parameter estimates. The maximisation (*M*) step estimates the model parameters by maximising the likelihood using standard procedures, given complete data. The procedure iterates through these two steps until convergence is obtained. Convergence occurs when the change in parameter estimates from iteration becomes negligible. An important part of the EM algorithm is restoring error variability to the imputed values during the *E*-step. Replacing a missing value by an imputed value using the EM algorithm results in EM single imputation (EMSI). DA (which resembles EM) follows the following process:

1. Initialise parameters randomly. Set $t = 0$.
2. *I*-step: Given a current estimate $\theta^{(t)}$, select a value of the missing data from the conditional predictive distribution of X_{miss}, $X_{miss}^{(t+1)} \sim P(X_{miss} \mid X_{obs}, \theta^{(t)})$.
3. *P*-step: Conditioning on $X_{miss}^{(t+1)}$, draw a new value of θ from its complete data posterior $\theta^{(t+1)} \sim P(\theta \mid X_{obs}, X_{miss}^{(t+1)})$. Through an iterative process, two distributions are obtained, $P(\theta \mid X_{obs})$ and $P(X_{miss} \mid X_{obs})$. For a suitable large t, we can implement a DA algorithm by Tanner and Wong [48], which iterates between sampling θ^{t+1} from $P(\theta \mid X_{obs})$ and sampling $X_{miss}^{(t)}$ from $P(X_{miss} \mid X_{obs})$.
4. Iterate steps 2 and 3 until convergence.

The imputation (I) step simulates a random imputation of missing data under assumed values of the parameters. The posterior (P) step draws new parameters from a Bayesian posterior distribution based on the observed and imputed data. The procedure of alternately simulating data and parameters creates a Markov chain (MC) $X_{\text{miss}}^{(1)}, \theta^{(1)}, X_{\text{miss}}^{(2)}, \theta^{(2)}, \ldots$ [19], which eventually stabilises or converges in distribution to $P(X_{\text{miss}}, \theta \mid X_{\text{obs}})$. The procedure iterates through these two steps until convergence is obtained. The rate of convergence is related to the fraction of missing information. DA can be thought of as a small-sample refinement of the EM algorithm using simulation, with the imputation step corresponding to the E-step and the posterior step corresponding to the M-step. This is the approach we follow in this chapter which we shall now call Bayesian multiple imputation (BAMI). MI has several desirable features: (1) introducing appropriate random error term into the imputation process which makes it possible for the method to get approximately unbiased estimates of all parameters, (2) repeated imputation allows one to get good estimates of standard errors, (3) MI can be used with any kind of data and any kind of analysis without specialised software, (4) MI saves money, since for the same statistical power, MI requires a smaller sample size than, say, listwise or case deletion and (5) once imputations have been generated by a knowledgeable user, researchers can use them for their own statistical analysis. However, certain requirements must be met for MI to have these desirable features. First, the data must be MAR. Second, the model used to generate the imputed values must be 'correct' in some sense. Lastly, the model used for the analysis must match up, in some sense, with the model used in the imputation. The reader is referred to Refs. [46] and [51] for a rigorous description of all these conditions.

11.4   SIMULATION EXPERIMENTS

In this section, the behaviour of the five proposed OOSM procedures against model-based imputation procedures is explored. The four methods selected are based on the following KFs: fixed sampling and random delay Kalman filter (FSRD-KF); measurement extrapolation Kalman filter (ME-KF); state augmentation for random delays Kalman filter (SARD-KF), minimum storage Kalman filter (MR-KF) and Bayesian framework Kalman filter (BF-KF). The tracking performance is characterised by the root mean square error (RMSE) over 1000 Monte Carlo run for each specific scenario. The root mean square deviation (RMSD) or RMSE is a measure of the differences between values predicted by a model or an estimator and the values actually observed from the thing being modelled or estimated. The RMSD of an estimator $\hat{\theta}$ with respect to the estimated parameter θ is defined as the square root of the mean square error (MSE):

$$\text{RMSD} = \text{RMSE}\,(\hat{\theta}) = \sqrt{\text{MSE}(\hat{\theta})} = \sqrt{E((\hat{\theta} - \theta)^2)}. \tag{11.16}$$

Following Bar Shalom [3], two cases (process noise $q = 0.1$ and 4) corresponding to $\lambda = 0.3$ and 2 are examined, that is, the underlying target performs in a straight line motion, or is highly manoeuvring. Data are generated randomly for each run starting with an initial state

$$x(0) = [200 \text{ km}, \quad 0.5 \text{ km/s}, \quad 100 \text{ km}, \quad -0.08 \text{ km/s}] \tag{11.17}$$

A two data-point method [43] is used to initialise the filters with

$$P(0 \mid 0) = \begin{pmatrix} P_0 & 0 \\ 0 & P_0 \end{pmatrix} \quad \text{where } P_0 = \begin{pmatrix} R & R/T \\ R/T & 2R/T^2 \end{pmatrix} \tag{11.18}$$

for *a priori* error covariance or to form the initial error covariance for augmented state. Like in Challa et al. [52], we assume that the OOSM can only have a maximum of one lag delay, and the data delay is uniformly distributed within the whole simulation period with probability P_r that the current measurement is delayed. All statistical tests were conducted using the MINITAB statistical software program [53]. Analyses of variance, using the general linear model (GLM) procedure [54], were used to examine the main effects and their respective interactions. This was done using a three-way repeated measures design (where each effect was tested against its interaction with the simulated dataset). The fixed effect factors were the OOSM and imputation methods, the probability of measurement delay and the manoeuvring index.

11.4.1 EXPERIMENT I AND THE RESULTS

In order to empirically evaluate the performance of the five OOSM methods and EMSI in terms of RMS error, an experiment on simulated data is used. This experiment is carried out in order to rank individual OOSM methods and also assess the impact of delayed measurements (at various time and distance intervals) on a single delay against single imputation in terms of position error. Experimental results on the effects of delayed measurements on one lag delay in terms of the RMS position error are described. The behaviour of these methods is explored for distance and time intervals. From these experiments the following results are observed.

All the main effects were found to be significant at the 5% level of significance ($F = 37.17$, df $= 5$ for OOSM methods and EMSI; $F = 6.195$, df $= 1$ for probability of measurement delay; $F = 9.39$, df $= 1$ for manoeuvring index; $p < 0.05$ for each effect). From Figure 11.3, BF-KF is the overall best technique for handling delayed measurements on a one lag with an excess error rate of 5.6%, closely followed by EMSI, FSRD-KF and MR-KF, with excess error rates of 6.1%, 8.2% and 8.5%, respectively. The worst technique is SARD-KF, which exhibits an error rate of 9.9%. Tukey's multiple comparison tests showed no significant differences between ME-KF and SARD-KF (on the one hand) and FSRD-KF and MR-KF (on the other hand). The significance level for all the comparison tests is 0.05. We found all interaction effects to be insignificant at the 5% level of significance. No interaction effects were found to be significant at the 5% level. Hence, they are not discussed in this chapter.

Figures 11.4 and 11.5 show simulation results where the performance of OOSM and imputation methods for single delay over 1000 runs are compared. We have the following observations:

1. SARD-KF and ME-KF have similar RMS error performance (on the one hand) with FSRD-KF, MR-KF and EMSI achieving similar performances (on the other hand). However, the latter methods always achieve lower RMS error rates. BF-KF achieves higher accuracy rates at all time levels. This is the case for non-manoeuvring target tracking (Figure 11.4).
2. In case of the manoeuvring target tracking, BF-KF (once again) outperforms all the methods with serious competition from EMSI. The differences in performance are mostly prominent at higher probabilities of measurement (Figure 11.5).

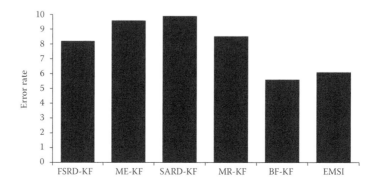

FIGURE 11.3   Results of OOSM and single imputation—overall means.

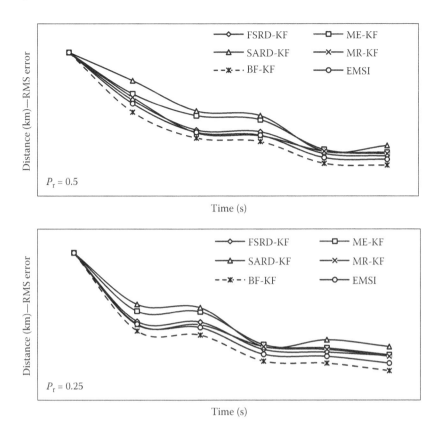

FIGURE 11.4 The performance of a straight line motion target with single delay OOSM (with $P_r = 0.5$ and 0.25 and manoeuvring index = 0.3).

3. For both manoeuvres, increase in the probability of measurement delay (P_r) is associated with increase in performance differences between methods. In fact, the performance by all the methods degrades with increase in the probability of measurement.

4. The accuracy of BF-KF and EMSI is achieved at a higher computational cost in terms of minutes (Table 11.1). Both methods take about twice (in some situations thrice) to compute compared to the others.

11.4.2 Experiment II and the Results

The main aim of this experiment is to compare the performance of OOSM and imputation methods for multiple delays, especially the top two OOSM methods that exhibited higher accuracy rates in the previous experiment. These are FSRD-KF and BF-KF. Also, we thought it would be interesting to test the effectiveness of multiple imputation (a procedure for handling incomplete data) against methods that have been proposed to deal with the delay measurement problem. All the main effects were found to be significant at the 5% level of significance ($F = 54.8$, df = 2 for OOSM and multiple imputations methods; $F = 11.62$, df = 1 for the probability of measurement delay; $F = 12.93$ df = 1 for manoeuvring index; $p < 0.05$ for each).

Figure 11.6 shows the average results of 12,000 experiments (3 OOSM and multiple imputation methods × 2 probability of measurement delay × 2 manoeuvring index) which summarise the accuracy of each method. Figure 11.6 further shows that BAMI has the best accuracy throughout the entire spectrum in terms of the probability of measurement and manoeuvring index. Tukey's multiple comparison tests showed significant differences between BAMI and the other individual OOSM methods at the 5% level.

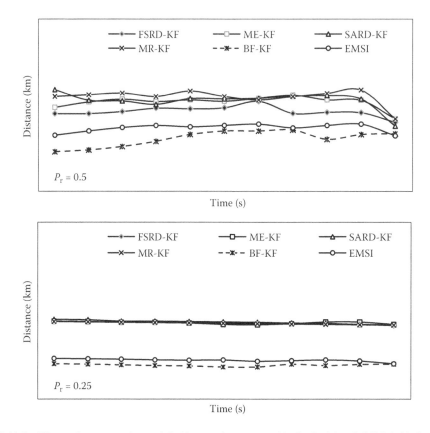

FIGURE 11.5 The performance of a straight line motion target with single delay OOSM (with $P_r = 0.5$ and 0.25 and manoeuvring index = 1).

Once again, no interaction effects were found to be statistically significant at the 5% level. From these experimental results, we have the following observations:

1. In case of non-manoeuvring tracking, all the three methods significantly reduce accuracy at all time and distance levels. Otherwise, all the methods show a very good fit when no measurements are delayed. In fact, at lower distance levels (between 20 and 90 s), BAMI and BF-KF compare favourably. Overall, BAMI achieves the highest accuracy rates as a method for handling delayed observations, followed by BF-KF and FSRD-KF, respectively (Figure 11.7).

TABLE 11.1

Comparison of OOSM and Single Imputation Methods (Single Delay)

	Methods					
P_r	FSRD-KF	ME-KF	SARD-KF	MR-KF	BF-KF	EMSI
0	2.67	2.78	3.71	2.01	6.54	6.78
0.25	3.15	3.40	4.43	2.57	6.54	6.78
0.5	3.74	3.97	4.78	2.64	6.55	6.78

Note: Computational times in minutes.

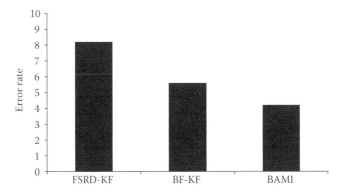

FIGURE 11.6 The results of OOSM and multiple imputation methods (multiple delays)—overall means.

2. In case of manoeuvring tracking, there appears to be no difference in performance between BAMI and BF-KF, especially when the probability of delay increases. For lower probabilities, the difference in performance is quite prominent as shown in Figure 11.8. Nonetheless, BAMI still outperforms FSRD-KF in terms of RMS error.

3. The computational cost of BAMI is about three times that of FSRD-KF and almost one-and-a-half times that of BF-KF (Table 11.2).

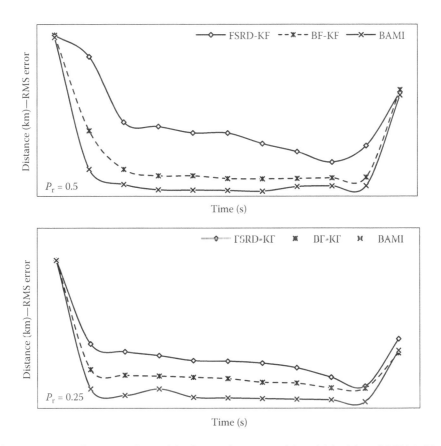

FIGURE 11.7 The performance of a straight line motion target with multiple delays OOSM (with $P_r = 0.5$ and 0.25).

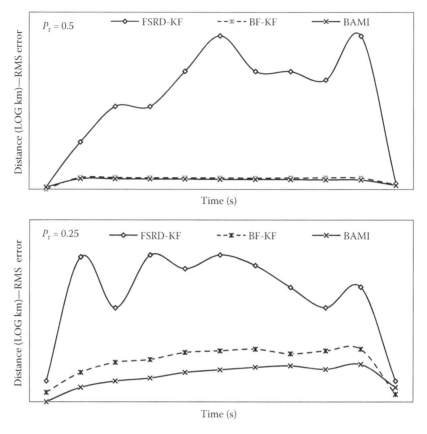

FIGURE 11.8 The performance in the case of highly manoeuvring with multiple delays OOSM (with $P_r = 0.5$ and 0.25).

TABLE 11.2
Comparison of OOSM and Multiple Imputation Methods (Multiple Delay)

	Methods		
P_r	FSRD-KF	BF-KF	BAMI
0	3.19	8.14	10.67
0.25	4.43	8.14	10.67
0.5	5.99	8.14	10.67

Note: Computational times in minutes.

11.5 DISCUSSION ON OOSM EXPERIMENTS AND CONCLUSIONS

The main contribution of the chapter is the use of simulation experiments to demonstrate the effectiveness of OOSM algorithms to handle delayed measurements. The referred techniques are well known but the extensive empirical evaluation of these methods is an original contribution. Furthermore, imputation procedures are not of widespread use in sensor data fusion, so showing the possibility of using the techniques on handling OOSM data is another contribution of this chapter for robotics learning.

The empirical data analysis study is based on simulated data, and the results suggest that imputation strategies can be successfully applied to deal with delayed measurements. Based on preliminary evidence, it has been found that EMSI performs comparable with BF-KF for single delay measurements data while BAMI achieves higher accuracy rates for multiple delayed measurements. The good performance of BAMI could be attributed to its variance averaging benefit even if it came at a high computational cost. The results further show the impact on the performance of methods is caused by the probability of measurement delays. Bigger positional error rates were achieved by methods for high-probability delays with bigger performance differences among methods. Also, given that the performance of each method varies by probability of measurement delay, it appears that the treatment of delayed measurements not only depends heavily on the probability of measurement delay but also on the range of manoeuvring target tracking. The worst performance achieved by these methods is for non-manoeuvring target tracking. This was rather a surprising result, which is not in accordance with statistical theory which considers missing completely at random (MCAR) as easier to deal with and IM data as very difficult to handle [46]. Potential limitations of the experimental studies include the use of simulated data, which could have involuntarily introduced biases, especially if those measurements considered as being delayed contained important information and they were not delayed as we assumed. However, the experimental results were carefully validated. For example, the experiments were conducted under the supervision of a domain expert who had a deep understanding for sensor data. This was a time-consuming exercise on ourselves and the expert.

The aspect of deciding to apply an imputation strategy to a given sensor dataset (with the fact that there are delayed measurements) must be considered. For the work described here, the data were simulated. Unfortunately, this type of information is rarely known for most 'real-world' applications. In some situations, it may be possible to use domain knowledge to determine the mechanism generating the delayed measurements. For situations where this knowledge is not available, the conservative nature of the consensus dictates that the measurements will be delayed randomly.

The two methods of OOSM and imputation were applied on only one dataset. This work could be extended by considering a more detailed simulation study using much more balanced additional types of datasets or even smaller datasets required to understand the merits of imputation. In addition, using as many datasets as possible in our comparative simulation study would enable a more sound generalisation of our results. This work could also be extended to a comparative evaluation of datasets with artificially simulated missingness against original datasets. The above issues should be investigated in the future. In summary, this chapter provides the beginnings of a better understanding of the relative strengths and weaknesses of model-based imputation strategies to handle delayed measurements. It is hoped that it will motivate future theoretical and empirical investigations into incomplete data and software prediction, and perhaps reassure those who are uneasy regarding the use of imputed data in software prediction.

ACKNOWLEDGEMENTS

This work was funded by the Department of Electrical and Electronic Engineering Science at the University of Johannesburg. The author would like to thank his colleagues at the University of Johannesburg for their helpful discussion and the anonymous reviewers for their useful comments.

REFERENCES

1. Kalman, R.E. 1960. A new approach to linear filtering and prediction problems, *Transaction of the ASME—Journal of Basic Engineering*, 82(D): 33–45.
2. Allison, P.D. 2001. *Missing Data*. Sage, Thousand Oaks, CA, USA.
3. Bar-Shalom, Y. 2000. Update with out-of-sequence measurements in tracking: Exact solution. *Proceedings of the SPIE Conference on Signal and Data Processing of Small Targets*, Orlando, FL, USA, pp. 51–556.

4. Wang, H., Kirubarajan, T., Li, Y. and Bar-Shalom, Y. 1999. Precision large scale air traffic surveillance using and IMM estimator with assignment, *IEEE Transactions in Aerospace, Electronic Systems, AES*, 35(91): 255–266.

5. Mallick, M., Coraluppi, S. and Carthel, C. 2001. Advances in asynchronous and decentralized estimation. *Proceedings of the 2001 IEEE Aerospace Conference*, Big Sky, MT, USA.

6. Mallick, M. and Marrs, A. 2002. Comparison of the KF and particle filter based out-of-sequence measurement filtering algorithms. *Proceedings of the 6th International Conference on Information Fusion*, July 8–10, Cairns, Australia.

7. Mallick, M., Krant, J. and Bar-Shalom, Y. 2002. Multi-sensor multi-target tracking using out-of-sequence measurements. *Proceedings of the 5th International Conference on Information Fusion*, August 8–11, Annapolis, MD, USA.

8. Mallick, M., Zhang, K. and Li, X.R. 2003. Comparative analysis of multiple-lag out-of-sequence measurement filtering algorithms. *Proceedings Signal and Data Processing of Small Targets*, San Diego, CA, USA.

9. Challa, S., Evans, R.H. and Wang, X. 2003. A Bayesian solution and its approximation to out-of-sequence measurement problems, *Information Fusion*, 4:185–199.

10. Bar-Shalom, Y., Chen, H. and Mallick, M. 2004. One-step solution for the multistep out-of-sequence measurements problem in tracking, *IEEE Transactions on Aerospace and Electronics Systems*, 40:27–37.

11. Bar-Shalom, Y., Chen, H., Mallick, M. and Washburn, R. 2002. One-step solution for the general out-of-sequence measurement problems in tracking. *Proceedings 2002 IEEE Aerospace Conference*, Big Sky, MT, USA.

12. Bar-Shalom, Y., Li, X.R. and Kirubarajan, T. 2001. *Estimation with Applications to Tracking and Navigation*. Wiley & Sons, New York, USA.

13. Bar-Shalom, Y. and Li, X-R. 1993. *Estimation and Tracking: Principles, Techniques and Software*. Artech House, MA, USA.

14. Blackman, S.S. and Ropoli, R. 1999. *Design and Analysis of Modern Tracking Systems*. Artech House, MA, USA.

15. Marcus, G.D. 1979. *Tracking with Measurements of Uncertain Origin and Random Arrival Times*, MS thesis, Department of Electrical Engineering and Computer Science, University of Connecticut, Storrs, USA.

16. Hilton, R.D., Martin, D.A. and Blair, W.D. 1993. Tracking with Time-Delayed Data in Multisensor Systems, NSWCDD/TR-93/351, Dahlgren, VA.

17. Thomopoulos, S.C.A. and Zhang, L. 1994. Decentralized filtering with random sampling and delay, *Information Sciences*, 81:117–131.

18. Ravn, O., Larson, T.D., Andersen, N.A. and Poulsen, N.K. 1998. Incorporation of time delayed measurements in a discrete time Kalman filter. *Conference on Decision and Control*, Tampa, FL, USA, *Proceedings of the 37th IEEE, National Bureau of Standards*, pp. 3972–3977.

19. Ray, A. 1994. Output feedback control under randomly varying distributed delays, *Journal of Guidance, Control Dynamics*, 17(4):701–711.

20. Lewis, R. 1986. *Optimal Estimation with an Introduction to Stochastic Control Theory*. John Wiley & Sons, Inc., New York, USA.

21. Challa, S., Evans, R. and Wang, X. 2001. Target tracking in clutter using time delayed out-of-sequence measurements. *Proceedings of Defence Applications of Signal Processing (DASP)*, Adelaide, Australia.

22. Tasoulis, D.K., Adams, N.M. and Hand, D.J. 2009. Selective fusions of out-of-sequence measurements, *Information Fusion*, 11(2):183–191.

23. Kwakernaak, H. 1967. Optimal filtering in linear systems with time delays, *IEEE Transactions on Automatic Control*, 12:169–173.

24. Richard, J.P. 2003. Time delay systems: A review of some recent advances and open problems, *Automatica*, 39:1667–1694.

25. Mallick, M. and Bar-Shalom, Y. 2002. Non-linear out-of-sequence measurement filtering with applications to GMTI tracking. *Proceedings of SPIE Conference Signal and Data Processing of Small Targets*, Orlando, FL, USA.

26. Zhang, H., Zhang, D. and Xie, L. 2003. Necessary and sufficient condition for finite horizon H_∞ estimation of time delay systems. *Proceedings of the 42nd IEEE Conference on Decision and Control*, 9–12 December 2003, Maui, Hawaii, USA, Vol. 6, pp. 5735–5740.

27. Zhang, H., Zhang, D. and Xie, L. 2003. An innovation approach to H_∞ prediction for continuous-time systems with application to systems with time delayed measurements, *Automatica*, 40:1253–1261.

28. Anderson, B.D.O. and Moore, J.B. 1979. *Optimal Filtering.* Prentice Hall, Englewood Cliffs, New Jersey, USA.

29. Hsiao, F-H. and Pan, S-T. 1996. Robust Kalman filter synthesis for uncertain multiple time-delay stochastic systems, *Journal of Dynamic Systems, Measurement, and Control,* 118(4): 803–808.

30. Kaszkurewicz, E. and Bhaya, A. 1996. Discrete time state estimation with two counters and measurement delay. *Proceedings of the 35th IEEE Conference on Decision and Control,* Kobe, Japan.

31. Matveev, A. and Savkin, A. 2003. The problem of state estimation via asynchronous communication channels with irregular transmission times, *IEEE Transactions on Automatic Control,* 48(4):670–676.

32. Larsen, T., Poulsen, N., Anderson, N. and Ravino, O. 1998. Incorporating of time delayed measurements in a discrete-time Kalman filter. *CDC'98,* Tampa, FL, USA.

33. Alexander, H.L. 1991. State estimation for distributed systems with sensing delay, In V. Libby (Ed.) *SPIE, Data Structures and Target Classification,* 1470:103–111.

34. Zhang, K., Li, X.R. and Zhu, Y. 2005. Optimal update with out-of-sequence measurements, *IEEE Transactions on Signal Processing,* 53(6):1992–2004.

35. Kailath, T., 1970. The innovations approach to detection and estimation theory. *Proceedings of the IEEE,* 58:680–695.

36. Maybeck, P.S. 1979. *Stochastic Models, Estimation and Control.* Vol. 1, Academic Press, New York, USA.

37. Orton, M. and Marrs, A.D. 2001. A bayesian approach to multi-target tracking and data fusion with out-of-sequence measurements. *IEEE International Workshop on Target Tracking Algorithms and Applications,* Enschede, the Netherlands.

38. Geld, A. 1974. *Applied Optimal Estimation.* The MIT Press, Cambridge, Massachusetts, USA.

39. Bar-Shalom, Y. and Li, X.R. 1995. *Multitarget-Multisensor Tracking: Principles and Techniques.* YBS Publishing, Connecticut, USA.

40. Julier, S.J. and Uhlmann, J.K. 2005. Fusion of time delayed measurements with uncertain time delays. *Proceedings of the American Control Conference 2005,* pp. 4028–4033.

41. Rhéaume, F. and Benaskeur, A. 2007. Out-of sequence measurements filtering using forward prediction. *Technical Report TR 2005-484,* Defence R & D Canada—Valcartier.

42. Lu, X., Zhang, H.S., Wang, W. and Teo, K.L. 2005. Kalman filtering for multiple time-delay systems, *Automatica,* 41(8):1455–1461.

43. Bar-Shalom, Y. and Fortmann, T.E. 1988. *Tracking and Data Association.* Academic Press, New York, USA.

44. Rubin, D.B. 1987. *Multiple Imputation for Nonresponse in Surveys.* John Wiley and Sons, New York, USA.

45. Rubin, D.B. 1996. Multiple imputation after 18+ years, *Journal of the American Statistical Association,* 91:473–489.

46. Schafer, J.L. 1997. *Analysis of Incomplete Multivariate Data.* Chapman and Hall, London.

47. Breiman, L., Friedman, J.H., Olshen, R.A. and Stone, C.J. 1984. *Classification and Regression Trees.* Chapman and Hall Inc., New York, USA.

48. Tanner, M.A. and Wong, W.H. 1987. The calculation of posterior distributions by data augmentation (with discussion), *Journal of the American Statistical Association,* 82:528–550.

49. Gilks, W.R., Richardson, S. and Spiegelhalter D.J. 1996. *Markov Chain Monte Carlo in Practice.* Chapman and Hall, London.

50. Dempster, A.P., Laird, N.M. and Rubin, D.B. 1977. Maximum likelihood estimation from incomplete data via the EM algorithm, *Journal of the Royal Statistical Society,* Series B, 39:1–38.

51. Allison, P.D. 2001. *Missing Data.* Sage, Thousand Oaks, CA.

52. Challa, S., Evans, R.J., Wang, X. and Legg, J. 2002. A fixed lag smoothing solution to out-of-sequence information fusion problems, *Communications in Information and Systems,* 2(4):327–350.

53. MINITAB. 2002. *MINITAB Statistical Software for Windows 9.0.* MINITAB, Inc., PA, USA.

54. Kirk, R.E. 1982. *Experimental Design* (2nd Ed). Brooks, Cole Publishing Company, Monterey, CA, USA.

12 Robotic Path-Planning in Dynamic and Uncertain Environment Using Genetic Algorithm

G. K. Singh and Ajith K. Gopal

CONTENTS

12.1 INTRODUCTION

Robotic navigation encompasses (1) motion-planning which includes dynamical modelling and (2) path-planning which restricts itself to spatial and geometrical modelling. Motion-planning is used mainly in real-time guidance applications and deals with generating a feedback control law to provide torques to manipulator joints and/or drive wheels, in order to track a supplied reference trajectory. The obstacle avoidance is primarily based on a 'sense and avoid' philosophy, utilizing the on-board sensors. On the other hand, path-planning finds application mainly in high-level off-line navigation tasks. The output of path-planning exercise is a feasible/optimal trajectory or path through the associated space of possible configurations of the robot and the known obstacles in the working area, from a given initial position to the desired final position of the robot. Genetic algorithms [1] are a powerful tool based on models of natural selection and evolution and facilitate an exhaustive search over large discontinuous spaces (Chapter 1).

The path-planning problem for robotic systems deals with finding a feasible path from a given starting point to a specified destination avoiding the known obstacles. Additionally, the path is also required to be optimal in terms of either minimum distance/time or fuel spent. These problems are normally posed as numerical optimization problems. Various optimization algorithms have been utilized for this purpose. Castillo and Trujillo [2] used multiple-objective genetic algorithms (MOGA) for the problem of off-line point-to-point path planning on a flat two-dimensional (2-D) terrain. The objectives minimized are the length of the path and the degree of difficulty. The motion-planning problem for the 'Khepera' robot is solved using GAs [3]. Most of these works consider only static obstacles in the domain of interest. In many practical problems, the obstacles are known to be moving dynamically, for example, in the case of multiple-robotic manipulators on an assembly line, the other robots which act as obstacles would have a defined dynamic behaviour. The presence of a moving obstacle introduces the 'time' dimension into the path-planning problem, which could otherwise be solved for in the configuration space only.

Tychonievich et al. [4] applied the manoeuvre-board approach for path-planning in the presence of moving circular obstacles. Fiorini and Shiller [5] proposed a 'velocity obstacle' concept, based on which local avoidance manoeuvres could be computed. A sequence of such avoidance manoeuvres computed at discrete time intervals then defines the 'trajectory'. Vadakkepat et al. [6] consider the case of moving goals and obstacles. The independent variables are used to define the individual potential fields for the obstacles and the goal and the robot motion is assumed along the resultant potential field. These works deal with the motion-planning and are based on the 'sense-and-avoid' principle. Hence, the paths generated by these are locally optimal and need not be globally optimal. This approach is obviously essential, for example, in the case of an autonomous robot navigating in an uncertain road traffic situation. In many cases, the path tracked by the other entity (obstacle) is known *a priori*, within some uncertainty while they are being executed. van den Berg and Overmars [7] start with the assumption that the only information about the obstacles is the maximum velocity with no information on the direction. This, however, is quite a pessimistic approach. Guibas et al. [8] assumed polygonal obstacles with uncertainty on the position of the vertices and used a probabilistic path-planning approach to reduce the collision-risk probability while navigating to the 'target'. The autonomous robot path-planning using a GA was also considered in Ref. [9], whereas Walther et al. [10] considered B-Splines for mobile path representation and motion control. Singh and Gopal [11] used genetic algorithms to address the robotic path-planning problem in the presence of moving elliptical obstacles with a bounded uncertainty in the position and orientation of the obstacles. The nominal movement trajectory of the obstacles in time was assumed to be known, albeit with some bounded uncertainty. A new analytical result for computing the axes-aligned bounding box for the ellipses with bounded uncertainty was presented. In most of the above references, the robot is assumed to be a 'point', since the dimensions of the robot can always be incorporated as an appropriate increase in the obstacle size. Finally, a brief discussion on motion-planning using genetic algorithm is given.

12.2   PATH REPRESENTATION IN 2-D

The first step in solving the path-planning problem is to have a parametric representation of a path from designated 'start' to 'stop' positions within a specified domain. This section is restricted to motion in a 2-D plane but the methods discussed here can be easily extended to higher dimensions. The domain over which a robot can move is typically restricted, without loss of generality to $[0,1] \times [0,1]$. Further, the starting point and destination represented as (x_0, y_0) and (x_d, y_d) are assumed to be (0,0) and (1,1), respectively. Singh and Gopal [11] discussed about some ways of representing a path in a 2-D plane in the available literature. In summary, the paths could be represented as a vector of $(\Delta x, \Delta y)_k$ increments, or in the polar form as a vector of $(R, \theta)_k$ values (e.g., Refs. [2] and [9], respectively). In these representations, it becomes difficult to ensure that a path will stay within the domain of interest and/or reach the specified goal location. Hence, Castillo and Trujillo [2] require a 'path-repair' mechanism to ensure that infeasible path can be converted into a feasible one. Tian and Collins [12] considered the trajectory planning for a two-joint two-link robot moving in a 2-D plane. It was modelled as a 2-dof manipulator, with the two angular rotations used for the path parameterization. They represented the time histories of the trajectory in the task space by a polynomial based on Hermite cubic interpolation. In this case, the joint kinematics ensure that the trajectories remain in the feasible domain. However, the issue with using angles as the parameterizing parameter is that the angles wrap around 360°. This can be resolved by using the normal out-vector (which will have two components in the 2-D plane (x_n, y_n)) for defining the position instead of the angular position. Thus, it is observed that the parameters for path representation should be carefully selected and it should be ensured that all possible paths could be adequately represented. Alternatively, B-Splines offer an attractive way of representing a path as a composition of piece-wise continuous segments defined locally [1,10,11]. Further, it is known that B-Spline fits can be generated to meet any specific smoothness requirements. The spline representation can be easily bounded to ensure that trajectories stay within the domain and also reach the target location. Hence, the B-Spline representation is used in this work. A path is represented as a linear combination of a set of local basis functions defined over some experimental time interval. For example, Figure 12.1 shows the linear and quadratic basis functions with knots located at $[0, t_1, t_2, \ldots, t_{expt}] = [0, 1, 2, \ldots, 10]$. Let B_J^k denote a kth-order basis function starting at the value J. It should be noted that the basis functions B_J^1 has non-zero values, that is, spans over 2 knots while B_J^2 spans over 3 knots; thus, there are $(N - k - 1)$ basis functions over the domain of interest. A path/trajectory $x_k(t)$ over time is then easily expressed as

$$x_k(t) = \sum_{J=0}^{J=N-k-1} a_J B_J^k(t) \tag{12.1}$$

where N is the number of knot locations and a_J are free parameters forming a linear combination of the basis functions, to be selected based as required by the user. It is also obvious that a linear combination of B_J^1 would result in C^0 continuous trajectories while B_J^2 would result in C^1 trajectories, that is, with possible discontinuities in acceleration. Higher order continuities, if required, can be implemented by using higher order B-Splines.

Note that the value of trajectory at the last knot location will always be $x_k(t_{expt}) = 0$, by definition. Thus, the knot locations should be appropriately selected to have t_{goal} less than t_{expt}. In the case of kth-order splines, at any given time instant $t_0 \geq t_J$, basis functions from only the previous $(k + 1)$ knots can have non-zero values. For the first-order spline, this implies that

$$x_k(t = t_0) = a_{J-1} B_{J-1}^1(t_0) + a_J B_J^1(t_0) \tag{12.2}$$

Since the free parameters a_J are seen to have a local effect on the curve, it is easy to enforce the 'goal-reaching' condition by an appropriate choice of the Jth control parameter, based on the previous

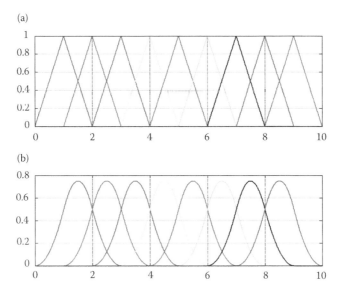

FIGURE 12.1   B-Spline basis functions of the first and second order. (a) Linear B-spline basis functions and (b) quadratic B-spline basis functions.

$(J-1)$th parameter. For the first-order spline we can only fix the initial and final positions. A second-order spline can be used to fix initial and final velocity and positions. The requirement for the curve to remain within the domain of interest $[0,1] \times [0,1]$ can be easily met by constraining the free parameters to lie within $[0,1)$. Using the same set of basis functions for the x and y trajectories over time ensures that both of them reach the goal $(1,1)$ at the same time. Thus, the set of possible trajectories in a 2-D plane is easily parameterized using $2*([(N-k-1)]$ free parameters constrained as above.

12.3   REPRESENTATION OF DYNAMIC OBSTACLES

Obstacles of various 2-D geometries have been considered in the literature. These include the polygonal (e.g., Ref. [8]) and circular obstacles among others. Circular obstacles have been widely used, since they are easy to represent in the 2-D plane, and also that the rotation of the obstacle becomes irrelevant; however, they offer a conservative representation if not used properly. For example, a polygonal obstacle could possibly be adequately represented as a collection of a number of circular obstacles [5]. In this chapter, we consider the class of elliptical obstacles since their rotation assumes significance. An ellipse in a 2-D plane is defined by its centre, (x_c, y_c), its semi-major and minor axis, (a, b), respectively and the rotation (θ) from a reference axis (X). Unlike van den Berg and Overmars [7], it is assumed that a nominal time-trajectory of the obstacle centre: both translation $[x_N(t), y_N(t)]$ and rotation $[\theta_N(t)]$ are known. It is assumed that the physical dimension of the obstacle specified by (a, b) is specified constants. Further, it is assumed that there is a bounded uncertainty associated with (1) the position of the centre and (2) the rotation, over and above the given nominal trajectory. The uncertain terms are indicated by the subscript 'U'. In particular, the actual trajectory of the obstacle, including the perturbation from the 'nominal' terms, becomes

$$x_{aC}(t) = x_N(t) \pm x_U(t)$$
$$y_{aC}(t) = y_N(t) \pm y_U(t) \qquad (12.3)$$
$$\varphi(t) = \theta_N(t) \pm \theta_U(t)$$

where the bounds on the uncertainty are given as

$$|x_U(t)| = h.$$
$$|y_U(t)| = k. \qquad (12.4)$$
$$|\theta_U(t)| = \Theta.$$

for specified values of $[h, k, \Theta]$.

12.4   OBSTACLES WITH AXIS-ALIGNED BOUNDING BOXES

With the time trajectory of obstacles as defined above, a bounding box for the set of uncertain ellipses is found as discussed below. The bounding rectangle is also parameterized by the same set of five parameters. The centre and rotation of the box are assumed to be the same as the nominal trajectory, that is, $[x_N(t), y_N(t), \theta_N(t)]$. The major and minor dimensions of the bounding rectangle (a_{RB}, b_{RB}) need to be computed for the obstacle for each time instant. It should be noted that the translation of the ellipse centre from the origin does not affect the dimension of the bounding box; hence, the following developments assume that the ellipse centre is at the origin $(0,0)$. At any time instant t, an elliptic obstacle centred at the origin can be easily represented in a parametric form as below. The time dependence has been omitted in the following for brevity:

$$x_e(p; [x_U, y_U, \phi]) = x_U + a*\cos(p)*\cos(\phi) - b*\sin(p)*\sin(\phi)$$
$$y_e(p; [x_U, y_U, \phi]) = y_U + b*\sin(p)*\cos(\phi) + a*\cos(p)*\sin(\phi) \qquad (12.5)$$

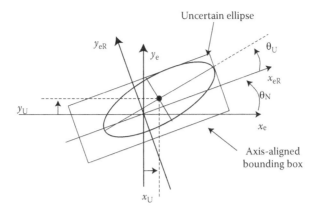

FIGURE 12.2   Geometry of the axis-aligned bounding box.

Note that p is a running variable from $[-\pi$ to $+\pi]$, and the angle $\phi = \theta_N + \theta_U$, where θ_N is the nominal rotation angle and θ_U represents the uncertainty. A schematic showing the geometry for the following developments is given in Figure 12.2. The bounding box is axis aligned, that is, its centre and rotation is specified as $[0, 0, \theta_N]$. The distance from any point of the ellipse (defined above) to the rotated axes (x_{eR}, y_{eR}) can be shown to be

$$x_{eR}(p;[x_U, y_U, \phi]) = x_e(p;[x_U, y_U, \phi]) * \cos(\theta_N) + y_e(p;[x_U, y_U, \phi]) * \sin(\theta_N)$$
$$y_{eR}(p;[x_U, y_U, \phi]) = -x_e(p;[x_U, y_U, \phi]) * \sin(\theta_N) + y_e(p;[x_U, y_U, \phi]) * \cos(\theta_N)$$
(12.6)

The dimensions of the bounding box are computed as follows:

$$a_{RB} = \max\{x_{eR}(p;[x_U, y_U, \phi])\} \text{ over } \{[-\pi \text{ to } +\pi]; [h, k, \theta_N \pm \Theta]\};$$
$$b_{RB} = \max\{y_{eR}(p;[x_U, y_U, \phi])\} \text{ over } \{[-\pi \text{ to } +\pi]; [h, k, \theta_N \pm \Theta]\};$$
(12.7)

The $x_{eR}(\cdot)$ and $y_{eR}(\cdot)$ in the earlier equation can be simplified to obtain

$$x_{eR}(p;[\cdot]) = x_U * \cos(\theta_N) + y_U * \sin(\theta_N) + a * \cos(p) * \cos(\theta_U) + b * \sin(p) * \sin(\theta_U)$$
$$y_{eR}(p;[\cdot]) = -x_U * \sin(\theta_N) + y_U * \cos(\theta_N) + a * \cos(p) * \sin(\theta_U) + b * \sin(p) * \cos(\theta_U)$$
(12.8)

Thus, it is seen from the above expansion of $x_{eR}(\cdot)$ and $y_{eR}(\cdot)$ that they can be partitioned into two terms, the first of which is seen to depend on the translational uncertainty of the centre (x_U, y_U) and the nominal rotation angle (θ_N), while the second term is a function of the rotational uncertainty (θ_U) only. It is well known that for two functions f_1 and f_2,

$$\max(f_1 + f_2) \le \max(f_1) + \max(f_2)$$
(12.9)

Thus, the maximum values of $x_{eR}(\cdot)$ and $y_{eR}(\cdot)$ can be computed using the maximum values of their component functions. It can be shown [11] that the dimensions of the bounding rectangle (a_{RB}, b_{RB}) can be obtained as

$$a_{RB} = d_{hk} * \max(|\cos(\Gamma_{12} - \theta_N)|) + \max\{a, \text{sqrt}\sqrt{[a^2 * \cos^2(\Theta) + b^2 * \sin^2(T)]}\},$$
$$b_{RB} = d_{hk} * \max(|\sin(\Gamma_{12} - \theta_N)|) + \max\{b, \text{sqrt}\sqrt{[a^2 * \sin^2(\Theta) + b^2 * \cos^2(\Theta)]}\}.$$
(12.10)

where

$$d_{hk} = \sqrt{(h^2 + k^2)}, \quad \text{and}$$
$$\Gamma_{12} = [\tan^{-1}(k/h),\ \tan^{-1}(-k/h)].$$

(12.11)

Note that Γ_{12} represents two values as above, and that the d_{hk} *mac($\cdot$) in the first terms needs to be evaluated over the two possible values as

$$d_{hk}*\max(|\cos(\Gamma_{12} - \theta_N)|) = d_{hk}*\max([|\cos(\Gamma_1 - \theta_N)|],\ [|\cos(\Gamma_2 - \theta_N)|]) \qquad (12.12)$$

The analytical result above can be used to find the bounding rectangles over time for each obstacle. The path-planning problem then can be solved treating these bounding boxes as the obstacles, instead of the elliptic obstacles.

12.5   IMPLEMENTATION OF PATH-PLANNING WITH GENETIC ALGORITHM

The example problem below uses first-order B-Splines with knots located at time instants [0, 2, 4, 6, 8, 10]. Thus, there are $N = 4$ basis spline functions and corresponding four free parameters $[a_1, a_2, a_3, a_4]$. The time taken to reach the goal was selected as $t_{goal} = 8$. The 'goal-reaching' condition in this case simplifies to $a_4 = 1$. Thus, for the current problem we are left with $(N - 1) = 3$ independent parameters, which should be constrained to lie in [0,1). A set of $[a_1, a_2, a_3]$ defines the x or y trajectories in time. The same spline configuration is used for both the x and y trajectories with independent parameters $\{\mathbf{X},\ \mathbf{Y} \in R^3\}$, and hence the path-planning problem requires a search over a six-dimensional space. A possible x–y path is thus represented as $P(\mathbf{X}, \mathbf{Y})$. A set of obstacles is assumed given for the problem as in Ref. [11]. The 'obstacle avoidance' is checked for each (x, y) point of the trajectory over time. If a particular point lies within any of the bounding rectangles, it implies an intersection with the obstacle. An obstacle function has been defined which gives a positive real number in case of a collision, and negative otherwise. The collision function (CF) is defined as the number of times the obstacle function is 'positive' along a path. It is also desired to have a minimum path length (PL). The velocity of the point object is computed as $v_{body} = \sqrt{v_x^2 + v_y^2}$. A velocity constraint is also introduced to ensure that $v_{body} \leq 0.5$. The velocity constraint function (VCF) is defined as the number of times the above constraint is violated along a given path. Additionally, a minimum norm $|[\mathbf{X}, \mathbf{Y}]|$ solution is being searched for. Thus, the minimization problem can be stated as follows.

Find $([\mathbf{X}, \mathbf{Y}])$ that minimizes

$$PL(P(\mathbf{X},\mathbf{Y})) + CF(P(\mathbf{X},\mathbf{Y})) + VCF(P(\mathbf{X},\mathbf{Y})) + |[\mathbf{X},\mathbf{Y}]| \qquad (12.13)$$

subject to

$$\mathbf{X}, \mathbf{Y} \in \left[0,1\right).$$

Genetic algorithms offer an effective and powerful method of solving such optimization problems over a large search spaces with no specific gradient or continuity requirements [12]. It employs a population-based search method, and hence has a good chance of overcoming the local minima problem faced in conventional search techniques. The search is implemented using genetic operators like cross-over, selection, and mutation on individuals, each of which are represented as a chromosome of independent variables. The initial population for the first generation is created using a

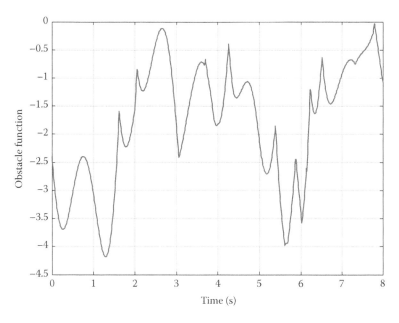

FIGURE 12.3   Obstacle function trajectory.

random number generator. Every individual in a population represents a possible 2-D path for which the objective function is evaluated. The members of the next generation are obtained by applying the genetic operations to the previous generation so as to improve the overall fitness of the population. The current work uses the 'Genetic Algorithm and Direct Search Toolbox' available with MATLAB®.

12.5.1 Numerical Simulation Example

As discussed earlier, an elliptic obstacle (E) can be described by the set of five parameters: position of the centre, the semi-major and minor axes, and the rotation from a reference axis as $E: = \{x_C, y_C, a, b, \theta\}$. The nominal parameters for five elliptic obstacles are defined as [11]

E1: $\{0.55 + \sin(t)/20,$ $0.45 + \sin(t)/20,$ $0.12, 0.06,$ $1.5\sin(1.5t)\}$
E2: $\{0.20 + \sin(2t)/20,$ $0.15 + \sin(t)/20,$ $0.04, 0.12,$ $1.5\sin(2t)\}$
E3: $\{0.80 + \sin(t)/20,$ $0.80 + \sin(t)/20,$ $0.04, 0.12,$ $\pi/4 + 1.5\sin(2.5t)\}$
E4: $\{0.20 + \sin(2t)/20,$ $0.80 + \sin(t)/20,$ $0.12, 0.04,$ $\pi/2 + 1.5\sin(1.25t)\}$
E5: $\{0.80 + \sin(2t)/20,$ $0.20 + \sin(t)/20,$ $0.12, 0.04,$ $\pi/3 + \sin(3t)\}$

A GA optimization call with 80 as the population size is used to obtain an optimal path. The body velocities were found to be within limits. The obstacle avoidance of the optimum path is shown in the obstacle function plot in Figure 12.3. A snapshot of the workspace at $t = 2.64$s is shown in Figure 12.4. The black line depicts the found optimum path. The black circle indicates the position of the robot at the current time instant. Also shown in Figure 12.4 are the uncertain elliptic obstacles and their bounding boxes (rectangles). It can be seen that two obstacles have actually coalesced at this point in time.

12.6   MOTION-PLANNING USING GA

As we have seen in Chapter 1, GAs have been found to be very effective methods for solving multi-criteria/multi-objective optimization problems in science and engineering. The GAs mimic models of natural evolution and possess an ability to adaptively search large spaces in near-optimal, often

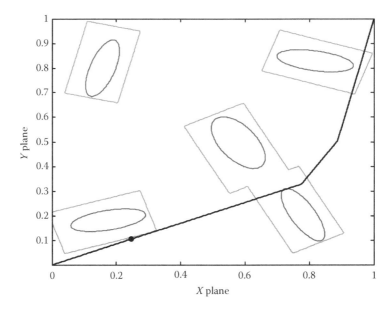

FIGURE 12.4   A snapshot of the workspace at $t = 2.64$ s.

global optimal, ways. One important application of the GA is in the area of evolutionary robotics. In this field, GA is used for designing behavioural controllers for robots and autonomous agents for autonomous mobile vehicles. Especially, GA can be used for path-planning (PPGA) that proposes the evolution of a chromosome attitudes structure to control a simulated mobile robot [13]. These attitudes specify the basic robot actions to reach a goal point. The PPGA performs straight motions and avoids obstacles. The fitness function employed to teach robot's movements is designed to achieve this type of behaviour in spite of any changes in the robot's goals and environment aspects. This learning process in this PPGA is not dependent on the obstacle environment distribution. Also, it does not depend on the initial and final points. Hence, this PPGA-based controller is not required to be retrained when these parameters change.

Due to its very interesting and promising value, a particular robot's simulator is described in some details here [13]. In the particular simulated mobile robot system, there are eight sensors which allow detection of the proximity of objects at its (robot's) four sides: front, back, left or right. For example, each sensor returns an integer value between 0 and 1023, the value of 0 meaning that there is no obstacle, whereas the value of say, 1023 means that an obstacle is very close to the sensor, that is, close to the robot itself [13]. The intermediate numbers can be taken to suggest intermediate distances between the sensor and the obstacle. The robot has two motors and each is able to achieve a speed integer value ranging from, say −10 to +10. The robot's simulator would also allow the readings of the position (x, y) and the direction angle (a) of the robot, the x and y coordinates ranging from 0 to 1000 and the direction angle (a) ranging from $-\pi$ to π degrees. When the robot is looking to the right, the direction angle is considered as 0 degrees. The PPGA-based robot, by utilizing the input information provided by the sensors and the distance to the arrival point, moves to the goal point without touching any obstacle. This approach is much different from the graph-based and other classical PMP algorithms (Chapter 3), where in addition to the description of the graph-based algorithms, the so-called heuristic methods are described, of which GA is one of them. However, GA is also considered to be a part of the soft computing paradigm. The evolutionary procedure would consist in programming a GA system to evolve the robot's controller. In PPGA, each chromosome represents a group of basic attitudes, which define the robot's movements in agreement with the feedback provided by the robot's environment. This feedback is used as input to the system and is obtained from the sensors readings and on the robot's direction to its goal point. The sensor system can be simplified to detect obstacles at the robot's left, right and back sides. The other inputs to this robot simulator system are the robot's direction to its goal location which is calculated as: consider a as the direction's angle of the robot, provided by the simulator at any location of the robot; define b as the angle of minimum distance between robot's position and its fixed arrival position; then the difference between b and a indicates whether the goal point is in the front of the robot, behind it or to its left or right sides; this then determines the robot's target direction. By using the sensors' readings and the robot's direction inputs, the attitude of the robot can be defined using the following rule [13]:

```
If ((Sleft > L) or (Sright > L) or (Sback > L));
Then
an obstacle is detected
the proximity-sensor = highest value (Sleft, Sright, Sback)
Else
An obstacle is not detected
target direction = b − a
END
```

In the above states rule, the constant L represents a collision threshold and may be determined between the sensor's range [0, 1023]. Each possible attitude defined by the states rule corresponds to a gene of the chromosome, and each gene is composed of the pair of the robot's motor speeds (M1, M2). The states attitude representation is divided into two parts: (i) the first part determines the basic robot's target direction movements when no obstacles are detected and (ii) the second part, the obstacle detected states, considers the motor speed actions that avoid collision, being independent from the goal's point location. Thus, if the robot's state is defined as collision free and its goal point is in its left side, then the robot's attitude to be carried out corresponds to a particular gene of the chromosome; on the other hand, if the robot detects the proximity of an object at its right-hand side, the robot would take the action indicated at the other gene without considering the target direction. As we know, GA requires a fitness function to monitor its progress to the global optimal. This fitness function F can be calculated at each robot's time step and could contain three components, say, V, D and A. The component V is maximized by speed of the robot, the component D by the straight motion and the component A by the robot's action. A depends on the attitude defined according to the states rule: (i) if the robot's state is collision free and the arrival point is at its front, then the GA should evolve the pair M1, M2 that minimizes this distance, (ii) if the robot is free but not oriented to the goal point, then the robot should find the pair M1, M2 that rotates its front in the direction of the target and (iii) if an object is located in the range of the robot's view, then the GA should forget the goal point during some steps and find the best way, a pair M1, M2 to avoid possible collision. Then the action component A is calculated by the mean of the total *number* of steps executed per attitude because of the random noise present in the data, especially in the motor speed and sensors' readings.

12.7 CONCLUDING REMARKS

An algorithm to compute the axes-aligned bounding box for the uncertain ellipses is presented in the context of robot path-planning. This has been applied to robotic path-planning problem in the presence of dynamically moving elliptic obstacles with velocity constraints in the 2-D plane. A solution to the bounded velocity shortest path problem while avoiding the moving obstacles has been obtained using genetic algorithm. The method is illustrated with a simulation result. Also, a brief description of a GA-based path-motion-planning algorithm that obtains a path solution based on the sensors' measurements and some well-defined fitness functions has been given.

REFERENCES

1. Kostaras A.N., Nikolos I.K., Tsourveloudis N.C. and Valvanis K.P. Evolutionary algorithm based on-line path planner for UAV navigation, *Proceedings of the 10th Mediterranean Conference on Control and Automation—MED2002*, Lisbon, Portugal, July 9–12, 2002.
2. Castillo O. and Trujillo L. Multiple objective optimization genetic algorithms for path-planning in autonomous mobile robots, *International Journal of Computers, Systems, and Signals*, 6(1), 2005, 48–63.
3. Thomaz C.E., Pacheco M.A.C. and Vellasco M.M.B.R. Mobile robot path planning using genetic algorithms, *Lecture Notes in Computer Science*, 1606, 1999, 671–679. (Proceedings of the International Work-Conference on Artificial and Natural Neural Networks: Foundations and Tools for Neural Modelling).
4. Tychonievich L., Zaret D., Mantegna J., Evans R., Muehle E. and Martin S. A maneuvering-board approach to path-planning with moving obstacles, *Eleventh International Joint Conference on Artificial Intelligence*, 2, 1989, 1017–1021.
5. Fiorini P. and Shiller Z. Motion planning in dynamic environments using velocity obstacles, *International Journal of Robotics Research*, 17(7), 1998, 760–772.
6. Vadakkepat P., Tan K.C. and Ming-Liang W. Evolutionary artificial potential fields and their application in real-time robot path-planning, *Proceedings of the 2000 Congress on Evolutionary Computation*, 1, 2000, 256–263.
7. van den Berg J. and Overmars M. Planning time-minimal safe paths amidst unpredictably moving obstacles, *International Journal of Robotics Research*, 27(11–12), 2008, 1274–1294.
8. Guibas L.J., Hsu D., Kurniawati H. and Rehman E. Bounded uncertainty roadmaps for path planning, *International Workshop on the Algorithmic Foundations of Robotics*, Guanajuato, Mexico, 2008.
9. Candido S. *Autonomous Robot Path Planning Using a Genetic Algorithm*, project report downloaded from http://www-cvr.ai.uiuc.edu/~scandido/pdf/GApathplan.pdf
10. Walther M., Steinhaus P. and Dillmann R. Using B-Splines for mobile path representation and motion control, *Proceedings of the 2nd European Conference on Mobile Robots (ECMR)*, Ancona, Italy, 2005.
11. Singh G.K. and Gopal A. Path planning in the presence of dynamically moving obstacles with uncertainty, In special issue on Mobile Intelligent Autonomous Systems of *Defence Science Journal*, 60(1), 2010, 55–60.
12. Tian L. and Collins C. An effective robot trajectory planning method using a genetic algorithm, *Mechatronics*, 14, 2004, 455–470.
13. Thomaz, C.E., Pacheco, M.A.C. and Vellasco, M.M.B.R. Mobile robot path planning using genetic algorithm. In *Foundations and Tools for Neural Modeling*, Lecture Notes in Computer Science, 1999, Vol. 1606/1999, pp. 671–679, DOI: 10.1007/BFb0098225, http://www.springerlink.com/content/krn12872x2367p55, July 2011.

13 Temporal Logic Motion Planning in Robotics

Motlatsi Seotsanyana

CONTENTS

13.1   INTRODUCTION

Robotic computer systems have become increasingly ubiquitous in everyday life and this has led to a need to develop safe and reliable robot systems. There are areas where robotic systems are required to perform some critical functions: (i) transport-related applications such as intra-logistics, automated parking garages and autonomous vehicles; (ii) mining-related applications such as automated mine vehicles and mine sensing; (iii) defense force-related applications such as autonomous vehicles and (iv) hospital-related applications such as surgical procedures. In such applications, any failure of a robotic system may result in more than just a mere inconvenience, such as incorrect information by a robotic receptionist, loss of time or even in a worse case may cause catastrophic loss of human life in the case of surgical and mining automation. To ensure safety and reliability of these systems, the four main verification techniques are usually considered: (a) theorem proving, (b) model checking, (c) peer reviews and (d) simulation and testing in the context of the practical robotic systems, for both hardware and software sub-systems as deemed appropriate.

Theorem proving is the process of using deductive methods to develop computer programmes that show that some statement (i.e., conjecture) is a logical consequence of a set of axioms and hypotheses. Unfortunately, the theorem-proving process is generally harder and requires considerable technical expertise and a deep understanding of the specification. It is also generally slower, more error-prone and labour intensive. Model checking, on the other hand, is an automatic verification technique for finite-state concurrent systems such as safety critical systems, communication protocols, sequential circuit design and so on. Model checking is an attractive alternative to simulation and testing to validate and verify systems. But model checking techniques are hindered by the state-space explosion problem (curse of dimensionality!), where the size of the representation of the behaviour of a system grows exponentially with the size of the system. The current practice is that the correctness of computer systems is achieved by human inspection: peer reviews and simulation and testing with little or no automation. Peer reviews refer to the inspection of software by a team of engineers that were preferably not involved in the design of the system (software), while simulation refers to a process whereby the model that describes the behaviour of the system is executed with some scenarios which can be provided by either the user or a tool and testing, on the other hand, refers to a process whereby software is executed with some inputs, called test cases, along different execution paths known as runs. However, peer reviews, simulation and testing are never complete: it is difficult to say when to stop as it is not feasible to check all the runs of a complex robotic system and/or model; therefore, it is easy to omit those runs (which might reveal subtle errors). They also have a drawback of showing the presence of errors, but not their absence! Around 1977, Pnueli introduced linear temporal logic (LTL) to specify requirements for reactive computer systems and since then the use of temporal logics in verifying critical systems has grown tremendously. In the early 1980s, two independent pairs of researchers, Clarke and Emerson [1] and Queille and Sifakis [2], introduced another type of temporal logic called computation tree logic (CTL). Therefore, there are currently two schools of temporal logic: linear temporal logic which assumes an implicit universal quantification over all the executions of a system and branching-time temporal logic which assumes an explicit existential and universal quantification over the executions. The details of some of the variants of these logics are presented in Section 13.4. The main hurdle in robotics is to be able to verify autonomous systems to ensure safety and reliability of these systems in the presence of people. Temporal logics can play an important role in ensuring that robotic systems operate safely and robustly in the presence of human beings and in this chapter, we review two approaches that are trying to address the problem: programme synthesis and runtime verification. Programme synthesis involves the construction of a reactive system (e.g., a controller) that assigns to every input the corresponding output that satisfies a certain system requirement. Runtime verification, on the other hand, refers to the process of detecting faults in a system during its normal execution.

13.2   THE IMPORTANCE OF AUTONOMY

The field of autonomous driving refers to a vehicle (or car) that can drive entirely on its own in a complex environment without human driver and remote control. This field has a rich history that spans over a decade and its benefits are evident. Modern advanced sensors are used to provide the vehicle with an internal world model, that is, a representation of local surrounding environment and positioning system of the vehicle in the environment. The introduction of these vehicles provides and promises a number of benefits to the society: the reduction of car accidents, reduction (or eradication) of traffic congestions, and also has a direct impact on areas such as mining, farming, health, constructions and so on. A lot of successful work in this area of research has been reported [3]. In South Africa, the document prepared by Road Traffic Management Corporation (RTMC) [4] reports the alarming road accidents despite the steady month-on-month decrease in the number of fatal crashes and fatalities since July 2006. From April 2007 to March 2008 the total number of vehicles involved in fatal crashes was 15,172 and the total cost of fatal crashes was approximately R113 billion (SA Currency Rand). The introduction of autonomous mobile vehicle has a potential to significantly reduce these road accidents and the corresponding costs. In addition to safety, autonomous mobile vehicle can also bring a number of benefits to communities: driving home human drivers (or passengers) who cannot drive themselves, as well as grandparents, schoolchildren, intoxicated drivers and so on. Despite the fact that other autonomous activities such as farming and construction vehicles have to some extent reached maturity, autonomous cars on public roads are still far from being commercialized because of the problems still being faced in this area. Public roads are inherently uncertain (and dynamically too busy), and accounting for moving obstacles where the intentions (of the moving obstacles, public, etc.) are not well known in advance—this being a challenging problem. However, the need for autonomous vehicle is considerable, especially in the mining industry where there are places which are hazardous to human beings. The main challenge of autonomous mobile vehicles is to develop computationally efficient frameworks and algorithms that allow the interaction of these vehicles with common people to accomplish their tasks. These tasks may be expressed in human-like languages such as temporal logics to allow high-level specifications such as 'start at stop A, go to stop B and wait for 30 min, and then go to stop C, wait for 30 min before going to stop A again' and so on. This approach will enable mobile robots to solve a large number of practical problems. Section 13.3 briefly details a general problem definition for the verification of autonomous systems.

13.3   GENERAL PROBLEM DEFINITION

Verification techniques such as simulation and testing do not scale up well to large robotic systems, which depend heavily on the complex environment and change over time. This behaviour makes it hard to predict and verify these systems prior to their execution and this is because the current *state* of a robot includes both its physical environment and configuration, which are referred to as a *state space*. State space captures all possible situations that could arise—the state could, for example, represent the position and orientation of a robot, the locations of an obstacle and so on. In motion planning, the current state is very important because it makes it possible to generate plans at real time, and thus makes the task of designing, developing and formally verifying the correctness of motion planning algorithms possible. Unfortunately, it is virtually impossible for a mobile robot to precisely know its state [5] and an appropriate approach—probably the only one—is to estimate its state from information provided by sensors, lasers, GPS and so on. Information sourced from these devices poses a number of challenges for designing and developing safe and reliable motion planning algorithms for autonomous robots. These challenges include uncertainty due to the failure of a device, unreliability due to the noisy data, unsafe if these robots operate in hazardous environments such as mining industry, unavailability of data due to the loss of connection to the GPS and so on. The only solution is to synthesize the state of the robot through probabilistic filtering techniques.

Motion planning architectures mostly consist of three components (i) the world which represents robot's physical environment and configuration, (ii) the estimation techniques to filter sensory data and (iii) motion planning. Motion planning through programme synthesis is the main focus of this chapter and it takes two inputs: the world model and formal specification, and produces one output (control strategies). The world model is the representation of ever-changing environment while formal specification refers to the formalized high-level specifications of static environment and tasks to be undertaken by the robot. Motion planning may consist of a number of levels (e.g., as in Ref. [6]) to eventually output control strategies to the robot. The works reviewed in Section 13.5.1 try to solve a formal specification problem stated in the form:

$$\varphi = \{ (\phi_{init}^{env} \wedge \square \; \phi_{safety}^{env} \wedge \square \; \lozenge \phi_{goal}^{env}) \Rightarrow (\phi_{init}^{sys} \wedge \square \; \phi_{safety}^{sys} \wedge \square \; \lozenge \phi_{goal}^{sys}) \} \qquad (13.1)$$

where ϕ_{init}^{env} and ϕ_{init}^{sys} are initial conditions of the environment and a robotic system, respectively; ϕ_{safety}^{env} and ϕ_{safety}^{sys} are the conditions of the environment and the robotic system that must always be true during the entire operation and ϕ_{goal}^{env} and ϕ_{goal}^{sys} represent the conditions that the goals of both the environment and the system that will often be achieved. The square and diamond symbols are explained in the next section; the temporal formula φ can be read as saying that if the vehicle and environment are in a given state and both of them always satisfy their requirements, then both of them will often reach their goals. The high-level specification tasks such as 'start at stop A, go to stop B and wait for 20 minutes, and then go to stop C, wait for 30 minutes before going to stop A again' can be naturally translated into temporal formulas φ such as linear temporal logic (LTL), computational tree logic (CTL), timed computation tree logic (TCTL), probabilistic computation tree logic (PCTL) and so on. Hence, in Section 13.5.1, only motion planning techniques that include tools of theory of computation and formal methods are considered. Apart from programme synthesis, runtime verification is an ideal approach to verify robotic systems. The traditional verification techniques such as peer reviews, simulation and testing do not often scale up well to large and open systems such as robotic systems, which depend heavily on the environment and changes over time. This behaviour makes it hard to predict and verify these dynamic systems prior to their execution. Therefore, runtime verification is an appropriate technique to complement these techniques. Leucker and Schallhart [7] define runtime verification as the 'discipline of computer science that deals with the study, development, and application of those verification techniques that allow checking whether a run of a system under scrutiny satisfies or violates a given correctness property'. Although not new, runtime verification is receiving increasing interest due to the advent of new verification techniques such as model checking techniques. In Section 13.5.2, we present some background and analysis of the current state of runtime verification.

13.4   AN OVERVIEW OF TEMPORAL LOGICS

The literature reviewed in Sections 13.5.1 and 13.5.2 use temporal logics for programme synthesis and runtime verification, respectively. Therefore, an overview of temporal logics is presented. These logics have been used to precisely describe the properties of concurrent systems (such as *safety* and *liveness* properties) and were first introduced by Pnueli around 1977 for the specification and verification of computer systems. The two most widely used types of temporal logics are LTL and CTL. More information on these logics is presented in Refs. [1,8,9].

13.4.1   LINEAR TEMPORAL LOGIC

The LTL is a modal temporal logic with modalities referring to time. In LTL, formulas are encoded about the future of paths (or runs) such that a condition will eventually be true, that a condition will be true until another fact becomes true and so on.

13.4.1.1 LTL Syntax

The syntax of LTL is defined in terms of atomic propositions, logical connectives and temporal operators. Atomic propositions are the most simple statements that can be made about the system in question and thus take the value *true* or *false*. Examples of atomic propositions are: *the door is closed, x is less than 2* and so on. Atomic propositions can be represented by alphabetic symbols such as p and q. The set of atomic propositions is referred to as *AP*. The Boolean operators that are used in the syntax of LTL are $\vee$, $\wedge$, $\neg$, $\Rightarrow$ and $\Leftrightarrow$ in addition, there are several temporal operators:

- $\square$ denotes 'always'
- $\lozenge$ denotes 'eventually'
- U denotes 'strong until'
- X denotes 'next'

The structure of a formula of LTL is given by the following grammar expressed in Backus–Naur Form (BNF) notation: $\alpha ::= p|\neg\alpha|\alpha \vee \beta|X\alpha|\alpha \ U \ \beta$

The operators $\wedge$, $\Rightarrow$, $\Leftrightarrow$, *true*, *false*, $\lozenge$ and $\square$ which are not mentioned in this syntax, can be thought of merely as abbreviations by using the rules:

$$
\begin{array}{llllll}
\alpha \wedge \beta & \equiv & \neg(\neg\alpha \vee \neg\beta) & \alpha \Rightarrow \beta & \equiv & \neg\alpha \vee \beta \\
\alpha \Leftrightarrow \beta & \equiv & (\alpha \Rightarrow \beta) \wedge (\beta \Rightarrow \alpha) & true & \equiv & \neg\alpha \vee \alpha \\
false & \equiv & \neg true & \lozenge\alpha & \equiv & true \ U \ \alpha \\
\square\alpha & \equiv & \neg\lozenge\neg\alpha & & &
\end{array}
$$

13.4.1.2 LTL Semantics

The syntax defines how LTL formulas are constructed, but does not provide an interpretation of the formulas or operators. Formally, LTL formulas are interpreted in terms of a model defined as a triple $M = (S,R,Label)$, where (i) S is a non-empty countable set of states, (ii) $R:S \rightarrow S$ is a function which assigns to each $s \in S$ a unique successor $R(s)$ and (iii) $Label:S \rightarrow 2^{AP}$ is a function which assigns to each state $s \in S$ the atomic propositions $Label(s)$ that are valid in s. The meaning of LTL formulas are defined in terms of a satisfaction relation, denoted by $\vDash$, between a model M, states $s \in S$ and the formulas α and β. Therefore, $M, s \vDash \alpha$ if and only if α is valid in the state s of the model M. If it is understood from the context that s is a state of the model M, then M is dropped and the satisfaction relation is mathematically defined as follows:

$$
\begin{array}{lll}
s \vDash p & \text{iff} & p \in Label(s) \\
s \vDash \neg\alpha & \text{iff} & \neg(s \vDash \alpha) \\
s \vDash \alpha \vee \beta & \text{iff} & (s \vDash \alpha) \vee (s \vDash \beta) \\
s \vDash X\alpha & \text{iff} & R(s) \vDash \alpha \\
s \vDash \alpha \ U \ \beta & \text{iff} & (\exists j \geq 0:R^j(s) \vDash \beta) \wedge (\forall \ 0 \leq k < j:R^k(s) \vDash \alpha)
\end{array}
$$

Here, R^i is used to denote i applications of the function R. For example, $R^3(s)$ is the same as $R(R(R(s)))$. The formal interpretation of the other connectives, *true, false*, $\wedge$, $\Rightarrow$, $\lozenge$ and $\square$ can be derived in a similar way from the definitions above.

13.4.2 COMPUTATION TREE LOGIC

The CTL is based on the concept that for each state there are many possible successors, unlike in LTL which is based on a model where each state s has only one successor s'. Because of this branching notion of time, CTL is also classified as a *branching temporal logic*. The interpretation of CTL is therefore based on a *tree* rather than a *sequence* as in LTL.

13.4.2.1 CTL Syntax

The formulas of CTL consist of atomic propositions, standard Boolean connectives of propositional logic and temporal operators. Each temporal operator is composed of two parts, a path quantifier (universal $\forall$ or existential $\exists$) followed by a temporal modality ($\Diamond$, $\square$, X, U). Note that some authors use G and F for $\forall$ and $\exists$, respectively. The temporal modalities have the same meanings as in Section 13.4.1. The syntax is given by the BNF:

$$\alpha ::= p \mid \neg\alpha \mid \alpha \vee \beta \mid \alpha \wedge \beta \mid \exists X\alpha \mid \exists[\alpha\ U\ \beta] \mid \forall[\alpha\ U\ \beta]$$

13.4.2.2 CTL Semantics

CTL semantics differs slightly from that of LTL defined in Section 13.4.1, that is, the notion of a *sequence* is replaced by a notion of a *tree*. The interpretation of CTL is defined by a satisfaction relation $\models$ between a model M, one of its states s and some formula. Let $AP = \{p,q,r\}$ be a set of atomic propositions, $M = (S,R,Label)$ be a CTL model, $s \in S$, and α and β be CTL formulas. In order to define the satisfaction relation ($\models$), the following definitions are first given: (i) A *path* is an infinite sequence of states $s_0, s_1, s_2, \ldots$ such that $(s_i, s_{i+1}) \in R$, (ii) Let $\rho \in S^w$ denote a path. For $i \geq 0$, $\rho[i]$ denotes the $(i + 1)$th element of ρ, that is, if $\rho = s_0, s_1, s_2, \ldots$, then $\rho[i] = s_i$ and (iii) $P_M(s) = \{\rho \in S^w \mid \rho[0] = s\}$ is a set of paths starting at s. Just like in LTL if it is understood from the context, M can be dropped in the satisfaction relation $\models$ defined as follows:

$s \models p$	iff	$p \in Label(s)$
$s \models \neg\alpha$	iff	$\neg\,(s \models \alpha)$
$s \models \alpha \vee \beta$	iff	$(s \models \alpha) \vee (s \models \beta)$
$s \models \exists\,X\alpha$	iff	$\exists\rho \in P(s){:}\rho[1] \models \alpha$
$s \models \exists\,[\alpha\ U\ \beta]$	iff	$\exists\rho \in P(s){:}\exists\,j \geq 0{:}(\rho[j] \models \beta \wedge \forall\,0 \leq k < j{:}\rho[k] \models \alpha)$
$s \models \forall\,[\alpha\ U\ \beta]$	iff	$\forall\rho \in P(s){:}\exists\,j \geq 0{:}(\rho[j] \models \beta \wedge \forall\,0 \leq k < j{:}\rho[k] \models \alpha)$

13.4.3 Timed Computation Tree Logic

The temporal logics presented in Sections 13.4.1 and 13.4.2 focus on the temporal order of events and do not explicitly state the actual time taken by these events. Time-critical robotic systems necessitate the consideration of quantitative time between the occurrence of events, that is, the correctness of most robotic systems does not only depend on the functional requirements but also on the time requirements. In this section, the syntax and semantics of TCTL is presented. But first, an overview of timed automata is given.

13.4.3.1 Timed Automata Syntax

Finite-state real-time systems are modelled with timed automata. A timed automaton is a standard finite-state automaton extended with a set of non-negative real-valued *clock variables* (or just *clocks* in short). Clocks are assumed to proceed at the same rate to measure the time elapsed since they were last reset. In order to formally define a timed automaton, *clocks* and *clock constraints* are first defined as follows: (i) A *clock* is a variable ranging over $\mathbb{R}^+$ (where $\mathbb{R}^+$ represents non-negative real numbers), (ii) For set C of clocks with $x,y,z \in C$, a *clock constraint* α over C is defined by $\alpha ::= x \prec c \mid x - y \prec c \mid \neg\alpha \mid (\alpha \wedge \alpha)$, where $\prec\, \in \{<, \leq\}$ and (iii) $\Psi(C)$ is the set of all possible clock constraints. Clocks are defined to range over the non-negative real numbers, that is, $x,y,z \in \mathbb{R}^+$. A state of a timed automaton consists of a *location* and values of clocks. Clock constraints are used to label the edges of a timed automaton and represent *guards* that are used to either enable or block transitions between locations. Clock constraints are also used to label locations and such constraints are then *invariants* that limit the amount of time to be spent in a location. Formally, a timed automaton A over set of actions Σ, set of atomic propositions AP and set of clocks C is defined as a tuple $(L, l_0, I, Label)$, where (i) L is a non-empty set of locations with the initial location $l_0 \in L$, (ii) $E \subseteq L \times \Psi(C) \times \Sigma \times 2^C \times L$ corresponds to a set of edges. $(l, g, a, r, l') \in E$ represents an edge from location l to location l' with clock constraint g (also known as enabling condition of the edge or guard) action a to be performed and the set of clocks r to be reset, (iii) $I : L \to \Psi(C)$ is a function which assigns a clock constraint (i.e., an invariant) for each location and (iv) $Label$: $L \to 2^{AP}$ is a function which assigns to each location $l \in L$ set of atomic propositions that hold in the location.

13.4.3.2 Timed Automaton Semantics

The interpretation of a timed automaton is defined in terms of an infinite transition system and in order to formally define the semantics of the timed automaton, the *clock assignment* function and *state* of a timed automaton are defined as follows:

- A *clock valuation (clock assignment)* u for the set of clocks C is a function $u : C \to \mathbb{R}^+$ assigning each clock $x \in C$ its value $u(x)$. Let the set of all clock valuations over C be denoted by $V(C)$. The clock evaluation has the following characteristics:
  - For $u \in V(C)$ and $d \in \mathbb{R}^+$, clock valuation $u + d$ over C means that all clocks are increased by d, that is, $u(x) + d$ for all $x \in C$.
  - For $C' \subseteq C$, $u[C' \to 0]$ means that all the clocks in C' are assigned to zero, that is, all assigned and zero clocks in C' are reset, so that $u[C' \to 0](x) = 0$ for all $x \in C'$ and $u[C' \to 0](x) = u(x)$ for all $x \notin C'$. If C' is the singleton set $\{z\}$, just $u[z \to 0]$ shall be written.
  - For a given clock valuation $u \in V(C)$ and a clock constraint $\alpha \in \Psi(C)$, $\alpha(u)$ is a Boolean value stating whether or not α is satisfied or not.
- A *state* is a pair (l, u) where l is a location of an automaton A and u is a clock valuation over C.

The operational semantics of a timed automaton $A = (L, E, I, Label)$ over the clock set C is therefore defined by an infinite-state transition system $M_A = (S, s_0, \to, Label)$, where

- $S = L \times V(C)$ is the set of states
- s_0 is the initial state of A (l_0, u_0)
- $\to$ is the transition relation with its members defined by the following two rules:
  - *action transition*: $(l, u) \xrightarrow{a} (l', u')$ if there is an edge $(l \xrightarrow{g, a, r} l')$ such that $g(u)$ holds and $u' = u[r \to 0]$, and $inv(u')$ holds for each $inv \in I(l')$
  - *delay transition*: $(l, u) \xrightarrow{d} (l, u')$ if, for $d \in \mathbb{R}^+$, $u' = u + d$ and $inv(u + d')$ holds for all $d' \leq d$ and $inv \in I(l)$
- $Label : S \to 2^{AP}$ is an atomic proposition function extended from $Label : L \to 2^{AP}$ simply by $Label(l, u) = Label(l)$.

13.4.3.3 TCTL Syntax

The syntax of TCTL is based on the syntax of CTL, extended with clock constraints. In order to clearly define the syntax, the following definitions are given:

- A *path* is an infinite sequence $s_0 a_0, s_1 a_1, \ldots$ states alternated by transition labels such that $s_i \xrightarrow{a_i} s_{i+1}$ for all $i \geq 0$, where a_i is either (g, a, r) or d.
- Let $\rho \in S^w$ denote a path. For $i \geq 0$, $\rho[i]$ denotes the $(i + 1)$th element of ρ (see Section 13.4.2).
- $P_M(s) = \{\rho \in S^w \mid \rho[0] = s\}$ is a set of paths starting at s (see Section 13.4.2).
- A *position* of a path is a pair (i, d) such that d equals 0 if $a_i = (g, a, r)$, and equals a_i otherwise.
- Let $Pos(\rho)$ be the set of positions in ρ. For convenience, the state $(l_i, v_i + d)$ can also be written as $\rho(i, d)$.

- A total order of positions is defined by
  - $(i,d) \ll (j,d')$ if and only if $(i < j) \vee (i = j \wedge d \leq d')$.
- Path ρ is called *time-divergent* if $\lim_{i \to \infty} \Delta(\rho, i) = \infty$, where $\Delta(\rho, i)$ denotes the time elapsed from s_0 to s_i, that is, $\Delta(\rho,0) = 0 \Delta(\rho,i) = \Delta(\rho,i) + \begin{cases} 0 & \text{if } a_i = (g,a,r) \\ a_i & \text{if } a_i \in \mathfrak{R} \end{cases}$, where $\mathfrak{R} = \mathbb{R}^+$.
- Let $P_M^\infty(s) = \{\rho \in S^w \mid \rho[0] = s\}$ denote the set of time-divergent paths starting at s.

Let $p \in AP$ and D be a non-empty set of clocks that is disjoint from the clocks of A (i.e., D is the set of clocks of the TCTL formulas and $C \cap D = \varnothing$), $z \in D$ and $\alpha \in \Psi(C \cap D)$. The TCTL formulas are then defined by the following BNF:

$$\beta ::= p \mid \alpha \mid \neg\beta \mid \beta \vee \beta \mid z \, in \, \beta \mid \exists[\beta \, U \, \beta] \mid \forall[\beta \, U \, \beta]$$

A clock constraint α is defined over formula clocks and timed automaton clocks and thus allows comparison of both formula and timed automaton clocks. Clock z is known as a *freeze identifier* and bounds formula clocks in β. For instance, $\forall[\beta \, U_{\leq 4} \, \phi]$ can be defined as $z \, in \, \forall[(\beta \wedge z \leq 4)U\phi]$.

13.4.3.4   TCTL Semantics

For $p \in AP$, $\alpha \in \Psi(C \cap D)$ is a clock constraint over $C \cup D$, model $M = (S, \to, L)$ is an infinite transition system, $s \in S$, $w \in V(D)$ and ψ, ϕ are TCTL formulas. The satisfaction relation $\vDash$ is defined as follows:

$s,w \vDash p$	iff	$p \in L(s)$
$s,w \vDash \alpha$	iff	$v \cup w \vDash \alpha$
$s,w \vDash \neg\phi$	iff	$\neg(s,w \vDash \phi)$
$s,w \vDash \phi \vee \psi$	iff	$(s,w \vDash \phi) \vee (s,w \vDash \psi)$
$s,w \vDash z \, in \, \phi$	iff	$s,w[z \to 0] \vDash \phi$
$s,w \vDash \exists[\phi \, U \, \psi]$	iff	$\exists\rho \in P_M^\infty(s) : \exists(i,d) \in Pos(\rho) : (\rho(i,d), w + \Delta(\rho,i) \vDash \psi \wedge$
		$\forall(j,d') \ll (i,d) : \rho(j,d'), w + \Delta(\rho,j) \vDash \phi \vee \psi))$
$s,w \vDash \forall[\phi \, U \, \psi]$	iff	$\forall\rho \in P_M^\infty(s) : \exists(i,d) \in Pos(\rho) : (\rho(i,d), w + \Delta(\rho,i) \vDash \psi \wedge$
		$\forall(j,d') \ll (i,d) : \rho(j,d'), w + \Delta(\rho,j) \vDash \phi \vee \psi))$

13.4.4   PROBABILISTIC TEMPORAL LOGICS

There are probabilistic temporal logics (PTLs) that are worth mentioning when talking about verification in autonomous robotic systems. These PTLs are important to specify probabilistic requirements of robotic systems. We present continuous stochastic logic (CSL) as it also covers PCTL. There are also other relevant temporal logics such as stochastic game logic (SGL) [10] which can also specify open systems like autonomous robotic system. The CSL syntax and semantics are briefly outlined as follows.

13.4.4.1   CSL Syntax

The formulas of CSL consist of atomic propositions, standard Boolean connectives of propositional logic and probabilistic operator $P_{\prec p}(.)$. The probabilistic operator replaces the path quantifiers: universal $\forall$ and existential $\exists$. Let $p \in [0,1]$, $\prec \in \{\leq, <, \geq, >\}$, and $t \in \mathbb{R}^+$, then the syntax is given by the BNF:

$$\varphi ::= p \mid \neg\varphi \mid \varphi \wedge \varphi \mid P_{\prec p}(X\varphi) \mid P_{\prec p}(X^{[0,t]}\varphi) \mid P_{\prec p}(\varphi \, U \, \varphi) \mid P_{\prec p}(\varphi \, U^{[0,t]} \, \varphi)$$

13.4.4.2   CSL Semantics

The interpretation of CSL is defined by a satisfaction relation $\models$ between a continuous (or discrete) time Markov chain model M, one of its states s and some formula. The semantics for temporal operators such as next, until and so on are similar to those of CTL explained in Section 13.4.2 with the exception of the bounded path temporal operators that require that the formula is satisfied within the specified time interval.

13.5   AN OVERVIEW OF ROBOTIC VERIFICATION

Autonomous robotic systems are classified as open system due to their intensive interaction with the environment and the common specification languages such as LTL and CTL cannot precisely specify their properties. This is because these logics are natural specification languages for closed systems. In order to deal with verification of autonomous robotic systems, new specification languages are needed. Fortunately, work has already been started in trying to find solution to the verification of open systems and in this chapter we present two of such works: programme synthesis and runtime verification. These approaches are not new, but they have gained a new momentum due to the advent of model checking techniques. Next, we review work done in Refs. [6,11–14]. These papers discuss programme synthesis in robot motion planning. Section 13.5.2 highlights some of the work done in runtime verification and we concentrate on Refs. [15–19] to show their relevance in the verification of open systems, without going into details.

13.5.1   PROGRAMME SYNTHESIS IN MOTION PLANNING

Motion planning of autonomous mobile vehicles is a fundamental problem in robotics. It is a complex and challenging problem and this is mainly because of the inherently unreliable, dynamic environments in which robotic applications operate. The ability of a motion planner to generate accurate and safe trajectories despite these conditions is fundamental to autonomous mobile vehicles to perform their tasks effectively and reliably. Hence, the use of programme synthesis is a promising approach to the generation of these accurate and safe trajectories (control strategies) in motion planning. We discuss the motion planning techniques that use programme synthesis to generate trajectories that conform to some formal specifications.

13.5.1.1   The Discrete Event Models Temporal Logic

In Ref. [11], the idea is to develop a robust compiler programme—similar to Silicon Compilers—to synthesize controller programmes of different robotic applications and manufacturing tasks, based on discrete event systems (DES) theory [20], Petri Nets [21,22] and temporal logic [23]. The synthesizer (or compiler) takes two inputs: (i) a model of a robotic problem and (ii) a set of high-level specifications expressed in temporal logic and outputs a synthesized controller that produces control commands that satisfy the synthesized controller. The running example that is used for testing and simulation of the synthesizer is the 'walking machine problem' with four legs. The model of the machine is divided into two layers: discrete and continuous. The discrete layer refers to the scheme used to synchronize states of the legs and this is modelled with a finite-state machine (FSM). There are six states: start, unload, recover, load, drive and slipped, of which each corresponds to different movements of a leg. The continuous layer, on the other hand, represents different kinematic equations at each state of FSM. Every leg of the 'walking machine' is modeled by an FSM, and a synchronized product FSM of the four legs has 1296 states and 5184 transitions. The transitions between these states are governed by the sensory data and only the position information was used for simulation purposes. The position p_r represents the rear position whereas p_f represents the front position of the legs. It is possible that the system may allow the steps that are longer than the leg and this is restricted by a graph traversal which maintains the minimum and maximum of difference between the rear and the front legs, that is, $\Delta(p_{feet}) = |p_f - p_r|$. For example, the equation $\Delta(p_{feet})[Load_1, Driver_2] = \Delta(p_{feet})[Recover_1, Driver_2] + \frac{1}{2} step$ assumes that the rear leg takes a very small step and the equation $\Delta(p_{feet})$ [$Load_1$, $Driver_2$] = $\Delta(p_{feet})$[$Recover_1$, $Driver_2$] + 2 $step$ assumes a full step distance. The desired constraint is that $\Delta(p_{feet}) < \ell$, where ℓ is derived from the mechanics of the walking machine. The authors of Ref. [11] follow a number of steps for the process of the controller synthesis. This includes first, the modification of the standard DES and the use of temporal logic for specification and verification properties. Second, a modified model checker is used to mark undesired states of the machine. For example, the system should not be in a state where both legs of a train are recovering, driving or slipping (i.e., these states should be avoided). The CTL formula used to mark these states is specified as

$$AG(\neg state([Driver_1, Driver_2]) \wedge \neg state([Recover_1, Recover_2]) \wedge \neg state([Slipping_1, Slipping_2]))$$

Third, the modified model checker is also used to verify the supervisory synthesizer to ensure that it maintains the required behaviour of the machine and this is the original intention of the authors—to verify the correctness of the synthesizer itself. More interesting properties are also verified, and the results are discussed in Section 13.6.

13.5.1.2 Symbolic Planning and Control of Robot Motion

In Ref. [6], the challenges of robot motion planning are stated. These challenges involve the development of computationally efficient framework which takes into account the constraints of the robot, and the complexity of environment while at the same time facilitating a detailed high-level specification of tasks. The framework is used to solve the motion planning problem which is usually divided into three layers: The first level (called specification level) is about dividing the configuration state space into cells and these are usually represented by a graph. The second level (called execution level) finds the shortest path that avoids obstacles from the initial state to the goal state. The last level (called implementation level) generates a reference trajectory and controllers are developed to follow the trajectory. The authors of Ref. [6] use theory of computation and formal methods tools to represent specification tasks, robot constraints and environment and they are the ones who coined the term 'symbolic' to refer to the use of these tools. Symbolic motion planning can be easily incorporated into the three aforementioned levels of the framework. Ref. [6] outlines the challenges and research directions of incorporating these tools in motion planning. Section 13.6 discusses the detailed analysis of these challenges and research directions.

13.5.1.3 Temporal Logic Motion Planning for Mobile Robot

A novel approach in Ref. [12] for linking discrete AI planning with motion planning is presented. The authors state that formal formulation of specification such as sequencing and so on in motion planning provides new challenges such as introducing computationally efficient methods that deal with the complexity of these approaches. These properties can be expressed in temporal logics such as LTL and CTL. Authors differentiate their approach from the earlier related approaches that used model checking to generate discrete paths that satisfy temporal logic specifications. The research has resulted in the following tools: model-based planning (MBP) [24], temporal logic planning (TLPLAN) [25] and universal planning for non-deterministic domains (UMOP) [26]. The goal is to generate continuous trajectories which satisfy temporal logic formulas. This is achieved through three steps: (i) the decomposition of workspace into cells [27,28], (ii) the use of model checking NuSMV (new symbolic model verifier) [29] to generate plans for discrete motion planning that satisfy LTL properties and (iii) the generation of continuous trajectories that satisfy the specified LTL properties.

The authors of Ref. [12] have chosen to model a robot that is operated in a polygonal environment P and the motion of the robot is expressed as

$$\dot{x}(t) = u(t) \quad x(t) \in P \subseteq \mathbb{R}^2 \quad u(t) \in U \subseteq \mathbb{R}^2 \tag{13.2}$$

where $x(t)$ is the position of the robot at time t and $u(t)$ is the control input. The objects of interest such as rooms, corridors and so on are atomic propositions represented by a set $\Pi = \{\pi_1, \pi_2, ..., \pi_n\}$ and the observation map, which is associated with Equation 13.2, is defined as

$$h_C : P \rightarrow \Pi \tag{13.3}$$

and this is an observation map that takes continuous states of the robot and maps it to the set of propositions. The proposition is a convex set of the form

$$P_i = \{x \in \mathbb{R}^2 \bigwedge_{1 \leq k \leq m} a_k^T x + b_k \leq 0, a_k \in \mathbb{R}^2, b_k \in \mathbb{R}\} \tag{13.4}$$

The relationship between the observation map $h_C : P \rightarrow \Pi$ and the set of atomic propositions (objects of interests) $\Pi = \{\pi_1, \pi_2, ..., \pi_n\}$ is defined as follows: $h_C(x) = \pi_i$ if and only if x belongs to some related set P_i. This is given in Equation 13.3. Given the robot model, Equation 13.1, observation map, Equation 13.2, initial state $x(0) \in P$ and LTL formula φ, the problem is to construct control input $u(t)$ such that the resulting robot trajectories satisfy the state $x(t)$. In order to solve this problem, let $x[t]$ define the robot trajectories starting at state $x(t)$. The meaning of LTL-formula φ is defined in terms of a satisfaction relation, denoted by $\models_C$, over continuous robot trajectories $x[t]$. Then, $x[t] \models_C \varphi$ states that the trajectory $x[t]$ starting at $x(t)$ satisfies the formula φ. Other LTL formulas can be constructed recursively. The process of generating continuous robot trajectories that satisfy LTL formulas involves three steps: (i) discrete abstraction of robot motion, (ii) temporal logic planning using model checking and (iii) continuous implementation of discrete plan. The first step is to partition the workspace P into triangles and the two reasons for the choice of the partitioning algorithm are: (i) there exist many efficient triangulation algorithms [30] and (ii) the controller used is proved to be efficiently computable on triangles [28]. Therefore, the map $T:P \rightarrow Q$ sends states $x \in P$ to the finite set of triangles $Q = \{q_1, ..., q_n\}$. Given a partitioned workspace of P, the robot motions are defined by the following transition system:

$$D = (Q, q_0, \rightarrow_D, h_D) \tag{13.5}$$

where

- Q is a set states
- $q_0 \in Q$ is a cell containing the initial state $x(0) \in P$
- $\rightarrow_D \in Q \times Q$ is a transition relation, defined as $q_i \rightarrow_D q_j$ if and only if the triangles q_i and q_j are topologically adjacent to each other
- $h_D : Q \rightarrow \Pi$ is an observation map, where $h_D(q) = \pi$ if there is a state $x \in T^{-1}(q)$ such that $h_C(x) = \pi$. And $T^{-1}(q)$ contains all the states $x \in P$ labelled by q.

The trajectory p of D is defined as a sequence $p[i] = p_i \rightarrow_D p_{i+1} \rightarrow_D p_{i+2} ...$, where $p_i = p(i) \in Q$. In step 2, the model checkers NuSMV [29] and 'spin' (SPIN) [31] are used to generate trajectories $p[i]$ of the system D explained in the first step. However, model checking tools are not meant to generate trajectories (i.e., witnesses), but to verify the system and output *yes* if the property is satisfied or a *counterexample* if it is not satisfied. To generate trajectories that satisfy the formula φ, the *counterexample* algorithms in these tools are used. Since these *counterexample* algorithms produce a computation $p[i]$ that satisfies $\neg\varphi$ (i.e., $p[i] \models \neg\varphi$), the original formula is negated and verified with one of the model checkers and the *counterexample* for $p[i] \models \neg(\neg\varphi)$ is generated. This trajectory satisfies $\varphi = \neg(\neg\varphi)$ and is used in the next step to guide the generation of a continuous trajectory. In the third step, to generate continuous trajectories that satisfy the formula φ, the following continuous transition system is defined:

$$C = (P, x(0), \rightarrow_C, h_C) \tag{13.6}$$

- P is a set of polygonals
- $x(0) \in T^{-1}(T(x))$ is the initial state
- $\rightarrow_C \subset P \times P$ is a transition relation, defined as $x \rightarrow_C x'$ between states in P, if and only if x and x' belong to adjacent triangles
- $h_C : P \rightarrow \Pi$ is an observation map, where $h_C(x) = \pi$ maps continuous state to areas of interest (i.e., the set Π)

In order for system C to implement trajectories that are generated by any of the model checkers, the triangulation of P must satisfy bi-simulation property [32]. That is, $T:P \rightarrow Q$ is called bi-simulation if the following conditions hold for $\forall x,y \in P$:

- If $T(x) = T(y)$, then $h_C(x) = h_C(y)$ (i.e., observation preserving)
- If $T(x) = T(y)$ and if $x \rightarrow_C x'$, then $y \rightarrow_C y'$ with $T(x') = T(y')$ (i.e., reachability preserving)

If bi-simulation property is satisfied by the triangulation of the environment, the controllers can be designed that satisfy this property. There are a number of frameworks that can be used, including Refs. [28,33].

13.5.1.4 Sensor-Based Temporal Logic Motion Planning

In Ref. [13], the authors draw a distinction between two approaches to motion planning: (i) bottom-up and (ii) top-down. In bottom-up, the emphasis is laid on generating control inputs to robot models that take a robot from one configuration space to another, while on the other hand, top-down approach focuses on finding discrete robot actions in order to achieve high-level complex tasks—including the interaction of robots in a multi-robot environment, sequencing of temporal actions and so on. High-level task planning and low-level motion planning were previously not possible until the advent of hybrid systems. Hybrid systems integrate discrete and continuous systems and now it is possible to integrate high-level task planning and low-level robot motion planning. This new paradigm of hybrid systems has made it possible to introduce new approaches in robot motion, such as Refs. [12,34,35]. The two novelties in motion planning are found: (i) the temporal logic used addresses sensor inputs directly and (ii) the use of fragment temporal logic called general reactivity (GR) [36] that is computationally polynomial. Its reduced complexity does not affect its expressiveness even though there are properties that cannot be addressed in this logic.

The goal has been to develop a framework that automatically and verifiably generates controllers that satisfy high-level specification tasks expressed in temporal logic. To achieve this, the following should be defined: the model of the robot, admissible environments and the desired system specification.

- *Robot model*: It is assumed that a robot is operating in a polygonal workspace P and the motion of the robot is expressed as

$$\dot{p}(t) = u(t) \quad p(t) \in P \subseteq \mathbb{R}^2 \quad u(t) \in U \subseteq \mathbb{R}^2 \tag{13.7}$$

where $p(t)$ is the position of the robot at time t and $u(t)$ is the control input. Also, it is assumed that P is partitioned into a number of cells $P_1, P_2, \ldots, P_n$, where $P = \bigcup_{i=1}^{n} P_i$ and

$P_i \cap P_j = \emptyset$ if $i \neq j$. It is also assumed that the partitions are convex polygon and each partition creates a set of propositions $\gamma = \{r_1, r_2, \ldots, r_n\}$. Therefore, the proposition r_1 is true if and only if $p \in P_1$, while other propositions are false.

- *Admissible environments*: A robot interacts with its environment through sensors. Authors of Ref. [13] assume that sensors are binary and m sensor variables $\chi = \{x_1, x_2, \ldots, x_m\}$ are not modelled explicitly; instead high-level assumptions are made so that variables can model admissible environments. These admissible environments are modelled in LTL formulas of the form φ_e.
- *System specification*: The desired behaviour of the robot is expressed in LTL formula φ_s and the specifications which can be expressed in LTL include coverage, sequencing and avoidance properties.

Given the robot model, Equation 13.7, initial state $t(0) \in P$ and some LTL formula φ, the problem is to construct a control input $u(t)$ such that the resulting robot trajectories satisfy the state $p(t)$. In order to solve this problem, let $x[t]$ define the robot trajectories starting at state $x(t)$. The meaning of LTL formula φ is defined in terms of a satisfaction relation, denoted by $\models_C$, over continuous robot trajectories $x[t]$. Then, $x[t] \models_C \varphi$ states that the trajectory $x[t]$ starting at $x(t)$ satisfies the formula φ. Other LTL formulas can be constructed recursively.

Given an LTL formula, an automaton that generates an acceptable behaviour by the LTL formula is synthesized. In Ref. [37], it is proved that the synthesis process is doubly exponential. However, the algorithm used is polynomial $O(n^3)$ time, where n is the number of valuations for sensor and state variables [36]. The synthesis process is compared with a game played between the robot and the environment. First, the environment makes a transition according to its transition relation and then the robot does the same. If the robot can satisfy the LTL formula φ, no matter what the environment does, the robot is winning; otherwise, the environment is winning and the desired behaviour cannot be achieved. Given the following winning condition (i.e., GR(1)) $\phi = \varphi_g^e \rightarrow \varphi_g^s$, the robot is winning if φ_g^s is true or φ_g^e is false. If the robot is winning, an automaton that represents desired behaviour is synthesized and it is formally defined as tuple

$$A = (\chi, Y, Q, q_0, \delta, \gamma) \qquad (13.8)$$

where

- χ is a set of input environment propositions
- Y is a set of output system propositions
- $Q \subset \mathbb{N}$ is a set of states
- $q_0 \in Q$ is the initial state
- $\delta: Q \times 2^\chi \rightarrow 2^Q$ is a transition relation, that is, $\delta(q, X) = Q' \subset Q$ where $q \in Q$ is a state and $X \subseteq \chi$ is the subset of sensor propositions that are true
- $\gamma: Q \rightarrow 2^Y$ is the set of state propositions that are true in state q

The importance of this automaton is to generate a path that a robot can follow under admissible inputs. Given admissible input sequence $X_1, X_2, X_3, \ldots, X_j \in 2^\chi$, the automaton generates a run $\sigma = q_0, q_1, \ldots$. This run σ is interpreted as sequence $y_0, y_1, \ldots$ where $\gamma(q_i) = y_i$ is the label of the ith state. The last step is to use the run σ to influence the continuous behaviour of the robot. There are a number of hybrid controllers [27,28] that can drive robots from region to region. In Ref. [13], a controller that satisfies the so-called bi-simulation property [32] is chosen and this controller is presented in Ref. [27].

13.5.1.5   Automatic Synthesis of Multi-Agent Motion Task

In Ref. [14], the authors state that there is an increasing interest in the control theory to develop automated controllers that satisfy complex desired requirements. The key issue in this direction of research is the use of formal specification. In Ref. [38], local controllers are synthesized based on specification represented in the form of graphs, while the authors of Ref. [39] use LTL specifications. The authors of Ref. [12] also mention that some studies use several motion description languages [40]. The authors of Ref. [14] use LTL specification due to its capability to express properties quantitatively and its similarity to natural languages. The following situation is assumed to test the methodology presented below: (i) there are m robots moving in the workspace $W \subset R^2$ and (ii) each robot $i = 1 \dots m$ occupies a disk in the workspace $R_i = \{q \in R^2 : \|q - x_i\| \le r_i\}$, where $x_i \in R^2$ is the centre of the disk and r_i is the radius. The configuration of each robot is presented by x_i and the configuration space by C. The kinematic model is

$$\dot{x} = u \tag{13.9}$$

and let $\phi(x, x_0, x_f)$ be a multi-robot navigation function, where robots do not overlap. Then the control law is

$$u = -\nabla \tag{13.10}$$

where $\nabla \circ = [(\partial/\partial x_1), (\partial/\partial x_2), \dots, (\partial/\partial x_n)]\circ$ and ϕ is a multi-robot navigation function that drives all robots from the feasible initial configuration x_0 to any feasible final state x_f. Two levels of motion controllers are defined. The global convergent controller manages the primary motion task and set of other controllers that lie within its range of convergence. The focus is to synthesis Büchi automaton that realizes the behaviour of the requirements of a robotic system. The steps for achieving this objective are outlined as follows:

1. Given an LTL formula ϕ, a Büchi automaton is constructed and accepts words that satisfy ϕ. After the construction, the largest non-blocking sub-automaton A_ϕ^{NB} is constructed from A_ϕ. If A_ϕ^{NB} is empty, then the LTL formula ϕ should be rewritten. The LTL formulas that are used to synthesize controllers are of the form $\Box G \wedge \phi$, where $\Box G$ refers to the global controller which must always be active.
2. Using A_ϕ^{NB} as a model, a function $\Delta : S \times O \rightarrow \{0,1\}^{|C|}$ is created to generate observation and controller predicates.
3. The last step is to define controllers that conforms to the following control law: $u = -k_1 \nabla \phi + \beta \cdot u_2^*$. The details of how to derive this equation are given in Ref. [14].

13.5.2   Runtime Verification

There is currently an increasing amount of work being done on runtime verification. Here, we only present a part of this work. Runtime verification is not a new topic, but recently it has taken a huge momentum and there is now a new international conference every year (started in 2010) that is dedicated to the work that involves verification of systems at runtime. We review the work done only in Refs. [15–19] since all these sources use formal methods techniques.

13.5.2.1   Monitoring and Checking

Monitoring and checking (MAC) [15] provides a general framework that makes sure that the target programme runs correctly with respect to a formal requirement specification. The framework consists of two specification languages: primitive event definition language (PEDL) and meta event definition language (MEDL). The former is used to define methods and objects to be monitored; a filter keeps a list of monitored local and global variables as well as addresses of monitored objects. The latter is used to write high specification requirement. The main reason for using two specification languages is to separate the implementation details from high-level requirement checking and thus makes the framework portable to different programming languages and specification formalisms.

13.5.2.2 Monitoring-Oriented Programming

Monitoring-oriented programming (MoP) [16] is a framework and methodology for building programme monitors. It allows formal property specification to be added to the target programme and does not place any restriction on a formalism to be used, as long as the corresponding translator of the specification language exists. The translated code must contain the following components: declaration, initialization, monitoring body, success condition and failure condition. The user puts annotations in the target programme at which the monitoring code must be inserted. Currently, the MoP supports three specification languages: past time and future time LTL as well as extended regular expressions.

13.5.2.3 Java with Assertion

Java with assertion (JASS/Jass) [17] is a general monitoring methodology implemented for sequential, concurrent and reactive systems written in Java. The tool Jass is a pre-compiler that translates annotations into a pure Java code in which a compliance with specification is tested dynamically at runtime. Assertions extend the design by contract [41] that allows specification of assertions in the form of pre- and post-conditions, class invariants, loop invariants and additional check to be inserted in any part of the programme code. Jass also offers *refinement check* and *trace assertion*. Refinement check is used to facilitate the specification of classes on different levels of abstraction, while trace assertion is used to monitor the correct behaviour of method invocations, ordering and timing of methods invocation.

13.5.2.4 Java PathExplorer

Java PathExplorer (JPaX) [18] is a general-purpose monitoring approach for sequential and concurrent programmes developed in Java. The tool offers two main facilities: logic-based monitoring and error pattern analysis. Formal specifications are written in LTL (both past and future) or in Maude [42,43]. JPaX instruments Java byte code to send stream of events to the observer that performs two functions: it checks events against a high-level specification (logic-based monitoring) and also checks low-level programming error (error pattern analysis).

13.5.2.5 Temporal Rover

Temporal rover (TR) [19] is a specification-based commercial tool for programmes written in C, C++, Java, Verilog and VHDL. In TR, the user annotates the sections of the target programme where a property needs to be checked at runtime. TR supports the following formalisms: linear-time temporal logic (LTL) and metric temporal logic (MTL) and the properties which can be specified with these logics include future time temporal properties as well as lower and upper bound properties, and relative-time and real-time properties. The tool takes the target programme as its input and its parser generates an identical programme to the properties inserted in the target programme, and during execution the generated code validates the executing programme against specified properties.

13.6 ANALYSIS

Ref. [11] introduces some interesting research directions in motion planning. However, there are some research questions that are not yet answered and some issues are not clearly explained. Some of these questions include the following: (i) The failure to achieve the goal of the paper, that is, to develop a compiler similar to Silicon Compilers. (ii) The ambiguity of some of the CTL formulas. That is, there are some of the CTL specifications in which supervisor synthesis could fail, for example,

$AX(AG(p_1)) \vee AX(AG(p_2))$, where the labels $A(s_2) = p_1$ and $A(s_4) = p_2$, and the states s_2 and s_4 do not have outgoing edges, but self-looping. The failure could be due to the maximal ambiguous controllable sub-language. (iii) Another open problem is to investigate the use of restricted CTL as in Ref. [44]. In addition, no real-time issues and obstacle avoidance algorithms of motion planning for the 'walking machine' are discussed. Apart from these unanswered research questions, there are some important issues that need some detailed explanation. This includes the modified model checker and the input specification language for the model checker. The detailed discussion of the modified model checker is important as this might affect the complexity of model checking algorithms. Other problems are the description of the specification language for the state machines, and the interpretation of the results. These are not yet clearly explained in detail. In Ref. [11], the usage of the system is tested with a train of two legs: the rear and front legs. The state machine which represents the behaviour of a leg was outlined and unregulated verification of the property similar to the one explained in Section 13.5.1 (i.e., states which should be avoided) gives NIL result, which states that the property is not satisfied. In the case that there are some states that are undesirable, the model checker outputs those states which are removed from the desired behaviour. The following shows an example of the results:

```
CMUCL 7> (omega-op K legs uncontrollable-events)
;;Debugging deleted...
>>OMEGA(0): removable states = ((D1 SL2) (SL1 D2))
-----------
;;Debugging deleted...
>>OMEGA(1): removable states = NIL
#<Representation for the approximation to K>
CMUCL 8>
```

Ref. [6] provides incomplete answers and also highlights problems and challenges to answer the following question: 'Can a computational framework that enables for specifying such a task in a high-level, human-like language, with automatic generation of provably correct robot control laws be developed?' These problems and challenges are centred around the concept of discretization which can either be environment-driven or be control-driven. In the latter, an environment is represented by LTL at implementation level. This representation poses many open questions. It is not clear as to which is the best specification language to use, that is, whether to use LTL or CTL? This is a problem since there are high-level specification tasks that cannot be expressed in LTL or vice versa CTL. In addition, a too expressive temporal logic might affect the performance of the analysis. For dynamic mobile robots, it might not be possible to execute strings over partitioned regions as in environment-driven discretization. The best approach is to do the discretization at a controller level. The idea behind control-driven discretization is to divide the system into sub-tasks, for example, sensing modality and the behaviour of each sub-task make up words in motion description languages (MDLs) [45]. The approaches in control-driven discretization include Control Quanta, Motion Primitives and Feedback Encoding. Other low-complexity methods use experimental data to mimic, for example, the behaviour of a human operator. In multi-robot systems, the control strategies could be influenced by studying the behaviour of flocks of birds or school of fish which can lead to some predictable behaviour. Alternatively, such communications or control strategies can be achieved through the use of embedded graph grammars [46].

However, most of the problems are not addressed in Ref. [6]; instead, it provides more questions than answers! In the case of environment-driven discretization, when methods such as model checking analysis and choice of specification languages are applied to a real-world problem, the following three questions need to be answered: (i) controllers guaranteeing robot transition from one region to another or making a region an invariant for a robot have not yet developed for robots with non-holonomic constraints, (ii) this approach should take into account constraints induced by digital controllers and sensors such as finite input and output spaces and (iii) given a team of locally interacting robots, and a high level specification over some environment, how can provably correct (local) control strategies be generated? What global (expressive) specifications can be efficiently distributed? How should local interactions (e.g., message passing vs. synchronization on common events) be modelled?

In the case of control-driven discretization, the following questions need to be answered: (i) what is the best choice of motion primitives for achieving a given class of tasks?, (ii) given an alphabet of motion primitives, what is the penalty associated with restricting the robots trajectories to those obtained through combination of those primitives with respect to a larger set of primitives? and (iii) can this symbolic approach to motion planning be extended to multi-robot environment? However, some of these questions could be answered through the proposed method presented in Section 13.7. In Ref. [12], the goal is to generate continuous trajectories that satisfy temporal logic formulas. This is achieved through three steps: (i) the decomposition of workspace into cells [27,28], (ii) the use of model checking NuSMV [29] to generate plans for discrete motion planning that satisfy LTL properties and (iii) the generation of continuous trajectories that satisfy the specified LTL properties. This is a good approach that gives some hope that one day people will be able to interact with mobile robots safely. However, there are some concerns that need to be addressed in order to achieve this goal and this includes the use of model checking algorithms without any modification. This is because when model checking tools such SPIN are used, they are likely to generate unnecessarily long paths (trajectories). For example, let us say that Figure 13.1 depicts an instance of a state space (of the environment) that is used to generate a path that satisfies some LTL property φ (i.e., $p[i] \models \neg$ $(\neg\varphi)$ as in Section 13.5). If model checkers that employ a nested depth first search algorithm are used to compute strongly connected components (SCC) with an accepting cycle [9] (as shown in Figure 13.1), the trajectory might be $I \rightarrow A \rightarrow B \rightarrow C \rightarrow D \rightarrow E \rightarrow F \rightarrow C$, where B represents some million of states from the initial state I. But the shortest trajectory is $I \rightarrow A \rightarrow C \rightarrow D \rightarrow E \rightarrow F \rightarrow C$, from the initial state I. It is clear that the use of model checking algorithms without any modification might affect the performance of the synthesis process. In Ref. [12], the problem formulation and method summarized in Section 13.5.1 facilitate the synthesis of the following properties:

1. The requirement is to visit rooms in no particular order, and it is formally defined as $(\Diamond r_1 \wedge \Diamond r_2 \wedge \Diamond r_3 \wedge \Diamond r_4 \wedge \Diamond r_5 \wedge \Diamond r_6)$. The generation of a trajectory by NuSMV is fast and the synthesis with MATLAB® takes not more than 15 s.

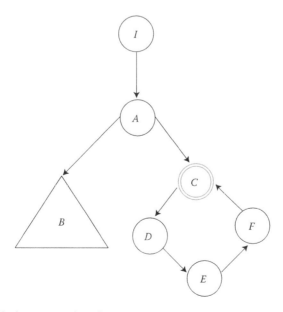

FIGURE 13.1 A graphical representation of a state space.

2. The requirement is to visit room r_2, then room r_1 and then cover rooms r_3, r_4, r_5 while avoiding obstacles o_1, o_2 and o_3. The path is generated with SPIN. It is not mentioned how long each process takes, but the figure showing the environment and the path is shown in Ref. [12].

3. The requirement is to start in a white room and go to both black rooms. The environment for this requirement consists of 1156 rooms and its discrete abstraction consists of 9250 triangles. The path generation takes about 55 s with NuSMV and the controller synthesis with MATLAB lasts for about 90 s.

Ref. [13] presents some promising results in using temporal logic in motion planning. Robots use sensors to gather information about their surroundings and GR(1) formulas are appropriate in representing this interaction. Since GR(1) is a class of LTL, if the same type of logic can be derived out of CTL*, a more robust and expressive logic can be found to synthesize many requirements in robot motion planning. The method presented in Ref. [13] is tested with two examples: (i) single robot—nursery scenario and (ii) multi-robot—search and rescue, which are outlined as follows:

1. The first example states that 'starting in region 1, keep checking whether a baby is crying in region 2 or 4. If you find a crying baby, go look for an adult in regions 6, 7 and 8. Keep looking until you find him. After finding the adult, go back to monitor the babies and so on …' It takes 2 s to synthesis an automaton that realizes the requirement and the automaton has 41 states.

2. The second example states that 'in this search and rescue scenario, we employ two UAVs that continuously search regions 1, 2, 3, 7 and 8 for injured people. Once an injured person is found, a ground vehicle (ambulance) goes to the person's location and helps, the ground vehicle does not move …' In this case, it takes about 60 s to synthesis an automaton which consists of 282 states.

In autonomous robotic systems, the main problem is noisy data which is derived from sensors. This problem necessitates the use of Bayesian filters. Bayesian filters are very useful and powerful statistical tools that are used to estimate variables of interest from sensory data and are also used to fuse data from different sensors. However, both programme synthesis and runtime verification methods presented in this chapter do not use these filters to compute probabilistically correct results. In our framework, we therefore use these filters to compute probabilistic constraints and properties.

13.7   SUGGESTED METHODOLOGY

The traditional verification techniques such as peer review, simulation and testing often do not scale up well to large systems such as robotic systems, which depend heavily on the environment and change over time. This behaviour makes it hard to predict and verify these systems prior to their execution. Therefore, stochastic runtime verification is an appropriate technique to complement these techniques and this is what we are proposing in this chapter. Although not new, runtime verification is receiving increasing interest due to the advent of new verification techniques such as model checking techniques. Despite the introduction of model checking techniques in runtime verification, robotic systems still pose a number of challenges that necessitate the need for a paradigm shift in verification, which should aim to achieve high-level expectations of autonomous operation, predictability and robustness. These challenges include among others environmental uncertainty, resource limitations and the use of library code with no accompanying source code. Currently, there are no verification techniques that address these problems. These challenges (i.e., environmental uncertainty, resource limitations and library programmes) call for the use of stochastic reasoning, discrete and continuous dynamics, qualitative and quantitative measures, goal-oriented approaches and so on, as well as online techniques such as machine learning, game theory and planning techniques. For instance, the use of stochastic reasoning will enable us to determine (or verify) the likelihood of a given situation occurring, or how long it can be expected to take place. This is different from normal model checking where the software only outputs a simple yes or no answer, instead of outlining the reasons for the output, that is, current modelling techniques concentrate more on closed systems and do not precisely address the requirements of open systems like robotic systems. On the other hand, online techniques such as machine learning will enable the continuous verification of these systems to ensure that they process their sensory inputs correctly and make the right decisions, and can also take appropriate steps if a problem occurs. This will reduce the need for a frequent human interaction and this has the benefits of good performance and efficiency.

The software platforms underlying the control of machines in mining robotics deal with a far more complex environment. Instead of finite, well-defined processing tasks, robotic machines might be engaged in a continuous series of negotiations with other machines (and people around them), as they use devices like GPS and scanners to sense the world, and routers to wirelessly exchange the information among themselves. These machines will also have to operate in an uncertain environment in which resources like bandwidth may become scarce. The idea is to let them autonomously adjust to deal with any situation that may arise. The ultimate goal of our work is to design and develop a runtime verification software platform that will learn, monitor and verify other systems at runtime to ensure that they work correctly all the time. In the words of Oliver G. Selfridge, '… show the program how to find and fix a bug, and the program will work forever'. To enable the development of such a software platform, we propose to use an observer design pattern [47]. The observer design pattern also known as Publish-Subscribe or Dependents defines a one-to-many dependency between interacting objects so that when one object (the subject) changes state, all its dependents (the observers) are notified and updated automatically. Although not new, the pattern is receiving increasing interest because of its usefulness in event-driven systems. It encompasses a well-established communications paradigm that allows any number of subjects (publishers) to communicate with any number of observers (subscribers) asynchronously and anonymously via event channels. We believe that the use of this design pattern will enable runtime verification that can easily be incorporated in a number of dynamic systems. In this case, a runtime verification module will be implemented as an observer while other modules will be implemented as subjects. This set-up will allow the encapsulation of interaction between components and promotes loose coupling of components. It allows components to refer to each other explicitly, and sends and receives messages to and from the components. Inspired by the success of formal verification and probabilistic robotics, the innovative challenge of the proposed platform does not only involve verification and statistical techniques at specification, design or implementation levels, but also incorporates these techniques during the runtime of the systems that are based on the platform, so that these systems can safely and reliably cope with the ever-changing complexity of the environment. Therefore, this approach will involve among others (i) the investigation of the fundamental principles of stochastic processes and runtime verification, (ii) formal specification of stochastic processes and (iii) the specification, design, verification and implementation of the proposed platform.

The target platform will provide an interface to robotic systems that need to be runtime verified. The interface will enable a seamless incorporation of these systems together with their corresponding formal requirements that are verified during the execution. For instance, a simultaneous localization and mapping (SLAM) system will have to implement the interface provided by the proposed platform, so that it (SLAM system) exposes its possible transitions and state variables that are verified against formally specified requirements to ensure safe and reliable operation. The following are some of the benefits that will be provided by the platform: (i) the platform will enable the runtime verification of systems that cannot be fully verified off-line, this includes complex third-party components such as software libraries, black box systems, general-purpose operating systems and so on; (ii) in the case where the behaviour of the system depends heavily on the complex environment, it is difficult (or almost impossible) to find a precise model of the environment and thus makes it difficult to apply classical verification techniques; therefore, the platform will enable an adequate verification of such systems as it

monitors the observed behaviour against the required one; (iii) the platform will also be used to complement classical verification techniques (i.e., theorem proving, model checking and testing), to double
check that what is proved or tested during an off-line verification also holds at runtime.

13.8   CASE STUDY

In this section, we present some of our preliminary work on the proposed platform. As we emphasized in the previous sections, the goal is to develop a framework that enables the verification of
dynamic robotic systems at runtime and the problem we are out to solve is to determine the probability that a robot is at a certain position on the map and/or it is approaching a certain area on the
map. The states of a system are then defined as a tuple of timestamps and positions, that is, $\langle t_i, p_i \rangle$,
where t_i is timestamps and p_i is xy-coordinates of the robot on the map at step i. The system requirements, on the other hand, describe the desired behaviour of objects (i.e., robots, people, etc.) interacting within the environment. These requirements could include avoidance, coverage, sequencing
and conditions. Avoidance refers to the use of sensors to avoid colliding with other objects, coverage
refers to those rules that specify regions of interest to traverse, sequencing refers to the traversal of
specific regions in a certain order and condition refers to those logical conditions that express a
function from truth values to truth values.

Problem: In order to test our verifier, let us consider a robot moving around the table as shown in
Figures 13.2 and 13.3. The question we want to answer is: what is the probability that a robot is far
from the position (x_{1k}, y_{1k})? Where (x_{1k}, y_{1k}) can be any xy-coordinates at timestamps k. In order to
answer the question, we filter the robot's odometry using a particle filter and then calculate the distance between two points using Euclidean distance. We use normal distribution density function to
estimate/approximate the distance between two points. In what follows, Equation 13.11 shows the
requirement that we want to verify, while Equations 13.12 and 13.13 are used to estimate distance
probability between two points:

$$P_{\geq 0.7}(\Diamond((-2.5 \leq x_k \leq -2.0) \wedge (-2.5 \leq y_k \leq -2.0))) \tag{13.11}$$

$$d_k = \sqrt{(x_{2k} - x_{1k})^2 + (y_{2k} - y_{1k})^2} \tag{13.12}$$

$$P(d_k) = \frac{1}{\sigma\sqrt{2\pi}} e^{-(d_d - \mu)^2 / 2\sigma^2} \tag{13.13}$$

FIGURE 13.2   CSIR-MIAS Vision LAB (the author's lab-work space).

Distance travelled by the robot

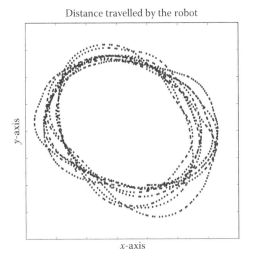

FIGURE 13.3   Robot's path.

Figures 13.3 and 13.4 show the trajectory of the robot and the normalized weights with respect to its position on the map, respectively. We use the *xy*-coordinates with high weights (i.e., high probabilities) to select positions of the robot so that we can estimate how far the robot is from its destination. We use a particle filter to compute these weights. Figures 13.5 and 13.6 show the probability estimates from the start and end goals. For example, in Figure 13.5, the graph shows probabilities of a start position set at the beginning of the robot trajectory to the end of the trajectory; at probability 0.40 the robot has reached its goal position. The negative distances reflect that the robot has past its goal position and the positive distances show that the robot is going towards its goal position. In Figure 13.6, we set the goal position at the beginning of robot trajectory and this graph shows us that the robot can only go away from its goal position. The results of these graphs are computed by a P_{-p} (φ) operator as explained in Section 13.4.4. Lastly, when we run the verifier, it produces the tuples of the form: (True, probability), where the first argument is a true value that tells whether a requirement is satisfied (true) or not (false). The second element of the tuple tells the

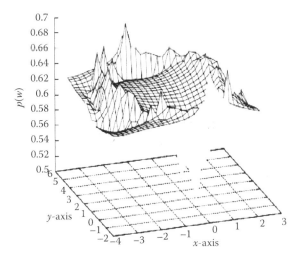

FIGURE 13.4   Estimated path.

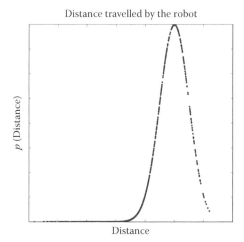

FIGURE 13.5   Go to the last position.

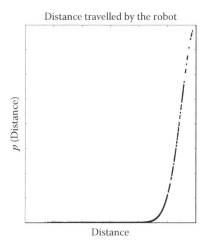

FIGURE 13.6   Go to the original position.

probability at which we must accept the result. An example of the verifier's results is presented as follows:

```
. . .
(False, 0.75803)
(True, 0.241970)
(True, 0.241970)
. . .
```

13.9   FURTHER POSSIBILITIES

Programme synthesis using temporal logic is a promising approach in designing and developing motion planning frameworks and algorithms for autonomous mobile vehicles. This chapter reviewed five research works that incorporate this approach (i.e., programme synthesis) using of theory of computation and formal methods tools and techniques. The aim of the first paper was to develop a compiler to synthesis controller programmes for robotic applications and manufacturing tasks. The second paper was a result of a workshop held in 2006 by the authors to address the issues, challenges and problems of using theory of computation and formal methods tools and techniques in motion planning. They are the ones who coined the term 'symbolic motion planning'. The third paper uses existing model checking tools such as SPIN and NUSMV to generate a discrete path that satisfies a certain requirement. The path is then translated into continuous trajectory to drive the robot from some initial state to the goal state. The fourth and the fifth papers also use temporal logics to synthesize automata that realize some complex requirements in robot motion planning. The procedure is similar to that one of the third paper with some differences in the methodology and the temporal logic used.

The main aim of this chapter has been to verify robotic systems at runtime and due to the stochastic nature of environment in which these systems operate, this calls for a robust verifier that can cope with the uncertainties of the environment. In this chapter, we only highlighted runtime verification frameworks that incorporate formal methods techniques. The preliminary results shown in Section 13.8 make us believe that *stochastic game theory* and *machine learning techniques* are ideal techniques for this kind of verification. We want to proceed by looking at the following issues:

1. Interaction of robots and the environment is very important and we want to proceed by modelling this interaction using game theory. We are going to look at the kind of requirements which are similar to the type of properties Piterman et al. [36] discussed, although they concentrate on LTL, which does not cover the type of requirements we are looking at. We can look at the requirements shown in Equations 13.14 and 13.15:

$$P_{\prec p}((\phi_{init}^{env} \wedge \Box\phi_{safety}^{env} \wedge \Box\Diamond\phi_{goal}^{env}) \Rightarrow (\phi_{init}^{sys} \wedge \Box\phi_{safety}^{sys} \wedge \Box\Diamond\phi_{goal}^{sys})) \tag{13.14}$$

$$P_{\prec p}(\phi_{init}^{env} \wedge \Box\phi_{safety}^{env} \wedge \Box\Diamond\phi_{goal}^{env}) \Rightarrow P_{\prec p}(\phi_{init}^{sys} \wedge \Box\phi_{safety}^{sys} \wedge \Box\Diamond\phi_{goal}^{sys}) \tag{13.15}$$

   where *env* and *sys* refer to the environment and a robotic system, respectively. On the other hand, *init*, *safety* and *goal* refer to initial condition, safety requirement and goal requirement of the formulas, respectively.
2. Inheritance is an important concept in our stochastic verification. Since a robotic system or even the environment may be designed in a hierarchical fashion to simplify the complexity of the system design and implementation, our probabilistic requirements can also be implemented in the same way, so that different components of both the environment and the system can inherit different formal properties and thus facilitate component reuse.
3. Machine learning techniques are also another important techniques that we intend to look at. Since our verifier relies on data from both the environment and robotic systems, we believe that the verifier can use these data to learn and react accordingly to the needs of the dynamic environment in order that our systems operate safely and reliably.

REFERENCES

1. E. M. Clarke and E. A. Emerson, Design and synthesis of synchronization skeletons using branching-time temporal logic, In *Logic of Programs, Workshop*, London, UK: Springer-Verlag, 1982, Vol. 131, pp. 52–71.
2. J. P. Queille and J. Sifakis, Specification and verification of concurrent systems in CESAR, In *Proceedings of the 5th Colloquium on International Symposium on Programming*, London, UK: Springer-Verlag, 1982, pp. 337–351.
3. M. Campbell, M. Egerstedt, J.P. How and R.M. Murray, Autonomous driving in urban environments: Approaches, lessons, and challenges, *Philosophical Transactions of the Royal Society A*, 368, 4649–4672, 2010.
4. R. T. M. Corporation, Road Traffic Report, March 2008, downloaded from http://www.arrivealive.co.za/documents/March_2008_-_Road_Traffic_Report_-_March_2008.pdf
5. S. M. LaValle, *Planning Algorithms*, University of Illinois: Cambridge University Press, 2006.

6. C. Belta, Antonio, M. Egerstedt, E. Frazzoli, E. Klavins and G. J. Pappas, Symbolic planning and control robot motion, *Robotics and Automation Magazine, IEEE*, 14(1), 61–70, 2007.

7. M. Leucker and C. Schallhart, A brief account of runtime verification, *Journal of Logic and Algebraic Programming*, 78(5), 293–303, 2009. The 1st Workshop on Formal Languages and Analysis of Contract-Oriented Software (FLACOS'07).

8. M. Seotsanyana, *Formal Specification, Development, and Verification of Safety Interlock Systems: Comparative Case Study*, 1st ed. Saarbrücken, Germany: VDM Verlag, September 2008.

9. C. Baier and J.-P. Katoen, *Principles of Model Checking*, 1st ed. Cambridge, MA: MIT Press, 2008.

10. B. Christel, B. Tomáš, G. Marcus and K. Antonín, Stochastic game logic, In *Fourth International Conference on the Quantitative Evaluation of Systems (QEST 2007)*, Los Alamitos, Washington, Tokyo: IEEE Computer Society, 2007, pp. 227–236, Edinburgh, Scotland.

11. M. Antoniotti and B. Mishra, Discrete event models + temporal logic = supervisory controller: Automatic synthesis of locomotion controllers, in *IEEE International Conference on Robotics and Automation*, 1995, pp. 1441–1446, Nagoya, Aichi, Japan.

12. G. E. Fainekos, H. Kress-Gazit and G. J. Pappas, Temporal logic motion planning for mobile robots, In *Proceedings of the 2005 IEEE International Conference on Robotics and Automation*, April 2005, pp. 2020–2025, Barcelona, Spain.

13. H. Kress-Gazit, G. E. Fainekos and G. J. Pappas, Where's Waldo? Sensor-based temporal logic motion planning, In *Proceedings of the IEEE Conference on Robotics and Automation*, Vol. 2324, 2003, pp. 3546–3551.

14. S. G. Loizou and K. J. Kyriakopoulos, Automatic synthesis of multi-agent motion tasks based on LTL specification, In *43rd IEEE Conference on Decision and Control*, 14–17, February 2004, Atlantis, Paradise Island, the Bahamas.

15. M. Kim, S. Kannan, I. Lee, O. Sokolsky and M. Viswanathan, Java- MaC: Run-time assurance tool for Java programs, In *Proceedings of the Fourth IEEE International High Assurance Systems Engineering Symposium*, 1999, pp. 115–132.

16. F. Chen and G. Roşu, Towards monitoring-oriented programming: A paradigm combining specification and implementation, *Electronic Notes in Theoretical Computer Science*, Elsevier, 89(2), 2003.

17. D. Bartetzko, Jass—Java with assertions, In *Proceedings of the First Workshop Runtime Verification (RV'01)*, K. Havelund and G. Roşu (eds.), *Electronic Notes in Theoretical Computer Science*, Paris, France: Elsevier Science, 55(2), 2001.

18. K. Havelund and G. Roşu, Java PathExplorer—A runtime verification tool, *Symposium on Artificial Intelligence, Robotics and Automation in Space*, Montreal, Canada, June 2001.

19. D. Drusinsky, The temporal rover and the ATG rover, *SPIN 2000*. Cupertino, CA: Time-Rover, Inc.

20. P. J. Ramadge and W. M. Wonham, The control of discrete event systems, In *Proceedings of the IEEE*, 77(1), 81–98, 1989.

21. L. E. Holloway and B. H. Krogh, Synthesis of feedback control logic for a controlled Petri nets, *IEEE Transactions on Automatic Control*, 35(5), 514–523, 1990.

22. B. J. McCarragher and H. Asada, A discrete approach to the control of robotic assembly tasks, In *IEEE International Conference on Robotics and Automation*, IEEE, 1993, pp. 331–336, Atlanta, Georgia, USA.

23. E. M. Clarke, E. A. Emerson and A. P. Sistla, Automatic verification of finite-state concurrent system using temporal logic specifications, *ACM Transactions on Programming Langauges and Systems*, 8(2), 244–263, 1986.

24. P. Bertoli, A. Cimatti, M. Pistore, M. Roveri and P. Traverso, MBP: A model based planner, In *Proceedings of the IJCAI'01 Workshop on Planning under Uncertainty and Incomplete Information*, Seattle, August 2001. [Online]. Available: citeseer.ist.psu.edu/bertoli01mbp.html.

25. F. Bacchus and F. Kabanza, Using temporal logics to express search control knowledge for planning, *Artificial Intelligence*, 116(1–2), 123–191, 2000.

26. R. M. Jensen and M. M. Veloso, OBDD-based universal planning for synchronized agents in non-deterministic domains, *Journal of Artificial Intelligence Research*, 13, 13–189, 2000.

27. D. Conner, A. Rizzi and H. Choset, Composition of local potential functions for global robot control and navigation, In *Proceedings of 2003 IEEE/RSJ International Conference on Intelligent Robots and Systems (IROS 2003)*, IEEE, 2007, pp. 3116–3121, Las Vegas, Nevada, USA.

28. C. Belta and L. Habets, Constructing decidable hybrid systems with velocity bounds, In *43rd IEEE Conference on Decision and Control*, Bahamas, 2004.

29. A. Cimatti, E. Clarke, F. Giunchiglia and M. Roveri, NUSMV: A new symbolic model verifier, In *Proceedings Eleventh Conference on Computer-Aided Verification (CAV'99)*, ser. Lecture Notes in Computer Science, N. Halbwachs and D. Peled, Eds., no. 1633. Trento, Italy: Springer, July 1999, pp. 495–499.

30. M. de Berg, M. van Kreveld, M. Overmars and O. Schwarzkopf, *Computational Geometry: Algorithms and Applications*, 2nd ed., Berlin, Heidelberg, New York: Springer-Verlag, 2000.
31. G. J. Holzmann, *The Spin Model Checker: Primer and Reference Manual*, Lucent Technologies Inc., Bell Laboratories: Addison-Wesley, 2004.
32. R. Alur, T. A. Henzinger, G. Lafferriere, George and G. J. Pappas, Discrete abstractions of hybrid systems, In *Proceedings of the IEEE*, 2000, pp. 971–984.
33. L. Habets and J. van Schuppen, A control problem for affine dynamical systems on a full-dimensional polytope, *Journal of Artificial Intelligence Research*, 40(1), 21–35, 2004.
34. M. Kloetzer and C. Belta, A fully automated framework for control of linear systems from temporal logic specifications, *IEEE Transactions on Automatic Control*, 53(1), 287–297, 2008.
35. G. E. Fainekos, H. Kress-Gazit and G. J. Pappas, Hybrid controllers for path planning: A temporal logic approach, In *Proceedings of the 44th IEEE Conference on Decision and Control*, December 2005, pp. 4885–4890, Seville, Spain.
36. N. Piterman, A. Pnueli and Y. Sa'ar, Synthesis of reactive(1) designs, *Lecture Notes In Computer Science*, Springer-Verlag, 3855, 364–380, 2006.
37. A. Pnueli and R. Rosner, On the synthesis of a reactive module, In *POPL '89: Proceedings of the 16th ACM SIGPLAN-SIGACT Symposium on Principles of Programming Languages*, New York, NY: ACM, 1989, pp. 179–190.
38. E. Klavins, Automatic synthesis of controllers for assembly and formation forming, In *Proceedings of the International Conference on Robotics and Automation*, 2002.
39. T. Tabuada and G. J. Pappas, Linear time logic control of linear systems, submitted, *IEEE Transactions on Automatic Control*, 51(12), 1862–1877, 2006.
40. M. Egerstedt, Motion description languages for multi-modal control in robotics, In *Control Problems in Robotics*, A. Bicchi, H. Cristensen and D. Prattichizzo (Eds.), Atlanta, GA: Springer-Verlag, pp. 75–90, 2002.
41. B. Meyer, Applying design by contract, *IEEE Computer*, 25(10), 40–51, 1992.
42. M. Clavel, S. Eker, P. Lincoln and J. Meseguer, *Principles of Maude*, Electronic Notes in Theoretical Computer Science, Vol. 4, 1996.
43. M. Clavel, F. Duran, S. Eker, P. Lincoln, N. Marti-Oliet, J. Meseguer and J. Quesada, Using maude, In *Proceedings of the Third International Conference on Fundamental Approaches to SE*, Lecture Notes in CS 1783, pp. 371–374, 2000, Berlin, Germany.
44. B. Mishra and E. M. Clarke, Hierarchical verification of asynchronous circuits using temporal logic, *Theoretical Computer Science*, 38, 269–291, 1985.
45. M. Egerstedt and R. Brockett, Feedback can reduce the specification complexity of motor programs, *IEEE Transactions on Automatic Control*, 48(2), 213–223, 2003.
46. E. Klavins, R. Ghrist and D. Lipsky, A grammatical approach to self-organizing robotic systems, *IEEE Transactions on Automatic Control*, 51(6), 949–962, 2006.
47. Judith Bishop. *C# 3.0 Design Patterns*, Ed. 1. O'Reilly, December 2007.

14 A Constraint Programming Solution for the Military Unit Path Finding Problem

Louise Leenen, Alexander Terlunen and Herman le Roux

CONTENTS

14.1 INTRODUCTION

The path/motion planning problem in the context of a mobile robot/vehicle was discussed in Chapter 3 and some results using A* algorithm were given. In this chapter, we present an algorithm to solve the Dynamic Military Unit Path Finding Problem (DMUPFP) which is based on Stentz's well-known D* algorithm to solve dynamic path finding problems [1]. The Military Unit Path Finding Problem (MUPFP) is the problem of finding a path from a starting point to a destination where a military unit has to move, or be moved, safely while avoiding threats and obstacles and minimising path cost in some digital representation of the actual terrain [2].

Path finding is the problem of moving an object from a starting point to a destination point, while avoiding obstacles and minimizing costs. It has many applications ranging from computer games, transportation, robotics, networks and others. Path finding algorithms can be divided into two different types, static (or global) and dynamic algorithms. For a static algorithm, the environment is assumed to be known and an optimal path is calculated before the object is moved. For a dynamic algorithm, the environment may not be completely known before the object starts moving, or it may change while the object is moving. In such a case, the path plan has to be updated while it is being executed. The path finding problem has been well studied and there is a vast body of research in the literature.

During military operations, military units often encounter the path finding problem; they aim to avoid threats or at least pass (or avoid) threats with a minimum separating distance while attempting to move as quickly as possible. In the MUPFP, the requirement is to maintain a balance between the two main criteria, route speed and safety [2]. Although there are methods to solve the MUPFP, the existing approaches combine the optimization of the various criteria in some form of an objective function. Our approach, on the other hand, is to minimise path costs while ensuring that certain criteria, such as safety requirements, are met. This means that our objective function is a pure cost function. We adopt a constraint-based approach with a clear distinction between the goal of obtaining an optimal cost and adherence to safety measures. Such an approach also allows for flexibility in terms of modelling different constraints. If the user has a new requirement in terms of adhering to safety measures or some other constraint on the problem, this simply entails the addition of constraints or modification of existing constraints, but the main algorithm stays the same. Section 14.3 gives an overview of Constraint Programming (CP) and Constraint Satisfaction Problems (CSPs). We present a new algorithm to solve the Dynamic MUPFP (DMUPFP) in Section 14.4. The objective function only represents the cost of traversing a particular path, that is, a representation of the route speed. We model the safety criteria as constraints which have to be satisfied. For example, a constraint can specify that obstacles must be avoided, or that an acceptable distance has to be observed when passing a threat such as a sniper. Previous work done revolved around a static constraint-based approach which was based on the A* algorithm [3].

14.2   OVERVIEW OF PATH FINDING METHODS

A significant volume of research has been done on path finding techniques, and the success of a particular technique relies on the environment and the constraints that are imposed on it. In this chapter, the scope is limited to a domain where a military unit has to move, or be moved, safely to a destination while avoiding obstacles such as threats and structures. We assume that the terrain can be presented as a two-dimensional grid or a graph. Tarapatta [4] gives an overview of techniques for terrain representation. The most common approaches to solving path finding problems are based on Dijkstra's shortest path algorithm and the A* search algorithm. The A* search algorithm is a graph-based algorithm that finds the least cost path between a source and destination node by minimising the estimated cost to the goal from the current node, as well as the cost of the path so far. This algorithm is regarded as one of the most efficient graph-based path finding algorithms. A* search-based algorithms are used in most electronic games for path finding. A* search is a heuristic, optimal algorithm and also optimally efficient, that is, at least as efficient in terms of node expansion as any other optimal algorithm [5]. The choice of a heuristic function is very important. A* search is optimal if the heuristic function never overestimates the minimum cost from a node n to the destination node. Common heuristic functions are the Manhattan or Euclidean distance. One of the disadvantages of the A* search algorithm is that it requires a large amount of memory space. Path finding on large maps can be problematic but many extensions of the algorithm address this deficiency by reducing search space, and time and memory requirements. The A* search algorithm is formulated for static environments but there is an extension, the D* algorithm, that is designed for dynamic environments. We discuss this algorithm in more depth later on.

Botea et al. [6] provide a survey of hierarchical path finding algorithms where the task is broken down into smaller components in terms of terrain presentation. The simplest instance is a two-level hierarchical approach [7]. The original map is abstracted into clusters, and a path is found from the one boundary of a cluster to the other. An example of a cluster can be a room in a building or a part of a field. This approach is fast but not necessarily optimal. Another approach is to abstract a map efficiently in terms of points of visibility [8]. The corners of convex obstacles are represented together with edges linking relevant nodes. This approach works well when the number of obstacles is relatively small and if obstacles do not have concave or curved shapes. Duc et al. [9] show how they successfully apply Hierarchical Path Finding, Points of Visibility and Reinforcement Learning in the strategy game and simulation environment. Quadtrees are used to decompose maps hierarchically by partitioning a map into square blocks of different sizes such that a block contains either walkable cells or block cells [10]. If this is not the case, a block is further decomposed. In this approach, a path is found between the centres of two adjacent blocks. In Ref. [11], the initial graph is decomposed into a set of sub-graphs and a global sub-graph to link the fragmented sub-graphs. Information is also stored for re-use in subsequent calculations. In Hierarchical A* search [12], a hierarchy of abstract spaces is built until a one-state space is obtained. The aim of the hierarchical representations of a space is to reduce the overall search time, as well as the size of the search space. Some of these algorithms have been modified for dynamic environments. Wichmann et al. [13] show how to reduce the size of the search space by computing a multi-resolution representation of the terrain. They start with a low-resolution representation of the terrain and find a route from a source to a destination node. They then proceed by using each pair of nodes in the low-resolution solution as a source and destination node in a higher resolution representation of the terrain. This process is repeated until a solution has been found for the original terrain representation. Experiments show that multi-resolution processes cuts down the search time and number of nodes visited significantly although the smoothness of the resulting path decreases as resolution decreases. This technique can be applied to search algorithms other than A* search as well. In an incremental search, information from previous searches is re-used to find a solution rather than solving a series of problems from scratch. These algorithms are useful in dynamic environments. Koenig et al. [14] provide an overview of incremental search techniques and addresses planning when the environment is dynamic. Lifelong Planning A* (LPA*) [15] is an incremental version of A* search where information of previous searches is re-used in cases where the environment suddenly changes. The first iteration of LPA* is similar to A* search. If the graph changes, subsequent searches can be a lot faster because it uses information from the earlier search which are based on identical parts of the search tree. LPA* is also equipped to deal with changes to the graph during its search. LPA* is an optimal algorithm.

The D* algorithm [1] is based on the A* algorithm but caters for path finding in unknown, partially known, and changing environments. It is optimal and is general enough to be applied to any cost path optimization problem where the cost values change during execution. In D* Lite (algorithm) properties of LPA* and D* are combined [16]. A complete planner can solve any solvable problem and indicate failure for non-solvable problems. However, completeness often leads to long running times. Weaker forms of completeness have been defined, such as probabilistic completeness. A planner is called probabilistically complete [17] if, given a solvable problem, the probability of solving it converges to 1 as the running time approaches infinity. Svestka and Overmars [17] give an overview of probabilistic path planning. Examples of successful probabilistic planners are the probabilistic path planner (PPP), planners using genetic algorithms and the randomised path planner (RPP) [18]. Ant Colony Optimisation (ACO) is a probabilistic technique for solving computational problems that can be applied for path finding. These algorithms are based on heuristic information formulated from ants' behaviour. Ants search for food and when they return to their colony they leave pheromones on their way back. Ants that find pheromone trails are likely to follow these paths and also leave pheromones. The pheromones evaporate and become weaker over time. Pheromones on a longer path back to the colony will disappear faster than those on a shorter path, and thus a shorter path is more likely to be followed and reinforced. Mora et al. [2,19] have adapted ACO algorithms for the MUPFP with great success. Very little work has been done in terms of formulating path finding as a CSP. Both Gualandi et al. [20] and Allo et al. [21] successfully used concurrent programming to address path finding for an aircraft.

14.2.1 D* Algorithm for Solving Dynamic Path Finding Problems

Stenz introduced the D* algorithm for solving dynamic path finding problems in 1994 [1]. We give a short overview of this algorithm here to make Section 14.2 (where we describe our modified D* algorithm to solve the DMUPFP) easier to read. For a detailed description of the algorithm, refer Ref. [1]. The algorithm initially calculates an optimal path from the destination node back to the start node, using a best first approach. For each node x, the algorithm maintains an estimate of the total sum of the edge costs from the node to the destination node, $h(x)$. Ideally, this estimate is equal to the minimum cost from the node x to the destination node. The algorithm also maintains a list of OPEN nodes, sorted according to the key value, $k(x)$, of each node. The key value of a node x is the minimum of its $h(x)$ value before possible modifications to map values, and all its $h(x)$ values since the node x was first placed on the OPEN list. The key value of a node x identifies the node either as a RAISE state or a LOWER state. A RAISE state is when $k(x)$ is smaller than $h(x)$, which means the path cost has increased and this information has to be propagated. On the other hand, a LOWER state, $k(x) \geq h(x)$, indicates that there may have been a path cost reduction. If $k(x)$ equals $h(x)$, it indicates that the path from the destination node to the node x is optimal and the neighbours of x are expanded. The algorithm keeps track of current best paths by setting back-pointers from each node 'back' to its current predecessor in the solution path. The main algorithm, *MUPFP_D Star* and the *Initialise* function are presented in Figures 14.1 and 14.2. The first phase of the algorithm halts when the start node is removed from the OPEN list. The back-pointers are then followed from the start node to the destination node while a sum of the actual costs of each edge in this path is calculated. If this sum of actual costs is in dispute with the h value of a node on the path, the algorithm halts at that particular node. Such a discrepancy indicates that there has been a modification to cost of an edge, and the algorithm re-calculates an optimal path from the node where the discrepancy has been identified to the destination node. The first phase of the D* algorithm is similar to the

```
MUPFP_DStar( )

1    Initialise ( );
2    ComputeOptimalPath( );
3    for ever do
4        ...wait for changes...
5      if the value of an edge (x, y) has changed then
6        ModifyCost(x, y); //Call this if the change was for a cost
7      else if the status of a node changed then
8        StatusChange(s, status_old);
9        for x ∈ PutOn do
10         if t(x) ≠ NEW then
11            InsertOnOpenQ(x, h(x));
12         PutOn = ∅;
13         for i ∈ TakeOff do
14           flag = true;//local Boolean variable
15           for x ∈ Neighbour(i) do
16             if t(x) ≠ NEW then
17                flag = false;
18                exit for loop;
19           if flag = true then
20              t(i) = NEW;
21              Remove(OpenQ, i);
22           TakeOff = ∅;
23    s = -1; //Initialise s
24    ComputeOptimalPath( );
```

FIGURE 14.1   The main algorithm, *MUPFP_D Star*.

```
Initialise( )

1    OpenQ = ∅;
2    TakeOff = ∅;
3    PutOn = ∅;
4    for  s ∈ N do
5       h(s) = ∞;
6       t(s) = NEW;
7       b(s) = NULL;
8    InsertOnOpenQ(s_goal, 0);
```

FIGURE 14.2   The *Initialise* function.

```
ComputeOptimalPath( )

1  while  s ≠ s_start AND  OpenQ ≠ empty do
2    s = Dequeue(OpenQ);
3    k_old = k(s);
4    NSet = Neighbour(s);
5    for all  x ∈ NSet do  //Remove any invalid nodes in loop
6      if ¬ ConstraintCheck(x) then
7          Remove(NSet, x);
8    if  k_old < h(s) then  //a RAISE state
9      for all  x ∈ NSet do
10       if  h(x) ≤ k_old AND  h(s) > h(x) + c(x, s) then  //path via x
         better than via s
11          b(s)  = x;
12          h(s) = h(x) + c(x, s);
13   if  k_old = h(s) then  //path via s is optimal
14     for all  x ∈ NSet do  //can path cost be lowered via x?
15       if  (t(x) = NEW)
         OR  (b(x) = s AND  h(x) ≠ h(s) + c(s, x))
         OR  (b(x) ≠ s AND  h(x) > h(s) + c(s, x)) then
16          b(x) = s;
17          InsertOnOpenQ(x, h(s) + c(s, x));
18   else  //s is a LOWER state
19     for all  x ∈ NSet do
20       if  (t(x) = NEW)
         OR  (b(x) = s AND  h(x) ≠ h(s) + c(s, x)) then
21          b(x) = s;
22          InsertOnOpenQ(x, h(s) + c(s, x));
23       else if  b(x) ≠ s AND  h(x) > h(s) + c(s, x) then
24          InsertOnOpenQ(s, h(s))
25       else if  b(x) ≠ s AND  h(s) > h(x) + c(x, s)  AND  h(x) > k_old
         AND  t(x) = CLOSED then
26          InsertOnOpenQ(x, h(x))
```

FIGURE 14.3   The *ComputeOptimalPath* function.

ComputeOptimalPath function shown in Figure 14.3 without the constraint checking code in lines 6–7. The second phase of D* replaces nodes on the OPEN list where edge cost has been changed.

14.3   OVERVIEW OF CONSTRAINT SATISFACTION PROBLEMS

The modelling of problems in terms of constraints has the advantage of a natural, declarative formulation. When a problem is defined as a Constraint Satisfaction Problem (CSP), it is a representation of what must be satisfied, without specifying how it should be satisfied. A CSP consists of a set of variables, a set of domains for the variables, and a set of constraints. Each constraint is defined over a subset of the set of variables. A constraint is a logical relation involving one or more variables, where each variable has a domain of possible values. A constraint thus restricts the possible values that variables can have. Constraints can specify partial information about variables, are declarative and may be non-linear. A solution to a CSP specifies values for all the variables such that all the constraints are satisfied. There are many general-purpose techniques that can be used to solve CSPs, for example, integer programming, local search, and neural network techniques, but there is a special-purpose technique that is widely used: tree search in conjunction with backtracking and consistency checking. Constraint Programming (CP) is a term that refers to the computational systems used to solve CSPs. CP emerged from a number of disciplines such as artificial intelligence, computational logic, programming languages and operations research. CP has proven to be effective at solving combinatorial and over-constrained problems, that is, sets of constraints that cannot be satisfied. There are several texts that provide an introduction to CP and CSPs [22,23].

14.4   MILITARY UNIT PATH FINDING CONSTRAINT SATISFACTION PROBLEM

In this section, we discuss a constraint programming approach to solve the MUPFP by formulating the problem as a CSP. In earlier work, we defined and implemented a modified A* search algorithm to solve our CSP formulation of the MUPFP for the static case. This algorithm was briefly discussed in Section 14.1. We introduce a modified D*-based algorithm to solve the DMUPFP in Section 14.2 and show some examples of its application in Section 14.3.

The advantage of following a constraint-based approach to our path finding problem is the flexibility this framework offers. The MUPFP involves an environment and goals that can effectively be represented by constraints, where new information can be added with ease. In our CSP formulation of the problem we opted for a graph-based approach for ease of implementation. The terrain map is divided into nodes and edges, where each edge has an associated cost. A node represents a geographical location and an edge represents a route between two nodes, that is, two geographical areas. In our formulation, the cost of an edge represents the difficulty of moving though a particular part of the actual route represented by that edge. For example, if there is a structure on a route through which a soldier cannot move (an obstacle), this edge will have a bounded cost (maximal) value. We represent the avoidance of danger through constraints. An example of a threat is a sniper whose presence at a certain node is known. In this formulation, we include constraints to disallow movement to a node where there is a known obstacle or a threat, and constraints that disallow movement within a certain distance of a node containing a known threat. Threats and obstacles are dynamic in the sense that they can move or disappear. Suppose the terrain map is divided into a graph with a set of nodes, $N = \{x_1, x_2, \ldots, x_k\}$ and a set of edges, $A = \{(x_i, x_j) \mid i, j \in N\}$. Every edge in the set A, (x_i, y_j), has an associated cost value, $c(x_i, y_j)$. The function $status: N \rightarrow \{O, T, U\}$ defines whether there is an obstacle (O) or a threat (T) present at a node, or if the node is unoccupied (U). In the latter case, there is no known obstruction. Let the set of variables be $V = \{V_1, V_2, \ldots, V_n\}$, where each variable represents one node in the solution path. A path is a sequence of nodes in a graph such that there is edge from each node in the sequence to the next node in the sequence. The domain of each variable is $Dom = \{x_1, x_2, \ldots, x_k\}$. The set of constraints, C, contains at least the ones listed below:

C_1: An all-different constraint, $V_1 \neq V_2 \neq \cdots \neq V_n$. This constraint ensures that each node in the solution path is unique.

C_2: $V_1 = s_{start}$ and $V_n = s_{goal}$, where s_{start} represents the starting node in the path and s_{goal} represents the destination node in the path.

C_3: For every V_{i+1}, $i = 1, \ldots, n - 1$, if $V_i = x_j$ then $V_{i+1} = x_m$ if and only if there exists an edge $(x_j, x_m) \in N$. This constraint ensures that the value assignment of the variables forms a path.

C_4: For every V_i, $i = 2, \ldots, n - 1$, $V_i \neq x_j$ if $status(x_j) = \{O,T\}$. This constraint disallows a node containing a threat or an obstacle to form part of the solution path.

C_5: For every V_i, $i = 2, \ldots, n - 1$, $V_i \neq x_j$ if there exists an edge (x_i, x_j) such that $status(x_j) = \{T\}$. This constraint ensures that the solution path does not venture too close to a threat. In this case we model a safe distance to be more than a single edge, that is, we do not allow a node (that is a neighbour of a threat) to be included in the optimal path.

The solution to the problem is an assignment of values for the variables in the set V such that

$$\min \sum_{s=1,\ldots,n-1} c(x_i, y_j), \text{ where } V_s = x_i \text{ and } V_{s+1} = y_j$$

Note that constraint C_5 can be alternatively modelled such that it limits the allowed cost of the edge between a node in the path and a neighbour node: for every V_i, $i = 2, \ldots, n - 1$, if $status(x_j) = \{T\}$, then $V_i \neq x_j$ there exists an edge (x_i, x_j) such that $c(x_i, x_j) < SafeDistance$, where $SafeDistance$ is some integer bound.

14.4.1 Solving the Static Military Unit Path Finding Constraint Satisfaction Problem

Leenen et al. defined and implemented a modified A* search algorithm to solve the static MUPFP formulated as a CSP [3]. This algorithm maintains a list of potential (partial) solutions from the designated start node to the designated destination node. From this list the best potential partial solution (in terms of total path cost while satisfying constraints) is selected and expanded: a new waypoint is added to the end of the chosen partial path for every possible extension. This algorithm does not use heuristic information. Any new waypoint has to satisfy all the constraints.

A CSP has a finite, known number of variables. In our application we do not know in advance what the length of an optimal or good solution path is. We thus have to decide what the value of n (number of decision variables in the CSP) is before solving our CSP. The D* algorithm calculates an optimal solution path and then retraces this path while testing for changes in cost values. If the initial solution path contains m edges, then we assign values to $n = m + 1$ variables. The value of n may change if there had been cost changes and another execution of the algorithm is required.

14.4.2 Solving the Dynamic Military Unit Path Finding Constraint Satisfaction Problem

The dynamic version of the MUPFP allows for changes in the graph at any time. The D* algorithm only considers possible changes to edge costs after it has calculated a potential optimal path. It then re-traces the potential optimal path and checks at each node if the path is indeed optimal or whether there have been edge cost value changes that may affect its optimality. Our Modified D* algorithm, MUPFP_D*, deviates somewhat in this regard: our algorithm computes an optimal path while satisfying all the constraints, and then it enters an event-driven phase where it checks constantly for any modifications that have been made to graph information which has not been considered during the calculation of the current best path. These deviations in our modified D* algorithm are loosely based on approaches followed in the D* Lite and LPA* algorithms [15,16].

In our application, we have modelled constraints to avoid obstacles and to keep a specified distance from known threats. For simplicity, we modelled this safe distance to be 'one node away', that is, our path will not include any node that is a neighbour of a threat node. When we consider the list of neighbours of a particular node, x, for expansion, we check if a neighbour node, y, is an obstacle or a threat (i.e., we perform constraint checking). Our algorithm calls a function, *ComputeOptimalPath*, to compute an optimal path for the current graph while satisfying safety constraints, and then it enters an infinite loop where it waits for either of the following events to occur: (a) the cost of an edge has been modified, or (b) the status of a node has been modified. The calculation of an initial optimal path is similar to that of the D* algorithm. It initially calculates an optimal path from the destination node back to the start node, using a best first approach. For each node x, the algorithm maintains an estimate of the total sum of the edge costs from the node to the destination node, $h(x)$. Ideally, this estimate is equal to the minimum cost from the node x to the destination node. The algorithm also maintains a list of OPEN nodes, sorted according to the key value, $k(x)$, of each node. The key value of a node x is the minimum of its $h(x)$ value before possible modifications to map values, and all its $h(x)$ values since the node x was first placed on the OPEN list. The key value of a node x identifies the node either as a RAISE state, a LOWER state or an optimal state. A RAISE state is when $k(x)$ is smaller than $h(x)$, that is, path cost increase information has to be propagated. On the other hand, a LOWER state, $k(x) \geq h(x)$, indicates that there may have been path cost reductions. If $k(x)$ equals $h(x)$, the path to the node x is optimal, and its neighbour node are expanded. The algorithm keeps track of current best paths by setting back-pointers from each node 'back' to its current predecessor in the solution path. For each of these above-mentioned events our algorithm executes a number of steps and then the *ComputeOptimalPath* function is called again to re-compute an optimal path for the changed graph. The algorithm waits for one of the events above to occur and then the cycle is repeated.

In the case of a cost change, we follow steps similar to such an occurrence in the D* algorithm by identifying nodes to be re-inserted on the OPEN list. We also adjust the h values of nodes that may be affected by the change in cost. In the case of a node whose status has changed, the algorithm does the necessary checks to identify nodes that have to be re-evaluated by calling the function *StatusChange* in which one of the following cases are identified:

- A node s's status changed from an obstacle or unoccupied to that of a threat. This node is made invalid for inclusion in an optimal path and for every one of its neighbours, x, the function, *RevertPath*, is called. The purpose of this function is to trace expanded paths through the node x and make adjustments to predecessor nodes. This is necessary because the node x is now the neighbour of a threat node and may not be included in a current best path to any node from the goal node. These paths need to be destroyed.
- If a node s's status changed from unoccupied to an obstacle, then *RevertPath* is called to ensure that s is not included in a best path from the goal node to any other node.
- If a node s's status changed from an obstacle to unoccupied, then the neighbour nodes of s are inserted on the OPEN list if they qualify for consideration.
- A node s's status changed from a threat to either unoccupied or an obstacle. In this case, the neighbours of s had not been considered for expansion prior to the status change, so the neighbour nodes are now considered for insertion on the OPEN list.

The main algorithm, *MUPFP_D Star* and other sub-algorithms, the *Initialise* function, the *ComputeOptimalPath* function, the *ConstraintCheck* function, the *InsertOnOpenQ* function, the *ModifyCost* function, the *StatusChange* function, the *RevertPath* function, and the *DangleNode* function, are given in Figures 14.1 through 14.9. We now describe this new algorithm in greater detail.

14.4.3   MUPFP_D* ALGORITHM

The algorithm assumes the existence of the following information and data structures:

- A CSP with n number of variables, a set of constraints C and a set of domain values *Dom*.
- A graph consisting of a set of nodes, N, and a set of edges, A. Every node $x \in N$ has a label, a status value, a heuristic value, $h(x)$, a key value, $k(x)$ and a back-pointer, $b(x)$, which can point to another node in N. Every edge $(x, y) \in A$ has an associated cost, $c(x, y)$.

- A function, *status: N → {U, O, T}*, where *U* represents an unoccupied node, *T* represents the presence of a threat and *O* represents an obstacle.
- A labelling function, *t: N → {NEW, OPEN, CLOSED}*. NEW indicates that a node has not yet been expanded, OPEN indicates that the node is a member of the OPEN list and CLOSED indicates that the node has been considered and removed from the OPEN list.
- A designated source node, s_{start}, and a designated destination node, s_{goal}.
- *OpenQ* is a priority queue that represents the OPEN list of nodes to be expanded and it is sorted in non-decreasing order in terms of the key values of its members.
- *PutOn* and *TakeOff* are two sets. The first set contains nodes that may have to be returned to *OpenQ* after a status change in a node, and the second set contains nodes that may have to be removed from *OpenQ* after a status change in a node.

The following functions are called in the algorithm:

- *Initialise()* initialises global variables and inserts the goal node into *OpenQ*.
- *ComputeOptimalPath()* computes an optimal path from the source node to the destination node while satisfying the threat and obstacle constraints.
- *ConstraintCheck(s)* performs the constraint checking: it checks whether any neighbour of the node *s* is an obstacle, a threat, or the neighbour of a threat. If this is the case, it returns a value of *false*, otherwise it returns *true*.
- *StatusChange(s,status_{old})* is called when a node *s* is identified to have received a new status value. This function makes the necessary modifications to various nodes' values and other collections such that a new optimal path can be computed. It calls the two functions *RevertPath* and *DangleNode* if necessary. Note that the second parameter of *RevertPath* is optional. These three functions are described in more detail later in this section.
- *ModifyCost(x,y)* is called when the cost of an edge (*x, y*) has been changed. It returns relevant nodes to *OpenQ*.
- *Neighbours(s)* returns a set that contains all the neighbour nodes of the node *s*.
- *InsertOnOpenQ(s, value)* insert a node *s* with new *h* value, *value*, onto the priority queue *OpenQ* after calculating the new *k* value of the node *s*.
- *Enqueue(Q,s,v)* is the operator that adds a node *s* with a value *v* to a priority queue *Q*.
- *Dequeue(Q)* removes and returns the first element (element with a minimal associated value) from a priority queue, *Q*.
- *Remove(Q,s)* removes a member *s* from a priority queue, *Q*.
- *Add(S,s)* inserts a member *s* into a set *S*.

The assumption is that neither the start node nor the destination node is an obstacle, a threat or the neighbour of a threat.

14.4.3.1 The Main Function, MUPFP_DStar

The main function of the MUPFP_D* algorithm is shown in Figure 14.1. It calls the function *Initialise* shown in Figure 14.2. After computing an optimal path in line 2 of the main function, there is a *for*-loop in lines 3–24 that runs until the user halts the algorithm. This algorithm is event driven and the *if*-statement starting in line 5 checks for any changes that have been made to the graph. The *ComputeOptimalPath* function is similar to first phase of Stenz's D* algorithm with the addition of constraint checking, and can be viewed in Figure 14.3. It calls the *ConstraintCheck* and *InsertOnOpenQ* functions shown in Figures 14.4 and 14.5.

If the cost of an edge (*x, y*) has been changed in the main part of the algorithm (lines 5–6), the function *ModifyCost* is called. This function is shown in Figure 14.6. The function checks if the nodes *x* and *y* are included in some best current path from to goal. In this case, the ancestor node is

```
ConstraintCheck(s)

1 if status(s) = O then //is s a 'Obstacle'
2   return false;
3 for x ∈ Neighbour(s) do
4   if status(x) = T then //is s a 'Threat'
5     return false;
6 return true; // All the other checks failed, so the node is safe
```

FIGURE 14.4   The *ConstraintCheck* function.

```
InsertOnOpenQ(s, h_new)

1  if  s = s_goal  then
2     h_new = 0;
3  if  t(s) = NEW then
4     k(s) = h_new;
5  else if  t(s) = OPEN then
6     k(s) = min(k(s), h_new);
7  else if  t(s) = CLOSED then
8     k(s) = min(h(s), h_new);
9  Enqueue(OpenQ, s, k(s))
10 h(s) = h_new;
11 t(s) = OPEN;
```

FIGURE 14.5   The *InsertOnOpenQ* function.

```
ModifyCost(x, y)

1  if  b(y) = x then
2     if t(x) = CLOSED then
3        Insert(OpenQ, x, h(x));
4  else if b(x) = y then
5     if t(y) = CLOSED then
6        InsertOnOpenQ(y, h(y));
7  else
8     if t(y) = CLOSED then
9        InsertOnOpenQ(y, h(y));
10    if t(x) = CLOSED then
11       InsertOnOpenQ(x, h(x));
```

FIGURE 14.6   The *ModifyCost* function.

re-inserted on *OpenQ* if it has a CLOSED status. Otherwise, if either *x* or *y* is CLOSED, it also has to be re-inserted on *OpenQ*.

Note that our implementation allows a user to make a map change, that is, change the status of a node or the cost of an edge, at any time. These changes are made immediately in the graph, but the implementation keeps a record of all changes such that the functions *StatusChange* or *ModifyCost* will be called in the main *for*-loop.

StatusChange (line 8 of the main algorithm) is called when any node *s* has had a change in its status. This function appears in Figure 14.7. If the new status of a node *s* is that of a threat (lines 3–10), then it cannot be considered for inclusion in any best path and neither can its neighbours. *RevertPath* is called for every neighbour *x* of *s* to destroy partial best paths that include *x*. If the

```
StatusChange(s, status_old)

1   status_new = status(s);
2   if status_new = T then //new status is a 'Threat'
3     h(s) = k(s) = ∞;
4     if t(s) = OPEN then
5       Remove(OpenQ, s);
6     t(s) = NEW;
7     for x ∈ Neighbours(s) do
8       if t(x) ≠ NEW then
9         RevertPath(x, s);
10    b(s) = NULL;
11  else if status_old = U then //changed from 'Unoccupied' ⇒ 'Obstacle'
12    if t(s) ≠ NEW then
13      RevertPath(s);
14  else if status_old = O then //changed from 'Obstacle' ⇒ 'Unoccupied'
15    for all x ∈ Neighbour(s) do
16      if t(x) ≠ NEW AND status(x) = U then
17        InsertOnOpenQ(x, h(x));
18  else if status_old = T then //was status 'Threat' before
19    for all x ∈ Neighbour(s) do
20      for all y ∈ Neighbour(x) do
21        if x ≠ s AND t(y) ≠ NEW AND status(y) = U then
22          InsertOnOpenQ(y, h(y));
```

FIGURE 14.7   The *StatusChange* function.

status of s changed from unoccupied to an obstacle (lines 11–13), *RevertPath* is called to destroy partial best paths that include s. In the case of a status change from an obstacle to unoccupied (lines 14–17), all unoccupied neighbours that are not NEW are re-inserted onto *OpenQ*. Finally, in the case of a change from a threat to a non-threat or an obstacle, the complete neighbourhood of s had been invalid nodes prior to the status change. Thus, we have to consider every node x which is a neighbour of s, and place all the neighbours of x on *OpenQ* unless they are NEW or unoccupied. *RevertPath* calls the function *DangleNode* to identify invalid best path created by the status change. These functions are shown in Figures 14.8 and 14.9, respectively.

Lines 9–22 of the main algorithm consider nodes that were placed in the sets *PutOn* and *TakeOff* by calls to *DangleNode*. Nodes in the first set have to be inserted on *OpenQ* if they have any neighbour that is either CLOSED or OPEN, whereas nodes in the latter set can be removed from *OpenQ* if they only have NEW neighbour nodes.

```
RevertPath(s, i)

1   h(s) = k(s) = ∞;
2   b(s) = NULL;
3   if t(s) = OPEN then
4     Remove(OpenQ, s);
5   t(s) = NEW;
6   if b(s) ≠ NULL AND t(b(s)) ≠ NEW then
7     InsertOnOpenQ(b(s), h(b(s)));
8   for x ∈ Neighbours(s) do
9     if ConstraintCheck(x) AND i ≠ x AND b(x) = s then
10      DangleNode(x);
```

FIGURE 14.8   The *RevertPath* function.

```
DangleNode(s)

1  if t(s) ≠ OPEN then
2    t(s) = NEW;
3  else
4    Add(TakeOff,s);
     //Nodes in TakeOff may later be removed from OpenQ
5  for all x∈ Neighbour(s) do
6    if x ≠ b(s) AND ConstraintCheck(x) then
7      if b(x) = s then
8        DangleNode(x);
9      else if t(x) = CLOSED then
10         Add(PutOn, x);
       //Nodes in PutOn will later be considered for insertion on OpenQ
11 b(s) = NULL;
```

FIGURE 14.9   The *DangleNode* function.

14.4.3.2   *ComputeOptimalPath*

This function removes a node s with a minimal k value from *OpenQ* (line 2) and removes all its invalid neighbours from its neighbour set (lines 4–7). An invalid neighbour node is a node that does not satisfy one of the safety constraints, that is, it is an obstacle, a threat or the neighbour of a threat.

It then identifies whether s is a RAISE state (its k value is smaller than its h value) in lines 8–12, or a LOWER state (its k value is larger than its h value) in lines 18–26. In these cases, it propagates the changed costs. If s is optimal (its k value is equal to its h value) it expands the neighbours of s (lines 13–17).

14.4.3.3   *StatusChange*

In line 2, we identify that the node s has changed its status to that of a threat. This means that neither s nor any of its neighbours may be included in the best path from the goal node to any other valid node. If s is currently on *OpenQ*, it is removed (lines 4–5). In lines 7–9, the function ensures that the invalidity of neighbour nodes of s is recognised. If s has become an obstacle node, then lines 11–13 ensure that it is made invalid. Lines 14–22 deal with a previously invalid node that has become valid and must now be considered for inclusion in the best paths from the goal to other nodes by inserting relevant node on *OpenQ*.

14.4.3.4   *RevertPath(s,i)*

This function identifies and destroys partial best paths that include the parameter s which has become an invalid node. The parameter i is optional and is used to avoid the re-checking of a known invalid node, say node a which is known to have become a threat. Lines 3–4 ensure that s is not on *OpenQ*. In lines 6–10, we check for the existence of best paths between the destination node and other nodes which include the newly invalid node s. In this case, the function *DangleNode* is called to destroy the path. Otherwise the neighbour node x is the ancestor of s in its current best path from the destination node and x is placed *OpenQ* for re-consideration.

14.4.3.5   *DangleNode(s)*

Node s forms part of a best path that has become invalid. If it is on *OpenQ*, it will later be considered for removal otherwise its label is changed to NEW (lines 1–4). The neighbours of s are checked for any paths through it which will also now be invalid (lines 6–8), or otherwise for a neighbour that may have to be re-inserted on *OpenQ* (lines 9–10).

We have implemented this algorithm in C++ (Visual Studio 2008) and show a few examples next.

14.4.3.6   Example 14.1: Calculate an Initial Optimal Path

Consider the graph in Figure 14.10. The start node is node number 2 (flagged node at top right between nodes 4 and 9) and the destination node is node number 16 (flagged node at the bottom between nodes 13 and 14). The nodes numbered 1 and 18 are obstacles (darkest shaded nodes), and all the other nodes are unoccupied (shaded light grey). We show how the algorithm calculates the initial optimal path.

The destination node is inserted on *OpenQ* by the *Initialise* function. In the first execution of *MUPFP_DStar*, $s = 16$, and its neighbours, nodes 13 and 14, are placed on *OpenQ*. The following steps are performed:

$s = 13$ and *OpenQ*: 14, 9, 5
$s = 14$ and *OpenQ*: 9, 6, 7, 5
$s = 9$ and *OpenQ*: 6, 7, 5, 12, 2
$s = 6$ and *OpenQ*: 7, 5, 12, 11, 2, 17
$s = 7$ and *OpenQ*: 5, 12, 17, 11, 2
$s = 5$ and *OpenQ*: 12, 17, 3, 11, 2
$s = 12$ and *OpenQ*: 17, 3, 11, 2, 15
$s = 17$ and *OpenQ*: 3, 11, 2, 15
$s = 3$ and *OpenQ*: 11, 2, 15, 0, 19
$s = 11$ and *OpenQ*: 2, 15, 0, 19, 8
$s = 2$ and *OpenQ*: 15, 0, 19, 8, 4

The optimal path, 2–9–13–16, has a total path cost of 3, and is shown in black in Figure 14.11.

14.4.3.7   Example 14.2: The Status of Nodes Changes from Unoccupied to an Obstacle

Suppose node 13 changes from unoccupied to an obstacle. The previous optimal path has to be re-calculated. The main algorithm will identify the status change of node 13 and call the function *StatusChange(13)*. *RevertPath* is called in line 13. The neighbours of node 13, nodes 5, 9, 16 and 18, are evaluated in lines 5–10 of *RevertPath*. Node 16 is inserted on *OpenQ* (line 7), node 18 is an obstacle so nothing is done, and *DangleNode* is called for nodes 5 and 9 (line 10). The best paths through nodes 5 and 9 are invalidated: the call *DangleNode(9)* generates calls for nodes 2 and 12 to *DangleNode* which results in nodes 4 and 15 being inserted in *TakeOff* and later removed from *OpenQ* by the main function. Similarly, the call *DangleNode(5)* results in the nodes 0 and 19 being removed from *OpenQ*.

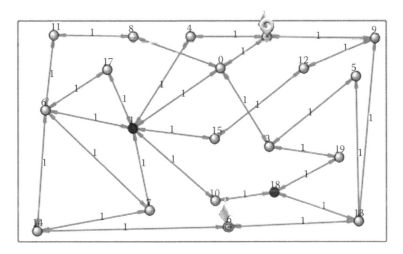

FIGURE 14.10   The initial graph in Example 14.1.

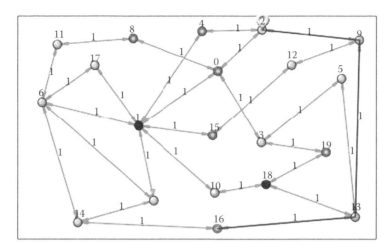

FIGURE 14.11 The optimal path calculated in Example 14.1.

OpenQ contains only nodes 16 and 8. A new optimal path, 2–0–8–11–6–14–16, is calculated by *ComputeOptimalPath* with a path cost of 6, and is shown in Figure 14.12. Note that the three obstacle nodes in Figure 14.12 are 1, 18 and 13, and the nodes 8, 4, 0, 15 and 19 are still on *OpenQ*.

14.4.3.8   Example 14.3: The Status of Nodes Changes from an Obstacle to a Threat

Now we change the status of node 1 from an obstacle to a threat and we change the status of node 13 back to unoccupied. Keep in mind that our application makes all map changes immediately and these changes are recorded such that *for*-loop in the main algorithm detects that such a change has occurred. In this case, these two status changes will result in two calls to *StatusChange*. *OpenQ* currently contains the nodes 3, 4, and 9. *StatusChange(1)* calls *RevertPath* for every neighbour of node 1. Nodes 3, 4, and 9 are removed from *OpenQ*, and node 14 is inserted. It also destroys the current best path via nodes 6 and 0. The call, *StatusChange(13)*, calls *RevertPath(13)* which re-inserts node 16 on *OpenQ*. Nodes 5 and 9 have a NEW status and node 18 is an obstacle.

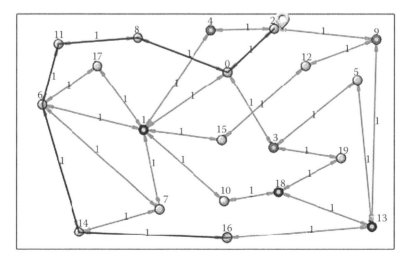

FIGURE 14.12 The optimal path calculated in Example 14.2.

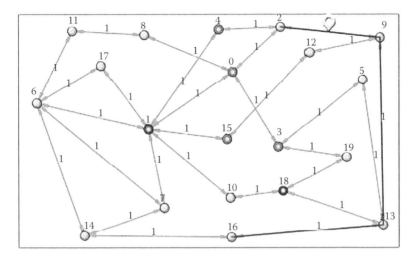

FIGURE 14.13   The optimal path calculated in Example 14.3.

OpenQ contains nodes 14 and 16 and then *CalculateOptimalPath* is called to recalculate the optimal path as 2–9–13–16. This is illustrated in Figure 14.13. Note that nodes 0, 3, 4, and 15 are still on *OpenQ*, node 18 is an obstacle and node 1 is a threat.

14.5   CONCLUDING REMARKS

In this chapter, we have introduced a constraint-based algorithm to solve a CSP formulation of the Dynamic Military Unit Path Finding Problem. Our algorithm is based on the well-known D* algorithm for dynamic path finding problems: we extended the D* algorithm by adding constraints that have to be satisfied. Our objective function represents only the path cost. The advantage of a CSP approach to path finding problems is the flexibility it provides in modelling problems with particular characteristics such as the military unit path finding where path costs have to be minimised while safety aspects have to be taken into account. This algorithm can be improved by including heuristics to direct the search. This should reduce computation considerably. The use of heuristic information will reduce computation considerably by directing the search.

REFERENCES

1. Stentz, A., 1994. Optimal and efficient path planning for partially-known environments, *IEEE International Conference on Robotics and Automation*, Carnegie Mellon University, Pittsburgh, USA.
2. Mora, A.M., Merelo, J.J., Laredo, L.L.J., Millan, C. and Torrecillas, J., 2009. *International Journal of Intelligent Systems*, Wiley Periodicals Inc., published online in Wiley InterScience, vol. 24, pp. 818–843 (www.interscience.wiley.com).
3. Leenen, L., Vorster, J.S. and Le Roux, W.H., 2010. A constraint-based solver for the military unit path finding problem, *SpringSim 2010*, April, Orlando, Florida, USA.
4. Tarapata, Z. 2003. Military route planning in battlefield simulation: Effectiveness problems and potential solutions. *Journal of Telecommunications and Information Technology*, **4**, 47–56.
5. Russel, S. and Norvig, P., 1995. *Artificial Intelligence: A Modern Approach*. 1st edn. New Jersey: Prentice Hall.
6. Botea, A., Müller, M. and Schaeffer, J., 2004. Near optimal hierarchical path-finding. *Journal of Game Development*, **1**(1), 7–28.
7. Rabin, S., 2000. A* aesthetic optimizations. In: M. Deloura, ed, *Game Programming Gems*. Hingham, MA, USA: Charles River Media, pp. 264–271.

8. Rabin, S., 2000. A* speed optimizations. In: M. Deloura, ed, *Game Programming Gems*. Hingham, MA, USA: Charles River Media, pp. 272–287.

9. Duc, L.M., Sidhu, A.S. and Chaudhari, N.S., 2008. Hierarchical path finding and AI-based learning approach in strategy game design. *International Journal of Computer Games Technology*, **2008** (Article ID 873913), http://www.hindawi.com/ journals/ijcgt/2008/873913/

10. Samet, H., 1988. *An Overview of Quadtrees, Octrees and Related Hierarchical Data Structures,* in NATO ASI Series, Vol. F40, pp. 51–68.

11. Shekhar, S., Fetterer, A., and Goyal, B., 1997. Materialization trade-offs in hierarchical shortest path algorithms, *5th International Symposium on Large Spatial Databases (SSD'97)*, Berlin, Germany.

12. Holte, R., Perez, M., Zimmer, R. and MacDonald, A., 1996. Hierarchical A*: Searching abstraction hierarchies efficiently, *Thirteenth National Conference on Artificial Intelligence (AAAI-96)*, Portland, Oregon, USA.

13. Wichmann, D.R. and Wuensche, B.C., 2004. Automated route finding on digital terrains, *International Image and Vision Computing New Zealand Conference*, University of Auckland, Auckland, New Zealand.

14. Koenig, S., Likhachev, M., Liu, Y. and Furcy, D., 2004. Incremental heuristic search in artificial intelligence. *Artificial Intelligence Magazine*, **25**(2), 99–112.

15. Koenig, S., Likhachev, M. and Furcy, D., 2004. Lifelong planning A*. *Artificial Intelligence*, **155**(1–2), 93–146.

16. Koenig, S. and Likhachev, M., 2002. D* lite, *Proceedings of the Eighteenth National Conference on Artificial Intelligence (AAAI-02)*, Edmonton, Alberta, Canada.

17. Svestka, P. and Overmars, M.H., 1998. Probabilistic path planning. In: J. Laumond, ed, *Robot Motion Planning and Control (LNCIS 229)*. Heidelberg: Springer-Verlag, pp. 255–304.

18. Barraquand, J. and Latombe, J.C., 1991. Robot motion planning: A distributed representation approach. *International Journal Robotics Research*, **10**(6), 628–649.

19. Mora, A.M., Merelo, J.J., Laredo, J.L.J., Castillo, P.A., Millan, C. and Torrecillas, J., 2007. Balancing safety and speed in the Military Path Finding problem: Analysis of different ACO algorithms, *Genetic and Evolutionary Computation Conference (GECCO'07)*, London, England, UK.

20. Gualandi, S. and Tranchero, B., 2004. Concurrent constraint programming-based path planning for uninhabited air vehicles, *Proceedings of SPIE's Defense and Security Symposium*, Orlando, Florida, USA.

21. Guettier, C., Allo, B., Legendre, V., Poncet, J.C., and Strady-Lécubin, N., 2002. Constraint model-based planning and scheduling with multiple resources and complex collaboration schema, *The Sixth International Conference on AI Planning and Scheduling (AIPS'02)*, Toulouse, France.

22. Dechter, R., 2003. *Constraint Processing*. 1st edn. San Francisco: Morgan Kaufmann.

23. Rossi, F., Van Beek, P. and Walsh, T. (eds), 2006. *Handbook of Constraint Programming* 1st edn. Amsterdam, The Netherlands: Elsevier.

15 Simultaneous Localization and Mapping for a Mobile Vehicle

Jitendra R. Raol

CONTENTS

15.1 INTRODUCTION

The simultaneous localization and mapping (SLAM) is a problem of robot localization and also of map building or upgrading the already available map (of the robot environment) simultaneously or in a boot-strap, or hand-in-hand manner. This problem is also known as concurrent mapping and localization (CML) within the map. The Visual SLAM (V-SLAM) is the combination of robotics and artificial vision with the aim of providing better solutions to the SLAM problem. Thus, with SLAM it is possible for a mobile robot to be placed at an unknown location in an unknown

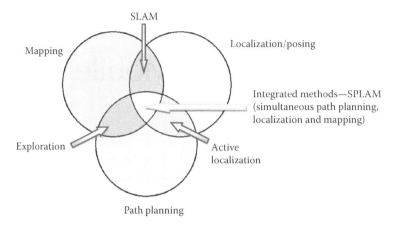

FIGURE 15.1 Integrated methods including SLAM and path planning.

environment. What this means is that the robot incrementally builds a consistent map of this environment while simultaneously determining its (robot's) location within this map, and hence a solution to the SLAM problem can/would provide the means to make a robot truly autonomous [1]. The SLAM problem is extensively studied and discussed in Refs. [1–5]. The SLAM is also very useful for unmanned aerial vehicles (UAVs), micro- and mini-air vehicles (MICAV/MAVs) and many other types of autonomous vehicles: from indoor (hospital service robots) to outdoor robots, underwater vehicles (UWV), unmanned ground vehicles (UGV) and airborne systems, for the latter it is called aerial SLAM. Thus, the SLAM is a process by which a mobile robot can build a map of an environment in which it is to move or traverse, and at the same time use this map to determine, decide, or find (in fact by process of state/parameter estimation) its own location. Both the trajectory of the robot's platform and the location of available landmarks (and thus map building) are estimated (online/real time) without any *a priori* knowledge of the robot's location. A domain-overview of the SLAM and the integrated problems of robot path planning are depicted in Figure 15.1 [3]. A mobile robot can traverse through an unknown environment by taking into consideration (meaning using these observation into onboard SLAM programme/algorithm), the relative observations of the landmarks (with respect to itself) provided. The estimates of these landmarks are most likely to be correlated with each other because of the common error in estimated vehicle location [2]. Thus, a good complete solution to the joint combined localization and mapping problem would require a joint state composed of the vehicle pose and every landmark position, to be updated following each landmark observation—joint parameter-state estimation problem, requiring to use a large state vector (if the number of landmarks maintained in the map is very large) [4]. This will be computationally burdensome.

15.2 SLAM PROBLEM

Let a mobile robot move through an environment taking relative observations of a number of unknown landmarks using a sensor located on the robot. At a time instant k, the quantities defined are [2]: (i) $\mathbf{x}_k$ is the state vector of the location (x, y positions, and here z is assumed to be non-significant) and orientation (angles) of the vehicle, (ii) $\mathbf{u}_k$ is the control vector, applied at time $k - 1$ to drive the vehicle to a state $\mathbf{x}_k$ at time k, what this means is that the robot's state will change from time $k - 1$ to time k, (iii) $\mathbf{m}_i$ is a vector that describes the location of the ith landmark, the true location of which is assumed time invariant and (iv) $\mathbf{z}_{ik}$ is an observation from the vehicle's sensor about the location (relative measurement with respect to the vehicle location) of the ith landmark at time index k. Also, some more relevant variable sets defined are: (a) $\mathbf{X}_{0:k} = \{\mathbf{x}_0, \mathbf{x}_1, \ldots, \mathbf{x}_k\} = \{\mathbf{X}_{0:k-1}, \mathbf{x}_k\}$ as the time history of the vehicle's locations, (b) $\mathbf{U}_{0:k} = \{\mathbf{u}_1, \mathbf{u}_2, \ldots, \mathbf{u}_k\} = \{\mathbf{U}_{0:k-1}, \mathbf{u}_k\}$ as the time history

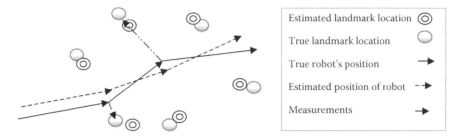

FIGURE 15.2   SLAM problem geometry.

of the applied control inputs, (c) $\mathbf{m} = \{\mathbf{m}_1, \mathbf{m}_2, \ldots, \mathbf{m}_n\}$ as the set of all landmarks and (d) $\mathbf{Z}_{0:k} = \{\mathbf{z}_1,$ $\mathbf{z}_2, \ldots, \mathbf{z}_k\} = \{\mathbf{Z}_{0:k-1}, \mathbf{z}_k\}$ as the set of all landmark observations. Figure 15.2 illustrates the SLAM problem's geometry [2]. In a nutshell, the problem of SLAM is to estimate the pose, x (robot's location, i.e., its position coordinates) and the map, m (the coordinates of the landmarks) of/for a mobile robot at the same time, $p(x, m|z, u)$ wherein the x is robot's poses, m is the map (represented by its coordinates with respect to some global coordinate/reference frame), z denotes the observations and movements of the robot and u denotes the control inputs. The SLAM problem is hard because (i) a map is needed to localize (to pose) the robot and (ii) pose (estimate) is needed to build a map. The map depends on the poses of the robot during data acquisition, if the poses are known, mapping is easy, and if the mapping is done accurately, the robot poses are accurate as well.

15.2.1 PROBABILISTIC FORM OF SLAM

In this representation, the simultaneous localization and map building problem requires that the following probability distribution be computed [2] for all times k.

$$P(\mathbf{x}_k, \mathbf{m} \mid \mathbf{Z}_{0:k}, \mathbf{U}_{0:k}, \mathbf{x}_0) \tag{15.1}$$

The above probability distribution specifies the joint posterior probability density function (pdf) of the landmark locations (m) and vehicle state, x, (at time k) given the measurements/observations (z) and control inputs (u) up to and including time k. The initial state of the vehicle, that is, the initial conditions are given. To solve the SLAM problem, we start with an estimate for the distribution $P(\mathbf{x}_{k-1}, \mathbf{m}|\mathbf{Z}_{0:k-1}, \mathbf{U}_{0:k-1})$ at time $k - 1$. Then the joint posterior probability distribution following a control input $\mathbf{u}_k$ and observations $\mathbf{z}_k$, is computed using Bayes theorem. This requires state transition and observation models to be specified. This measurement/observation model describes the probability of making observation $\mathbf{z}_k$ when the vehicle location (x) and landmark locations (m) are known and is described as

$$P(\mathbf{z}_k \mid \mathbf{x}_k, \mathbf{m}) \tag{15.2}$$

It is assumed that, once the vehicle location/state and map (coordinates) are defined, the observations are conditionally independent. The robot motion model is given in terms of a probability distribution on state transitions as

$$P(\mathbf{x}_k \mid \mathbf{x}_{k-1}, \mathbf{u}_k) \tag{15.3}$$

The state transition process in the above equation is assumed to be a Markov process in which the next state $\mathbf{x}_k$ depends only on the immediately preceding state $\mathbf{x}_{k-1}$ and the applied control $\mathbf{u}_k$.

The SLAM algorithm is then implemented in a standard two-step recursive prediction, time-propagation, and correction, measurement/data-update form as follows [2]:

Time-propagation

$$P(\mathbf{x}_k, \mathbf{m} \mid \mathbf{z}_{0:k-1}, \mathbf{U}_{0:k}, \mathbf{x}_0) = \int P(\mathbf{x}_k \mid \mathbf{x}_{k-1}, \mathbf{u}_k) \times P(\mathbf{x}_{k-1}, \mathbf{m} \mid \mathbf{Z}_{0:k-1}, \mathbf{x}_0) \, d\mathbf{x}_{k-1} \qquad (15.4)$$

Measurement/data-update

$$P(\mathbf{x}_k, \mathbf{m} \mid \mathbf{Z}_{0:k}, \mathbf{U}_{0:k}, \mathbf{x}_0) = P(\mathbf{z}_k \mid \mathbf{x}_k, \mathbf{m}) P(\mathbf{x}_k, \mathbf{m} \mid \mathbf{Z}_{0:k-1}, \mathbf{U}_{0:k}, \mathbf{x}_0) / P(\mathbf{z}_k \mid \mathbf{Z}_{0:k-1}, \mathbf{U}_{0:k}) \quad (15.5)$$

The above two equations for the time propagation, or time update and the measurement/data update of the joint estimation vector provide a recursive procedure for calculating the joint posterior for the robot state $\mathbf{x}_k$ and the map $\mathbf{m}$ at a time k based on all observations $\mathbf{Z}_{0:k}$ and all control inputs $\mathbf{U}_{0:k}$ up to and including the time k. These computations are performed by Kalman filter in linear cases, and by extended Kalman filter (EKF) for non-linear cases. For the latter even derivative-free Kalman filter and particle filters are used.

15.2.2 Observation of the Structure of the SLAM

In general, much of the error between estimated and true landmark locations is found to be common between landmarks [2,4]. It is due to the errors in knowledge of where the robot is (x) when landmark observations (z) are made, implying that the errors in landmark location estimates are highly correlated. This means that the relative location between any two landmarks, $\mathbf{m}_i - \mathbf{m}_j$, may be known with high accuracy (common errors getting nearly cancelled out), even when the absolute location of a landmark $\mathbf{m}_i$ might be uncertain. An important observation is that the correlations between landmark estimates increase as more observations/measurements are made. This means that the knowledge of the relative location of landmarks improves, that is, the joint pdf on all landmarks $P(\mathbf{m})$ becomes highly peaked as more measurements are made. This situation occurs because the observations made by the mobile robot can be considered as 'nearly independent' measurements of relative location between the landmarks.

15.3 SLAM PROBLEM SOLUTION APPROACH

For solving the probabilistic SLAM problem we need appropriate representations for the observation/measurement and motion/state models. The most appropriate form is that of a state-space model (and the associated measurement model, often in algebraic discrete-time form) with additive Gaussian noise. For the linear systems this is a Gauss–Markov model process. For non-linear systems this entails the use of the extended Kalman filter (EKF). However, there are many alternative solutions available: derivative-free Kalman filter (unscented Kalman filter), particle filter and H-infinity (H-∞) filters. The work on the application of the H-∞ filters to SLAM problem is very little, or almost nil. In this chapter, we present some results of use of the H-∞ posteriori filter in SLAM problem. The EKF described in Chapter 31 can be easily modified to solve the SLAM problem, with appropriate definitions of the state and measurement variables.

15.3.1 Convergence of SLAM

The convergence, for the EKF-based SLAM, of the map is implicit in the monotonic convergence to zero of the determinant of the map covariance matrix $\mathbf{P}_{\mathrm{mm},k}$ and all landmark pair sub-matrices [2,4]. The individual landmark variances converge towards a lower bound determined by initial uncertainties in robot position/observations where these bounds can be determined by extending the usual Cramer–Rao bounds (CRB of maximum likelihood method) to the EKF.

15.3.2 Computational Aspects

In the observation/data update cycle all landmarks and the joint covariance matrix are updated every time an observation is made and incorporated into filter computation. The computations would grow in quadratic manner with the number of landmarks. Some other efficient methods can be explored to reduce the computations. Hence, the SLAM computations span in quadratic manner with the number of landmarks in a given map. This is a limitation if the SLAM algorithm is to be implemented in real time. To handle this complexity there are several approaches [2,4]: (i) linear time state augmentation, (ii) sparsification in information form, (iii) partitioned updates and (iv) sub-mapping methods. The algorithm based on state-space approach requires the joint estimation of state that is an augmentation of a robot pose (x) and the locations of the observed landmarks (m). Since, the process-state mathematical model only affects the vehicle pose states and the measurements model only makes reference to a single vehicle–landmark pair, many approaches are available to exploit this structure and develop on this situation to reduce the computational complexity. The optimal SLAM algorithms reduce the computations, yet obtain the estimates very close to the original algorithm, whereas the conservative algorithms might have larger uncertainty than the optimal ones, but can be more efficient and useful for real-time implementation.

15.3.2.1 State Augmentation

To reduce the computational complexity one can (i) limit the time-propagation computation by using the state-augmentation method and (ii) limit the measurement-update computation by using a partitioned form of the update equations, because at any time the joint algorithm's state vector has two parts: the robot pose-state, $\mathbf{x}$, and the set of map-landmark locations, $\mathbf{m}$. The vehicle model can propagate only the pose-state according to the given control inputs, $\mathbf{u}$. The map states are left unaffected. Since, only the pose-states (x) are affected (by the robot model), the covariance prediction matrix can be re-written such that it has now linear complexity in the number of landmarks [2]. A new landmark (for/in the map) is initialized as a function of the robot pose-state and an observation $\mathbf{z}_k$. Then the augmented states (the joint state vector) are a function of a small number of existing states. As a result, the EKF prediction step and the process of adding new landmarks the computations are such that they are linear in the number of landmarks.

15.3.2.2 Updates by Partitioning

Similarly, in the original implementation of the SLAM the measurement-data update is carried out by updating all robot and map states every time a new measurement is made. One method to reduce this (quadratic) burden is to confine sensor-rate updates to a small local region and update the global map only at a much lower rate. One approach is to generate a short-term sub-map with its own local coordinate frame—it avoids very large global covariance and is numerically more stable and less affected by the linearization errors. In this local sub-map algorithm the two independent SLAM estimates are maintained [2]: (i) a map composed of a set of globally referenced landmarks (along with the global reference pose of a sub-map coordinate frame) and (ii) the local sub-map (along with a locally referenced robot pose-state) with locally referenced landmarks. As the measurements are made and incorporated into the filtering/estimation process the updates are carried out within the local sub-map and with only those landmarks held in the local sub-map. Actually it is possible to obtain a global robot pose estimate at any time by combination of the locally referenced pose-state and the global estimate of the sub-map coordinate frame. An optimal global estimate is obtained periodically by registering the sub-map with the global map, and a new sub-map is created and the estimate process continues further. This sub-map approach has some merits: (i) the number of landmarks are limited to only those that are described in the local sub-map frame, (ii) there is lower uncertainty in a locally referenced frame and (iii) sub-map registration can use batch-validation gating.

15.3.2.3   Sparsification

The reformulation of the state-space SLAM into information form renders the information matrix (IM) for reducing the computational burden. An alternative representation is in information form using the information vector and the associated IM, which is the inverse of the covariance matrix (arising from the KF computations). For large-scale maps, the advantage of the information form is that several off-diagonal components of the normalized IM are very close to zero. This renders a sparsification procedure feasible and we can set near-zero elements of the normalized IM to actual zero values [2]. With the IM being sparse, efficient update procedures for information estimates can be constructed to obtain the maps. Now several consistent sparse solutions are available. Thus, in the IM form of the SLAM problem the state augmentation is a sparse operation.

15.3.2.4   Submaps

In this approach the idea is to break the map into the sub-regions with the local coordinate systems and then arrange (the map) in a hierarchical manner. The local updates occur and then are refined periodically by inter frame updates. These techniques provide generally the conservative estimates in the global frame. The submap methods are means of addressing the issue of quadratic-computation complexity during measurement updates. Submap methods are [2]: (i) globally referenced and (ii) locally referenced. A submap defines local coordinate frame and nearby landmarks are estimated with respect to the local frame. These submap estimates are obtained using the standard/optimal SLAM algorithm using only the locally referenced landmarks. These submap structures are arranged in a hierarchy leading to a computationally efficient suboptimal global map. The global submap methods estimate the global locations of the submap coordinate frames relative to a common base frame. These methods reduce computations to linear or constant time dependence by maintaining a conservative estimate of the global map. Since, as the submap frames are located relative to a common base coordinate frame, global submaps do not alleviate the linearization problem that would arise from large pose uncertainties. In the relative submap methods there is no common coordinate frame. The location of any given submap is recorded only by its neighbouring submaps that are connected in a graphical network. Global estimates are obtained by summation along a path in the network. The relative submap method: (i) produces locally optimal maps with computational complexity independent of the size of the complete map, (ii) by treating updates locally it is numerically very stable, (iii) allows batch association between frames and (iv) minimizes problems arising from linearization in a global frame.

15.3.3   DATA ASSOCIATION

The standard EKF-SLAM process is weak to incorrect association of measurements to landmarks. The data association (DA) solutions are needed to be incorporated in the SLAM process. This data association situation is compounded in environs when the landmarks are not simple points and indeed look different from different viewpoints. Thus, the important issue is to correctly associate observations of the landmarks with the landmarks held in the map. That is to say that given an environment map, and a set of sensor observations, associate observations with the map elements. This is in the realm of data association (DA). The new measurements are associated with existing map landmarks, before fusing data into the map. After the fusion, these associations cannot be revised. A single incorrect data association can induce divergence into the map estimate. Some important aspects in data association in continuous SLAM are [2,4]: (i) nearest neighbour v/s joint compatibility, (ii) SLAM with laser and sonar, (iii) map joining and (iv) the loop closing problem. One should consider the influence of the type, density, precision and robustness of features in addressing the DA problem. The approaches to DA are: (i) search in configuration space: find robot–vehicle location with maximal data to map overlapping, this can be done either with raw data or with features; (ii) search in correspondence space: find a consistent correspondence hypothesis and compute robot–vehicle location, extract features from data (if the data are sparse move and build a local map), obtain feature-based map, search for data feature to map feature correspondences.

15.3.3.1 Batch Validation

To avoid unlikely associations (of the measurement data to the track/landmark) most SLAM implementations perform DA using only statistical validation gating a method from the target-tracking literature. Some early implementations of SLAM algorithms considered each measurement-to-landmark association individually by checking if an observed landmark were close to a predicted location or not. This type of individual gating is very unreliable if the vehicle pose (x) is very uncertain (state-error covariance matrix is large) and fails in all but the most sparsely populated and structured environs. The concept of batch gating, wherein multiple associations are considered simultaneously would be more advantageous. The mutual association compatibility utilizes the geometric relationships between landmarks. The two forms of batch gating are [2,4]: (i) the joint compatibility branch and bound (JCBB) method which is a tree-search and (ii) the combined constraint data association (CCDA) which is a graph search. The batch gating process alone is often sufficient to achieve reliable DA: (i) if the gate is sufficiently constrained, association errors have minimal effect and (ii) if a false association is made with an incorrect landmark which is physically close to the right one, then the inconsistency is minor.

15.3.3.2 Multi-Hypothesis Data Association

For robust target tracking in cluttered environments, the multi-hypothesis data association (MHDA) is essential [2], because it resolves association ambiguities by generating a separate track estimate for each association hypothesis and thereby creating, over time, an ever-branching tree of tracks. The low likelihood tracks are pruned from the hypothesis tree. Particularly in large complex environments, multi-hypothesis tracking (MHT) is also important for robust SLAM implementation. In loop closure in SLAM, a robot should ideally maintain separate hypotheses for suspected loops. It should also maintain a 'no-loop' hypothesis for cases where the perceived environment is structurally similar. In SLAM application of MHT, a major bottleneck is the computational overhead of maintaining separate map estimates for each hypothesis. The FastSLAM algorithm is actually the MHT solution, with each particle having its own map estimate. An attribute of the FastSLAM algorithm is its ability to perform per-particle data association [2,4].

15.3.4 Non-Linearity

The EKF-based SLAM employs linearized models of non-linear motion and observation models and hence inherits these approximations which might cause inaccuracies as well as divergence of the SLAM algorithm. It might lead to inconsistency in solutions. The convergence and consistency are only guaranteed in the linear case. The SLAM problem is inherently a non-linear problem. It must be very clear from the foregoing that the SLAM process provides a solution to the competency of mapping and localization for any autonomous robot. Some of the areas where still some work can be done are [3]: (i) the information (state/matrix)-oriented approach for large-scale mapping, problems of coordinating many vehicles in mixed environments with sensor networks and dynamic landmarks, (ii) the handling of out-of-sequence measurement data problem for the SLAM in order not to sacrifice quality and robustness, (iii) appearance- and pose-based SLAM methods for mapping and location estimation without the need for strong geometric landmark descriptions, (iv) larger and more persuasive implementations and demonstrations, (v) to demonstrate SLAM approach to large problems, (vi) mapping an entire wide area without using GPS and (vii) to demonstrate true autonomous localization and mapping of structures.

15.4   MATHEMATICAL FORMULATIONS FOR SLAM

We will mainly focus on the use of Kalman filtering algorithm for the solution of the SLAM problem. A mobile robot used the observations (z) which are made relative to the locations to the landmarks in the robot's environment [5]. This environment is the one that the mobile robot is going to traverse through in its total journey. These measurement data are used by the robot's computer to estimate the robot's states and the landmarks locations (x). The state vector can contain the robot's position, velocity and acceleration in any or more directions: x, y, z. In addition to these it can contain the robot's orientations. Thus, the KF would be using three mathematical models: (i) the vehicle model, (ii) landmarks model and (iii) the sensor model. The vehicle model describes the kinematics and dynamics of the robot model. The sensor model relates the measurements with the state vector to be estimated.

15.4.1   MATHEMATICAL MODELS

The mathematical models can be considered to be linear for the sake of the simplicity of the exposition and to clearly understand the problem of estimation in SLAM. However, in general, these models are non-linear and asynchronous in reality.

15.4.1.1   Vehicle Mathematical Model

The vehicle's mathematical model is given as

$$x_v(k+1) = F_v\{x_v(k), u_v(k+1), k+1\} + w_v(k+1) \tag{15.6}$$

The u is control vector comprising velocity inputs and steering angles and x is the state vector of the model comprising of the vehicles' states (position, velocity, etc.). The matrix F captures the mobility, kinematics and dynamics of the robot, including orientation dynamics. Any unmodelled behaviour is represented by the so-called process or state noise which is assumed to be zero mean white noise with Q as its covariance matrix. The vehicle state vector can be expanded to include other higher states like acceleration and orientations.

15.4.1.2   Landmarks' Mathematical Model

The landmarks' locations are generally considered as fixed with some uncertainty or without it. The landmarks can be considered as a point model or with some additional features, and shapes, the latter complicating the landmark mathematical model, adding higher-order dimensions. The landmark model considered is simple one with two parameters with respect to some global reference coordinate frame. Thus, the landmark model is a time invariant. The ith point landmark in the environment is defined as

$$p_i = \begin{bmatrix} x_i \\ y_i \end{bmatrix} \tag{15.7}$$

The landmark is considered stationary and hence it is represented as

$$p_i(k+1) = p_i(k) = p_i \tag{15.8}$$

In the landmark model no additive uncertainty is considered since the landmark location is assumed to be known precisely; however, in reality this may not be true. The assumption here implies that if there is any uncertainty, it is assumed to be constant with time.

15.4.1.3 Measurement Model

The measurement model is given as

$$z_i(k) = H_i\{x_v(k), p_i, k\} + v_i(k) \tag{15.9}$$

The variables have usual meaning. The variable z denotes the observation of the landmark location relative to the location of the robot. The unmodelled uncertainties are lumped into the measurement noise v that is assumed to be a zero mean white noise with its covariance matrix as R. For the purpose of the work presented in this chapter, primarily the models assumed are linear; however, for the generation of the simulated results by the software modules 'ekfslam_sim' and 'hislamsim', the linearization is used to compute the covariance propagation/updates and Kalman gains in the appropriate EKF and H-∞ filters.

15.4.2 Representation of Maps

The robot's environment map representation can be done in two ways [5]: (i) absolute form and (ii) relative form. In absolute form the landmarks are represented as locations registered in a common global coordinate frame of reference. The absolute map is given as a vector and is called the absolute map vector. However, the relative map represents the relationships between these fixed landmarks positions, and these relationships are considered as the states of the relative map. Basically if there are two landmarks with known locations (with respect to some global frame of reference), then the relative- (map-) state is the difference between the two location coordinates.

15.5 H-∞ FILTER-BASED SLAM OF A MOBILE ROBOT

As we have seen the above traditionally, Kalman filtering algorithm and its many variants have been used. In this section the SLAM problem is solved in the domain of joint state/parameter estimation using H-∞ filtering approach. The H-∞ filter is based on the so-called H-∞ norm which is minimized or brought under a bound to obtain robust filtering algorithm. Thus, we present a new solution to the SLAM problem.

15.5.1 Robust Filtering

Recently, the H-∞-based concept has gained importance and now a good number of H-∞ norm-based filtering algorithms are available. We discuss one such H-∞ filtering algorithm which is based on the so-called Krein space. We then use this algorithm into the SLAM structure. We study the performance of this filtering algorithm using numerical simulation data. In the Kalman filtering approach the signal processing system is state-space model driven by a white-noise process with known statistical properties. The measurement signals are assumed to be corrupted by white noise with known statistical properties. The aim of the filtering is to minimize the variance of the terminal state estimation error. The H-∞ filtering problem differs from Kalman filtering in two respects [6]: (a) the white noise is replaced by unknown deterministic input-disturbance of finite energy and (b) a pre-specified positive real number (gamma, a scalar parameter) is defined. Then the aim of the filter is to ensure that the energy gain from the input-disturbance to the (output-) estimation error is less than this number. This number can be called a threshold for the magnitude of the transfer function between output estimation error and the input disturbance energies. One important aspect is that the Kalman filter evolves from the H-∞ filter as this threshold number tends to infinity. From the point of view of robustness we see that the H-∞ concept, at least in theory, would yield a robust filtering algorithm, if not the optimal one. In many applications robustness is more important to achieve than mere optimality.

15.5.2 H-∞ NORM

The concept of H-∞ norm and H-∞ filter emerges from the theory of optimal control synthesis in frequency domain [6]. Interestingly, H-∞ (optimal) control is a frequency-domain optimization and synthesis theory, and the theory of H-∞ explicitly addresses the crucial question of modelling of errors. The basic tenet of the H-∞ concept and norm and hence H-∞ filtering philosophy or paradigm is to treat the worst-case scenario: plan for the worst situation and then optimize. The idea is to minimize the maximum of the output error. This is called the min–max problem. Thus, the framework has the following properties: (i) capable of dealing with plant modelling errors and unknown disturbances, (ii) it must represent a natural extension to the existing theory, say H-2-based Kalman filtering theory, thereby H-∞-based theory being more general one, (iii) it must be amenable to meaningful optimization and (iv) it must be applicable to multivariable problem. The H-∞ norm involves RMS value of a signal, that is, a measure or a metric of a signal that reflects eventual average size of root-mean-square (RMS) value. This is a classical notion of the size of a signal, used in many areas of engineering. The H-∞ norm [6] used in deriving the robust filtering algorithm is given as

$$\frac{\sum_{i=0}^{N} (\dot{x}_f(k) - x(k))^t (\dot{x}_f(k) - x(k))}{(\dot{x}_{0f} - x_{0f})^t P_{0f}(\dot{x}_{0f} - x_{0f}) + \sum_{k=0}^{N} w^t(k)w(k) + \sum_{i=1}^{m} \sum_{k=0}^{N} v_i^t(k)v_i(k)} \tag{15.10}$$

It can be readily seen from the above expression (15.10) of the H-∞ norm that the input to the filter consists of energies (represented as variances in the denominator) due to the errors in: (a) the initial condition (of the state error), (b) the state disturbance (process noise) and (c) the measurement noise (here in this chapter considered for both the sensors). The output energy of the filter (given in the numerator) is due to the error in the fused state. Basically this ratio, H-∞ norm, should be less than square of gamma, which can be considered as an upper bound on the maximum energy gain from the input to the output, that is, the worst case. It must be emphasized here that no statistical assumptions on the noise processes are required to be made. So, by class the H-∞ filters are deterministic filters where robustness in emphasized more than the randomness or stochastic aspects of the signals.

15.5.3 H-∞ FILTER

It has been recently shown that H-∞ estimation and control problems related to risk-sensitive estimation, and adaptive filtering can be studied in a simple and unified manner in the indefinite metric space called a Krein space [7,8]. In the conventional H-2 framework, on which the celebrated Kalman filter is based [9], the unknown parameter (or state) vector and the additive disturbance (noises) are assumed to be random variables, or stochastic processes. The optimal, rather robust solutions to the filtering problems in the H-∞ space are found by minimizing expected prediction error energy. In the H-∞ framework no statistical assumptions are made on the unknown vector and the disturbances, however, the robustness of the solution/estimates is ensured by minimizing (or atleast bounding) the maximum energy gain from the disturbances to the estimation errors, via H-∞ norm defined in expression (15.10).

15.5.4 MATHEMATICAL MODEL

The dynamic model of a mobile robot is given by

$$x(k + 1) = Fx(k) + Gw(k) \tag{15.11}$$

Here, F is the state coefficient matrix, G is the process disturbance gain matrix and T is the sampling interval (represented by discrete-time index k). The landmarks' relative measurements are modelled by

$$z(k) = Hx(k) + v(k) \tag{15.12}$$

Here, the H is the measurement model/matrix, basically it is a sensor's mathematical model. The process measurement noises are considered as unknown deterministic disturbances.

15.5.5 H-∞ A POSTERIORI FILTER

The state estimates are obtained by H-∞ *a posteriori* filter (HPOF) [7,8]. The time-propagation is given as

$$P_i(k + 1) = FP_i(k)F' + GQG' - FP_i(k)[H_i^t L_i^t]R_i^{-1}\begin{bmatrix} H_i \\ L_i \end{bmatrix}P_i(k)F'$$

$$R_i = \begin{bmatrix} I & o \\ 0 & -\gamma^2 I \end{bmatrix} + \begin{bmatrix} H_i \\ L_i \end{bmatrix}P_i(k)[H_i^t L_i^t] \tag{15.13}$$

The expression for R includes the gamma-constant as a scalar threshold parameter and L denotes the linear combination of state estimates (used in Krein space).

The H-∞ filter gain equation is given by

$$K_i = P_i(k + 1)H_i^t(I + H_iP_i(k + 1)H_i^t)^{-1} \tag{15.14}$$

Measurement update of states is obtained by

$$\hat{x}_i(k + 1) = F\hat{x}_i(k) + K_i(z_i(k + 1) - H_iF\hat{x}_i(k)) \tag{15.15}$$

The above update combines the time propagation of the state estimates. Although certain quantities in the HI filters are called grammarian [7] we continue to use the conventional notations as covariance for the sake of simplicity and to remain connected with the classical and conventional notations of Kalman filter and stochastic theories.

15.6 NUMERICAL SIMULATION RESULTS

Since the SLAM is non-linear filtering problem we have used a linearization process similar to that used in the extended Kalman filtering algorithm. Due to this linearization process the H-∞ filter was directly applicable to be incorporated into the SLAM structure. The vehicle states are: x-axis, y-axis and heading (phi) angle. This is defined as the vehicle-pose and defines the vehicle model. It also includes the vehicle velocity, V. The math model as represented in the HISLAMSIM programme is given as [10]

```
xv = [xv(1) + V * dt * cos(G + xv(3,:));
      xv(2) + V * dt * sin(G + xv(3,:));
      pi_to_pi(xv(3) + V * dt * sin(G)/WB)]
```

Here, G is the steering angle and WB is the vehicle base. The Q matrix is chosen as diagonal with elements as $Q = [0.09\ 0;\ 0;\ 0.003]$. Similarly the measurement covariance matrix $R = [0.01\ 0;\ 0$

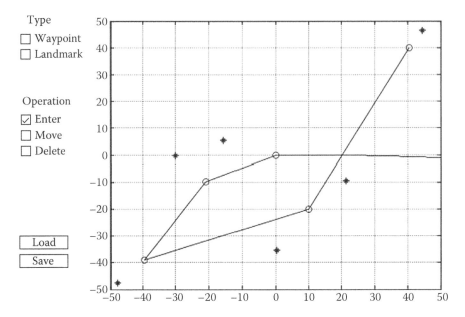

FIGURE 15.3 Specification of waypoints (circles joined by lines) and landmarks (starred points) for the ekfslamver2.0 interactive programme for generation of SLAM results.

0.0003]. The maximum range chosen is 30 m. The maximum steering angle is 30°, and the maximum steering rate is 20°/s. For the data association the usual method of using the covariance matrix of the innovations has been used. The disposition layout of the waypoints and landmarks is shown in Figure 15.3 and the actual and estimated slam trajectories obtained with the ekfslamver2.0 are shown in Figure 15.4. As such, Figure 15.3 shows the specification of way points (circles joined by lines) and landmarks (starred points) for the ekfslamver2.0 interactive programme for generation of SLAM results [10]. Figure 15.5 depicts the vehicle true data (left subplot) and vehicle path estimates (right subplot) obtained using the ekfslamver2.0 interactive programme for generation of SLAM

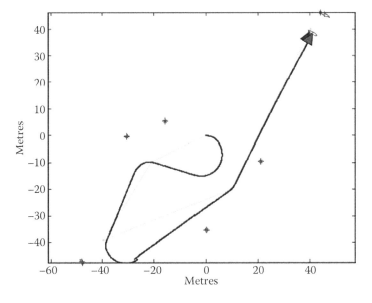

FIGURE 15.4 The trajectories using for the ekfslamver2.0 interactive programme for generation of SLAM results.

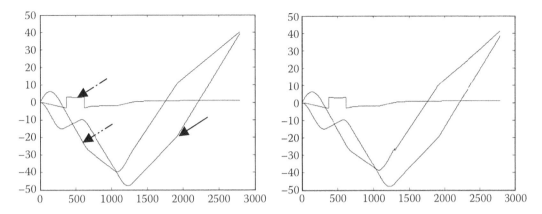

FIGURE 15.5 Vehicle true data (left subplot) and vehicle path estimates (right subplot) obtained using the ekfslamver2.0 interactive programme for generation of SLAM results (heading ---; x-position ----; y-position --).

results [10]. Although the plots are not superimposed it can be readily seen that the time history match of the x-, y-, and heading trajectories is very good.

Another layout of waypoints and landmarks is shown in Figure 15.6. The results of SLAM simulation obtained with the H-∞ filter with layout of Figure 15.6 are shown in Figures 15.7 through 15.14. The actual and estimated slam trajectories obtained with the HISLAMSIM programme are shown in Figure 15.7. The time history match of the actual and estimated slam trajectories (i.e., the three states described above) obtained with the HISLAMSIM programme is depicted in Figure 15.8. In Figure 15.9, the actual and estimated slam trajectories obtained with the HISLAMSIM programme (landmarks: *—given, +—estimated; trajectory: actual—thick line, estimated—thin line merged with the thick line; only thin line → the waypoints of Figure 15.6) with noise index = 0 for Q, R, and lambda = 1.05 in the H-∞ filter are shown. Figure 15.10 shows in left subplot the actual and estimated slam trajectories obtained with the HISLAMSIM programme; that is, the time histories of the three

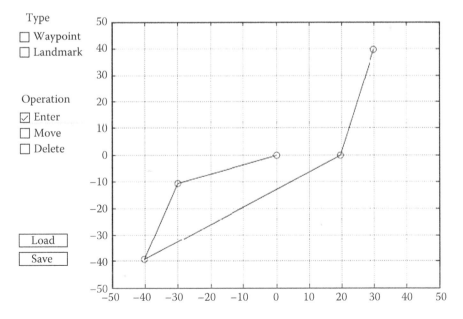

FIGURE 15.6 Specifications of way points (circles joined by lines) and landmarks (starred points) for the hislamsimver1.0 interactive programme for generation of SLAM results.

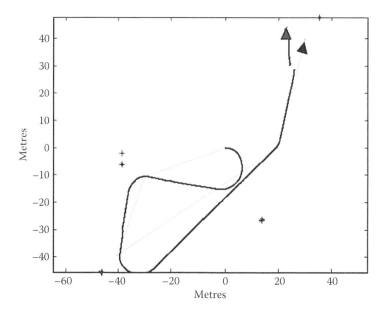

FIGURE 15.7 Actual and estimated slam trajectories obtained with the HISLAMSIM programme. (Landmarks: *—given, +—estimated; trajectory: actual—thick line, estimated—thin line merged with the thick line; only thin line → the waypoints of Figure 15.6; noise index = 0 for Q, R.)

trajectories: actual dashed line, estimated continuous line. In the right subplot, the state-error plots: x-axis thick line, y-axis faint line, and heading error middle line are depicted for the noise index = 0 for Q, R, with lambda = 1.05 in the H-∞ filter. Similar results for the noise index = 1 for $2*Q$, $2*R$, that is, covariance matrices of magnitude double of the previous ones, with lambda = 1.5 and 1.3 (in H-∞ filters) are shown in Figures 15.11 through 15.14. The percentage fit errors for the x-axis, y-axis and heading are given in Table 15.1 for various noise indices and lambda values. We see that reasonable

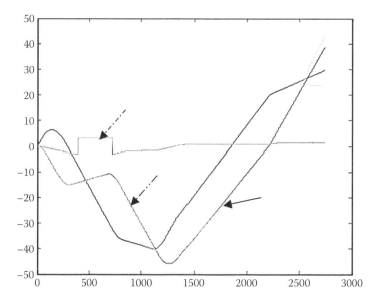

FIGURE 15.8 Time history match of the actual and estimated slam trajectories obtained with the HISLAMSIM programme. (Time histories of the three trajectories → : actual—dashed line, estimated—continuous line; heading ---; x-position ----; y-position --.)

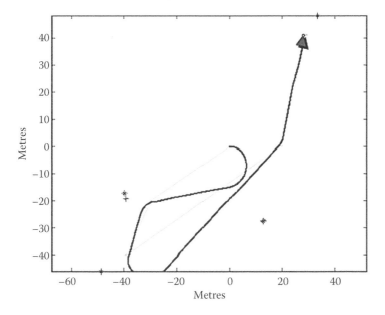

FIGURE 15.9 Actual and estimated slam trajectories obtained with the HISLAMSIM programme. (Landmarks: *—given, +—estimated; trajectory: actual—thick line, estimated—thin line merged with the thick line; only thin line → the waypoints of Figure 15.6; noise index = 0 for Q, R, lambda = 1.05 in HI filter.)

low values have been achieved for the H-∞ SLAM problem. From these simulated results we can infer that the performance of the H-∞ filter for SLAM is satisfactory and encouraging. Further, it is planned to evaluate the HISLAMSIM for more data and in the possible sensor data fusion scenario [11]. In this section, we have presented a new solution to the SLAM problem in the area of navigation and guidance for a mobile robot. We have specifically used an H-∞ filter for joint state and parameter estimation.

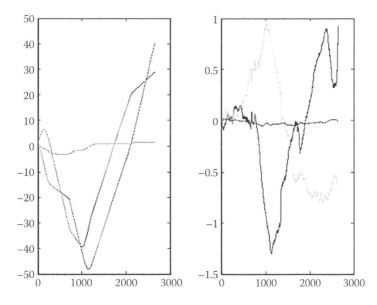

FIGURE 15.10 Left subplot: Actual and estimated slam trajectories obtained with the HISLAMSIM programme. (Time histories of the three trajectories: actual—dashed line, estimated—continuous line; Right subplot: The state-error plots: x-axis—thick line, y-axis—faint line, and heading error—middle line. Noise index = 0 for Q, R, lambda = 1.05 in HI filter.)

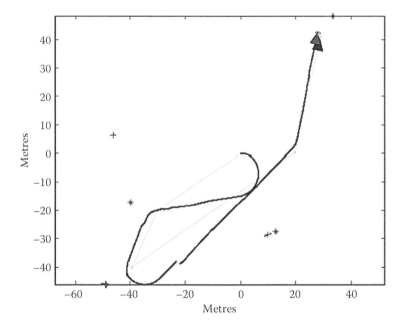

FIGURE 15.11 Actual and estimated slam trajectories obtained with the HISLAMSIM programme. (Landmarks: *—given, +—estimated; trajectory: actual—thick line, estimated—thin line merged with the thick line; only thin line → the waypoints of Figure 15.6; noise index = 1 for 2*Q, 2*R, lambda = 1.5 in HI filter.)

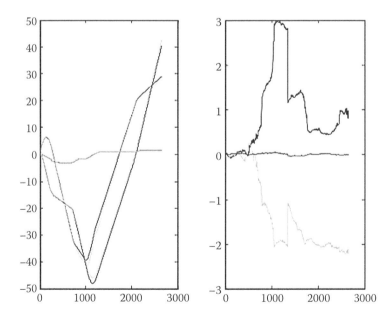

FIGURE 15.12 Left subplot: Actual and estimated slam trajectories obtained with the HISLAMSIM programme. (Time histories of the three trajectories: actual—dashed line, estimated—continuous line; Right subplot: The state-error plots: x-axis—thick line, y-axis—faint line, and heading error—middle line. Noise index = 0 for Q, R, lambda = 1.05 in HI filter.)

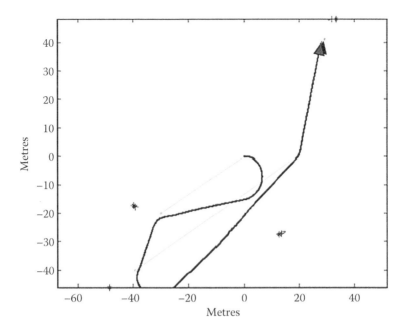

FIGURE 15.13 Actual and estimated slam trajectories obtained with the HISLAMSIM programme. (Landmarks: *—given, +—estimated; trajectory; actual—thick line, estimated—thin line merged with the thick line; only thin line → the waypoints of Figure 15.6; noise index = 1 for 2*Q, 2*R, lambda = 1.3 in HI filter.)

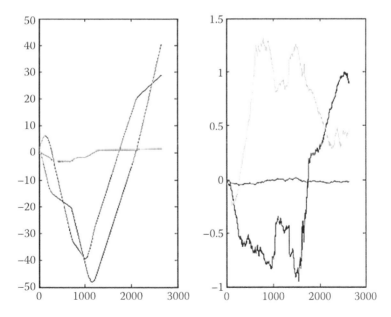

FIGURE 15.14 Left subplot: Actual and estimated slam trajectories obtained with the HISLAMSIM programme. (Time histories of the three trajectories: actual—dashed line, estimated—continuous line; Right subplot: The state-error plots: x-axis—thick line, y-axis—faint line, and heading error—middle line. Noise index = 1 for 2*Q, 2*R, lambda = 1.3 in HI filter.)

TABLE 15.1

The Percentage Fit Errors for the H-∞ Filter for SLAM

Noise Index NI	Noise Variances	Lambda in the HI Filter	x-Axis	y-Axis	Heading
0	Q, R	1.05	2.5	2.06	1.78
1	$2*Q, 2*R$	1.5	5.8	5.9	1.344
1	$2*Q, 2*R$	1.3	2.77	3.0	1.4

REFERENCES

1. Aulinas, J. *3D Visual SLAM Applied to Large-Scale Underwater Scenarios*. MSc thesis, Institute of Informatics and Applications, University of Girons, Girons, Spain, 2008.
2. Hugh, D.-W. and Tim, B. Simultaneous localization and mapping: Part I. *IEEE Robotics & Automation Magazine*, 99–108, June 2006.
3. Giorgio, G., Cyrill, S. and Wolfram, B. Improved techniques for grid mapping with Rao–Blackwellized particle filters, *Transactions on Robotics*, 23, 34–46, 2007.
4. Tim, B. and Hugh, D-W. Simultaneous localization and mapping: Part II. *IEEE Robotics & Automation Magazine*, 108–117, September 2006.
5. Newman, P.M. *On the Structure and Solution of the Simultaneous Localization and Map Building Problem*. PhD thesis, Australian Center for Field Robotics, The University of Sydney, 1999.
6. Green, M. and Limebeer, D.J.N. *Linear Robust Control*. Prentice-Hall, Englewood Cliffs, NJ, 1995.
7. Hassibi, B., Sayed, A.H. and Kailath, T. Recursive linear estimation in Krein spaces—Part I: Theory. *Proceedings of the 32nd IEEE Conference on Decision and Control*, San Antonio, TX, Dec. 1993.
8. Hassibi, B., Sayad, A.H. and Kailath, T. Linear estimation in Krein spaces—Part II: Applications. *IEEE Trans. on Autom. Contrl.*, 41(1), 34–49, 1996.
9. Raol, J.R., Girija, G. and Jatinder, S. *Modelling and Parameter Estimation of Dynamic Systems*. IEE (/IET) Control Series Vol. 65, IEE London, UK, August, 2004.
10. Bailey, T. ekfslamver2.0, a MATLAB program for EKF based SLAM. http://www-ersonal.acfr.usyd.edu.au/tbailey/software/slam_simulations, accessed July 2011.
11. Raol, J.R. *Multi-Sensor Data Fusion with MATLAB*. CRC Press, Taylor & Francis, FL, USA, 2009.

16 Robotic and Moving Vehicle Architectures

Jitendra R. Raol

CONTENTS

16.1 INTRODUCTION

In any robotic system as well as for any mobile autonomous system, the hardware (HW) and software (SW) architecture both play a very crucial role in showing all the interconnecting sub-systems. These HW/SW architectures also show functionality and controlling aspects of the entire active/reactive system. In this chapter, we briefly discuss different types of robot (control) architectures which with some more modifications could be useful for developing such architectures for any mobile vehicles/systems [1–6]. We discuss certain architectural design properties also. Some classification and comparison of these approaches is attempted. The attributes and properties of these different architectures are highlighted.

Complex mobile autonomous robots and systems are better designed to be intelligent and with a good level of autonomy. Such systems are supposed to perform various assigned tasks concurrently and/or asynchronously, adding, of course, to the systems' complexity. This requires a definition of an appropriate HW/SW structure of how all sub-systems should interact. A robotic architecture is a collection of some HW components and SW modules that operate in the onboard computing systems/processors. These SW operating-building blocks (algorithms/computational processes/numerical computations/control algorithms/embedded neural networks and fuzzy inference systems-related computations) facilitate highly specific and individual tasks for the mobile robot. The HW architecture for a robot at the basic level has sensors, signal processing/conditioning systems, actuators and the physical (mechanical platform/mechanical arms/any other type of extensions) which interact with each other to perform a specific activity and task. Here, in this chapter, we only concentrate on the sensing–planning–actuation abstraction of a system's architecture. The field of autonomous systems and robotics is a challenging one for the study and development of robust and intelligent architectures which could include many capabilities relating to (i) acting in real time, RTC (real-time control); (ii) sensors/actuators and their control, including smart sensors/systems, (iii) concurrency, (iv) detection and reaction to extraneous and exceptional situations, (v) dealing with uncertainty and (vi) integrating higher level planning with lower level control tasks.

16.2   ARCHITECTURAL DESIGN ASPECTS

The design of architecture for a mobile vehicle is an important aspect of mobile intelligent autonomous system (MIAS)/robotics [1]. Possible design criteria are outlined next.

16.2.1   SYSTEM RELIABILITY

The reliability of functioning of the sub-systems as well as the entire MIAS/robotic system is very important to achieve the designated tasks without fail or any hindrance. The reliability should be ensured even at the (slight if not very high) cost of performance reduction in time (some functions take more time), cost (higher cost), or accuracy (slightly reduced accuracy). The reliability should be evaluated by using some function/aspects of sensors, actuators, and communications channels and measuring effects on task achievement and performance metrics. It can exploit the redundancy of various processing functions. The overall reliability of the robotic system is enhanced when there are fault-tolerant mechanisms at the HW, SW or both HW/SW levels in the robotic system. Thus, the reliability and dependability of the MIAS/robotic system are very important aspects for achieving fault tolerance (Chapters 29, 31 and 32) of such autonomous systems.

16.2.2   GENERALIZATION FEATURE

It is the ability of the MIAS system to act appropriately, even if sub-optimally, in situations that the system has not encountered earlier. The generalization is assessed, as in machine learning, by separating training, testing and demonstration performance in any new situation. However, one must remember that if a system is highly generalized, then it might not perform very well in some other routine/normal tasks.

16.2.3   SYSTEM'S ADAPTABILITY

This is the ability to modify system's behaviour to perform better in new situations. Any such system/robot should have the ability to refine the current task and its behaviour according to current goal and execution. The adaptability could be easily achieved if the system's architecture is modular and flexible. This adaptability can be achieved in a better way in behaviour-based architecture because the basic concept suits here. The adaptability can also be linked to automatic fault detection, isolation and management, including the aspect of reconfigurability.

16.2.4 SYSTEM'S MODULARITY

The system should be built up of several modules and components that are able to connect and interact with each other with correct functionality. The main merit of modularity is that its components can be easily replaced when any of them shows some error or fault. The modularity adds to the flexibility of the system's functioning. It can reduce downtime of the system, thereby enhancing the time reliability of the MIAS/robotic system.

16.2.5 AUTONOMOUS BEHAVIOUR

It is the system's ability to carry out its own action independently without the user's involvement. Often, these tasks are planned in advance, and the system is programmed/asked to carry out these tasks when called upon to perform. In a truly autonomous system, it is achieved online/in real-time learning from the previous or just the current observations of the behaviour of the system. The autonomous behaviour is very important for some mobile vehicle systems since these might be called upon to operate in some hazardous situations where the direct involvement of the operator would be very risky. Sometimes, a very long duration of operations might be required of these systems, for example, mining robots. In some experimental situations autonomy would also be very advantageous.

16.2.6 ROBUSTNESS PROPERTIES

Any MIAS/robotic system is supposed to perform satisfactorily in the presence of imperfect inputs, unexpected events and certain kind of uncertainty. Thus, the system is required to be built such that it has certain robustness properties. The main aspect of robustness is required for the control sub-systems. Robustness means that the system (including sub-systems) and/or its various functions should continue to perform at the level of expected/designated accuracy and precision despite the presence of uncertainty in the mathematical models used for simulation and controller design.

16.2.7 EXTENSIBILITY OF TASK PERFORMANCE

The system's certain learning capabilities should be possible so that the MIAS/robotic system could be used for certain tasks in an extended mode, beyond well-specified ones. This is like an extended memory of a computer or an extended arm of a robot itself supporting human's functional movements. Sometimes a mobile vehicle might be required to perform an additional task that was not envisaged earlier, and hence the system architecture should be such that the system should be able to extend its capability, even temporarily for that additional specified task. This requires that the system is originally designed for extended task, but in normal situations it operates for a limited task.

16.2.8 REACTIVITY

A system is supposed to respond to situations with adequate responses, and to any new situations with some appropriate responses such that the system's overheads are not more. This reactivity is the common feature of all types of robotic architectures.

16.2.9 Artificial Intelligence

Most autonomous systems and vehicles should have the so-called artificial intelligence (Chapters 1 and 28). The system would have a computer to command its functions—the robot's inherent brain-power. However, to cope with unusual circumstances, the robot should be provided with some features of artificial intelligence (AI) with sufficient adaptability. This automatically leads to the behaviour-based learning capability for a robot.

16.2.10 Flexibility

It is an ability to introduce new features such as learning and adaptation methods, besides the usual flexibility to operate with several interconnected HW/SW modules, which add to the flexibility of robot architecture. Modularity and flexibility are complementary functions.

In summary, these design requirements properties can be used as measures while evaluating the performance of the robotic architecture. It should be kept in mind that a single system might not have all the properties or features, for example, generalization may be more desirable than extensibility for certain tasks. Hence, a compromise is always required to be exercised while designing these MIAS/robotic architectures. A system that incorporates all the above features would be highly sophisticated and might turn out to be costly. Also, the requirements of combination of some functionality might not be feasible due to the fact that these might be contradictory or might require some trade-offs between the systems requirements and performance of the overall system.

16.3 ROBOTIC SENSE–PLAN–ACT PARADIGM

In the open literature, the most common and well-accepted robotic primitives are: (i) sense, (ii) plan and (iii) act in terms of how sensory data are acquired and processed and how the vehicles' move/motions are planned in advance and generated/propagated through the MIAS/robotic system [1,2]. In the following sections, we give a brief description for each of these primitives in robotic architecture.

16.3.1 Sensing

Any MIAS/robotic system cannot function without using some sensors (Chapter 8). The sensors are the 'eyes' for such systems, including all the aerospace vehicles. The measurements from these sensors provide continually the information on the internal as well as external states of the vehicle. This information in turn is assessed, processed and further analysed and the required results are obtained which are used for communication, command and control of the vehicles. The decision making squarely depends on the sensor's output data/responses. Thus, 'sense' (sensing) paradigm represents the sensor that a robot uses to perceive and 'feel' the world around it. The robot gets the 'sense' of its belonging, its own fixation, its own pose and then after the decisions are made by some algorithm the robot is commanded to its next move to achieve the specified goal. Thus, these sensed data are very crucial for the planning process.

16.3.2 Planning

Many robotic functions are planned in advanced for the task-oriented systems. Thus, the 'plan' (planning paradigm) represents a planner, normally an algorithm residing in the onboard computer. This might usually be a complex planner that uses some sophisticated problem-solving approach. Planning involves various tasks to be carried out by the vehicle, including path and motion planning. Planning process could be quite complex if the vehicle is on a difficult task or mission. Planning process requires a good model of the robot's environment, in addition to its own mathematical model, an optimal path planning algorithm and an associated decision-making logic system.

16.3.3 ACTING–ACTION

The act–acting–action (AAA) paradigm represents actuators, the acting mechanism, with which the robot can act upon against its environment to perform a given task. The AAA paradigm also requires a sophisticated planning and it is an integral part of the planning process. It also requires the mathematical models of the actuators and in conjunction with the mathematical model of the robot's dynamics, the AAA process requires to draw a decision tree that would then command the vehicle to its desired goal. The commanded inputs are carefully monitored and the responses of the robot are also monitored and used as feedback to improve the stability properties of the vehicle. This AAA and decision tree (ADT) can be designed using fuzzy logic and genetic algorithms. In summary, all the autonomous systems including robots need to pass through the above three phases to complete a given mission successfully.

16.4 ROBOTIC ARCHITECTURES

Based on these basic paradigms of sense–plan–act, the three classes of the robot architectures defined are: (i) hierarchical/function-based architecture (which is also called deliberative architecture), (ii) the behaviour-based (called reactive) architecture and (iii) a hybrid architecture that combines the first two architectures. The concepts of these architectures are depicted in Figure 16.1.

16.4.1 FUNCTION-BASED ARCHITECTURE

The function-based architecture, sense–plan–act–feedback-to-sense, is a classical approach which was the dominant paradigm in the early days of (AI) robotics (Figure 16.1a). Then, much focus was on robot (path/motion) planning and higher level reasoning. The architecture follows a top-down approach sense→plan→act with an appropriate feedback mechanism, and this cycle is repeated for every task/mission. The emphasis then was on constructing a detailed environmental model and then carefully planning what functions/steps are required to be carried out by the robot. The sensing module would translate sensor measurements into an internal world model; of course, this mathematical modelling would require the use of some system identification/parameter estimation algorithm which can be either implemented in the onboard processor (for online/real-time planning) or might have been used in an off-line mode. This is an involved and complex task indeed. Then the

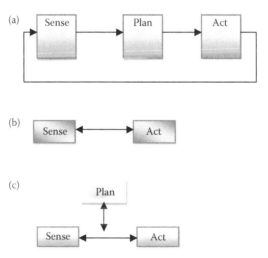

FIGURE 16.1 Three main robotic architectures. (a) Function-based robotic architecture paradigm. (b) Behaviour-based robotic architecture paradigm. (c) Hybrid architecture paradigm.

planner (algorithm/software) would take this internal world model into account along with specified goal to generate a plan of series of actions for the vehicle to follow. The executor then follows the plan and sends actions to the robot actuators. The merit of this architecture is its ability to utilize past experience and expert knowledge to accomplish a task. However, this expert knowledge could be incorporated into fuzzy If…Then rules and the traditional function-based architecture can be converted into an expert system architecture.

16.4.2 Behaviour-Based Architectures

This architecture does not involve planning block (Figure 16.1b). This refers to the fact that such a system exhibits various behaviours, some of which are of urgent nature and hence there is no time to plan. Hence, such robotic systems are characterized by tight/direct coupling between sensors and actuators with minimal computation. However, since there is no explicit planning block, it is prudent to assume that some kind of planning is absorbed or dovetailed into either sense or act block, or is distributed into both the blocks. This has to be done very carefully, otherwise it will not obviate the need of the planning block. What is meant by this is that the sense part and the act part can be made very 'smart' so that the plan block is totally avoided. The robot then consists of behaviour modules and the feedback control for various behaviours closely couples the robot to the real world. This architecture is more suitable for the robots that behave in a dynamic environment as the robots react reasonably well to changes in the environment. Since the planning block is absent, the behaviour-based system's main challenge is that it might lose its appeal if the environmental task's complexity increases. Since, with this, the number of behaviours that a robot may need to exhibit would also increase, making the prediction of ultimate behaviour very difficult for the robot. The architecture has low level of intelligence since there is no planning and there is also a lack of representation. The features of the two architectures are given in Table 16.1.

16.4.3 Hybrid Architecture

The hybrid architecture tries to blend the features of the two architectures and retains planning in some way with direct coupling as well (Figure 16.1c). This architectural paradigm has been very useful and successful as it combines the qualities of both the classical architectures, enabling the autonomous robots to have both decision-making and reactive flexibility. The hybrid architecture is more suitable for complex environment which can be both static and dynamic for a short time. However, there are a good number of variations of this architecture depending on the design details, for example, what techniques are used and how they are combined to obtain the new variation [5,6].

TABLE 16.1
Features of Robotic Architectures

Function-Based/Deliberative Architecture	Behaviour-Based/Reactive Architecture
• Planning is required/planner-based	• No planning is required
• Heavy computational burden	• Less computational burden
• Has slow response	• Has fast response
• World representation is required and used	• No world representation is used
• Suitable for static environment	• More suitable for dynamic environment
• Relatively high level of intelligence is used	• Relatively low level of intelligence used

Source: Adapted from Mtshali M and Engelbrecht A, Robotic architectures (review paper). In Raol JR and Ajith G. (Eds), *Mobile Intelligent Autonomous Systems*, Sp. issue of the *Def. Sc. Jl.*, 60, 1, 15–22, 2010.

Some hybrid architectures might inherit the weaknesses of both the functional and the behavioural architectures; the challenge lies in finding a reasonable balance between the two. Several specific robotic architectures can be found in Refs. [7–9]. An interesting approach that is closely related to a hybrid architecture [4] is the integration of the two distinct architectures to come up with a dual architecture rather than developing one hybrid architecture.

16.5 COORDINATION FUNCTIONS IN BEHAVIOUR-BASED ARCHITECTURE

In this architecture, the behaviours, which are of different types but are required to be carried out for some complex and interconnected task, need to be properly and meaningfully coordinated to have one behavioural command sent to the actuators. The question is how should the system arbitrate/coordinate/cooperate its behaviours and actions so that there is no conflict in the commanded signal to be sent to the actuators. The coordination mechanisms avoid this conflict between two or more behaviours that are active to carry out any task, this mechanisms being (i) competitive and (ii) cooperative. In competitive coordination, the behaviours compete with each other and only one succeeding behaviour is selected and activated. The coordination concepts are depicted in Figure 16.2. The behaviours can be selected based on priority with the use of suppression and inhibition functions [2]. Yet, another way of competition-based approach is to use action–selection coordination. This method lets behaviours generate votes for actions and the action with most weighted votes is chosen for action. The merits of this competitive coordination are (i) modularity, (ii) robustness and (iii) tuning time. The demerits are (i) reduced performance, (ii) increased development time and (iii) increased complexity. These demerits are because of the mechanisms used to select the optimal behaviour among the various competing behaviours. A cooperative coordination function differs in a sense that the different behaviours are all considered and appropriately fused together. It provides

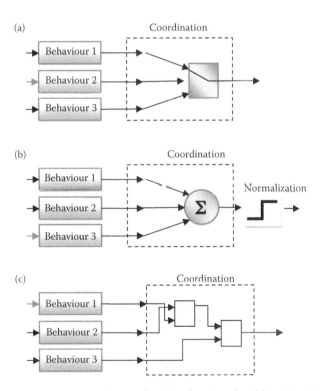

FIGURE 16.2 Command coordination schemes for behaviour-based architecture. (a, b) Competitive command generation scheme. (c) Hybrid command coordination scheme.

the ability to use concurrently the output of more than one behaviour at a time. The main aspect is the combination mechanism for all behavioural outputs to one that is the fused behaviour. The resultant behaviour is then normalized. The merits are (i) increased performance, (ii) less development time and (iii) simplicity.

Subsequently, hybrid coordination has come out which is a combination of a competitive and a cooperative coordination to overcome their respective demerits. The structure and the functionality of this hybrid coordination are very apparent from Figure 16.2. However, the hybrid approach would increase latency (time) due to multi-processing of different behaviours.

16.6   LAYERED APPROACH

It is prudent to make a design of architecture more efficient for robots to perform complex tasks in a dynamic environment. One can use a layered architecture with separate layers for each functionality [10,11]. Thus, three-layered architecture has become very common in robot control architectures; of course, there is no fixed limit to the number of layers (Figure 16.3). The top layer adopts a more goal-oriented view. It plans over a longer/larger scope using information obtained from the sensory data. The lower layer provides fast and short horizon decision about quickly executed actions that are based on sensory data inputs. The upper layer is deliberative whereas the middle one is reactive. The lower layer controls the robot architecture components, like the sensors and actuators. These layers differ between architectures as well as the mechanisms for communicating state information and coordinating activity for a mobile robot.

16.7   CENTRALIZED AND DISTRIBUTED ARCHITECTURES

Like in a multi-sensory data fusion system/approach, in a robotic system, it is crucial to decide on choosing a centralized or a distributed approach. The centralized approach coordinates multiple goals and constraints in a complex environment. A purely centralized architecture is not very appropriate for a real-time system where environment is dynamic or uncertain [9]. The distributed architecture offers reactivity to dynamic environments and is very suitable for such tasks, since the flexibility and robustness are enhanced. Communication among modules is a challenge in the distributed systems. In some architecture, there is direct communication between modules. This then provides the system designer with a high degree of control (flexibility) over the operation of the system, which may be desirable when modules are engineered to interact with each other.

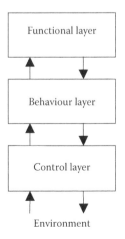

FIGURE 16.3   A three-layered robotic architecture.

16.8 DEVELOPMENT TOOLS FOR THE ARCHITECTURES

In the literature on robotic architectures, a number of robot programming languages are available. A review on these tools can be found in Refs. [12,13] wherein the authors also propose a programming language for functional robotics (FROB). However, there is still no uniform language. This is basically due to the fact that a large number of issues are pending to be resolved for the development of robots and its various features, and these are (i) variety of the tasks that a robot has to perform, (ii) the environmental changes, that is, structured versus unstructured environment, (iii) the static/dynamic environments and (iv) recent beginning of the use of AI components in robotic developments. Some further developmental and architectural aspects about robotic systems can be found in Refs. [14–17].

16.9 FOUR-DIMENSIONAL/REAL-TIME CONTROL MODEL ARCHITECTURE

This architecture (4DRTC) is the latest reference model architecture for intelligent control system design and is suitable for robot as well as many other mobile systems. It has [18] (a) sensing element to obtain perception, (b) cognitive ability, (c) decision-making ability and (d) usual planning and control aspects. It also incorporates many diverse concepts. The 4DRTC architecture is a unique one having several features like (i) hierarchical structure, (ii) distributed architecture, (iii) deliberative and (iv) yet reactive. It connects the cognitive, reflective, planning and feedback control. A variety of planning algorithms are used: (a) case-based reasoning, (b) search-based optimization method and (c) schema-based scripting. Each node in the distributed architecture has its own planner that can be performed autonomously onboard the UGV (unmanned ground vehicle) or can use a computer-assisted planner. In this architecture, the declarative knowledge about the condition of the external world is represented in a format that can be manipulated, decomposed and analysed by reasoning engine. This knowledge in turn describes size, shape, position, orientation, velocity and class of entities. It also enables the system to know the present position of the environment as well as the system's own status. In 4DRTC architecture, the symbolic and iconic knowledge can be incorporated. This symbolic knowledge is about the abstract data structures and is used to represent actions, entities and events. The iconic knowledge is about the information on objects and situations in space/time. This information is about images, maps and state time histories.

16.10 AN ALTERNATIVE HYBRID ARCHITECTURE

An alternative hybrid architecture is that a conceptual architectural model with three layers is shown in Figure 16.4. This architecture includes sensor data fusion explicitly. Also, it is suggested that fuzzy logic be used to incorporate heuristic knowledge to support behaviour-based learning and planning, the factors that were missing in the architecture proposed in Ref. [19].

It has both deliberative and behavioural layers. The deliberative layer is supposed to ensure good planning mechanism and high-level reasoning to be carried out by the robotic system. The control layer represents the HW units that control the robot's sensing and actuation mechanism. The behavioural layer is to ensure that the robot reacts quickly to unknown environment/situations by creating appropriate behaviours. The proposed modified architecture includes the main components: (i) world model, (ii) planner, (iii) learning component and (vi) a command generator. The world model defines the external environment with which the robot would interact. It can represent both the dynamic and the static environments. The planner–processor unit interacts with the world model and decides if any planning is necessary; otherwise, it reactively responds by creating suitable behaviours, which can be supported by incorporating heuristic knowledge by using fuzzy logic. In case the planning is required, the planner derives a plan in terms of sequential tasks that the system has to undertake to achieve a goal.

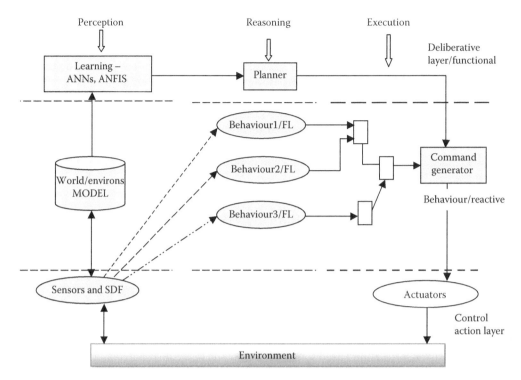

FIGURE 16.4   A revised conceptual hybrid architecture for a robotic and a mobile system with sensor data fusion (SDF) included explicitly for assisting building of the world model. The dotted lines indicate the choice of one or more behaviours, wherein heuristic knowledge can be incorporated using fuzzy logic.

Various behaviours are coordinated together using a hybrid coordinator that uses both competitive and cooperative mechanism to combine the different behaviours. The command generator would create required commands for the robot's actuator to perform. To bring in the AI component into the proposed modified architectural system, a learning component is introduced that would incorporate new knowledge, facts, behaviours and rules into the system. Some of the learning mechanisms that can be used are (a) reinforcement learning, (b) learning by imitation and (c) artificial neural networks. One can use switching operations to choose one behaviour in preference to another one. Multi-sensor data fusion can be advantageously used in generating accurate world model. This alternative hybrid architecture with some AI components can be easily built, tested and validated.

16.11   CONCLUDING REMARKS

Several concepts of robotic architectures were discussed briefly in this chapter. Some features of the function-based, behaviour-based, hybrid, layered and 4DRTC architectures have been discussed. Some features of a modified architecture that include (a) sense–plan–act cycle, (b) reactive element, (c) hybrid command coordination, (d) learning element, (e) layered behavioural part and (f) sensor data fusion to assist in building world model have been discussed. It can be taken up as an example architecture and tested for a given robotic system task.

ACKNOWLEDGEMENTS

The author acknowledges that Ms. Mtshali Mbali (earlier with CSIR, SA) put in her initial efforts and time to study the robotic architectures under the guidance of the author. Her efforts and contributions are gratefully appreciated. The author is also very grateful to Professor Andries P. Engelbrecht (University of Pretoria) for encouraging us to pursue this study.

REFERENCES

1. Arkin R, *Behavior-Based Robotics*, MIT Press, USA, 1998.
2. Murphy R, *Introduction to AI Robotics*, A Bradford Book, The MIT Press, Cambridge, Massachusetts, London, England, 2000.
3. Gat E, Integrating planning and reacting in a heterogenous asynchronous architecture for controlling real-world mobile robots, *Tenth National Conference on Artificial Intelligence (AAAI)*, San Jose Convention Center, San Jose, California, pp. 809–815, 1992.
4. Langland B, Jansky O, Byrd J, and Pettus R, Integration of dissimilar control architectures for mobile robot applications, *Journal of Robotic Systems* 14(4), 251–262, 1997.
5. Connell J, SSS: A hybrid architecture applied to robot navigation, *Proceedings of the IEEE Conference on Robotics and Automation (ICRA-92)*, Nice, France, pp. 2719–2724, 1992.
6. Mithun S. and Marie desJardins, Data persistence: A design principle for hybrid robot control architectures, Paper presented at *International Conference on Knowledge Based Computer Systems*, Mumbai, India, 2002.
7. Volpe R, Nesnas I, Estlin T, Muts D, Petras R, and Das H. The CLARAty architecture for robotic autonomy, *Aerospace Conference, IEEE Proceedings*, Big Sky, MT, USA, vol. 1, pp. 121–132, March 2001.
8. Arkin R and Balch T, AuRA: Principles and practise in review, *Journal of Experimental & Theoretical Artificial Intelligence* 9(2–3), 175–189, 1997.
9. Rosenblatt J, DAMN: A distributed architecture for mobile navigation, *Journal of Experimental & Theoretical Artificial Intelligence* 9(2–3), 339–360, 1997.
10. Brooks R, A robust layered control system for a mobile robot, *IEEE Journal of Robotics and Automation* 2(1), 14–23, 1986.
11. Gat E, *On Three-Layer Architectures*, Artificial Intelligence and Mobile Robots: 195–210. http://www.flownet.com/gat/papers/tla.pdf, 1998.
12. Pembeci I and Hager G, *A Comparative Review of Robot Programming Languages*, CIRL Lab Technical Report, University of Oregon, Eugene, OR, USA, 2001.
13. Biggs G and MacDonald B, A survey of robot programming systems, In: *Proceedings of the 2003 Australasian Conference on Robotics and Automation (ACRA)*, Auckland, 2003.
14. Brooks, RA, How to build complete creatures rather than isolated cognitive simulators, *Architectures for Intelligence*, Lawrence Erlbaum Associates, Hillsdale, NJ, 1991, pp. 225–239.
15. Konolige K, Myers K, and Ruspini E, The saphira architecture: A design for autonomy, *Journal of Experimental and Theoretical Artificial Intelligence*, 9, 215–235, 1997.
16. Bonasso R, Firby R, Gat E, Kortenkamp D, Miller D, and Slack M, Experiences with an architecture for intelligent, reactive agents, *Journal of Experimental and Theoretical Artificial Intelligence*, 9(2–3), 237–256, 1997.
17. Simmons R, Structured control for autonomous robots, *IEEE Transactions on Robotics and Automation*, 10(1), 34–43, 1994.
18. Madhavan R, Messina ER, and Albus JS (Eds.). *Intelligent Vehicle Systems: A 4D/RCS Approach*. Nova Science Publishers, Inc., New York.
19. Mtshali M and Engelbrecht A, Robotic architectures (review paper). In Raol JR and Ajith G. (Eds), *Mobile Intelligent Autonomous Systems*, Sp. issue of the *Def. Sc. Jl.*, 60, 1, 15–22, 2010.

17 Multi-Robot Coordination

Isaac O. Osunmakinde

CONTENTS

17.1 INTRODUCTION

The problem of multi-robot coordination lies at the heart of many robotics applications. A multi-robot system (MRS) is a group of robots that are organized into a multi-agent architecture to cooperatively carry out a common task. The concept of such systems is emerging as an important model for designing intelligent and complex software applications in cooperative robotics research because they possess some peculiar capabilities, such as cooperative localization, cooperative behaviour, cooperative planning and cooperative control [1,2]. The coordination problem can be seen as the problem of avoiding conflicts in a group of robots to optimize their common goal. It can be addressed using several approaches, such as wireless communication technologies, learning, defining a common knowledge or role among the robots, and so on [3]. Wireless communication technologies provide the basic capabilities, which are information sharing and explicit coordination between the robots to support sophisticated cooperation and coordination algorithms development. Within a group of robots learning over time, every robot would be aware of or be able to predict the subtasks carried out by the other robots in the group and successfully coordinates its action to achieve a common complex goal.

In this chapter, we present a survey of recent work within the cooperative robotics, focusing specifically on multi-robot coordination. We segment the multi-robot coordination problem into five fundamental phases—MRS, coordination, controllers, software architecture and control a group of robots. The major contributions of this work are (i) surveying and establishing a categorization structure for current multi-robot coordination approaches which concretely place work within a larger cooperative robotics community, and (ii) identifying open areas for future research and contributing knowledge on laying the foundation for such structures.

17.2 FROM SINGLE TO MULTI-ROBOT CONTROL

Regarding a movement from a single to multi-robot control, we note that control architecture and developing tools which are not platform dependent are needed for handling the platform movement.

We pause to place one of the intents of this chapter within the context of previous control architecture literature. An open-source robotic architecture based on the Sense-Plan-Act paradigm developed for Shakey robot which is introduced in Ref. [4] features three distinct stages of operation. Their strict decomposition is not very suitable for dynamic environments. Another set of control architectures which include reactive planners [5] were also developed. They include a three-layer behaviour-based system that incorporates the low-level control routines currently used on some rovers. The middle behaviour layer uses Multi-Objective Behavior Control action selection mechanisms. Layered control robot architectures, such as OROCOS (open robot control software) [6] and Player/Stage [7], are currently used for the development because of the challenges of managing behaviours to achieve long-range goals in the previous efforts. For specifics of implementation, Figure 17.1 shows control architecture for two robots that can be extended to multiple robot systems [8]. Figure 17.1 is a centralized MRS, where robots and users are connected to a central server (computer), which executes the off-board planning, scheduling and resource sharing and so on. It requires spatio-temporal information which will be used in various ways by the on-board local control and the off-board remote control. The on-board local control is mostly a reactive one, which deals with safety tasks, such as obstacle avoidance according to the directives of remote global planning.

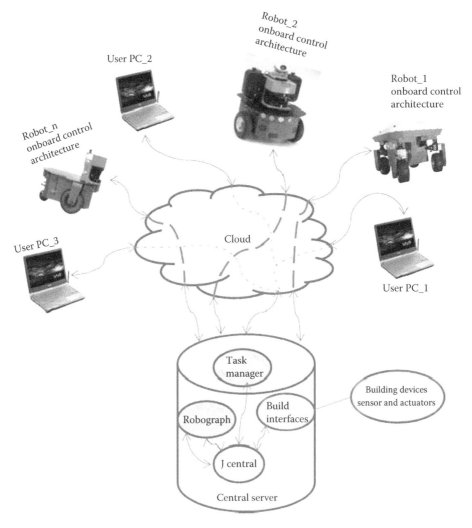

FIGURE 17.1   Multi-robot control architecture.

The off-board remote control is mostly a deliberative one, which deals mainly with planning tasks for multiple robots. It implies that it handles deliberative computing, but it needs some information about accomplishment of tasks so as to change plans when necessary. Also sensory information, such as camera images are required for the user interfaces. The onboard control architecture on a robot in the system is organized into four modules: (a) hardware server governing hardware interaction, (b) control integrating sensor and motion information, (c) executive for application development and (d) interface modules for debugging and tracing.

17.3 CHALLENGES OF MRS

The challenges of multi-robots can be seen in a system where multiple robots have to coordinate their actions, and it is infeasible to model all possible joint actions since this number grows exponentially with the number of robots. For instance, the problem of coordination and balancing between reactivity and deliberation in software architectures are some of the scientific challenges that have been presented in RoboCup competitions [9]. Wireless communication among robot players is allowed, and it can be exploited to achieve good coordination. However, the robots may not fully depend on communication or information provided by other robots due to frequent communication failures. Moreover, coordinating an MRS, where there is a fully distributed processing of sensory input sometimes leads to many possible sources of errors, which also appear to be a challenging problem. For succinctness, some of the important research problems in MRS are multi-robot inspection in underground mines and multi-robot object transportation. In the former, a team of robots could be deployed to inspect loose rocks in the underground mines to improve mine safety. For a team of two robots, one could be deployed to a stope area of the mine while another could be deployed to the gallery region of the mine as shown in Figure 17.2. Their cooperative actions allow for a decomposition of complex inspection tasks into simpler subtasks. In the latter, a group of autonomous robots move cooperatively to transport an object from a position to a goal location and orientation in a static or dynamic environment [10]. It is a challenging task to transport an object which may be more heavier than what a single robot can handle alone and the environment may allow fixed and moving obstacles. Determination of an appropriate cooperative strategy is a major challenge of MRS in transporting an object successfully. This implies deriving optimal amplitudes and positions of the applied forces of the robots, while avoiding obstacles during the transportation.

17.4 NECESSITY OF COORDINATION IN MRS

The presence of coordination in MRS within different areas of human activities is becoming ever more prevalent. Overcoming the difficulties of avoiding obstacles effectively in dynamic environments is achieved, if a learning capability which possesses characteristics of adaptation, fault tolerance and self-organization is integrated into the coordination system. A natural and practical

(a)   (b)

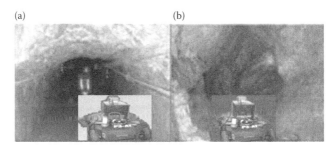

FIGURE 17.2 A team of two pioneer robots for inspecting an underground mine. (a) Gallery region. (b) Stope region.

extension of this is that behaviour conflicts and cooperation failures may be increasingly generated among robots when they select actions concurrently, and it is reasonable to expect an effective coordination about what every robot should do to avoid this. In particular, when two robots select the same action to achieve a common task, they will compete with each other, thereby violating a set-up of coordination strategy. Considering these challenges, good coordination strategy in MRS would ensure that one robot does consider the actions of the other robots before it selects its own action or makes its own decision. From a system engineering viewpoint, coordination typically makes MRS to carry out tasks faster and facilitates inherent redundancy, as robots may still work as a group when a group member is malfunctioning. These necessities of coordination make multi-robots useful in a variety of domains, such as exploration [11], patrolling [12] and transportation [10]. In RoboCup [9], coordination improves team performance and can be measured in terms of the goals scored against the opponents, gaining ball possession and defending their own goal. In exploration, coordination increases system performance in MRS by avoiding interference with one another and a robot avoids exploring same places that other robots have already searched.

17.5   FROM LOCAL INTERACTION TO GLOBAL COORDINATION

There are certain aspects of multi-robot coordination which are becoming popular among all applications to date. One is the fact that a local interaction among some robots could be designed to emerge global coordination of specified behaviours. However, it remains elusive on how to design local interaction rules to achieve certain global behaviours in a group of robots. One approach of handling this in applications is the use of divide-and-conquer design [13] to achieve a specified global coordination. This idea decomposes a global specification into sub-specifications, which are held true by individual robots. A local controller is then designed for each robot to satisfy the local specifications so as to achieve good global behaviours. The following are some research questions that are required to be answered in order to develop similar designs: (i) how to describe the global specification and subtasks in a succinct and formal way?, (ii) how to decompose the global specification?, (iii) is it always possible to decompose? and (iv) what are the necessary and sufficient conditions for decomposability? Another approach involves the use of a coordination graph (CG) [14]. Here a global payoff function can be decomposed into a linear combination of local payoff functions. To illustrate this, Figure 17.3 shows a CG describing the local interactions and a global coordination issue. Similar to Bayesian Network models in Chapter 34, Nodes R_1–R_4 represent robots while edges are depicted as dependencies f_1–f_3 between the robots. Only robots that are directly connected interact locally at any time. From the local viewpoint, robot R_2 interacts with R_1, R_4 interacts with R_3, R_3 interacts with both R_4 and R_1, and R_1 interacts with both R_2 and R_3. This implies that global coordination of multi-robot is achieved when replaced by a number of local interactions of robots pairs. This helps to handle exponential growth of actions as the number of robots increases in an MRS.

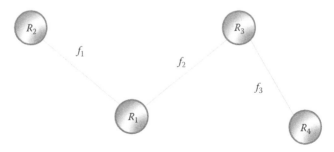

FIGURE 17.3   A CG for local interaction to global coordination.

17.6   INTERACTION THROUGH THE ENVIRONMENT

Although robots are already being used in assistive technology, surgery and therapy [15], the question of continuity of robots interacting through the environments occupied by humans has not been well studied in the robotics community [16]. Unlike the industrial rovers and mining robots, service robots perform service activities directly for human beings for enhancing quality of life, such as addressing the caregiver-shortage problem, rehabilitation [17] and robots interacting with children with disability within an educational context. Safety is a critical issue since a service robot interacts directly with humans. In designing a domain for a service robot, the interacting space may be determined by many rules, such as [18] (i) being a social activity that can interact with humans in a friendly approach; (ii) must tolerate ambiguity—when the robot does not know the actual actions to take to achieve a given task, it should have a human-like intelligence to determine what to do; and (iii) the applications of the service robots must be tested with proven technologies without allowing trial and errors. Communication between robots and humans are of various methods, which involves the use of cameras for visual recognition, audio for sound recognition, through touching and patterns of behaviours. For instance, infrared sensors are being used to communicate to the security systems in elderly homes [19]. These identify challenging research and development areas within interaction of robots through the environment that have not been properly explored. Within this context, robots should be designed to adapt to human by learning complex behaviours of humans and minimize machine factors.

17.7   DESIGN AND ANALYSIS OF MRS

Various studies have been undertaken on the applications of MRS in a common workplace as it increases flexibility and productivity. This section presents the analysis of MRS and potential future work. For an MRS to be analysed, the following characteristics are considered [20]: (i) decentralized data system; (ii) capability for solving a problem by each robot team member; (iii) no global control; (iv) asynchronous computation; (v) errors in sensing, actuating and communicating devices and (vi) collision avoidance by robot cooperation. Since cooperation is an important aspect of multi-robot that can be used to measure a team performance, the following are key research areas that examine the mechanism of cooperation in a team: (i) group architecture, (ii) resource conflict, origin of cooperation, (iii) learning and (iv) geometry problems. Also, performance evaluation of an MRS is based on a collective individual performance of every robot in the team. Particularly in RoboCup [9], team performance can be measured in terms of the goals scored against the opponents, individual gaining ball possession and defending own goal.

   To avoid collision between robots, mutual exclusion method is popularly used and on the basis of the concept of the multi-processor system, an embedded Markov chain model for the MRS that has a common workspace can be constructed as an option. The state space of the Markovian model becomes intractably large as the system size increases where this may introduce some configuration issues. MRS configuration maps multi-robot motions into simultaneous motion, coordinated motion and overlap motion [21]. The simultaneous motion and the coordinated motion are applicable to complex tasks that a single robot cannot accomplish. An example is the object transportation problem given in Section 17.3. In contrast, the overlap motion does not need to operate robots tightly together. The overlap motion increases productivity because robots accomplish various tasks without changing software programmes. MRS approaches to date have focused on performance in terms of flexibility and tolerance for errors, but need more work in efficiency from the viewpoint of engineering. The concept of Dwarf intelligence [22] and Mobiligence derived from biologist and engineers [23] would be appropriate in this regard. Another research problem area is the MRS design issue, which is currently unsolved. The design has to do with the determination of the number of robots and their working environments. This looks more like optimization problems.

17.8   PRINCIPLED SYNTHESIS OF MRS CONTROLLERS

The principled synthesis of MRS controllers in the presence of uncertainties and inaccuracies of sensorial data and environmental noise is a challenging task [24]. With this challenge noted, the navigation controller of a team member of an MRS often addresses three basic problems, such as tracking a reference trajectory, point stabilization and following a path [25,26]. For instance, looking at the trajectory tracking problem, a wheel of a mobile robot in the team is to follow a pre-specified trajectory. Kinematics and vehicle dynamics are sometimes neglected or considered in the tracking problems. In the cases of when dynamics are neglected, vehicle control inputs are calculated by a control algorithm as it assumes that there is a perfect velocity tracking [27]. When a perfect velocity tracking is considered, two assumptions formulate the basis of control algorithms: (i) complete kinematics and dynamics of the robot is known and the dynamics of the robot in not included in the controller design; and (ii) a robot follows a required trajectory without any velocity error. In this case, the following are examples of the control algorithms without the kinematic controllers: (i) behaviour-based controllers [28], and (ii) fuzzy logic controllers [29]. On the other, when dynamics is considered, the following are examples of the control algorithms based on kinematic controllers: (i) adaptive controllers [30] and (ii) neural network controllers [31]. Based on these, there is need for more research and development of an elaborate control system for the poor performance of perfect velocity tracking. The deficiency in sensor data actually contributes to the poor performance when the dynamics in the tracking is neglected. It will be interesting to see a controller of a robot in a team to also track on an open trajectory.

17.9   MULTI-ROBOT SOFTWARE ARCHITECTURE

The software processes in software architecture of an MRS must have an efficient set of low-level skills and must be able to coordinate themselves to act as a team. For instance, in Figure 17.4, software architecture, the main processing node (laptop) of every CAMBADA robot [32] in a team, runs several software processes, such as image acquisition, image analysis, integration and communication with the low-level modules. The scheduling of the processes is arranged by a process manager, which stores the characteristics of each process to activate. The shared data structure of a Real-Time Database (RTDB) stores all information from perception processes, odometry from the holonomic base, presence of obstacles and so on gathered locally by every robot. The shared area of the RTDB of each robot is communicated to the other robots using a multicast communication protocol. It is important that the software architecture of every robot in a team must allow a successful intra-robot and inter-robot integration of different activities by supporting communication and cooperation [33]. ETHNOS IV is a programming environment for the design of real-time systems for a multi-robot, which addresses the communication issue.

17.10   REAL-TIME FRAMEWORK FOR MULTI-ROBOT COORDINATION

Multi-robot frameworks categorize their coordination into different approaches, thereby grouping similar coordination methods together. A framework to coordinate multiple robots uses the coordination graph described earlier in Section 17.5. It assumes that a group of robots is embedded in a continuous and dynamic domain and that the robots are able to perceive their surroundings with sensors.

In the context of position selection behaviour (PSB) framework [34], it is often used in the RoboCup team where it examines the current view of the pitch and uses it to suggest a good place to move to. The factors considered for a position selection in a game includes position of teammates, position of opponents, position of the ball and abilities of the robot players. All team members are assigned different roles, which in turn define home positions for them. Since the environment of soccer pitch and robot motion are dynamic and unpredictable, the coordination framework assists

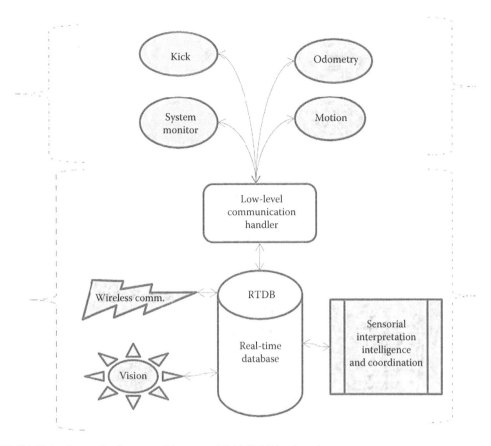

FIGURE 17.4 Layered software architecture of CAMBADA robots in a team.

in knowing a good position to be selected to receive the ball from an opponent. A real-time multi-agent framework for coordinating the activities between robots is also described in ref. [10]. The framework comprises four software agents, including vision agent, learning agent, two robot assistant agents and two physical robots. Based on the information from their own sensors and internal states, the agents cooperatively determine a cooperative strategy to achieve a common goal. The four software agents constitute a high-level coordination subsystem, which assign commands to the low-level control and execution subsystem of the physical robots. The vision agent is in charge of acquiring and processing images using the cameras. The learning agent is used as a learner, a monitor and an advisor. The assistant agents serve as intermediary between the physical robots and the software agents by forwarding information, such as the robot positions to agents at the high level, and vice versa.

17.11 CONTROL OF GROUPS OF ROBOTS

The control and coordination of a group of robots in a distributed and unsupervised mode is a decentralized control problem since only local information is available for the control of every robot member [35]. A decentralized control law would be required to trigger the group of autonomous robots to move at the same time, along a given path and at a given speed in order to resolve collisions between robots. An approach of addressing such laws is based on developing consensus algorithms using nearest-neighbour rules. This implies reaching a consensus among robots. For instance, the motion of each autonomous robot can be controlled using information about the velocities of several other robots, which are the closest neighbours of a robot at a given time. Formation control approaches [36], such as potential-based and behaviour-based, are also used to control groups of robots. Although potential-based has a drawback of local minima such that a desired pattern of formation may not always be guaranteed, it uses some artificial potential functions and the negative gradient of the function provides the control of the robots. It carries out the inter-robot interactions and the interactions with the environment. In the context of a behaviour-based approach, it is difficult to guarantee a convergence to the desired formation pattern mathematically. Some other approaches assume that communication network between robots in a group is constant at all times while others allow disconnection for some periods of time. The leader–follower approach is the most studied formation control strategy, which uses a hierarchical arrangement of individual controllers. This implies that the problem of formation control for a group of robots is reduced to individual tracking problems [37]. Simulations of improving these control approaches in various scenarios would be good for illustrations.

17.12   CONCLUSIONS

In this chapter, we have presented a good survey of MRS techniques employed to address the multi-robot coordination problems. Multi-robot coordination has the attractive characteristics of being an intuitive communication medium to non-robotics experts. Additionally, multi-robot coordination complements many traditional policies of single-robot coordination techniques, offering a solution to some of the weaknesses in traditional approaches. Consequently, multi-robot coordination has been successfully applied to many robotics applications. The multi-robot coordination robotics community, however, suffers from the lack of an established structure in which to organize approaches. In this survey, we have contributed such a structure, through categorization of multi-robot coordination techniques as described in the previous sections. Although multi-robot coordination has proven a successful concept for robot policy development, there still exist many open areas for research, several of which we have identified in most sections above.

REFERENCES

1. R. Rocha, J. Dias, A. Carvalho, Cooperative multi-robot systems: A study of vision-based 3-D mapping using information theory, *Journal of Robotics and Autonomous Systems*, 53(3–4), 282–311, 2005.
2. G. A. S. Pereira, V. Kumar, M. F. M. Campos, Closed loop motion planning of cooperating mobile robots using graph connectivity. *Journal of Robotics and Autonomous Systems*, 56(4), 373–384, 2008.
3. C. Boutilier, Planning, learning and coordination in multiagent decision processes. In: *Proceedings of the 6th Conference on Theoretical Aspects of Rationality and Knowledge*, pp. 195–210, Renesse, Holland, 1996.
4. V. Ng-Thow-Hing, K. R. Thórisson, R. K. Sarvadevabhatla, J. Wormer, T. List, Cognitive map architecture: Facilitation of human–robot interaction in humanoid robots. *IEEE Robotics & Automation Magazine*, 16(1), 55–66, 2009.
5. P. Pirjanian, T. L. Huntsberger, A. Trebi-Ollennu, H. Aghazarian, H. Das, S. S. Joshi, P. S. Schenker, CAMPOUT: A control architecture for multi-robot planetary outposts, In: *Proceedings of SPIE Conference on Sensor Fusion and Decentralized Control in Robotic Systems III*, pp. 221–230, Boston, MA, November 2000.
6. Bruyninckx, H., Open robot control software: the OROCOS project, In: *Proceedings of the IEEE International Conference on Robotics and Automation*, ICRA, ISBN: 0-7803-6576-3, pp. 2523–2528, Seoul, Korea, 2001.
7. B. P. Gerkey, R. T. Vaughan, A. Howard, The Player/Stage Project: Tools for multi-robot and distributed sensor systems. In: *Proceedings of the International Conference on Advanced Robotics*, pp. 317–323, Coimbra, Portugal, 2003.
8. J. López, D. Pérez, E. Zalama, A framework for building mobile single and multi-robot applications. *Journal of Robotics and Autonomous Systems* 59, 151–162, 2011.
9. M. Asada, H. Kitano, The RoboCup challenge. *Journal of Robotics and Autonomous Systems* 29, 3–12, 1999.

10. N. Miyata, J. Ota, T. Arai, H. Asama, Cooperative transport by multiple mobile robots in unknown static environments associated with real-time task assignment. *IEEE Transactions on Robotics and Automation* 18(5), 769–780, 2002.

11. K. H. Low, G. J. Gordon, J. M. Dolan, P. Khosla, Adaptive sampling for multi-robot wide-area exploration, *2007 IEEE International Conference on Robotics and Automation*, Roma, Italy, 10–14 April 2007.

12. N. Agmon, S. Kraus, G.A. Kaminka, Multi-robot perimeter patrol in adversarial settings, *IEEE International Conference on Robotics and Automation, ICRA 2008*, Pasadena, California, pp. 2339–2345, 2008.

13. M. Karimadini, H. Lin, Guaranteed global performance through local coordinations. *Journal of Automatica* 47, 890–898, 2011.

14. C. Guestrin, D. Koller, R. Parr, Multiagent planning with factored MDPs. In: *Advances in Neural Information Processing Systems*, Vol. 14, MIT Press, Cambridge, MA, pp. 1523–1530, 2002.

15. M. Hans, B. Graf, R. D. Schraft, Robotic home assistant care-o-bot: Past–present–future. In: *IEEE Ro-man, 11th InternationalWorkshop on Robot and Human Interactive Communication*, Berlin, Germany, pp. 380–385, 2002.

16. S. Haddadin, S. Parusel, R. Belder, J. Vogel, T. Rokahr, A. Albu-Schaffer, G. Hirzinger, Holistic design and analysis for the human-friendly robotic coworker, *IEEE/RSJ International Conference on Intelligent Robots and Systems*, Taipei, Taiwan, pp. 4735–4742, 2010.

17. M. Hillman, Rehabilitation robotics from past to present—A historical perspective. In: *IRCORR, The Eighth International Conference on Rehabilitation Robotics*, KAIST, Daejeon, Korea, 2003.

18. L. Leifer, Tele-service robots: Integrating the Socio-Technical Framework of Human Service through the Internet-WWW. In: *Proc. of International Workshop on Biorobotics: Human–Robot Symbiosys*, Japan, 1995.

19. K. Haigh, L. Kiff, J. Myers, V. Guralnik, K. Krichbaum, J. Phelps, T. Plocher, D. Toms, The independent lifestyle assistant: Lessons Learned, Tech. Rep., Honeywell Laboratories, 3660 Technology Drive, Minneapolis, 2003.

20. J. Ota, Multi-agent robot systems as distributed autonomous systems. *Journal of Advanced Engineering Informatics*, 20, 59–70, 2006.

21. C.-K. Tsai, Multiple robot coordination and programming. *IEEE International Conference on Robotics and Automation*, Sacramento, CA, pp. 978–985, 1991.

22. T. Arai, J. Ota, Dwarf intelligence—A large object carried by seven dwarves. *Journal of Robotics and Autonomous Systems*, 18(1–2), 149–55, 1996.

23. H. Asama, M. Yano, K. Tsuchiya, K. Ito, H. Yuasa, J. Ota, A. Ishiguro, T. Kondo, System principle on emergence of mobiligence and its engineering realization. In: *Proceedings of the IEEE/RSJ International Conference of Intelligent Robots System*, pp. 1715–1720, 2003.

24. P. Coelho, U. Nunes, Path following control of mobile robots in presence of uncertainties. *IEEE Transactions on Robotics and Automation*, 21(2), 252–261, 2005.

25. M. S. Kim, J. H. Shin, S. G. Hong, J. J. Lee, Designing a robust adaptive dynamic controller for nonholonomic mobile robots under modelling uncertainty and disturbances. *Journal of Mechatronics*, 13(5), 507–519, 2003.

26. E. Maalouf, M. Saad, H. Saliah, A higher level path tracking controller for a four-wheel differentially steered mobile robot. *Journal of Robotics and Autonomous Systems*, 54(1), 23–33, 2006.

27. R. Fierro, F. L. Lewis, Control of a nonholonomic mobile robot using neural networks. *IEEE Transactions on Neural Networks*, 9(4), 589–600, 1998.

28. M. Egerstedt, X. Hu, A hybrid control approach to action coordination for mobile robots. *Journal of Automatica*, 38(1), 125–130, 2002.

29. F. M. Raimondi, M. Melluso, A new fuzzy robust dynamic controller for autonomous vehicles with nonholonomic constraints. *Journal of Robotics and Autonomous Systems*, 52(2–3), 115–131, 2005.

30. W. Dong, K. D. Kuhnert, Robust adaptive control of nonholonomic mobile robot with parameter and non-parameter uncertainties. *IEEE Transactions on Robotics and Automation*, 21(2), 261–266, 2005.

31. D. Gu, H. Hu, Neural predictive control for a car like mobile robot. *Journal of Robotics and Autonomous Systems*, 39(2), 73–86, 2002.

32. Azevedo, J, M. Cunha, L. Almeida Hierarchical distributed architectures for autonomous mobile robots: A case study. In: *Proceedings of the 12th IEEE Conference on Emerging Technologies and Factory Automation*, Greece, pp. 973–80, 2007.

33. D. Nardi, G. Adorni, A. Bonarini, A. Chella, G. Clemente, E. Pagello, M. Piaggio, ART'99: Azzurra robot team, In: M. Veloso, E. Pagello, H. Kitano (Eds.), *RoboCup'99: Robot SoccerWorld Cup III*, *Lecture Notes on Artificial Intelligence*, Vol. 1856, Springer, Berlin, pp. 695–698, 2000.

34. M. Hunter, K. Kostiadis, H. Hu, A behaviour-based approach to position selection for simulated soccer agents. In: *Proceedings of the RoboCup Euro 2000 Workshop*, 2000.

35. W. Ren, R. W. Beard, *Distributed Consensus in Multi-Vehicle Cooperative Control*, Springer, London, 2008.
36. V. Gazi, B. Fidan, Coordination and control of multi-agent dynamic systems: Models and approaches. In: *Swarms Robotics*. Lecture Notes in Computer Science, Springer, Berlin, pp. 71–102, 2007.
37. A. Fujimori, T. Fujimoto, G. Bohacs, Distributed leader follower navigation of mobile robots. In: *Proceedings of International Conference on Control and Automation* (ICCA '05), 2, pp. 960–965, Budapest, Hungary, June 2005.

18 Autonomous Mobile Robot with a Biological Brain

Kevin Warwick, Dimitris Xydas, Slawomir J. Nasuto,
Victor M. Becerra, Mark W. Hammond, Julia H. Downes,
Simon Marshall and Benjamin J. Whalley

CONTENTS

18.1 INTRODUCTION

It is usual for an autonomous mobile robot to be controlled by a computer system, which is either embedded or linked via a wireless or umbilical connection. However, it is now possible for biological neurons to be cultured and trained to act as the brain of such a robot—in doing so, these neurons either completely replace or operate in a cooperative manner with a computer system. Hybrid systems of this type, when embodied in a robot platform, can provide insights into the general operation of biological neural structures and therefore research in this field has immediate medical implications as well as considerable potential in terms of new robotic structures. The aims of the research described in this chapter are to assess the computational and learning capacity of dissociated, cultured neuronal networks. A hybrid system incorporating closed-loop control of an autonomous robot by a culture of neurons has been created. The chapter contains an overview of the problem area, describes the culturing process, gives an idea of the breadth of ongoing research, details operable system architecture and, as an example, reports the results of conducted experiments with 'real-life' robots.

The human brain is a complex computational platform with the ability to rapidly process vast amounts of information, adapt to noise and tolerate faults. Recently, progress has been made towards hybrid systems that integrate biological neurons and electronic components. Reger et al. [1] demonstrated that it was possible to use the brain of a lamprey to control the trajectory of a robot while others were successfully able to send control commands to the nervous system of cockroaches [2] or rats [3], as if they were robots. Although such studies can inform us about information processing and encoding in the brains of living animals [4], they do pose ethical questions and can be technically problematic as access to the brain is limited by barriers such as the skin and skull, and data interpretation is confounded by many factors including the sheer number of neurons present in the brain of even the neurophysiologically simplest animal. Moreover, whole animal approaches capable of recording the activity of individual neurons or their small populations are limited by the invasive, and hence destructive, nature of such techniques. For these reasons, neurons cultured under laboratory conditions on a planar array of noninvasive electrodes provide a far more attractive platform for probing the operation of biological neuronal networks.

This area of research is vital for a number of reasons. Firstly, understanding neural behaviour is important in establishing better bidirectional interactions between the brain and external devices. Secondly, in dealing with numerous neurological disorders, establishing an improved understanding of the fundamental basis for the manifestation of neuronal activity as meaningful behaviours are critical. A robot body can potentially move around a defined area and the effects within a biological brain, which is controlling the body, can be witnessed. This opens up the possibility of gaining a fundamental appreciation and understanding of the cellular correlates of memory and considered actions based on learning and habit.

Recent research has focused on culturing networks of some tens of thousands of brain cells grown *in vitro* [5]. These cultures are created by dissociating neurons obtained from foetal rodent cortical tissue using enzymes and then culturing them in a specialized chamber by providing suitable environmental conditions and nutrients. An array of electrodes is embedded in the base of the chamber (MEA, a multielectrode array) providing an electrical interface to the neuronal culture [6–9]. The neurons in such cultures begin to spontaneously branch out and within an hour of placement, even without external stimulation they begin to reconnect with other nearby neurons and commence both chemical and electrical communication. This propensity to spontaneously connect and communicate demonstrate an innate tendency to network; studies of neural cultures demonstrate distinct periods of development defined by changes in activity which appear to stabilize after 30 days and last for at least 2–3 months [10,11]. The neuronal cultures form a monolayer upon the MEA on the base of the chamber, making them particularly amenable to optical microscopy and accessible to both physical and chemical manipulation [9]. Lewicki [12] provided a review of methods for spike sorting, the detection and classification of neural action potentials.

The objective of the project and experiments described in this chapter is to investigate the use of cultured neurons for the control of mobile robots. However, to produce useful processing, we postulate that disembodied biological networks must develop in the presence of meaningful input/output relationships as part of closed-loop sensory interaction with the environment. This is supported by animal and human studies which show that development in a sensory-deprived environment results in poor or dysfunctional neural circuitry [13,14]. The overall closed-loop hybrid system involving a primary cortical culture on an MEA and a mobile robot ensures a sufficiently rich and consistent environment for the culture and hence constitutes an interesting and novel approach to examining the computational capabilities of biological networks [15]. Typically, *in vitro* neuronal cultures consist of thousands of neurons generating highly variable and multidimensional signals. To extract components and features representative of the network's overall state from such data, appropriate preprocessing and dimensionality reduction techniques must be applied. Several schemes reported in the literature have till now been constructed to investigate the capacity of hybrid systems. Of note, Shkolnik [16] created a very interesting control scheme for a simulated robot. Two channels of an MEA were selected and an electrical stimulus consisting of a pulse (of ± 600 mV, and 400 μs biphasic) was delivered at varying interstimulus intervals. The concept of information coding was formed by testing the effect of electrically induced neuronal excitation with a given time delay termed the interprobe interval between two stimulus probes. This technique gives rise to a characteristic response curve that forms the basis for deciding the robot's direction of movement using basic commands (forward, backward, left and right). In other experiments [16], physical robots such as 'Koala' and 'Khepera' robots were used in an experiment wherein one of the robots (the Koala) was able to maintain a constant distance from the Khepera robot, which was moving under random control. It was reported that the Koala robot managed to successfully approach the Khepera and maintain a fixed distance from it. It is important to stress here that spontaneous activity of the culture was sent to a computer which then made a binary decision as to what action the Koala should take. It is important to note that the culture itself was not directly controlling the Koala through a feedback loop and no learning effect was reportedly exploited. In contrast, both closed-loop control and learning are central aims in our study. In a well-publicized experiment, DeMarse and Dockendorf [17] also investigated the computational capacity of cultured networks by introducing the idea of implementing the control of a 'real-life' problem, such as controlling a simulated aircraft's flight path (e.g., altitude and roll adjustments). Meanwhile, recent developments have focused on the application of learning techniques in neuronal cultures. Shahaf and Marom [18] reported one of the first experiments to achieve desired discrete output computations by applying a simple form of supervised learning to disembodied neuronal cultures. More recently, Bull and Uruokov [19] successfully applied a learning classifier system to manipulate culture activity towards a goal level using simple input signals. However, in both cases the desired result was only achieved in about one-third of experiments, indicating the underlying complexity of neuronal networks, the influence of experimental variability and the difficulties in achieving repeatability in these systems.

Nevertheless, it is clear that even at such an early stage, such re-embodiments (real or virtual) have a prevailing role in the study of biological learning mechanisms. Our proposed physical robots provide the starting point for creating a proof-of-concept control-loop embedding the neuronal culture and a basic platform for future—more specific—reinforcement learning experiments. As the fundamental problem is the coupling of the robot's goals to the culture's input–output mapping, the design of the robot's architecture discussed in this chapter emphasizes the need for flexibility and the use of machine learning (ML) techniques in the search of such coupling.

18.2 PREPARATION FOR CULTURED NEURAL NETWORK

To create the cultured neural network, cortical tissue is dissected from the brains of embryonic rats and neuronal cells enzymatically dissociated before seeding onto planar MEAs. Cells are restricted to lie within the recording horizon of the electrode array by means of an inverse template constructed from adhesive tape placed on the MEA before seeding and removed immediately after cells have settled (~1 h). The MEA is also filled with a conventional cell culture medium containing nutrients, growth hormones and antibiotics of which 50% is replaced twice weekly. Within 1 h of seeding, neurons appear to extend connections to nearby cells and within 24 h, a thick matt of neuronal extensions is visible across the seeded area. This connectivity increases rapidly over subsequent days. After 7 days, initial electrical signals are observed in the form of single action potentials that, in the 'disembodied culture' (not connected within the closed loop), over the following week transform into dense bursts of almost simultaneous electrical activity across the entire network that continues through to maturity (30 days *in vitro* and onwards). However, such continued bursting behaviour, after this initial development phase, may subsequently be representative of an underlying pathological state resulting from impoverished sensory input and may differ from activity of a culture developing within a closed loop [20]. On average, cultures remain highly active until ~3 months of age. During this time, they are sealed with Potter rings [21] to maintain sterility and osmolarity, and are maintained in a humidified, 37°C, 5% CO_2 incubator. Recordings are undertaken in a non-humidified 37°C, 5% CO_2 incubator for between 30 min and 8 h dependent on environmental humidity and the resulting stability of activity.

18.3 EXPERIMENTAL SET-UP

The MEA enables voltage fluctuations (relative to a reference ground electrode outside the network) to be recorded at 59 sites out of 64 in an '8 × 8' array (Figure 18.1), allowing the detection of neuronal action potentials within 100-m radius around an individual electrode. Using spike sorting algorithms [12], it is possible (although nontrivial) to separate the firing of multiple individual

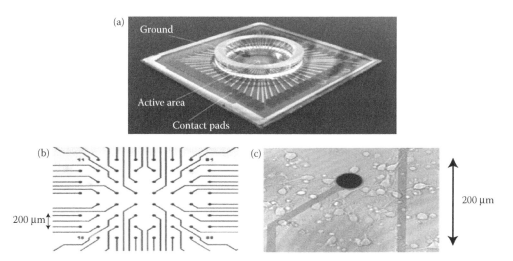

FIGURE 18.1   (a) The MC200/30iR-gr MEA shows the 30-μm electrodes which led to the electrode column–row arrangement; (b) seen under the optical microscope, with 4× magnification, the electrode arrays in the centre of the MEA; and (c) an MEA at 40× magnification, showing neuronal cells in close proximity with visible extensions and interconnections.

neurons, or small groups of neurons, from a single electrode. Consequently, multielectrode recordings permit a picture of the global activity of the entire neuronal network to be formed. It is also possible to electrically stimulate through any of the electrodes to induce focused neural activity. The MEA therefore forms a functional and nondestructive bidirectional interface to the cultured neurons. The electrically evoked responses and spontaneous activity of the culture (the neuronal network) are coupled to the robot architecture through a ML interface that maps the features of interest to specific actuator commands. Associating sensory data feedback from the robot with a set of appropriate stimulation protocols delivered to the culture closes the robot–culture loop. Thus, signal processing can be broken down into two discrete parts: (a) 'culture to robot', in which an output ML procedure processes live neuronal activity and (b) 'robot to culture', which involves an input mapping process, from robot sensor to stimulus. Our overall system has been designed as a closed-loop, modular architecture. Neuronal networks exhibit spatiotemporal patterns with millisecond precision [22], the processing of which necessitates a very rapid response from neurophysiological recording and robot control systems. The software developed for this project runs on Linux-based workstations communicating over the ethernet via fast server–client modules, thus providing the necessary speed and flexibility required when working with biological systems. The study of cultured biological neurons has in recent years been greatly facilitated by commercially available planar MEA systems. These consist of a glass specimen chamber lined with an 8 × 8 array of electrodes (as shown in Figure 18.1).

A standard MEA (Figure 18.1a) measures 49 × 49 × 1 mm and its electrodes provide a bidirectional link between the culture and the rest of the system. The associated data-acquisition hardware includes a head-stage (MEA connecting interface), 60-channel amplifier (1200× gain; 10–3200-Hz bandpass filter), stimulus generator and PC data-acquisition card.

So far, we have successfully created a modular closed-loop system between a (physical) mobile robotic platform and a cultured neuronal network using a MEA, allowing for bidirectional communication between the culture and the robot. It is estimated that the cultures employed in our studies consist of ~100,000 neurons, the actual number depending on natural density variations in proliferation postseeding and experimental aim. The spontaneous electrochemical activity of the culture is used as input to the robot's actuators and the robot's (ultrasonic) sensor readings are (proportionally) converted into stimulation signals received by the culture, effectively closing the loop. For the robotic framework, we selected the Miabot, a commercially available robotic platform, from Merlin Robotics, UK, which exhibits very accurate motor encoder precision (~0.5 mm) and has a maximum speed of ~3.5 m/s. Recording and stimulation hardware is controlled through open-source MEABench software [23]. We have also developed custom stimulator control software that interfaces with the commercially available stimulation hardware, with no need for hardware modification [23].

A simulated counterpart for the real-life robot and its environment has also been developed. The simulation can interface with the culture software in exactly the same manner as the real robot system, thereby extending the modular capabilities of the system. It is expected that this simulation will be particularly helpful in long-running experiments where a real robot could face issues such as battery depletion, as well as in the deployment of various ML experiments explained below. It must be stressed here that our key drive is for the culture to directly control the physical Miabot (Figure 18.2). The simulation has been created as an auxiliary tool to this end and is principally useful for system setup and to ensure that systems are running appropriately.

The overall closed-loop system therefore consists of several modules including the real-life or simulated robot, the MEA and stimulating hardware, a directly linked workstation for conducting computationally expensive neuronal data analyses and a separate machine running the robot control interface; a network manager routing signals directly between the culture and the robot body. The various components of the architecture communicate through TCP/IP sockets, allowing for the distribution of processing loads to multiple machines throughout the University of Reading's internal network. The modular approach to the problem is shown in more detail in Figure 18.3.

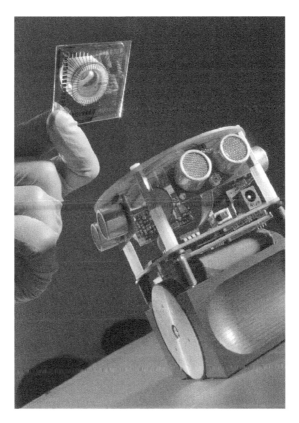

FIGURE 18.2 The Miabot robot with its cultured neural network.

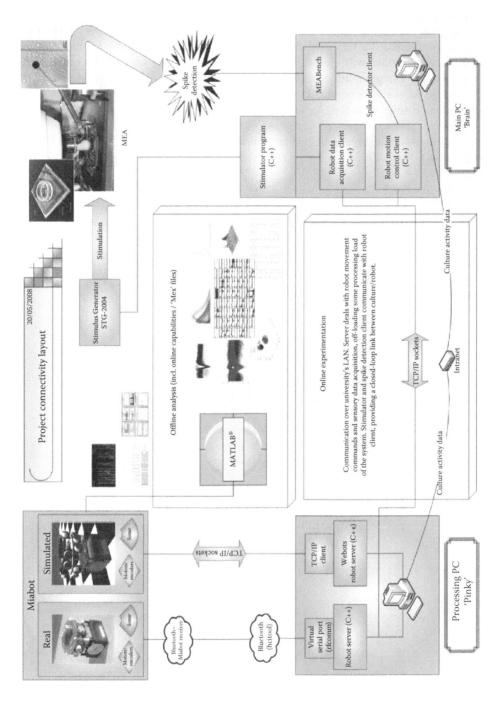

FIGURE 18.3 The robot–MEA system's modular layout.

The Miabot is wirelessly controlled through Bluetooth. Communication and control are performed through custom C++ server code and TCP/IP sockets and clients running on the acquisition PC that has direct control of the MEA recording and stimulating software. The server sends motor commands and receives sensory data through a virtual serial port over the Bluetooth connection, whereas the client programmes contain the closed-loop code that communicates with and stimulates the MEA culture. The client code also performs text logging of all important data during an experiment run, which can then be analysed offline. This modular approach to the architecture has resulted in a system with easily reconfigurable components. The obtained closed-loop system can efficiently handle the information-rich data that is streamed through the recording software. A typical sampling frequency of 25 kHz of the culture activity demands large network, processing and storage resources. Consequently, on-the-fly streaming of spike-detected data is the preferred method when investigating real-time closed-loop learning techniques.

18.4 RESULTS

Initially, we tested all system components and operation of the entire closed loop in experiments utilizing the custom stimulation software [24]. We then conducted the same experiment with a live culture. To this end, an existing appropriate neuronal pathway was identified by searching for strong input/output relationships between pairs of electrodes. Suitable input/output pairs were defined as those electrode combinations in which neurons proximal to one electrode responded to stimulation of the other electrode at which the stimulus was applied (at least one action potential within 100 ms of stimulation) more than 60% of the time and responded no more than 20% of the time to stimulation on any other electrode.

An input–output response map was then created by cyclic stimulation of all preselected electrodes individually with a positive-first biphasic waveform (600 mV; 100 μs each phase, repeated 16 times). By averaging over 16 attempts, this ensures that the majority of stimulation events fall outside any inherent culture bursting that may occur. In this way, a suitable input/output pair could be chosen, dependent on how the cultures had developed, to provide an initial decision-making mechanism for the robot. The robot followed a forward path within its corral confines until it reached a wall, at which point the front sonar value decreased below a threshold (set at ~30 cm), triggering a stimulating pulse (as shown in Figure 18.4). If the responding/output electrode registered activity following the input pulse then the robot turned to avoid the wall. Essentially, activity on the responding electrode was interpreted as a command for the robot to turn to avoid the wall. As a result, it was apparent that the robot also turned spontaneously whenever activity was registered on the response/output electrode. However, the most interesting (and relevant) result was the occurrence of the chain of events: wall detection–stimulation–response.

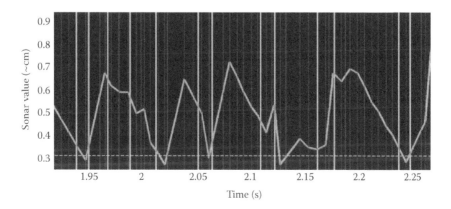

FIGURE 18.4   The robot's activity during a simple wall-detection and right-turn trial experiment.

Such a study opens up the possibility of investigating response times of different cultures under different conditions and how they might be affected by external influences such as electrical fields and pharmacological stimulants [24]. At any one time, we have typically 25 different cultures available and hence such comparative developmental studies are now being conducted. With the sonar threshold set at ~30 cm from a wall, a stimulation pulse was applied to the culture, through its sensory input, each time this threshold was breached effectively when the robot's position was sufficiently close to a wall. An indication of the robot's typical activity during a simple wall-detection/right turn experiment is shown in Figure 18.4. The main trace indicates the front sonar value. Vertical bars indicate stimulus pulse times and sonar timing/actuator command timings as appropriate. Response events (single detected spikes) may occur spontaneously or due to electric stimulation as a result of the sensor threshold being breached. These events are deemed 'meaningful' only in the case when the delay between stimulation and response is less than 100 ms. In other words, this event is a strong indicator that the electric stimulation on one electrode caused a neural response on the recording electrode. When rotation commands are sent to the robot, these events are always coupled (the first one starts the right-turn rotation and the second simply ends the rotation). The results of these rotations can be clearly seen here as the response of the robot, to a sensor threshold breach, causes the sonar value being recorded to increase dramatically. This is as a direct consequence of electrode stimulation (as a result of electrode firing which instantly initiates a rotation command). A 'meaningful' event chain would be, for example, at 1.95 s, where sonar value drops below threshold (30 cm) and a stimulation–response occurs.

Table 18.1 contain typical results from both a model cell and live culture test. If the live culture acted 'perfectly', making no mistakes, then the two columns (Model Cell—showing 'ideal' performance and live culture) would be identical. In Table 18.1, 'total closed-loop time' refers to the time between wall detection and a response signal witnessed from the culture. 'Meaningful turns' refer to the robot turning because of a 'wall detection–stimulation–response' chain of events. A 'wall-to-stimulation' event corresponds to the 30-cm threshold being breached on the sensor such that a stimulating pulse is transmitted to the culture. Meanwhile, a 'stimulation to response' event corresponds to a motor command signal, originating in the culture, being transmitted to the wheels of the robot to cause it to change direction. It follows that for the culture some of the 'stimulation to response' events will be in 'considered' response to a recent stimulus—termed as meaningful—whereas other such events—termed as spontaneous—will be either spurious or in 'considered' response to some thought in the culture, about which we are unaware.

In fact, by totaling all of the trials carried out in this experiment (over 100), considerable differences (as typically indicated in Table 18.1) are observed between the ratio of expected and spontaneous turns between the 'ideal' performance as directed by the model cell and the live culture. Under control of the model cell $95 \pm 4\%$ (mean $\pm$ SD) meaningful turns were observed while the remaining spontaneous turns ($5 \pm 4\%$) were easily attributable to aspects of thresholding spike activity. In contrast, the live culture displayed a relatively low number of meaningful turns ($46 \pm 15\%$) and a

TABLE 18.1

Basic Statistics Resulting from Wall-Avoidance Experiment

Events/Results	Live Culture	Model Cell
Wall → Stimulation event	100%	100%
Stimulation → Response event	67%	100%
Closed-loop time (total)	0.2–0.5 s	0.075 s
Run time	140 s	240 s
Meaningful turns	22	41
Spontaneous turns	16	41

large number of spontaneous turns (54 ± 19%) as a result of intrinsic spontaneous neuronal activity. Such a large number of spontaneous turns were to be expected in an uncharacterized system and the current work aims to both reduce the level of ongoing spontaneous, reminiscent of epileptiform, activity present in such cultures and to discover more appropriate input sites and stimulation patterns. This experiment has 'closed the loop' with the ability to apply custom stimulation protocols and has set the basis for subsequent experiments which will focus on characterizing the culture responses using machine learning techniques for performing more complex robot control. As a follow-up closed-loop experiment, the robot's individual (right and left separately) wheel speeds were controlled through the spike firing frequency recorded from the two chosen motor/output electrodes. The frequency is actually calculated by means of the following simple principle: a running mean of spike rate from both the output electrodes was computed from the spike detector. The detected spikes for each electrode were also separated and divided by the signal acquisition time to give a frequency value. These frequencies were then linearly mapped (from their typical range of 0–100 Hz) to a range of 0–0.2 m/s for the individual wheel linear velocities. Meanwhile, the received sonar information was used to directly control (proportionally) the stimulating frequency of the two sensory/input electrodes. The typical sonar range of 0–100 cm was linearly rescaled into the range 0.2–0.4 Hz for electrode stimulation frequencies (600 mV voltage pulses). The overall set-up is reminiscent of a simple Braitenberg model [25]; however, in this case the sensor-to-speed control is mediated by the cultured network within the overall feedback loop. Certain experiments investigate methods of inducing an appropriate receptive state in the culture that may allow control over the formation of memories [26] involving learning in a cortical associative memory source. For comparative purposes, the experiment was performed with both real and simulated robots with runtime of ~30 min. It might be felt that such a test time is not long enough to evoke long-term potentiation, that is, directed neural pathway changes in the culture, thereby effecting plasticity between the stimulating-recording electrodes. Although this was not a major target in carrying out this part of the experiment, it has been noted elsewhere that a high-frequency burst time can induce plasticity very quickly [27,28]. As a result, we are now investigating spike timing-dependent plasticity based on the coincidence of spike and stimulus.

18.5   MACHINE LEARNING PARADIGM

Initially, the inherent operating characteristics of the cultured neural network have been taken as a starting point to enable the physical robot body to respond in an appropriate manner. The culture then operates over a period of time within the robot body in its corral area. Experimental duration, for example, how long the culture is operational within its robot body, is merely down to experimental design. Several experiments can therefore be completed within a day, whether on the same or differing cultures. The physical robot body of course can operate 24/7.

Learning and memory investigations are at an early stage. However, we were surprised to see that, during the system tests with the live culture, the robot appeared to improve its performance over time in terms of its wall-avoidance ability. We are currently investigating this promising initial observation and examining whether it can be repeated robustly and subsequently quantified. What we have witnessed could mean that neuronal structures/pathways that bring about a satisfactory action tend to strengthen purely through a process being habitually performed. Such plasticity has been reported elsewhere, for example, [29], and experimentation has been carried out to investigate the effects of sensory deprivation on subsequent culture development. In our case, we are now monitoring changes and attempting to provide a quantitative characterization relating plasticity to experience and time. The potential number of confounding variables is considerable, as the subsequent plasticity process is (most likely) dependent on such factors as initial seeding and growth near electrodes as well as environmental transients such as feed rate, temperature and humidity. After the completion of these first phases of the infrastructure set-up, a significant research contribution, it is felt, lies in the application of ML techniques to the hybrid system's closed-loop experiments. These techniques may be applied in various areas, such as the spike-sorting process (dimensionality reduction of spike data profiles, clustering of neuronal units), the mapping process between sensory data and culture stimulation as well as the mapping between the culture activity and motor commands, and the application of learning techniques on the controlled electrical stimulation of the culture, in an attempt to exploit the cultured networks' computational capacity.

18.6   CONCLUDING REMARKS

At this stage, we can conclude that we have successfully realized a closed-loop adaptive feedback system involving a (physical) mobile robotic platform and a cultured neuronal network using an MEA and employing using electrophysiological methods. This necessitates real-time bidirectional communication between the culture and the robot. A culture being employed consists of ~100,000 neurons, although at any one time only a relatively small proportion of these neurons are actively firing. Initial trial runs have been carried out with the overall robot and comparisons have been made with an 'ideal' performance that responds as expected to stimuli. It has been observed that the culture on many occasions responds as expected, on other occasions; however, it does not, and in some cases it provides a motor signal when it is (some might feel) not expected to do so.

In fact, in these circumstances, the mere concept of an 'ideal' response is difficult to address as a biological network is involved, and it should perhaps not be seen as a negative when the culture does not adhere to or achieve such an ideal. As we still know very little about the fundamental neuronal processes that give rise to meaningful behaviours, particularly where learning is involved, we should perhaps retain more of an open mind as to a culture's performance. The culture preparation techniques employed are constantly being refined and have led to stable cultures that exhibit both spontaneous and induced spiking/bursting activity that develops; this is very much in line with the findings of other groups, for example, [15,21]. The stable robotic infrastructure has been extensively tested and is now in place for future ML and culture behaviour experiments. The embodiment module can be instantiated via either a robotic hardware platform or as a software simulation for comparative purposes. The existing, successfully tested infrastructure could be easily modified to investigate culture-mediated control of a wide array of alternative robotic devices, such as a robot head, an 'autonomous' vehicle, robotic arms/grippers, mobile robot swarms and multilegged walkers.

In terms of robotics, this study and others, show that a robot can have merely a biological brain to make its 'decisions'. The 100,000 neuron basis is merely due to present-day limitations—clearly this will increase. The whole area of research is therefore a rapidly expanding one as the range of sensory inputs is expanded and the number of cultured neurons encapsulated rises. The potential capabilities of such robots, including the range of tasks they can perform, therefore need to be investigated.

18.7   FUTURE RESEARCH

There are a number of ways in which the current research programme can be taken forward. Firstly, the Miabot can be extended to include additional sensory devices such as extra sonar arrays, audio input, mobile cameras and other range-finding hardware such as an on-board infrared sensor. This could provide an opportunity to investigate sensory fusion in the culture and to perform more complex behavioural experiments, possibly even attempting to demonstrate links between behaviour and culture plasticity, along the lines of [29], as different sensory inputs are marshaled. Provision of a powered-floor for the robot's corral is also important, to provide the robot with relative autonomy for a longer period of time whereas the suggested ML techniques are applied and the culture's behavioural responses are monitored. For this, the Miabot must be adapted to operate on an in-house powered floor, providing the robot with an unlimited power supply. This feature, which is based on an original design for displays in museums [30], is necessary as ML and culture behaviour tests will be carried out for many minutes and even hours at a time. Other groups have used a simulated rat [31] that moved inside a four-wall environment including barrier objects. It is worth pointing out, however, that the robotic simulation provides an alternative solution to continuous operation of the closed-loop avoiding current hardware limitations. Current hardcoded mapping between the robot goals and the culture input/output relationships can be extended by using ML techniques to reduce, or even eliminate, the need for an *a priori* choice of the mapping. In particular, modern reinforcement learning techniques can be applied to various mobile robot tasks such as wall following and maze navigation, in an attempt to provide a formal framework within which the actual learning capabilities of the neuronal culture will be introduced.

To increase the effectiveness of culture training beyond the ~30% success rate seen in previous work, biological experiments are currently being performed to identify physiological features that may play a role in cellular correlates of learning processes. These experiments also investigate possible methods of inducing an appropriate receptive state in the culture that may allow greater control over its processing abilities and the formation of memories [26] involving specific network activity changes (switch between input and feedback states), which may allow identification of the function of given network ensembles. A further area of research is to identify the most suitable stage of development at which to place cultures within the closed loop and whether a less pathological (epileptiform), and therefore more effectively manipulated, state of activity is achieved when cultures are allowed to undergo initial development in the presence of sensory input. In addition, progression of the project requires benchmarking both the ML techniques and the results obtained by the culture. To achieve this, it is necessary to develop a model of the cultured neural network, based on experimental data about culture density and activity. This behavioural evaluation model is likely to provide great insight into the workings of the neuronal network by comparing the model's and culture's performance. In particular, we hope to gain a better understanding of the contribution of culture plasticity and learning capacity to the observed control proficiency.

There are also possibilities to extend the culture size into a three-dimensional version, consisting of ~30 million neurons at today's standards. Coupled with this is the use of human, as opposed to rat, neurons as the basis for the culture itself. However, the embedding of 30 million human neurons to control a robot body, while it opens up medical issues in a pertinent manner, raises numerous ethical questions regarding life and rights [32]. The authors are aware that these issues need to be discussed thoroughly.

ACKNOWLEDGEMENTS

This work is funded by the UK Engineering and Physical Sciences Research Council (EPSRC) under Grant No. EP/D080134/1. The team wishes to acknowledge the Science Museum (London), and in particular Louis Buckley, for its housed display explicitly on this work from October 2008 to August 2009. We also wish to thank *New Scientist* for its general popular coverage of our robot system [33].

REFERENCES

1. Reger, B., Fleming, K., Sanguineti, V., Simon Alford, S. and Mussa-Ivaldi, F. Connecting brains to robots: An artificial body for studying the computational properties of neural tissues. *Artificial Life*, 2000, **6**, 307–324.

2. Holzer, R., Shimoyama, I. and Miura, H. Locomotion control of a bio-robotic system via electric stimulation. *Proceedings of International Conference on Intelligent Robots and Systems*, Grenoble, France, 1997.

3. Talwar, S., Xu, S., Hawley, E., Weiss, S., Moxon, K. and Chapin, J. Rat navigation guided by remote control. *Nature*, 2002, **417**, 37–38.

4. Chapin, J., Moxon, K., Markowitz, R. and Nicolelis, M. Real-time control of a robot arm using simultaneously recorded neurons in the motor cortex. *Nature Neuroscience*, 1999, **2**, 664–670.

5. Bakkum, D.J., Shkolnik, A., Ben-Ary, G., DeMarse, T. and Potter, S. *Removing Some 'A' from AI: Embodied Cultured Networks*, Lecture Notes In Computer Science, 2004, 130–145.

6. Thomas, C., Springer, P., Loeb, G., Berwald-Netter, Y. and Okun, L. A miniature microelectrode array to monitor the bioelectric activity of cultured cells. *Experimental Cell Research*, 1972, **74**, 61–66.

7. Gross, G. Simultaneous single unit recording in vitro with a photoetched laser deinsulated gold multimicroelectrode surface. *IEEE Transactions on Biomedical Engineering*, 1979, **26**, 273–279.

8. Pine, J. Recording action potentials from cultured neurons with extracellular microcircuit electrodes. *Journal of Neuroscience Methods*, 1980, **2**, 19–31.

9. Potter, S., Lukina, N., Longmuir, K. and Wu, Y. Multi-site two-photon imaging of neurons on multielectrode arrays. *In SPIE Proceedings*, 2001, **4262**, 104–110.

10. Gross, G., Rhoades, B. and Kowalski, J. Dynamics of burst patterns generated by monolayer networks in culture. In: Bothe, H.-W., Samii, M. and Eckmiller, R. (eds), *Neurobionics: An Interdisciplinary Approach to Substitute Impaired Functions of the Human Nervous System*, 1993, 89–121, Elsevier, Amsterdam.

11. Kamioka, H., Maeda, E., Jimbo, Y., Robinson, H. and Kawana, A. Spontaneous periodic synchronized bursting during the formation of mature patterns of connections in cortical neurons. *Neuroscience Letters*, 1996, **206**, 109–112.

12. Lewicki, M. A review of methods for spike sorting: The detection and classification of neural action potentials. *Network (Bristol)*, 1998, **9**(4), R53.

13. Saito, S., Kobayashik, S., Ohashio, Y., Igarashi, M., Komiya, Y. and Ando, S. Decreased synaptic density in aged brains and its prevention by rearing under enriched environment as revealed by synaptophysin contents. *Journal of Neuroscience Research*, 1994, **39**, 57–62.

14. Ramakers, G.J., Corner, M.A. and Habets, A.M. Development in the absence of spontaneous bioelectric activity results in increased stereotyped burst firing in cultures of dissociated cerebral cortex. *Experimental Brain Research*, 1990, **79**, 157–166.

15. Chiappalone, M., Vato, A., Berdondini, L., Koudelka-Hep, M. and Martinoia, S. Network dynamics and synchronous activity in cultured cortical neurons. *International Journal of Neural Systems*, 2007, **17**(2), 87–103.

16. Shkolnik, A.C. Neurally controlled simulated robot: Applying cultured neurons to handle an approach/avoidance task in real time, and a framework for studying learning *in vitro. Mathematics and Computer Science*, 2003, Masters thesis, Department of Computer Science, Emory University, Georgia.

17. DeMarse, T.B. and Dockendorf, K.P. Adaptive flight control with living neuronal networks on microelectrode arrays. *Proceedings of the 2005 IEEE International Joint Conference on Neural Networks*, 2005, **3**, 1548–1551.

18. Shahaf, G. and Marom, S. Learning in networks of cortical neurons. *Journal of Neuroscience*, 2001, **21**(22), 8782–8788.

19. Bull, L. and Uruokov, I. Initial results from the use of learning classifier systems to control in vitro neuronal networks. *Proceedings of the 9th Annual Conference on Genetic and Evolutionary Computation (GECCO)*, pp. 369–376, 2007, ACM, London, England.

20. Hammond, M., Marshall, S., Downes, J., Xydas, D., Nasuto, S., Becerra, V., Warwick, K. and Whalley, B.J. Robust methodology for the study of cultured neuronal networks on MEAs. *Proceedings 6th International Meeting on Substrate-Integrated Micro Electrode Arrays*, 2008, pp. 293–294.

21. Potter, S.M. and DeMarse, T.B. A new approach to neural cell culture for long-term studies *Journal of Neuroscience Methods*, 2001, **110**, 17–24.

22. Rolston, J.D., Wagenaar, D.A. and Potter, S.M. Precisely timed spatiotemporal patterns of neural activity in dissociated cortical cultures. *Neuroscience*, 2007, **148**, 294–303.

23. Wagenaar, D., Demarse, T.B. and Potter, S.M. MEABench: A toolset for multi-electrode data acquisition and on-line analysis. *Proceedings of the 2nd International IEEE EMBS Conference on Neural Engineering*, 2005, pp. 518–521, IEEE, Piscataway, NJ.

24. Xydas, D., Warwick, K., Whalley, B., Nasuto, S., Becerra, V., Hammond, M. and Downes, J. Architecture for living neuronal cell control of a mobile robot. *Proceedings of European Robotics Symposium EUROS08*, 2008, pp. 23–31, Prague.

25. Hutt, B., Warwick, K. and Goodhew, I. Emergent behaviour in autonomous robots. Chapter 14. *Information Transfer in Biological Systems* (Design in Nature Series, vol. 2), Bryant, J., Atherton, M. and Collins, M. (eds.), 2005, WIT Press.

26. Hasselmo, M.E. Acetycholine and learning in a cortical associative memory source. *Neural Computation Archive*, 1993, **5**, 32–44.

27. Cozzi, L., Chiappalone, M., Ide, A., Novellino, A., Martinoia, S. and Sanguineti, V. Coding and decoding of information in a bi-directional neural interface. *Neurocomputing*, 2005, **65/66**, 783–792.

28. Novellino, A., Cozzi, L., Chiappalone, M., Sanguinetti, V. and Martinoia, S. Connecting neurons to a mobile robot: An *in vitro* bi-directional neural interface. *Computational Intelligence and Neuroscience*, 2007, 13, doi: 10.1155/2007/12725.

29. Karniel, A., Kositsky, M., Fleming, K., Chiappalone, M., Sanguinetti, V., Alford, T. and Mussa-Ivaldi, A. Computational analysis *in vitro*: Dynamics and plasticity of a neuro-robotic system. *Journal of Neural Engineering*, 2005, **2**, S250–S265.

30. Hutt, B. and Warwick, K. Museum robots: Multi-robot systems for public exhibition. *Proceedings of 35th International Symposium on Robotics*, 2004, p. 52, Paris.

31. DeMarse, T., Wagenaar, D., Blau, A. and Potter, S. The neurally controlled animal: Biological brains acting with simulated bodies. *Autonomous Robots*, 2001, **11**, 305–310.

32. Warwick, K. Implications and consequences of robots with biological brains. *Ethics and Information Technology*, 2010, **12**, 223–234.

33. Marks, P. Rat-brained robots take their first steps. *New Scientist*, 2008, **199**(2669), 22–23.

Safe and Effective Autonomous Decision Making in Mobile Robots

Rahee Walambe

CONTENTS

19.1   INTRODUCTION

This chapter discusses research carried out towards the development of safe and effective autonomous decision making in intelligent robotic systems, specifically for mobile robots which are employed in hazardous applications, such as landmine clearance, nuclear decommissioning, excavation for construction sites, archaeological excavations and so on.

In the last few decades, owing to a large number of accidents and considerable loss of human life in hazardous environments [1–6], a need to replace human workers by autonomous robots has arisen. The work discussed in this chapter focusses primarily on two areas: (1) to investigate how to integrate safety into the autonomous decision-making process of the autonomous robot and (2) implementation of control architecture (taking into account modularity and re-use) to a mobile robot so that it can perform the task of navigation and path planning autonomously and safely. Combination of methods in robotics and AI are used to bring about the necessary achievements.

In view of the aforementioned scope, RCS-RMA (real-time control system—reference model architecture) has been employed as an architectural framework. RCS is a hybrid hierarchical architecture and facilitates modularity, re-use of components and flexibility of use of various control algorithms at different levels of hierarchy. A probabilistic reasoning model, partially observable Markov decision process (POMDP), has been employed for safety decision making and for dealing with the uncertainty introduced due to partial observability of the environment. It was observed that the POMDP model maps well into the basic operational unit of RCS-RMA and the POMDP process can be developed within this RCS node which facilitates safety decision making.

19.2   ROBOTIC ARCHITECTURES: CLASSIFICATION AND SELECTION

System 'architecture' primarily refers to the software and hardware framework for controlling the system. Intelligent systems are complex and difficult to develop. They have various subsystems performing their own functions while obtaining real-time information and propagating it throughout other subsystems [7,8].

Most of the architectural styles described in the technical literature and developed over the last few decades can be classified into three categories: deliberative, reactive and hybrid. Figure 19.1 depicts the spectrum of these architectural strategies: the left side represents methods that employ

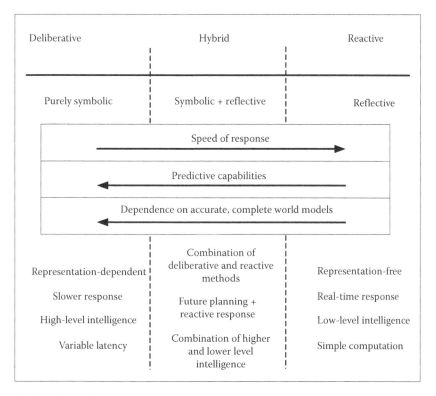

FIGURE 19.1 Modified robot architecture spectrum.

deliberative reasoning, the right side represents reactive control and midway between them is the hybrid style [9]. There are other architectures that one has to consider separately: Sekiguchi et al. [10], for instance, proposed controlling a mobile robot by a structured neural network. Along with this, there are biologically inspired architectures which are not considered in the following classification [11–15].

Deliberate reasoning often requires strong assumptions about the world model; primarily that the knowledge upon which reasoning is based is consistent, reliable and certain. If the information, which the reasoner uses, is inaccurate or has changed since it was obtained, the outcome of reasoning may err seriously. NASREM (NASA/NBS Standard Reference Model) [16,17] is an example of deliberative architecture. The right side of the spectrum depicted in Figure 19.1 represents reactive systems. Reactive control is a technique for tightly coupling perception and action, typically in the context of motor behaviours, to effectively produce timely robotic response in complex, dynamic and unstructured domains. Reactive control for autonomous robots grew out of the recognition that planning, no matter how well intentioned, is often a waste of time. Subsumption is a reactive architecture developed by Brookes [18,19].

Reactive behaviour-based robotic control can effectively produce robust performance in complex and dynamic domains. In some ways, however, the strong assumptions that a purely reactive system makes, serve as a disadvantage at times. Introducing various forms of knowledge into a robotic architecture can often make behaviour-based navigation more flexible and general. Deliberative systems permit representational knowledge to be used for planning purposes in advance of execution. Hybrid architectures combine both deliberative and reactive strategies for better behaviour.

The hybrid style combines both reactive and deliberative control in a heterogeneous architecture. It facilitates the design of efficient low-level control with a connection to high-level reasoning. However, the connection between the two levels must be carefully designed and implemented to provide the right mix of reactivity and deliberation [17]. Hybrid deliberative/reactive robotic architectures have recently emerged, combining aspects of traditional AI symbolic methods and their use of abstract representational knowledge, but maintaining the goal of providing the responsiveness, robustness and flexibility of purely reactive systems. Hybrid architectures permit reconfiguration of reactive control systems based on available world knowledge through their ability to reason over the underlying behavioural components.

The hybrid system's architect contends that neither approach is entirely satisfactory in isolation but that both must be taken into account. Each of these methods addresses different subsets of the complexities inherent in intelligent robotics. Human behaviour is also a combination of deliberative planning and reactive responses, weighted by circumstances and experience [20,21]. Examples of hybrid architectures are AuRA (autonomous robot architecture) [22,23], BERRA (behaviour-based robot research architecture) [24] and RCS-RMA [25].

RCS-RMA is a hybrid architecture developed by the National Institute of Standards and Technology (NIST). RCS-RMA differs from other hybrid architectures (such as AuRA, BERRA) in that it does not relegate planning to the upper levels and reactive behaviour to the lower levels of a hierarchical control system [17,25]. RCS combines deliberative with reactive components at every hierarchical level. Every level of the RCS hierarchy contains a planner and reactive executor. Planners at higher levels in RCS have longer term planning horizons with lower resolution planning space. Figure 19.2 shows the RCS architectural model for automated guided fleet of vehicles [26].

The basic operational block of RCS is called an RCS node (as shown in Figure 19.3) [26]. Within the RCS node(s) at every level of RCS hierarchy, a reactive and deliberative component is present. Along with the sensor, actuators and other components, the RCS node is responsible for the basic functioning of RCS hierarchy.

In summary, the reasons for choosing RCS as reference model architecture for this development are listed below:

1. RCS combines many concepts from AI with control theory. RCS supports a rich dynamic world model at many different levels of resolution of space and time. It combines deliberative with reactive behaviour at many different levels of resolution.

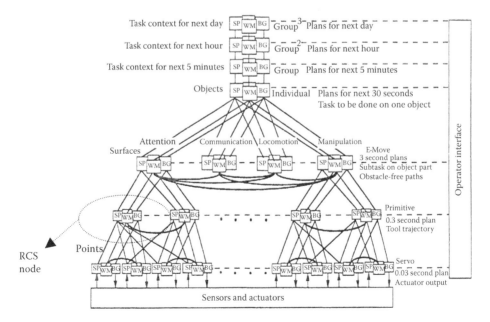

FIGURE 19.2   RCS architecture showing RCS node.

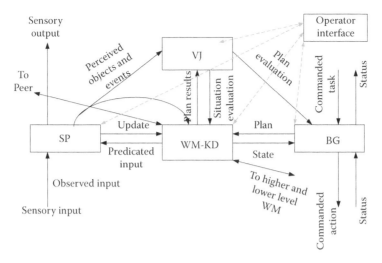

FIGURE 19.3   RCS computational node.

2. The operator interface (OI) provides the capability for the operator to interact with the system at any time at a number of different levels. The OI functionality can be defined according to application.
3. In RCS, the inherent complexity of intelligent systems is dealt with by several means, for example, hierarchical layering, focus of attention and software re-use.
4. RCS uses the principle of hierarchical levelling to facilitate software re-use.
5. It is easier to divide a system into modules and to establish relations between them, and thus it satisfies the criteria for modularity. When going up the hierarchy, RCS follows the IIDP (increasing intelligence while decreasing precision) principle [27,28].

RCS is a flexible architecture design (i.e., the freedom of the RCS designer in selecting the hierarchy of the controller). The designer can choose any structure of controller that will meet the control objectives.

The RCS architecture has been used in various applications [25] such as autonomous submarine vehicles, post office automation, coal mining and so on over the last two decades successfully. It is a fairly generalised reference architecture and hence highly developed among the available system architectures. RCS is a mature system with a number of software engineering tools and software libraries that are available to potential users. NIST provides support for RCS architecture and RCS library [25] implementation.

19.3   COMPONENTS OF RCS-RMA

RCS is a hierarchical architecture. RCS partitions the control problem into four basic elements called functional elements [29]. The RCS node shown in Figure 19.3 is the fundamental building block of the RCS architecture.

19.3.1   RCS COMPUTATIONAL NODE

The functional elements of the RCS reference model architecture are *sensory processing* (SP), *world model* (WM)—knowledge database (KD), *value judgement* (VJ) and *behaviour generation* (BG) [29]. When a particular application is designed using RCS architectural framework, it is divided into smaller subtasks or subsystems. One node is assigned to one or more of these subtasks. The fundamental function of each node is to complete this subtask(s). RCS clusters the four functional elements into computational nodes according to these various subsystems and arranges these nodes in hierarchical layers such that each layer has characteristic functionality and timing. There can be more than one node at any level of RCS hierarchy. Each layer provides a mechanism for integration of deliberative (planning) and reactive (feedback) control.

Koestler [30] explains through different examples that the hierarchical organisation of systems is an inbuilt feature of not only the biological and sociological life but also of any complex evolving system. Not only is the time needed for the development greatly shortened when hierarchical methods are used, primarily because of the decomposition of the task into subtasks, various control algorithms can be used which are suitable for each subsystem. Also, the hierarchical methods employ the principle of reusability that reduces the development time to a certain extent. Hierarchical methods also offer inherent benefits in terms of maintenance, regulation and restoration [31].

Each RCS node, if not at the highest level, looks upward to a higher level node from which it takes commands, for which it provides sensory information and to which it reports status. Each non-leaf RCS node (nodes at levels other than the lowest of the hierarchy) also looks downwards to one or more lower level nodes to which it issues commands and from which it accepts sensory information and status. Every node may also communicate with nodes at the same level (peer nodes) with which it exchanges information. Each node (Figure 19.3) in the architecture acts as an operational unit in an intelligent system.

The SP element consists of a set of processes by which sensory data interact with the prior knowledge to detect and recognise useful information about the world. Perception occurs in an SP system element that compares sensory observations with expectations generated by an internal world model. WM is a functional process that constructs, maintains and uses a world model knowledge database (KD) in support of BG and SP. WM is an internal representation of the external world. WM at any given level of RCS hierarchy is defined and constructed using a control algorithm appropriate at that level. WM also generates and maintains a best estimate of the state of the world that can be used for controlling the current actions and planning the future behaviour. It simulates the results of the possible future plans based on the estimated state of the world and planned actions. Simulated results are evaluated by the VJ system to select the best plan for execution. VJ computes the cost, risk and benefits of actions and plans. VJ also evaluates perceived and planned actions and assesses the reliability of information. The ability of VJ to calculate the rewarding or punishing effects of the perceived states and events enables BG to select goals and set priorities. BG undertakes the planning and control of actions intended to achieve or maintain behavioural goals. Behaviour goal is a desired result that behaviour is intended to achieve or maintain. BG accepts commands and formulates and/or selects plans using information from the WM.

In Figure 19.3, SP, VJ, WM-KD and BG represent four functional elements. Each of these functional elements can communicate with the operator interface if needed. Various links between the functional elements show the data flows between and functions carried out within them. The SP–WM–BG loop completes the feedback loop between predicted and observed sensory data, updates the knowledge and plans accordingly. Within SP, observed input from sensors and lower level nodes is compared with predictions generated by WM. Differences between observations and predictions are used by WM to update the knowledge database. The SP–VJ–WM connection is responsible for estimating and planning and represents the deliberative component of RCS. Within BG, goals from higher levels are compared with the state of the world as estimated in the knowledge database. BG typically involves planning and execution functions. Differences between goals and estimated states are used to generate action. Information in the knowledge database of each node can be exchanged with peer nodes for purposes of synchronisation and information sharing. WM–VJ–BG connection represents the reactive component of RCS at the basic level and is responsible for evaluating the plans, learning and generating actual physical behaviour.

A collection of RCS computational nodes, such as that illustrated in Figure 19.3, can be used to construct distributed hierarchical reference model architecture, such as that shown in Figure 19.2.

A similar architecture could be developed for an autonomous land vehicle, a construction machine on a construction site, an undersea vehicle carrying out cooperative operations with other undersea vehicles and so on.

19.3.2 Control Algorithms within RCS Hierarchy

This section describes a generalised approach for designing the robotic applications by implementing RCS-RMA. Each RCS node (Figure 19.3) is responsible for different subtasks [32]. Hence, each level has its appropriate time–space resolution and behaviour (deliberative or reactive). Consequently, the control methods required at each level need to be different. If the same control method is used at all levels of hierarchy, a well-featured method could be wasteful at lower levels due to more than required functionality it could offer and a simpler method could be insufficient at higher levels due to lack of sufficient features. In summary, the hierarchical nature of the architecture can be visualised by considering the upper-level WM to offer a world-centric representation, low in resolution but large in scope, providing a general picture of the robot workspace. At the lower levels, the WM will take the form of an ego-centric representation allowing the system to evaluate its current state and interaction with its immediate environment such that the robotic system can be considered stationary within a dynamic world. Given the variation in control requirements and scope within the architectural hierarchy, the mode of control at each RCS level (and also at intra-level nodes) must adapt according to the characteristics defining that specific level of decision making.

For the aforesaid reasons, it is proposed that the use of a common control method at all levels of such a hierarchical architectural scheme may not be appropriate. Diverse control methods are required within RCS to achieve the desired operational goal and to provide the required level of autonomy. The need for supervision of controller functions, the distributed nature of many systems and the need in many systems for an interface to a human user (e.g., for monitoring and specifying system goals) leads to a hierarchical structure for a controller. An example of a hierarchical structure is shown in Figure 19.2 (this is sometimes called the *organisational hierarchy* of the controller). In Figure 19.2, the boxes represent different modules of the controller, each assigned a different task (or subtask) and the lines represent communication links.

19.4 AUTONOMOUS SAFETY MANAGEMENT FOR ROBOTIC SYSTEMS

The important challenges to the development of autonomous mobile robots to perform useful tasks in real work environments are effective operation, commercial viability and, above all, operational safety [33]. It is relatively easy to make safety arguments for situations where mobile robots replace humans in hazardous environments, such as landmine-filled terrains and nuclear waste clearance. In these circumstances, it is important that the robot is intelligent enough to take care of its own safety. It would not serve its purpose if the robot stops working or gets trapped due to some unsafe action-decision choice and the operator is required to venture into the unsafe territory to rescue it. However, in other situations, such as construction sites or hospitals, the robot itself can be the main safety concern due to the fact that these locations are crowded. Here, an unsafe robot means increasing the possibility of collision accidents with humans.

Safety is a relative term. No system is absolutely safe. The system which one perceives as safe can be unsafe from someone else's perspective and it is impossible to ensure complete safety. Due to this inability in assuring complete safety, it is necessary to secure a tolerable level of risk, which fits the circumstances in which the application under development is being used. In this respect, it is essential to understand various terms related to the system, viz., safety, reliability and uncertainty. Uncertainty often gives rise to unsafe behaviour as already explained above. Reliability concerns the ability of a system to continue performing according to its specification. An uncertain system may be reliable but at the same time unsafe. Although reliability and safety are related in many respects, they do not correspond to the same objective and hence the work on system reliability falls short of defining a system which can operate safely [34]. Pace points out the following two aspects which give rise to this difference of operation of safety and reliability:

1. The system can still be fully reliable but operate in an unsafe manner if it encounters situations for which it was not designed to operate in.
2. The consequences of an unreliable system are not considered adequately in the context of system risks, and therefore, it is not possible to adequately assess how reliability, or the lack of it, influences the system's safety.

It is also important to distinguish between safety and the ability to complete a given task in an effective manner. The latter requires availability and possibly flexibility and freedom of access. In other words, the traditional means of making the robots safe in the factories by providing clear separation and barriers is often not an option. Likewise, the option to simply stop and wait for further instructions whenever uncertainty arises is also not conducive to effective task completion. It is clear that if mobile robots are to remain effective while operating in unstructured environments, they will need to reason about their own actions and do this in a manner which is tolerably safe.

19.4.1 System Safety Analysis and Uncertainty

The selection of a specific action affects the safety management of a robotic system to a great extent. Unsafe operating conditions can arise from scenarios not explicitly related to an internal fault, but rather from the robot environment. The following three main hazards are identified [35]:

1. Impact—This involves being struck by a moving part of the robot or by a part or tool carried or manipulated by the robot. It can be caused by the unexpected movement of the robot or by the robot ejecting or dropping workpieces or molten metal.
2. Trapping—This can be caused by the movement of the robot in close proximity to fixed objects like machines, equipment, fences and so on. Trapping points can also be caused by the movement of the work carriages, pallets, shuttles or other transfer mechanisms. They can also be presented on the robot itself, on the arm or mechanism of the robot.
3. Other—This would include hazards inherent to the application itself like electric shock, arc flash, burns, fume, radiation, toxic substances, noise and so on.

Dhillon and Fashandi presented an overview [36] of the various robot safety and reliability assessment techniques. Various assessment methods, for example, failure mode and effects analysis, fault tree analysis and Markovian analysis, which had been developed for robot safety and reliability are discussed in detail. Both the robot and the environment are at risk due to various hazards such as system faults, environmental factors and sensor unreliability, giving rise to accidents such as collision, toppling and so on. If the robot is capable of reasoning about its own safety and taking the required safe decisions, the surroundings and the robot itself will be automatically protected. It is critically important that the robot working in such environment(s) is well equipped with the required decision-making capacity. These decisions are strongly influenced by the availability of knowledge about the robot's surroundings and how fast the system reacts in real time to sudden changes in the surroundings.

Moreover, robots are capable of delivering large amounts of energy on collision and performing a wide range of quick and erratic movements [37] which make them hazardous even in a closed environment.

Incidences of accidents are not uncommon in industry [38–40]. The primary reasons for such accidents is often related to robot design and autonomy (e.g., robot not capable of detecting the sudden and unexpected objects in its surroundings, robots unable to make quick safety decisions and robot being too slow to respond). Safety reasoning and subsequently safe behaviour generation comes from within the robotic system design. Margrave et al. [41] suggested the use of a separate module mounted on the robot called the 'safety manager' for ensuring the safe operation of the mobile robot. However, this approach is not advantageous (and hence not employed here) for the following reasons:

1. Concentrating too much effort in terms of safety management into one single module— Although this reduces the criticality of the other elements, it puts excessive burden on the safety manager in terms of guaranteeing safety. For such an accurate operation, the safety manager would have to be designed to an extremely high integrity level.
2. The complexity of the safety manager—Given that a robotic system operating in an unstructured environment has to deal with complex behaviour, it would make the safety manager deal with a myriad of different problems.
3. Difficulties in obtaining the right compromise between safe behaviour and effective task completion—The decision-making process necessitates appropriate evaluation of the safety as well as goal achievement consequences of the system's planned actions. Therefore, it makes little or no sense to have a whole plan generated by one module, only to be rejected by the safety manager.

Spreading safety-related activities over the whole system and possibly over different subsystems could reduce the integrity level requirements. Using safety decision making at different levels of control also makes the implementation manageable. Spreading the safety management throughout the system also ensures that the action selection achieves a balance between both safety and task achievement perspectives.

The above discussion does not explicitly discard the safety manager approach, but rather points in the direction that it may be better to have several safety managing modules with the primary objective being that of constraining the decision-making process at its various levels and operational nature to take safety into account.

Owing to the above-mentioned reasons, the robot's architecture must embed the safety reasoning within its decision-making process. Instead of considering safety as a separate entity, safety management should be considered as an intrinsic part of the decision-making process [34]. By introducing safety reasoning within the robot's decision-making process, operational safety can be improved. In other words, safety consideration should be included in the action-selection/decision-making process of the robotic architecture. As this research employs RCS as an architectural framework, the following discussion is based on various operational aspects of RCS, and more importantly, how safety can be made intrinsic to the decision-making process of RCS-RMA.

The deployment of robots in unstructured surroundings such as construction and nuclear decommissioning will require them to become more responsive to their local environments and to assume responsibility for decision making in uncertain environmental conditions. A significant cause of hazards is uncertainty. The ability to handle underlying operational uncertainty is highlighted by Albus [29], Moravec [42] and Brooks [19] among many others who relate it to the definition of intelligent behaviour. Uncertainty is the root cause of unsafe or risky behaviour. The two major sources of uncertainty in terms of system operation are

1. The understanding and representation of the robot's own state and the state of the external world
2. The limitation on the system resources and hence on the representation of the world

The identification of the cause of uncertainty, the way in which it influences the decision-making process and the degree of uncertainty is crucial for finding a satisfactory solution. From a robot's perspective, the following types of uncertainties are required to be managed [34]:

1. *Uncertainty about the current state of the system and current state of the world*—The interpretations of the current internal system state and external environment state are never fully accurate. This is mainly owing to the fact that the sensors cannot perceive and sense all the parameters of the external world to the highest degree of reliability. Sensors cannot be considered as 100% reliable and some tolerance has to be assigned to the sensory outputs. Uncertainty also always arises from inadequate understandable interpretation of the sensory data.
2. *Uncertainty about the outcome of the action/behaviour which a system chooses to perform*—This type of uncertainty originates from the wrong interpretation of the environment. The uncertainty that is associated with the sensory data interpretation gives rise to uncertainty about the outcomes of the action decisions. If the system is unsure about the outcome of action it decides to perform, it might make a wrong choice. This is the uncertainty about the identification of the future state following a certain action execution, that is, expected or estimated future state.
3. *Uncertainty resulting from internal faults such as sensor faults or system faults*—Sensor faults are due to unreliable sensor behaviour resulting into uncertain or wrong sensory data being processed. System faults include the faults introduced due to system failure to interpret sensor data in good time.

Section 19.2 discussed the adoption of RCS for this research. The next section explains how safety can be incorporated into the decision-making levels of RCS-RMA.

19.4.2 SAFETY INTEGRATION WITHIN RCS

RCS is a much-generalised architectural framework [25] and allows various control methods to be incorporated within its hierarchical layout. However, there is still room for modifications and improvement. One aspect of this improvement is an increase in robustness in the presence of uncertainty and a need for safety. Recommendations for increasing the robustness and reliability of RCS are enlisted next:

1. RCS does not take into consideration safety in general but it is possible to introduce safety behaviours and management through the reasoning. There are several issues such as partial observability (and hence uncertainty), incomplete and inconsistent data, hazards, causes of hazards and so on which need to be taken into account when analysing a system's safety, particularly if the system operates autonomously in an unpredictable and uncontrolled environment.
2. The RCS framework on its own does not deal with uncertainty. It is the RCS designer's job to implement suitable algorithms, but RCS does support the use of various algorithms within its framework. For a faithful representation of reality, uncertainty should be taken into consideration. The system has to model the external world effectively. The system should represent not only what it is certain about but also what it is uncertain about, along with the level of uncertainty (as well as how this uncertainty alters decisions). RCS provides a working architectural framework. But incorporating the uncertainties can make RCS more robust.

The full integration of safety is important at many levels of RCS so that it is considered throughout planning as well as execution. This departs from most other approaches which generally only consider safety at the lowest level of control at the point of direct interaction with the environment. The following issues in relation to embedding safety within RCS are considered [43] here:

1. Should safety be an additional value in the value judgement of the RCS?—The value judgement parameter can take the form of a risk measurement (in the field of risk analysis, measurable risk values are considered, so theoretically there are ways of quantifying the risk).
2. How would a safety value judgement be processed?—Potentially, this could be through the evaluation of possible accident risks for the proposed plans (i.e., the value judgement process is nothing more than a risk assessment process for the proposed plans with the resulting 'value' being the risk measure).
3. What are the implications of including safety in decision making?—There will need to be some form of evaluation of the dependability of the world model created as well as the sensory processing (i.e., how sure can one be that the world model and the system perceptions are a true reflection of reality). What is 'safe' is required to be defined in the context of system goals. The 'safe' actions might not always be 'task effective' and vice versa. Hence, for a 'safe' and 'effective' decision making, a mechanism to balance these goals has to be integrated within the decision-making algorithm.
4. How are risk measures integrated with decision making? How can other values such as task achievement be considered by VJ and how does BG employ these values to generate actions?
5. How do the architectural layers influence safety management? Essentially, this means operational risks can be considered on various time and spatial scales. The advantage is that the architecture layers act as a form of 'safety firewall' as decisions and actions are passed down the architecture, where each level has a role to play in eliminating hazards to which the robot is exposed. The safety firewall works as an arbiter and/or barrier depending on the circumstances and the context. This is a very important issue and is integral to the RCS as well as to the philosophy of considering safety throughout planning and execution.

The different hierarchical levels of RCS are assigned different subtasks. As one goes down the hierarchy, the focus becomes more concentrated on execution of safe decisions, whereas the upper levels are more concerned about actually making these safety decisions. The upper levels are also responsible for long-term planning for task completion. It can be further said that every subsystem at every layer, along with the main task for which it is designed, also plays its role to keep the system safe. In this case, the underlying objective is to ensure that the system remains safe from hazards during travel, for example, the higher levels are more active in planning the safe path and so on, but the lower levels are more reactive and respond in a systematic way to ensure safety. It is assumed that, for example, at the lowest level, there are still electromechanical interlocks, such as limit switches, bumpers and so on, present to ensure that the system will ultimately stop rather than harm itself or others. The act of stopping could create an unsafe condition but if there is no other alternative (such as moving to a safer position because of the appearance of a sudden and unexpected obstacle inside the robot's safety perimeter which might trigger the bumper switches), the robot should decide to stop operation. The electromechanical interlocks work as a final barrier at the lowest level and prevent a system from harming its surrounding and if possible itself. Of course, the user does not want the robot to stop as this clearly prevents task achievement, so in more concrete terms all the higher levels should work together so as to ensure that the lowest level does not need to intervene.

Hence, the whole of the RCS-RMA can be said to be a safety management system. Each module (some more than others) of the RCS caters for safety, and safety is not considered as one of the subtasks of the system, which is allocated to one of the modules of the hierarchical architecture. This is in exact opposition to the concept of 'safety manager' suggested by Seward et al. [44–46] as discussed earlier in this section. For a case of mobile robot, such an RCS 'safety hierarchy' is feasible. One more thing to be considered at this stage is that some applications may require the robot to enter an hazardous environment, such as a nuclear decommissioning site. The possibility of the lowest level intervening is reduced if each and every level considers safety. By distributing safety there is no burden on any particular level and as it is distributed at all levels, it becomes more manageable.

19.4.3   USE OF PROBABILITIES FOR REDUCTION OF UNCERTAINTY

In uncertain and unstructured environments, uncertainty quantification is the major issue. Probabilities provide a way to quantify and measure uncertainty. The first and foremost reason for the use of probabilities in this research is that various variables of the system can be modelled and quantified using probabilities. After modelling the uncertainty problem using the probabilistic reasoning technique, it becomes easier to deal and handle them. The various uncertainties are converted from an measurable and non-structural form to a more recognisable and definitive structure. This quantification provides a method to define, measure and respond to the uncertainties.

The human mind works/thinks/comes to a decision/selects actions on the basis of probabilities for each of the choices available. The use of probabilities for dealing with uncertainty also agrees with this analogy of mapping human thinking to machines. The human brain considers and weighs each given action (option) and chooses the one that is the best in the given circumstances and related to the focus of the current thinking process. Human reasoning is based on probability evaluation, albeit a fuzzy version. Pearl [47] calls probabilistic reasoning a faithful guardian of common sense.

Utility theory [48] argues that humans make decisions in order to maximise benefits (pleasure) and minimise costs (pain). As stated by the eighteenth-century utilitarian philosopher Jeremy Bentham, the purpose of any activity should be to 'maximise the greatest happiness of the greatest number' [49]. Inherent within this decision process is an, often subconscious, consideration of the probability of reaching the desired outcome. Also, different humans can reach different conclusions about the most appropriate action to choose depending upon their current 'objectives' or 'beliefs' or personality. Thus, a cautious and safety-conscious individual would make different decisions to one who is bold and task focussed.

The POMDP approach to decision making utilises probabilities to deal with uncertainties [50]. The probabilities are calculated by the Bayesian network technique. The Bayesian network calculates the probability distribution for a state variable while taking into consideration the partial observations obtained from sensors, the conditional probabilities between the state variables and the information learnt from prior observations. Actions that maximise benefits as perceived by the 'objectives' of the system are assigned highest rewards. The integration of POMDP within RCS enhances its scope for dealing with decision making in the presence of uncertainty. Furthermore, the fact that this POMDP decision making can be incorporated into several hierarchical levels means that the system provides 'defence in depth' against accidents.

The second advantage of using probabilities is that they can be modelled robustly and precisely using mathematical models [51,52]. A probabilistic approach, such as the Markov decision process model, has been used for localization of mobile robots and the mapping of the environment by Thrun et al. [53,54] and Olson [55]. Simmons and Koenig [56] have proposed the use of POMDP for probabilistic navigation. Probabilistic techniques to deal with uncertainty are also implemented in other fields such as telemedicine [57,58] and so on.

Alternative approaches, such as Kalman filters and Bayesian networks, are used for localization by Roumeliotis [59,60] among other authors. A Bayesian model has also been used to model and keep track of dynamic environmental characteristics [61]. Kalman filtering is also used in fault detection and diagnosis by various authors [62]. Nehmzow make use of neural networks on the other hand for learning in mobile robots [63]. Zurada et al. [64] used a combination of neural networks and fuzzy logic to deal with system safety. Although these various control strategies such as fuzzy logic, Kalman filtering and neural networks are available, the probabilistic model has been selected for this development.

The underlying working principle on which humans, natural intelligent systems, animals and others base their decisions, make choices and decide what to choose from a vast array of options is probability. The reasons for using POMDP are discussed in the next section.

19.4.4 WHY POMDP?

Uncertainties are introduced primarily because the state (of system and of environment) is not fully observable due to various reasons mentioned in Section 19.3.1. As the system operates in unstructured domain which is partially observable, a method which can specifically deal with this problem should be employed. The POMDP model provides an effective way of dealing and handling the unknowns and uncertainties using probabilities.

In a POMDP, a given system problem introduced due to partial observability is modelled using probabilities. The model uses the information gained from these partial observations obtained from sensors. Sensory observations consisting of uncertain and incomplete information about the surroundings and the robot itself form the most important part of POMDP model. The uncertainties are modelled using a probabilistic reasoning technique and then the POMDP model develops the reward/value function using these probabilities and the information gained in the past (learning). This value/reward function is responsible for decision making and choosing the action which would generate safe behaviours.

The POMDP deals with the various probabilities present and then assigns values to each of them and makes the system decide the action that provides the highest reward. The POMDP approach was employed for probabilistic navigation [56], for decision-making and the action-selection uncertainties [34]. The various reasons for choosing POMDP are listed below:

- Hazardous environments are by nature partially observable. This introduces the uncertainties. The POMDP approach to decision making utilises probabilities to deal with uncertainties [50]. Uncertainty can be quantified using probabilities. The POMDP approach can effectively handle such uncertainties primarily because the uncertainties can be quantified by using probabilistic reasoning. Using POMDP, uncertainties can be modelled mathematically and robustly.
- POMDP has already been successfully applied on various robotic mechanisms, such as navigation, localisation [53,54,65,66], safety against toppling [34] and so on. Simmons et al. also developed a POMDP-based robot architecture called 'Xavier' [67]. POMDP has also been modified and employed for a conversational robot [68]. In this research, their use is being extended to ensure safety in navigation and ensure obstacle avoidance to minimise the likelihood of collision accidents.
- Retrospectively, as demonstrated earlier, the POMDP model maps very well onto RCS-RMA framework.

19.5 CASE STUDY: MOBILE ROBOT APPLICATION

This and the next few sections discuss the implementation of the RCS methodology for a case study of a mobile robotic application [25]. The RCS methodology supports the development of a robotic system which can continue to operate even in the event of partial or complete failure of some of its components or subsystems. It supports incremental implementation and allows the extensibility of the system. A typical RCS design starts with the study of the tasks to be performed by the system and proceeds by defining a controller structure composed of hierarchically organised RCS nodes, which are alternatively called as *modules* or *control modules*. A control module is a generic processing structure consisting of an RCS node that could implement control algorithm. The inter-level (between different layers) operation is normally asynchronous but the intra-level (within a layer) operations are synchronous. The inter-level communication is achieved by a special protocol

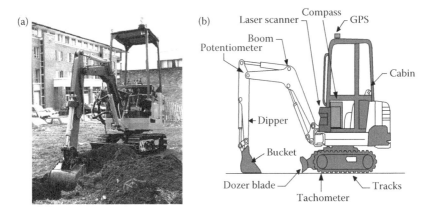

FIGURE 19.4   (a) LUCIE. (b) Main elements of LUCIE.

termed as NML (neutral messaging language) [25] used for RCS development. There are three basic steps in a typical RCS design [25]:

1. Decomposition of a given application into tasks and subtasks
2. Defining the controller architecture
3. Deciding the control algorithms and data flows for each control

The Lancaster University (LU) Engineering Department is active in the development of autonomous and safe mobile robots for two decades (late 1980s to date). As a part of this research, Lancaster University Computerised Intelligent Excavator (LUCIE) [69,70] was developed. For the purpose of demonstration and implementation of the concept development, a case study of a mobile robot whose design is based on LUCIE is considered. The application under consideration here involves safe navigation of this mobile robot excavator while avoiding the collision with surrounding obstacles in an unstructured environment. From an analytical perspective, system hardware aspects are important for understanding the design of LUCIE. Figure 19.4 illustrates the autonomous excavator LUCIE (a) and its principal hardware components (b). The excavator is equipped with a sensory suite (shown in Figure 19.4b), including

1. Potentiometers on arms for arm position feedback
2. A two-axis tilt sensor for monitoring excavator tilt
3. Two laser range scanners (one each on the front and the rear) with 180° field-of-view
4. Differential global positioning system (DGPS) device for location and navigation
5. Two tachometers (one per track) for speed measurements
6. An electronic compass for orientation purposes

19.5.1 TASK DECOMPOSITION

Figure 19.5 shows the task decomposition for a mobile robot excavator. The task is to travel from the start position on the worksite to the goal position and to start digging. This task is decomposed into smaller tasks as shown in Figure 19.5.

Figure 19.6 shows the controller hierarchy for the excavator application. The area of interest here is the development of the mobile robot navigation, with associated path planning and motion control [71,72], obstacle detection and related safety issues. Other aspects such as digging, pick-and-place and so on are not considered at this point. Hence, only the part indicated inside the dotted square in

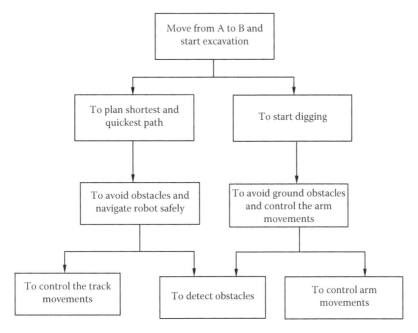

FIGURE 19.5 Task decomposition analysis.

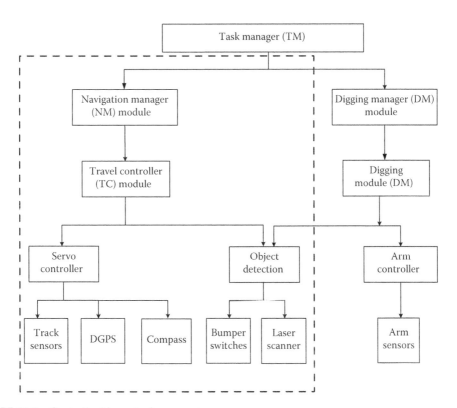

FIGURE 19.6 Controller hierarchy for excavator.

the controller hierarchy of Figure 19.6 is considered and developed in this research. This controller employing different control methods is described in greater detail in the next section.

19.5.2 Implementation of Control Algorithms

The reasons for the use of different control methods within a hierarchical architecture such as RCS-RMA are explained in Section 19.2. Although RCS is a generalised architecture and provides the framework, the RCS designer is required to define and write the control algorithm(s) for the different levels of the RCS hierarchy. Furthermore, the control method applied at each control level must suit the representation and management of uncertainty, given the subtask objective, resolution and scope of that level. This section discusses the control and decision-making strategies which are used for integration within the four-layer mobile robot RCS hierarchy (as shown in Figure 19.6).

The abbreviations NM, TC, SC and OD are used alternatively throughout this chapter for navigation manager module, travel controller module, servo controller module and object detection module, respectively. Also, robot, vehicle and excavator all refer to the mobile robot. The following control and decision-making strategies are proposed for integration within RCS framework as shown in Figure 19.7.

19.5.3 Four-Layer Mobile Robot Application

The control strategies implemented at each level are explained in detail in this section. The suffix used after functional elements (BG, VJ, WM and SP) refers to the name of the module these functional elements reside in.

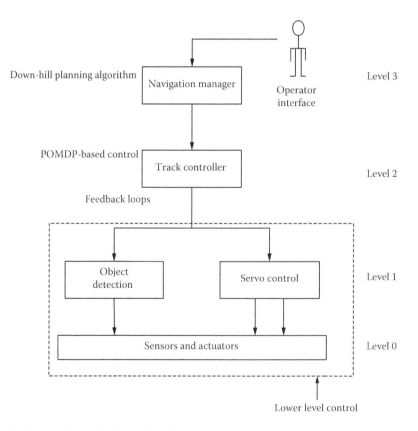

FIGURE 19.7 Control hierarchy of mobile robot architecture.

19.5.3.1   Level 3: Navigation Manager Module

The task of this level is to plan the path from start position to goal position. Hence, planning algorithms such as Refs. [71–73] are appropriate at this level. This implementation uses a downhill planning algorithm-based method [73] for path planning. This level is responsible for 'strategic planning'. It is the most deliberative in nature and the least reactive of all the layers. It consists of a grid-based dynamic path-planning algorithm. This level ensures that the task gets completed in the most cost-effective and time-effective manner. The algorithm plans the path which is optimally safe. It avoids the obstacles but does not take into account the local and possibly changing terrain conditions while planning.

Initially, the NM module generates a map of an environment. The world and obstacles are represented in a two-dimensional occupancy grid of the size corresponding to the area of the worksite. Each cell in the grid corresponds to a small section of the real world. The squares on the grid corresponding to the obstacles in the world are marked 'FULL', whereas the open areas are marked 'EMPTY'. As the obstacles are unknown to the robot at the start of the operation, the whole area is marked as 'EMPTY'. If the robot discovers an obstacle, the TC module computes the size and position of the obstacle and sends this data to NM. NM then updates the corresponding area on the occupancy grid and marks it as 'FULL'. Changes in the occupancy grid require a re-planning step since a new obstacle might obstruct the planned route or the newly discovered open space might offer a short cut.

The planned path is in the form of a cost grid. Each cell in the grid is an estimate of the shortest travel distance from that point to the goal. The cost at the goal cell is 0.0 and the cost at other cells increases the further they are from the goal. Cost grid is computed mathematically within the algorithm. To get a better idea of the cost grid, one can view a three-dimensional surface generated by plotting the cost at each point as a height. The high spots on the 3D plot are obstacles. The goal is at the low point. It can be visualised that following the plan is just like rolling a ball down hill, hence down-hill planning. The path is given by drawing a line through the minimum cost cells starting from the initial position in the direction of the goal position. Thus, path planning algorithm in NM follows the following steps:

1. Initialise the occupancy grid to 'EMPTY'
2. Plan
3. Check sensory information giving robot position, goal and nearby obstacles
4. Update the occupancy grid for corresponding areas to 'FULL'
5. Recalculate the plan
6. Check plan and initiate movement along the shortest path
7. Go to step 3

After describing the algorithm, it is important to discuss how the functional elements participate in the running of this algorithm. The SP_{NM} (SP in NM module—refer to Figures 19.3 and 19.7) receives information such as the total number and type of detected obstacles and current robot position from level 2 (travel controller module). World model (WM_{NM}) consists of a dynamic map of whole area (grid size = 100) and way-point data from the start to the target position. The WM_{NM} is world-centric in nature. The NM module has the broadest (wider scope, less details) world model with the longest time horizon but is general in scope with relatively low resolution. It lacks the details. The WM_{NM} is updated at a relatively slow rate (in minutes); hence, it must be made certain that the WM_{NM} remains relatively valid between updates from the point of view of the NM operational objective.

The WM_{NM} contains the information of the planned path (way-points) of the robot. It can be imagined that the robot is stationary with a constantly moving world around it. BG_{NM} sends the command PLAN_GLOBAL_PATH consisting of the next way-point to TC which then sends the command to set the reference to SC and OD. The commands for NM module are shown in Table 19.1.

TABLE 19.1

Commands for NM

Command	Description
INIT	Performed when the human operator switches from manual to automatic operation. Initialises the system: Actuates the excavator. Initialises any needed variables and sends INIT to the travel controller (TC).
HALT	Performed when the excavator reaches the goal point or the human operator switches from automatic to manual operation. Halts operations: saves any needed data and sends HALT to the TC.
PLAN_GLOBAL_PATH	Normal operation: Runs as long as the automatic controller is on, that is, when the user tells the onboard computer to control. Plans the path from current position to goal position considering obstacle information received from level below.

19.5.3.2   Level 2: Travel Controller Module

This middle level deals with the safety decision making in the presence of uncertainty. In situations where safety is an issue of concern and where potentially unsafe operation arises from perceptual and cognitive uncertainties [34], it is particularly important to implement an algorithm that can deal with the uncertainties. POMDP provides an effective way of dealing with and handling the unknowns and the uncertainties using probabilities [74]. At an intermediate level of control, it is argued that POMDP offers a suitable structure for developing the desired control strategies allowing the mapping of the functional RCS elements, viz., SP, WM, VJ and BG, onto specific components of the POMDP itself [32,75]. This mapping provides the means of ensuring the appropriate interaction between these elements of the control level, providing a coherent controller behaviour and decision-making process which succeeds in managing the uncertainty present at such a level of control.

TC is responsible for confirming that the obstacle data received from the lower level (OD) is reliable and then sending these data to NM. It is also responsible for initialising the lower level modules, commanding them to perform their tasks and demanding the obstacle and robot status data from them. This level performs decision making to ensure the safety of operation and precise movement of the robot. This level receives commands from the level above (NM module). It can be said that this level makes the decisions and plans 'tactically' within the limits of 'strategic planning' made by level 3 (NM module). It is less reactive than level 1 and level 0, but more reactive than level 3. It is also less deliberative in nature than level 3, but more deliberative than the levels below. In short, this module is responsible for safety decision making and safely performing the manoeuvres that the NM module instructs.

The SP_{TC} receives the processed sensory input about the robot state and obstacles from lower level (level 1) modules. SC sends the robot data (e.g., orientation, position and speed) and OD sends the laser scanner (object data) in each cycle to TC. The WM_{TC} is less world-centric in nature and more ego-centric. It has detailed information regarding its surroundings, but it lacks the information outside its sensor range. The information is centred on the robot and it has a time horizon of seconds. The WM_{TC} consists of the next way-point position data sent previously by NM module. BG_{TC} commands the system to perform the action.

The main reason for assigning the task of decision making to TC is that it explicitly deals with uncertainty. TC has the capacity to override the commands sent by NM. Although NM computes the path, TC is responsible for deciding whether re-planning of the path is required or not. NM on its own is not capable to decide whether an unexpected and sudden obstacle has arrived and deal with it. That responsibility lies with TC. If it receives the sensory data from OD and determines that an obstacle(s) is in the vicinity of the vehicle, it decides what action to take. Although it does not actually decide the next way-point, it asks the NM to recalculate the path. Subsequently, the

TABLE 19.2

Commands for TC

Command	Description
INIT	Initialises any needed variables and sends INIT to the SC and OD.
HALT	Saves any needed data and sends HALT to the SC and OD.
PROCEED	Normal operation of the excavator. This action decision is taken until the OD encounters obstacle. This command is sent from TC to SC and OD.
CHANGE_DIR	Change the direction of motion. This action decision is taken if the NM sends TC new way-points based on obstacle data earlier sent by TC. This command is sent from TC to SC.
WAIT	Wait till the NM calculates and sends the new way-point. This happens in case of obstacle detection. This command is sent from TC to SC.
SPEED_UP/ DOWN	Increase or decrease the speed of the robot. This command is sent from TC to SC depending on the obstacle position and orientation w.r.t. robot.

command to set the reference for position, orientation and so on, against which the actual sensory data is compared from within the algorithm, is sent to the SC and OD modules. The commands for the travel controller module are shown in Table 19.2.

19.5.3.3 Level 1

As shown in Figure 19.7, level 1 consists of two modules, viz., OD and SC. Both these modules are explained in this section. It is important to note that this level does not take part in high-level decision making or planning. The control method used at this level is feedback control. Both these modules at level 1 help keep the system safe by reducing the uncertainty about (1) the current state of the robot, for example, position, orientation, speed and tilt angles and (2) the current state of the world, for example, obstacle presence, obstacle speed and type.

19.5.3.3.1 Level 1: Object Detection Module

This module deals with obstacle detection. The WM_{OD} contains mainly expected obstacle data. OD is designed to act reactively when an obstacle comes within the safe operating perimeter. This module is also capable of halting the system if an unexpected obstacle suddenly comes into close proximity (<5 m) of the robot. The time horizon is in milliseconds. Simple closed feedback loops are considered to be the most appropriate at the lowest level consisting of sensors, actuators and their servo-feedback loops, providing a minimum capacity for handling uncertainty.

SP_{OD} receives the observations from sensors, viz., laser scanner and bumper switches. It processes these data and decides the position and nature of various obstacles within the scanner range. The WM_{OD} consists of the data about robot position, goal position and obstacle data. BG_{OD} receives command from TC to start the OD (object detection) process, sends the status of obstacle detection to TC (travel controller) and actuates sensors. Commands for the OD are shown in Table 19.3. The process of object detection and collision avoidance is discussed in Section 19.6.

TABLE 19.3

Commands for OD

Command	Description
INIT	Initialises any required variables, turns on any needed devices.
HALT	Saves any needed data, notifies the TC and turns off any needed devices.
SET_REF	Transmits the data to TC. Usually data contain information about robot and its surroundings.

TABLE 19.4

Commands for SC

Command	Description
INIT	Initialises any needed variables, turns on any needed devices.
HALT	Saves any needed data, notifies the TC and turns off any needed devices.
SET_REF	Transmits the data to TC. Usually data contain information about robot and its surroundings.

19.5.3.3.2   Level 1: Servo Controller Module

This module ensures accurate movement of the robot towards the target position. It ensures smooth operation of the tracks. It is also capable of halting the system if required. WM_{SC} contains robot data, such as position, speed and orientation. It receives the robot data from sensors and sends the processed sensory data to TC. It receives the command to set references (about speed, position and orientation of the vehicle) from TC. It also receives the data about the next action to choose from TC. Based on this action, it calculates the valve control signals for the left and right valves of the excavator. After calculating the valve signals, it then sends them to the vehicle. Commands for the SC module are shown in Table 19.4.

19.5.3.4   Level 0: Sensors and Actuators

Different types of intelligent sensors are used at the lowest level of the controller hierarchy, along with actuators (proportional control valves). Both OD and SC modules at Level 1 interact with these sensors and actuators. The sensors that the excavator is equipped with are capable of determining the orientation of the vehicle, existence of objects in the surroundings and determining whether a detected object is likely to collide. The sensors participate to ensure safety by forming the final barrier that can stop the system if required. In an ideal scenario, the system would not be called upon to stop. The higher levels should operate in such a way to ensure that the lowest levels do not need to act reactively.

Having described the RCS hierarchy and having identified the algorithms suitable for each level of this hierarchy, the next step is the development of these algorithms. To define more fully how the modules interact with the system, the operator and with each other, it is necessary to define the communications that can occur (e.g., commands sent to and status received from, lower level nodes).

- INIT and HALT are sent from higher level nodes to lower level nodes. INIT initialises while HALT stops the operation.
- Command PLAN_GLOBAL_PATH corresponds to a request 'Manoeuvre the robot excavator to the required position'. This command is sent by the supervisor or human operator and accepted by NM. This command should carry information about the numeric value of the position (in terms of the coordinates or parameters of GPS) to which the excavator is expected to reach (i.e., the 'GOAL POSITION'). This command initialises the NM; the NM then sends the 'INIT' command to TC.
- When TC receives 'INIT' it sends the 'INIT' to OD and SC. Once all the modules are initialised, the NM receives request from TC for a new command.
- The NM must then send the command PLAN_PATH to TC to search for the obstacle(s) in the robot vicinity and make the next action decision.
- TC then sends the command/signal 'SET_REF' (set reference) that prompts the lower level module OD to start the object detection process. This command is sent to find the obstacle status in the surrounding of the robot.

For the purpose of this research, an 'object' is an entity present in the robot surroundings. An 'obstacle' is a subset of an object. When an object is such as there is possibility of collision with the robot, it is considered an 'obstacle'. The OD module is responsible for detecting objects in the vicinity of the robot. TC is responsible for deciding whether a detected object might collide with the robot considering the position and path of the robot. If TC perceives a possibility of collision, it sends the information about obstacles to NM. At the same time, TC also requests the robot data from SC.

- Once NM receives the obstacle data, it starts re-planning the path from the current robot position to the goal position.
- Meanwhile, TC sends the command WAIT (corresponding to 'wait till the path is recalculated by NM') to SC. TC then generates the appropriate control valve signals to 'CRUISE' with the same speed and orientation and send to SC. Once TC receives the next position (way-point) from NM (based on re-planned path), it computes the desirable speed and orientation and sends the corresponding control signals to SC.
- If TC does not perceive a possibility of collision with an obstacle, it sends the PROCEED command to SC which corresponds to 'no obstacle' and hence the robot can proceed. TC then asks for 'next way-point' to NM, computes control signals corresponding to the desirable speed and orientation and sends these data to SC.
- When robot position = goal position, the NM sends HALT to TC which turns HALT to OD and SC. In the case of unforeseen and unavoidable possibility of collision, the bumper switches send the corresponding signal to OD which immediately stops the system operation.

Figure 19.8 shows the RCS hierarchy with the function of each functional element within every module discussed so far.

19.5.4 SAFETY HIERARCHY IN RCS

Each layer of RCS caters for safety, and safety is not just the responsibility of one element/component of the system but of all the layers. In this respect, it is possible to represent safety as a parameter to be mapped at various levels of the RCS architecture for the excavator as shown below. Figure 19.9 shows how safety management is spread over the whole architecture and how each of the modules play an important role in keeping the system safe. Hence, the whole of the RCS-RMA can be said to be a safety management system.

19.6 OBJECT DETECTION AND COLLISION AVOIDANCE ALGORITHM

An algorithm for object recognition and obstacle avoidance was developed which makes use of the limited range of sensors available on the robot. The existing algorithms [76–81] for this purpose were not suitable as the sensory suite present on the robot was not sufficient and developed enough to provide sensory input to these algorithms. Another reason for this independent and novel development was that the dynamic map (consisting of obstacle positions) drawn by the system from available sensory data has to be updated in every time step so as to make safe decisions in order to avoid collisions. When the OD module is initiated, it detects the objects from laser scanner data, determines the obstacle position with respect to the robot and sends the object position data to the TC (in that sequence). This section explains the object detection and collision avoidance algorithm in detail.

19.6.1 OBJECT DETECTION

When the controller is initiated, all the modules of the hierarchy are initiated. After initiation, level 1 modules (OD and SC) start receiving sensory data from the simulated robot. The laser scanner data are specifically important for obstacle detection and collision avoidance purposes. Laser scanner data are in the form of the distance from which the laser beam is reflected. If there is no obstruction, the beam distance is 15 m. If the beam is reflected from an object, the distance is less

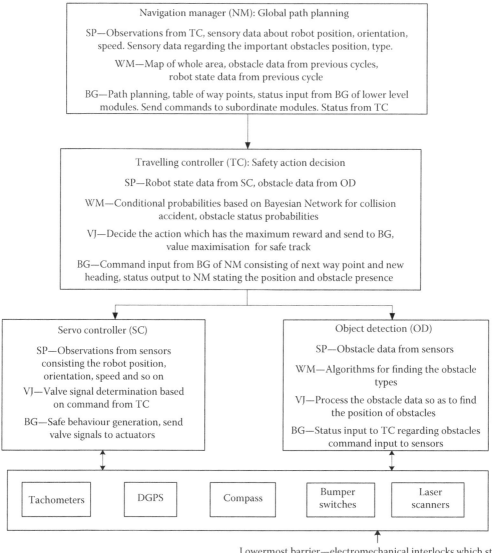

FIGURE 19.8 RCS-RMA for mobile robotic systems.

than 15 m. There is one beam for every two degrees. The laser scanner total range is 180°. Hence, there are a total of 90 beams per laser scanner (front and rear). If the scanner distance for two consecutive beams is less than 15 m, the OD increments the number of detected objects by one. After this, the start angle is calculated. The start angle is equal to the angle of the laser scanner corresponding to the first <15 m reading. It is denoted by 's'. The beam angle is then incremented till the next reading when scanner distance = 15 m. For the last reading < 15 m for the object under consideration, the end angle is calculated. End angle is denoted by 'e'. Next, the point on obstacle corresponding to the minimum scanner distance is determined. This is the 'tip' of the part of the object detected by the scanner. The beam angle corresponding to this point is termed as middle angle and is denoted by 'm'. The scanner distance corresponding to the start angle is D_curr_s, corresponding to the middle angle is D_curr_m and corresponding to the end angle is D_curr_e. This can be seen in Figure 19.10.

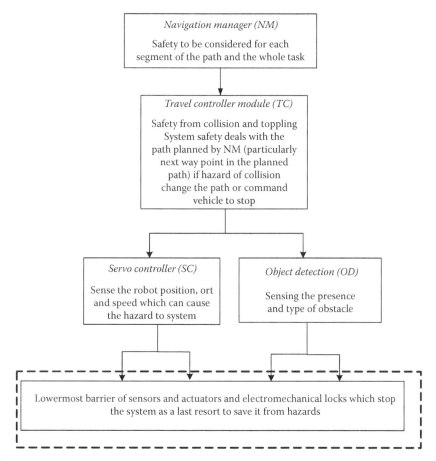

FIGURE 19.9 Safety hierarchy.

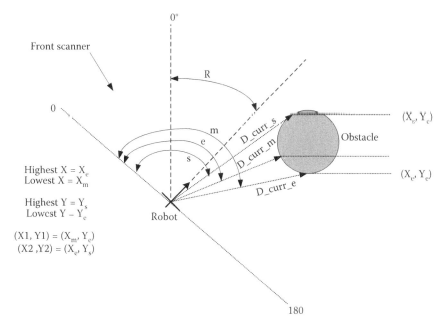

FIGURE 19.10 Object detection.

19.6.2 Computing Obstacle Position Coordinates

The scanner angles in 0–180° range when received by the OD are converted to the map quadrant system. All the angles, that is, start angle 's', middle angle 'm' and end angle 'e' are first converted to the map quadrant system. These converted angles are denoted by s_in_map, e_in_map and m_in_map. Based on the s, e, m, D_curr_s, D_curr_m and D_curr_e, the coordinates of the start position, end position and middle position of obstacle width are determined by using trigonometric functions. Figure 19.11 shows the case when the orientation is in the first quadrant. In Figure 19.11, only the start position coordinates are shown. The end position and middle position coordinates are calculated in exactly the similar fashion, only by replacing the start angle 's_in_map' by 'e_in_map' and 'm_in_map', respectively.

Depending on the orientation of the robot, the cosine and sine terms can be either negative or positive which accounts for either adding or subtracting W and H from the robot position coordinates X and Y, respectively. Once all three position coordinates corresponding to start angle, end angle and middle angle are determined, the lowest (X1, Y1) point and highest (X2, Y2) point are decided. This is demonstrated in Figure 19.10. The data related to the object: X1, Y1 and X2, Y2 are sent to travel controller via the status channel.

19.6.3 Dynamic Map Updating

The dynamic map updating procedure is shown in Figure 19.12. The TC module is responsible for deciding whether the object detected by the OD is an obstacle or not. If the TC determines a certain object to be an obstacle, it sends the position data for that obstacle to NM. In each cycle, if the TC detects an obstacle, the circle with a radius of 15 m and robot at its centre is cleared on the map. If no obstacle is detected, the map is not updated. Once this area is cleared, the NM marks the rectangle with width (X2–X1) and height (Y2–Y1) as an obstacle (area with BIG Cost) on the map.

It is important to note that the whole map is not updated in every cycle even if obstacles are detected. Only the circle with a radius of 15 m and robot at its centre is cleared. Next, the path from current robot position to the goal is recalculated. When the cell has big cost, the robot steers away from it to the adjacent cell with lowest cost. This facilitates obstacle avoidance while planning the path effectively. This procedure is repeated in every cycle. The flowchart shown in Figure 19.13 explains this in detail.

This concludes the overall development of four-level RCS hierarchy for a mobile robot application. Figure 19.14 shows the flowchart for the complete RCS process and how the data flow between levels of hierarchy. Having discussed the architectural design of the controller for the given application, the next step is to implement this controller. The RCS software library has been developed for this purpose [25].

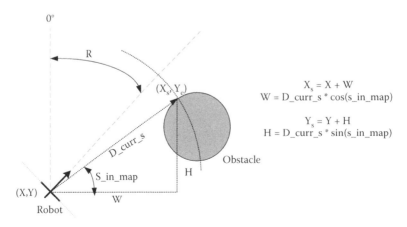

FIGURE 19.11 Obstacle in the first quadrant w.r.t. robot.

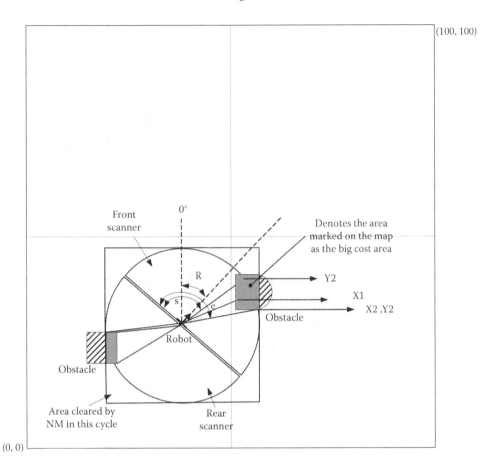

FIGURE 19.12   Map updating.

The next section establishes the relationship between RCS and POMDP and to show that the functional elements within RCS can be used to represent the computational elements of the POMDP model. This section is particularly focussed on the development of the POMDP process within the RCS node. Fundamental POMDP model aspects are given, outlining the basis for the applied computational model.

POMDP is used for safety decision making at the intermediate (travel controller module) level of the four-level mobile robot RCS hierarchy. As discussed in the previous sections, POMDP provides an effective way of dealing with uncertainty through probabilities. The integration of POMDP within RCS enhances its scope for dealing with decision making in uncertain domains. As shown in the safety hierarchy of Figure 19.9, all levels of RCS participate and contribute to the safe operation of the system.

19.7   DEVELOPMENT OF POMDP PROCESS IN RCS NODE

19.7.1   POMDP MODEL

This section describes the POMDP model. It first provides an insight into Markov decision processes (MDP), then considers decision making and action policy selection and finally considers the presence of uncertainty and partial state observability and how the MDP is modified to accommodate this partial observability.

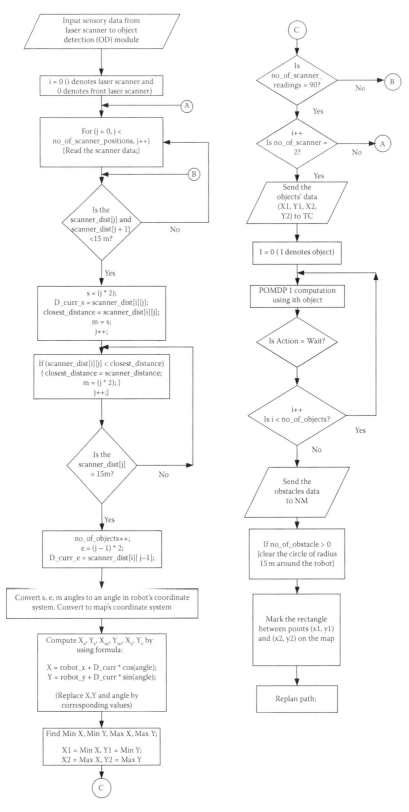

FIGURE 19.13 Flowchart for OD and collision avoidance algorithm.

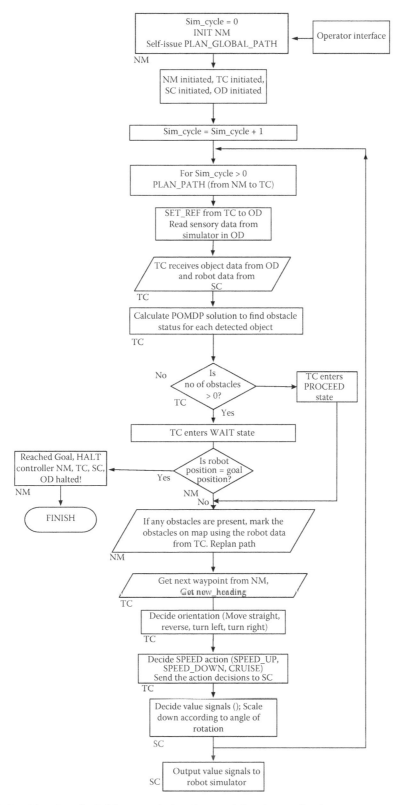

FIGURE 19.14 Flowchart for RCS process in four-layer mobile robot application.

Mathematically, the Markov process can be defined as a tuple $<S, T>$, where S is a finite set of states and T is the transition function. This transition function defines the transition of the system from one state to another. The transition conditional probability depends on the current system state and the action decision that is taken in that specific state.

From decision theory [48], decisions need to be based on the achievement of some objective, which in mathematical terms can be represented by the *reward/value* (i.e., the degree to which an objective is achieved). This *value* can be associated with a specific state or action. Defining MDP requires the identification of a set of possible actions (these actions will then influence transition probability).

This *value* is generally derived through some utility function, outlining the decision maker's preferences. The system dynamically adjusts the values of this utility function as a part of the learning process. In the case of an MDP, the *value* of an action or state can be represented by a set of rewards that the system would achieve by taking a specific action in a specific state. The rewards in this case can be defined in various ways. For example, a reward may be allocated to the system state or alternatively to the transition from one state to another state when a specific action is taken. Given the rewards that a system can achieve in each time step, the objective of the system is to select action decision, which maximises the total expected reward or utility value to be obtained over a specific number of time steps.

The MDP can be represented [74] in mathematical terms as a tuple $<S, A_d, T, R>$, where

- S is a finite set of states of the world
- A_d is a finite set of actions which can be taken
- $T:S \times A_d \rightarrow \Pi(S)$ is the state-transition function (TF), where for each state s_i and action a, a probability distribution over the world states is given, where $T(s_i, a, s_j)$ represents the probability of ending in state s_j. $T(S_i, a, s_j) = \Pr(S_{t+1} = s_j) \mid S_t = s_i, A_d = a)$
- $R:S \times A_d \rightarrow R$ is the reward function (RF), giving the expected immediate reward gained by the system for taking each action a in each state s, $R(s,a)$

In this model, the next state (s_j) and the expected reward depend only on the previous state and the action taken; even if we were to condition on additional previous states, the transition probabilities and the expected rewards would remain the same. This is known as the Markov property—the state and reward at time $t + 1$ is dependent only on the state at time t and the action at time t.

The above MDP model assumes that the system state can be fully observed, that is, the system knows exactly in which state it is. This assumption is unrealistic, especially in unstructured and uncertain domains. It will also be unable to recognise the state in which it will end up in each time step following the actions that will be taken. The uncertainty thus lies in the knowledge of the action outcome. In order to behave truly effectively in a partially observable world, it is necessary to use the memory of previous actions and observations to aid in the disambiguation of the states of the world. The POMDP framework provides a systematic method for this. One of the problems that has been identified with regard to safety management is the system's inability to precisely perceive the current operational conditions, be it the system's state of integrity or the environmental conditions, owing to limits on sensory and perceptual resources. Consequently, the system will only have partial knowledge of its current operational conditions.

A POMDP is an MDP in which the agent is unable to observe the current state. Instead, it makes an observation based on the action and resulting state. The agent's goal remains to maximise expected discounted future reward. POMDP describes a set of observations which represent the information that the system can receive. This information is related in a probabilistic sense to the system states and thus, the system can define the likelihood of being in a specific state, given the observations it receives. A mathematical representation of a POMDP as defined in Ref. [74] is based on a tuple $\langle S, A_d, T, R, \Omega, O \rangle$ where

- S, A_d, T and R describe an MDP
- Ω is a finite set of observations that the system can make of its world

- $O{:}S \times A_d \to \Pi(\Omega)$ is an observation function (OF), giving, for each action and resulting state, a probability distribution over possible observations. Here $O(s_j, a, o)$ signifies the probability of making an observation o at time $t + 1$, given that the system took action a at time t and landed in state s_j at time $t + 1$:

$$O(s_j, a, o) = \Pr(\Omega_{t+1} = o \mid A_{d,t} = a, S_{t+1} = s_j)$$

The above model represents the system state identification uncertainty problem realistically, since in reality, the system would be able to receive specific sensory observations and would have to come to conclusions on the likelihood of its operational state, given such observations. Observations thus form the most important part of this model. Having defined such a representation, it is possible to determine an action-decision process for the system that would give the highest expected reward. Unfortunately, this action-decision problem becomes much harder, primarily due to the lack of certainty of being in a specific system state [74]. Determining the effect of observational uncertainty on the action-selection policy thus becomes necessary.

19.7.2 THE DEVELOPMENT OF POMDP PROCESS WITHIN RCS COMPUTATIONAL NODE

The approximate mapping of various RCS functional elements with POMDP components is shown in Figure 19.15. In order to exhibit a relationship between RCS and POMDP, each of the functional elements of the RCS is explained in detail with respect to POMDP.

As mentioned in the RCS reference architecture, the functional elements, BG, WM, VJ and SP, are organised into an operational unit called the 'RCS node'. Each level has one or more nodes and each module has one node [29]. Although for this research, POMDP is employed within the TC module of the mobile robot RCS hierarchy, in the following section, a generalised POMDP process model is developed in the RCS node. Since an RCS node is the basic organisational building block of RCS-RMA, it is only appropriate to develop a POMDP process within a node, which can then be extended to various nodes within an RCS application. This development involves replacing the RCS functional elements and the links within them by appropriate POMDP components. This is shown

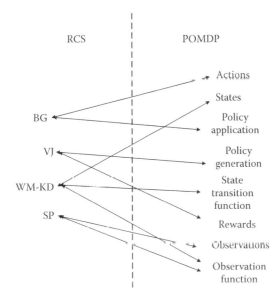

FIGURE 19.15 RCS and POMDP mapping.

POMDP process as expressed in RCS node

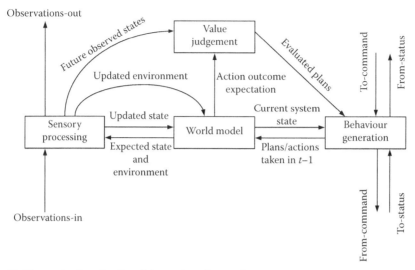

SP: Observation function: Set of observations Ω, state S_{t+1}

WM: Transition function: Current state S_t, set of actions available A, expected action outcome, pruning

VJ: Reward function: Pruning, evaluated plans

BG: Action selection: Policy Π

FIGURE 19.16 POMDP modification in an RCS computational node.

in Figures 19.3 and 19.16. The functional elements and flows in an original RCS node shown in Figure 19.3 are replaced by the corresponding POMDP components in the POMDP-modified node in Figure 19.16.

19.7.3 VALUE JUDGEMENT IN A POMDP-MODIFIED RCS NODE

The reward/value function in POMDP is the counterpart of VJ in RCS. Assessing and knowing the expected cost of a decision, the expected penalties for risky decision making or the expected rewards for safe decision making is vital for making good behavioural choices. The 'good' behavioural choice for the system under consideration is based on the ultimate goal of the system. Every system is designed to complete the application, but the design considerations depend significantly on other objectives such as safety, economy and so on. The most appropriate choice of action selection fulfils all these parameters (safety, speed of completion, economy and above all task completion) in combination.

VJ evaluates (reward assignment) the perceived and planned actions and situation, thereby enabling BG to select goals and generate suitable behaviour. Estimated plans/actions which are evaluated as more safe, less risky and less costly will be selected for execution over the others which are evaluated as unsafe, more risky and more costly. Further to the discussion carried out in Section 19.3, it is important to see how safety can be used as a significant value within VJ and how a value measurement process can be developed in relation to safety.

The goals or objectives (such as safe task achievement) of the system are divided into sub-goals. Safety management is spread over the RCS hierarchy and each module contributes to safety to a certain extent (some more than others). Once VJ receives a set of possible actions, they are assigned a cost (either reward or penalty depending on whether the action is 'good' or 'bad' for achieving this objective). Then, an action which is expected to incur maximum reward (action perceived as 'good') is chosen. This action is then passed on to the WM, which in turn sends the control signals corresponding to this action to BG in order to create this physical behaviour.

VJ selects a single action with the highest *value*, evaluates the situation, makes the plan and sends this information to WM and BG. Here, it is important to state what is meant by positive or negative reward for a particular sub-goal. Positive reward helps the system achieve the set goal/objective, whereas negative reward prohibits safe behaviour. Positive values are assigned to goal achievement reflecting hope (hope to gain) and negative value is assigned to lack of safety reflecting fear (fear to loose or expose to hazards). In more anthropomorphic terms, one way to model risk and risk-taking behaviour in a robot is by modelling the emotions of fear and hope [82]. Fear is essentially the biological brain's evaluation of risk. Hope is the evaluation of expected benefit. Balancing fear of loss against hope of gain is how the natural brain approaches risky behaviour. If the expected cost (fear) is greater than the expected benefit (hope), then the robot will avoid the risky behaviour. By adjusting the threshold where the hoped for benefit exceeds the fear of failure, one can change the robot's behaviour from cautious to aggressive. POMDP uses the probability distribution over the state-space and calculates the rewards for the available actions. The actions which incur positive rewards are considered to be good or hopeful for achievement of the system objective. The actions which incur negative rewards (penalties) are considered as bad or fearful for objective achievement. The method by which POMDP calculates the rewards is discussed in detail in later sections.

19.7.3.1 Value Judgement from a POMDP Perspective

The reward function defines the VJ process through the allocation of rewards/values to actions in terms of the predefined and identified system values. Table 19.5 shows the flows from/to VJ in a POMDP-modified RCS node (refer to Figures 19.3 and 19.16).

19.7.4 WORLD MODEL IN A POMDP-MODIFIED RCS NODE

The counterpart of WM in POMDP is the set of discrete states S, set of observations Ω, set of actions A and transition function TF. The WM model in each node contains knowledge of the world with range and resolution that are appropriate for control functions in the behaviour-generation process in that node. WM provides information to VJ to enable it to assign rewards for actions/plans. The POMDP model enables the learning process, but RCS framework provides the method to incorporate POMDP.

It is critical that the WM not only represents what the system is certain about but also what it is uncertain about. Uncertainty of what the system is doing, the current state of the system and where it will end after performing a particular action has to be modelled well. Uncertainty can be represented by probabilistic reasoning techniques. For the purpose of this research, the uncertainty is represented by probabilities using the Bayesian network formalism.

TABLE 19.5
Flows from/to VJ in Terms of POMDP

Flows	Original Link	Replacement Link	Functionality of the Link
VJ-BG	Plan evaluation	Evaluated plans/actions	Sending the available plans/actions along with their assigned values (rewards or penalties). Consideration is given in terms of the sequence of actions (policy) and not the latest action only.

It is important that the WM also includes an interpretation of sensory integrity. Having said that, it is also important understanding how sensory integrity evaluation arises and how such sensory integrity will influence the information gained from that sensor. The VJ based on safety regards the identification of conditions that will give rise to an accident. The identification of such conditions arises from the information in the WM and therefore, whatever those conditions are, it is necessary to put an appropriate evaluation of whether that condition has arisen or will arise. The role of sensory integrity and its effect on the information gained should be viewed in this respect.

Further to Section 19.3.2, WM is an important safety critical element of RCS (not the only one but definitely an important element—VJ also contributes significantly since it is the element that embeds what constitutes safety and what does not). Creating a reward/value is dependent on the WM and hence it is very important to have a WM with a high level of integrity as it will affect the performance of the whole system and all the other elements of the RCS. It should not be a false image of reality but as close to reality as possible.

19.7.4.1 World Model from POMDP Perspective

This element stores the current belief state. It also computes the transition probabilities given the available actions using the transition function. It is responsible for identification of future expected states and environment changes. The transition function estimates the next state (at $t = t + 1$) in which the system ends up following the selection of a specific action in step $t = t$. Table 19.6 shows the flows from/to WM in a POMDP-modified RCS node (refer to Figures 19.3 and 19.16).

19.7.5 Behaviour Generation in a POMDP-Modified RCS Node

The POMDP counterpart of BG is policy application. Depending on the focus of the task (task achievement or safety) the policies are generated in WM. The action decision is chosen by VJ after performing *value* maximisation and fed to BG to actually carry out the actions according to that policy. The commanded action may include parameters that specify how, where, when, how much, how fast and on what.

BG only performs physical actions based on the policy generated by VJ and state input it receives from WM. No or very little intelligence is assigned to the BG. BG executes the instructions to maintain a safe condition. BG may have embedded safety features (bumper switches, etc.), but the primary decision as to whether a particular action is safe or not is taken within VJ.

Goals are selected and plans are generated by a looping interaction between BG, WM and VJ elements. BG system element also monitors the execution of plans and sends the status to a higher level, which then modifies the policies if required. BG develops or selects plans by using *a priori* task knowledge and value judgement functions combined with real-time information provided by

TABLE 19.6
Flows from/to WM in Terms of POMDP

Flows	Original Link	Replacement Link	Functionality of the Link
WM-SP	Predicted input	Expected state and environment	This link is used only in case of more than one-step look ahead. In such a situation, it is responsible for sending the expected future state and environment following action transition estimation along with previously estimated state and environment given the action that was executed in time step t.
WM-VJ	Plan results	Action outcome expectation	The expectation of the chosen action outcome, which is based on expected state and environment estimation in WM.
WM-BG	State	Current state	The evaluated probable current state belief. It is sent from WM to BG with actions available for decision making.

sensory processing and world modelling to find the best assignment of tools and resources to agents and to find the best schedule of actions (i.e., the most efficient plan to get from an anticipated starting state to a goal state).

Policy selection process is performed in BG so as not just to find ways to achieve the goal but also to reduce uncertain and risky behaviour (or to increase safety). The process of selecting the action with the maximum reward is based on the reward function computations carried out in VJ. The action which is suitable for safe behaviour generation may not be best suitable for task completion. It is important to note that these two objectives (task completion and safety behaviour generation) have contrasting interests and may produce different behaviour choices. It is essential to obtain a balance between them, and BG is responsible for achieving this balance. Section 19.7.3 discusses the method to integrate safety and task achievement rewards. BG is responsible for action execution and physical behaviour generation.

19.7.5.1 Behaviour Generation from POMDP Perspective

This element is responsible for action selection and generation of behaviour, which reduces uncertainty and ensures safe operation and task achievement. Table 19.7 shows the flows from/to BG in a POMDP-modified RCS node (refer to Figures 19.3 and 19.16).

19.7.6 Sensory Processing in a POMDP-Modified RCS Node

The POMDP counterpart of SP is the set of observations (Ω) and the OF. Observations are the most important aspect of POMDP as those provide the basis for updating the WM and maximising the value in VJ and then in turn generating the policy, which gives maximum reward in BG. However, there is overlap between the role of WM and SP. In most cases, sensors cannot directly measure the state of the world. Sensors typically can only measure phenomena that depend on the state of the world. Thus, SP must infer the state of the world from input signals and *a priori* knowledge.

As mentioned in Section 19.3.2, RCS levels play important roles in safety management. RCS is a hierarchical architecture and the layers of architecture can be usefully employed to implement POMDP.

19.7.6.1 Sensory Processing from a POMDP Perspective

SP determines the belief (probability) of the current state (vehicle state and environment) based on expectations (from WM) and on actual observations (from sensors). The OF is responsible for current belief state estimation, future observed state estimation and pruning of future observed states. Pruning is an essential part of POMDP as it reduces the calculation complexity of the system and

TABLE 19.7
Flows from/to BG in Terms of POMDP

Flows	Original Link	Replacement Link	Functionality of the Link
BG-WM	Plan	Action	The action or command that had just been executed in $(t-1)$ is sent from BG to WM
Behaviour generation (at level $n-1$)–Behaviour generation (at level n)	Status	From-status	Current system status, for example, halt, execution and initialization
	Status	To-status	Status info of the lower level
	Command	To-command	Task to be executed by the level in consideration
	Command	From-command	Task to be executed by the level below, that is, the action selection by BG on lower level

TABLE 19.8

Flows from/to SP in Terms of POMDP

Flows	Original Link	Replacement Link	Functionality of the Link
SP-WM	Updates	'Updated state and environment'	It is the current belief distribution about the vehicle and the environment state
Sensors–SP	Sensory input	'Observations-in'	Depending on the position (level) of a particular node in the RCS hierarchy, this link is responsible for passing either observation variable values (calculated from a lower level module) or observed sensory input (directly from sensors) regarding the vehicle state and environment
SP (at level $n-1$)–SP (at level n)	Sensory output	'Observations-out'	Sends the summary of updated state and environment (to higher level module)
SP-VJ	Perceived objects and events	Future observed state	Sends the observed future states from SP to VJ. The computation of future observed state is carried out by SP and is based on the observation function (in SP) and input from WM

the subsequent processing time [83]. Table 19.8 shows the flows from/to SP in a POMDP-modified RCS node (refer to Figures 19.3 and 19.16).

Figure 19.17 shows the interconnections between all the functional elements of RCS node in terms of POMDP components. Figure 19.17 summarises the discussion so far. The flows between different functional elements of RCS node are replaced by their counterparts in a POMDP-modified node.

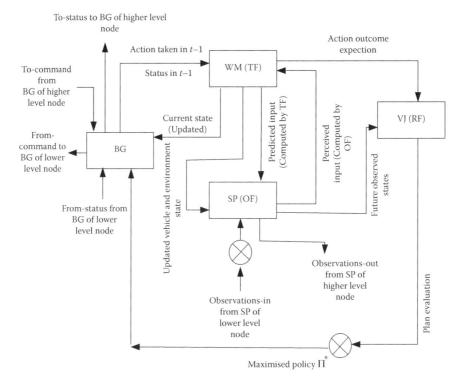

FIGURE 19.17 Relationship between SP_VJ_WM_BG in a POMDP-modified node.

19.8   DEVELOPMENT OF POMDP-BASED DECISION-MAKING ALGORITHM

The POMDP model is employed for safety decision making while navigating. The main concern is to avoid collisions with moving and static obstacles. As the sensory data are incomplete owing to the partial observability and hence uncertainty, the probability distribution over states is considered. The POMDP model is described in Section 19.6.1. Having defined such a representation, it is possible to determine an action-decision process for the system that would give the highest expected reward. The POMDP model is employed for safety decision making while navigating. The main concern is to avoid collisions with moving and static obstacles. The safest and effective action decision balances objectives of the task completion and safe behaviour generation.

Thus, the state-estimation component of a POMDP controller can be constructed from a given model. Bayesian networks provide formalism for reasoning about partial beliefs under conditions of uncertainty. The POMDP solution process developed as part of this research consists of the following steps:

1. Fault tree analysis for collision accidents
2. Development of observation belief network (OBN) for belief state computation
3. Computation of offline POMDP solution

The problem considered in this research deals with collision accidents. To decide the probability of the obstacle state, a Bayesian network is developed. An OBN is used to estimate the current belief state of an obstacle around the robot. From Figure 19.18, the following set of state variables is considered as possible contributors (parameters of system state vector) to collision accidents:

1. Robot speed
2. Robot orientation
3. Robot position
4. Obstacle presence
5. Obstacle position
6. Obstacle speed
7. Obstacle orientation
8. Fault state of the sensors

The following assumptions/modifications are considered:

1. Obstacle status is considered as a single state variable, which is associated with five different states. These states take into account all the data related to obstacles. This variable includes obstacle presence, obstacle type, obstacle position and obstacle orientation. For the development of OBN, other state variables are considered as parent (root) nodes providing information to the process to compute the obstacle status probability.
2. It is assumed that obstacle speed is not directly measurable with the use of available sensors and hence it is not considered in further calculations.
3. Terrain roughness can cause the collision with hidden ground obstacles. However, as mentioned earlier, this research investigates the collision accident with obstacles above ground (dynamic or static) and hence terrain roughness is not considered.

The following observation variables are taken into account:

• Laser scanner reading (front and rear)

The observation variables serve as the child node, which also provide the information to the process to compute the belief state of 'obstacle status'. Figure 19.19 shows the OBN for a collision

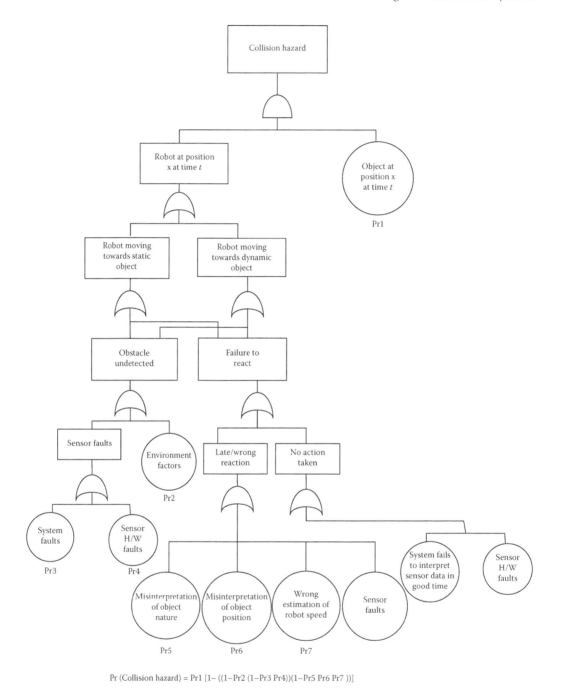

$$\text{Pr (Collision hazard)} = \text{Pr1} [1 - ((1 - \text{Pr2} (1 - \text{Pr3 Pr4}))(1 - \text{Pr5 Pr6 Pr7}))]$$

FIGURE 19.18   Simplified fault tree for collision accident.

accident. The obstacle status variable is affected by robot position and orientation. The obstacle status states are based on the relationship between obstacle and robot and the obstacle position or orientation is always considered with respect to the robot position and orientation. This information is integrated with the prior knowledge of the obstacles. The laser scanner (front and rear) provides the data regarding the obstacle position and orientation to compute the probability of obstacle status.

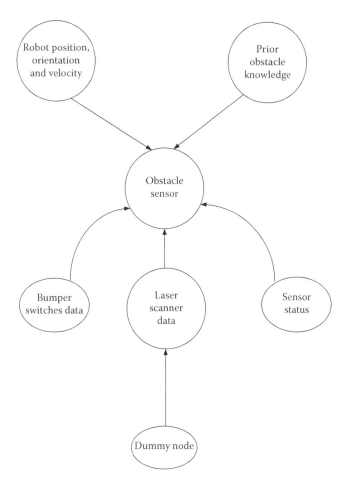

FIGURE 19.19 OBN for collision accident.

The sensor status node provides the information about the status of the LS to the obstacle status node. The dummy node is introduced in order to deal with knowledge about the LS data from consecutive multiple readings. Dummy node represents the crude data—the integration and interpretation of this data are shown in the LS data node. The OBN shown in Figure 19.19 is further simplified by considering the following assumptions. The simplified OBN is shown in Figure 19.20.

- Bumper switches (BS) are not considered because BS halt the system when they detect objects. There is no need to calculate the OBN and subsequent POMDP solution once the system comes to a halt.
- Robot speed, position and orientation are not affected by network links. Enough knowledge about them is available from various sensors. Hence, the network links for these state variables are not shown in the above diagram.
- Obstacle status and sensor status causes are grouped together to reduce the computational complexity. They form a single node. In this case, the node value probability corresponds to the probability of the obstacle status and sensor status variable taking corresponding specific values, that is,

$$\text{Probability} = \text{Pr (obstacle status)} * \text{Pr (sensor status)}$$

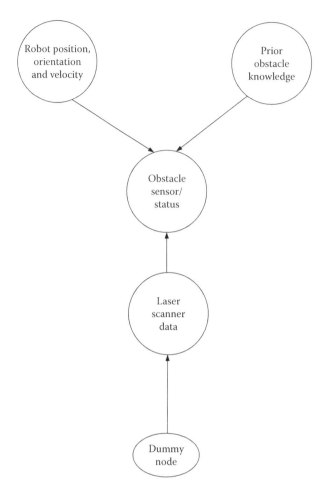

FIGURE 19.20 Simplified OBN for collision accident.

The OB/St node receives information about prior obstacle knowledge and robot data from parent nodes as shown. It receives sensory data from the child node. The simplified OBN of Figure 19.20 can be expressed in the form of acyclic circular graph [47] (as shown in Figure 19.21). This representation is used for computational definitions.

The solid nodes represent the state variables (parent nodes), whereas the clear nodes represent the observation variables (child nodes). The grey node in the centre represents the obstacle status/ sensor causes node for which the belief state is computed. This node receives information from the child and parent nodes.

19.8.1 Obstacle Status/Sensor Status Estimation

It should be noted that LS readings are influenced by the obstacle and robot relationship in terms of position, orientation, direction of motion and so on. The OB/St node receives messages from five parent nodes and one child node. The information received from the parent nodes is represented in the Π form whereas the information received from the child node is represented in the λ form [47]. The belief state of 'obstacle status' is computed from the messages it receives from parent and child nodes. The OB/St node receives information from five parent nodes as described below:

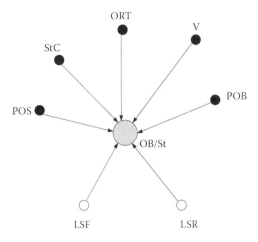

FIGURE 19.21 Belief network representations.

1. Parent node StC

$$\pi_{StC}(StC) = Pr(StC)$$
$$\pi_{StC}(OB/St) = Pr(St \mid StC)\pi_{StC}(StC)$$
(19.1)

2. Parent node ORT

$$\pi_{ORT}(ORT) = Pr(ORT)$$
$$\pi_{ORT}(OB) = Pr(OB \mid ORT)\pi_{ORT}(ORT)$$
(19.2)

3. Parent node POB

$$\pi_{POB}(POB) = Pr(POB)$$
$$\pi_{POB}(OB) = Pr(OB \mid POB)\pi_{POB}(POB)$$
(19.3)

4. Parent node POS

$$\pi_{POS}(POS) = Pr(POS)$$
$$\pi_{POS}(OB) = Pr(OB \mid POS)\pi_{POS}(POS)$$
(19.4)

5. Parent node V

$$\pi_{V}(V) = Pr(V)$$
$$\pi_{V}(OB) = Pr(OB \mid V)\pi_{V}(V)$$
(19.5)

The conditional probability matrix provides the relationship between the above five nodes and the 'obstacle status' variable. The OB/St node also receives information from a child node as below:

6. Child Node LS (Front and Rear)

$$\lambda_{LS}(OB \mid St) = \sum_{LS} Pr(LS \mid OB, St)Pr(LS)$$
$$\lambda_{LS}(OB, St \mid ORT, POS, POS, V) = \sum_{LS} Pr(LS \mid OB, St, ORT, POB, POS, V)Pr(LS)$$
(19.6)

The belief in OB/St can be computed as

$$
\begin{aligned}
\Pr(\text{OB/St}) = {}& \alpha\,\lambda_{\text{LSR}}(\text{OB/St})\lambda_{\text{LSF}}(\text{OB/St})\pi_{\text{StC}}(\text{OB/St})\pi_{\text{ORT}}(\text{OB/St}) \\
& \pi_{\text{POB}}(\text{OB/St})\pi_{\text{POS}}(\text{OB/St})\pi_{\text{V}}(\text{OB/St})
\end{aligned}
\tag{19.7}
$$

where α is the normalising factor which ensures that the addition of all the individual probabilities is 1. Substituting for π_{StC}, π_{ORT}, π_{POB}, π_{POS} and π_{V} gives

$$
\begin{aligned}
\Pr(\text{OB} \mid \text{St}) = {}& \alpha\,\lambda_{\text{LSR}}(\text{OB/St})\lambda_{\text{LSF}}(\text{OB/St})\Pr(\text{St} \mid \text{StC})\pi_{\text{StC}}(\text{StC}) \\
& \Pr(\text{OB} \mid \text{ORT})\pi_{\text{ORT}}(\text{ORT})\Pr(\text{OB} \mid \text{POB})\pi_{\text{POB}}(\text{POB}) \\
& \Pr(\text{OB} \mid \text{POS})\pi_{\text{POS}}(\text{POS})\Pr(\text{OB} \mid \text{V})\pi_{\text{V}}(\text{V})
\end{aligned}
\tag{19.8}
$$

Further substitution then gives

$$
\begin{aligned}
\Pr(\text{OB} \mid \text{St}) = {}& \alpha\,\lambda_{\text{LSR}}(\text{OB/St})\lambda_{\text{LSF}}(\text{OB/St})\Pr(\text{St} \mid \text{StC})\Pr(\text{StC}) \\
& \Pr(\text{OB} \mid \text{ORT})\Pr(\text{ORT})\Pr(\text{OB} \mid \text{POB})\Pr(\text{POB}) \\
& \Pr(\text{OB} \mid \text{POS})\Pr(\text{POS})\Pr(\text{OB} \mid \text{V})\Pr(\text{V})
\end{aligned}
\tag{19.9}
$$

Given the interpretation of the network, it is necessary to compute the belief in obstacle status and sensor status for each respective possible value of robot orientation (ORT) value, prior obstacle knowledge (POB) and each possible robot position (POS), velocity (V) value as follows:

$$
\lambda_{\text{LSF}}(\text{OB/St}) = \sum_{\text{LS}} \Pr(\text{LSF} \mid \text{OB, St, ORT, POS, POB, V})\Pr(\text{LSF})
\tag{19.10}
$$

Similarly for rear LS reading

$$
\lambda_{\text{LSR}}(\text{OB/St}) = \sum_{\text{LS}} \Pr(\text{LSR} \mid \text{OB, St, ORT, POS, POB, V})\Pr(\text{LSR})
\tag{19.11}
$$

Equations 19.9 through 19.11 are computed for every value of POS, POB, ORT and V. Once the values are obtained, the OB/St value conditional on each specific combination of POB, POS, ORT and V state can be calculated as

$$
\begin{aligned}
\Pr(\text{OB, St} \mid \text{ORT, POB, POS, V}) = {}& \alpha\,\lambda_{\text{LSR}}(\text{OB, St} \mid \text{ORT, POS, POB, V}) \\
& \lambda_{\text{LSF}}(\text{OB, St} \mid \text{ORT, POS, POB, V})\Pr(\text{St} \mid \text{StC}) \\
& \Pr(\text{OB} \mid \text{ORT})\Pr(\text{OB} \mid \text{POB})\Pr(\text{OB} \mid \text{POS})\Pr(\text{OB} \mid \text{V})
\end{aligned}
\tag{19.12}
$$

By taking into consideration various POB, POS, ORT and V states, the belief in the OB/St states can be evaluated as

$$
\Pr(\text{OB, St}) = \alpha \sum_{\text{ORT}} \sum_{\text{POS}} \sum_{\text{POB}} \sum_{\text{V}} \left\{ \begin{bmatrix} [\lambda_{\text{LSR}}(\text{OB, St} \mid \text{ORT, POS, POB, V})] \\ [\lambda_{\text{LSR}}(\text{OB, St} \mid \text{ORT, POS, POB, V})] \\ \Pr(\text{St} \mid \text{StC})\Pr(\text{OB} \mid \text{ORT}) \\ \Pr(\text{OB} \mid \text{POB})\Pr(\text{OB} \mid \text{POS})\Pr(\text{OB} \mid \text{V}) \end{bmatrix} \right\}
\tag{19.13}
$$

19.8.2 Obstacle Status Estimation

To determine individual obstacle status and sensor status, marginal probabilities need to be calculated as follows:

$$\Pr(OB) = \sum_{St} \Pr(OB/St)$$
$$\Pr(St) = \sum_{OB} \Pr(OB/St)$$

(19.14)

From Equation 19.14, the marginal probabilities can be derived. The conditional probability relation matrix mentioned in the above document is developed as a part of this research.

As mentioned in Section 19.6.2, POMDP action decisions need to be based on the achievement of some objective. A brief description of the POMDP problem computation is presented next. In a POMDP, the agent (in this case a robot controller) must select appropriate actions when operating in a stochastic environment where the state of the system is only partially observable. The value of an action or state can be represented by a set of rewards that the system can achieve by taking a specific action in a specific state. The rewards in this case can be defined in various ways. For example, a reward may be allocated to the system state or alternatively, to the transition from one state to another state when a specific action is taken or to a specific action irrespective of the state. Given the rewards that a system can achieve in each time step, the objective of the system is to select action decisions which maximise the total expected reward or utility value to be obtained over a specific number of time steps. According to Kaelbling et al. [74], the POMDP solution finds a policy (or action) that maximises the reward obtained through the controller's choice of actions. Policy is a description of the behaviour of an agent. In simpler terms, an action is nothing but a policy. As mentioned earlier, the POMDP solution can be found over any number of time steps look ahead. It is generally called the horizon. Thus, the 'infinite horizon' refers to an infinite lifetime of the agent. In an infinite horizon POMDP model, the rewards are added over the infinite lifetime of the agent. In order to represent the importance of rewards received earlier in its lifetime, the discount factor γ $(0 < \gamma < 1)$ is used. The agent should act to optimize:

$$E\left[\sum_{t=0}^{\infty} \gamma^t r_t \right]$$

(19.15)

where t is the number of time step (t varies between 0 and ∞), r_t is the reward received at time step t, E is expected sum of rewards and γ is the discount factor.

In this model, rewards received earlier in its lifetime have more value to the agent. The infinite lifetime is considered, but the discount factor ensures that the sum is finite. The larger the discount factor (closer to 1), the more effect the future rewards have on current decision making. For the purpose of this research, a three-step look ahead is considered.

The policy generated in an infinite horizon model is termed as 'stationary policy'. On the other hand, the policy generated in a finite horizon model is termed as 'non-stationary policy'. The stationary policy does not depend upon time. The non-stationary policy depends on the state and is indexed by time. The way an agent chooses its actions on the last few steps of the finite horizon model is generally very different from the way it chooses them when it has a long life ahead of it (in the case of the infinite horizon model).

In a finite horizon case, let $V_{\pi,t}(s)$ be the value function giving the expected sum of rewards obtained from starting in state s and executing non-stationary policy π for t steps. At the last step, the value function is the expected reward for taking the action specified by the final element of the policy (i.e., $V_{\pi,1}(s) = R(s, \pi_1 (s))$). To evaluate the future, one must consider all possible resulting states s', the likelihood of their occurrence (transition function $T(s, a, s')$) and their $(t - 1)$ step value under policy π given by $V_{\pi,t-1}(s')$. $V_{\pi,t}(s)$ for the finite horizon model can be defined as

$$V_{\pi,t}(s) = R(s,\pi_t(s)) + \gamma \sum_{s' \in S} T(s,\pi_t(s),s')V_{\pi,t-1}(s') \qquad (19.16)$$

The stationary policy is not dependent on time and hence for an infinite horizon problem, the value can be computed by

$$V_\pi(s) = R(s,\pi(s)) + \gamma \sum_{s' \in S} T(s,\pi(s),s')V_\pi(s') \qquad (19.17)$$

The value function V_π for policy π is the unique simultaneous solution of this set of linear equations, one equation for each state. The above discussion is based on the assumption that the policy is known to the designer and the value function can be calculated from the policy. But in reality, in fact, solving a POMDP problem is finding the optimal policy. If one knows the value function, computation of policy can be performed in the opposite direction from Equation 19.16 or 19.17 depending on the problem horizon. For an infinite horizon model, the policy obtained by taking the action that maximises expected immediate reward at every step plus the discounted value of the next state measured by V is given by

$$\pi_V(s) = \arg\max_a \left[R(s,a) + \gamma \sum_{s' \in S} T(s,a,s')V(s') \right] \qquad (19.18)$$

$R(s,a)$ is the expected immediate reward for taking action a, in state s. In case of POMDP, the state s is not completely observable and hence a 'belief state' is considered instead. The policy component of a POMDP agent must map the current belief state into action. A policy tree is a tree of depth t that specifies a complete t step non-stationary policy. The top node determines the first action to be taken. Then, depending on the resulting observation, a link is traced to a node on the next level, which determines the next action. This is a complete summary of the t steps. The expected discounted value to be gained from executing a policy tree p depends on the state of the world when the agent starts. In the single-horizon case, p is a one-step policy tree (a single action). Let $a(p)$ be the action specified in the top node of policy tree p. Substituting policy $\pi_t(s)$ in Equation 19.16 by $a(p)$, the value function is expressed as

$$V_{p,1}(s) = R(s,a(p)) \qquad (19.19)$$

More generally, if p is a t step policy tree, then

$$V_p(S) = R(s,a(p)) + \gamma \cdot (\text{expected value of the future}) \qquad (19.20)$$

Because the agent is not sure of the exact current state of the world it must be able to determine the value of executing a policy tree p from some initial belief state $b(s)$. This is an expectation over states of executing p in each state:

$$V_p(b) = \sum_{s \in S} b(s)V_p(s) \qquad (19.21)$$

Let $$\alpha_p = \langle V_p(s_1) \ldots V_p(s_n) \rangle, \quad \text{then } V_p(b) = b \cdot \alpha_p$$

α_p is also called the *alpha vector*. This gives the value of executing the policy tree p in every possible belief state. To construct an optimal t step non-stationary policy, however, is not exactly the same. It will generally be necessary to execute different policy trees from different initial belief states. Let P be the finite set of all t step policy trees. Then

$$V_t(b) = \max_{p \in P} b \cdot \alpha_p \qquad (19.22)$$

That is, the optimal t step value of starting in belief state b is the value of executing the best policy tree in that belief state. Each policy tree p induces a value function V_p that is linear in belief state b and the value function V_t is the upper surface of this collection of functions. V_t is piecewise linear and convex (PWLC) function. The optimal value function can be projected back down onto the belief space, yielding a partition into polyhedral regions. Within each region, there is some single policy tree p such that the scalar product of b and *value* (of the policy tree p) is maximal over the entire region. The optimal action for each belief state in this region is $a(p)$, the action in the root node of policy tree p. A single policy tree can be executed to maximise expected reward [74]. The value function can be calculated by using various algorithms such as value iteration, one-pass algorithm, enumeration algorithm, linear support, incremental pruning and witness algorithm [50,84–87].

19.8.3 POMDP SOLUTIONS: INTEGRATING GOAL- AND SAFETY-RELATED DECISIONS

The POMDP reward function or value function is generally derived through some utility function, outlining the decision maker's preferences. The analysis given in this section has, until now, focussed on an action-decision process based on safe operation. However, a practical and more precise action-decision selection system needs to cater to the achievement of the operational task together with maintaining system safety. It should be noted that the set of actions available for decision making has to be equivalent for both safety and operational task achievement goals. This is necessary because whichever action gets selected, it eventually influences the system state with respect to both safety and task achievement objectives. If the dominant goal is 'safety', the safest but not necessarily the quickest action is assigned maximum reward, and on the other hand, if 'task achievement' is the system focus, then the quickest or most effective but not necessarily the safest action is assigned the maximum reward.

A value function or reward function has been already defined for a POMDP problem solution. Since the discussion until now focussed on safety criteria, this reward function was considered to assign maximum rewards to the actions, which change the system state to safer system state. A similar approach to that applied for safety-related action selection could also be followed for the operational task-related decision making. In the latter case, rewards are assigned through a value function that reflects the action-decision preferences in terms of task achievement only. Pace [34] states that the 'action decision preferences and the respective attributed reward values for safety management and task achievement are unlikely to be compatible at all times and antagonistic preferences are likely to occur during the robotic system's operation'. It is therefore necessary to balance these two objectives and reach an optimum compromise between safety and task achievement preferences, which still ensures that the right level of safety is maintained.

19.8.3.1 Integrating Safety and Operational Task Achievement Preferences

Pace [34] suggested a concept of a weighting factor scale. Based on utility theory [48], two utility functions are defined: a safety-related utility function V_{Safe} and a task achievement utility function V_{Task} which maps the preferences of the decision-making process from both a safety and task achievement perspective. The total utility (value) for an action a_t taken at time t can be expressed as

$$V_{\text{Tot}}(a_t) = W_S V_{\text{Safe}}(a_t) + (1 - W_S) V_{\text{Task}}(a_t) \qquad (19.23)$$

In the above, W_S is simply a weighting factor ranging between 0 and 1, indicating the relative weight that is given between safety maintenance and task achievement. The concept of using the weighting factor-based utility factor has been implemented for this research in the further sections.

19.8.4 Action-Decision Selections from POMDP Solutions

POMDP solution is computed by implementing the incremental pruning algorithm [86,88]. Incremental pruning algorithm initially generates sets of *alpha vectors* for each action individually and then focusses on each observation at a time. To construct a vector requires selecting an action a and a vector V for each observation z. For a given action, one can construct all of $S(a, z)$ sets (one for each observation). Adding the immediate rewards is an easy step. The main problem is in finding all the different combinations of future strategies. Incremental pruning algorithm does this incrementally observation by observation. For example, let the action be a and the set of observations be $(z_1, z_2, z_3, \ldots, z_n)$. Initially, all the combinations of the $S(a, z_1)$ and $S(a, z_2)$ vectors are carried out which gives a value function. This value function gives rise to numerous vectors. Then the vectors which are completely dominated by other vectors over the entire belief space are eliminated. This is called pruning. Now, the vectors which are left after elimination are combined with $S(a, z_3)$ and the same process is carried out until all the vectors for a particular action (a) and all observations (z_1 to z_n) are combined together. The same process is carried out for other actions. This algorithm prunes the unnecessary vectors and generates a list of a few vectors which can be used for action decision.

19.8.4.1 Offline POMDP Solution Computation

The optimal policy calculation discussed so far provides direct mapping from the current belief state space to the action state space. The POMDP solution is found as discussed in Section 19.7.2 and finds the *alpha vector* consisting of alpha value for each action-state combination. The *alpha vector* gives the value function for carrying out each action in every possible state. In other words it gives the value of executing the particular action in every possible belief state over the entire belief space.

The method presented in Section 19.7.2 provides direct mapping from the current belief state space to the action-state space. Thus, the system in operation will only be required to estimate the current belief state through observations and prior beliefs and simply map that belief state into an action. It is argued that the real-time computation of the value-iteration algorithm is not necessary if all possible solutions to the POMDP model for the various belief states are computed beforehand. In run-time, only the belief state is calculated using the OBN discussed in Section 19.7.1. OBN is used as the means of estimating the current state of the system variables (Table 19.9). When a particular belief state is calculated based on the sensory observations, the POMDP solution (value function) is referred to and using the value function, the action with maximum rewards corresponding to the

TABLE 19.9
OBN Abbreviations

Symbol	Variable
StC	L.S. sensor status causes
OB/St	Obstacle status/L.S. sensor status
POB	Prior obstacle knowledge
POS	Robot position
ORT	Robot orientation
LSF	Observed laser scanner (front) reading
LSR	Observed laser scanner (rear) reading
V	Velocity of the vehicle

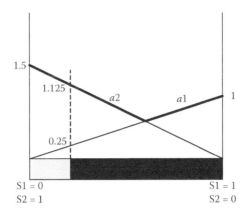

FIGURE 19.22 PWLC value function for a two-state problem.

belief state is decided from the POMDP solution. Rewriting Equation 19.22, $V_t(b) = \max_{p \in P} b \cdot \alpha_p$, it can be seen that, the value V_t for a belief state b at time t is equal to the maximum of the scalar product of belief state and the alpha value for a particular action (policy p) taken from the *alpha vector*.

For a two-state problem, let action $a1$ incur reward = 1 in state $s1$ and 0 in state $s2$ and let action $a2$ gain a reward = 0 in state $s1$ and 1.5 in state $s2$. Let the belief state be [0.25, 0.75], then the value of taking action $a1$ in this belief state is $0.25 \times 1 + 0.75 \times 0 = 0.25$. Similarly, action $a2$ has value $0.25 \times 0 + 0.75 \times 1.5 = 1.125$. This can be shown diagrammatically on a value function (as shown in Figure 19.22).

Similar analogy is applied to higher state POMDP problems, although it becomes difficult to express such problems due to the increase in belief space dimension. Figure 19.23 presents the complete POMDP solution process.

POMDP is used as a safety decision-making algorithm at level 2 (travel controller) of RCS hierarchy discussed in the previous sections. POMDP is responsible for choosing the safest action based on the observations and prior knowledge. The best action decision is associated with the state of mainly three parameters (causal factors):

1. Speed actions: reduction or increase in speed
2. Orientation actions: move straight, turn right, turn left, reverse choosing the suitable orientation to move
3. Obstacle status actions: waiting to update map or proceeding

Initially, a POMDP model was developed which consisted of a set of state vectors. A set of state vectors consisting of the above three causal factors is considered. The state variables forming the state vector are shown in Table 19.10.

The state vector consisting of the above three elements can be represented as

$$\begin{bmatrix} \text{speed state} \\ \text{orientation state} \\ \text{obstacle status state} \end{bmatrix}$$

Every possible combination of these three state variable states gave rise to a tremendously large set of states, actions and observations as shown below.

- Speed states (2)*orientation states (4)*obstacle status states (5) = 40 states
- Actions for speed (3)*actions for orientation (4)*actions for obstacle status (2) = 24 actions
- Obs. for speed (2)*obs. for orientation (4)*obs. for obstacle status (5) = 40 observations

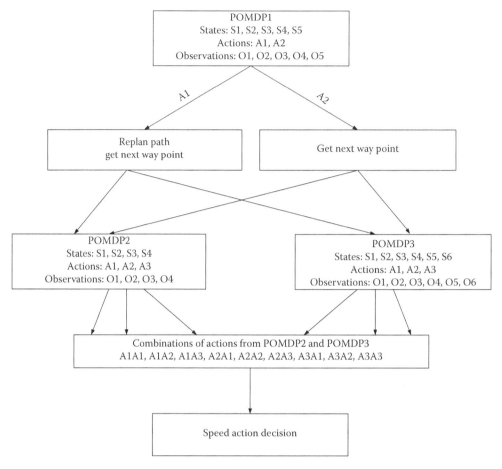

FIGURE 19.23 Split POMDP.

Such a huge POMDP model (40 states, 24 actions, 40 observables) proved impossible to compute using the existing solution methods. Partial observability takes into account the uncertainty, but at the same time introduces complexity in the POMDP computations. Despite the representational power of POMDPs, their use is significantly limited by the great computational cost of finding an optimal policy for the controller (decision maker).

19.8.5 REDUCTION OF COMPLEXITY IN POMDP SOLUTION COMPUTATIONS

Several means of complexity reduction are proposed [68,86,89–92] by various researchers which make the POMDP much more manageable and usable. It is not only important to reduce the computational complexity but also the time to find the solution. Current algorithms however are still generally unable to handle large and complex POMDP problems involving a large number of states. In this research, another method of reducing the complexity in solving the POMDP problem is developed. The theoretical concept and experimental results are discussed in the next few sections.

19.8.5.1 POMDP Problem Decomposition-Split POMDP

In various computer architectures, a task is divided into a hierarchy of subtasks and then each is solved by a different subsystem. This was the basis for developing the strategy for solving extremely big POMDP problems such as discussed above. The split POMDP method is a combination of a

TABLE 19.10

State Vector Components of POMDP

State Variable	Number of States	States	Actions Available
Speed	2	Slow	Speed_up
		Fast	Speed_down
			Cruise
Orientation	4	0–90	Go_forward
		90–180	Go_reverse
		180–270	Turn_right
		270–360	Turn_left
Obstacle status[a]	5	S1	
		S2	Wait
		S3	Proceed
		S4	
		S5	

[a] The obstacle status variable is associated with five different states, viz., S1 (dynamic obstacle within safety perimeter (SP) and moving towards the vehicle), S2 (dynamic obstacle present within SP and moving away from the vehicle), S3 (static obstacle present within SP and on the path of vehicle), S4 (static obstacle present within SP but not on the path of vehicle), S5 (no obstacle in the laser scanner range).

rule-based approach and POMDP problem solving. This method reduces the computation time for solving the POMDP problem by splitting a large POMDP problem into multiple smaller POMDP problems. It is important to note that split POMDP promises to be a highly effective method of complexity reduction, but it is not a generalised method yet. Although it is employed and used effectively for this research, in order to make it more generalised, it has to be proved effective by testing it (using it) for more applications.

The bigger and complex POMDP problem consisting of 40 states, 24 actions and 40 observations is broken down into three smaller POMDP problems:

1. Obstacle status POMDP
2. Speed_obs POMDP
3. Speed_ort POMDP

19.8.5.1.1 Obstacle Status POMDP

When an object is detected, the robot computes the belief state for each of the obstacles and generates a separate POMDP model for each of these obstacles. If the POMDP decision is such as to 'Proceed', the obstacle is neglected. If the decision is 'Wait', the data about that particular object is saved to update the map dynamically and to recalculate the next way-point. When the next way-point is computed, two smaller POMDPs (Speed_obs POMDP and Speed_ort POMDP) are computed. These two POMDPs are for deciding the speed of the robot in the next state. These two POMDPs are dependent upon the relationship between the robot and the obstacles around it.

19.8.5.1.2 Speed_obs POMDP

This POMDP is responsible for deciding whether there is an obstacle present near the robot. If there is an obstacle present, the robot has to take that into consideration while deciding whether to speed_up, speed_down or cruise.

TABLE 19.11

Split POMDP Components

POMDP Title	States	Actions	Observations
Obstacle status POMDP	5 (S1–S5)	Wait proceed	5 (corresponding to five states)
Speed_ort POMDP	Obstacle within SP and robot going fast, obstacle not within SP and robot going fast, obstacle within SP and robot going slow, obstacle not within SP, robot going slow	Speed_up, speed_down, cruise	4 (corresponding to four states)
Speed_obs POMDP	Obstacle moving towards robot and robot moving fast, obstacle moving away from robot and robot moving fast, robot speedily rotating about itself, obstacle moving towards robot and robot moving slow, obstacle moving away from robot and robot moving slow, robot slowly rotating about itself	Speed_up, speed_down, cruise	6 (corresponding to six states)

19.8.5.1.3   Speed_ort POMDP

If the robot is moving fast towards the obstacle or if the static obstacle is in the path of the robot, then this POMDP is responsible for speed reduction. If the obstacle is not on the path or moving away from the robot, this POMDP chooses the action which gives the highest reward, that is, speed-up. If there is no obstacle on the path, but there is still an obstacle present in the robot vicinity, then the POMDP chooses the action 'Cruise'.

The three smaller POMDP model components are shown in Table 19.11.

It can be noted that the actions chosen by the last two POMDP may differ. It is also important to note that the actions chosen when the system focus is safety might be different when the system focus is task achievement. These two objectives might be in opposition and may produce different combined actions for the same situation. This is demonstrated in Table 19.12.

When an object is detected by the object detection module, the object detection module decides the type of the object (i.e., if it is moving towards/away from the robot, if it is static on the robot path or not on the robot path) the OD sends the data for all the objects to the TC module. The TC module

TABLE 19.12

Combined Action Decision Based on the Actions Chosen for Speed_obs and Speed_ort POMDPs

Speed_obs Action Decision	Speed_ort Action Decision	Safety Combined Action Decision	Task Achievement Combined Action Decision
Speed_up	Speed_up	Speed_up	Speed_up
Speed_up	Speed_down	Speed_down	Speed_up
Speed_up	Cruise	Cruise	Speed_down
Speed_down	Speed_up	Speed_down	Speed_up
Speed_down	Speed_down	Speed_down	Speed_down
Speed_down	Cruise	Speed_down	Cruise
Cruise	Speed_up	Speed_down	Cruise
Cruise	Speed_down	Speed_down	Cruise
Cruise	Cruise	Cruise	Cruise

computes the belief state for each of the obstacles and generates a separate POMDP model for each of these obstacles. If the POMDP decision is such as to 'Proceed', the obstacle is neglected. If the decision is 'Wait', the data about that particular object is sent to the NM module which updates the map dynamically and recalculates the next way-point. When the TC receives the next way-point, another two smaller POMDPs are computed. These two POMDPs are for deciding the speed of the robot in the next state. These two POMDPs are dependent upon the relationship between the robot and the obstacles around it. The POMDP computational process model is shown in Figure 19.24.

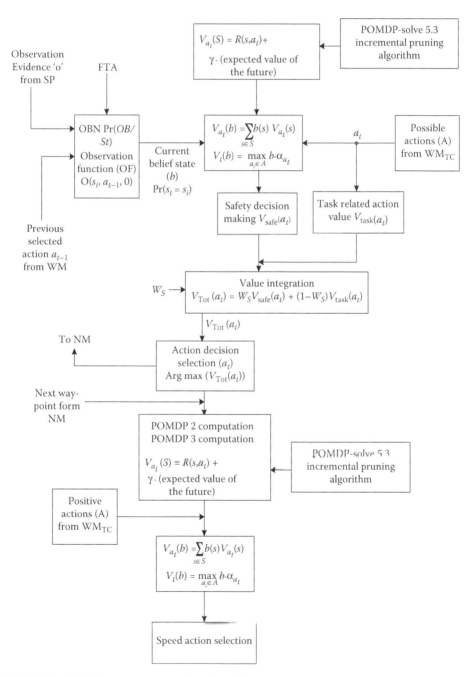

FIGURE 19.24 POMDP process computational model.

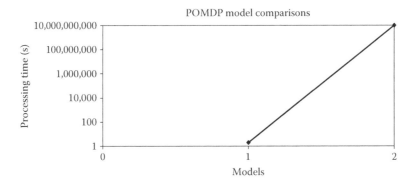

FIGURE 19.25   Normal POMDP and split POMDP comparison.

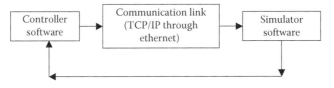

FIGURE 19.26   Experiment setup.

The POMDP process, as carried out in the travel controller module, is shown diagrammatically in Figure 19.25. The concept of split POMDP for safety DM is as shown in Figure 19.26. The three small POMDPs are solved in a particular sequence (as shown in Figure 19.26). This group of smaller POMDPs are solved individually by the process explained in Section 19.7. The action decisions are chosen based on the above-mentioned process. The objective of the system can be balanced as explained in Section 19.7. The experimental results are shown in the next section.

19.9   TESTING AND EVALUATION

This section focusses on the testing and evaluation of the development of safety decision-making process within the RCS architectural framework discussed so far. The two main areas of testing are

1. Evaluating POMDP model development
2. Evaluating the entire approach for safe and effective decision making in the presence of uncertainty for path planning and collision avoidance

19.9.1   TEST RESULTS AND EVALUATION OF THE POMDP MODEL DEVELOPMENT

Evaluating the POMDP model development consists of different sets of tests. The test set-up for all the following tests consisted only of a computer with Linux operating system which is used as a controller station.

The following objectives were considered from a test and evaluation point of view:

• Evaluate the split POMDP method for reducing the computational complexity
• Demonstration of obstacle avoidance and effective navigation of a mobile robot case study

Please note that for all the tests or experiments presented in the next few sections, unless otherwise stated, the POMDP model is for obstacle status problem. Please note that 'obstacle status' is considered as single-state variable, which is associated with five different states. The actions corresponding to this POMDP model are A0—Wait and A1—Proceed.

19.9.1.1 Split POMDP Model Evaluation

Splitting the given POMDP model into a number of smaller POMDPs makes the problem simpler, more understandable and easier to compute. In a split POMDP, a given POMDP task is divided into smaller subtasks and for each subtask, a smaller POMDP problem is developed. After solving these smaller POMDPs individually, a combined action decision is chosen by the agent as the ultimate action decision in that time step.

For an autonomous mobile robot, POMDP is employed for safety decision making for navigating in uncertain environments. POMDP is responsible for taking into consideration the obstacle data and safely command the robot to navigate. The OBN is used to find the belief state of the obstacle status variable which is associated with five states (S1–S5).

The bigger POMDP problem consisting of a huge number of states proved extremely difficult to compute. The memory and processing power requirements for such a big POMDP model are very high, an unrealistic criteria for the hardware on which these techniques are most likely to be deployed in the real world. For a large POMDP, it also takes a very long time (in terms of days) to find the solution. The split POMDP approach, as the name suggests, splits such a big POMDP into smaller POMDPs. These smaller POMDPs are then solved separately in a particular sequence as discussed in Section 19.7.

Primary evidence regarding the effectiveness of this approach can be obtained by comparing the processing time required for computing the bigger POMDP model solution and the smaller POMDPs derived after application of the split POMDP approach.

The experiment shows that for split POMDPs, the processing time reduces significantly. The results show that the split POMDP approach reduces the processing time and computation power significantly as compared to the big single POMDP. As the POMDP size goes on increasing, the time and computation power escalates exponentially. The processing time required by the controller for calculating one complete iteration of solving three split POMDPs (model 1) and then finding the combined action decisions is approximately 200 ms. On the other hand, for a bigger POMDP (model 2—consisting of 40 states, 40 observations (obs), 24 actions), the current processor could not compute the solution owing to the large size of the model for 3 days from start. Later, the process was terminated manually without completion of computation. Figure 19.25 shows the comparison of processing times for these two models.

19.9.2 Results and Evaluation of the Navigation and Collision Avoidance: Safety Decision-Making Process Evaluation

Having assessed the split POMDP method, the integration of this model development as a whole within the RCS framework for autonomous safety decision-making process can be evaluated.

Such an evaluation is based on a set of number of simulation runs of the simulated mobile robot, where the robot is exposed to the risk of collision accident while performing a travelling task. These tests also take into consideration the capability of the mobile robot for path planning and navigation in the presence of obstacles. Various types of dynamic and static obstacles are considered. The experimental setup consists of a controller and a simulator connected through a TCP/IP communication link as shown in the schematic of Figure 19.26 and the photograph of Figure 19.27a.

In the simulator programme, the sensory behaviour of bumper switches, laser scanners, tachometers, compass and so on is simulated [93] and these sensors produce sensory information based on the interaction between the simulated environment and robot. Figure 19.27b shows the simulator software interface. The laser scanners are assumed to detect the obstacles within 15 m and the terrain slopes and hence the laser scanner's tilt angles are not considered. The controller software is developed in C++ on Linux (Debian) OS [94,95].

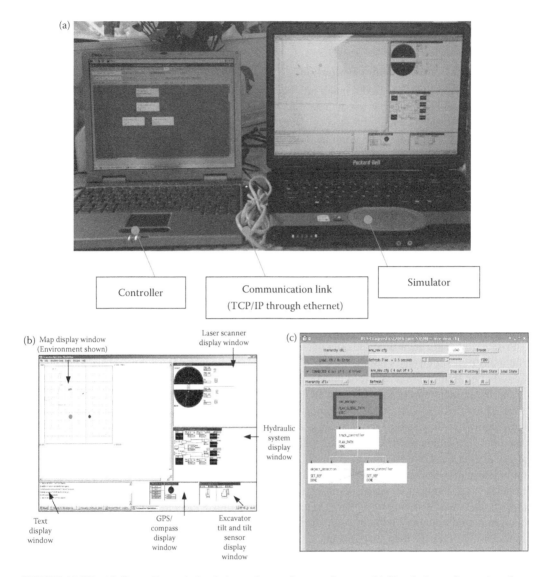

FIGURE 19.27 (a) Controller and simulation software for experiments. (b) Simulation software interface. (c) Controller software interface.

Figure 19.27c shows the controller interface using the RCS diagnostic tool [25]. Controller and simulator software run on two separate computers and operating systems (as shown in Figure 19.27a, b and c). The following tests were carried out to test the validity of the controller development. They demonstrate the effectiveness of the safety decision-making process during navigation, as discussed so far. With this, they also confirm the effectiveness of the obstacle-detection algorithm and path-planning algorithm. This evaluation also suggests some interesting points for future development.

For all the tests carried out for this evaluation, the environment is simulated by the simulator software. The map area is 100×100 m. The data were carefully recorded and plotted in each simulation run. In each time step, the position of the robot (x, y coordinate) is recorded and then plotted against each other. The robot (and sensory behaviour) is also simulated using the simulator software. The sensory range of detection for sensor (15 m) maps onto the grid (100×100 m).

The term 'processing time' refers to the time taken by the controller to process the data received from the simulator, compute the next action decision and send control signals to the simulator from the start of the operation until the robot reaches the goal. The term 'travelling time' is the superset of processing time. The travelling time consists of the processing time plus the time required by the simulated robot to move from start to goal. Although in the simulated system tested here, the difference between the travelling time and processing time is not significant, in a real robotic system, this difference will be considerable. The mechanical factors such as mechanical friction, roughness of ground, time required for the robot tracks to move and so on will add to the travelling time significantly.

19.9.2.1 Test 1

Aim: To test the effect of dynamic and static obstacles on robot behaviour.

This test examines how the robot behaves in the presence of static and moving obstacles. The robot is kept at the same starting position. The static and dynamic obstacles are present at the same position between the robot start and goal position. The goal position is at (50, 50) at the centre of the map (100 × 100 m). The travelling time and paths travelled in the presence of dynamic and corresponding static obstacles are considered. The obstacles are simulated by the simulator software. Dynamic obstacles available using the simulator are truck or person. The speeds and path of movement of dynamic obstacles is specified using the simulator. The static obstacles available using simulator are tree, fence, pole, mound and trench. The dimensions of these obstacles are specified using the simulator. To compare the behaviour of the robot in the presence of the dynamic (d) and static (s) obstacle, the following four different scenarios are considered:

1. A truck (d) or a trench (s) at (32,80), robot start position (30,84) and orientation 128°
2. A person (d) or a pole (s) at (68,69), robot start position (70,80) and orientation 0°
3. A person (d) or a fence (s) at (25,52), robot start position (45,14) and orientation 90°
4. A truck (d) or a tree (s) at (58,33), robot start position (65,28) and orientation 308°

As the starting position and the orientation of robot is same for both static and dynamic obstacle cases, a comparison can be carried out for each of the above scenario. Figure 19.28a and b shows the paths followed by the robot in the presence of dynamic and corresponding static obstacle from various start positions.

Table 19.13 shows the comparative travelling times for a particular starting position and orientation in the presence of a static or a dynamic obstacle. It can be noted that although the trench and a mound are considered as ground obstacles, they are not considered as collision obstacles and hence the robot can cross the trench or pass over the mound.

Conclusion: As expected, the travelling time required for the robot in the presence of the dynamic obstacle is higher than the corresponding static obstacle for the same starting position and orientation. This is due to the excess time the robot needs for safety decision making in the presence of continuously moving obstacles in the surrounding area.

19.9.2.2 Test 2

Aim: To check how the robot behaviour gets affected in the presence of increasing number of dynamic obstacles.

The test setup is the same as for the above tests. The robot is kept at the starting position (80,80). Initially, no obstacle is present. The obstacle number is then increased by one for every run up to the maximum of five obstacles. Every obstacle marked in the grid in Figure 19.28 is given an obstacle number corresponding to the test run.

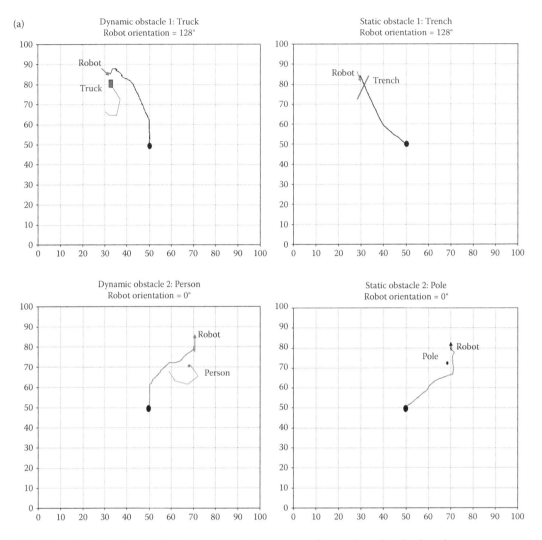

FIGURE 19.28   (a, b) Paths followed by robot in the presence of dynamic and static obstacle.

In each test run, all the obstacles present in the previous test runs are also present. Each mobile obstacle followed the same path in each test run with a pre-decided speed assigned at the environment creation stage. When the mobile obstacle finishes its one cycle from start to end, it toggles back to the initial starting position and follows the same path again. This process continues until the robot reaches the goal. All the mobile objects move into the view as the robot and the obstacle move.

Conclusion: The travelling time increases as the number of dynamic obstacles goes on increasing. Interestingly, the paths followed by the robot in the presence of increasing number of dynamic obstacles are not exactly as expected. It is important to note that the expected paths would be longer as the number of obstacles increases, but as is evident from the paths shown, it is not the case. As the number of obstacles goes on increasing, the robot may not necessarily follow a longer path. It can be seen (Figure 19.29) that the path followed by the robot in the presence of three obstacles is longer than the path followed by the robot in the presence of four or five obstacles. At a first glance, this may seem unexpected behaviour, but when the corresponding action decisions for the robot simulation runs for three and four obstacles are compared, it became evident that the action decisions required for the robot to take in the presence of four obstacles (1042 iteration cycles) are far

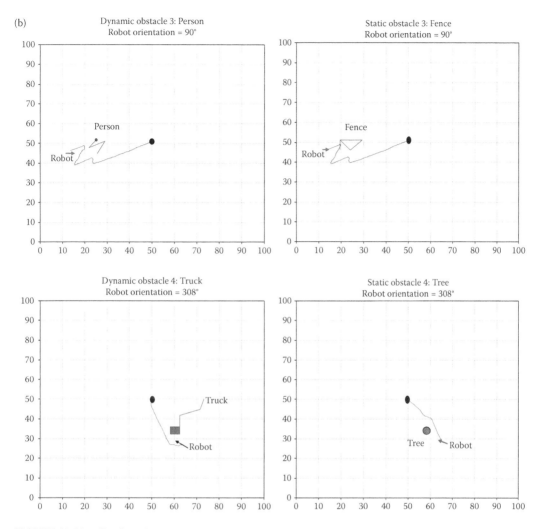

FIGURE 19.28 Continued.

TABLE 19.13

Comparison between Travelling Time for Various Starting Positions in the Presence of Static or Dynamic Obstacle

Start Position	Type of Obstacle	Processing Time
(30,84)	Dynamic (truck)	3 min 17 s
	Static (trench)	2 min 28 s
(70,80)	Dynamic (person)	3 min 50 s
	Static (pole)	2 min 3 s
(45,14)	Dynamic (person)	5 min 24 s
	Static (fence)	3 min
(65,28)	Dynamic (truck)	2 min 5 s
	Static (tree)	1 min 27 s

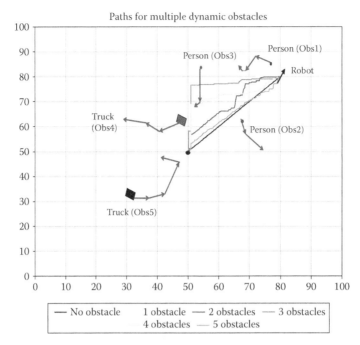

FIGURE 19.29 Paths followed by robot in the presence of multiple dynamic obstacles.

more than the action decisions required in the case of three obstacles (278 iteration cycles). One iteration cycle corresponds to one action decision. The robot becomes more cautious in the presence of a higher number of obstacles and needs more time to decide about what action to choose next. The path in the presence of five obstacles is straighter than the paths followed in the presence of three or four obstacles. This is because of the fact that as the time taken by robot in the presence of five obstacles is significantly more, the robot moves slowly and it sees the obstacles at different positions than the previous runs. As shown in Figure 19.30, when more obstacles are present in the surrounding area, the robot takes a slightly zig-zag path, taking very small but noticeable deviations from the straight line path. Also, the number of action decisions it needs to reach the goal are more and it becomes slower taking more time deciding on the best course of action at each time step. On the other hand, when lesser obstacles are present, the path becomes straighter and lesser action decisions are required. The travelling times and number of action decisions for each of these runs demonstrate this more accurately (Table 19.14).

19.9.2.3 Test 3

Aim: To test the effects of change in Weighting Factor on the robot behaviour.

The test setup is as for previous tests. The weighting factor is introduced in order to balance the objectives of task achievement and safety. The following equation is utilised for the purpose of calculating the value function (Equation 19.23): $V_{\text{Tot}}(a_t) = W_S V_{\text{Safe}}(a_t) + (1 - W_S)V_{\text{Task}}(a_t)$.

For this test, the experimental design consisted of varying the weighting factor keeping the start position, goal position, start orientation, obstacle positions and so on constant. For this test, the following two different conditions are considered:

- In the presence of static obstacles labelled Obs1 and Obs2 shown in the map (Figure 19.30), the robot is kept at start position $\cong (11,32)$.
- In the presence of a dynamic obstacle labelled Obs3 (as shown in Figure 19.31), the robot is kept at the start position $\cong (70,80)$.

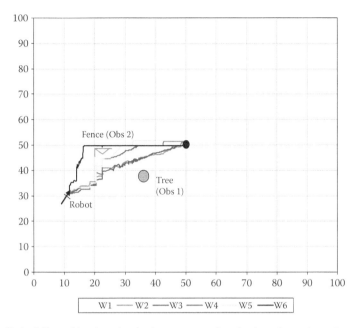

FIGURE 19.30 Paths followed by the robot in the presence of static obstacles and varying weighting factor.

The goal position is at (50, 50). The weighting factor is varied from 0 to 1 in increments of 0.2. A high weighting factor produces a more cautious robot. The paths followed by the robot for each weighting factor setting are observed. They are as shown in Figure 19.30 for static obstacles and as shown in Figure 19.31 for the dynamic obstacle.

Conclusion: As the weighting factor increases, the system becomes more safety conscious and less task oriented. For weighting factor 0, the system goal is entirely task oriented as demonstrated by Equation 19.23. As can be seen from this equation, when the weighting factor is shifted towards 1, the safety-related rewards go on increasing and the system becomes more safe behaviour oriented. As expected, increase in safety means reduction in task achievement effectiveness. The actions which lead to safer behaviour are generally less task effective (as in this case, the action 'Wait' which is chosen when the system senses an obstacle close by inevitably makes the system behaviour slow and increases the decision-making time and in effect also the travelling

TABLE 19.14

Travelling Time Goes on Increasing in the Presence of Increasing Number of Dynamic Obstacles

Number of Obstacles	Travelling Time	Action Decision Cycles
0	40 s	156
1	1 min 58 s	192
2	2 min 20 s	215
3	2 min 47 s	278
4	3 min	1042
5	3 min 11 s	2012

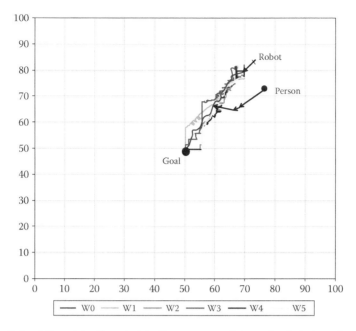

FIGURE 19.31   Paths followed by the robot in the presence of dynamic obstacle and varying weighting factor.

time). This can be evident from Table 19.15. It is important to note that the travelling time for the robot in the presence of static obstacles is higher than the travelling time for the same weighting factor in the presence of the dynamic obstacle. This is due to the fact that the starting positions and the distance required to be travelled are different for both the cases. Also, there are two static obstacles and only one dynamic obstacle which make the process of decision making slightly easier in the latter case.

For both sets of tests, exactly identical operating conditions such as starting position, obstacle position, distance and so on are deliberately not considered. This is because the earlier tests prove that for the same starting position, the robot needs higher travelling time in the presence of dynamic obstacle than static obstacle. Deliberately random starting positions and obstacle positions are considered in this test to examine the robot behaviour and observe the effectiveness of the development in various random operating conditions.

TABLE 19.15
Weighting Factor versus Travelling Time

Weighting Factor	Travelling Time in Presence of Dynamic Obstacle	Travelling Time in Presence of Static Obstacle
0.0	3 min 28 s	2 min 12 s
0.2	4 min 15 s	3 min 3 s
0.4	4 min 32 s	3 min 43 s
0.6	4 min 41 s	3 min 49 s
0.8	4 min 49 s	3 min 59 s
1	5 min 12 s	4 min 15 s

19.10   CONCLUSION AND FUTURE WORK

The following conclusions can be drawn from the results:

1. The RCS framework used for the entire controller design and implementation is effective in contributing to safety at all hierarchical levels. POMDP model maps effectively within the RCS node and POMDP process can be developed within an RCS node. The combination of using a highly powerful probabilistic reasoning model (POMDP) within a flexible architecture such as RCS improves the robustness of the architecture while at the same time providing a powerful method to deal with partial observability.
2. Computing the POMDP solution offline and beforehand proves to be an easier and far simpler method of POMDP solving than solving it in real time. The agent only has to calculate the belief state in real time which saves significant amount of processing time.
3. The split POMDP method reduces the computational processing time to a large extent and facilitates effective decision making by finding the combined action decision in reasonable time.
4. The integration of safety decision making within the architectural framework provides the means for safety-conscious behaviour generation. As the weighting factor increases, the focus of the system varies from task oriented to safety oriented, thus making the system more risk-averse and cautious.

The above aspects give rise to a system that not only tries to avoid the risk of collision by steering away from the obstacles but also tries to assess the environment for the path that offers least uncertainty and resistance.

In this way, system safety is integrated and improved while making certain that the system effectively completes the task set in front of it. The effectiveness of the development is judged primarily by the completion of task and by considering the time required to complete the task successfully.

Suggestions for future work:

1. The first and foremost important step forward is implementing/porting the work in this thesis onto real hardware and testing it by deploying it first in a controlled environment and then in the field, for example, for a specific application of a mobile robot, such as a robot for intelligent handling of unexploded shells. Work is currently underway to port the controller software to the Linux-based single-board computer which can be used on an actual robot.
2. Demonstration as to how the architecture can be used as part of a formal safety case, that is, as a way of proving that the robot is safe.

REFERENCES

1. Mine Safety and Health Administration, http://www.msha.gov/MSHAINFO/FactSheets/MSHAFCT2.HTM [cited August 2007].
2. Labour, U.M.O. http://www.msha.gov/stats/charts/allstatesnew.asp
3. Coal mining: Most deadly job in China Zhao Xiaohui & Jiang Xueli, Xinhua News Agency, Updated: 2004-11-13 15:01 [cited August 2007].
4. Watts, S., *BBC Newsnight*, http://news.bbc.co.uk/2/hi/programmes/newsnight/4330469.stm [cited August 2007].
5. Blackeye, D., *BBC- H2G2*, http://www.bbc.co.uk/dna/h2g2/A2922103 [cited August 2007].
6. BBC News, http://news.bbc.co.uk/1/hi/world/asia-pacific/6952519.stm [cited August 2007].
7. Simmons R. and Coste-Maniere E. Architecture, the Backbone of Robotic Systems, in IEEE International Conference on Robotics & Automation, San Francisco, 2000.
8. Medeiros A.A.D. A survey of control architectures for autonomous mobile robots, *Journal of the Brazilian Computer Society*, 4(3), 1998.

9. Arkin R.C. *Behaviour-Based Robotics*, MIT Press, Cambridge, MA, 1998.

10. Sekiguchi M. Nagata S. and Asakawa K. Behavior control for a mobile robot by a structured neural network, *Advanced Robotics*, 1992. 6(2): 215–230.

11. Aleksander I. *Neural Computing Architectures: The Design of Brain-like Machines*, MIT Press, Cambridge, MA, 1989.

12. Klein H., *General & Strategic Management Department, Temple University*, http://www.howhy.com/ucs2006/Abstracts/Klein.html [cited March 2008].

13. University of York, Department of Electronics, http://www.elec.york.ac.uk/intsys/projects/inspired.html [cited March 2008].

14. Darpa neural targeting, https://kat021zen.wordpress.com/2008/02/20/i-have-nothing-better-to-do/ [cited March 2008].

15. Ortega C. and Tyrrell A. Biologically inspired fault-tolerant architectures for real-time control applications, *Control Engineering Practice*, 1999. 7(5): 673–678.

16. Albus J.S. *Brains, Behaviour and Robotics*, Byte Books, Peterborough, NH, 1981.

17. Albus J.S. and Meystel A.M. A reference model architecture for design and implementation of intelligent control in large and complex systems, *International Journal of Intelligent Control and Systems*, 1996. 1(1): 15–30.

18. Brooks R.A. A robust layered control system for a mobile robot, *IEEE Journal of Robotics and Automation*, 1986. (2): 14–23.

19. Brooks R.A. *Planning Is Just a Way of Avoiding Figuring Out What to Do Next*, MIT Artificial Intelligence Laboratory, USA, 1987.

20. Rasmussen J. Skills, rules, knowledge; signals, signs, and symbols, and other distinctions in human performance models, *IEEE Transactions on Systems, Man and Cybernetics*, 1983. 13: 257–266.

21. Rasmussen J. The role of hierarchical knowledge representation in decision making and system management, *IEEE Transactions on Systems, Man and Cybernetics*, 1985. 15: 234–243.

22. Arkin R.C. Riseman E.M. and Hanson A.R. AuRA: An architecture for vision-based robot navigation, in DARPA Image Understanding Workshop, 1987.

23. Arkin R.C. and Balch T.R. AuRA: Principles and practice in review, *Journal of Experimental & Theoretical Artificial Intelligence*, 1997. 9: 175–189.

24. Orebick A. and Lindstrsm M. BERRA: A research architecture for service robots, in IEEE International Conference on Robotics & Automation, San Francisco, 2000.

25. Gazi V. et al. *The RCS Handbook—Tools for Real-Time Control Systems Software Development*, in Wiley Series on Intelligent Systems, ed. J.S. Albus, A.M. Meystel, and L.A. Zadeh, John Wiley and Sons, New York, 2001.

26. National Institute of Standards and Technology, http://www.isd.mel.nist.gov/projects/rcs/ref_model/TOF.htm [cited January 2008].

27. Albus J.S. Outline for a theory of intelligence, *IEEE Transactions on Systems, Man and Cybernetics*, 1991. 21(3).

28. Saridis G.N. Intelligent robotic control, *IEEE Transactions on Automatic Control*, 1983. 28(5): 547–557.

29. Albus J.S. and Meystel A.M. *Engineering of Mind: An Introduction to the Science of Intelligent Systems*, ed. J.S. Albus, A.M. Meystel, and L.A. Zadeh, John Wiley and Sons, New York, 2001.

30. Koestler A. *The Ghost in the Machine*, Penguin Group, London, England, 1967 (1990 reprint edition).

31. Edwards M. http://www.integralworld.net/edwards13.html [cited January 2008].

32. Agate R. et al. Control architecture characteristics for intelligence in autonomous mobile construction robots, in 23rd International Symposium on Automation and Robotics in Construction, ISARC2006, Japan, 2006.

33. Dhillon B. Fashandi A.R.M. and Liu K.L. Robot systems reliability and safety: A review, *Quality in Maintenance Engineering*, 2002. 8(3): 170–212.

34. Pace C. *Autonomous Safety Management for Mobile Robots*, Lancaster University, Lancaster, 2004.

35. Division, T.I.W. *Robot Safety*, Department of Labour, 1987.

36. Dhillon B. and Flynn A. Safety and reliability assessment techniques in robotics, *Robotica (Cambridge University Press)*, 1997. 15: 701–708.

37. Ramirez C.A. ed. *Safety of Robot International Encyclopedia of Robotics Applications and Automation*, ed. R.C. Dorf. John Wiley and Sons, New York, 1988.

38. Robotics online, http://www.roboticsonline.com/public/articles/archivedetails.cfm?id=1574 [cited February 2008].

39. Graham J.A.E. *Safety, Reliability and Human Factors in Robotic Systems*, Van Nortsand Reinhold, New York, 1991.

40. Dhillon B. *Robot Reliability and Safety*, Springer Verlag, New York, 1991.
41. Bradley D. Seward D.W. and Margrave F. Hazard analysis techniques for mobile construction robots, in 11th International Symposium on Robotics in Construction (ISARC), Brighton, UK, 1994.
42. Moravec H. Autonomous Mobile Robots Annual Report, Mobile Robot Laboratory, The Robotics Institute, Carnegie Mellon University, 1985.
43. Agate R.Y. Seward D.W. and Pace C.M. Emotions modelling for safe behaviour generation in robotic systems, in Safety and Reliability Conference ESREL2006, Portugal, 2006.
44. Seward D.W. et al. Safety Analysis of Autonomous Excavator Functionality, *in Proceedings of Reliability Engineering and Systems Safety,* 2000. 70: 29–39.
45. Seward D.W. et al. Developing the safety case for large mobile robots, in International Conference on Safety and Reliability, Lisbon, 1997.
46. Pace C. *Development of a Safety Manager for an Autonomous Mobile Robot*, Lancaster University, Lancaster, 1997.
47. Pearl J. *Probabilistic Reasoning in Intelligent Systems*, Morgan Kaufman, San Francisco, CA, 1998.
48. Joyce J.M. *The Foundations of Causal Decision Theory*, Cambridge University Press, UK, 1999.
49. Bentham J. *An Introduction to the Principles of Morals and Legislation, in Utilitarianism*, Fontana, 1789.
50. Sondik E. *The Optimal Control of Partially Observable Markov Decision Processes*, Stanford University, 1971.
51. Bertsekas D.P. *Dynamic Programming and Optimal Control*, Athena Scientific: Optimization and Computation, USA, Vol. 1. 2001.
52. Bertsekas D.P. *Dynamic Programming and Optimal Control*, Athena Scientific: Optimization and Computation, USA, Vol. 2. 2001.
53. Thrun S. Burgard W. and Fox D. A probabilistic approach to concurrent mapping and localisation for mobile robots, *Machine Learning*, 1998. 31: 29–53.
54. Fox D. Burgard W. and Thrun S. Active Markov localisation for mobile robots, *Robotics and Autonomous Systems*, 1998. 25: 195–207.
55. Olson C.F. Probabilistic self-localisation for mobile robots, *IEEE Transactions on Robotics and Automation*, 2000. 16(1): 55–66.
56. Simmons R. and Koenig S. Probabilistic navigation in partially observable environments, in International Joint Conference on Artificial Intelligence (IJCAI–95), 1995, pp. 1080–1087.
57. Rose C. Smaili C. and Charpillet F. A dynamic Bayesian network for handling uncertainty in a decision support system adapted to the monitoring of patients treated by hemodialysis, in IEEE International Conference on Tools with Artificial Intelligence (ICTAI 05), Hongkong, China, 2005.
58. Moghadasi M.N. Haghighat A.T. and Ghidary S.S. Evaluating Markov decision process as a model for decision making under uncertainty environment, in International Conference on Machine Learning and Cybernetics, Oregon State University, Corvallis, USA, 2007.
59. Roumeliotis S.I. Sukhatme G.S. and Bekey G.A. Circumnavigating dynamic modelling: Evaluation of the error-state Kalman filter applied to mobile robot localization, in IEEE International Conference on Robotics and Automation (ICRA1999), Detroit, 1999.
60. Roumeliotis S.I. and Bekey G.A. Bayesian estimation and Kalman filtering: A unified framework for mobile robot localisation in IEEE International Conference on Robotics and Automation (ICRA2000), San Francisco, 2000.
61. Cox I.J. and Leonard J.J. Modelling a dynamic environment using a Bayesian multiple hypothesis approach, *Artificial Intelligence*, 1994. 66: 311–344.
62. Goel P. et al. Fault detection and identification in a mobile robot using multiple model estimation and neural network, in IEEE International Conference on Robotics and Automation (ICRA2000), San Francisco, 2000.
63. Nehzmow U. *Mobile Robotics: A Practical Introduction*, Springer Verlag, London, 2000.
64. Zurada J. Wright A.L. and Graham J.H. A neuro-fuzzy approach for robot system safety, *IEEE Transactions on Systems, Man and Cybernetics—Part C: Applications and Reviews*, 2001. 31(1): 49–64.
65. Roy N. et al. Coastal navigation: Robot navigation under uncertainty in dynamic environments, in IEEE International Conference on Robotics and Automation (ICRA), 1999.
66. Thrun S. et al. *Simultaneous Mapping and Localization with Sparse Extended Information Filters: Theory and Initial Results*, CMU-CS-01–112, 2002.
67. Simmons R. and Koenig S. *Xavier: A Robot Navigation Architecture Based on Partially Observable Markov Decision Process Models*, Carnegie Mellon University, School of Computer Science, 1998.
68. Pineau J. and Thrun S. Hierarchical POMDP decomposition for a conversational robot, in Workshop on Hierarchy and Memory in Reinforcement Learning (ICML), Williams College, MA, USA, 2001.

69. Bradley D. and Seward D. Developing real-time autonomous excavation—The LUCIE story, in 34th IEEE Conference on Decision and Control, New Orleans, 1995.

70. Bradley D. and Seward D.W. The development, control and operation of an autonomous robotic excavator, *Springer Journal of Intelligent and Robotic Systems*, 1998. 21(1):73–97.

71. Latombe J.C. *Robot Motion Planning*, Vol. 124, Kluwer International Series in Engineering and Computer Science, Boston, USA, 2000.

72. Brady M. *Robot Motion: Planning and Control*, ed. Brady M. et al. MIT Press, Cambridge, MA, 1982.

73. Balch T.R. Grid-based navigation for mobile robots, *The Robotics Practitioner*, 1996. 2(1): 7–10.

74. Kaelbling L.P. Littman M.P. and Cassandra A.R. Planning and acting in partially observable stochastic domains, *Artificial Intelligence*, 1998. 101: 99–134.

75. Seward D.W. Pace C. and Agate R.Y. *Safe and Effective Navigation of Autonomous Robots in Hazardous Environments*, Springer—Construction Robotics—Special Issue, 2006.

76. Rankin A. Huertas A. and Matthies. L. *Evaluation of Stereo Vision Obstacle Detection Algorithms for Off-Road Autonomous Navigation*, Jet Propulsion Laboratory, 2005.

77. Tsai-Hong Hong. Legowik S. and Nashman M. *Obstacle Detection and Mapping System*, Intelligent Systems Division, National Institute of Standards and Technology (NIST), 1998.

78. Badal S. et al. A practical obstacle detection and avoidance system, in Proceedings of the Second IEEE Workshop on Applications of Computer Vision, Sarasota, Florida, 1994.

79. Soumare S. Ohya A. and Yuta S. Real-time obstacle avoidance by an autonomous mobile robot using an active vision sensor and a vertically emitted laser slit, in The 7th International Conference on Intelligent Autonomous Systems (IAS-7), USA, 2002.

80. Karuppuswamy J. *Detection and Avoidance of Simulated Potholes in Autonomous Vehicle Navigation in an Unstructured Environment*, University of Cincinnati, 2000.

81. Borenstein J. and Koren Y. Tele-autonomous obstacle avoidance, http://www-personal.umich.edu/~johannb/teleauto.htm, January 2006 [cited January 2008].

82. Albus J.S. Personal Email, R.Y. Agate, Editor, 2005.

83. Agate R.Y. *Annual Review 2004-05*, Lancaster University, Lancaster, 2005.

84. Monahan G.E. A survey of POMDP: Theory, models and algorithms, *Management Science*, 1982. 28(1): 1–16.

85. Cheng H.T. *Algorithms for Partially Observable Markov Decision Processes*, University of British Columbia, Canada, 1988.

86. Cassandra A.R. Littman M.L. and Zhang N.L. Incremental pruning: A simple, fast, exact method for partially observable Markov decision processes, in Proceedings of Uncertainty in Artificial Intelligence (UAI), Rhode Island, USA, 1997.

87. Littman M.L. The witness algorithm: Solving partially observable Markov decision processes, in *Technical Report CS-94-40*, Brown University, 1994.

88. Cassandra A.R. http://www.pomdp.org/pomdp/code/index.shtml. 2003 [cited 2006].

89. Russell S. and Parr R. Approximating optimal policies for partially observable stochastic domains, in *IJCAI*, Morgan Kouffmann, Quebec, 1995.

90. Parr R. and Russell S. *Reinforcement Learning with Hierarchies of Machines in Advances in Neural Information Processing Systems: Proceedings of the 1997 Conference*, MIT Press, Cambridge, MA, 1998.

91. Pineau J. Roy N. and Thrun S. A hierarchical approach to POMDP planning and execution, in Workshop on Hierarchy and Memory in Reinforcement Learning (ICML), Williams College, USA, 2001.

92. Theocharous G. and Mahadevan S. Learning the hierarchical structure of spatial environments using multiresolution statistical models, in IEEE International Conference on Intelligent Robots and Systems, Lausanne, Switzerland, 2002, pp. 1038–1043.

93. Agate R.Y. Safe and Effective Decision Making in Autonomous Mobile Robots, PhD thesis, Lancaster University, Lancaster, 2008.

94. Stevens R. and Rago S.A. *Advanced Programming in the UNIX(R) Environment*, Addison-Wesley Professional Computing Series, USA, 1992.

95. SS64, http://www.ss64.com/bashsyntax/vi.html [cited September 2006].

20 Partially Integrated Guidance and Control of Unmanned Aerial Vehicles for Reactive Obstacle Avoidance

Radhakant Padhi and Charu Chawla

CONTENTS

20.1 INTRODUCTION

Rapid advance of various technologies associated with unmanned aerial vehicles (UAVs) has enabled many complex tasks to be carried out with minimal human intervention. UAVs can be deployed for numerous applications such as reconnaissance and surveillance, targeted attacks with less collateral damage, battle damage assessment, traffic monitoring for crime prevention, detection and containment of hazardous leakages in industries, assessment and rehabilitation in case of natural calamities and so on [1]. However, for successful autonomous missions, it is quite obvious that UAVs should have a good built-in mechanism and associated guidance laws to avoid collisions in its flight path. This is in fact a problem of 'path planning', which typically consists of two layers: (i) a global path planner and (ii) a local path planner. The global planner usually attempts to find a longer and preferably optimal path (usually from offline calculations) such that known obstacles can be avoided and the destination is reached. For local path planning, however, the main aim is to avoid collisions with obstacles in the near vicinity, especially for the pop-up ones which have not been accounted earlier in the global path planner. In such a situation, collision avoidance is the major concern and optimality of the path from the fuel and control minimization point of view is not quite important.

Whenever an onboard sensor detects an obstacle on its flight path in close vicinity, the reactive collision avoidance guidance algorithm attempts to manoeuvre the vehicle away from the danger as soon as possible so that the impending collision is averted. Such a guidance is also called as a 'reactive guidance', since decisions are quickly taken within the available small time-to-go resulting in high manoeuvre of the vehicle for a small duration of time. Another usual restriction on such a guidance law is that it should be computationally quite efficient, preferably available in closed form, since the time to react is typically quite low. Other than rotating wing vehicles (e.g., quad-rotor UAVs), which can reduce their speed with relative ease and even hover if necessary, fixed wing vehicles usually cannot reduce their speed quickly and they must continue to move sufficiently fast to sustain their lift. While avoiding the obstacles, the guidance law should also ensure that the vehicle should not deviate too much from its original intended path either. This is both because other obstacles should not come on its new flight path and it should not deviate too much away from seeking its intended destination. Otherwise, putting it back into its original path becomes a difficult task. The problem of reactive collision avoidance for UAVs and associated innovated solutions has been heavily reported in recent literature. For an extensive list of literature and associated discussions on advantages and drawbacks of various attempted approaches, an interested reader can refer to Ref. [2]. To name a few, the 'artificial potential field method' is a popular approach due to its intuitive nature without compromising on sound mathematical features [3,4]. In this approach, a potential function to be optimized is selected in such a way that obstacles have a repulsive field while the destination has an attractive field. The optimized field leads to a safe direction for the UAV to move along. Even though this algorithm is quite suitable for offline global path planning and has been tried out for limited local path planning as well, it is not quite a well-suited reactive guidance since it relies on an iterative optimization algorithm and hence becomes computationally intensive. Another popular approach is the philosophy of rapidly exploring random tree (RRT), which is a randomized search algorithm having many attractive features including computational efficiency [5]. However, RRT is primarily a heuristic approach, which leads to several concerns. For example, the algorithm usually generates several extraneous branches that are eventually not useful. Moreover, it does not lead to an optimal solution either. Although reactive collision avoidance permits manoeuvres that are non-optimal, the wastage in the path found by RRT (and hence computational wastage) is usually significant. For a comprehensive overview of path planning and collision avoidance guidance philosophies, one can see Ref. [2].

Even though obstacle sensing and an effective reactive collision avoidance guidance is an important feature and must be considered with due importance, one must not forget that the airframe of an UAV reacts only to its control inputs in the form of fin deflections through altering its body rates. Hence the task of control (autopilot) design for the vehicle is also important, for which one must consider the full non-linear six-degree-of-freedom (six-DOF) dynamics and the associated involved physics to design an effective control law. The usual practice in aerospace industry is to first derive a guidance law from the geometric (kinematic) relationship or, at the maximum, from the point mass equation. Next, an autopilot design is carried out which tracks the guidance commands as closely as possible. This essentially operates in a three-loop structure: (i) first, the guidance commands are generated from an outer loop, (ii) next, these guidance commands are transformed to the necessary pitch and yaw rates in an intermediate loop and (iii) the commanded body rates are tracked by generating the necessary fin deflections in the inner loop. Here it is called 'conventional approach'. Unfortunately, the conventional three-loop structure introduces time delays in each loop and hence the overall delay becomes large. Such a structure also introduces large overall transients in general, as the transient of an inner loop acts over the transient of its outer loop. For reactive collision avoidance purpose, obviously these are not welcome features as these may lead to collision before the transients decay down and/or before the errors settle down. Quite interestingly, the low time availability, sharp turning requirement and so on are quite relevant in the terminal phase missile guidance and control as well. Because of these concerns, several interesting 'integrated guidance and control (IGC)' approaches have been proposed in the missile guidance and control literature (see, e.g., [6–8]). The basic philosophy here is to formulate a control design problem from a larger-dimensional state space problem embedded in the six-DOF dynamics directly so that the guidance objective is achieved. Quite obviously, unlike the conventional approach, this philosophy offers several advantages such as minimization of overall settling time, optimality of the overall design and so forth.

Unfortunately, however, the IGC approach does not take into account the inherent time-scale separation property that exists in aerospace vehicles. In general, it tries to ignore the large moment generation capability of the control surfaces as compared to their force generation capability. Since guidance problem is essentially to translate the whole vehicle, it rather attempts to do that directly by altering the states of the system. However, the design tuning must indirectly ensure that the body rates generated by deflection of control surfaces should give meaningful turning the vehicle while achieving the guidance objective of translating the vehicle. Unless the tuning is done with extreme care, it leads to the instability of rotational dynamics in general. In other words, tuning of the design becomes quite difficult and to a large extent becomes initial condition dependent as well, which is highly undesirable. This happens primarily because of the lack of explicit information about the 'desired body rates' in an IGC formulation. To overcome the above difficulty, Pasha et al. [9] have proposed a partial integrated guidance and control (PIGC) philosophy [10] in the context of missile guidance and have explicitly shown that the PIGC performs better than both the conventional as well as the IGC approaches. The PIGC philosophy essentially works in two loops, both of which directly reply on the six-DOF dynamics of the vehicle and hence there is no need of a separate compatible point mass model for the guidance purpose. While retaining the benefits of the IGC, the PIGC approach does not get trapped in its difficulties (including tuning difficulties). In this chapter, inspired by the collision cone-based collision detection and avoidance philosophy [11,12] and the aiming point guidance [13], a PIGC design approach is presented for reactive collision avoidance of UAVs. In this approach, the necessary angular correction commands in horizontal and the vertical planes are first computed from algebraic relationships. Using these angular commands, the necessary body rates are computed from an outer loop using dynamic inversion [14]. Next, the necessary fin deflection commands are computed from an inner loop (again using dynamic inversion) to track these desired body rates. To demonstrate the usefulness of the proposed PIGC approach, numerical experiment has been carried out using the six-DOF model of a fixed wing UAV, which also has actuator models for aerodynamic as well as thrust controls. The details of the vehicle as well as its mathematical model are discussed next.

20.2   SIX-DOF MODEL OF A PROTOTYPE UAV

The six-DOF dynamics of a prototype UAV, named as AE-2 (see Figure 20.1), has been used to demonstrate the PIGC development with numerical experiment. AE-2 is designed and developed at the UAV Lab of Department of Aerospace Engineering, Indian Institute of Science, Bangalore, India. It is a fixed wing airplane designed for long endurance. The thrust generating unit is an electric motor with propeller. It has a pusher configuration of thrust, so that onboard sensors like cameras can be mounted at the nose.

FIGURE 20.1   AE-2 (all electric Airplane-2).

Under the assumption that the airplane is a rigid body as well as the earth to be flat, the set of equations that describe the equation of motion (in body and inertial frames) are given by the following differential equations [15]:

Force equations

$$\dot{U} = RV - QW - g\sin\theta + \frac{1}{m}(X_a + X_t) \tag{20.1}$$

$$\dot{V} = PW - RU + g\sin\phi\cos\theta + \frac{1}{m}(Y_a) \tag{20.2}$$

$$\dot{W} = QU - PV + g\cos\phi\cos\theta + \frac{1}{m}(Z_a) \tag{20.3}$$

Moment equations

$$\dot{P} = c_1 RQ + c_2 PQ + c_3 L_a + c_4 N_a \tag{20.4}$$

$$\dot{Q} = c_5 PR + c_6 (R^2 - P^2) + c_7 (M_a - M_t) \tag{20.5}$$

$$\dot{R} = c_8 PQ - c_2 RQ + c_4 L_a + c_9 N_a \tag{20.6}$$

Kinematic equations

$$\dot{\phi} = P + Q\sin\phi\tan\theta + R\cos\phi\tan\theta \tag{20.7}$$

$$\dot{\theta} = Q\cos\phi - R\sin\phi \tag{20.8}$$

$$\dot{\psi} = Q\sin\phi\sec\theta + R\cos\phi\sec\theta \tag{20.9}$$

Navigation equations

$$\dot{x}_i = U\cos\theta\cos\psi + V(\sin\phi\sin\theta\cos\psi - \cos\phi\sin\psi)$$
$$+ W(\cos\phi\sin\theta\cos\psi + \sin\phi\sin\psi) \tag{20.10}$$

$$\dot{y}_i = U\cos\theta\sin\psi + V(\sin\phi\sin\theta\sin\psi + \cos\phi\cos\psi)$$
$$+ W(\cos\phi\sin\theta\sin\psi - \sin\phi\cos\psi) \tag{20.11}$$

$$\dot{h}_i = U\sin\theta - V\sin\phi\cos\theta - W\cos\phi\cos\theta \tag{20.12}$$

Note that U, V, W can be computed from V_T, α and β in the following manner:

$$U = V_T\cos\alpha\cos\beta$$
$$V = V_T\sin\beta \tag{20.13}$$
$$W = V_T\sin\alpha\cos\beta$$

where $V_T = \sqrt{U^2 + V^2 + W^2}$ is the total velocity and the angle of attack (α) and side slip angle (β) are defined as $\alpha = \tan^{-1}(W/U)$ and $\beta = \sin^{-1}(V/V_T)$. The coefficients c_1–c_9 in Equations 20.4 through 20.6 are function of moment of inertias associated with the UAV in the body axes system and are given as [15]

$$\begin{bmatrix} c_1 \\ c_2 \\ c_3 \\ c_4 \\ c_8 \\ c_9 \end{bmatrix} \triangleq \frac{1}{I_{xx}I_{yy} - I_{xz}^2} \begin{bmatrix} I_{zz}(I_{yy} - I_{zz}) - I_{xz}I_{xz} \\ I_{xz}(I_{xx} - I_{yy} + I_{zz}) \\ I_{zz} \\ I_{xz} \\ I_{xz}I_{xz} + I_{xx}(I_{xx} - I_{yy}) \\ I_{xx} \end{bmatrix}, \quad \begin{bmatrix} c_5 \\ c_6 \\ c_7 \end{bmatrix} \triangleq \frac{1}{I_{yy}} \begin{bmatrix} I_{zz} - I_{xx} \\ I_{xz} \\ 1 \end{bmatrix}$$

The aerodynamic and thrust forces and moments are given by

$$[X_a \quad Y_a \quad Z_a] = \frac{\bar{q}S}{m}[-C_X \quad C_Y \quad -C_Z]$$
$$[L_a \quad M_a \quad N_a] = \bar{q}S[bC_l \quad cC_m \quad bC_n]$$
$$X_t = \frac{1}{m}(T_{max}\,\sigma_t)$$
$$M_t - d\,(T_{max}\,\upsilon_t)$$

The geometric and inertia parameters of AE-2 are given in Table 20.1.

The aerodynamic force and moment coefficients are found from curve fitting on wind tunnel data of the UAV, t_{max} is the maximum thrust value (15 N) that can be produced by the electric motor and

TABLE 20.1

Geometrical and Inertia Parameters of AE-2 UAV

b (m)	c (m)	d (m)	m (kg)	I_{xx} (kg-m²)	I_{yy} (kg-m²)	I_{zz} (kg-m²)	I_{xz} (kg-m²)
2	0.3	0.26	6	0.5062	0.89	0.91	0.0015

propeller assembly and $\sigma_t \in [0,1]$ is the percentage of maximum thrust applied to the system (a control variable). The various aerodynamic force and moment coefficients are given as

$$C_X = C_{X_0} + C_{X_\alpha}(\alpha)\alpha + C_{X_{\delta_e}}(\alpha)\delta_e + C_{X_Q}(\alpha)\bar{Q}$$

$$C_Y = C_{Y_\beta}(\alpha)\beta + C_{Y_{\delta_a}}(\alpha)\delta_a + C_{Y_{\delta_r}}(\alpha)\delta_r + C_{Y_P}(\alpha)\bar{P} + C_{Y_R}(\alpha)\bar{R}$$

$$C_Z = C_{Z_0} + C_{Z_\alpha}(\alpha)\alpha + C_{Z_\beta}\beta + C_{Z_{\delta_e}}\delta_e + C_{Z_Q}(\alpha)\bar{Q}$$

$$C_l = C_{l_\beta}(\alpha)\beta + C_{l_{\delta_a}}(\alpha)\delta_a + C_{l_P}(\alpha)\bar{P} + C_{l_R}(\alpha)\bar{R}$$

$$C_m = C_{m_0} + C_{m_\alpha}(\alpha)\alpha + C_{m_\beta}(\alpha,\beta)\beta + C_{m_{\delta_e}}(\alpha)\delta_e + C_{m_Q}(\alpha)\bar{Q}$$

$$C_n = C_{n_\beta}(\alpha)\beta + C_{n_{\delta_r}}(\alpha)\delta_r + C_{n_P}(\alpha)\bar{P} + C_{n_R}(\alpha)\bar{R}$$

where $[\bar{P} \quad \bar{Q} \quad \bar{R}] = 1/2V_T[bP \quad cQ \quad bR]$. Various static and dynamic derivatives are functions of α and β. They are obtained from curve fitting on wind tunnel data, the details of which are given in the Appendix of this chapter. The models for aerodynamic and thrust actuators are represented by first-order systems, which are given as follows:

$$\dot{\delta} = -9.5\delta + 9.5u_\delta \tag{20.14}$$

$$\dot{\sigma}_t = -4.5\sigma_t + 4.5u_{\sigma_t} \tag{20.15}$$

where Equation 20.14 represent the aerodynamic actuator model for the aerodynamic control surface deflections (u_δ is the aerodynamic actuator input) and Equation 20.15 represent the thrust control actuator model (u_{σ_t} is the thrust actuator input). For more details about the vehicle model, one can refer to Ref. [16]. For the original wind tunnel aerodynamic data, one can refer to Ref. [17].

20.3   AIMING POINT SELECTION FOR OBSTACLE AVOIDANCE

After a pop-up obstacle detected in the close vicinity, the collision avoidance algorithm presented here has two key components: (i) to detect whether the obstacle can be a threat to the UAV and (ii) to steer it sufficiently in case the obstacle is a critical one. For carrying out these two tasks, one can rely on the 'collision cone' approach [11]. In the approach, an artificial safety ball is first put around the obstacle and, if the extended velocity vector enters this safety ball, an appropriate 'aiming point' is selected on the surface of the boundary of the ball towards which the velocity vector is reoriented. Details of these logics are outlined below.

20.3.1   COLLISION CONE CONSTRUCTION

The 'collision cone' is an effective tool for (i) detecting collision and (ii) finding an alternate direction of motion that will avert the collision [11]. The construction of the collision cone is shown in Figure 20.2.

First, a safety ball of radius r is constructed around the obstacle's geometrical centre (it is assumed that the shape of the obstacle is known from appropriate onboard sensor package) with appropriate safety margin from its geometric boarders such that even if the UAV's centre of gravity (CG) touches any of the points on the surface of the ball, it still does not result in any physical collision with the obstacle. Next, from the current position of the UAV (position of its CG, to be exact), a cone is constructed by collecting all tangential lines from the CG of the vehicle to the safety ball,

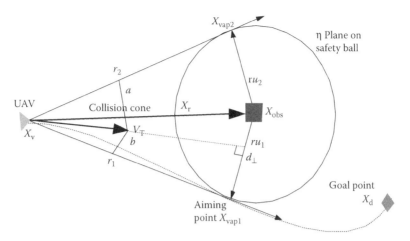

FIGURE 20.2 Collision cone representation.

which is called as the 'collision cone'. Quite obviously, if the velocity vector of the UAV lies within this cone, it is on a collision course, as in that case the vehicle is guaranteed to enter inside the safety ball and in that case the threat is said to be 'critical'. In such critical cases, corrective measures should be taken to divert it away. However, if the vehicle is diverted too much away, it can again come in collision course with other obstacles and, moreover, getting it back to its original intended path will become difficult. Hence, the idea is to divert it just enough so that the vehicle just touches the collision cone and moves forward, thereby successfully avoiding the obstacle.

20.3.2 AIMING POINT COMPUTATION

In critical cases, the 'aiming point' on the surface of the sphere (towards which the velocity vector should be aimed at) is found as follows. From Figure 20.2, it is obvious that the relative distance between the UAV and the obstacle is given by $X_r = X_{obs} - X_v$, where X_{obs} and X_v are the positions of the obstacle and vehicle, respectively, in some plane that is parallel to inertial plane. This non-rotating frame is assumed to be centred at the CG of the vehicle (i.e., $X_v = 0$). Since X_r and V_T share the common origin, one can visualize a plane spanned by these two vectors. This plane, when cuts the safety sphere will generate a 'great circle' as it is guaranteed to contain the centre of the sphere. The geometrical relationships in this 3-D scenario (containing the great circle) are given in Figure 20.3.

In Figure 20.3 (which is fairly self-explanatory), after some algebra, one can derive the following relationships [18]:

$$d_\perp = X_r - \left(\frac{X_r \cdot V_T}{\|V_T\|^2} \right) V_T \tag{20.16}$$

After obtaining a numerical value for $d_\perp$, it is obvious that in case $d_\perp < r$, the velocity vector V_T lies within the collision cone and hence the obstacle is critical. In this case, the velocity vector needs to be diverted away so as to make it parallel to one of the tangent lines to the circle. Obviously one should aim for that particular tangent line which is closer to the velocity vector and hence will demand minimal correction (which will ultimately result in minimum control effort). Towards this objective, one needs to first compute the tangential vectors r_1 and r_2, which can be done as follows.

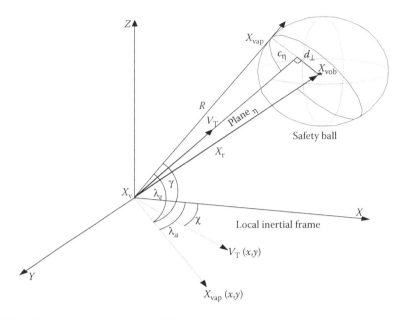

FIGURE 20.3 3D vectorial representation of the guidance logic.

$$r_1 = X_r + ru_1 \tag{20.17}$$

$$r_2 = X_r + ru_2 \tag{20.18}$$

where u_1 and u_2 are the unit vectors of the radius of the ball perpendicular to the tangents. These unit vectors are given by

$$u_1 = -\frac{1}{\|X_r\|^2}(c(X_r \cdot V) + r)X_r + cV$$
$$u_2 = \frac{1}{\|X_r\|^2}(c(X_r \cdot V) - r)X_r - cV \tag{20.19}$$

where c is computed as

$$c = \sqrt{\frac{\|X_r\|^2 - r^2}{\|X_r\|^2 \|V\|^2 - (X_r \cdot V)^2}} \tag{20.20}$$

Next, the components of the velocity vector on r_2 and r_1, that is, a and b are computed as

$$a = \frac{1}{2}\left(\frac{X_r \cdot V_T}{\|X_r\|^2 - r^2} + \frac{1}{cr}\right) \tag{20.21}$$

$$b = \frac{1}{2}\left(\frac{X_r \cdot V_T}{\|X_r\|^2 - r^2} - \frac{1}{cr}\right) \tag{20.22}$$

After computing a and b, if $a > 0$ and $b > 0$, then the obstacle under consideration is critical as it results in safety ball incursion in case the vehicle continues to move in the same direction. This happens only when $d_\perp < r$. Finally, in critical cases, the desired aiming point is determined in the following manner:

$$\text{If } a > b, \quad X_{ap} = X_v + r_1 \tag{20.23}$$

$$\text{If } b > a, \quad X_{ap} = X_v + r_2 \tag{20.24}$$

Note that, when no obstacle is critical, the goal point becomes the aiming point, that is, $X_{ap} = X_g$, where X_g is the goal point co-ordinates. Hence, the guidance problem addressed here has sufficient generality both to avoid obstacles as well as to reach the goal point. Note that in all calculations above wherever required the standard Euclidean (second) norm is used. For more details about the derivation of the above relationships, one can refer to Refs. [11,12].

20.3.3 TIME-TO-GO (t_{go}) COMPUTATION

Some times it is necessary to compute the time-to-go (i.e., time to reach the aiming point or the goal point), t_{go}, atleast in an approximate way. This information can be very useful in several ways, including tuning the guidance loop. This can be done in the following manner. First, the relative distance between the UAV and the aiming point is given by

$$X_{vap} = X_{ap} - X_v \tag{20.25}$$

The X_{vap} vector can be any one of r_1 or r_2 direction depending on the condition specified in Equation 20.23 and in Equation 20.24. Next, assuming that the velocity vector is instantaneously aligned and its magnitude remains constant, t_{go} can be computed as follows:

$$t_{go} = \frac{(X_{vap} \cdot V_T)}{\|V_T\|^2} \tag{20.26}$$

This t_{go} information can be useful in tuning the gains, which are selected in such a way that the error settles down within a fraction of the available t_{go}.

20.4 VELOCITY VECTOR ORIENTATION AND DYNAMICS OF ORIENTATION ANGLES

Before formulating the problem in the partial IGC framework, concepts about velocity vector orientation as well as the dynamics of the orientation angles are also necessary, which are discussed in this section.

20.4.1 CURRENT AND DESIRED ORIENTATION OF THE VELOCITY VECTOR

The first task is to find the current and desired orientation of the velocity vector so that an appropriate guidance problem can be formulated. In Figure 20.2, the XYZ frame is coordinate frame whose origin is at the CG of the vehicle, but orientation of the axes are assumed to be parallel to an inertial frame. For convenience, this coordinate frame can be called as a *virtual inertial frame (VI frame)*, even though it is not an inertial frame. Note that for collision avoidance purpose, the real inertial frame is not required as the information about the target as well as velocity component in the VI frame as well as the information about body attitude (i.e., the body frame) is good enough to compute the required control action. However, for information about the vehicle trajectory, one should also know the real inertial frame as well. Note that onboard accelerometers usually give the velocity component information in the inertial frame. Hence, the components v_x, v_y, v_z of the total velocity vector V_T in the VI frame are known. From this, the the two flight path angles γ and χ that define the orientation of the instantaneous velocity vector (see Figure 20.2), that is, the flight path angle and the heading angle, respectively, can be found from the components of the velocity vector as follows:

$$\gamma = \tan^{-1}\left(\frac{v_z}{\sqrt{v_x^2 + v_y^2}}\right) \tag{20.27}$$

$$\chi = \tan^{-1}\left(\frac{v_y}{v_x}\right) \tag{20.28}$$

Next, it is assumed that the safety envelope (one can say that as 'obstacle position') $X_{obs} = [x_{obs}\, y_{obs}\, z_{obs}]^T$ is available in the VI frame (in reality, this can be obtained after doing the necessary information transformation from sensor frame to VI frame). With this information as well as the information about velocity vector orientation, one can infer whether the obstacle is critical and, if so, the necessary aiming point position $X_{vap} = [x_{vap}\, y_{vap}\, z_{vap}]^T$ can be computed as discussed in this section. With this information, the desired orientation of the velocity vector (which along the X_{vap} vector), that is, the desied flight path angle and the desired heading angle, respectively, can be computed as follows:

$$\gamma^c = \tan^{-1}\left(\frac{z_{vap}}{\sqrt{x_{vap}^2 + y_{vap}^2}}\right) \tag{20.29}$$

$$\chi^c = \tan^{-1}\left(\frac{y_{vap}}{x_{vap}}\right) \tag{20.30}$$

Next, quite logically, the objective of the guidance and control design is to make sure that $\gamma \to \gamma^c$ and $\chi \to \chi^c$ as soon as possible. One can note that here the velocity vector magnitude is assumed to be maintained constant (which is justified because of the small time-to-go), whereas its directions are manipulated to achieve the guidance objective.

20.4.2 Dynamics of Flight Path and Heading Angles

To achieve the objective that $\gamma \to \gamma^c$ and $\chi \to \chi^c$ (which is essentially a tracking problem), one should first have the dynamics of γ and χ. First, the expression for $\dot{\gamma}$ can be derived as follows.

From Figure 20.2, one can notice that the rate of coordinates of the inertial frame can be given as

$$\dot{x}_i = V_T \cos\gamma \cos\chi \tag{20.31}$$

$$\dot{y}_i = V_T \cos\gamma \sin\chi \tag{20.32}$$

$$\dot{h}_i = V_T \sin \gamma \tag{20.33}$$

However, the rate of change of height can also be obtained from Equation 20.12 in Section 20.2. Hence, equating the two expressions for $\dot{h}$ from Equations 20.12 and 20.33, one can write

$$\sin \gamma = \frac{U}{V_T} \sin \theta - \frac{V}{V_T} \sin \phi \cos \theta - \frac{W}{V_T} \cos \phi \cos \theta \tag{20.34}$$

Equation 20.34 can simplified further by using the definitions of α, β. After carrying out the necessary algebra, one can write

$$\sin \gamma = \cos \alpha \cos \beta \sin \theta - \sin \beta \sin \phi \cos \theta - \sin \alpha \cos \beta \cos \phi \cos \theta \tag{20.35}$$

Here one can notice that it is necessary in flight control to ensure that side slip angle β remains as small as possible throughout the flight to minimize the drag. Ensuring side slip angle to remain small through 'coordinated turn' [15] (which is done by generating necessary bank angle as discussed later) and assuming $\sin \beta \approx \beta$, $\cos \beta \approx 1$, Equation 20.35 simplifies to

$$\sin \gamma = \sin \theta \cos \alpha - (\sin \phi \cos \theta)\beta - \sin \alpha \cos \phi \cos \theta \tag{20.36}$$

Taking time derivatives on both sides of Equation 20.36 and carrying out the necessary algebra, the dynamics of the flight path angle can be written as

$$\dot{\gamma} = \frac{1}{\cos \gamma} (C_1 \dot{\theta} + C_2 \dot{\phi} + C_3) \tag{20.37}$$

where C_1, C_2, C_3 are given by

$$\begin{aligned}
C_1 &= \cos \theta \cos \alpha + (\sin \phi \sin \theta)\beta + \sin \alpha \cos \phi \sin \theta \\
C_2 &= (\cos \phi \cos \theta)\beta + \sin \alpha \sin \phi \cos \theta \\
C_3 &= (-\sin \theta \sin \alpha - \cos \alpha \cos \phi \cos \theta)\dot{\alpha} + (\sin \phi \cos \theta)\dot{\beta}
\end{aligned} \tag{20.38}$$

In Equation 20.38, the expressions for $\dot{\alpha}$, $\dot{\beta}$ are given as

$$\dot{\alpha} = \frac{U\dot{W} - W\dot{U}}{U^2 + W^2} \tag{20.39}$$

$$\dot{\beta} = \frac{(V_T \dot{V} - V\dot{V}_T)\cos \beta}{U^2 + W^2} \tag{20.40}$$

where

$$\dot{V}_T = \frac{(U\dot{U} + V\dot{V} + W\dot{W})}{V_T}$$

The rate at which side slip angle changes is assumed to be zero, ($\dot{\beta} \approx 0$), since β is almost constant and its value is nearly zero in C_3. The expressions for $\dot{\alpha}$ and $\dot{\beta}$ can be expanded further by substituting Equations 20.1 through 20.3 into Equations 20.39 and 20.40. Next, the dynamics of heading angle can be derived as follows. From Equations 20.31 and 20.32, it is obvious that

$$\frac{\dot{y}}{\dot{x}} = \tan \chi \tag{20.41}$$

However, from the navigation equation of six-DOF equations, that is, Equations 20.10 and 20.11, it is clear that

$$\frac{\dot{y}}{\dot{x}} = \left(\frac{\begin{array}{l} \cos\alpha\cos\theta\cos\beta\sin\psi + \sin\beta(\sin\phi\sin\theta\sin\psi + \cos\phi\cos\psi) \\ + \sin\alpha\cos\beta(\cos\phi\sin\theta\sin\psi - \sin\phi\cos\psi) \end{array}}{\begin{array}{l} \cos\alpha\cos\theta\cos\beta\cos\psi + \sin\beta(\sin\phi\sin\theta\sin\psi - \cos\phi\sin\psi) \\ + \sin\alpha\cos\beta(\cos\phi\sin\theta\cos\psi + \sin\phi\sin\psi) \end{array}} \right) \tag{20.42}$$

Hence, from Equations 20.41 and 20.20 one obtains

$$\tan\chi = \left(\frac{\begin{array}{l} \cos\alpha\cos\theta\cos\beta\sin\psi + \sin\beta(\sin\phi\sin\theta\sin\psi + \cos\phi\cos\psi) \\ + \sin\alpha\cos\beta(\cos\phi\sin\theta\sin\psi - \sin\phi\cos\psi) \end{array}}{\begin{array}{l} \cos\alpha\cos\theta\cos\beta\cos\psi + \sin\beta(\sin\phi\sin\theta\sin\psi - \cos\phi\sin\psi) \\ + \sin\alpha\cos\beta(\cos\phi\sin\theta\cos\psi + \sin\phi\sin\psi) \end{array}} \right) \tag{20.43}$$

Here, assuming α and β to be 'small' and substituting $\sin \alpha = \sin \beta = 0$, Equation 20.43 simplified to

$$\tan\chi = \left(\frac{\cos\alpha\cos\theta\sin\psi}{\cos\alpha\cos\theta\cos\psi} \right) = \tan\psi \tag{20.44}$$

Since tangent is a one-to-one function in $[-\pi, \pi]$, from Equation 20.44 it can be inferred that

$$\chi = \psi \tag{20.45}$$

within $[-\pi, \pi]$, provided α and β are 'small'. Assuming Equation 20.45 to be valid for all time, the dynamics of the heading angle can be written as

$$\dot{\chi} = \dot{\psi} = Q \sin\phi \sec\theta + R \cos\phi \sec\theta \tag{20.46}$$

20.5   PARTIAL IGC FORMULATION

As pointed out in Section 20.1, the PIGC formulation operates in two loops (see Figure 20.4) and combines the benefits of both the conventional three-loop approach as well as the one-loop integrated guidance and control approach. It retains and utilizes the inherent separation existing between the faster and slower dynamics of aerospace vehicles by manipulating its six-DOF model in two steps intelligently. The outer loop generates the desired body rates in pitch and yaw for achieving the desired orientation of the velocity vector V_T, while the desired roll rate is generated to

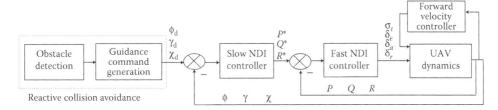

FIGURE 20.4   Block diagram of PIGC design.

ensure the necessary roll angle that is necessary for turn coordination. On the other hand, the inner loop tracks the desired body rates sufficiently fast (in all three roll, pitch and yaw channels). The details of this approach are discussed in this section.

20.5.1   Roll Angle for Turn Coordination

In fixed-wing aircrafts, it is quite essential to have coordinated turn as there is a heavy amount of drag penalty otherwise. This is ensured by making sure that the side slip angle β remains as small as possible, or equivalently, by ensuring $V \to 0$ as soon as possible and thereafter $V \approx 0$ for all time. To ensure this requirement, one can aim to enforce $\dot{V} = -k_V V$, where the gain $k_V > 0$. Substituting for $\dot{V}$ from Equation 20.2, it leads to

$$PW - RU + Y_a + g\cos\theta\sin\phi = -k_V V \tag{20.47}$$

Solving for ϕ from Equation 20.47, the desired role angle necessary for coordinated turn is given as

$$\phi_d = \sin^{-1}\left(\frac{-k_V V - (PW - RU + Y_a)}{g\cos\theta}\right) \tag{20.48}$$

Note that computation of force Y_a requires the information about the control surface deflection, which is assumed to be held at their previous values in any update cycle (initially the deflection is assumed to be zero). This is very much justifiable since the force components generated by control surface deflections are usually very small.

20.5.2   Generation of Body Rates from the Outer Loop

In the outer loop, the objective is to generate the necessary body rates (i.e., roll, pitch and yaw rates) such that the physical angles ϕ, γ and χ track the desired values ϕ_d, γ_d and χ_d, respectively. To do this, following the philosophy of dynamic inversion [14], the following error dynamics is enforced:

$$\begin{bmatrix} \dot{\phi} - \dot{\phi}_d \\ \dot{\gamma} - \dot{\gamma}_d \\ \dot{\chi} - \dot{\chi}_d \end{bmatrix} + \begin{bmatrix} k_\phi & 0 & 0 \\ 0 & k_\gamma & 0 \\ 0 & 0 & k_\chi \end{bmatrix} \begin{bmatrix} \phi - \phi_d \\ \gamma - \gamma_d \\ \chi - \chi_d \end{bmatrix} = 0 \tag{20.49}$$

Assuming the obstacles to be stationary $\dot{\gamma}_d = \dot{\chi}_d = 0$. It is also assumed that $\dot{\phi}_d = 0$. Note that even though these angles may keep changing, one can bring in the 'quasi-steady assumption', whereby they are updated at every time step, but its value is assumed to be constant until next time step. Next, substituting for $\dot{\phi}, \dot{\gamma}$ and $\dot{\chi}$ from Equations 20.7, 20.37 and 20.46, respectively, and carrying out the necessary algebra, the desired body rates are solved in a closed form as follows:

$$\begin{bmatrix} P^* \\ Q^* \\ R^* \end{bmatrix} = g_A^{-1}(b_A - f_A) \tag{20.50}$$

where

$$f_A \triangleq \begin{bmatrix} 0 \\ C_3 \sec\gamma \\ 0 \end{bmatrix}, \quad b_A \triangleq \begin{bmatrix} -k_\phi(\phi - \phi_d) \\ -k_\gamma(\gamma - \gamma_d) \\ -k_\chi(\chi - \chi_d) \end{bmatrix}$$

$$g_A \triangleq \begin{bmatrix} 1 & \sin\phi\tan\theta & \cos\phi\tan\theta \\ C_2\sec\gamma & \sec\gamma(C_1\cos\phi + C_2\sin\phi\tan\theta) & \sec\gamma(C_2\cos\phi + \tan\theta - C_2\sin\phi) \\ 0 & \sin\phi\sec\theta & \cos\phi\sec\theta \end{bmatrix}$$

20.5.3 Generation of Control Surface Deflections from the Inner Loop

After generating the necessary body rates, it is required that the actual body rates of the vehicle should track these desired values. Once again, following the philosophy of dynamic inversion [14], this can be done by enforcing the following first-order error dynamics

$$\begin{bmatrix} \dot{P} - \dot{P}^* \\ \dot{Q} - \dot{Q}^* \\ \dot{R} - \dot{R}^* \end{bmatrix} + \begin{bmatrix} k_P & 0 & 0 \\ 0 & k_Q & 0 \\ 0 & 0 & k_R \end{bmatrix}\begin{bmatrix} P - P^* \\ Q - Q^* \\ R - R^* \end{bmatrix} = 0 \tag{20.51}$$

Once again, following the 'quasi-steady assumption', one can assume $\dot{P}^* = \dot{Q}^* = \dot{R}^* = 0$. Next, substituting the moment equations $\dot{P}, \dot{Q}, \dot{R}$ from Equations 20.4 through 20.6, respectively into Equation 20.51 and carrying out the necessary simplification, it leads to

$$f_R + g_R U_c = b_R \tag{20.52}$$

By carrying out the necessary algebra, the closed-form solution obtained by inverting the body rates dynamics is

$$U_c = g_R^{-1}(b_R - f_R) \tag{20.53}$$

where $U_c = [\delta_a\ \delta_e\ \delta_r]^T$ and other terms are defined as follows:

$$f_R \triangleq \begin{bmatrix} c_1 RQ + c_2 PQ + c_3 L_{a_x} + c_4 N_{a_x} \\ c_5 PR + c_6(P^2 - R^2) + c_7(M_{a_x} - M_t) \\ c_8 PQ - c_2 RQ + c_4 L_{a_x} + c_9 N_{a_x} \end{bmatrix}, \quad g_R \triangleq \begin{bmatrix} c_3 L_{a_u} & 0 & c_4 N_{a_u} \\ 0 & c_7 M_{a_u} & 0 \\ c_4 L_{a_u} & 0 & c_9 N_{a_u} \end{bmatrix},$$

$$b_R \triangleq \begin{bmatrix} \dot{P}^* - k_P(P - P^*) \\ \dot{Q}^* - k_Q(Q - Q^*) \\ \dot{R}^* - k_R(R - R^*) \end{bmatrix}$$

where

$$L_{a_x} \triangleq \bar{q}Sb[C_{l_\beta}(\alpha)\beta + C_{l_P}(\alpha)P + C_{l_R}(\alpha)R]$$

$$M_{a_x} \triangleq \bar{q}Sc[C_{m_0} + C_{m_\alpha}(\alpha)\ \alpha + C_{m_\beta}(\alpha,\beta)\beta + C_{m_Q}(\alpha)Q]$$

$$N_{a_x} \triangleq \bar{q}Sb[C_{n_\beta}(\alpha)\beta + C_{n_P}(\alpha)P + C_{n_R}(\alpha)R]$$

$$[L_{a_u} \quad M_{a_u} \quad N_{a_u}] \triangleq \bar{q}Sb[C_{l_{\delta a}} \quad C_{m_{\delta e}} \quad C_{n_{\delta r}}]$$

Note that the thrust is assumed to be a slow-varying parameter, which is updated at a lesser frequency, as discussed in Section 20.5.4. Also note that the gain selection in inner loop is typically done higher than that of the outer loop because the body rate dynamics are usually much faster than the angular rate dynamics used in the outer loop.

20.5.4 VELOCITY CONTROL

While manoeuvring away from an obstacle by using aerodynamic control, it should be remembered that the vehicle thrust is also a control parameter. However, since it can only be varied slowly (and hence demands a longer time to be effective) and since the time availability for reactive manoeuvres are usually small, it cannot be used as a very effective controller. However, it should at least be used to prevent vehicle acceleration (e.g., in case of an altitude drop) so that the available time to go is not curtailed further. In terms of mathematics, it means that the forward velocity U should track the desired velocity U^*, which can be assumed as the initial value at the start of the manoeuvre (and hence a constant value). To enforce this, one can again take the help of the same dynamic inversion philosophy [14] and enforce the following error dynamics:

$$(\dot{U} - \dot{U}^*) + k_U(U - U^*) = 0 \tag{20.54}$$

Separating the state and control terms in $\dot{U}$ from Equation 20.54 and carrying out the necessary algebra, the throttle control can be written as

$$\sigma_t = g_U^{-1}(b_U - f_U) \tag{20.55}$$

where

$$f_U \triangleq RV - QW - g\sin\theta + X_a, \quad g_U \triangleq \frac{T_{max}}{m}, \quad b_U \triangleq \dot{U}^* - k_U(U - U^*)$$

20.5.5 ACTUATOR DYNAMICS AND DESIGN OF THE ACTUATOR INPUT

AE-2 employs electromechanical servos for the control surface deflection, which are all similar. The control deflections generated by the inner loop are passed to the first-order actuator, system. The actuator dynamics for the elevator servo is given by Equation 20.14 and the actuator dynamics for the throttle servo is given by Equation 20.15. The actual aerodynamic control deflections given as a control to the plant dynamics are obtained through the actuator. The tracking error introduced by the actuator model due to the first-order delay may adversely affect the system performance.

Therefore, in order to compensate for the lag either a fast actuator should be used or a controller for the actuator should be designed based on the tracking error. In this study, a controller has been designed with the assumption that the actuator states (control deflections) are available for the feedback. The controller is designed based on the error of the actual states of the actuator $\sigma_t, \delta_e, \delta_a, \delta_r$ and the desired state of the actuator $\sigma_t^*, \delta_e^*, \delta_a^*, \delta_r^*$, respectively. This is done by enforcing the following first-order error dynamics for the elevator deflection (other channels are very much similar to this).

$$(\dot{\delta}_e - \dot{\delta}_e^*) + k_{\delta_e}(\delta_e - \delta_e^*) = 0 \tag{20.56}$$

Next, substituting the actuator dynamics from Equation 20.14 into Equation 20.56 and solving for u_{δ_e}, one gets

$$u_{\delta_e} = \frac{1}{9.5}[-9.5\delta_e + \dot{\delta}_e^* - k_{\delta_e}(\delta_e - \delta_e^*)] \tag{20.57}$$

Note that the actuator dynamics is also subjected to position and rate limits, which have been accounted in the simulation studies. The details, however, are omitted here for brevity. One may consult Ref. [19] for further details on it.

20.6 NUMERICAL RESULTS

In PIGC framework, the reactive nonlinear guidance algorithm is validated with full nonlinear six-DOF model of UAV. For better accuracy, UAV model is integrated with Runge–Kutta method [20] throughout the simulations. Scenarios with multiple obstacles and with varied radius of the safety ball around them have been considered on the path to the goal point. It is implemented because in practice, the obstacles of different sizes will be encountered, which should be sensed with appropriate size of safety ball around them. In the present study, the radius of the safety ball varies from 5 to 20 m. In the PIGC framework, the outer and the inner loop work on the NDI technique. In NDI, the gain selection corresponding to the outer loop and the inner loop is dictated by the settling time of the system dynamics and required time-to-go to reach the aiming point. In all the scenarios, the experiment was conducted with the first-order actuator model for the control surface deflections. The control surface deflections after passing through the actuator cause first-order delay in tracking. Therefore, to compensate for the tracking errors, actuator controller is designed with the assumption that the control deflections are available as a state for feedback. The actuator controller observes the position and the rate limit as posed by the system. All the results have been exhibited with the effect of the actuator controller.

20.6.1 TRIM CONDITION

Trim condition is calculated for steady level flight at a given velocity and altitude [16]. The state vector $X = [U \, V \, W \, P \, Q \, R \, \phi \, \theta \, \psi \, x_i \, y_i \, h_i]^T$ representing the UAV is initialized with trim values as stated in Table 20.2.

The following constraints were imposed by the actuator on the control variables associated with the UAV as shown in Table 20.3.

It is assumed that the instantaneous position of the obstacles are obtained through the aboard passive sensors. Since the attitude of the vehicle is changing so the sensor direction will change and it may cause the motion of the static view in its frame. Keeping in view, the delicacy of the sensor frame the obstacle locations have been considered in the inertial frame to perform the guidance command computation. Two scenarios are considered, one with the single obstacle and other with the multiple obstacles (two obstacles in the present case). In all the scenarios, the effect of the actuator and the vehicle constraints are taken into account.

TABLE 20.2
Trim Values of the State and Control Variables

Velocity	$V_T = 20$ m/s
Position	$x_{trim} = 0$ m, $y_{trim} = 0$ m, $h_{trim} = 50$ m
Body Angular rates	$P_{trim} = 0$ deg/s, $Q_{trim} = 0$ deg/s, $R_{trim} = 0$ deg/s
Euler angles	$\phi_{trim} = 0°$, $\theta_{trim} = 3.1339°$, $\psi_{trim} = 0°$
Aerodynamic angles	$\alpha_{trim} = 3.1339°$, $\beta_{trim} = 0°$
Control surface deflections	$\sigma_{trim} = 0.3708$, $\delta_{e_{trim}} = -3.2673°$, $\delta_{a_{trim}} = 0°$, $\delta_{r_{trim}} = 0°$

TABLE 20.3
Actuator Constraints Description

	Throttle (σ_t)	Elevator (δ_e)	Aileron (δ_a)	Rudder (δ_r)	ϕ_d
Upper-level limit	1	+5 deg	+15 deg	+15 deg	+45 deg
Lower-level limit	0	−25 deg	−15 deg	−15 deg	−45 deg
Upper-rate limit	N/A	+45deg/s	+45 deg/s	+45 deg/s	N/A
Lower-rate limit	N/A	−45 deg/s	−45 deg/s	−45 deg/s	N/A

Note: N/A, Not applicable.

20.6.2 Scenario 1: Single Obstacle

The starting point of the simulation is assumed to be where it senses the obstacle first, which was assumed to be $X_i = [0\ 0\ 50]^T$. The goal point was assumed to be $X_d = [200\ 15\ 52]^T$. The position of the obstacle in the inertial frame is $X_{obs} = [80\ 7\ 49]^T$ and the radius of the ball is 15 m around it.

Figure 20.5 shows the effectiveness of the guidance algorithm in avoiding the obstacle and finally reaching the goal point within the available time to go.

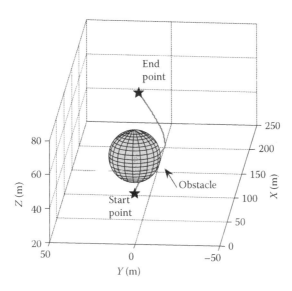

FIGURE 20.5 3D view of the obstacle avoidance trajectory for a single obstacle.

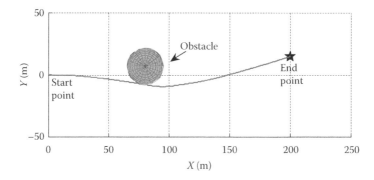

FIGURE 20.6 2D view of the obstacle avoidance trajectory in X–Y plane for a single obstacle.

Figure 20.6 shows the 2D view of the avoidance scenario, in the X–Y plane which gives better insight about the turning capability of UAV.

Figure 20.7 represents the control effort required by the vehicle in the longitudinal plane. It represents the control profiles of throttle and the elevator. The reference command in Figure 20.7 represents the input to the actuator (i.e., the control generated from the inner loop) and the output of the actuator controller is tracking the reference value. In Figure 20.7, throttle control is used to keep the forward velocity U = constant because the velocity will tend to decrease due to drag being more than the total lift. It happens during turning that only one component of the lift vector is available to balance the weight of the UAV.

Figure 20.8 shows the response of the control effort required in the lateral plane. Figure 20.8 represents two control profiles with the actuator output following its reference value (i.e., input to the actuator). It can be seen that both the aileron and rudder control responds to the avoidance manoeuvre within the prescribed limits. It can be seen that both the deflections settles down to their trim values when the avoidance is over.

It can be seen from Figure 20.9 that the judicious selection of the gain values for the inner loop causes the body angular rates to track their commanded values efficiently. Figure 20.9 shows the tracking of the commanded body angular rates by the actual body rates.

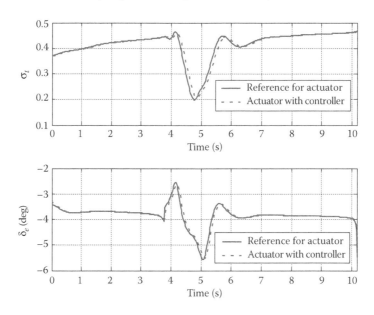

FIGURE 20.7 Longitudinal control surface deflections for a single obstacle.

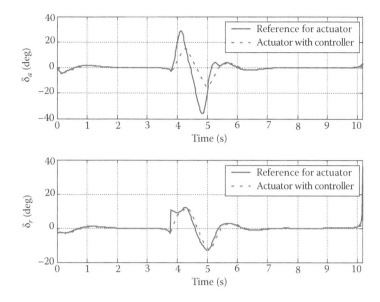

FIGURE 20.8 Lateral control surface deflections for a single obstacle.

In Figure 20.10, it can be seen that in the outer loop, flight path angles γ and χ tracks the guidance command which shows the successful achievement of the aiming point and the goal point. It can be inferred from Figure 20.10 that the total velocity V_T of the UAV maintains its constant value and gets deviated only when the obstacle is avoided.

Figure 20.11 shows the forward velocity profile and the corresponding aerodynamic angles α and β which gets deviated from their trim values only when the obstacle avoidance takes place. It can be seen that the side slip angle is near to zero.

Figure 20.12 represents the attitude of the UAV, where the roll angle tracks its commanded value quickly even with actuator in the loop without violating the turning constraint. It can be inferred from Figure 20.12 that the attitude defining angles settles down to their steady-state values when the obstacle is averted.

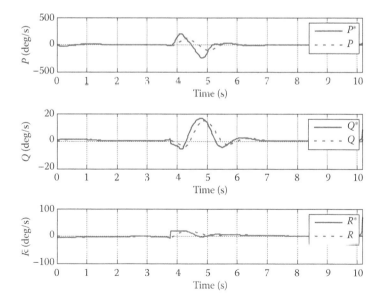

FIGURE 20.9 Tracking of commanded body angular rates for a single obstacle.

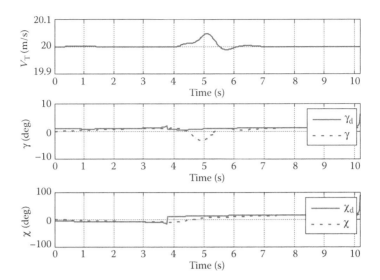

FIGURE 20.10 Total velocity and its directions for a single obstacle.

20.6.3 SCENARIO 2: TWO OBSTACLES

It has been observed through simulations, that the separation between the UAV and the obstacle should be at least five times the radius r of the ball with which the obstacle is surrounded. Even the distance between the obstacles should also be 50 m. These constraints on the guidance algorithm are imposed by the vehicle capability considered for the current problem. In case of the multiple obstacles, two obstacles have been considered. The position of the obstacles in the inertial frame are $X_{obs1} = [80\ 7\ 49]^T$ and $X_{obs2} = [180\ -27\ 52]^T$ and the goal point $X_d = [300\ -34\ 54]^T$.

Figure 20.13 shows the turning trajectory of the UAV, where guidance algorithm is efficient enough to avoid the two obstacles in the environment on the way to the destination point. It can be inferred that the irrespective of the number of the obstacles, the obstacle avoidance algorithm has

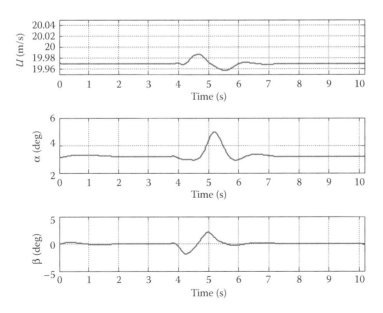

FIGURE 20.11 Forward velocity and aerodynamic angles for a single obstacle.

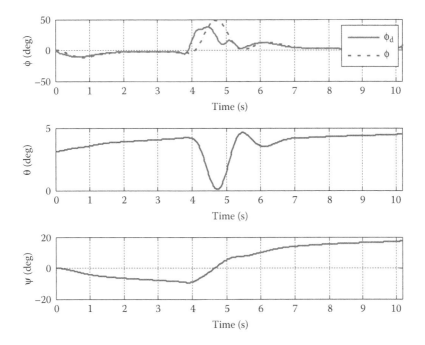

FIGURE 20.12   Tracking of roll angle and Euler angles for a single obstacle.

the capability to work in the cluttered environment. Figure 20.14 gives a better view in the 2D X–Y plane of the avoidance manoeuvre executed by the UAV.

Figure 20.15 represents the control profiles of throttle and elevator in the longitudinal plane which responds two times to the demand of avoidance manoeuvre for two obstacles in the scenario. Similarly, Figure 20.16 represents the control profiles of aileron and rudder in the lateral plane which responds twice to the demand of avoidance of obstacles. In Figures 20.15 and 20.16, the control deflections are smooth in both the planes due to the presence of the actuator. Moreover, the actuator output follows its reference value (i.e., the input to the actuator).

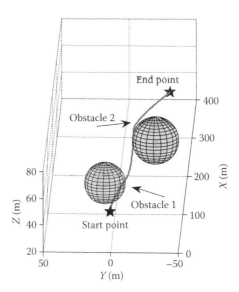

FIGURE 20.13   3D view of the obstacle avoidance trajectory for multiple obstacles.

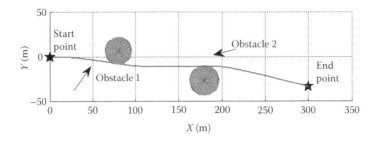

FIGURE 20.14 2D view of the obstacle avoidance trajectory in X–Y plane for multiple obstacles.

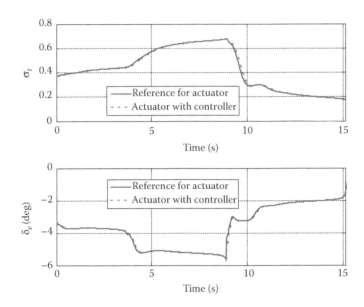

FIGURE 20.15 Longitudinal control surface deflections for multiple obstacles.

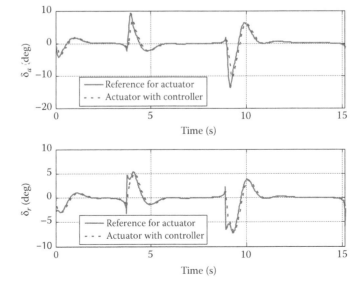

FIGURE 20.16 Lateral control surface deflections for multiple obstacles.

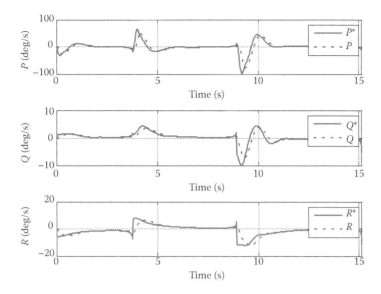

FIGURE 20.17   Tracking of commanded body angular rates for multiple obstacles.

It can be inferred from Figure 20.17 that the efficient tracking of commanded body rates is not affected by the number of obstacles present in the scenario even with the actuator in the loop.

Figure 20.18 shows the profile of total velocity which remains almost constant to its trim value. The tracking of the guidance commands in Figure 20.18 are executed efficiently, till the goal point is reached which shows the successful achievement of the aiming point.

Figure 20.19 shows the efficiency of the throttle control in keeping the forward velocity constant. The constancy of the aerodynamic angles α and β at their trim values can be assured from Figure 20.19. The aerodynamic angles only responds when the obstacle avoidance takes place and during the task of reaching the goal point.

Figure 20.20 shows the tracking of the commanded roll angle for efficient coordinated flight in the presence of the actuator. It can be seen in Figure 20.20 that the number of the obstacles does not

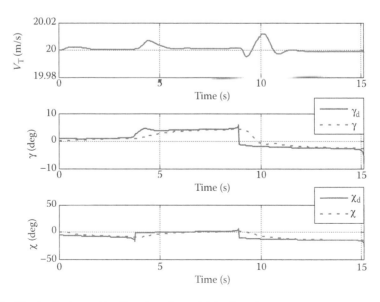

FIGURE 20.18   Total velocity and its directions for multiple obstacles.

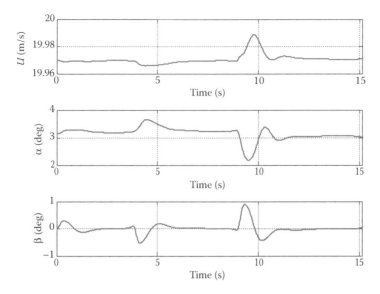

FIGURE 20.19   Forward velocity and aerodynamic angles for multiple obstacles.

prevent the attitude of the UAV to get settled to its steady-state value even with the actuator controller. Note that similar promising results were observed in a number of simulation studies carried out. These results clearly demonstrate the proposed PIGC technique, a very good design approach for collision avoidance of UAVs, especially for pop-up obstacles.

20.7   CONCLUDING REMARKS

In this work, the problem of reactive collision avoidance for UAVs has been addressed with an innovative technique known as Partially Integrated Guidance and Control (PIGC). Unlike conventional approach, it executes in two-loop structure and minimizes the transients due to time lags in multiple loop tracking. Thus PIGC algorithm performs faster in avoiding the susceptible collisions in small time span available.

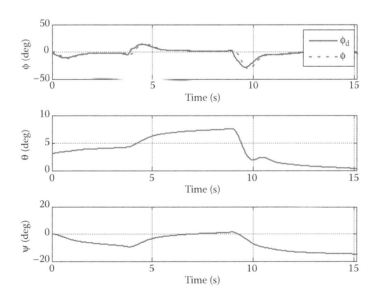

FIGURE 20.20   Tracking of roll angle and Euler angles for multiple obstacles.

Moreover, PIGC does not execute in a single-loop structure like IGC approaches which fail to take into account the inherent time-scale separation property that exists in aerospace vehicles. Thus, PIGC algorithm overcomes the disadvantages of both the conventional and IGC approaches while retaining their benefits. PIGC explicitly uses full nonlinear six DOF model of UAV to design its guidance and control loop structures. In this chapter, inspired by the collision cone-based detection and avoidance philosophy, aiming point guidance algorithm is designed in PIGC framework. In this approach, the necessary angular correction commands in horizontal and the vertical planes are first computed from algebraic relationships. Using these angular commands, the necessary body rates are computed from an outer loop using dynamic inversion. Next, the necessary control surface deflection commands are computed from an inner loop (again using dynamic inversion) to track these desired body rates. The usefulness of the proposed PIGC approach is supported by the numerical results, using the six-DOF model of a fixed wing UAV, which also has actuator models for aerodynamic as well as thrust controls. The tracking error introduced by the actuator model due to the first-order delay may adversely affect the system performance. Therefore, in order to compensate for the lag, a controller has been designed with the assumption that the actuator states (control deflections) are available for the feedback.

APPENDIX: AERODYNAMIC COEFFICIENTS

The aerodynamic model of the AE-2 UAV is given in Section 20.2. Various force and moment coefficients of the model are given in this Appendix for completeness. Some of the derivatives are partitioned with respect to angle of attack into linear region and nonlinear region. From observation of wind tunnel data, the value of α for partition was found to be $10°$. In expanded derivative functions, α_{10} is defined as $\alpha_{10} = 0°$ if $\alpha \le 10°$ and $\alpha_{10} = \alpha - 10°$ if $\alpha \ge 10°$. With this definition, various aerodynamic derivatives are computed as follows:

$$C_{X_\alpha}\alpha = x_{10}\alpha + x_{11}\alpha^2 + x_{12}\alpha_{10}^2 + x_{13}\alpha_{10}^3 + x_{14}\alpha_{10}^4$$
$$C_{X_{\delta e}} = x_{20} + x_{21}\alpha + x_{22}\alpha_{10}^2$$
$$C_{X_Q} = x_{30} + x_{31}\alpha$$
$$C_{Y_\beta} = y_{10} + y_{11}\alpha + y_{12}\alpha^2 + y_{13}\alpha_{10}^2 + y_{14}\alpha_{10}^3$$
$$C_{Y_{\delta a}} = y_{20} + y_{21}\alpha + y_{22}\alpha^2 + y_{23}\alpha^3 + y_{24}\alpha_{10}^2 + y_{25}\alpha_{10}^3 + y_{26}\alpha_{10}^4$$
$$C_{Y_{\delta r}} = y_{30} + y_{31}\alpha_{10}^2 + y_{32}\alpha_{10}^3$$
$$C_{Y_P} = y_{40} + y_{41}\alpha$$
$$C_{Y_R} = y_{50} + y_{51}\alpha$$
$$C_{Z_\alpha}\alpha = z_{10}\alpha + z_{11}\alpha_{10}^2 + z_{12}\alpha_{10}^3$$
$$C_{Z_Q} = z_{20} + z_{21}\alpha$$
$$C_{l_\beta} = l_{10} + l_{11}\alpha + l_{12}\alpha^2 + l_{13}\alpha_{10}^2 + l_{14}\alpha_{10}^3$$
$$C_{l_{\delta a}} = l_{20} + l_{21}\alpha + l_{22}\alpha^2 + l_{23}\alpha^3 + l_{24}\alpha_{10}^2 + l_{25}\alpha_{10}^3 + l_{26}\alpha_{10}^4$$
$$C_{l_p} = l_{30} + l_{31}\alpha + l_{32}\alpha^2$$
$$C_{l_R} = l_{40} + l_{41}\alpha + l_{42}\alpha^2$$
$$C_{m_\alpha} = m_{10} + m_{11}\alpha + m_{12}\alpha^2 + m_{13}\alpha^3 + m_{14}\alpha^4 + m_{22}\alpha\beta$$
$$C_{m_\beta} = m_{20}\beta + m_{21}\alpha + m_{23}\alpha^2\beta$$
$$C_{m_{\delta e}} = m_{30} + m_{31}\alpha + m_{32}\alpha^2$$
$$C_{m_Q} = m_{40} + m_{41}\alpha + m_{42}\alpha^2$$
$$C_{n_\beta} = n_{10} + n_{11}\alpha + n_{12}\alpha^2 + n_{13}\alpha_{10}^2$$
$$C_{n_{\delta r}} = n_{20} + n_{21}\alpha_{10}^2 + n_{22}\alpha_{10}^3$$
$$C_{n_p} = n_{30} + n_{31}\alpha + n_{32}\alpha^2$$
$$C_{n_R} = n_{40} + n_{41}\alpha + n_{42}\alpha^2$$

TABLE 20.4
Numerical Values of Various Constants

C_{x_0}	0.0386	y_{51}	0.0035514	m_{11}	−0.00026206
x_{10}	−0.0040376	C_{z_0}	0.1653	m_{12}	−1.7853e−005
x_{11}	−0.0010525	z_{10}	0.087138	m_{13}	−2.1109e−006
x_{12}	0.0027887	z_{11}	−0.0091867	m_{14}	1.1346e−007
x_{13}	0.00010917	z_{12}	0.00024242	m_{20}	0.00024049
x_{14}	−5.3586e−6	C_{z_β}	−0.0020001	m_{21}	−7.8566e−006
x_{20}	−0.00035832	$C_{z_{\delta e}}$	0.0039823	m_{22}	1.0663e−006
x_{21}	−2.2061e−5	z_{20}	6.9303	m_{23}	−7.8866e−007
x_{22}	−5.7342e−6	z_{11}	−0.047657	m_{30}	−0.0145
x_{30}	−0.18476	l_{10}	0.0022856	m_{31}	9.2552e−006
x_{31}	−0.10227	l_{11}	6.4827e − 005	m_{32}	9.0437e−006
y_{10}	0.0099319	l_{12}	−3.0529e − 006	m_{40}	−13.954
y_{11}	0.00029462	l_{13}	−2.7687e − 005	m_{41}	0.0017379
y_{12}	1.7831e−005	l_{14}	1.7713e − 006	m_{42}	0.0016743
y_{13}	−0.00030969	l_{20}	0.0029091	n_{10}	−0.0015474
y_{14}	1.6759e − 005	l_{21}	9.0047e − 006	n_{11}	6.1309e−005
y_{20}	0.0022145	l_{22}	−7.4562e − 006	n_{12}	−1.8989e−006
y_{21}	0.00041878	l_{23}	3.0423e − 007	n_{13}	−5.5706e−006
y_{22}	1.3117e − 5	l_{24}	−2.5531e − 005	n_{20}	0.00077238
y_{23}	−1.1549e − 6	l_{25}	4.1263e − 006	n_{21}	1.1379e−006
y_{24}	−5.2196e − 5	l_{26}	−2.0918e − 007	n_{22}	−4.1705e−008
y_{25}	8.8682e − 6	l_{30}	−0.44336	n_{30}	−0.015512
y_{26}	−3.2717e − 7	l_{31}	0.00075577	n_{31}	−0.011325
y_{30}	−0.0016884	l_{32}	−0.00013921	n_{32}	9.8251e−005
y_{31}	−1.3637e − 05	l_{40}	0.076582	n_{40}	−0.085307
y_{32}	1.3214e − 06	l_{41}	0.010019	n_{41}	0.00080338
y_{40}	−0.14504	l_{42}	1.1783e − 005	n_{42}	−0.00026197
y_{41}	0.013516	C_{m_0}	0.0346		
y_{50}	0.13784	m_{10}	−0.013841		

The numerical values of the various constants in the above expressions are given in Table 20.4. For more details on the model and the origin of the aerodynamic coefficients, one can consult Ref. [17].

REFERENCES

1. DeGarmo, M. and Nelson, G. M., Prospective unmanned aerial vehicle operations in the future national airspace system, *Proceedings of the 4th Aviation Technology, Integration and Operations (ATIO) Forum*, AIAA, Chicago, IL, 20–22 Sept. 2004.
2. Mujumdar, A. and Padhi, R., Evolving philosophies on autonomous obstacle/collision avoidance of unmanned aerial vehicles, *Journal of Aerospace Computing, Information, and Communication*, 2011, 8(2), 17–41.
3. Scherer, S., Singh, S., Chamberlain, L. and Elgersma, M., Flying fast and low among obstacles: Methodology and experiments, *The International Journal of Robotics Research*, 2008, 27(5), 549–574.
4. Paul, T., Krogstad, T. R. and Gravdahl, J. T., Modeling of UAV formation flight using 3D potential field, *Simulation Modeling Practice and Theory*, 2008, 16(9), 1453–1462.
5. LaValle, S. M. and Kuffner, J. J., Rapidly-exploring random trees: Progress and prospects, *Algorithmic and Computational Robotics: New Directions*, 2001.

6. Palumbo, N. F., Reardon, B. E. and Blauwkampand, R. A., Integrated guidance and control for homing missiles, *Johns Hopkins APL Technical Digest*, 2004, 25(2), 121–139.
7. Xin, M., Balakrishnan, S. N. and Ohlmeyer, E. J., Integrated guidance and control of missiles with $\theta - D$ method, *IEEE Transactions on Control Systems Technology*, 2006, 14(6), 981–992.
8. Mingzhe, H. and Guangren, D., Integrated guidance and control of homing missiles against ground fixed targets, *Chinese Journal of Aeronautics*, 2008, 21, 162–168.
9. Padhi, R., Chawla, C., Das, P. G. and Venkatesh, A., Partial integrated guidance and control of surface-to-air interceptors for high speed targets, *American Control Conference*, 10–12 June 2009, St. Louis, USA.
10. Chawla, C. and Padhi, R., Reactive obstacle avoidance of UAVs with dynamic inversion based partial integrated guidance and control, *AIAA Guidance, Navigation, and Control Conference and Exhibit*, 2–5 August 2010, Toronto, Canada.
11. Watanabe, Y., Calise, A. J. and Johnson, E. N., Minimum effort guidance for vision-based collision avoidance, *AIAA Atmospheric Flight Mechanics Conference and Exhibit*, 21–24 August 2006, Keystone, Colorado.
12. Chakravarthy, A. and Ghose, D., Obstacle avoidance in a dynamic environment: A collision cone approach, *IEEE Transactions on Systems, Man and Cybernetics-Part A: Systems and Humans*, 1998, 28(1), 562–574.
13. Tsao, P. L., Chou, C. L., Chen, C. M. and Chen, C. T., Aiming point guidance law for air-to-air missiles, *International Journal of Systems Science*, 1998, 29(2), 95–102.
14. Enns, D., Bugajski, D., Hendrick, R. and Stein, G., Dynamic inversion: An evolving methodology for flight control design, *International Journal of Control*, 1994, 59(1), 71–91.
15. Stevens, B. and Lewis, F., *Aircraft Control and Simulation*, 2nd Edition, John Wiley & Sons, Hoboken, NJ, 2003.
16. Singh, S. P. and Padhi, R., Automatic path planning and control design for autonomous landing of UAVs using dynamic inversion, *American Control Conference*, 10–12 June 2009, St. Louis, USA.
17. Surendranath, V., Govindaraju, S. P., Bhat, M. S. and Rao, C. S. N., Configuration Development of All Electric Mini Airplane, ADE/DRDO Project, 2004, Project Ref. No: ADEO/MAE/VSU/001, Department of Aerospace Engineering, Indian Institute of Science, Bangalore.
18. Carbone, C., Ciniglio, U., Corraro, F. and Luongo, S., A Novel 3D Geometric Algorithm for aircraft autonomous collision avoidance, *Proceedings of the 45th IEEE Conference on Decision and Control*, 13–15 December 2006, San Diego, CA, USA.
19. Chawla, C., Robust Partial Integrated Guidance and Control of UAVs for Reactive Obstacle Avoidance, Thesis, Department of Aerospace Engineering, Indian Institute of Science, Bangalore, December 2010.
20. Atkinson, K. E., *An Introduction to Numerical Analysis*, John Wiley & Sons, New York, 2001.

21 Impedance-Controlled Mechatronic Systems for Autonomous Intelligent Devices

Kostadin Kostadinov and Dimitar Chakarov

CONTENTS

21.1 INTRODUCTION

The development of modern technologies results in reaching the limits of the capabilities of the existing mechatronic systems (MS). On other hand, the development of new trends in engineering and intelligent autonomous systems requires new capabilities for obtaining new functional or better qualitative results. In most cases, mechatronic and robot systems perform the contact tasks with their technological environment during the process. Various force control methods have been developed to meet the requirement of regulating the contact force within a specific range. The force control is based on two distinct methodologies: (i) pure force control and (ii) impedance control (Figure 21.1). Pure force control can be applied only when the end-effecter is in contact with the environment. While in impedance control the force is regulated by controlling the position and its

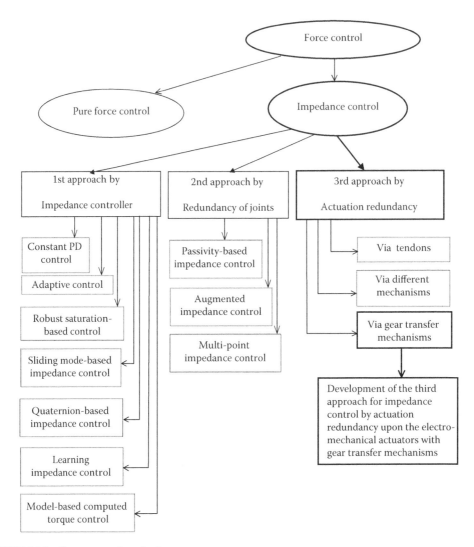

FIGURE 21.1   Force control methods.

relationship with the force, that is, mechanical impedance of the mechatronic system [1]. The mechanical impedance $Z(s)$ expresses the common resistance of an MS to the applied forces $\delta F(s)$, that is, the dynamic properties of the MS considered [2]. It yields a deviation from the desired position δx, that is [1]

$$Z(s) = \frac{\delta F(s)}{\delta x} = Ms^2 + Bs + K \tag{21.1}$$

where M, B and K are dynamic system parameters, mass, damping and stiffness, respectively. The control of the dynamic properties of MS is based on the assumption that the MS accepts the technological environment (TE) as a physical system. At the interaction between two physical systems, one of them could be expressed as impedance, while the second one is characterized with admittance and vice versa. To control the mechanical impedance, it is necessary to set up two control sets [1]: (i) the flow source which defines the requirements to the output link motion and (ii) the control set which expresses the mechanical impedance of the actuator Z. The impedance control can be realized by three approaches [1]. A lot of work has been carried out to develop the first approach for impedance control (by impedance controller) using one of the known approaches [3]: (i) constant PD control, (ii) adaptive control [4], (iii) model-based computed torque control [5], (iv) sliding mode-based impedance control [6], (v) learning impedance control [7], (vi) robust saturation-based control [8] and/or (vii) quaternion-based impedance controller [9]. They are characterized with the different shortcomings such as (a) the desired impedance cannot be maintained because of changes in the robot configuration and velocity; (b) model-based computed torque control is too sensitive to uncertainties in the MS dynamic model; (c) the measurement noise decreases the accuracy of the estimation of the dynamic parameters and (d) the requirement of extensive computation and high gains. The second approach for impedance control, that is, by redundancy of joints, can be realized for the specified application task where the impedance in the necessary directions has to be controlled [10]. Here, the manipulator possesses degrees of freedom (DoF) that outnumber the dimensions of the task space and the redundancy thus obtained is called 'kinematic redundancy'. The redundancy of joints is at least that direction where the desired impedance control has to be achieved. The shortcoming is the realization of the impedance control for complex plane where it is necessary to switch the impedance control between different joints of the MS or robot.

The second approach is reduced for one joint into the third approach for impedance control realized by redundancy of drives for each MS or robot joint. Many researchers who developed the third approach for impedance control used antagonistically driven joints by two actuators via tendons [11,12]. By this approach, the number of the actuators is more than that of the DoF; thus, the resulting redundancy is called 'actuation redundancy' [13]. Additional actuators significantly enhance the load-handling capacity of the MS. However, actuation redundancy is used to minimize joint torques or to satisfy actuator constrains [14–16]. At the same time, it enables the system to modulate the end-effecter impedance by realizing an internal load distribution [17,18]. Parallel manipulators [19,20] as special mechatronic systems, are designed with actuation redundancy, while kinematic redundancy is a characteristic of serial manipulators where the second approach for impedance control is used. Parallel structures are also created by serial manipulators contacting the environment, mobile robots, multi-fingered end-effectors or coordinated multiple space manipulators [21]. The hybrid (serial–parallel) manipulators combine the properties of both types and admit kinematic as well as actuation redundancy. Here, the second and the third approaches for impedance control can be applied.

Manipulation against kinematic constraints will be considerably simplified if it is possible to specify the motion of the end-effecter in response to arbitrary disturbance forces. Control strategies, most generally known as stiffness control by redundant actuation [22–24], are developed for this purpose. The shortcomings here consist of redundancy of drives, complicated servo mechanisms, increased energy consumption and so on. All these shortcomings of the known approaches for impedance control realization specify the spectra of practical applications, defining in this way the approach for their design. By analysing the dynamic interaction of MS with TE ideas, the design approach is generated for some subsystems—drive, mechanic, sensor, control and information as well as the whole MS. This could allow the desired quality parameters and functionality of advance MS to be achieved.

This chapter considers impedance-controlled mechatronic systems based on the redundancy of actuators.

21.2   MECHANICAL IMPEDANCE CONTROL BY REDUNDANCY ACTUATION

The actuators with drive redundancy are a simplified model of the antagonistic drive in the (living) nature. The synthesis of electromechanical actuators for MS is based on the motion transition method [25]. It consists of two-zone dynamic controlled gearing on the actuator output link through the introduction of an antagonistic drive unit that is identical to the existing one. In this way, the kinematic chain is symmetrical and closed due to the antagonistic force/torque developed by the

TABLE 21.1

Kinematic Redundancy Schemes with Internal Gearing

	Disk	Cylinder	Cone	Toroid
Disk				
Cylinder				
Cone				
Toroid				

additional drive unit. Redundancy actuation causes internal force/torque in the transfer mechanisms, which does not perform any effective work to the external world, but the actuator joint stiffness depends on this internal torque [14]. On that basis are synthesized [26] 16 variants of kinematic redundancy schemes of the transfer mechanisms with internal (Table 21.1) and 10 with external gearing (Table 21.2).

21.2.1 METHOD AND SCHEMES FOR MOTION TRANSITION WITH MECHANICAL IMPEDANCE CONTROL

The above-mentioned kinematic schemes can be realized through non-self-stopping and self-stopping gearings. Some of the possible variants of these kinematic redundancy schemes of the transfer mechanisms are used in feeding and positioning mechatronic autonomous intelligent devices in robotized complexes for dimensional quality control, in an MS (Figure 21.1) for precise feeding operations used in robot-assisted material removal, and as a barrier actuator (Figure 21.2) in a Langmuir–Blodgett film deposition mechatronic system. The mechanical impedance consists of two components: knot and non-knot components. The knot mechanical impedance Z_0 is determined by the stiffness K and damping B. The non-knot component Z_1 reflects the inertia properties. As it was mentioned above, the actuator joint stiffness depends on this internal force/torque T_a. For the double worm transfer mechanism (Figure 21.3), the stiffness is obtained by the expression

$$K = \frac{0,8184.10^6 n_0}{\cos \lambda} \sqrt{\frac{T_a}{m_0 y_k d_w W(\alpha) \cos \alpha}} \qquad (21.2)$$

TABLE 21.2

Kinematic Redundancy Schemes with External Gearing

	Disk-1	Cylinder-2	Cone-3	Toroid-4
Disk-1	1-1	1-2	1-3	1-4
Cylinder-2	2-1	2-2	2-3	2-4
Cone-3	3-1	3-2	3-2	3-4
Toroid-4	4-1	4-2	4-3	4-4

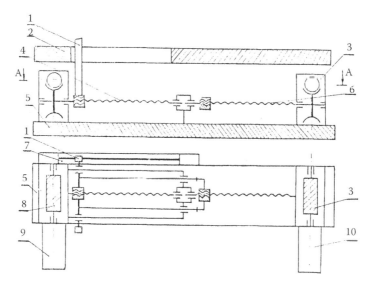

FIGURE 21.2 Kinematic chain of the barrier drive. 1: Barrier; 2: upper plate; 3: worm; 4: ball screw-nut system; 5: base; 6: ball screw-nut system; 7: linear displacement sensor; 8: worm; 9,10: DC motors.

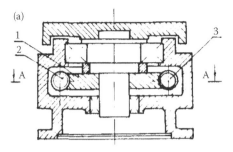

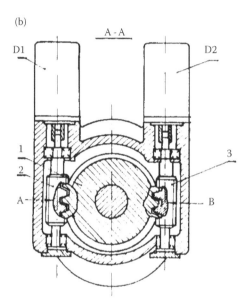

FIGURE 21.3 (a,b) Cross-section of gearing zones 1-2 and 1-3. Drawings of the rotary table. 1: Worm wheel; 2,3: worms; D1,D2: DC motors; A,B: gearing zones.

If the damping of double worm gearing B_w is considered caused by the friction in the two gearing zones A and B shown in Figure 21.3, then it could be expressed by

$$B_w = B_c + \frac{2\mu . f_T(a)}{\dot{\varphi}_{OL} d_k}\left[1 + \frac{\sin(\alpha_1 - \rho_1 - \rho_0)\cos\rho_2}{\sin(\alpha_3 + \rho_1 + \rho_2)\cos\rho_0}\right]T_a = B_c + B_a(T_a) \qquad (21.3)$$

where B_c is the damping of the conventional worm gearing mechanism, $\dot{\varphi}_{OL}$ the output link velocity and $B_a(T_a)$ the additional term caused by the internal torque T_a. The other parameters in Equations 21.2 and 21.3 are constructive parameters of the worm gear used. So, by controlling the internal torque T_a, the knot mechanical impedance can be controlled according to Equations 21.2 and 21.3 since the output link motion of the transfer mechanism can be performed at different levels of the motor activation [26].

21.2.2 MODELLING OF THE ACTUATORS WITH DRIVE REDUNDANCY

The actuators discussed can be treated in turn as a component of a complex system consisting of mechanical and electrical subsystems and a servo controller. The structure of impedance-controlled MS is shown in Figure 21.4 (with transfer mechanism, Figure 21.3). The synthesized kinematic

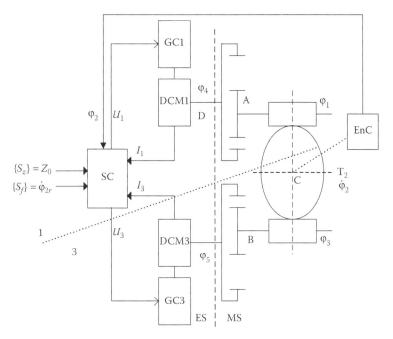

FIGURE 21.4 Structure of impedance-controlled actuators. GC1,GC3: gait converters; DCM1,DCM3: DC motors; EnC: encoder; SC: servo controller; 1: drive unit; 3: antagonistic drive unit; ES: electrical subsystem; MS: mechanical subsystem.

schemes with redundancy are defined with two inputs and one output. The functional conditions of the drive motors are considered depending of the type of gear mechanisms: non-self-stopping or self-stopping.

21.2.2.1 Impedance-Controlled Mechatronic Actuators with Self-Stopping Gearing

The antagonistic motor voltage U_3 sets a reference output link velocity $\dot{\phi}_{2d}(t)$. The difference between the control voltages of the drive U_1 and antagonistic U_3 motors setting up their virtual $\dot{\phi}_1^v$ and reference $\dot{\phi}_3^r$ velocities set up the desired joint mechanical impedance Z_0 according to

$$T_a = T_1 - T_3 = k_e k_T (\dot{\phi}_1^v - \dot{\phi}_3^r)/R_a = k_T (U_1 - U_3) R_a \tag{21.4}$$

To close the kinematic chain with internal torque T_a, it is necessary to fulfil the following condition:

$$\dot{\phi}_1^v > \dot{\phi}_3^r \tag{21.5}$$

So, the two control sets of impedance control $\{S_f\}$ and $\{S_z\}$ are fed to the antagonistic 3 and drive 1 unit, respectively.

21.2.2.2 Impedance-Controlled Mechatronic Actuators with Non-Self-Stopping Gearing

The internal torque T_a is caused by the antagonistic motor T_3. The drive motor DCM1 defines the desired output link velocity $\dot{\phi}_{2d}(t)$. Two control sets $\{S_f\}$ and $\{S_z\}$ of impedance control are fed to the drive 1 and antagonistic 3 units, respectively. When the direction of motion is changed, the antagonistic unit will be driven, and vice versa for the drive unit in both cases of the transfer mechanisms with non self-stopping and self-stopping gearings. Therefore, in both cases of non-self-stopping and self-stopping gearing, the kinematic chain is always closed without backlash and with the knot mechanical impedance control.

21.3   CONTROL STRATEGY

An approach for accommodation control of the dynamic accuracy [2] is used. The scheme for accommodation impedance control consists of two parts. The first part is a feed-forward controller constructed off-line using the dynamic model of the impedance-controlled MS. The open-loop impedance control involves off-line planning of the control sets $\{S_f\}$ and $\{S_z\}$ for the desired output link velocity $\dot{\phi}_{2r}(t)$, smoothness motion $\Delta\dot{\phi}_2(t)$ and expected impacts and disturbances T_a. This allows open-loop disturbances and impacts rejection. The torque of interaction T_i between the actuator output link and TE can be expressed by the equation

$$T_i = T(\phi,\dot{\phi}) - J\frac{d\dot{\phi}}{dt} = K(\phi_r - \phi_a) + B(\dot{\phi}_r - \dot{\phi}_a) - J\frac{d\dot{\phi}}{dt} \tag{21.6}$$

where $T(\phi,\dot{\phi})$ is a non-inertial component of the actuator impedance and J the inertia tensor in task space. The dynamics of the TE end-effecter when it is on an ideal rigid body is expressed by

$$J_\Sigma \frac{d\dot{\phi}}{dt} = T_e + T_i \tag{21.7}$$

where J_Σ is the inertia tensor of the TE's end-effecter and T_e are unknown torques and impacts.

The motion equation of the system 'actuator shaft—TE' is

$$T_i = T(\phi,\dot{\phi}) - J\frac{d\dot{\phi}}{dt} = K(\phi_r - \phi_a) + B(\dot{\phi}_r - \dot{\phi}_a) - J\frac{d\dot{\phi}}{dt} \tag{21.8}$$

It is most important to assure smooth motion for the feeding operation, that is, $d\dot{\phi}/dt \approx 0$. Equation 21.6 will be

$$(J_\Sigma + J)\frac{d\dot{\phi}}{dt} = K(\phi_r - \phi_a) + B(\dot{\phi}_r - \dot{\phi}_a) + T_e - J\frac{d\dot{\phi}}{dt} \tag{21.9}$$

Equation 21.9 means to reject the disturbances at some impacts by variation of the actuator knot mechanical impedance Z_0. The desired actuator shaft response $\phi_a(t)$ and $\dot{\phi}_a$ to the reference motion ϕ_{2r} and $\dot{\phi}_{2r}$ and the external torques T_i are defined by the desired mechanical impedance (Equation 21.6). In this way, the control strategy of the actuator with drive redundancy is based on adjusting the actuator knot mechanical impedance parameters with given disturbances or impact T_i and reference parameters of motion–position error $\Delta\phi_2$ and velocity unevenness $\Delta\dot{\phi}_2$:

$$\begin{aligned}
\Delta\phi_2 &= \phi_{2r} - \phi_{2a}(T_i) \\
\Delta\dot{\phi}_2 &= \dot{\phi}_{2r} - \dot{\phi}_{2a}(T_i)
\end{aligned} \tag{21.10}$$

The open-loop impedance control involves the off-line planning of antagonistic motor voltages U_1 and U_3 such that the desired actuator knot mechanical impedance characteristics and the net effective load at some preliminary known impacts have been obtained. This allows open-loop disturbances and impacts rejection by general accommodation impedance control scheme shown in Figure 21.5.

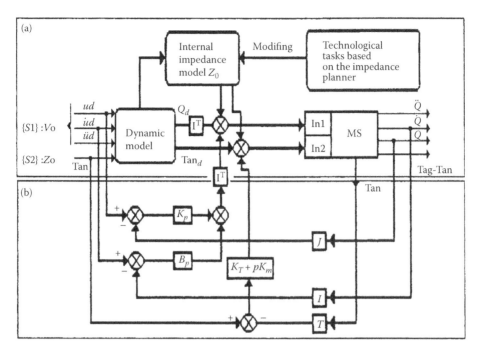

FIGURE 21.5   General accommodation impedance control scheme. (a) Feedforward controller. (b) Feedback controller. $Q, \dot{Q}, \ddot{Q}$: mechatronic system task coordinates.

The second part is a feedback controller used to compensate on-line small perturbation and non-modelled dynamics. Let us consider the equation of motion of the mechatronic feeding device with impedance-controlled actuators [26], which is characterized by the following parameters:

$$L = 1.37 \times 10^{-3}\,\text{H}; \quad R = 1.14\,\Omega; \quad m = 52; \quad n = 62/30;$$

$$T_{vT} = 0.1\,\text{ms}; \quad J_2 = 2.2136 \times 10^{-3}\,\text{kgm}^2; \quad J_r = 2.429 \times 10^{-3}\,\text{kgm}^2;$$

$$J_{2R} = 1.321 \times 10^{-3}\,\text{kgm}^2; \quad J_{1R} = 2.415 \times 10^{-5}\,\text{kgm}^2; \quad J = 4.11 \times 10^{-3}\,\text{kgm}^2;$$

$$B_2 = B_r = B_{2R} = B_{1R} = 0.1; \quad C = 9.6962 \times 10^6\,\text{Nm/rad}; \quad \bar{C} = 2.424 \times 10^5\,\text{Nm/rad};$$

$$k_{p1} = k_{p2} = 0.1; \quad k_{g1} = k_{g2} = 0.05; \quad k_{vT} = 1.5; \quad k = k_e = 0.172\,\text{Vs/rad}$$

A control synthesis is made by a program using the optimization method ARSTI (adaptive random search with translating intervals [27]). The optimal profiles of control voltages U_1 and U_3 can be obtained for desired actuator output link velocity and smoothness of motion. The obtained control voltage of the drive motor U_1, setting up the reference velocity $\dot{\varphi}_{2r}(t)$ of the actuator output link is shown in Figure 21.6. The corresponding velocity $\dot{\varphi}_{2a}(t)$ of the output link for the case with applied impact T_2 shown in Figure 21.7 is characterized with unevenness less than the reference for the reference angle φ_{2r} as it could be seen from Figure 21.8. Hence, the proposed control strategy for impedance control of mechatronic systems with drive redundancy does not need large computational resources and time in on-line control, while adjusting the knot mechanical impedance of the actuator allows open-loop impacts and disturbances rejection.

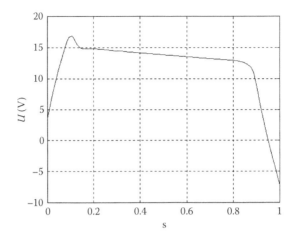

FIGURE 21.6   Control voltage of the drive motor.

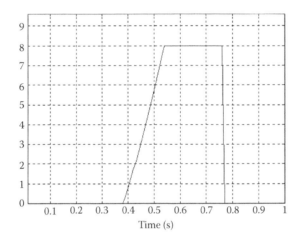

FIGURE 21.7   Expected impact T_2 (Nm).

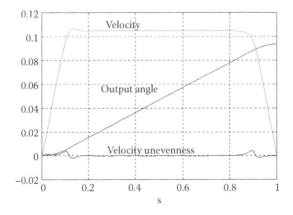

FIGURE 21.8   Parameters of output link motion.

21.4 EXPERIMENTAL RESULTS AND DISCUSSION

Three experimental test beds with actuator redundancy have been designed to investigate the possibilities of the MS with impedance control realized by the third approach:

- A drive unit for the barrier of a Langmuir–Blodgett monomolecular film deposition mechatronic system, which is a kinematic scheme (as shown in Figure 21.2)
- A rotary table for feeding operations in robot-assisted material removal with its transfer mechanism (as shown in Figure 21.3)
- Harmonic joint actuator (HawIC) for robots and peripheral mechatronic devices (Figure 21.9)

Experimental investigations on the rotary table for smooth feeding operations independent of the technological cutting forces are made using Bruel & Kjaer Dual Channel Signal Analyzer BK2034, accelerometer B12 for capturing the actuator response due to excitation, amplifier BK2635 with linear characteristic in the ratio 1 Hz to 10 kHz and with non-linearity of 10% for frequency less than 1 Hz, and impact hammer with force sensor BK8200 for excitation. The frequency response function $H_2 = F/\ddot{\phi}_2$ has been obtained for output link velocity $\dot{\phi}_2 = 0.105$ rad/s. The results for two different values of the actuator mechanical impedance are shown in Figure 21.10. The resonance frequency for the case of low impedance is 72.5 Hz and for the case of higher impedance is 107 Hz. For this case, the impedance ratio can be obtained, dependent on the displacement, as follows:

$$D_k = \frac{K_{max}}{K_{min}} = \frac{\omega_{max}^2}{\omega_{min}^2} = \frac{f_{max}^2}{f_{min}^2} = 2.18$$

For the same output link velocity, the impedance ratio, dependent on the velocity, is obtained for the same values of the actuator mechanical impedance. The damping is determined by the measuring procedure in the time domain consisting of measuring the response due to the excitation (Figure 21.11). For every measurement, the magnitudes A_1 and A_2 are registered. The damping of the impedance-controlled actuator is determined as a logarithm decrement: $B = \lg(A_2/A_1)$.

Hence, the impedance ratio dependent on the velocity is a ratio between the two values of damping for the cases with lower and higher impedance mentioned above, that is

$$D_B = \frac{B_{max}}{B_{min}} = \frac{\lg(A_2/A_1)_{max}}{\lg(A_2/A_1)_{min}} = 2.06$$

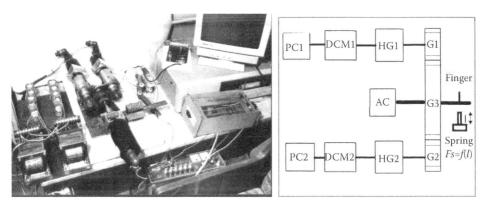

FIGURE 21.9 Harmonic actuator with impedance control (photo and kinematic scheme). PC: pulse coder; DCM: DC motor; HG: harmonic gear; G: gear; AC: axes encoder.

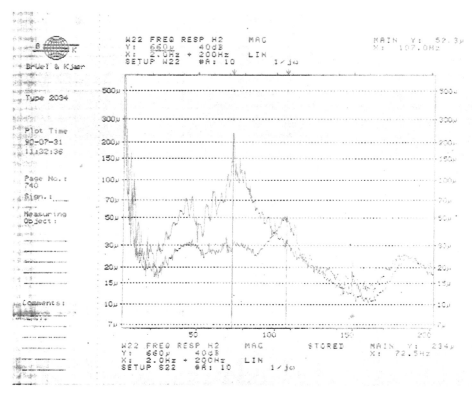

FIGURE 21.10 Spectra of frequency response function. Impedance ratio, dependent on the displacement, is $D_k = K_1/K_2 = f_1^2/f_2^2 = 107^2/72.5^2 = 2.18$. (The figure is obtained from the author's printer.)

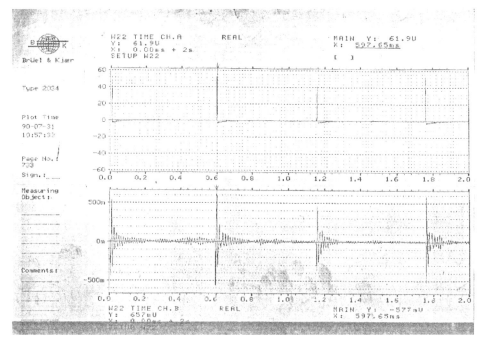

FIGURE 21.11 Experimental determination of damping. Impedance ratio, dependent on the velocity, is $D_B = B_1/B_2 = \lg(A_i/A_{ii})_1/\lg(A_i/A_{ii})_2 = 2.06$. (The figure is obtained from the author's printer.)

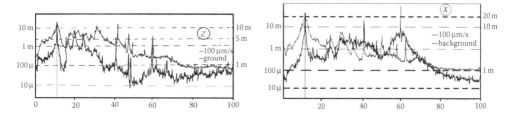

FIGURE 21.12   Acceleration spectra of the background and barrier motion with velocity 100 μm/s in z and x direction. MR: 10 m:10^{-4} m/s^2.

The redundancy actuator of the barrier of Langmuir–Blodgett monomolecular film deposition system is also experimental, investigated in regard of micro-motions, velocity ratio and the generated vibrations. The last is an important requirement for such devices since the vibrations generated could destroy the surface layer. The micro-motion of the barrier achieved is 1 μm, which particularly depends on the position sensor used at a range of motion (350 mm). A wide technological max-to-min velocity ratio (1:10,000) with smooth motion is also achieved. Figure 21.12 shows the results of the acceleration spectra of the background and barrier motion at a velocity of 100 μm/s in the z and x direction. The experimental results obtained show that the generated vibrations are negligible, less than the references. For the barrier motion, this difference between the backgrounds is less than 10 m (10^{-4} m/s^2) at 60 Hz. If the magnitudes for these frequencies around 60 Hz are high, then by the control of the actuator mechanical impedance, these magnitudes can be decreased. The results for x direction express the unevenness of barrier motion. For the three different values (low, medium and high), the knot mechanical impedance of the HAwIC experimental results are obtained. The relative displacement Δ between the drive motor and the antagonistic motor for these three values and two turnovers of joint shaft ϕ_2 with an impact during every turnover are shown in Figure 21.13a, b and c, respectively. The results obtained show the influence of the impacts on the correspondence of the

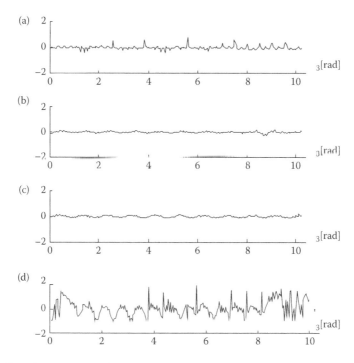

FIGURE 21.13   Experimental results for low (a), medium (b) and high (c) knot mechanical impedance compared with the conventional harmonic actuator (d).

position of the joint shaft with the antagonistic motor shafts is becoming negligible. This means that the HAwIC is characterized with significant improvement of the correspondence of the position information of the joint shaft absolute encoder with the information of the motor shafts encoders through the knot mechanical impedance modification, independent of the impact applied on the actuator shaft. In Figure 21.13d, this correspondence is shown for the conventional harmonic actuator. Accommodation control approach was adopted for the HAwIC and improvement in system dynamic response was observed under the influence of impacts and actuator mechanical impedance. Hence, the HAwIC can work under some impacts with high accuracy adjusting its knot mechanical impedance. In this case, the correspondence of the position information of the joint shaft absolute encoder with the information of the motor shafts encoders also is improved.

In conclusion, the main possibilities of impedance-controlled MS with drive redundancy consist of (i) controlled modification of the actuator mechanical impedance, (ii) micro-motions of the output link—particularly for the barrier—1 μm, at a range of motion (350 mm) and a big range of loading of the output link and (iii) smooth motion in a wide technological max-to-min velocity ratio (1:10,000) regardless of the force interaction at adjusting the actuator impedance with the TE's impedance.

21.5   SERIAL–PARALLEL MANIPULATORS WITH ANTAGONISTIC ACTUATORS

The studied manipulators [28] are serial–parallel structures including basic link 0 and some other links 1, …, n connected in between in a serial chain. The end-effecter M is situated in the end link n of this chain, which moves in a v operation space. The driving chains A_1, …, A_m, with number m, are attached to the basic link 0 and to the end link n, forming parallel chains (as shown in Figure 21.14).

21.5.1   KINEMATIC MODELLING OF SERIAL–PARALLEL MANIPULATORS WITH ACTUATION REDUNDANCY

These chains consist of similar type drive modules with one single drive joint for a linear motion and two passive rotational joints. The number of the DoF of each parallel chain is equal to three in the planar case and six in the spatial one. In this way, the number of the manipulator DoF is defined by the number of the DoF of the basic chain h. The number of DoF h of the studied manipulators with actuation redundancy should be less than the number of the drive modules m ($h < m$). The parameters of the relative motion in the drive linear joints of the parallel chains

$$\lambda = \left[\lambda_1, \ldots, \lambda_m\right]^T \tag{21.11}$$

and the parameters of the relative motion in the joints of the basic chain

$$q = \left[q_1, \ldots, q_h\right]^T \tag{21.12}$$

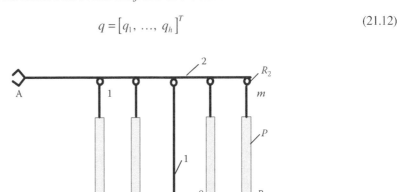

FIGURE 21.14   Generalized kinematic scheme of a serial–parallel manipulator.

are selected as generalized parameters for the kinematic system definition. Every closed loop in the parallel manipulator structure produces links between the generalized parameters in Equations 21.11 and 21.12. These links can be defined by means of the vector function $\lambda = \Phi(q)$ which allows to calculate the parameters in Equation 21.11 from parameters in Equation 21.12. Part of the parameters in Equation 21.11 with number h, $\lambda_u = [\lambda_1, \ldots, \lambda_h]^T$, which are sufficient for the kinematics system definition, are called independent parameters, and the remaining part of the parameters with number $(m - h)$, $\lambda_d = [\lambda_{h+1}, \ldots, \lambda_m]^T$ are called dependent parameters. Then, the vector of Equation 21.11 will be

$$\lambda = \left[\lambda_u; \lambda_d\right]^T \tag{21.13}$$

The differentiation of $\lambda = \Phi(q)$ with respect to time is

$$\dot{\lambda} = L\dot{q} \tag{21.14}$$

and defines $(m \times h)$ matrix of the first partial derivatives

$$L = \left[\frac{\partial \lambda}{\partial q}\right] = \left[L_u; L_d\right]^T \tag{21.15}$$

comprising of $(h \times h)$ matrix $L_u = [\partial \lambda_u/\partial q]$ and $(m - h) \times h$ matrix $L_d = [\partial \lambda_d/\partial q]$.

The matrix of the partial derivatives of all the parameters in the drive joints (see Equation 21.11) with respect to the independent parameters λ_u is defined with the help of the above matrixes as

$$\left[\frac{\partial \lambda}{\partial \lambda_u}\right] = \left[\begin{array}{c} E \\ L_d L_u^{-1} \end{array}\right] \tag{21.16}$$

where E is the unity $(h \times h)$ matrix.

The double differentiation of Equation 21.16

$$\ddot{\lambda} = L\ddot{q} + \dot{q}^T H \dot{q} \tag{21.17}$$

produces $(m \times h \times h)$ matrix of the second-order partial derivatives

$$H = \left[\frac{\partial L}{\partial q}\right] = \left[H_u, H_d\right]^T \tag{21.18}$$

Each plane above, H_i, $i = 1, \ldots, m$ is a $(h \times h)$ matrix with elements:

$$H_{i,j,k} = \frac{\partial^2 \lambda_i}{\partial q_k \partial q_j} = \frac{\partial}{\partial q_k}\left(\frac{\partial \lambda_i}{\partial q_j}\right) \quad j = 1, \ldots, h; \quad k = 1, \ldots, h$$

The link between the parameters of the basic chain, Equation 21.12 and the end-effector coordinates

$$X = \left[X_1, \ldots, X_v\right]^T, \quad v \leq 6 \tag{21.19}$$

is defined by means of the direct problem of the kinematics of the serial manipulators. This problem on the level of displacements, velocities and accelerations is presented by means of the equations $X = \Psi(q)$, $\dot{X} = J\dot{q}$ and $\ddot{X} = J\ddot{q} + \dot{q}^T G\dot{q}$, where

$$J = \left[\frac{\partial X}{\partial q} \right] \tag{21.20}$$

is the $(v \times h)$ matrix of Jacoby and

$$G = \left[\frac{\partial J}{\partial q} \right] \tag{21.21}$$

is the $(v \times h \times h)$ matrix of the second-order partial derivatives.

21.5.2 STIFFNESS MODEL AT REDUNDANT ACTUATION

The manipulation system which is in contact with the environment is considered to be in a static balance without consideration of the gravitation forces [29]. We can define the driving forces in the linear joints of the parallel chains with

$$F = \left[F_u ; F_d \right]^T \tag{2.22}$$

where F_u and F_d are vectors of the driving forces corresponding to the independent parameters λ_u and the dependent parameters λ_d. The forces $F_u = [F_1, \ldots, F_h]^T$ are sufficient for the manipulator motion along a given trajectory and are considered as basic forces, while the forces

$$F_d = \left[F_{h+1}, \ldots, F_m \right]^T \tag{21.23}$$

are considered as additional driving forces.

The effective generalized forces in the linear joints corresponding to the independent parameters λ_u can be defined by means of the $(h \times 1)$ vector:

$$U = \left[U_1, \ldots, U_h \right]^T \tag{21.24}$$

The effective generalized torques in the joints in the basic serial chain corresponding to the parameters in Equation 21.12 can be defined by means of the $(h \times 1)$ vector:

$$Q = \left[Q_1, \ldots, Q_h \right]^T \tag{21.25}$$

The external forces and torques applied at the end-effecter and corresponding to the coordinates in Equation 21.19 can be defined by means of the $(v \times 1)$ vector:

$$P = \left[P_1, \ldots, P_v \right]^T, \quad v \le 6 \tag{21.26}$$

The relations between the forces cited above can be defined according to the principle of the virtual work. In this way, from Equation 21.16 follows the relation between the driving forces, Equation 21.22 and the effective generalized forces Equation 21.24:

$$U = \left[\begin{array}{c} E \\ L_d L_u^{-1} \end{array} \right]^T F \tag{21.27}$$

or

$$U = F_u + \left[L_d L_u^{-1}\right]^T F_d \tag{21.28}$$

The relation between the generalized effective torques, Equation 21.25 in the joints of the basic chain and driving forces in the linear joints of the parallel chains can be defined with the help of Equation 21.14:

$$Q = L^T F \tag{21.29}$$

The link between the external forces and torques (Equation 21.26) and the effective generalized torques, Equation 21.25, can be defined using Equation 21.20:

$$Q = J^T P \tag{21.30}$$

Equations 21.29 and 21.30 define the link between the external and driving forces:

$$J^T P = L^T F \tag{21.31}$$

Differentiation of Equation 21.31 with respect to parameters in Equation 21.12 after consideration of Equations 21.20 and 21.21 as well as Equations 21.18 and 21.15 produces the following equation:

$$G^T P + J^T \frac{\partial P}{\partial X} J = H^T F + L^T \frac{\partial F}{\partial \lambda} L \tag{21.32}$$

Equation 21.32 defines the manipulator as an elastic system in which the end-effector stiffness $K = \partial P / \partial X$ is defined by the axes stiffness in the linear drive joints $K_{sh} = \partial F / \partial \lambda$ and the latent stiffness described by the expressions $G^T P$ and $H^T F$. The latent stiffness is a result of the static equilibrium in the contact process, Equation 21.31 between the external forces, Equation 21.26 and the driving forces, Equation 21.22 on the one hand, and a result of the balance, Equation 21.28 between the basic and the additional driving forces F_u and F_d. The generated stiffness as a result of the antagonistic force equilibrium is called the antagonistic stiffness. The equality in Equation 21.32 can be defined using Equations 21.18, 21.22 and 21.28:

$$-H_u^T I_u^{-T} J^T P + G^T P + J^T K J = -H_u^T L_u^T L_d^T F_d + H_d^T F_d + L^T K_{sh} L \tag{21.33}$$

Actuation-redundant manipulators allow specification of the end-effecter stiffness by distribution of the magnitudes of the additional driving forces, Equation 21.23 according to Equation 21.33. It is convenient in the process of simulations and control to use the reverse stiffness matrix K

$$B = K^{-1} \tag{21.34}$$

called the compliance matrix of the end-effector. The compliance of the end-effector which is specified in the process of control is investigated without consideration of the external forces. According to Equations 21.33 and 21.34 and $P = 0$, the end-effector compliance defined in the operation space can be expressed as

$$B = J[-H_u^T L_u^{-T} L_d^T F_d + H_d^T F_d + L^T K_{sh} L]^{-1} J^T \tag{21.35}$$

21.5.3 Conditions for Specification of Desired End-Effector Compliance

The stiffness control realization for control of manipulators with actuation redundancy is linked with specification of a suitable end-effecter stiffness or compliance. This stiffness is a reason for generation of restoring forces, as an expected response to the external excitation due to a contact.

During the performance of feedback stiffness control [23], the end-effector compliance (Equation 21.35) is specified by means of stiffness selection in the drive joints $K_{sh} = \partial F/\partial \lambda$ or else by means of proper selection of the magnitudes of the additional driving forces F_d (Equation 21.23) in the realization of antagonistic actuation stiffness control [22,24]. It is necessary that the partial derivative matrixes J, L and H not to be singular in the process of stiffness specification [17]. Also, it has to be considered that the symmetric compliance matrix, Equation 21.34, is positively defined and also with positive eigenvalues or

$$\det B > 0 \text{ and } b_{ii} > 0 \qquad (21.36)$$

where b_{ii} are the diagonal matrix components.

The symmetrical compliance $(v \times v)$ matrix (Equation 21.34) has $\mu = v \times (v + 1)/2$ independent components. The manipulator must have μ additional driving joints theoretically for the full compliance matrix specification according to Equation 21.35. Each drive joint is considered a generator of an antagonistic acting force (element of F_d) (Equation 21.23). The general number of the drive joints must be $\mu + h$ because for the manipulator motion control with h DoF are necessary h number of drive joints. The parallel manipulator must possess $m = \mu + h$ number of parallel chains if every parallel chain possesses one drive joint.

The desired end-effecter compliance can be represented by the vector

$$b^0 = \left[b_1, ..., b_j, ..., b_\mu \right]^T \qquad (21.37)$$

including the independent components of the matrix of Equation 21.34.

The desired compliance, Equation 21.37, can be specified by μ number of components of the vector of the additional driving forces F_d. In this case, the linear equation system derived from Equation 21.35 is

$$\begin{vmatrix} b_1 = f_1(F_d) \\ b_j = f_j(F_d) \\ b_\mu = f_\mu(F_d') \end{vmatrix} \qquad (21.38)$$

The upper system solutions can be derived using unrealistically high values of the driving forces, Equation 21.23. The possibilities to find solutions of system of Equation 21.38 are narrowed due to the existence of practical restrictions in the magnitudes of these forces. Solving the problem for the specification of a given compliance using restricted parameters can be performed by means of optimization. In this case, there are difficulties in defining the real forces F_d, which generate the compliance matrix satisfying limits in Equation 21.36. Computer experiments are carried out in order to investigate the correlation of the generated compliance with respect to inequality in Equation 21.36. The carried-out experiments are two-dimensional (2D), where the compliance in the operation space is defined by a 2D linear compliance matrix.

$$B^L = \begin{bmatrix} b_{xx} & b_{xy} \\ b_{xy} & b_{yy} \end{bmatrix} \qquad (21.39)$$

The next coefficient is derived on the basis of Equation 21.36 for the 2D case from Equation 21.39:

$$\kappa = \frac{b_{xy}}{\sqrt{b_{xx} b_{yy}}} \tag{21.40}$$

The values of this coefficient, which are among the limits $-1 < \kappa < 1$ define stiffness generation. Besides, this coefficient can be used indirectly for qualitative evaluation of this stiffness. If compliance is represented in a graphic way by means of compliance ellipse, the values of κ close to 1 define the biggest angle of deviation of the main ellipse axes of the coordinate axes, while at $\kappa = 0$, the ellipse axes coincide with the coordinate axes. Four areas of orientation of the main compliance ellipse axes are formed for the generated stiffness evaluation assigned with A, B, C and D (as shown in Figure 21.15). The four areas are identified by means of the coefficient magnitude, Equation 21.40, and the relation of the elements b_{xx} and b_{yy} as cited below:

$$\begin{array}{l} \text{A: } -0.5 \le \kappa \le 0.5 \quad \text{and} \quad b_{xx} < b_{yy}\text{; } \text{B: } 0.5 \le \kappa \le 1 \\ \text{C: } -0.5 \le \kappa \le 0.5 \quad \text{and} \quad b_{xx} > b_{yy}\text{; } \text{D: } -1 \le \kappa \le -0.5 \end{array} \tag{21.41}$$

The experiments are performed using a 2D parallel manipulator (as shown in Figure 21.16). The manipulator consists of a basic chain with one immovable and two movable links l_0, l_1, l_2 ($l_1 = 0.4$ [m]) linked by means of two rotational joints J_1 and J_2 ($h = 2$). The manipulator can include in the structure 1–6 parallel drive chains attached to the immovable base l_0 and to link l_2 (as shown in Figure 21.16). The parallel chains are shown in Figure 21.16 with numbers 1, 2, 3, 4, 5, 6. The end-effector compliance is evaluated with a discrete variation of the magnitudes and the direction of the driving forces in the limits:

$$-300 \le F_i \le 300, \quad i = 3, \dots, 6, \ [N] \tag{21.42}$$

The stiffness generation is defined by the values of the coefficient, Equation 21.40, which are situated in the areas, Equation 21.41. The stiffness generation at a given force distribution is graphically interpreted by one of the signs — \ | / corresponding to the areas A, B, C and D. Several structural variants are considered varying the number of the manipulator driving chains. First of

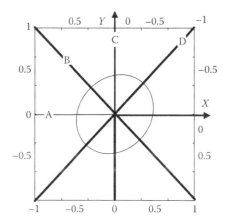

FIGURE 21.15 Areas of orientation of the main axes of the compliance ellipses.

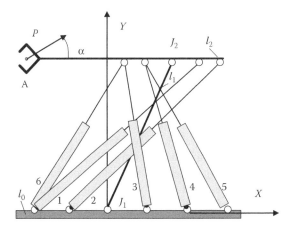

FIGURE 21.16 Parallel manipulator with six driving chains.

all a manipulator shown in Figure 21.16 with four parallel chains marked with 1, 2, 4 and 5 is considered. Discrete values within the limits, Equation 21.42, are assigned to the additional forces F_d in two driving joints (4 and 5). The experimental results are shown in the graphics in Figure 21.17a. The values of the additional forces F_{d4} and F_{d5} are pointed out along the coordinate axes. A solution of a variant of a manipulator with five parallel chains (1, 2, 3, 4, 5 in Figure 21.16) is considered. The additional driving forces F_{d3}, F_{d4}, F_{d5} in the three joints 4, 5, 6 vary within the limits, Equation 21.42. The results for stiffness generation are shown in Figure 21.17b, where the same coordinate axes (F_{d4} and F_{d5}) are used. There exist cases of different orientation of the compliance ellipse axes for one value of the forces F_{d4} and F_{d5} on the graphics shown in Figure 21.17b, which are derived by variation of the third force F_{d3}. The last structural solution revealed is of a manipulator variant with six parallel chains (1, 2, 3, 4, 5, 6 in Figure 21.16). The forces F_{d3}, F_{d4}, F_{d5}, F_{d6} receive discrete values within the limits (Equation 21.42). The results are shown in a similar way in Figure 21.17c. The basic conclusions of the results of the carried-out experiments are: (1) the effect of the antagonistic stiffness does not exist for all values of the driving forces in spite of the existence of an antagonistic equilibrium among them; (2) the possibility for orientation variation of the compliance ellipse axes by means of variation of the magnitudes and orientations of the additional forces is limited and (3) increasing the number of the additional driving chains gives possibilities for axes orientation variation of the compliance ellipse and widens the range where the forces generate stiffness.

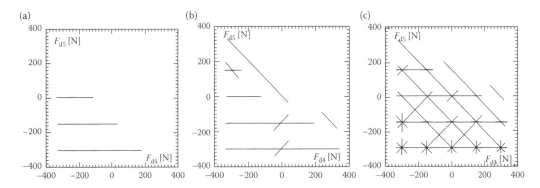

FIGURE 21.17 Variation of the main axis orientation of the compliance ellipse by means of variation of the magnitudes and orientations of the additional forces by different number of active parallel chains: (a) four parallel chains marked with 1, 2, 4, 5; (b) five parallel chains marked with 1, 2, 3, 4, 5; (c) six parallel chains marked with 1, 2, 3, 4, 5, 6.

21.5.4 SOME APPROACHES FOR COMPLIANCE SPECIFICATION

Specification of a random compliance matrix of a manipulator by means of antagonist driving forces distribution is not always possible according to the above-cited conditions. On the other hand, the full compliance matrix specification is not necessary for many case problems. For series technological tasks specification is sufficient only for certain components of the compliance matrix or specification is enough only along a single direction in the operation space or guaranteeing is needed for the upper compliance limits in the operation space. Some approaches are developed for these reasons which do not include specification of the full compliance matrix. The following optimization procedures are carried out for these approaches and verified by computer experiments.

21.5.4.1 Specification of the Upper Compliance Limits in the Operation Space

In this way, the compliance is restricted in all directions up to a defined upper limit. The biggest compliance value in the operation space b_{max} is selected from the bigger one among the main compliances b_1 and b_2 (if $b_1 > b_2$ then $b_{max} = b_1$ and vice versa). The main compliances are linear compliance matrix eigenvalues and can be defined by the equation

$$\det\left[B^L - B^0\right] = 0$$

where

$$B^0 = \begin{bmatrix} b_1 & 0 \\ 0 & b_2 \end{bmatrix}$$

If the desired upper compliance limit is defined as b_d, then the reference function is

$$G_A = \left(b_{max} - b_d\right)^2 \tag{21.43}$$

Assuming that the joint stiffness k_{sh} are constant, then a zero minimum of Equation 21.43 is to be defined with respect to the additional driving forces F_d

$$\min G_A\left(F_d\right) \Rightarrow 0 \tag{21.44}$$

The value $b_{max} - b_d \geq 0$ is satisfied in the optimization performed.

21.5.4.2 Compliance Specification along a Given Direction in the Operation Space

The desired direction in the 2D solution is defined by the unity vector: $P = [p_x; p_y]^T$, where $P^T P = 1$. The compliance in the same direction is defined by the equation $b_p = P^T B^L P$ where B^L is the linear compliance matrix, Equation 21.39. The compliance in the given direction P is defined by the magnitudes of the additional driving forces, Equation 21.23. If we mark the desired compliance in the given direction with b_{pd}, then the following aim function can be defined:

$$G_B = \left(b_P - b_{Pd}\right)^2 \tag{21.45}$$

and we are looking for a zero minimum of Equation 21.45 with respect to the driving forces F_d:

$$\min G_B\left(F_d\right) \Rightarrow 0 \tag{21.46}$$

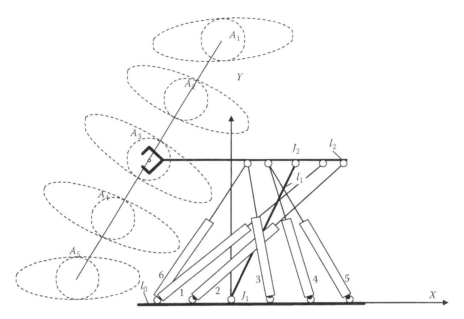

FIGURE 21.18 Compliance ellipses in the motion trajectory points corresponding to the forces presented in Table 21.3.

The value $b_p - b_{pd} \geq 0$ is satisfied in the optimization performed. Computer experiments are carried out for the manipulator shown in Figure 21.16 using the above-cited approach A. The biggest end-effecter compliance value is specified at a defined upper limit $b_d = 0.005$ [m/N]. The compliance in several points $(A_1, A_2, A_3, A_4, A_5)$ is specified from a single trajectory S in the operation space as shown in Figure 21.18. The distribution is to be defined of four additional driving forces assigned as F_3, F_4, F_5, F_6. The function, Equation 21.44, is minimized with respect to the forces using a computer method of adaptive random search [28]. The calculated optimal values of the additional forces F_3, F_4, F_5, F_6 for each trajectory point are presented in Table 21.3. This table shows the magnitudes of the forces F_1 and F_2 according to Equation 21.28 as well as the averaged value of all the forces marked with F. Figure 21.18 shows the compliance ellipses in the trajectory points $(A1, A2, A3, A4, A5)$ generated by driving forces presented in Table 21.3. In Figure 21.18, the circular ellipse of the limit compliance $(b_1 = b_2 = b_d)$ is shown. The experiment performed show that the approaches accepted allow easy and effective definition of the desired stiffness without full compliance matrix specification. The first approach is a more common one restricting the compliance along all directions due to the higher driving forces values. The dimensions of these forces is diminished at a higher drive number. The previously cited experiments are carried out using zero stiffness values in the joints k_{sh}. The estimation of the constant

TABLE 21.3

Optimal Values of the Driving Forces for Each Trajectory Point

	F_1	F_2	F_3	F_4	F_5	F_6	F
A_1	−225	−985	−247	−244	−246	−247	−366
A_2	−821	−507	−324	−317	−324	−326	−436
A_3	−1117	−404	−433	−446	−452	−439	−548
A_4	−628	−612	−442	−452	−459	−444	−506
A_5	−49	−1015	−387	−389	−397	−387	−437

values of the joint stiffness in the performed optimization results in diminishing the values of the necessary driving forces and appearing as the joint elastic deformations.

21.6 CONCLUDING REMARKS

Development of mechanical impedance-controlled mechatronic systems by actuation redundancy implements the third approach for mechanical impedance control. Based on this approach, intelligent autonomous mechatronic systems with such actuators can accomplish very fine motion, regardless of their dynamic interaction with remaining technological equipment or environment. The developed kinematic model of parallel manipulators and a stiffness model at actuation redundancy are a basis for investigation on the mechanical properties of parallel manipulators and for developing control strategies. The presented investigations of the stiffness show that the existence of the antagonist equilibrium is not always sufficient for specification of desired stiffness of the end-effector. The possibilities for this effect are bigger with a higher number of the redundant actuators. Approaches which do not require specification of the full compliance matrix are developed for solving this problem. The investigations carried out on the antagonist stiffness can be used in autonomous intelligent mechatronic system fulfilling contact tasks where the generated antagonist stiffness gives the system response to the arbitrary disturbance forces.

REFERENCES

1. Hogan, N., Impedance control: An approach to manipulation, *Transactions of ASME Journal*, DSMS, 1, 1985, 1–23.
2. Kostadinov, K.Gr., Accommodation control in the drive dynamic accuracy for positioning robot, 38 *Int. wissenschaftliches kolloquium*, Tagungsband, ss. 100–108, Ilmenau, 20-23.09, 1993.
3. Lu, Z. and Goldenberg, A., Robust impedance control and force regulation: Theory and experiments, *The International Journal of Robotics Research*, 14(3), 1995, 225–254.
4. Lee, S., Yi, S.-Y., Park, J.-O., Lee C.-W., Reference adaptive impedance control and its application to obstacle avoidance trajectory planning, *Intelligent Robots and Systems. IROS '97, Proceedings of the IEEE/RSJ International Conference*, 7–11 September 1997, Grenoble, France, vol. 2, pp. 1158–1162.
5. Chan, S.P. and Liaw, H.C., Experimental implementation of impedance based control schemes for assembly task, *Journal of Intelligent & Robotic Systems*, 29, Nr.1, 2000, 93–110.
6. García-Valdovinos, L.-G., Parra-Vega, V. and Arteaga, M.A., Observer-based sliding mode impedance control of bilateral teleoperation under constant unknown time delay, *Robotics and Autonomous Systems*, 55(8,31), 2007, 609–617.
7. Cheah, C.-C. and Wang, D., Learning impedance control for robotic manipulators, *IEEE Transactions on Robotics and Automation*, 14(3), 1998, 452–465.
8. Liu, G. and Andrew A., Goldenberg: Comparative study of robust saturation-based control of robot manipulators: Analysis and experiments, *International Journal of Robotic Research*, 15(5), 1996, 473–491.
9. Caccavale, F., Natale, C., Sicoliano, B. and Villani, L., Six-DOF impedance control based on angle/axis representation, *IEEE Transactions on Robotics and Automation*, 15(2), 1999, 289–300.
10. Tsuji, T., Jazidie, A. and Kaneko, M., Hierarchical control of end-point impedance and joint impedance for redundant manipulators, *Systems, Man and Cybernetics, Part A: Systems and Humans, IEEE Transactions on*, 29(6), 1999, 627–636.
11. Jacobsen, S.C., Ko, H., Iversen, E.K. and Davis C.C., Antagonistic control of a tendon driven manipulator, *Proceedings of the IEEE International Conference on "Robotics and Automation"*, May 14–19, 1989, Scottsdale, Arizona, USA, pp. 1334–1339.
12. Mittal, S. Tasch, U. and Wang, Y., A redundant actuation scheme for independent modulations of stiffness and position of a robotic joint: Design, implementation and experimental evaluation, DSC-Vol.49, *Advances in Robotics, Mechatronics, and Haptic Interfaces*, ASME, 1993, pp. 247–256.
13. Dasgupta, Bh. and Mruthyunjaya, T.S., Forse redundancy in parallel manipulators: Theoretical and practical issues. *Mechanism and Machine Theory*, 33(6), 1998, 727–742.
14. Tadokoro, S., Control of parallel mechanisms, *Journal of Advanced Robotics*, 8(6), 1994, 559–571.
15. Gardner, J.F., Kumar, V. and Ho, J.H., Kinematics and control of redundantly actuated closed chains, *IEEE International Conference on Robot and Automation*, Scottsdale, 1989, pp. 418–424.

16. Nahon, M.A. and Angeles, J., Force optimization in redundantly—Actuated closed kinematic chains, *IEEE International Conference on Robot and Automation*, Scottsdale, 1989, pp. 951–956.

17. Byung-Ju Yi and Freeman, R., Geometric characteristics of antagonistic stiffness in redundantly actuated mechanisms, *IEEE International Conference on Robot and Automation*, Atlanta, 1993, pp. 654–661.

18. Byung-Ju Yi, Il Hong Suh and Sang-Rok Oh, Analysis of a 5-bar finger mechanism having redundant actuators with applications to stiffness, *IEEE International Conference on Robot and Automation*, Albuquerque, 1997, pp. 759–765.

19. Merlet, J.-P., *Les Robots paralleles*, Hermes, Paris, 1990.

20. Gosselin, C., Stiffness mapping for parallel manipulators, *IEEE Transactions on Robotics and Automation*, 6(3), 1990, 377–382.

21. Tarn, T., Bejczy, A. and Yun, X., Design of dynamic control of two cooperating robot arms: Closed chain formulation, *Proceedings of the 1987 IEEE International Conference on Robotics and Automation*, Raleigh, North Carolina, March 31–April 3, 1987, vol. 4, pp. 7–13.

22. Byung-Ji Yi, Freeman, R. and Tesar, D., Open-loop stiffness control of over constrained mechanisms/robotic linkage systems, *IEEE International Conference on Robotics and Automation*, Scottsdale, 1989, pp. 1340–1345.

23. Yokoi, K., Kaneko, M. and Tanie, K., Direct compliance control of parallel link manipulators, *8th CISM—IFToMM Symposium, Ro.man.sy'90*, Cracow, 1990, pp. 224–251.

24. Kock, S. and Schumacher, W., A parallel *x-y* manipulator with actuation redundancy for high-speed and active-stiffness applications, *IEEE International Conference on Robot and Automation*, Leuven, Belgium, 1998, pp. 2295–2300.

25. Kostadinov, K.Gr. and Parushev, P.R., *Method of Motion Transition*, Bulg. Patent No.44365, 30.06.1987.

26. Kostadinov, K., *Synthesis and Investigation of the Electromechanical Drives for Positioning Robots*, PhD thesis, Institute of Mechanics and Biomechanics, 1994, Sofia (in Bulgarian).

27. Edissonov, I., The new ARSTI optimization method: Adaptive random search with translating intervals, *American Journal of Mathematical and Management Sciences*, 14, 3&4, 1994, 143–166.

28. Chakarov, D., Study of the passive compliance of parallel manipulators, *Mechanism and Machine Theory*, 34(3), 1999, 373–389.

29. Chakarov, D., Study of the antagonistic stiffness of parallel manipulators with actuation redundancy, *Mechanism and Machine Theory*, 39/6, 2004, 583–601.

22 Hydro-MiNa Robot Technology for Micro- and Nano-Manipulation in Manufacturing of Micro-Products

Kostadin Kostadinov

CONTENTS

22.1 INTRODUCTION

Many scientific and industrial tasks in the domain of micro- and nano-technology require micro-/nano-operations such as pick-and-place, assembly, feeding and so on, combined with large up to several centimetres robotized manipulations (for some field robotics and medical robotics). Such complex tasks need one large-range robot and one or more low-range manipulators with high precision. Depending on the end-effector of the robot system, different applications are possible for biological and industrial micro- and nano-technological operations. To meet these requirements, the robot technology for precise manufacturing and operation has to offer a robot able to be designed with high resolution of motion, high repeatability and high bandwidth capabilities, while the robot's dimensions could be in the macro-scale range with sizes of tens to hundreds of millimetres [1]. Development of such micro-robot meeting all these requirements involves kinematic schemes, mechanical joints, materials, fabrication, actuators, sensors and control schemes. Compliant parallel micro-manipulators featuring with parallel kinematic structure with various types of motion and flexure hinge-based joints are preferred for micro- and nano-applications [1,2]. Some micro-manipulators utilize piezo- (PZT) actuators or bimorph piezo-ceramic plates [2–7] which offer substantial advantages for biological cell manipulation such as large force, high frequency and small displacements with high accuracy [4,8]. In the area of micro- and nano-technologies the tele-operation method of control is widely used. It is accompanied by attempts to entirely automate the control of some micro- and nano-operations [3,7,9–11]. Most of the robotic systems have a specialized vision module providing real-time control of the process. The number of cameras, their orientation, objective's magnification and resolution are of great importance. The optical module is the feedback of many robotic systems providing visual control of the position and the orientation of the working tool and objects [2,12]. In the hybrid assembly approach, the aim of the European integrated FP6 project #026622 Hydromel the self-assembly technology [13] is combining with high-performance robotic tools such as precise manipulators with submicron resolution and mechatronic handling or feeding devices, innovative vision to detect either the micro-objects or the tip of the technological end-effectors, force sensing and robot system control. This chapter presents enabling robot technology including web application for automation of the synthesis of closed structures for micro- and nano-applications, utilizing the advantages of tense piezo-actuators and closed robot kinematical structures and developed automated and teleoperated control approaches based on the different scaling approaches and techniques and real-time control applications utilizing the optical sensing approach providing pipette tip detection and tracking with high submicron resolution.

22.2   HYDRO-MiNa ROBOT TECHNOLOGY

The enabling robot technology subject of this chapter includes web application for automation of the synthesis of closed structures for micro- and nano-applications, utilizing the advantages of tense piezo-actuators and closed robot kinematical structures. The algorithm, integrated into the developed web-based application [14], offers a synthesis of robot kinematic chains without extensive knowledge in this domain of the possible users from the domain of micro-products manufacturing to meet the user requirements. This technology also facilitates a synthesis of such kind of kinematic chains from specialists who can generate optimal solutions for automation and robotization of the requested micro- and nano-process with necessary specifications as over all dimensions, position accuracy and resolution, range of motion, force measurement or control. Design methodology based on precise functional task formulation, and development of appropriated regional and local robot systems as parts of modular robot system able to fulfil the desired functional task, is shown in Figure 22.1.

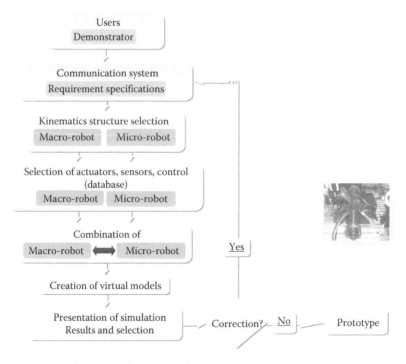

FIGURE 22.1   Hydro-MiNa robot design methodology.

22.2.1 UNIFIED APPROACH FOR FUNCTIONAL TASK FORMULATION IN THE DOMAIN OF MICRO-/NANO-MANIPULATIONS

Hydro-MiNa communication system is developed to gather necessary information from users utilizing micro- and nano-manipulation processes in order to formulate precisely their functional task. It is developed as an interactive programme for specifying the user requirements to the robotization of the precise manufacturing process [14] utilizing any micro- and nano-manipulation or operation tasks (Figure 22.2). The interactive simulation module developed allows specifying parameters and/or reference requirements for manipulation or operation tasks and illustrates the user about the features of the robot virtual model. The robot performing the requested/required technology task is designed as a combination of a regional and local robot structures realizing micro- and nano-motions. Regional robot structure or large-range robotic system is designed based on the modular concept [15]

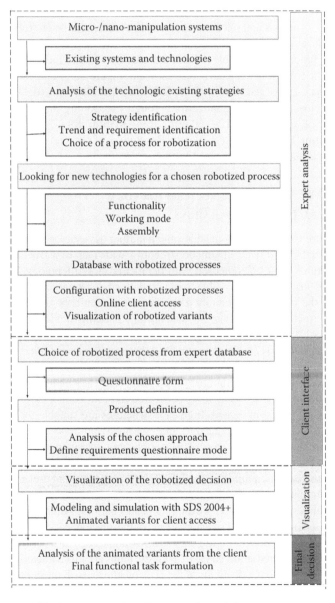

FIGURE 22.2 Algorithm block scheme for functional task formulation.

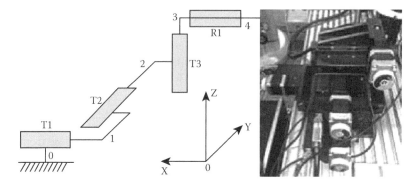

FIGURE 22.3   Kinematic scheme and prototype of the regional structure of the robot with 4 DoF.

and the procedure for synthesis and selection of optimal structures for macro robots [16]. As an example—macro robot developed for the automation of manufacturing of micro-products and for cell micro-manipulation functional tasks able to fulfil the necessary macro motion requirements has been characterized with the robot kinematic structure shown in Figure 22.3. The corresponding prototype (Figure 22.3) developed is characterized with the following specifications: (i) 3 translations (T1:T3), 1 rotation (R1) with closed loop; (ii) application—rough positioning and orientation of the glass pipette; (iii) range of motion—T1:T2—50 mm; R3—30 mm; R4—360°; (iv) working space–XYZΘ–50 mm/50 mm/30 mm/360°; (v) 0.1 μm Resolution of the end-effector in XYZ; and (vi) Max. XYZ speed—25 mm/s; 25 mm/s and 12 mm/s, respectively.

22.2.2   SYNTHESIS OF TENSE PIEZO STRUCTURES FOR LOCAL MICRO- AND NANO-MANIPULATIONS

A local robot structure is developed as mechatronic handling devices for micro- and nano-operation tasks using synthesized closed tense kinematic structures with 1 up to 3 DoF. In order to be tensed, piezo-ceramic structures must be composed with parallel or closed topology. To achieve tension in closed piezo-ceramic structures it is possible to use predefined deformation in elastic joints or antagonistic interaction of the redundant actuators. For this purpose, an approach has been developed for synthesis of closed structures appropriate for mechatronic handling devices (MHD) based on structured piezo-ceramics [2] or for MHD using piezo-stack actuators and single ceramic actuators for micro- and nano-applications [17]. The developed synthesis approach considers three cases: (a) when the structure includes basic links connected in between, only by means of the actuators, (b) when the structure includes basic links connected in between a serial chain to which are connected the actuators and (c) when the structure includes basic links connected in between in a parallel chain to which are connected the actuators. The synthesized three case structures are suitable for devices based on the structured piezo-ceramics, including polarized and non-polarized areas or for devices using piezo-stack actuators and single ceramic actuators. A web application for automation of the synthesis procedure is created [16] to facilitate the technology application. Using input parameters: degrees of freedom, number (DoF) of actuators and number of basic links, one user can easily select the suitable solution of closed structures for micro- and nano-applications, utilizing piezo-stack actuators or structured tense integrated piezo-ceramics. It is shown with examples of mechatronic handling devices for micro- and nano-applications based on synthesized structures. One of the developed local structures with 3 DoF for cell manipulations [18] shown here in Figure 22.4 as an example has the specifications: (i) stack piezo-actuators (A1:A3) with integrated strain gauges, (ii) elastic joints (J1:J3) characterized with no-backlash, (iii) application—fine positioning, orientation of the glass pipette and cell injection, (iv) motion Range—A1:A2—30 μm; A3—60 μm, (v) size of the working space: orientation around XY— 180 μm/180 μm and translation in Z—60 μm and (vi) resolution in XYZ—3.6 nm/3.6 nm/0.6 nm. Another example

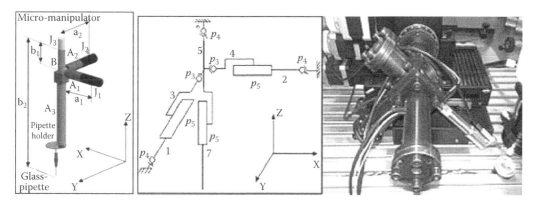

FIGURE 22.4   Local robot structure for cell micro-manipulations (simulation model, kinematic scheme and prototype).

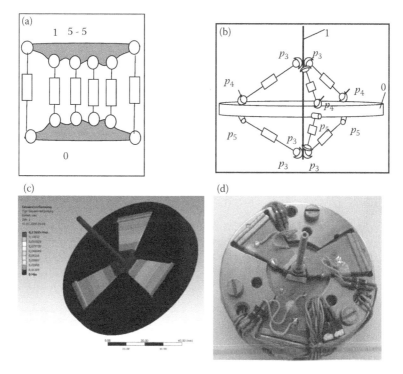

FIGURE 22.5   MHD with piezo-structured piezo-ceramics as a local robot structure with $n^0 = 2$ (type A) and $h = 3$, $m = 6$. (a) Structure, (b) kinematic schemes, (c) FE simulation model and (d) the optimized prototype developed).

is mechatronic device for cell handling operation with 3 DoF which is based on the kinematic structures type A with two links of the basic serial chain $n^0 = 2$ and three Dof $h = 3$. The number of the actuators **m** are bigger than the number DoF of the device $m = h + r$, or $m = 3 + 3 = 6$. The chosen structure **6–6** corresponds to the above parameters according to the introduced condition and is developed with the synthesized variants [5]. The structure is symmetric and each basic link can be chosen as an immovable one as it is shown in Figure 22.5a. There are different possible distributions of the revolution joints with 3, 4 or 5 DoF expressed by p_3, p_4 and p_5, respectively, such as

$$p_3 = 3, p_4 = 9$$
$$p_3 = 4, p_4 = 7, p_5 = 1$$

$p_3 = 5, p_4 = 5, p_5 = 2$
$\boldsymbol{p_3 = 6, p_4 = 3, p_5 = 3}$
$p_3 = 7, p_4 = 1, p_5 = 4$

On the basis of the above, a kinematics scheme with joints $\boldsymbol{p_3 = 6}$, $\boldsymbol{p_4 = 3}$ and $\boldsymbol{p_5 = 3}$ is built as shown in Figure 22.5b. The basic two links are: 0 and 1, while link 0 is chosen as an immovable one. A simulation of a device prototype with structured piezo-ceramics is shown in Figure 22.5c.

22.3 HYDRO-MiNa ROBOTIC SYSTEMS FOR MICRO- AND NANO-MANIPULATIONS

To exploit all possibilities offered by the technology of structured piezo-ceramics, a two-stage design method is developed [19]. The first step involves kinematic design of the actuator system specifying the kinematic structure and raw geometric data from application requirements like actuator force and motion. During the second step, optimization and fine tuning of geometric and material parameters could be performed in an iterative process based on a detailed FE-simulation of the

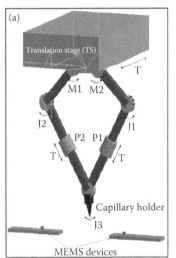
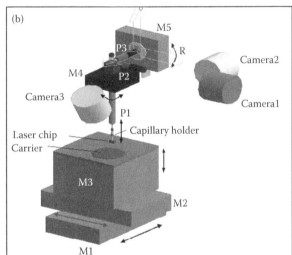

FIGURE 22.6 Virtual models of Hydro-MiNa robotic system (a) for micro-assembly of MEMS micro-grippers and force sensors; (b) laser chip inspection.

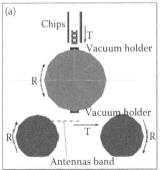
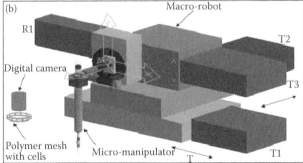

FIGURE 22.7 Virtual models of Hydro-MiNa robotic system for (a) micro-assembly of RFID tags and (b) cell injections.

mechatronic system. Virtual models of a Hydro-MiNa robotic system developed for other applications of micro-manufacturing such as micro-assembly of MEMS micro-grippers and force sensors, and for laser chip inspection, are shown in Figure 22.6a,b; and for micro-assembly of RFID tags, and for cell injections, are shown in Figure 22.7a,b. They are based on the synthesis and selection of optimal regional structures for macro robot motions [20]. An example of Hydro-MiNa robot with 7 DoF for cell injection with force sensing is presented in Figure 22.7 utilizing the enabling robot technologies to perform the requested cell operation tasks.

22.4 ROBOT SENSING AND CONTROL WITH HIGH-RESOLUTION IMAGING AND FORCE FEEDBACK

Appropriate optical system developed (Figure 22.8) provides high-resolution imaging of the injection pipette over the working area defined by the cells holder dimensions. Numerical algorithms for tip pipette detection and tracking during the working process are developed. The sub-pixel accuracy of these algorithms and high-precision linear measuring system integrated in the large-range robot allow precise calibration of the image space. In this way, the visual feedback is used for autonomous control of the whole Hydro-MiNa robot system and the tip pipette position with respect to the target cell. Optical system provides a resolution of 4 µm and a field of view 1×1 mm^2 while an integrated optical fibre attached to the glass pipette allows precise tracking of the pipette tip. Integrated red LED background illumination improves the pipette tip separation in the color space. Two different approaches for pipette tracking in the micro-scene are realized, that is, the pipette point tracking based on the normalized cross-correlation and pipette point tracking through joint transform correlation (Figure 22.8b). Software application integrates the developed image processing routines and the robot control. Numerical algorithms for pipette tip detection and autonomous tracking during the working process are developed with 0.1 µm resolution. The sub-pixel accuracy algorithms and high-precision linear measuring system integrated in the large-range robot allow precise calibration of the image space. In this way, the robot control system tracks the pipette with respect to the target cell. Once the cell's position is detected and defined in the image space, the injection process could be autonomously automated. Many scientific and industrial tasks in the domain of micro- and nano-technology require force sensors in µN/nN range. These sensors have to be integrated within manipulators and/or end-effectors. Such a complex device needs proper size/force scale matching of all system components. Depending on manipulated objects and end-effectors, different applications are possible for biological and industrial micro- and nano-technological operations. For cell injection it was shown that the cell-membrane reaction forces are in the range of 1–30 µN. Hence, a force sensor for sub-µN force range has been developed (Figure 22.9) providing the robot control system with sub-µN resolution feedback.

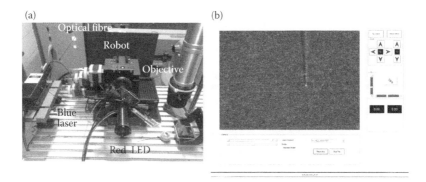

FIGURE 22.8 (a) Optical robot system for high-resolution imaging of the injection pipette. (b) Pipette point tracking through joint transform-correlation.

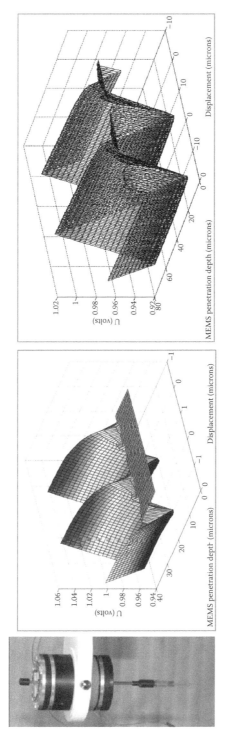

FIGURE 22.9   Prototyped force sensor and MΞMS output signal during tip penetration into soft layer and transversal displacement in ±1 μm and ±10 μm.

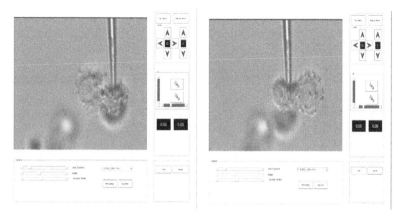

FIGURE 22.10   Screenshots of the injection process of single HTC using the developed Hydro-MiNa robotic system.

Automated and teleoperated control approaches [9,21] for the robots are developed based on the different scaling approaches and techniques and real-time control applications. This technology is based on the two software applications for telemanipulation and virtual operator self-learning. The first application realizing the method for self-learning based on the impedance method and impedance scaling approach [22] provides an improvement of the teleoperation operator control in the domain of micro-/nano-manipulations. The developed teleoperated control approach improves operator skills in teleoperated robotized micro-/nano-manipulation process. With the impedance scaling technique it is possible to sense the working micro-/nano-environment that is unknown and unnatural for humans. Thus, the operator can feel the boundaries of the working space and can control the velocity of the end-effectors. This approach is verified with the developed two micromanipulators with 3 DoF based on the structured piezo-ceramics actuators. Demonstrator systems with accent on the assembly, processing and manufacturing of micro-parts and products with dimensions less than 1 mm such as MEMS parts, RFID tags, bio-cells, optical systems and nanowires are chosen to prove the robot enabling technologies developed as a part of the hybrid assembly technology developed [13]. The robotized injection of two types of cells—*Xenopus* oocyte and hard to transfer cells (HTC) is achieved by the developed Hydro-MiNa robot. The cell injection of HTC with a size of 20–25 μm is shown in Figure 22.10.

22.5   CONCLUDING REMARKS

The enabling robot technologies developed includes web application for formulation of user functional task in terms of robotics and automation, automation of the synthesis of closed structures for micro- and nano-applications, utilizing the advantages of tense piezo-actuators and closed robot kinematical structures. The robot performing the requested technology task has been designed as a combination of a regional and local robot structures realizing the necessary micro and micro motions. The kinematic schemes for local robot structure are synthesized based on the stack piezo-actuators or on the structured piezo-ceramics with a tense closed kinematic structure. The algorithm, integrated into the developed web-based application, offers a synthesis of robot kinematic chains without extensive knowledge in this domain of the possible users from micro-products manufacturing field. Automated and teleoperated control approaches for the robots are developed based on the different scaling approaches and techniques and real-time control applications. This technology is based on the two software applications for telemanipulation and virtual operator self-learning. The first application realizing the method for self-learning based on the impedance method and impedance scaling approach provides the improvement of the teleoperation operator control in domain of micro-/nano-manipulations. In the frame of the EC FP6 project HYDROMEL a Hydro-MiNa robot with 7 DoF has been developed for cell manipulations and demonstrated with injections of single *Xenopus* oocyte and HTC. It is a modular robotic system with a large working range for automation of cell injection process, high precision and integrated two-dimensional vision control. The large-range robot performs regional position and orientation motions of the glass pipette in a working space with dimensions up to $50 \times 50 \times 50$ mm and realizes rough positioning to the cell membrane with accuracy of 1 μm. The integrated linear measuring system provides the large-range robot control system with 0.1 μm resolution feedback. The technological motions as fine positioning, orientation and cell penetration are realized by a micro-manipulator designed as a local robot structure, actuated by 3 piezo-stack actuators with integrated strain gauges having a resolution of 0.6 nm and stroke up to 100 μm, and second prototype with 3 DoF actuated by structured piezo-ceramics as a result of DFG project KA-1161. A glass pipette, handled and manipulated by the Hydro-MiNa robot, is used for cell penetration and injecting cells from 10 up to 800 μm in a diameter. Sophisticated numerical algorithms allow pipette tip tracking with resolution of 100 nm. Software application integrates the developed image processing routines and the robot control.

ACKNOWLEDGEMENTS

The investigations in this chapter are partially supported by the European Commission through the FP6 Integrated Project HYDROMEL with contract no. 026622-2 and DFG through the KA-1161 Project.

REFERENCES

1. Q. Xu, Y. Li, N. Xi, Design, fabrication, and visual servo control of an XY parallel micromanipulator with piezo-actuation, *IEEE Trans. on Automation Science and Engineering*, 6(4), 710–719, 2009.
2. R. Kasper, M. Al-Wahab, W. Heinemann, K. Kostadinov, D. Chakarov, Mechatronic handling device based on piezo ceramic structures for micro and nano applications, *Proceedings of 10th International Conference on New Actuators*, Bremen, Germany; pp.154–158, 2006.
3. A. Kortschack, A. Shirinov, T. Trüper, S. Fatikow, Development of mobile versatile nanohandling micro-robots: Design, driving principles, haptic control, *Journal of Robotica*, 23, 419–434, 2005.
4. K. Gr. Kostadinov, F. Ionescu, R. Hradynarski, T. Tiankov, Robot based assembly and processing micro/nano operations, in W. Menz and St. Dimov (Eds) *1st International Conference on Multi- Material Micro Manufacture 4M2005 (Karlsruhe, 29.06.–01.07.05)*, Elsevier, Karlsruhe, Germany, pp. 295–298, 2005.
5. Y. Irie, H. Aoyama, J. Kubo, T. Jujioka, T. Usuda, Piezo-impact-Driven X-Y stage and precise sample holder for accurate microlens alignment, *Journal of Robotics and Mechatronics*, 21(5), 635–641, 2009.
6. H. Kawasaki, M. Yashima, Piezo-driven 3 DoF actuator for robot hands, *Journal of Robotics and Mechatronics*, 2(2), 129–134, 1990.
7. N. Ando, M. Ohta, K. Gonda, H. Hashimoto, Workspace analysis of parallel manipulator for telemicro-manipulation systems, *Journal of Robotics and Mechatronics*, 13(5), 488–496, 2001.
8. H. Maruyama, F. Arai, T. Fukuda, On-chip microparticle manipulation using disposable magnetically driven microdevices, *Journal of Robotics and Mechatronics*, 18(3), 264–270, 2006.
9. K. Kostadinov, R. Kasper, T. Tiankov, M. Al-Wahab, D. Gotseva, Telemanipulation control of mechatronic handling devices for micro/nano operations, In St. Dimov, W. Menz and Y. Toshev (Eds.), *4M2007 3rd Int. Conf. "Multi-Material Micro Manufacture" (Borovetz, 03.10.-05.10.2007)*, Whittles Publishing CRC Press, ISBN 978-1904445-52-1, pp. 241–244, 2007.
10. P.T. Szemes, N. Ando, P. Korondi, H. Hashimoto, Telemanipulation in the virtual nano reality, *Transaction on Automatic Control and Computer Science CONTI, Romania*, 45(49)(1), 117–122, 2000.
11. J. Unger, R. Klatzky, R. Hollis, A telemanipulation system for psychophysical investigation of haptic interaction, *Proceedings of the International Conference on Robotics and Automation, Taipei*, Taiwan, pp. 1253–1259, 2003.
12. A. Otieno, Ch. Pedapati, X. Wan et al., Imaging and wear analysis of micro-tools using machine vision, *06 IJME–INTERTECH Proceedings* IT301:Paper071, 2006.

13. Al. Steinecker, Hybrid assembly for ultra-precise manufacturing, in *Precision Assembly Technologies and Systems: 5th IFIP Wg 5.5 International Precision Assembly Seminar*, IPAS 2010, Chamonix, France, 14–17 February, 2010, Proceedings, Vol. 315 of IFIP Advances in Information and Communication Technology Series, Ed. Sv. Ratchev, Springer, Berlin, Heidelberg, New York, Germany, pp. 89–96, 2010.

14. K. Kostadinov, R. Kasper, T. Tiankov, M. Al-Wahab, D. Chakarov, D. Gotseva, Unified approach for functional task formulation in domain of micro/nano handling manipulations, in W. Menz and St. Dimov (Eds.), *4M2005 2nd International Conference on Multi-Material Micro Manufacture (Grenoble, 20.09.–22.09.2006)*, Elsevier, Oxford, pp. 255–258, 2006.

15. A. Burisch, A. Raatz, J. Hesselbach, Design of modular reconfigurable micro-assembly systems, In *Micro-Assembly Technologies and Applications*, Springer, Boston, Vol. 260, pp. 337–344, 2009.

16. D. Chakarov, K. Kostadinov, D. Gotseva, T. Tiankov, Web-based synthesis of robot structures for micro and nano manipulations, *Journal of Solid State Phenomena*, 147–149, 25–30, 2009.

17. D. Chakarov, M. Abed Al-Wahab, R. Kasper, K. Kostadinov, Synthesis of tense piezo structures for local micro- & nano- manipulations, *Proceedings of the "8. Magdeburger Maschinenbau-Tage"*, Otto-von-Guericke-Universität Magdeburg, Magdeburg, pp. 173–180, 2007.

18. K. Kostadinov, D. Chakarov, T. Tiankov, Fl. Ionescu, *Robot for Micro and Nano Manipulations*, BG Patent application Nr. 110432/28.07.2009.

19. K. Kostadinov, R. Kasper, M.A. Al-Wahab, D. Chakarov, T. Tiankov, D. Gotseva, Mechatronic handling device for cell micro operations with human assisted automation control, *Presented at the International Conference "Motion and Vibration Control: MOVIC 2008"*, Munich, Germany, 12pp., 2008.

20. P. Genova, K. Kostadinov, Vl. Kotev, Synthesis and selection of optimal structures for macro robots. *Journal of ICMaS*, 3, 087–092, 2008.

21. T. Tiankov, P. Genova, Vl. Kotev, K. Kostadinov, Strategy for control of a hybrid macro–micro robot with a 5-link closed structure—An inverse problem of kinematics, *Journal of ICMaS*, 4, 113–118, 2009.

22. K. Kostadinov, Impedance scaling approach for teleoperation robot control. *Journal of ICMaS*, ISSN 1842-3183, 1, 059–062, 2006.

Part III

Allied Technologies for MIAS/Robotics

23 Real-Time System Identification of an Unstable Vehicle within a Closed Loop

Jitendra R. Raol, C. Kamali and Abhay A. Pashilkar

CONTENTS

23.1 INTRODUCTION

The system identification, Kalman filtering and state/parameter estimation for aerospace—and other related systems like micro air vehicles (MAVs), and unmanned aerial vehicles (UAVs), underwater and ground vehicles (UWGVs) and even robotic systems—have several potential applications. These range from modelling for flight control, rapid flight-test envelope expansion, aerodynamic data update, assessment of flight safety, mathematical modelling of the dynamics of these systems and target tracking. These techniques are also used for building world models of environments of the mobile robots [1–5]. The development of online estimation algorithms is very essential for reconfigurable/restructurable flight control and related applications [1,5–7]. Among many algorithms, which have been used for real-time parameter estimation, the computational simplicity of equation error methods makes them more suitable for real-time applications [8]. In this chapter, an evaluation of a time domain algorithm based on equation error-recursive least squares (RLS) principle is carried out for several applications: (i) application of online parameter identification (PID) for post-failure model estimation, (ii) a method to perform online estimation of model parameters of a new aircraft undergoing developmental tests, assuming the absence of calibrated flow angles and (iii) a method to evaluate real-time stability margins, for an unstable aircraft. Thus, the emphasis here is to show the suitability of one signal-processing tool (RLS) for three important problems. Although the applications considered in this chapter are related directly to aircraft, the approaches can be very well and easily extended to determine mathematical models of various robotic sub-systems. These empirically determined math models would then be used for simulation studies for robot path/motion planning, multi-robot coordination, and other mobile intelligent autonomous system (MIAS) related modelling and simulation studies.

Many fault-tolerant control systems in an aircraft and other aerospace vehicles have the potential of detection, identification and accommodation of sensor and actuator failures. Some such systems might have even reconfiguration capability built into them. A recursive estimation method based on discrete Fourier transform (DFT) was evaluated in Ref. [5] for online estimation of a mathematical model of aircraft that was supposed to have sustained damage to a primary control surface. The results presented in Ref. [5] had shown the potential of online PID within a fault-tolerant flight control system. The DFT technique performed well; however, it showed a relatively oscillatory/transient response until sufficient frequency-domain information was obtained. Its convergence rate was slow and required higher floating point operations/cycle (FLOPS). The DFT technique was compared with a Bayesian method for the same application and the Bayesian technique outperformed the DFT technique [7]. The Bayesian method is iterative and yet computationally intensive. In this chapter, we show that the RLS [1,5] method can provide less oscillatory/transient response, faster, monotonic convergence and have lesser FLOPS than the DFT technique. The simulated data are used to demonstrate the concept/procedure. The RLS algorithm with the forgetting factor has the tendency to become unstable under insufficiently excited conditions. A solution to this problem was proposed in Ref. [9] by including a parameter that stabilized RLS (SRLS). However, the SRLS does not differ from RLS with respect to the asymptotic convergence properties. A comparison of SRLS and DFT is carried out with respect to PID for reconfigurable/restructurable control. The RLS algorithm can be started with zero initial values of the unknown parameters which are to be estimated. The requirement of specific noise covariance is compensated with the forgetting factor. The choice of the forgetting factor and the stabilizing factor does not affect the accuracy of the final estimates; they only influence the convergence (faster and monotonic/less oscillatory) properties of the estimates with respect to time and hence the filter tuning is not very critical. For an aircraft undergoing developmental flight tests, measurements of calibrated angle of attack (α) and sideslip angle (β) might not be available. The RLS/SRLS algorithms are found to be very suitable for estimation of these flow angles. Simulated data from 6 DOF non-linear flight simulation are used to demonstrate this concept.

Modern fighter aircraft are designed to be longitudinally inherently unstable to gain good manoeuverability. Subsequently, these aircraft are provided with highly augmented full-authority control laws. A method of assessing safety is by evaluating the phase and gain margins. These margins serve as a basis for the flight-test ground support team to give clearance to the pilot to fly from one test point to the next, and flight clearance is given if the margins are satisfactory at the present test point. Often, the flight clearance analysis is performed off-line from the recorded flight-test data, and hence real-time stability margin estimation is an attractive option to improve flight-test efficiency. A technique achieving near real-time estimation of stability margins using frequency response of aircraft data was proposed in Ref. [10] with a comparative study of three techniques based on FFT with increasing order of complexity for near real-time stability margin estimation. The best technique for stability margin estimation relied on parameter estimation using the frequency response of short-period flight-test data. Off-line optimization was performed to fit the short-period frequency response data with known structure of short-period transfer function. Once the model parameters are derived from optimization, the loop transfer function is obtained by computing the product of the estimated aircraft transfer function and the known controller transfer function [10]. In this chapter, instead of frequency response model fitting, we use the RLS algorithm to perform real-time estimation of the short-period model for the purpose of stability margin computations. The measurements of angle of attack α and pitch rate q are assumed to be available. In case the angle of attack is not directly available, a combination of other signals including aircraft normal acceleration can be used to estimate the same. The angle of sideslip is estimated using mainly the lateral acceleration signal.

23.2 APPLICATION TO POST-FAILURE ESTIMATION

Online PID has a potential application in fault-tolerant flight control (FTFC) systems analysis. An FTFC system is required to perform failure detection, identification and accommodation (FDIA) with the following objectives: (1) minimum handling quality degradation following a failure, (2) in the event of a failure, providing higher priority for the safe continuation of flight without aborting the mission and (3) ensuring lower rate of aircraft loss. A block diagram of FTFC system highlighting the role of online PID is shown in Figure 23.1. The algorithms of DFT and RLS can be found in Ref. [6].

23.2.1 Stabilized RLS

The development in this chapter closely follows that in Ref. [6]. The SRLS algorithm is obtained by minimizing the cost function [6]:

$$J(\kappa(N)) = \sum_{i=1}^{N} \lambda^{N-i} \left| \varepsilon(i) \right|^2 + \delta \left| \kappa(N) - \kappa(N-1) \right|^2 \quad i = 1, 2, ..., N \tag{23.1}$$

Here, N represents the number of data points. The batch solution of Equation 23.1 is given by

$$\kappa(N) = \left[\sum_{i=1}^{N} X(i)X^T(i)\lambda^{N-i} + \delta I \right]^{-1} \left[\sum_{i=1}^{N} X(i)Y^T(i)\lambda^{N-i} + \delta\kappa(N-1) \right] \tag{23.2}$$

The state variable X and the measurement variable Y are as in a standard least squares (LS) problem. The recursive form of Equation 23.2 results in SRLS. The covariance matrix [1,6] of this algorithm is given by

$$P(N) = \left[\sum_{i=1}^{N} X(i)X^T(i)\lambda^{N-i} + \delta I \right]^{-1} \tag{23.3}$$

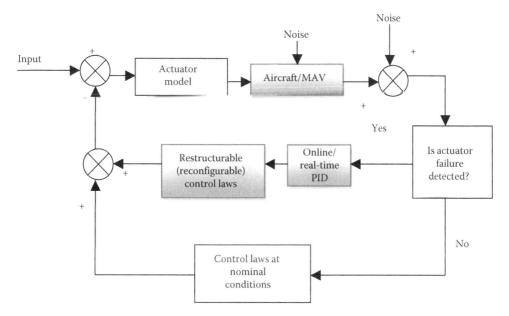

FIGURE 23.1 Fault-tolerant flight control schematic.

The $\delta|\kappa(N) - \kappa(N-1)|^2$ in Equation 23.1 ensures the stability of the algorithm. This additional term penalizes (may be large) changes in the parameter vector κ. The extra term $\delta\kappa(N-1)$ in Equation 23.2 prevents the parameter estimates to deviate rapidly from the previous value, which could otherwise diverge when the covariance matrix is ill conditioned. The recursive algorithm is summarized as follows [6]:

Initialize the algorithm with $P(0) = (1/\delta)I$.
The covariance matrix is updated by

$$P(n) = (1/\lambda)P(n-1) - (1/\lambda)P(n-1)C(n)[\lambda I + C(n)^T P(n-1)C(n)]^{-1}C(n)^T P(n-1) \qquad (23.4)$$

Here, the matrix $C(n)$ is given by

$$C(n) = \left[X^T(n)\sqrt{n_p\delta(1-\lambda)}e(n) \right] \qquad (23.5)$$

The $e(n)$ is a sequence of vectors as follows:

$$e(1) = \begin{bmatrix} 1 \\ 0 \\ . \\ . \\ . \\ 0 \end{bmatrix}, \quad e(2) = \begin{bmatrix} 0 \\ 1 \\ . \\ . \\ . \\ 0 \end{bmatrix}, \quad e(n_p) = \begin{bmatrix} 0 \\ 0 \\ . \\ . \\ . \\ 1 \end{bmatrix}, \quad e(n_p + 1) = \begin{bmatrix} 1 \\ 0 \\ . \\ . \\ . \\ 0 \end{bmatrix} \qquad (23.6)$$

Here, n_p is the number of parameters to be estimated in each of the state equations. The parameter updates are given by

$$\kappa(n) = \kappa(n-1) + P(n)X(n)^T[Y(n) - X(n)\kappa(n-1)] + \delta\lambda P(n)[\kappa(n-1) - \kappa(n-2)] \qquad (23.7)$$

23.2.2 MODEL FOR POST-FAILURE ANALYSIS

The aircraft model described in Ref. [5] is considered in the present study. It is assumed that the left elevator gets stuck at 15 s and there is also a loss of 67% in control surface effectiveness. It has also been assumed that the failure is detected in 1 s and the healthy right elevator is applied to perform PID. The short-period aircraft model [1,5,6] is

$$\begin{bmatrix} \dot{\alpha} \\ \dot{q} \end{bmatrix} = \begin{bmatrix} Z_\alpha & Z_q \\ M_\alpha & M_q \end{bmatrix}\begin{bmatrix} \alpha \\ q \end{bmatrix} + \begin{bmatrix} Z_{\delta el} & Z_{\delta er} \\ M_{\delta el} & M_{\delta er} \end{bmatrix}\begin{bmatrix} \delta_{el} \\ \delta_{er} \end{bmatrix} \qquad (23.8)$$

23.2.2.1 Modelling of Post-Failure

We assume actuator failure and/or battle damage. The elevator failure is considered to be more critical than aileron and rudder failures because of the unavoidable coupling between the longitudinal and lateral dynamics. The characteristics of a control surface can be modelled in terms of normal force, axial force and moment around some fixed points or axes. It is assumed that the axial forces exerted by longitudinal control surface deflections are negligible and only the normal forces will undergo changes. To derive a post-failure model, we need to obtain closed-form expressions of the non-dimensional aerodynamic stability and control derivatives as a function of the normal force coefficient relative to the control surface failure and/or battle damage [6]. The closed-form expressions for the derivatives in terms of $C_{L\delta}$ of the left and right side of the longitudinal control surface are to be obtained: $C_{L\alpha}, C_{m\alpha}, C_{L\dot{\alpha}}, C_{m\dot{\alpha}}, C_{Lq}, C_{mq}$. A study of post-failure modelling of an aircraft involving a stuck left elevator with a 67% effectiveness reduction is covered in Ref. [5], from where the numerical values of nominal aircraft model and post-failure model are taken for this work. The nominal model [4,5] is given as

$$\begin{bmatrix} \dot{\alpha} \\ \dot{q} \end{bmatrix} = \begin{bmatrix} -0.53 & 0.99 \\ -7.74 & -0.717 \end{bmatrix} \begin{bmatrix} \alpha \\ q \end{bmatrix} + \begin{bmatrix} -0.03 & -0.028 \\ -5.7 & -5.7 \end{bmatrix} \begin{bmatrix} \delta_{el} \\ \delta_{er} \end{bmatrix} \tag{23.9}$$

The failure condition model is given as

$$\begin{bmatrix} \dot{\alpha} \\ \dot{q} \end{bmatrix} = \begin{bmatrix} -0.534 & 0.99 \\ -4.72 & -0.38 \end{bmatrix} \begin{bmatrix} \alpha \\ q \end{bmatrix} + \begin{bmatrix} -0.0094 & -0.028 \\ -1.9 & -5.7 \end{bmatrix} \begin{bmatrix} \delta_{el} \\ \delta_{er} \end{bmatrix} \tag{23.10}$$

23.2.3 RESULTS AND DISCUSSION OF POST-FAILURE MODEL ESTIMATION

The PID exercise was carried out featuring the nominal and post-failure conditions in a single simulation process, considering clean and noisy simulated data. It was mentioned in Ref. [8] that the DFT technique could not start with left and right elevators separately during PID under nominal conditions as the control surface inputs were correlated. The PID implementation scheme reported here models the left and right elevators separately for both the nominal and post-failure conditions. Comparison of RLS with DFT (with respect to FLOPS carried out in Ref. [6]) shows that RLS is superior to DFT in terms of minimum onboard memory. The LS algorithm with the forgetting factor has a tendency to become unstable under insufficiently excited conditions, the solution being to persistently excite the regression vector by adding a small perturbation to the controls by white noise. But it will hamper the flight of the aircraft. An alternative solution is to use covariance resetting to induce sharp discontinuities and transients in the responses of the algorithm. The SRLS is the modified RLS, which avoids the need of covariance resetting during insufficiently excited conditions. The control surface inputs showing the normal left and right elevators (0–15 s) and normal right elevator with failed left elevator (15–30 s) are shown in Figure 23.2. The aircraft responses (α, q) are corrupted by Gaussian and white measurement noise with SNR = 10. The convergence results of parameters along with their true values for SRLS and DFT schemes are shown in Figure 23.3. The convergence of SRLS is found to be less oscillatory, monotonic and faster than DFT. The smoother convergence of SRLS is due to the use of stabilizing parameter δ that penalizes the estimates deviating from their previous values. This regularizing parameter δ was chosen to be 10. The use of δ is equivalent to introducing damping effect. The larger the value of δ, the slower will be the convergence. However, the choice of δ does not hamper the accuracy. The forgetting factor λ is 0.999. A Monte Carlo simulation of 500 runs was performed for SRLS with noisy data (SNR = 10). The results are presented in Table 23.1. The estimation results are found satisfactory. In Tables 23.1 through 23.3, the PEEN represents the parameter estimation error norm given as

$$\text{PEEN} = \frac{\text{norm}(\kappa_t - \hat{\kappa})}{\text{norm}(\kappa_t)} * 100 \tag{23.11}$$

Here, κ_t is the vector of true parameters and $\hat{\kappa}$ is the vector of estimated parameters.

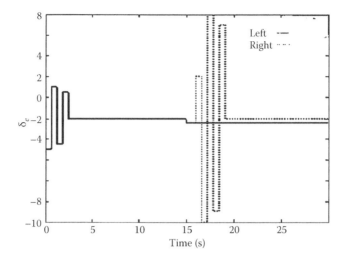

FIGURE 23.2   Doublet-like command inputs to aircraft control surface.

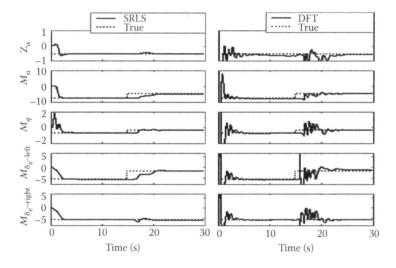

FIGURE 23.3   Results of SRLS and DFT algorithms (SNR = 10).

TABLE 23.1
SRLS Monte Carlo Analysis (SNR = 10)

Aircraft Parameters	True Values (Nominal Conditions)	Ensemble Average of Estimates	True Values (Post-Failure Conditions)	Ensemble Average of Estimates
Z_α	−0.534	−0.534	−0.534	−0.531
M_α	−7.740	−7.559	−4.720	−5.035
M_q	−0.717	−0.676	−0.380	−0.403
$M_{\delta e-\text{left}}$	−5.700	−5.589	−1.899	−1.948
$M_{\delta e-\text{right}}$	−5.700	−5.589	−5.700	−5.973
PEEN %	2.168		5.473	

23.3 ESTIMATION IN THE ABSENCE OF CALIBRATED ANGLES OF INCIDENCE AND SIDESLIP

The RLS used assumes the measurement of α for the estimation of aircraft short-period data and β for the estimation of Dutch roll data. However, when the calibrated measurements of α and β are not available, these should be reconstructed online using normal acceleration $\left(N_{z_{CG}}\right)$ and lateral acceleration $\left(N_{y_{CG}}\right)$ measured at the centre of gravity (CG) of the aircraft, respectively. The onboard inertial platform senses the local acceleration, which is a combination of acceleration at CG as well as the angular rates. These accelerations are measured at the sensor location, which has a definite relation with acceleration at CG:

$$N_{z_{\text{sensor}}} = N_{z_{CG}} + \dot{q}x_s \tag{23.12}$$

Here, x_s is the forward displacement (in metres) of the accelerometer sensor from the CG in the x direction. If the measurement of $N_{z_{CG}}$ is not available, then using $N_{z_{\text{sensor}}}$ and filtered differentiation of pitch rate, one can obtain $N_{z_{CG}}$. Similar approach applies to $N_{y_{CG}}$.

23.3.1 DEVELOPMENT OF MATHEMATICAL MODELS

AoA and AoS [1,4,11] can be constructed online by integrating the following equations:

$$\dot{\alpha} = q + \frac{g}{V}(\cos\theta\cos\phi + N_{z_{CG}}), \quad \dot{\beta} = p\sin\alpha_0 - r\cos\alpha_0 + \frac{g}{V}(N_{Y_{CG}} + \sin\phi) \tag{23.13}$$

Here, α_0 is the trim value of angle of attack. Equation 23.13 has been derived from the basic flight mechanics equations [11]. These reconstructed values of α and β can be utilized by the proposed algorithm to yield real-time parameter estimates. The block diagrams showing the methodology of conducting online PID in the absence of calibrated air data are shown in Figure 23.4. The accelerations $N_{z_{CG}}$ and $N_{y_{CG}}$ are noisy and integration of those variables will introduce drift in the derived α and β. The drift in the derived α and β affects estimation accuracy. Hence, the aircraft state space

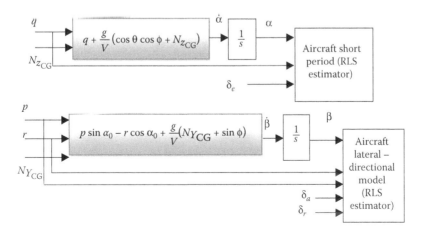

FIGURE 23.4 Estimation in the absence of calibrated flow angles.

model to be estimated can include a constant term for modelling the drift. The short-period model to be estimated can be augmented with drift constants k_1 and k_2 as follows:

$$\begin{bmatrix} \dot{\alpha} \\ \dot{q} \end{bmatrix} = \begin{bmatrix} Z_\alpha & Z_q \\ M_\alpha & M_q \end{bmatrix} \begin{bmatrix} \alpha \\ q \end{bmatrix} + \begin{bmatrix} Z_{\delta e} & k_1 \\ M_{\delta e} & k_2 \end{bmatrix} \begin{bmatrix} \delta_{el} \\ u_s \end{bmatrix} \tag{23.14}$$

The second-order Dutch roll lateral model can be augmented with drift constants k_1 and k_2 as follows:

$$\begin{bmatrix} \dot{p} \\ \dot{r} \end{bmatrix} = \begin{bmatrix} L_p & L_r \\ N_p & N_r \end{bmatrix} \begin{bmatrix} p \\ r \end{bmatrix} + \begin{bmatrix} L_\beta & L_{\delta a} & L_{\delta r} & k_1 \\ N_\beta & N_{\delta a} & N_{\delta r} & k_2 \end{bmatrix} \begin{bmatrix} \beta \\ \delta a \\ \delta r \\ u_s \end{bmatrix} \tag{23.15}$$

Here, u_s represents discrete unit step signal. Modelling the aircraft dynamics with augmented drift constants and estimating the parameters using RLS algorithm will yield the best LS solution while using derived α and β from the noisy acceleration signals.

23.3.2 Short-Period Results

The data for the short-period real-time parameter estimation were generated using 6 DOF flight simulation software. The data for a subsonic flight condition—the landing phase of aircraft—were generated and the aircraft short-period model is unstable. The convergence of important short-period parameters is shown for clean data in Figure 23.5. In order to test the robustness of algorithm to measurement noise, the estimation was repeated with responses (α, q) corrupted by Gaussian random noise with SNR = 10. The noisy $N_{z_{CG}}$ and q signals are shown in Figure 23.6. The δ_e is assumed uncorrupted by noise. The comparison of constructed α with simulated α is shown in Figure 23.6. The constructed α shows a drift. This drift is taken care of in the modelling so that we could still get reasonably good estimates. The convergence for important short-period parameters is shown in Figure 23.7. Then the algorithm is tested for a transonic and supersonic simulated data with and without measurement noise. The supersonic data were corrupted with Gaussian noise with SNR = 100, since δ_e and α amplitudes were significantly low at supersonic flight conditions. The results for noisy flight data are presented in Table 23.2. The assessment of estimates in Table 23.2 is

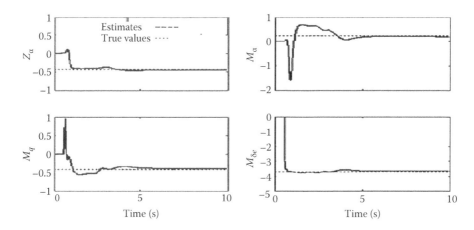

FIGURE 23.5 Short-period estimates: convergence to true values (data w/o noise).

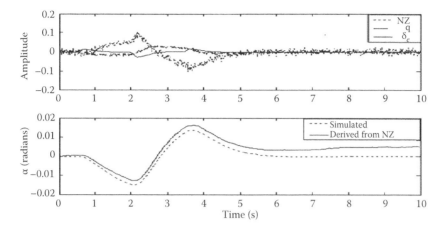

FIGURE 23.6   Comparison of constructed and simulated alpha (data with noise).

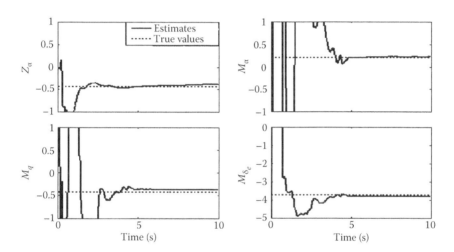

FIGURE 23.7   Short-period estimates: convergence to true values (data with noise).

TABLE 23.2
Short-Period Estimated Parameters (in the Absence of Calibrated Alpha)

Aircraft SP Parameters Estimates	Subsonic FC (SNR = 10)		Transonic FC (SNR = 10)		Supersonic FC (SNR = 100)	
	True	RLS Estimates	True	RLS Estimates	True	RLS Estimates
Z_α	−0.445	−0.389	−0.842	−0.785	−1.060	−1.010
Z_q	0.972	0.921	0.987	0.969	0.989	0.989
$Z_{\delta e}$	−0.181	−0.229	−0.331	−0.287	−0.199	−0.126
M_α	0.199	0.226	−4.917	−4.689	−25.086	−23.299
M_q	−0.428	−0.376	−1.616	−1.701	−0.707	−0.662
$M_{\delta e}$	−3.756	−3.815	−34.997	−34.911	−31.648	−30.056
PEEN (%)	2.53		0.758		5.93	

computed using PEEN. If the augmentation proposed for drift terms in Equation 23.14 were not applied, the estimates are found to deviate significantly from the true values.

23.3.2.1 Performance under Turbulent Conditions

The RLS/SRLS discussed is not modified to perform under turbulent wind conditions. However, a mild turbulence profile of intensity 0.5 ft/s in the forward velocity and vertical velocity is invoked in the simulation apart from sensor measurement noise of SNR = 10 to evaluate longitudinal performance. The reconstructed α is shown in Figure 23.8 that diverges beyond 5 s. Despite accounting for drift through a constant in estimation, it can be noticed that Z_α shows some tendency to diverge in Figure 23.9 compared to Figure 23.7. Nevertheless, for this wind profile, the PEEN value is found to be 5.177 which is satisfactory.

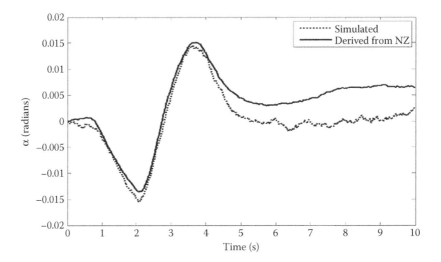

FIGURE 23.8 Constructed and simulated alpha: in the presence of turbulence.

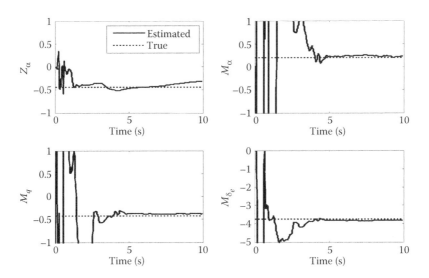

FIGURE 23.9 Short-period parameters: convergence to true values in the presence of turbulence.

23.3.3 Dutch Roll Results

The data for Dutch roll were simulated for the same aircraft. We tested the estimator with a simulated data set representing subsonic flight condition. The convergence of some lateral Dutch roll parameters for clean data is shown in Figure 23.10. To test the robustness of the technique, the subsonic data were corrupted by Gaussian measurement noise with SNR = 10. The reconstructed β was found to match with simulated β. The convergence of some lateral Dutch roll parameters is seen in Figure 23.11. The algorithm was tested for transonic and supersonic simulated data with and without measurement noise. The results for noisy flight data are presented in Table 23.3. The RLS algorithm performs well. If the augmentation proposed for drift terms in Equation 23.15 were not applied, the parameter estimates are found to deviate significantly from the true values.

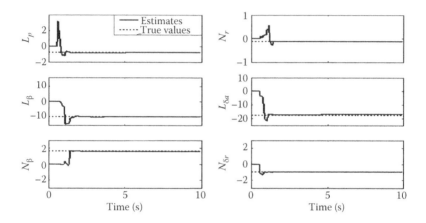

FIGURE 23.10   Dutch roll parameters: convergence to true values (clean data).

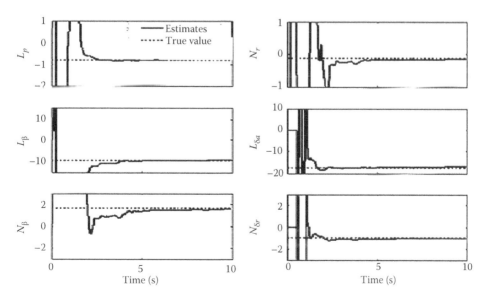

FIGURE 23.11   Dutch roll parameters: convergence to true values (data with noise).

TABLE 23.3
Dutch Roll Estimates (in the Absence of Calibrated Beta)

Aircraft Lateral DR Parameters Estimates	Subsonic FC (SNR = 10)		Transonic FC (SNR = 10)		Supersonic FC (SNR = 10)	
	True	RLS Estimates	True	RLS Estimates	True	RLS Estimates
L_p	−0.823	−0.743	−1.924	−1.807	−2.230	−12.056
L_r	0.755	0.787	1.629	1.364	0.798	0.549
L_β	−10.525	−10.065	−34.174	−34.58	−68.584	−67.56
$L_{\delta a}$	−17.56	−17.012	−103.99	−102.12	−102.793	−99.74
$L_{\delta r}$	1.79	1.813	12.256	13.327	6.149	7.284
N_p	−0.041	−0.067	−0.067	−0.036	−0.071	−0.079
N_r	−0.137	−0.155	−0.344	−0.378	−0.398	−0.372
N_β	1.603	1.552	9.212	9.558	13.702	12.735
$N_{\delta a}$	−1.39	−1.277	−10.762	−10.005	−14.524	−14.13
$N_{\delta r}$	−1.005	−1.079	−5.297	−4.919	−3.565	−3.262
PEEN (%)	3.52		2.136		2.78	

23.3.3.1 Performance under Turbulent Conditions

A turbulence of intensity 0.5 ft/s in the side velocity was added in the simulation apart from sensor measurement noise such that SNR = 10 to evaluate lateral estimation performance. The reconstructed β time history is shown in Figure 23.12 that is seen to deviate after 4 s. Even though the drift was accounted for using a constant in estimation, it is noticed that the estimates of the parameters L_p, L_β, N_β, $L_{\delta a}$ deviate from the true values as seen in Figure 23.13. These large deviations in the estimated parameters from their true values increase the PEEN to 36.64%, which is rather large. This calls for further study to improve the results. It is suggested that a mathematical model of the turbulence along with the measurement noise covariance matrix R be incorporated in the SRLS scheme to improve the estimation results.

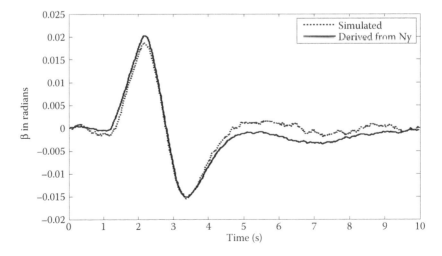

FIGURE 23.12 Constructed beta with simulated beta: under turbulence.

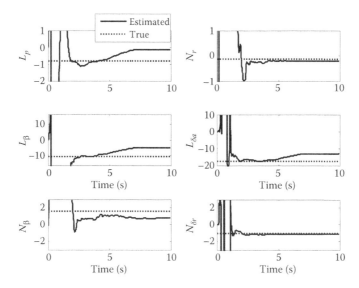

FIGURE 23.13   Dutch roll parameters: convergence in the presence of turbulence.

23.4   A METHODOLOGY FOR ESTIMATION OF STABILITY MARGIN IN REAL TIME

The two steps and phases required to perform stability margin estimation in real time are shown in Figure 23.14. The first phase (a) uses a real-time parameter estimation algorithm for estimating short-period model of the aircraft, and in the second phase (b), the product of estimated short-period model and (controller + actuator) transfer function (TF) is computed. This process yields the loop TF. The phase (b) can be made real time by pre-computing the controller and actuator TFs jointly. Then the phase and gain margins are computed from the open-loop TF.

23.4.1   UNSTABLE AIRCRAFT MODEL: STABILITY MARGIN ESTIMATION

The test aircraft is inherently longitudinally unstable. The model of the unstable aircraft is given in Equation 23.16. The longitudinal dynamics of the aircraft has two modes: short-period and phugoid modes. The left top corner of A matrix and top two elements of B matrix (the block matrices in Equation 23.16) represent the aircraft short-period parameters. The right bottom corner of A matrix and bottom two elements of B matrix of the same block matrix shown in Equation 23.16 represent the aircraft phugoid parameters [1,4,6,11]:

$$
\begin{bmatrix} \dot{\alpha} \\ \dot{q} \\ \dot{\theta} \\ \dot{u}/u_0 \end{bmatrix} = \begin{bmatrix} Z_\alpha & 1 & 0 & Z_{\dot{u}/u_0} \\ M_\alpha & M_q & 0 & M_{\dot{u}/u_0} \\ 0 & 1 & 0 & 0 \\ X_\alpha & 0 & X_\theta & X_{\dot{u}/u_0} \end{bmatrix} \begin{bmatrix} \alpha \\ q \\ \theta \\ u/u_0 \end{bmatrix} + \begin{bmatrix} Z_{\delta e} \\ M_{\delta e} \\ 0 \\ X_{\delta e} \end{bmatrix} \delta_e \tag{23.16}
$$

The numerical values [10] in the above state model are

$$
\begin{bmatrix} \dot{\alpha} \\ \dot{q} \\ \dot{\theta} \\ \dot{u}/u_0 \end{bmatrix} = \begin{bmatrix} -0.7771 & 1 & 0 & -0.1905 \\ 0.3794 & -0.8329 & 0 & 0.0116 \\ 0 & 1 & 0 & 0 \\ -0.9371 & 0 & -0.0960 & -0.0296 \end{bmatrix} \begin{bmatrix} \alpha \\ q \\ \theta \\ u/u_0 \end{bmatrix} + \begin{bmatrix} -0.2960 \\ -9.6952 \\ 0 \\ -0.0422 \end{bmatrix} \delta_e \tag{23.17}
$$

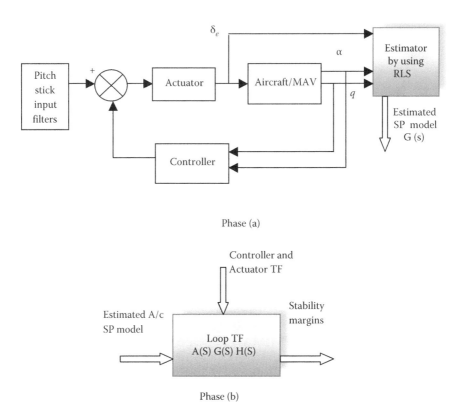

FIGURE 23.14 Proposed scheme for real-time stability margin estimation (TF, transfer function).

The eigenvalues from/of Equation 23.17 are −1.4917, 0.2506, −0.3523, −0.0461, with one unstable pole at 0.2506. This instability is mainly caused by the short-period dynamics. We estimate only the short-period parameters for the purpose of computing the stability margin.

23.4.2 STABILITY MARGIN ESTIMATION RESULTS

To demonstrate the concept of stability margin estimation, the simulated closed-loop responses (α, q) of the aircraft are considered without any measurement noise. The online estimates of short-period parameters converged as soon as the manoeuvre was over. As a result, the computation of stability margins around 3.5 s itself was convenient. The results are calculated for data up to the end of 10 s. The stability margin results along with the true values are given in Table 23.4. The accuracy of the margins' estimates is assessed as per the criteria in Ref. [10]. The robustness of RLS estimator to the

TABLE 23.4
Stability Margins

	Estimates of Margins		True Values		Error (True-Estimated)	
	Phase Margin in Degree	Gain Margin in dB	Phase Margin in Degree	Gain Margin in dB	Phase Margin in Degree	Gain Margin in dB
Data w/o noise	58.2	∞	57.9	∞	−0.27	0
Data with SNR = 10	57.9	∞	57.9	∞	−0.047	0

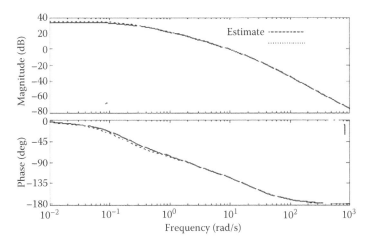

FIGURE 23.15   Bode diagrams for the true and estimated models: noisy data.

noisy data was assessed by adding white Gaussian measurement noise (with SNR = 10) to the responses α and q. The control surface input is kept free from measurement noise. In spite of noisy data, the estimates converged soon as the manoeuvre was over. The true open-loop TF and estimated loop TF Bode diagrams are shown in Figure 23.15. This matching looks satisfactory for noisy data. The stability margin estimates with noisy data along with the true values are compared in Table 23.4.

23.5   CONCLUDING REMARKS

In this chapter, we have presented a practical application of RLS algorithm in flight control and flight testing of an aircraft. These results can be easily extended to any kind of mobile vehicle and even robots depending upon the test situations. The SRLS performed better than DFT for post-failure aircraft model estimation in terms of transient, convergence characteristics and FLOPS. A novel approach for real-time estimation using RLS in the absence of calibrated angle of attack and sideslip has been utilized, and it was demonstrated that the RLS technique worked very well with reconstructed alpha and beta from normal and lateral accelerations, respectively. A novel application of RLS to evaluate the aircraft stability margin in real time has also been successfully demonstrated. It should be mentioned that these estimation schemes and approaches can be easily flight tested to evaluate the suitability of RLS onboard, for all the important applications discussed in this chapter. The demonstrated procedures of using RLS/SRLS estimation technique can be easily extended to micro/mini air vehicles (MAVs), UAVs, helicopter robots, and many similar autonomous ground vehicles (AGVs), including field robots where online/real-time system identification and parameter estimation might be required for evaluating the system performance as well as for determining the mathematical models of these systems from real data for simulation model updates.

APPENDIX: SYMBOLS AND NOMENCLATURE

α	Angle of attack (AoA) in rad
β	Angle of sideslip (AoSS) in rad
p	Roll rate in rad/s
q	Pitch rate in rad/s
r	Yaw rate in rad/s
κ	Vector of parameters (linear model)
J	Cost function

P	State-covariance matrix
δ_{el}	Left elevator control surface deflection in rad
δ_{er}	Right elevator control surface deflection in rad
δa	Aileron control surface deflection in rad
δr	Rudder control surface deflection in rad
g	Acceleration due to gravity in m/s^2
V	True air speed in m/s
θ	Pitch angle in rad
ϕ	Roll angle in rad
ψ	Yaw angle in rad
N_{zCG}	Normal acceleration at CG
N_{YCG}	Lateral acceleration at CG
$u\ wind$	Turbulence in forward velocity in ft/s
$v\ wind$	Turbulence in lateral velocity in ft/s
$w\ wind$	Turbulence in vertical velocity in ft/s
λ	Forgetting factor in the estimator
δ	Stabilizing parameter
T	Transposed variable
E	Mathematical expectation operator
$\wedge$	Estimated variable
PID	Parameter identification
6 DOF	Six degrees of freedom

REFERENCES

1. Raol J.R., Girija G. and Singh J. *Modelling and Parameter Estimation of Dynamic Systems*, IEE Control Engineering Book Series, Vol. 65, IEE/IET, London, UK, 2004.
2. Klein V. and Morelli E.A. *Aircraft System Identification: Theory and Practice*, AIAA, USA, 2006.
3. Morelli E.A. Real-time parameter estimation in the frequency domain, *AIAA Guidance, Navigation and Control Conference and Exhibit*, USA, 1999, Paper No. AIAA-99-4043.
4. Raol J.R. and Singh, J. *Flight Mechanics Modeling and Analysis*, CRC Press, Boca Raton, FL, USA, 2009.
5. Napolitano M.R., Song Y. and Seanor B. On-line parameter estimation for restructurable flight control systems, *Aircraft Design*, 4(2001), 19–50, 2001.
6. Kamali C., Pashilkar A.A. and Raol J.R. Real time parameter estimation for reconfigurable control of unstable aircraft, *Defense Science Journal*, 57(4), 527–537, 2007.
7. Han Y. et al. Frequency and time domain online parameter estimation for reconfigurable flight control system, AIAA Paper No 2009–2040, USA.
8. Morelli E.A. Practical aspects of the equation error method for aircraft parameter estimation, AIAA Paper No. 2006–6144, USA.
9. Shore D. and Bodson M. Flight testing of a reconfigurable control system on an unmanned aircraft, *Journal of Guidance, Control and Dynamics*, 28(4), 698–707, 2005.
10. Patel V.V., Deodhare G. and Chetty S., Near Real Time Stability Margin Estimation from Piloted 3–2–1–1 Inputs, AIAA Aircraft Technology Integration and Operations (ATIO2002), Technical Forum, October 2002.
11. Nelson R.C. *Flight Stability and Automatic Control*, Aerospace Series, McGraw-Hill International Editions, Singapore, 1990.

24 Smart Antennas for Mobile Autonomous Systems

Adaptive Algorithms for Acceleration

Hema Singh and R. M. Jha

CONTENTS

24.1   INTRODUCTION

In mobile intelligent autonomous systems, there are often dynamic scenarios, which require obstacle negotiation. For example, a robot while moving is often required to observe objects around it and avoid them. Such an obstacle negotiation in robotics can be achieved with the help of adaptive algorithms implemented in the sensors embedded within the robot. Yet another application is *miniature aerial vehicles* (MAVs) and *unmanned aerial vehicles* (UAVs). Here, the aim is to provide a system where the onboard computer system can take autonomous decision for mission-related activities, for example, reconnaissance, surveillance and tracking of targets. The onboard decision-making is based on the sensor data [1]. The system decides the 3-D theta-phi (elevation, azimuth) direction of movement that can be adapted once again, using the adaptive algorithms on the received sensor data. Next, the application considered may be of interplanetary land rover vehicle with its obstacle negotiation capability, while moving on the steep, uneven terrain of the moon and other planets of the solar system.

Such *artificial intelligence* (AI)-based autonomous systems are necessarily integrated with an antenna system as the front-end. It is advantageous to have a smart system, for example, multiple beam antenna (MBA) with the main beams simultaneously oriented towards the numerous desired directions. More importantly, however, it has to be adaptive to insulate it from the objects and targets, which must be avoided. In the state-of-the-art parlance, this would involve *smart* or *adaptive antenna*, which is capable of reconfiguring the pattern online in real time, and these are done using the adaptive algorithms. Although adaptive antennas have been known to exist for some time, it is important to accelerate the algorithms, which is the focus of this chapter. The efficiency of the algorithm lies not only in correct decision but also the rate of convergence and hence the speed of the decision-making capability. In an arbitrary real-time situation based on the signals received, the antenna and hence the processing block are required to have optimized distributions subject to certain circumstances, which might include external interference or receiver noise. The smart alternative choice is an adaptive array, which utilizes arbitrary element patterns, polarizations and spacing. In adaptive arrays, the pattern optimization is obtained by real-time active weighting of the received signal, and reconfigured according to the conditions under which it is operating. The optimum weights are obtained by employing an efficient adaptive algorithm. The difference between the signal (from desired directions) and the externally generated interference is used to suppress the interference and enhance the desired signal. These adaptive array techniques are useful because they not only suppress the probing sources but also have potential applications in mobile communication and broadcasting [2].

24.2 ACTIVE CANCELLATION IN SIDELOBE CANCELLERS

Generalized sidelobe canceller (GSC) is one of the adaptive beamformers that nullify the jammer (an interference source) and simultaneously maintain high-output *signal-to-interference-noise ratio* (SINR). However, GSC is quite sensitive to *direction of arrival* (DOA) mismatch. In such circumstances, the array tends to treat the desired signal like an interference source and tries to suppress it, leading to a phenomenon called *signal cancellation*. Moreover, in the case of GSC, the input signal is present in the stochastic gradient, which makes the gradient large, hence requiring a very small step size. This further reduces the speed of convergence. In order to get rid of such problems, one of the possible remedies is to use notch filter in conventional GSC. However, the blind equalizer can be included in the GSC for improving the robustness against various mismatch errors. Such a design [3] is collectively known as *decision-feedback generalized sidelobe canceller* (DF-GSC). Blind equalizers are in general used for digital signals in order to avoid *inter-symbol interference* (ISI). Another approach is to include different types of constraints in sidelobe cancellers for better efficiency and robustness. Mainly, three types of constraints, viz., point constraint, directional constraints and derivative constraints are popularly employed [4]. However, in the present simulations, only two constraints, point constraint and first derivative constraint, are included. It has been observed that due to the addition of constraints in GSC, robustness is increased but output SINR gets degraded.

24.2.1 CONVENTIONAL GENERALIZED SIDELOBE CANCELLER

The structure of GSC consists of two weights and a blocking matrix, B, as shown in Figure 24.1. The weight in the upper branch is known as quiescent matched filter w_q and the one in the lower branch is called interference cancelling filter w_a. The main element is the blocking matrix which has a peculiar property that each of its independent row sums to zero [5]. Many matrices can be formed with this property but *Walsh Hadamard equations* are most commonly used for the generation of the blocking matrix. *Walsh* functions are of three types: the *Walsh-ordered*, the *dyadic-ordered* and the *Hadamard-ordered Walsh function*. The Hadamard-ordered Walsh function, also called the Hadamard function, has been used in this report. These are a set of orthogonal functions consisting only of 1 and −1 [6]. In order to generate these functions, for example, for $N = 16$, a (16×4) matrix is generated by writing the binary codes of decimal numbers from 0 to 15.

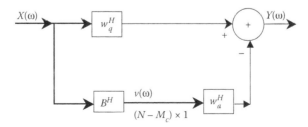

FIGURE 24.1 Generalized sidelobe canceller (GSC).

Thus, all the values of $b_k(i)$ [this is not related to the transmitted symbol $b_0(k)$] are obtained such that

$$b_3(0) = b_2(0) = b_1(0) = b_0(0) = 0$$
$$b_3(1) = b_2(1) = b_1(1) = 0, b_0(1) = 1$$
$$b_3(0) = b_2(0) = 0, \; b_1(0) = 1, \; b_0(0) = 0$$

and so on.

Then the (16×16) *Hadamard* matrix Had(i, j) can be obtained using the formula

$$\text{Had}(i, j) = (-1)^{\sum_{k=0}^{3} b_k(i) b_k(j)} \tag{24.1}$$

The *Hadamard* matrix thus obtained is *Hermitian*. All its columns are linearly independent. Since a point distortionless constraint is being used, one of the columns is removed to obtain a (16×15) blocking matrix. All the columns of this matrix sum up to zero. The *Hadamard* matrix can also be obtained in a recursive manner by the formula

$$\text{Had}_{n+1} = \begin{bmatrix} \text{Had}_n & \text{Had}_n \\ \text{Had}_n & -\text{Had}_n \end{bmatrix} \tag{24.2}$$

The $N \times 1$ steering vector is given as

$$S(\theta_m) = [1, e^{i\tau\theta_0}, e^{2i\tau\theta_0}, \ldots, e^{i(N-1)\tau\theta_0}]^T \tag{24.3}$$

where $\tau\theta_0 = 2\pi d/\lambda \sin\theta$ in which d is spacing between elements, λ is the wavelength of incoming signal and θ_m is the DOA of the mth source.

Then, the kth snapshot of the received signal is expressed as

$$x(k) = S(\theta_0)s_0(k) + \sum_{m=1}^{3} S(\theta_m)s_m(k) + n(k) = s(k) + i(k) + n(k) \tag{24.4}$$

where $s(k)$ is the desired signal, $s_0(k)$ is the transmitted signal, $l(k)$ is the interfering signal and $n(k)$ is the noise. In the present work, the optimization problem is

$$J = w^H R_x w \quad \text{subject to} \quad C^H w = f \tag{24.5}$$

Here, C is the $N \times P$ constraint matrix and f is the $P \times 1$ response vector, P being the number of constraints. J is *mean square error* (MSE), w is the weight vector and R_x is the input correlation matrix given by $R_x = E\{x(k)x^H(k)\}$.

The output of GSC can be expressed as

$$y(k) = (w_q - Bw_a)^H x(k) \tag{24.6}$$

where w_q is a vector of dimension $N \times 1$, B is $N \times (N-P)$, and w_a is $(N-P) \times 1$. The weight w_q can be determined using either an analytic or an iterative approach.

24.2.1.1 Analytic Approach

For simplicity, *quadrature phase-shift keying* (QPSK) modulation with *flat-fading* channel environments [7] is considered for the present work. In this approach, we have

$$w_q = C(C^H C)^{-1} f \tag{24.7}$$

Equation 24.5 can be re-written as

$$\min J = \min(w_q - Bw_a)^H R_x (w_q - Bw_a) \tag{24.8}$$

Then, one can find the optimum w_a as

$$w_{a,\text{opt}} = (B^H R_x B)^{-1} B^H R_x w_q \tag{24.9}$$

Let the optimum weight $w_{\text{opt}} = w_q - Bw_{a,\text{opt}}$. Then the *minimum mean-squared error* (MMSE) for Equation 24.8, denoted as $J_{\min}$, is given by

$$J_{\min} = w_{\text{opt}}^H R_x w_{\text{opt}} = w_q^H R_x w_{\text{opt}} \tag{24.10}$$

where R_x is the input correlation matrix and is given by

$$R_x = \sigma_{s_0}^2 S(\theta_0) S^H(\theta_0) + \sigma_{s_1}^2 S(\theta_1) S^H(\theta_1) + \sigma_{s_2}^2 S(\theta_2) S^H(\theta_2) + \sigma_{s_3}^2 S(\theta_3) S^H(\theta_3) + \sigma_n^2 I \tag{24.11}$$

Here, σ_{sj}^2 represents the variance of the jth signal.

R_{i+n} is the correlation matrix of interference plus noise and is given by

$$R_{i+n} = \sigma_{s_1}^2 S(\theta_1) S^H(\theta_1) + \sigma_{s_2}^2 S(\theta_2) S^H(\theta_2) + \sigma_{s_3}^2 S(\theta_3) S^H(\theta_3) + \sigma_n^2 I_d \tag{24.12}$$

where I_d is the identity matrix.

In case of GSC, the minimum output power, denoted by $P_{o,\min}$ is equal to $J_{\min}$

$$P_{o,\min} = J_{\min} \tag{24.13}$$

The output desired signal power, denoted by P_s, is expressed as

$$P_s = \sigma_{s_o}^2 \left| w_q^H S(\theta_o) \right|^2 \tag{24.14}$$

where $\sigma_{s_o}^2$ is the input desired signal power.

The general expression for optimum SINR [8] is

$$\text{SINR}_{\text{opt}} = \frac{P_s}{P_{o,\min} - P_s} \tag{24.15}$$

The gain for the beam pattern is expressed as

$$\text{Gain} = w_{\text{opt}}^H S(\theta) \tag{24.16}$$

24.2.1.2 Iterative Approach

In this approach, w_a is calculated iteratively using the standard least mean square (LMS) algorithm. The weight, $w_{a,\text{opt}}$, is calculated as

$$w_a(k+1) = w_a(k) + \mu_a v(k) e^*(k) \tag{24.17}$$

where μ_a is the step size controlling the convergence and $v(k)$ is the output of the blocking matrix given by

$$v(k) = B^H x(k) \tag{24.18}$$

Since the GSC implementation reuses the array output signal as the error signal $e(k)$, we have

$$e(k) = y(k) \tag{24.19}$$

The conventional GSC is sensitive to a DOA mismatch that can easily occur in practical situations. It can be shown that the stochastic gradient in Equation 24.17, that is, $v(k)y^*(k)$, will not be close to zero even for optimum weights. As a result, the excess MSE induced by the LMS algorithm will be large. To reduce the MSE, small step size should be used but this will result in slower convergence.

24.2.2 Decision-Feedback Generalized Sidelobe Canceller

A decision-feedback (DF) block is added to the conventional GSC in order to enhance the robustness and performance of conventional GSC. Moreover, a blind equalizer is introduced which is trained with error signal between output and input signal. It also contains a feedback filter, which is trained with the error signal between GSC and feedback filter output [3]. All these together constitute DF-GSC. The block diagram of DF-GSC is shown in Figure 24.2. The blind equalizer equalizes the channel and the DOA mismatch effect. The role of feedback filter is to cancel any desired signal component in the LMS error signal. This inclusion of decision-feedback blind equalizer removes the coupling between the signals. The blind equalizer and the feedback filter are trained by different error signals and they have different weights w_m and w_b, respectively. The blind equalizer can recover the transmitted bits, $b_0(k)$ with a phase ambiguity which is acceptable. The blind equalizer

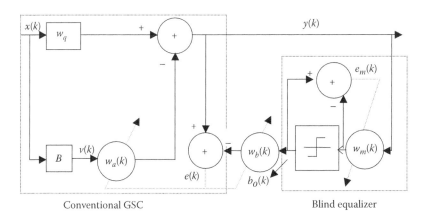

FIGURE 24.2 Decision-feedback generalized sidelobe canceller (DF-GSC).

is trained by the error signal $e_m(k) = \hat{b}_0(k) - w_m^* y(k)$, $\hat{b}_0(k)$ being the detected symbol and w_m the equalizer tap weight. The updated equation for w_m is expressed as

$$w_m(k+1) = w_m(k) + \mu_m y(k) e_m^*(k) \tag{24.20}$$

where μ_m is the step size controlling the convergence behaviour of w_m.

The minimum power denoted by $J_{\min}$ is given as

$$J_{\min} = w_q^H R_{i+n} w_{\mathrm{opt}} \tag{24.21}$$

Here, w_q and R_{i+n} are same as in conventional GSC and w_{opt} is given by

$$w_{\mathrm{opt}} = \begin{bmatrix} w_a \\ w_b \end{bmatrix} \tag{24.22}$$

where w_b is the feedback weight with a dimension of 1×1. It is assumed that the blind equalizer is able to detect the symbol correctly and hence

$$s_0(k) = \hat{s}_0(k) = b_0(k) = \hat{b}_0(k).$$

24.2.2.1 Analytic Approach

It can be shown that w_{opt} given in Equation 24.22 can be decomposed back in two weights as

$$w_{a,\mathrm{opt}} = (B^H R_x B)^{-1} B^H R_x w_q \tag{24.23}$$

and

$$w_{b,\mathrm{opt}} = S^H(\theta_o) w_q \tag{24.24}$$

These two formulas were used in the analytic weight calculation and numbers of snapshots were varied. The output power $P_{o,\min}$ and the desired signal power P_s are calculated using Equations 24.13 and 24.14, respectively.

24.2.2.2 Iterative Approach

In this approach, instead of using Equations 24.23 and 24.24 for weight calculation, a weight update equation is used. Both w_a and w_b are calculated recursively as

$$\begin{aligned} w_a(k+1) &= w_a(k) + \mu_a v(k) e^*(k) \\ w_b(k+1) &= w_b(k) + \mu_b b_0(k) e^*(k) \end{aligned} \tag{24.25}$$

where μ_a is the step size for w_a, μ_b is the step size for w_b, and $v(k)$ is the filter input vector given by Equation 24.18.

The error signal used to feed the above two weights is given as

$$e(k) = (w_q - B w_a)^H x(k) - w_b^* b_o(k) \tag{24.26}$$

The SINR equations remain the same for the DF-GSC. The equations for P_s and $P_{o,min}$ also remain the same. The difference is in the method of computation of w_{opt}. It is important to note that $P_{o,min}$ is not equal to J_{min}, which is given as

$$J_{min} = w_q^H R_{i+n} w_{opt} \tag{24.27}$$

Thus, $P_{o,min}$ is calculated using Equation 24.21.

24.2.3 Constraints in Adaptive Array Processing

In conventional GSC, in order to separate the desired signal component from the interfering signals, a notch filter is added. This is done to increase the robustness of the GSC, which is then guaranteed because of signal-free operation. It also lessens the deviation between the optimum and adaptive weights, and thus improves the convergence rate [9]. The main disadvantage of this approach is the need of the notch filter and a slave array for recovering the desired signal. There are several other approaches that can be used to increase the robustness of adaptive arrays. One of them is the use of constraints [10]. Optimum beamforming with multiple linear constraints is now a well-established technique in array processing. In the simplest case, a single constraint is imposed. The weight vector is then calculated by minimizing the beamformer output power subject to this constraint. There are many methods of constraining the main beam response of an adaptive array. These methods [11] attempt to (i) constrain the adaptive processor from responding to the signals in the main beam, (ii) constrain the processor to maintain constant gain in the desired look direction and the shape of response pattern in the vicinity of that direction and (iii) be capable of achieving a desired quiescent pattern for search when there is no interference present.

24.2.3.1 Point Constraints

This is the minimum constraint that must be applied to the adapted weight vector. In this case, we have $C = E, F = N$ (scalar), where C is the constraint vector, E is an $N \times 1$ vector of ones and N is the number of elements [12]. Thus, the constraint equation becomes

$$W^* E = N \tag{24.28}$$

which constraints the response of the weight vector, W, in the look direction. With this constraint, the weight vector minimizes the output power in all directions, except in the look direction. Since the useful signal (target) and the look directions usually do not coincide exactly, it is important to study the reaction of the weight vector to in-beam but off-axis signals. However, in practical applications, where perturbations exist, and where close-in target and interference lines dominate over the background, it is advantageous to utilize multiple constraints to maintain an angular sector of the main lobe.

24.2.3.2 Derivative Constraints

One knows that greater the desired signal power, the worse the signal suppression and the narrower the effective beamwidth becomes. A remedy to this problem is to impose the derivative constraints on the main beam in the look direction. Thus, by controlling the first few derivatives of the main lobe [13], adaptive arrays can avoid placing the nulls in the main lobe. In the present work, the anti-jamming performance of the first-order derivative constraint array with perfect pointing is considered. If these derivatives are set to zero [14], these constraint systems are of the form given below:

$$\tilde{V} = \begin{bmatrix} v(\theta) & \dot{v}(\theta) & \ddot{v}(\theta)... \end{bmatrix} \tag{24.29}$$

where $\tilde{V}$ denotes a matrix whose columns are the constraint vectors, $v(\theta)$ is the first point constraint column, $\dot{v}(\theta)$ is the first derivative and $\ddot{v}(\theta)$ is the second derivative.

These constraints allow direct control of the sensitivity of the beam pattern to mismatch between steer and actual arrival angle of the desired signal. The adaptive portion of the modified GSC structure consists of using a simple unconstrained LMS algorithm on a reduced set of steered element signals. The number of weights, which must be adapted, is therefore smaller than in the direct form version of the constrained beamformers. The performance of such a derivative constraint system is independent of the phase origin and can achieve the maximum output *signal-to-noise ratio* (SNR) when the pre-steering is perfect. It is observed that higher the orders of derivatives, the broader the beam in the look direction normally becomes. The broader beam is useful when the actual signal direction and known direction of signal are not precisely the same.

24.2.3.3   Directional Constraints

Adaptive arrays with a look-direction constraint are very effective in suppressing interferences and can achieve the maximum SNR. However, an error in the steering angle, called the pointing error, will cause the adaptive array to null out the desired signal as if it were a jammer. In order to make optimum beamformers robust against beam steering angle errors, the use of either *multiple directional constraints* or *multiple derivative constraints* on the polar response of the beamformer have been suggested. For a given multiple-directional-constraint system with the constraints separated in azimuth by an amount Δ, the beamformer performance is found to be affected by the magnitude of Δ [15]. It can be seen that in some practically useful situations, directional constraints are liable to problems with *ill conditioning* of correlation matrix, which derivative constraints avoid. Since the signal suppression for a derivative constraint approaches that of a directional constraint system as the angle between the directional constraints decreases, the use of derivative constraints may be preferable in these circumstances. However, it does not follow that derivative constraints are always *well conditioned*. The choice of which form to use is just a matter of fact that which of the above gives a better-conditioned system [16].

24.2.4   SIMULATION RESULTS

For a *uniform linear array* (ULA) of $N = 16$ antenna elements, spaced half a wavelength apart, let the signal environment consist of one desired source and three uncorrelated interfering sources probing from $0°$, $20°$, $50°$, and $-35°$, respectively. The SNR of the desired signal is 0 dB and the *interference noise ratio* (INR) is 20 dB per interference. The desired source and the interfering sources are taken randomly generated QPSK signal. A flat fading channel environment is assumed so that $s_0(k) = h(k) \times b_0(k)$, $h(k)$ being the channel coefficient. Here, $h(k) = 1$, in order to avoid complex computations. Noise is modelled as white Gaussian noise with variance 0.35. It is assumed that the desired signal's DOA is exactly known; thus, no training is required for DF-GSC. Here, only point distortionless constraint is considered. Figure 24.3 presents the learning curves of GSC and DF-GSC schemes. It shows the variation in output SINR (dB) with snapshots. The algorithm used for weight adaptation is the standard LMS algorithm. Since the generation of random signals is involved, the results are shown after averaging 200 simulations. The corresponding adapted beam pattern for GSC and DF-GSC is shown in Figure 24.4. The arrows denote the directions of adapted nulls/main lobe of desired source. This convention is followed in this chapter. It can be observed that adaptive DF-GSC capabilities are better in nulling the interference than GSC. The quiescent pattern (pattern without interfering sources) is also included. It is apparent that the adaptive DF-GSC can achieve higher SINR than the conventional GSC. The difference between the two schemes is almost 10 dB for each interfering source.

In order to compare the convergence rate for both algorithms, the learning curves for GSC and DF-GSC with the same SINR target are shown in Figure 24.5. It shows that DF-GSC converges around 150 snapshots whereas GSC converges around 350 snapshots. DF-GSC takes lesser time to reach the optimum value of output SINR, as compared to GSC. Thus, one can infer that the GSC implementation is of importance as it has less computational complexity. The addition of blind

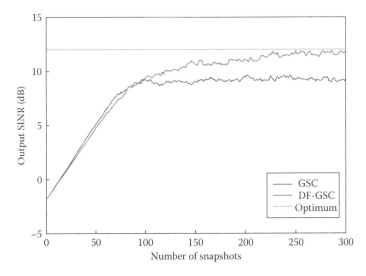

FIGURE 24.3 Learning curves for GSC and DF-GSC. (Adapted from Y. Lee and W.-R. Wu. *IEEE Transactions on Antennas and Propagation*, 53, 3822–3832, 2005.)

equalizer with decision-feedback in conventional GSC provides better robustness, faster convergence and high-output SINR. The resultant design, referred to as the DF-GSC scheme, can achieve higher SINR value for the same convergence rate. Moreover, on increasing the desired signal power, the SINR output is enhanced. In case of DF-GSC, deeper nulls are obtained which shows that DF-GSC has better suppression capabilities. The suppression capabilities of sidelobe cancellers can also be exploited in stealth techniques. Mounting these cancellers in suitable configurations on aircraft and missiles will allow the reception of only the desired signal, simultaneously suppressing the unwanted components. This will make them invisible for the radars at the enemy base or in their aircraft. Similarly, this capability of adaptive arrays can be used for smart behaviour of space probes and robotics.

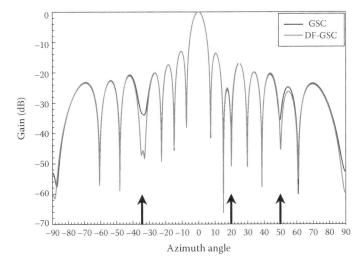

FIGURE 24.4 Beam pattern of GSC and DF-GSC after 200 snapshots. (Adapted from Y. Lee and W.-R. Wu. *IEEE Transactions on Antennas and Propagation*, 53, 3822–3832, 2005.)

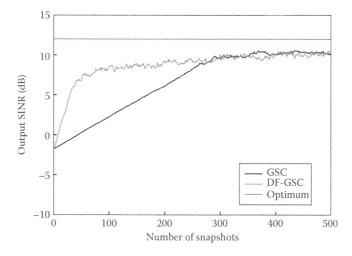

FIGURE 24.5 Learning curves for GSC and DF-GSC with the same SINR target. (Adapted from Y. Lee and W.-R. Wu. *IEEE Transactions on Antennas and Propagation*, 53, 3822–3832, 2005.)

24.3 DIRECTION OF ARRIVAL MISMATCH ANALYSIS

The conventional beamformers assume that the desired signal is absent in the training period. In such a case, the beamformer is known to be sufficiently robust against mismatch errors in the array response. Unfortunately, in typical practical applications, the signal-free training snapshots are difficult to obtain. In order to surmount the problem of array performance degradation due to the DOA mismatch, one of the remedies is to impose additional constraints such as point constraints, derivative constraints and linear constraints on the array weight vector. However, imposing additional constraints deteriorates the array's capability in suppressing interference and noise, that is, it consumes some of the array's degrees of freedom.

Both the point constraints and the first-derivative constraints are included in the formulation in this chapter, and the simulations are carried out to analyse the enhancement in the robustness of GSC and DF-GSC against the DOA mismatch errors. The DF-GSC scheme requires some initial training snapshots, in order to acquire the required robustness. In practice, however, it becomes difficult to decide this initial snapshot number. Thus, some *blind* and *semi-blind* methods are used in practice to overcome this problem. The performance analysis of DF-GSC is carried out using the *blind approach*, which has three stages of weight estimation. A novel *semi-blind canceller* scheme is also introduced here to enhance the robustness of the system, which has only two stages of computation. Simulations are done using different LMS algorithms. Steady-state analysis is performed for the blind DF-GSC using the standard LMS and the structured LMS algorithms.

24.3.1 SIGNAL MODEL WITH MISMATCH

If there is a mismatch between the actual and presumed the desired signal's DOA, that is,

$$\theta_o = \hat{\theta}_o + \Delta \tag{24.30}$$

where $\hat{\theta}_o$ is the estimated DOA and Δ is the mismatch value, the steering vector will be

$$S(\theta_o) = S(\hat{\theta}_o + \Delta) = \left[1,\, e^{i\alpha \sin(\hat{\theta}_0 + \Delta)},\, e^{i2\alpha \sin(\hat{\theta}_0 + \Delta)}, ...,\, e^{i(N-1)\alpha \sin(\hat{\theta}_0 + \Delta)} \right]^T \tag{24.31}$$

with $\alpha = 2\pi d/\lambda$. Since Δ is generally small

$$\sin(\hat{\theta}_o + \Delta) \approx \sin(\hat{\theta}_o) + \Delta\cos(\hat{\theta}_o) \tag{24.32}$$

Assuming $\hat{\theta}_o = 0°$ and using Equations 24.31 and 24.32, the mismatch steering vector is given by

$$S(\theta_o) = S(\Delta) = \left[1, e^{i\alpha\Delta}, e^{i2\alpha\Delta}, ..., e^{i(N-1)\alpha\Delta}\right]^T \tag{24.33}$$

Now, $s(k)$ is re-written as $s(k) = S(\Delta)s_o(k)$.

24.3.2 DOA Mismatch in GSC

With an estimation error of the desired signal's DOA, w_q is not matched to the desired signal's steering vector, and thus the blocking matrix cannot obstruct the desired signal entering the input of the interference-cancelling filter. If there is enough degree of freedom (DOF), the filter will cancel the desired signal from the quiescent filter output leading to signal cancellation. In case of GSC, even with optimum weight vector, it does not provide the ability to maximize the SINR [3]. The expression of the optimum output SINR for conventional GSC is given as

$$\text{SINR}_{\text{opt}} = \frac{P_s}{P_{o,\text{min}} - P_s} \tag{24.34}$$

where P_s is the output desired signal power. $P_{o,\text{min}}$ is the minimum power and is equal to J expressed as

$$\min J = \min(w_q - Bw_a)^H R_x (w_q - Bw_a) \tag{24.35}$$

where B is the blocking matrix and R_x is the correlation matrix.

Now, the output desired signal power in the conventional GSC becomes

$$P_s = \sigma_{s_o}^2 \left|(w_q - Bw_{a,\text{opt}})^H S(\Delta)\right|^2 \tag{24.36}$$

which is not the same as $P_s = \sigma_{s_o}^2 \left|w_q^H S(\theta_o)\right|^2$ because $w_{a,\text{opt}}^H B^H S(\Delta)$ is not zero now.

The actual amount of signal attenuation depends upon the power of signal and the amount of error. The minimum output power $P_{o,\text{min}}$ for calculating the optimum output SINR of the conventional GSC in mismatch is the same as Equation 24.35.

24.3.3 DOA Mismatch in DF-GSC

In the case of DF-GSC, the signal cancellation phenomenon is avoided. Since a correlation exists between two paths of GSC, a coupling takes place between $w_{a,\text{opt}}$ and $w_{b,\text{opt}}$ if there is a mismatch. Thus, the coupled $w_{a,\text{opt}}$ and $w_{b,\text{opt}}$ can be expressed as [3]

$$w_{a,\text{opt}} = (B^H R_{i+n} B)^{-1} B^H R_{i+n} w_q \tag{24.37}$$

$$w_{b,\mathrm{opt}} = S^H(\Delta)(w_q - Bw_{a,\mathrm{opt}}) \tag{24.38}$$

The MMSE of the DF-GSC with mismatch can be expressed as

$$J_{\min} = w_q^H R_{i+n} w_{\mathrm{opt}} \tag{24.39}$$

The power of the desired signal will be the same as Equation 24.36, but for the DF-GSC with mismatch, the minimum output power is calculated as

$$P_{o,\min} = w_{\mathrm{opt}}^H R_x w_{\mathrm{opt}} \tag{24.40}$$

The output SINR will be estimated using similar expression (Equation 24.34) as in the case of GSC.

24.3.4 CONVERGENCE ANALYSIS

The parameters considered for the convergence analysis of GSC and DF-GSC are MSE and SINR in steady state. The eigenvalues are used in the calculation of SINR.

24.3.4.1 MSE in Steady State

Let the MSE in steady state be denoted as $J(\infty)$. Then

$$J(\infty) = J_{\min} + J_{\mathrm{ex}}(\infty) \tag{24.41}$$

where $J_{\min}$ is MMSE and in case of DF-GSC, it is calculated as

$$J_{\min} = w_q^H R_{i+n} w_{\mathrm{opt}} \tag{24.42}$$

while

$$J_{\min} = w_{\mathrm{opt}}^H R_x w_{\mathrm{opt}} = w_q^H R_x w_{\mathrm{opt}} \tag{24.43}$$

in case of GSC and J_{ex} is excess MSE calculated as

$$J_{\mathrm{ex}}(\infty) = J_{\min} \sum_{l=1}^{N-P} \frac{\mu_a \lambda_l (B^H R_x B)}{2 - \mu_a \lambda_l (B^H R_x B)} \tag{24.44}$$

where $\lambda_l(B^H R_x B)$ is the lth eigenvalue of $B^H R_x B$. It can be observed that the excess MSE is roughly proportional to the resultant MMSE and the step size used.

24.3.4.2 SINR in Steady State

The output SINR in steady state is used as a performance index for adaptive GSC and DF-GSC schemes. The steady-state SINR with the LMS algorithm is given by

$$\mathrm{SINR}_{\mathrm{LMS}} = \frac{P_s}{P_{o,\min} + J_{\mathrm{ex}(\infty)} - P_s} \tag{24.45}$$

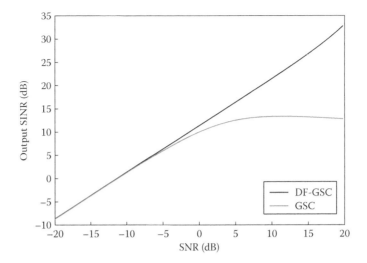

FIGURE 24.6 Steady-state SINR performance in different SNR environments. (Adapted from Y. Lee and W.-R. Wu. *IEEE Transactions on Antennas and Propagation*, 53, 3822–3832, 2005.)

It can be observed that the smaller the excess MSE value, larger the steady-state SINR becomes. The SINR performance in different SNR environments for GSC and DF-GSC is presented in Figure 24.6. Here, no averaging is carried out for the simulations. It can be seen that after a certain value of SINR, the SINR for GSC saturates while there is no such trend observed in DF-GSC.

24.3.4.3 Desired Signal's DOA Mismatch

So far, it is assumed that the DOA of the desired signal is perfectly known. However, in practical situations, there exists a difference between the received and the estimated direction of the desired signal. Considering a mismatch of 2°, the variation of SINR with snapshots is plotted for GSC and DF-GSC in Figures 24.7 and 24.8, respectively.

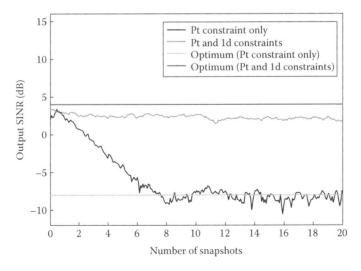

FIGURE 24.7 Learning curves for GSC with DOA mismatch utilizing point constraint only and first-order derivative constraint. (Adapted from Y. Lee and W.-R. Wu. *IEEE Transactions on Antennas and Propagation*, 53, 3822–3832, 2005.)

GSC with point constraint only: On inclusion of point constraint in the simulation of GSC, when the output SINR is computed over 20,000 snapshots, it is observed that output SINR degrades drastically, as shown in Figure 24.7. The convergence is achieved around SINR level of −8 dB.

GSC with first derivative and point constraint: Further including derivative constraints along with the point constraint, a comparative analysis is done for the performance of GSC. The corresponding learning curve is shown in Figure 24.7. It can be observed that additional constraint results in enhanced robustness. On using this additional first derivative constraint, the SINR is increased considerably. The SINR stabilizes after 1000 snapshots and around a value of 2.5 dB.

DF-GSC with point constraint only: Similarly, in Figure 24.8a, the learning curve for DF-GSC with DOA mismatch is presented. This result is simulated over 20,000 snapshots. The plot converges after 3000 snapshots and stabilizes to an approximate value of 9 dB.

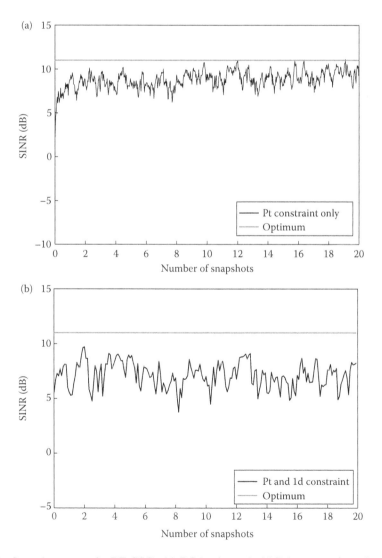

FIGURE 24.8 Learning curves for DF-GSC with DOA mismatch. (a) Point constraint only. (b) First-order derivative constraint. (Adapted from Y. Lee and W.-R. Wu. *IEEE Transactions on Antennas and Propagation*, 53, 3822–3832, 2005.)

DF-GSC with first derivative and point constraint: The learning curve for DF-GSC with DOA mismatch is shown in Figure 24.8b as in the case of GSC, including both first derivative constraint and point constraint. *There was no training required in this case.* Addition of the constraint lowers the SINR for the DF-GSC. The reason behind this is lesser number of available degrees of freedom. The two curves of point constraint and derivative constraint are shown separately for clarity. These simulations show that GSC with decision-feedback provides better robustness as compared to the conventional GSC structure.

24.3.5 Dependence of Output SINR on the Step Size of Weight Adaptation

There are two approaches, viz., analytic approach and simulated approach, for the estimation of SINR [3].

A. *Analytic approach*: The weights are calculated using the equation

$$w_{a,\text{opt}} = (B^H R_x B)^{-1} B^H R_x w_q \tag{24.46}$$

in which a specific formula is used for w_a. It should be noted that no iteration is done for weight estimation. The noise variance is set to 0.5. The desired signal power is taken as 0.68 while the interfering signal of power levels of 100 each is assumed. Figure 24.9 demonstrates the dependence of SINR of GSC and DF-GSC on the step size. Slow but continuous degradation in output SINR is observed. For DF-GSC, w_a and w_b are calculated using $w_{a,\text{opt}} = (B^H R_x B)^{-1} B^H R_x w_q$ and $w_{b,\text{opt}} = S^H(\theta_o) w_q$, respectively. It is noted that SINR stabilizes in this case and no degradation is obtained. This proves that in case of steady-state computation, DF-GSC is again far better than conventional GSC.

B. *Simulated approach*: In this approach, weights are calculated iteratively for GSC using

$$w_a(k+1) = w_a(k) + \mu_a v(k) e^*(k) \tag{24.47}$$

where μ_a is the step size, $v(k)$ is the array output and $e(k)$ is the error signal.

For DF-GSC, the weight adaptation equation used is given by

$$\begin{aligned} w_a(k+1) &= w_a(k) + \mu_a v(k) e^*(k) \\ w_b(k+1) &= w_b(k) + \mu_b b_0(k) e^*(k) \end{aligned} \tag{24.48}$$

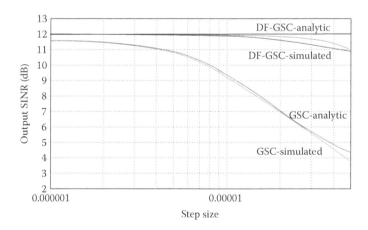

FIGURE 24.9 Steady-state SINR performance with different step sizes. (Adapted from Y. Lee and W.-R. Wu. *IEEE Transactions on Antennas and Propagation*, 53, 3822–3832, 2005.)

For uniformity, noise variance and power of interference sources are kept the same as in the above case. It is to be noted that both the analytic and simulated curves match satisfactorily with each other in both GSC and DF-GSC. Thus, the performance analysis carried out for GSC and DF-GSC using two performance indices, viz., SINR and MSE proves that DF-GSC has faster convergence and higher output SINR, even in the case of DOA mismatch. Its robustness is further enhanced by the use of first derivative constraint along with point constraints. It has been observed that after a certain value of SNR, the output SINR degrades for GSC whereas there is no such trend observed for DF-GSC. The role of step size in the performance of GSC and DF-GSC is demonstrated by the trend followed by output SINR w.r.t. the step size of the weight adaptation used. It can be observed that the SINR deteriorates in the case of conventional GSC whereas for DF-GSC it stabilizes after a certain value of step size. The results obtained are similar for both analytic and simulated approaches.

24.4 ADAPTIVE ALGORITHMS IN SIDELOBE CANCELLERS

Adaptive array processing involves the manipulation of signals induced on the elements of an array. Over the decades, many researchers have shown their interest in this area because of its widespread applicability. Many algorithms have been suggested for better performance and speed of convergence. Among these algorithms, LMS algorithm is one of the simplest algorithms that adjust the array sensors to respond to a desired signal while suppressing the noises and interferences [4]. The LMS algorithm is an important member of the family of stochastic gradient algorithms. The most significant feature of the LMS algorithm is its simplicity. It does not require any calculation of correlation matrix and matrix inversion. There are various forms of LMS algorithms such as *standard LMS*, *structured gradient LMS*, *improved LMS* and *recursive LMS algorithm*. Standard LMS algorithm is the most widely used algorithm since it has less complexity. Moreover, standard LMS can be easily implemented in the hardware. The improved LMS algorithm enhances the performance even more by reusing all the values of auto-correlation matrix (ACM) generated at previous iterations. In order to analyse the interference suppression capabilities of sidelobe cancellers, different forms of LMS algorithms have been employed for weight adaptation in different signal scenario. The output SINR and MSE are taken as performance indices. Further, the effect of step size and power level of the desired signal on the performance of GSC and DF-GSC scheme are analysed.

24.4.1 WEIGHT ADAPTATION USING LMS ALGORITHM

The LMS algorithm works on the principle of iterative calculation of weights. It is done using the difference between the array output and the reference signal. After each iteration, the weights are changed in the negative direction of the gradient by a small amount. The constant that determines this amount is referred to as step size [17]. This gradient is estimated in different ways in different forms of LMS algorithms [4]. The optimum values of weights are determined in order to minimize the mean output power. Although various forms of the LMS algorithm employ different method of gradient estimation, there is one feature that is common. In each algorithm, the weights in the $(n + 1)$th iteration are calculated using samples obtained after the nth iteration. There is no role of the previous samples. All algorithms do filtering and weight updating at every iteration.

Filtering:

$$y(n) = W^H(n)x(n) \qquad (24.49)$$

Weight updating:

$$W(n+1) = P[W(n) - \mu \, g(W(n))] + \frac{S_o}{S_o^H S_o} \qquad (24.50)$$

where $S_o = S(\theta_o)$, steering vector of the desired signal incident at an angle of θ_o. Similarly, P, the projection operator is given by

$$P = I - \frac{S_o S_o^H}{S_o^H S_o}$$ (24.51)

$g(W(n))$ is an unbiased estimate of the gradient of $W^H(n)RW(n)$ with respect to $W(n)$.

   Equation 24.50 consists of both the adaptive and non-adaptive parts. The non-adaptive term S_o contains the constraint in the look direction. The adaptive term contains the projection matrix P.

24.4.1.1   Standard LMS Algorithm

This algorithm involves the simplest estimate of the gradient. It tries to minimize the output power subject to $W^H(n)S_o = 1$, while leaving signal in look direction unaffected. The iterative estimation of optimum weight vector is done using Equation 24.50. The gradient is given by

$$g[W(n)] = 2X(n+1)X^H(n+1)W(n) = 2X(n+1)y^*[W(n)]$$ (24.51a)

24.4.1.2   Structured Gradient LMS Algorithm

The structured algorithm is the algorithm in which the gradient is estimated from ACM that is constrained to have *Toeplitz* structure [18], that is, the elements along the diagonal are equal. The procedure for the generation of the ACM is given below [19]: (a) The matrix $x(k) \cdot x^H(k)$ is calculated; (b) Consider only the N diagonals of the upper triangular matrix. For each diagonal, the diagonal elements are averaged. The new value for each diagonal element is the mean value thus calculated; (c) Now consider the $N-1$ diagonals of the lower triangular matrix. In each diagonal, every element is taken as the complex conjugate of the value in the corresponding diagonal of upper triangular matrix so that the new matrix (ACM) is Hermitian. The structured gradient LMS algorithm replaces the product $x(k) \cdot x^H(k)$ by the newly generated $\hat{R}(k)$ to obtain better performance. The gradient estimate is defined as

$$\hat{g}[W(n)] = 2\hat{R}(n+1)W(n)$$ (24.51b)

where $\hat{R}(n)$ is the structured estimate of the array correlation matrix at the nth instant of time and is given by

$$\hat{R}(n) = \begin{bmatrix} \hat{r}_o(n) & \hat{r}_1(n) & ... & \hat{r}_{L-1}(n) \\ \hat{r}_1^*(n) & . & & \\ ... & ... & & \\ ... & ... & & \\ \hat{r}_{L-1}^*(n) & & & \hat{r}_o(n) \end{bmatrix}$$ (24.52)

with

$$\hat{r}_l(n) = \frac{1}{N_l} \sum_i x_i(n)x_{i+1}^*(n), \quad l = 0, 1,..., L-1$$ (24.53)

where N_l denotes the number of possible combinations of elements with lag. For a linear array of equi-spaced elements, $N_l = L - l$. Each element of $\hat{R}(n)$ is a mean value of all the elements of $R(n)$ with the same spatial correlation lags.

24.4.1.3 Improved LMS Algorithm

This algorithm also uses the Toeplitz structure [19], in which the elements along the diagonal are equal, along with the earlier samples. The gradient [20] is calculated as

$$g_I[W(n)] = 2\tilde{R}(n+1)W(n) \tag{24.54}$$

where

$$\tilde{R}(n+1) = \frac{1}{n+1}\left[n\tilde{R}(n) + \hat{R}(n+1)\right] \tag{24.55}$$

The expression of $\hat{R}(n)$ is the same as given by Equation 24.52.

24.4.1.4 Recursive LMS Algorithm

This algorithm uses the following equation for gradient estimation:

$$g_R[W(n)] = 2R(n+1)W(n) \tag{24.56}$$

where

$$R(n+1) = \frac{1}{n+1}\left[nR(n) + X(n+1)X^H(n+1)\right] \tag{24.57}$$

24.4.1.5 Recursive Least Square Algorithm

Here, step size is replaced by gain matrix $R^{-1}(n)$ which is calculated iteratively as

$$R^{-1}(n) = R^{-1}(n-1) - \frac{R^{-1}(n-1)X(n)X^H(n)R^{-1}(n-1)}{1 + X^H(n)R^{-1}(n-1)X(n)} \tag{24.58}$$

with

$$R^{-1}(0) = \frac{I}{\varepsilon}, \quad \varepsilon > 0 \tag{24.59}$$

The required inverse $R^{-1}(n)$ is updated using the previous inverse and the present samples and thus the weight is calculated as

$$W(n) = \frac{R^{-1}(n)S_o}{S_o^H R^{-1}(n)S_o} \tag{24.60}$$

24.4.2 Novel Simulations

The performance of GSC and DF-GSC schemes of phased array is studied using different forms of LMS algorithms.

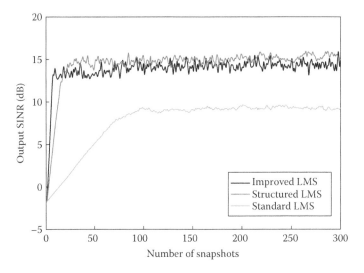

FIGURE 24.10 Learning curves of GSC using different forms of the LMS algorithm.

A. *Generalized sidelobe canceller:* The learning curves of GSC using standard LMS, structured gradient LMS and improved LMS algorithms are shown in Figure 24.10. The simulations are carried out for 300 snapshots. It can be seen that the structured gradient and improved LMS algorithms have faster convergence rate along with higher output SINR. The adapted beam pattern for GSC using structured LMS algorithm is shown in Figure 24.11. The plot includes the quiescent pattern, which is shown with dotted line. The overall performance of GSC is found to be much better as compared to the one obtained using the standard LMS algorithm.

The effect of changing the desired signal level on output SINR is presented in Figure 24.12. The signal levels considered are 1, 10 and 100. The algorithm used for weight adaptation is the structured LMS algorithm. It is observed that as the desired signal power is increased, the SINR output also increases which is according to the expectations. Calculations are averaged over 200 snapshots. The learning curves of GSC for three step sizes are shown in Figure 24.13. The step sizes used are 0.2×10^{-5}, 0.5×10^{-5} and 0.7×10^{-5}. It can be seen that as the step size is reduced, the convergence

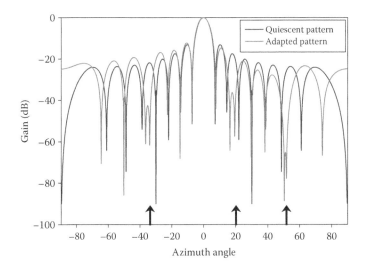

FIGURE 24.11 Beam pattern of GSC using the structured LMS algorithm.

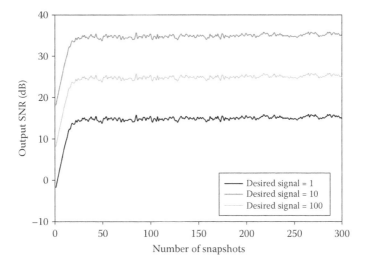

FIGURE 24.12   Learning curves of GSC using the structured LMS algorithm for different desired signal levels.

time is increased but the final SINR level remains almost same. In Figure 24.14, the beam pattern of GSC has been plotted *after varying the power ratio of interference sources* to 10, 1000 and 100. The reason for this is that if probing sources have equal powers, then the corresponding eigenvalues are also the same. On the other hand, if the sources have different power, the eigenvalue of one probing source will be much larger as compared to the eigenvalue of other probing source and thus the overall performance is better.

B. *Decision-feedback generalized sidelobe canceller:* Similar computations are performed for DF-GSC in Figure 24.15, using standard LMS, structured gradient LMS and improved LMS algorithms. It can be inferred that the improved LMS is best among the three LMS algorithms. The improved LMS gives the best convergence rate along with high SINR. Although the computation complexity in standard LMS is less, it gives less SINR with poor convergence rate.

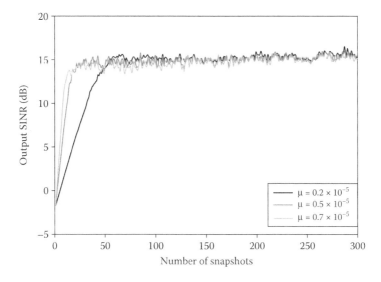

FIGURE 24.13   Learning curves of GSC using the structured LMS algorithm for different step sizes.

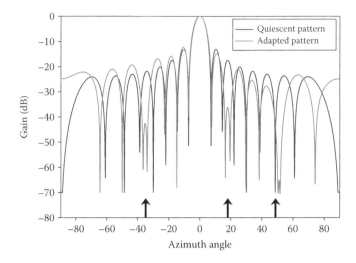

FIGURE 24.14 Beam pattern of GSC for interfering sources with different power levels, that is, 10, 1000, and 100 using the structured LMS algorithm.

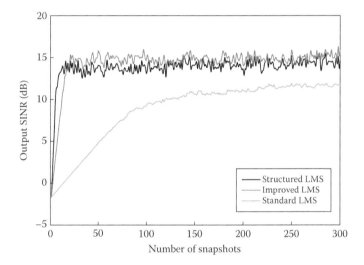

FIGURE 24.15 Learning curves of DF-GSC using three forms of the LMS algorithm.

In order to demonstrate *the effect of closely spaced interfering sources*, the beam pattern for DF-GSC is shown in Figure 24.16, after reducing the spacing between the interfering sources. The new directions of arrival (DOAs) of the sources are 36°, 48° and 66°, respectively. The power levels of interfering sources are kept constant at 100. The step size used is $\mu_a = 10.0 \times 10^{-7}$ and $\mu_b = 0.01$. It is observed that the pattern gets distorted and deeper nulls are obtained. This is because *the array treats closely spaced jammers as a continuous source and hence its performance improves.*

24.5 BLIND EQUALIZER AND FEEDBACK FILTER IN SIDELOBE CANCELLERS

The fact that the DF-GSC scheme has better suppression capabilities than the conventional GSC scheme is known. Its robustness can be further enhanced by the use of additional constraints such as derivative constraints and directional constraints. However, the output SINR deteriorates due to these additional constraints [3].

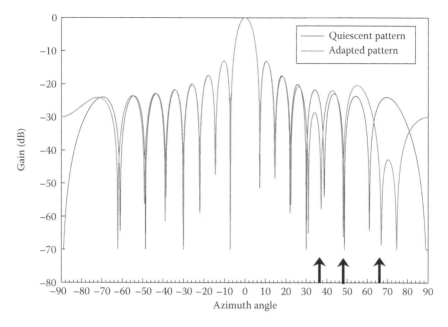

FIGURE 24.16 Beam pattern of DF-GSC for closely spaced interfering sources using the structured LMS algorithm.

In digital communications, an equalizer that can fully cancel the interference from a linear dispersive medium within a finite time span while not amplifying the noise is bound to be popular. Another point of consideration is that it can be implemented with ease. Equalization is important for the receiver of communication systems to correctly recover the symbols sent by the transmitter, because the received signals may contain interference, noises and so on. Many real communication devices contain equalization modules, such as modems, cellular phones and digital TVs. Blind equalization is useful to cancel the repeatedly transmitted training signals so as to improve system output. It is still a challenging work to find an effective, robust, computationally efficient algorithm to do blind equalization [5].

In order to improve spectral efficiency, several blind and semi-blind methods for channel identification, equalization and demodulation have been proposed by the researchers. Moreover, fading channels may deteriorate the known sequence of transmitted symbols severely, so that a conventional trained equalizer, which requires some training snapshots initially, becomes unreliable. The term 'blindness' means that the receiver has no knowledge of either the transmitted/received sequence or the channel impulse response. Only some of the statistical or structural properties of the transmitted and received signal are exploited in the process of adapting the equalizer [4].

24.5.1 BLIND APPROACH FOR DF-GSC WITH MISMATCH

Blind methods allow for tracking fast variations in the channel, and re-acquiring operational conditions using only information symbols. However, if any explicit prior knowledge is available about transmitted sequences, such information should be exploited. In most of the situations, the desired signal may become too weak to initiate blind equalizer. Thus, it is difficult to acquire correct decision initially when DOA mismatch occurs. With the blind approach, DF-GSC can skip the use of training symbols in the initial phase and reach the required SINR performance through adaptation. Conventional equalizers employ a pre-assigned time slot during which a training signal, which is known in advance, is transmitted. In the receiver, the coefficients are then changed or adapted by using some algorithm (e.g., LMS and RLS) so that the output of the equalizer closely matches the

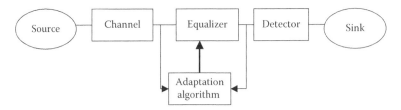

FIGURE 24.17 Blind adaptation scheme.

training sequence. However, inclusion of these training signals reduces the output of the system. Therefore, to overcome this problem, adaptation techniques are used that does not require training, that is, *blind adaptation schemes*. Figure 24.17 shows that blind adaptation uses signal samples from the input and output of the equalizing filter and thus blind adaptation is not affected by erroneously detected data symbols due to noise.

Principle: The principle of this method is to use some rough number of derivative constraints for the DF-GSC initially and then release the constraints gradually. Initially, the main beam gets widened with a certain amount of DOA mismatch value but it ensures that there is enough desired signal strength in the quiescent filter output. Then the equalizer can actively converge without training. After the convergence, GSC releases constraints one by one, due to which the columns in blocking matrix are reduced gradually. This causes increase in DOFs, leading to the enhancement in interference and noise suppression ability.

Steps of algorithm: The algorithm used for blind approach consists of three parts. The first part is called *initialization* in which the projection matrix is calculated and the dimension of the weight vector is increased. The second part is called *transition* in which the new blocking matrix and the new quiescent weight are calculated iteratively. We assume that the number of point-plus-derivative constraints is changed from p to $p - 1$ and thus DOFs are increased from $N - p$ to $N - (p - 1)$. The third part has a usual way of iterative calculation of weights. The detailed steps are described below [3]:

A. *Initialization:* The projection matrix P is formed as

$$P_{N-(P-1)} = B_{N-(P-1)}B^{H}_{N-(P-1)} \tag{24.61}$$

The projection vector, b, in terms of new constraint vector c_p is calculated as

$$b_{N-(P-1)} = P_{N-(P-1)}c_P \tag{24.62}$$

where c_P is the new constraint vector.

A difference vector, d, is calculated which is equal to the difference between the new and the old quiescent weight vector as

$$d_{N-(P-1)} = w_{q,N-(P-1)} - w_{q,N-p} \tag{24.63}$$

A zero vector is tapped onto w_a in order to avoid the mismatch between the new blocking matrix and the new weight vector

$$w_{a,N-(P-1)} = \begin{bmatrix} w_{a,N-P} & 0 \end{bmatrix} \tag{24.64}$$

B. *Transition:* The new weight vector $w_{q.N-(p-1)}$ and the new blocking matrix are calculated as

$$
B_{N-(p-1)} = \left[B_{N-p} \quad \gamma_k b_{N-(p-1)} \right]
$$
$$
w_{q.N-(p-1)} = w_{q.N-p} + \gamma_k d_{N-(p-1)}
$$
(24.65)

where $\gamma_k = 1 - \{(T-1)-k\}/(T-1),\ 0 \le k \le (T-1)$.

Repeat the above step till $\gamma_k = 1$ and weights are calculated iteratively in the last part.

Thus, following the above procedure, DF-GSC can skip the use of training without degradation in performance. Figure 24.18 shows the computation methodology in three stages where the first stage is the initialization stage where weights are calculated iteratively. In this stage, a projection matrix is calculated using Equation 24.61. Then, a difference vector is calculated using Equation 24.63, which is used in the transition stage for weight calculation. The second stage is the transition stage where, along with the weights, the blocking matrix is also changed recursively. In the last stage, weights are again calculated iteratively with a fixed dimension of blocking matrix.

The learning curve of the blind DF-GSC scheme of 10-element ULA is shown in Figure 24.19. The power of the desired signal level is 0.2 whereas three interference sources are kept 100 each. The DOA of three jammers is 20°, 50° and −35°, respectively. The DOA mismatch of 2° is considered. There are three stages involved in this approach. In the first stage, point and first derivative constraints are used and weights are calculated iteratively. After 115 snapshots, the second stage begins in which Equation 24.65 is used for weight adaptation. This stage continues till the 190th snapshot after which the last stage starts. Now, the dimension of the blocking matrix is kept fixed and simulations are performed. The power of the desired signal is kept at 0.2. The length of the transition period is 84. The results are compared for standard LMS, structured LMS and improved LMS algorithms. It can be seen that the standard LMS algorithm gives the lowest output SINR whereas higher output SINR is achieved in the case of the improved LMS algorithm. The fastest rate of convergence is, however, observed in the case of the standard LMS algorithm.

The corresponding normalized adapted beam pattern of blind DF-GSC using three forms of the LMS algorithm is shown in Figure 24.20. Quiescent pattern, as shown by the dotted curve, is included in each plot for the sake of comparison. Deeper nulls are obtained at the location of hostile sources for the case of the improved LMS as compared to the structured and standard LMS algorithm. The power level of each source is kept constant at 100 in order to maintain uniformity. The desired signal power is only 1. Considering another signal scenario with the equal-powered

Stage 1	Stage 2	Stage 3
Initialization	Transition	Iteration
Blocking matrix $N \times N-(P-1)$	Blocking matrix $N \times (N-1)$	Blocking matrix $N \times (N-1)$
w_a, w_b (Iterative calculation)	w_a, w_b (Iterative calculation)	w_a, w_b (Iterative calculation)
w_q (Calculated offline)	w_q (Iterative calculation)	w_q (Calculated offline)
Blocking matrix's dimension is fixed	Blocking matrix is calculated iteratively	Blocking matrix's dimension is fixed

FIGURE 24.18 Three stages of computation in blind DF-GSC.

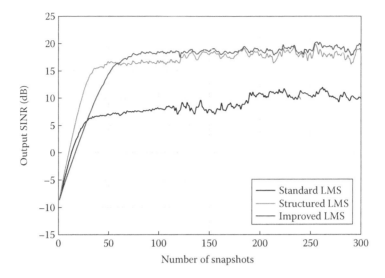

FIGURE 24.19   Learning curves of blind DF-GSC with three forms of the LMS algorithm.

probes located at 38°, 45° and 66°, the adapted pattern of blind DF-GSC using the structured LMS algorithm is shown in Figure 24.21. It can be seen that the nulls up to −70 dB are obtained, signifying efficient suppression of hostile sources.

Semi-blind approach: In this approach, only two stages of computations are done, excluding the last stage of iteration. First is the initialization stage while the second is the transition stage that involves the iterative calculation of the blocking matrix. It is termed the *semi-blind approach*. Figure 24.22 shows the learning curve of semi-blind DF-GSC using three forms of LMS

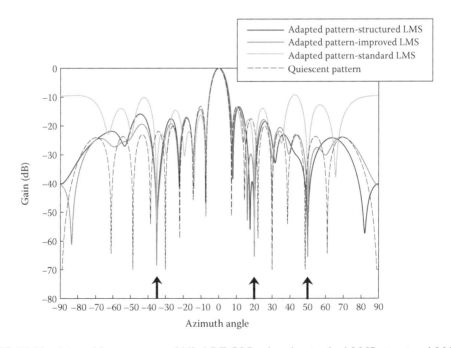

FIGURE 24.20   Adapted beam pattern of blind DF-GSC using the standard LMS, structured LMS and improved LMS algorithm.

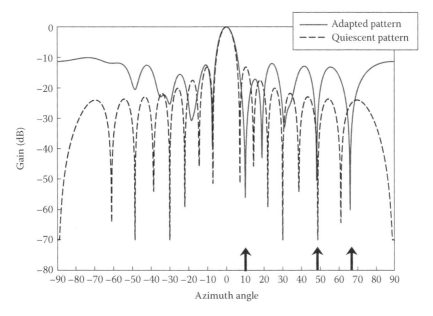

FIGURE 24.21 Adapted beam pattern of blind DF-GSC using the structured LMS algorithm for probing sources (10°, 48°, 66°; 100 each).

algorithms. It can be observed that higher output SINR is obtained using the improved LMS algorithm with a better convergence rate.

A comparative analysis was carried out between normal DF-GSC and the one that uses semiblind DF-GSC. The adapted pattern is shown in Figure 24.23. It can be seen that deeper nulls at each probing direction (20°, 50° and −35°) are obtained in the case of semi-blind DF-GSC. Figure 24.24 shows the beam pattern of blind and semi-blind DF-GSC for the same signal scenario using the structured LMS algorithm. It can be seen that semi-blind approach has better probe suppression than the *blind DF-GSC* approach. In the *blind DF-GSC*, the pattern is distorted a bit. However, this is not the case with *semi-blind DF-GSC*. The performance of semi-blind DF-GSC is compared for

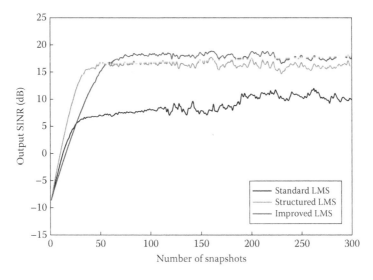

FIGURE 24.22 Learning curves of semi-blind DF-GSC with three forms of the LMS algorithm.

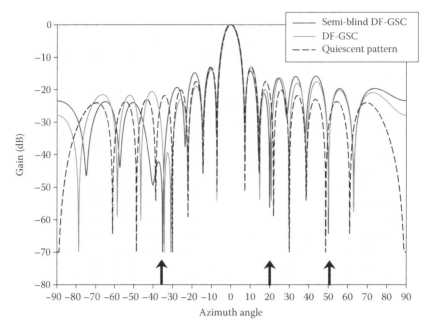

FIGURE 24.23   Adapted beam pattern of semi-blind DF-GSC and DF-GSC for probing sources (50°, 20°, −35°; 100 each).

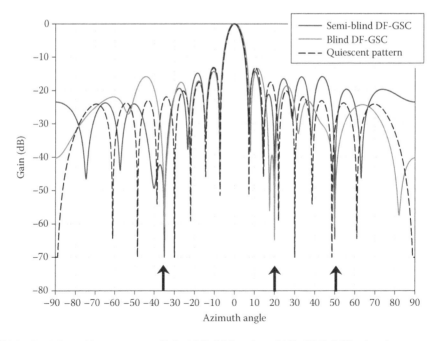

FIGURE 24.24   Adapted beam pattern of blind DF-GSC and semi-blind DF-GSC using the structured LMS algorithm.

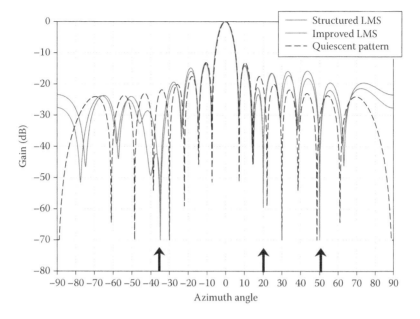

FIGURE 24.25 Adapted beam pattern of semi-blind DF-GSC using the structured and improved LMS algorithm.

the improved LMS and structured LMS algorithms in Figure 24.25. It can be seen that the depth of nulls is better in case of the improved LMS as compared to the structured LMS algorithm.

24.5.2 STEADY-STATE ANALYSIS OF DF-GSC WITH BLIND EQUALIZATION

A steady-state analysis is carried out using the standard and structured LMS algorithms. All the three stages of blind DF-GSC are considered for the simulations. The length of the training period is kept at 200. The step sizes used are $\mu_a = 5 \times 10^{-7}$ and $\mu_b = 0.01$. Computations are carried out for different values of SNR with a number of snapshots being fixed at only 1. The corresponding plot is shown in Figure 24.26. It can be observed that higher output SINR is achieved in case of blind

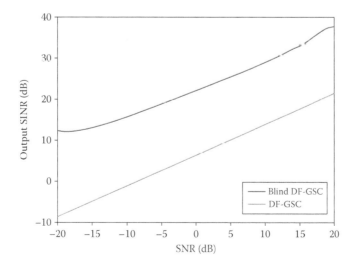

FIGURE 24.26 Steady-state analysis for DF-GSC and blind DF-GSC using the standard LMS algorithm.

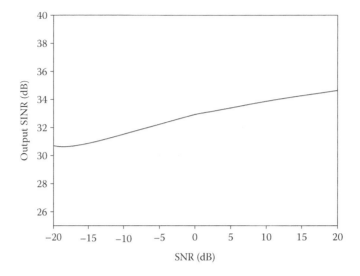

FIGURE 24.27   Steady-state analysis for blind DF-GSC using the structured LMS algorithm.

DF-GSC as compared to normal DF-GSC. Similarly, the steady-state analysis for blind DF-GSC is shown in Figure 24.27 using the structured LMS algorithm. The step sizes used are $\mu_a = 2.5 \times 10^{-7}$ and $\mu_b = 0.01$. For the sake of consistency, the length of the training period is kept the same as above. It is observed that a much higher output SINR is achieved in this case along with better convergence rate. Moreover, an output SINR as high as 34 dB is achieved in case of the structured LMS algorithm. The convergence rate is found to be faster as compared to the standard LMS algorithm.

24.6   WIDEBAND PROBE SUPPRESSION

The interference suppression capability of phased array is known to be frequency sensitive. Thus, the bandwidth of the probing source affects the array performance significantly. Defining f_o as the centre frequency, and an offset Δf, the phase factor u_i can be written as

$$u_i = \left(1 + \frac{\Delta f}{f_o}\right)\frac{\pi}{2}\sin\theta_i \qquad (24.66)$$

where θ_i is the azimuth direction of the incoming interfering signal and i is the number of hostile sources. Using this frequency-dependent expression for u_i, the interference source with bandwidth can be handled by dividing the source power according to a number of spectral lines. For computation of eigenvalues, the covariance matrix is used which is given by

$$M = M_q + \sum_{r=1}^{I} P_r M_r \qquad (24.67)$$

where P_r is the power ratio and M_r is the covariance matrix of the rth interfering source. For the receiver noise power equal to unity, M_q, the quiescent covariance matrix will be an identity matrix. Thus

$$M = I_d + \sum_{r=1}^{I} P_r M_r \qquad (24.68)$$

For the case of wideband sources, M may be re-written as [21]

$$M = I_d + \sum_{r=1}^{K} \sum_{l=1}^{L} P_{rl} M_{rl} \tag{24.69}$$

where M_{rl} is the covariance matrix of lth spectral line of rth interfering source.

The mnth component of M_{rl} is given as

$$(M_{rl})_{mn} = e^{j2u_{rl}(n-m)} \tag{24.70}$$

where $u_{rl} = (1 + (\Delta f_l / f_0)) \dfrac{\pi}{2} \sin \theta_r$.

The extent of the frequency spread, that is, $\Delta f / f_0$ for a particular hostile source of lth spectral line is computed using the expression

$$\frac{\Delta f}{f_0} = \left(\frac{B_r}{100} \right) \left[\frac{-1}{2} + \left(\frac{l-1}{L_r - 1} \right) \right] \tag{24.71}$$

where B_r is the bandwidth and L_r is the number of spectral lines.

24.6.1 CONVERGENCE BEHAVIOUR OF SIDELOBE CANCELLERS WITH WIDEBAND SOURCES

In this section, the role of various parameters, viz., the number of spectral lines, the power level, the number of sources, the arrival angles and the bandwidth on the suppression capabilities of antenna arrays are analysed. Output SINR is estimated and the rates of convergence for different algorithms are analysed in the presence of wideband probing sources of different bandwidths.

Single wideband probing source: A wideband probing source at $42°$ is assumed to probe the antenna array of 16 antenna elements, spaced half a wavelength apart. Figure 24.28 presents the learning curves of GSC scheme using the improved LMS algorithm. The results are shown for different bandwidths, viz., 2%, 4% and 5%. The power ratio is taken to be 100. The step size, $\mu_a = 1.0 \times 10^{-6}$ is kept constant for all the cases. It can be observed that output SINR increases with

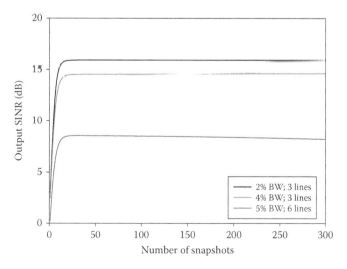

FIGURE 24.28 Learning curves of GSC for suppression of single wideband probing source at $42°$ with different bandwidths. The improved LMS algorithm is employed for weight adaptation.

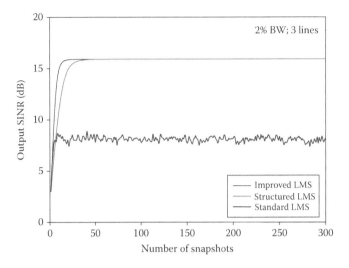

FIGURE 24.29 Comparative analysis of convergence rate of standard, structured and improved LMS algorithms in conventional GSC scheme.

the decrease in the bandwidth of the source. This is because a single wideband hostile source is equivalent to a set of narrowband sources depending on the spectral lines, spread around the angle of arrival [22]. Hence, with the increase in bandwidth, the number of narrowband sources probing the antenna array increases, resulting in decreased value of output SINR.

In Figure 24.29, the rate of convergence of the standard, structured and improved LMS algorithm is compared when a single probing source (42°) with 2% BW and power ratio of 100 is considered to probe the antenna array with the GSC scheme. It can be seen that when the structured and improved LMS algorithms are used, the output SINR of conventional GSC converges at a higher value of 16 dB as compared to the case of the standard LMS algorithm, which converges at 8 dB only. The performance of the improved LMS algorithm is again found to be the best owing to its faster rate of convergence. Table 24.1 shows that with the increase in bandwidth of the probing source, the performance of GSC using the standard LMS algorithm deteriorates. However, this is not the case with the structured LMS algorithm. Nulls up to −60 dB were obtained in all the three cases (Figure 24.30). The performance of the improved LMS algorithm for a single probing source (42°) with 2% BW and a power ratio of 100 is compared with the standard and structured LMS algorithms in Figure 24.31. The null placed towards the probing source is up to −80 dB in case of

TABLE 24.1

Comparative Analysis of Performance of a Conventional GSC Scheme with a Standard and Structured LMS Algorithm for a Single Probing Source with Different Bandwidth

Probing Direction (°)	BW (%)	Number of Spectral Lines	Algorithm	Extent of Suppression (dB)
42	2	3	Standard LMS	−38
			Structured gradient LMS	−60
42	5	6	Standard LMS	−34
			Structured gradient LMS	−60
42	15	11	Standard LMS	−25
			Structured gradient LMS	−60

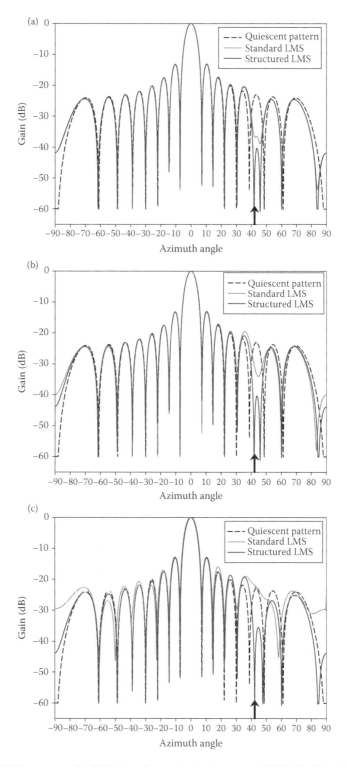

FIGURE 24.30 (a) Beam pattern for GSC using the standard and structured LMS algorithm. Source DOA = 42°, 2% BW, 3 spectral lines; power ratio = 100. (b) Beam pattern for GSC using the standard and structured LMS algorithm. Source DOA = 42°, 5% BW, 6 spectral lines; power ratio = 100. (c) Beam pattern for GSC using the standard and structured LMS algorithm. Source DOA = 42°, 15% BW, 16 spectral lines; power ratio = 100.

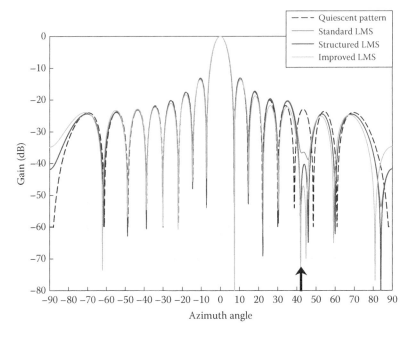

FIGURE 24.31 Beam pattern of GSC using the standard, structured and improved LMS algorithm. Single probing source at 42°, BW = 2%, 3 lines; power ratio = 100.

the improved LMS algorithm. This demonstrates the efficiency of the improved LMS algorithm in accurate and efficient suppression of probing. Table 24.2 compares the performance of three algorithms for a single wideband probing source. It is apparent that among the three forms of LMS algorithms used in GSC scheme, the improved LMS algorithm gives the best results.

Two wideband sources: It is assumed that two probing sources arrive at −35° and 25°, having 2% and 5% bandwidth, respectively. Unequal power ratios of 10 and 100 have been considered. In Figure 24.32, the corresponding adapted beam pattern of GSC scheme employing the standard and structured LMS algorithm is shown. It can be observed that the structured LMS algorithm performed better than the standard LMS algorithm in active cancellation of both the probing sources. Although the sources had different bandwidths and different power ratios, they were equally suppressed by the structured gradient algorithm unlike the standard LMS algorithm in GSC scheme. Table 24.3 demonstrates that GSC scheme with the structured LMS algorithm gives good nulls even with the less-powered probing source whereas the performance of the standard LMS algorithm degrades with decrease in power of the probing source.

Three sources with equal and unequal powers: Next, it is assumed that three wideband sources at −25°, 42° and 70° are trying to probe the antenna array. Two cases with equal and unequal power

TABLE 24.2

Comparative Analysis of Performance of Standard, Structured and Improved LMS Algorithms in GSC Scheme

Probing Direction (°)	BW (%), Lines	Power Ratio	Algorithm	Extent of Suppression (dB)
42	2, 3	100	Standard LMS	−38
			Structured LMS	−61
			Improved LMS	−79

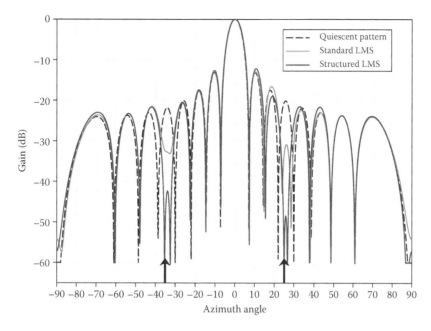

FIGURE 24.32 Beam pattern of GSC using the standard and structured LMS algorithm. Two probing sources with DOA = −35°, 25°; BW = 2%, 5%; number of lines = 3, 6; power ratio = 10, 100, respectively.

ratios have been considered to compare the efficiency of the standard and structured LMS algorithm in the GSC scheme. The adapted beam patterns for both the cases are shown in Figures 24.33 and 24.34, respectively. It can be observed that the performance of the standard LMS algorithm deteriorates when all the sources are considered to have equal power ratios whereas the structured LMS algorithm is seen to perform consistently better even for active cancellation of equal-powered sources.

Continuously distributed sources: In Figure 24.35, the adapted beam pattern for the GSC scheme using the structured and improved LMS algorithm is shown when three continuously spaced wideband probing sources with equal bandwidth of 2% were considered to probe the antenna array. The sources were assumed to have different power ratios of 1, 10 and 100. It can be observed that not only are nulls placed towards the direction of the sources but also the nearest lobes are suppressed to a great extent. The depth of the nulls is greater for the improved LMS algorithm.

So far, the suppression capabilities of the GSC scheme for wideband probing sources are discussed. Let us analyse the DF-GSC scheme for wideband signals. Figure 24.36 presents the

TABLE 24.3

Comparison of the Suppression Capability of GSC Scheme with Standard and Structured LMS Algorithms for Two Wideband Probing Sources with Different Bandwidths and Powers

Number of Sources	Probing Direction (°)	BW (%), Lines	Power Ratio	Algorithm	Extent of Suppression (dB)
2	−35	2, 3	10	Standard LMS	−35
				Structured gradient LMS	−60
	25	5, 6	100	Standard LMS	−43
				Structured gradient LMS	−60

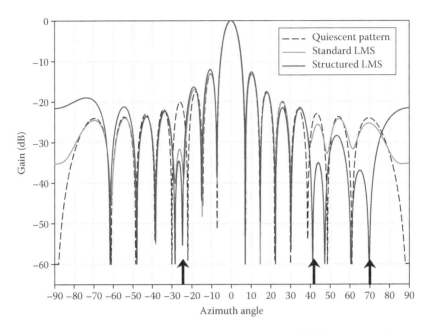

FIGURE 24.33 Beam pattern of GSC using the standard and structured LMS algorithm. Three sources with DOA = −25°, 42°, 70°; BW = 2%, 2%, 5%; number of lines = 3, 3, 6; power ratio = 10, 5, 100, respectively.

adapted pattern of the DF-GSC scheme for two equal-powered (100) probing sources at −35° and 25° with 2% and 5% bandwidth, respectively. It is clearly visible in the graph that broader nulls are obtained at the hostile source locations with no distortion in the rest of the pattern. Next, the role of spectral lines in wideband probe suppression is studied. The performance of the GSC scheme has been analysed (Table 24.4) for active cancellation of a single wideband probing

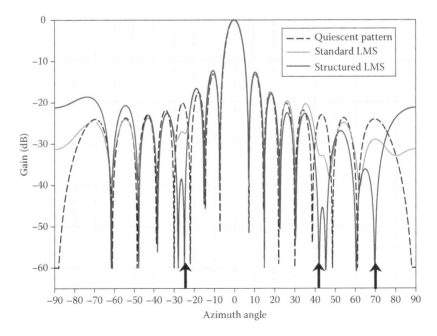

FIGURE 24.34 Beam pattern of GSC using the standard and structured LMS algorithm. Three probing sources with DOA = −25°, 42°, 70°; BW = 2%, 2%, 5%; number of lines = 3, 3, 6; power ratio = 100 each.

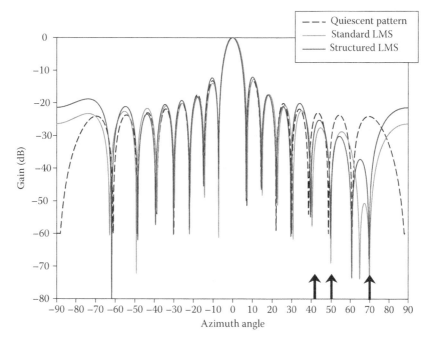

FIGURE 24.35 Beam pattern of GSC using the structured and improved LMS algorithms. Multiple continuously distributed sources with DOA = 42°, 50°, 70°; BW = 2%, 3 lines each; power ratio = 1, 10, 100.

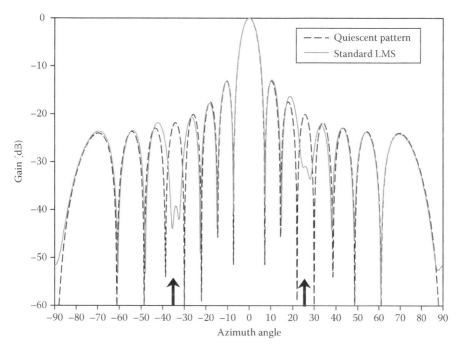

FIGURE 24.36 Adapted pattern of DF-GSC for two probing sources of equal power (100 each) at −35° and 25° with 2% and 5% bandwidth, respectively. The standard LMS algorithm is employed for weight adaptation.

TABLE 24.4

Performance of GSC Scheme with the Change in Number of Spectral Lines of Bandwidth of Probing Source

Probing Direction (°)	BW (%)	Algorithm	Number of Spectral Lines	Extent of Suppression (dB)
42	15	Standard LMS	3	−32
			9	−23
		Structured LMS	3	−60
			9	−42
		Improved LMS	3	−78
			9	−63

source having the same bandwidth but different number of spectral lines. It can be seen that when the GSC scheme uses the standard, structured or improved LMS algorithm for weight adaptation, wideband probing source having three spectral lines is suppressed to a greater extent, that is, deeper nulls are placed compared to the nulls obtained for the source with 9 spectral lines.

In Figure 24.37, learning curves of the DF-GSC scheme using the standard LMS algorithm are shown for multiple wideband probing sources. All the sources are assumed to have 2% bandwidth but unequal power level. The desired signal level is taken to be 0 dB for each case. It can be seen that both the output SINR and the rate of convergence decreases with the increase in number of sources.

Next, a single wideband probing source is considered for analysing the performance of the blind DF-GSC. The DOA of the hostile source is kept at 25° and the power at 100. The source is assumed to have a bandwidth of 2% with 3 spectral lines. The adapted pattern obtained is presented in Figure 24.38. The quiescent pattern is shown as the dotted curve. It can be seen that there is no distortion in the main beam and an accurate null is placed.

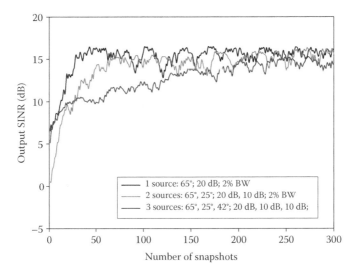

FIGURE 24.37 Learning curves of DF-GSC scheme for wideband sources using the standard LMS algorithm.

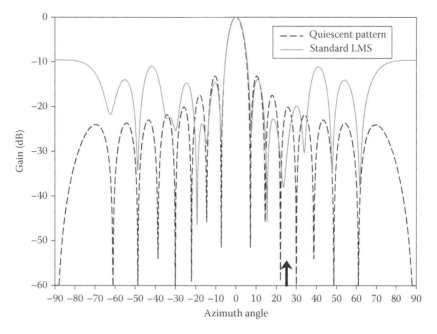

FIGURE 24.38 Adapted pattern of blind DF-GSC scheme for a single wideband source (25°; 2% BW, 3 lines; 100) using the standard LMS algorithm.

24.7 INTERFERENCE SUPPRESSION IN PLANAR PHASED ARRAYS

From the geometrical prospective, a planar antenna array may be viewed as a array of linear arrays. Thus, a planar radiating surface is obtained when a number of linear arrays are placed next to each other in the orthogonal directions themselves [23]. For example, a planar, say rectangular, array may be obtained by placing antenna elements along a rectangular grid, that is, elements are present along two orthogonal directions, for example, x and y. This arrangement provides additional variables for the antenna design and thus facilitates controlling and shaping of the array pattern. It is well known that the planar arrays are more versatile than linear arrays since they provide symmetrical patterns with lower sidelobes [24]. A planar antenna array is capable of scanning the main beam of the antenna towards any arbitrary point in the space. In phased antenna array, linear and circular arrangement of antenna elements provide beam steering in any arbitrary direction of the azimuth plane; however, the elevation radiation pattern of such an antenna array depends on the radiation pattern of the individual radiators. In contrast to this, a planar antenna array can also steer the beam in the elevation plane and generates a narrow beam [25]. Since planar arrays can scan the beam in 3-D space, these arrays are widely used in tracking radar, search radar, remote sensing and wireless communication systems. These arrays are very useful for portable devices due to the ability to scan the beam in any direction (θ, φ). Printed planar antenna arrays have added advantage of simple structure, low cost, low profile, small size, high polarization purity and broad bandwidth [26].

24.7.1 GENERATION OF RADIATION PATTERN

The geometry of planar array having rectangular cross section is shown in Figure 24.39. The rectangular array ($d_x \times d_y$) consist of discrete elements which are progressively phase shifted in $x-y$ plane with the centre chosen as the origin. The elements are arranged in two directions (say x and y), with different number of elements in either direction, forming the rectangular grid. There are additional variables available for the analysis due to the two-dimensional array geometry. This also

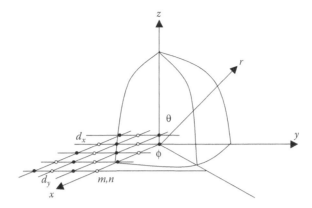

FIGURE 24.39   Geometry of a rectangular array of antenna elements.

provides a better control for the design of antenna array for specific requirements. In planar array, the directivity does not depend on the choice of the coordinate system [5]. In this special case of rectangular boundary, there are separable weightings (w_{nm}) corresponding to the two axes. The beam pattern is the product of two individual array factors [27], that is,

$$B(\psi_x, \psi_y) = B(\psi_x) \times B(\psi_y) \tag{24.72}$$

The beam pattern [5] of a rectangular array of uniform cross section is given by

$$B(\psi_x, \psi_y) = e^{-j([(N-1)/2]\psi_x + [(M-1)/2]\psi_y)} \sum_{n=0}^{N-1} \sum_{m=0}^{M-1} w_{nm}^* e^{j(n\psi_x + m\psi_y)} \tag{24.73}$$

where

$$\psi_x = \frac{2\pi}{\lambda} d_x \sin\theta \cos\phi$$
$$\psi_y = \frac{2\pi}{\lambda} d_y \sin\theta \sin\phi \tag{24.74}$$

Here, $B(\psi_x, \psi_y)$ is the beam pattern, d_x and d_y are the spacing between the elements in x and y directions, respectively, (θ, ϕ) is the angular distribution in azimuth and elevation planes, N and M are the number of elements along x and y directions, w_{nm}^* is the amplitude excitation, λ is the wavelength and $j = \sqrt{-1}$. In terms of (ψ_x, ψ_y), the visible region is

$$\sqrt{\left(\frac{\psi_x}{d_x}\right)^2 + \left(\frac{\psi_y}{d_y}\right)^2} \leq \frac{2\pi}{\lambda} \tag{24.75}$$

In u-space, Equation 24.74 can be expressed as

$$u_x = \sin\theta \cos\phi$$
$$u_y = \sin\theta \sin\phi \tag{24.76}$$

The visible region is

$$u_r \cong \sqrt{u_x^2 + u_y^2} \tag{24.77}$$

24.7.2 Performance of Planar Array in Hostile Environment

In this section, simulation results for generating the adapted pattern of planar arrays in different signal environments are discussed. Various distributions of the interfering signals such as those from (i) separated narrowband probing sources, (ii) sources with bandwidth, (iii) low-power probing sources and (iv) continuously distributed sources are considered for the computations. The results are presented for an adaptive array excited by different aperture distributions like the uniform and Dolph–Chebyshev distribution. Here, the improved LMS algorithm is employed for the generation of adapted pattern. The array geometry is taken to be a 16×10 rectangular array. The spacing between the elements are taken as be $d_x = 0.484\lambda$ and $d_y = 0.770\lambda$.

There are $N - 1$ DOFs available for N element antenna arrays. Thus, the antenna pattern will have $N - 1$ nulls or beam maxima signifying the cancellation of $N - 1$ probing sources. The available DOF are dependent on the number of probing sources. This is apparent from Figure 24.40, demonstrating the suppression for three probes ($\theta_1 = 20°$, $\theta_2 = 30°$, $\theta_3 = 40°$) with powers ($P_1 = 10$, $P_2 = 10$, $P_3 = 100$). The nulls are placed accurately towards each probing direction. The amplitude excitations for the antenna elements are done using uniform distribution. Further simulations are carried out for the suppression of five probing sources ($\theta_1 = -60°$, $\theta_2 = -45°$, $\theta_3 = -25°$, $\theta_4 = 25°$, $\theta_5 = 45°$; $P_1 = 10$, $P_2 = 10$, $P_3 = 100$, $P_4 = 10$, $P_5 = 100$) in azimuth plane for a 16×10 planar array. Figure 24.41 represents the corresponding adapted pattern. The suppression of pattern up to -35 dB is observed at each source location. Keeping the geometry of the array unchanged, computations are performed for the Dolph–Chebyshev distribution. Figure 24.42 represents the probe suppression of two hostile sources ($\theta_1 = 20°$, $\theta_2 = 60°$) in azimuth plane for a 16×10 element planar array with

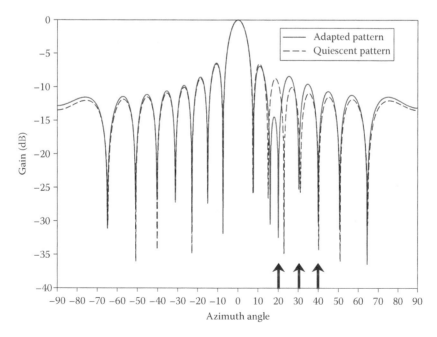

FIGURE 24.40 Probe suppression in azimuth plane of 16–10 planar array for three hostile sources ($\theta_1 = 20°$, $\theta_2 = 30°$, $\theta_3 = 40°$; $P_1 = 10$, $P_2 = 10$, $P_3 = 100$). Uniform excitation is employed here.

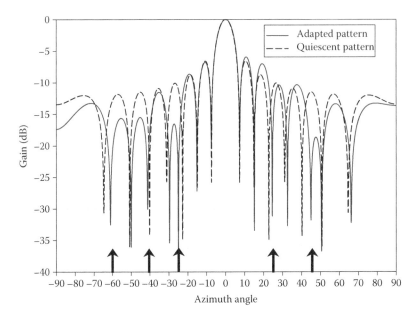

FIGURE 24.41 Suppression of five hostile sources in azimuth plane of 16×10 planar array ($\theta_1 = -60°$, $\theta_2 = -45°$, $\theta_3 = -25°$, $\theta_4 = 25°$, $\theta_5 = 45°$; $P_1 = 10$, $P_2 = 10$, $P_3 = 100$, $P_4 = 10$, $P_5 = 100$). Uniform excitation is employed here.

unequal powers ($P_1 = 10$, $P_2 = 100$) using the improved LMS algorithm. The sidelobe level is kept at -20 dB. Deep nulls are placed towards the probing directions. Further, Figure 24.43 represents the adapted patterns for two hostile sources probing at ($\theta_1 = -45°$, $\theta_2 = 55°$; $P_1 = 100$, $P_2 = 10$) for a 16×10 element planar array in the azimuth plane. In this case, Dolph–Chebyshev distribution is used for the pattern synthesis of 16×10 planar array for -20 dB sidelobe level. It is apparent that

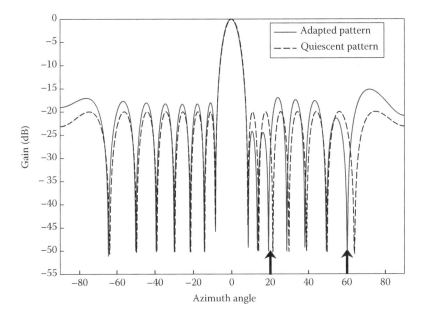

FIGURE 24.42 Probe suppression for two hostile sources ($\theta_1 = 20°$, $\theta_2 = 60°$; $P_1 = 10$, $P_2 = 100$) in azimuth plane of 16×10 planar array. Dolph–Chebyshev distribution with -20 dB sidelobe level is considered.

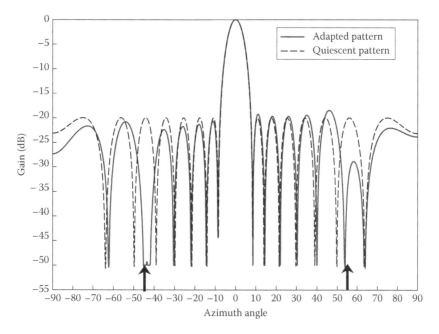

FIGURE 24.43 Probe suppression in azimuth plane of 16×10 planar array for two hostile sources ($\theta_1 = -45°$, $\theta_2 = 55°$; $P_1 = 100$, $P_2 = 10$) using Dolph–Chebyshev distribution with -20 dB sidelobe level.

the probe at $\theta_1 = -45°$ ($P_1 = 100$) is suppressed more than the probe at $\theta_2 = 55°$ ($P_2 = 10$). This observation emphasizes the fact that sources with high power level as compared to the desired signal can be more effectively suppressed by the antenna array. However, it also depends on the efficacy of the adaptive algorithm used for weight updation.

24.8 ACTIVE CANCELLATION OF PROBING EFFECT WITH SIMULTANEOUS MULTIPLE DESIRED SIGNALS

The overall perspective is provided by the fact that in robotics, mobile and satellite communications, many targets have to be monitored simultaneously. Hence, the antenna array must necessarily possess beamforming capability to receive simultaneous multiple signals with selective gains in desired directions, while at the same time minimizing output noise power. To summarize, a technique for adaptive beamforming with the capability of providing multiple beam constraints along with coherent probing suppression is presented. In this approach, an adaptive array beamformer is synthesized with multiple beam constraints. It essentially involves the construction of a steering matrix and a constraint vector according to the signal environment. The entries of the constraint vector denote the gain of the antenna array towards the signals. The signal model and the adaptive algorithm employed are capable of taking care of (partial or full) the correlation between the signals for accelerated convergence.

24.8.1 Weight Adaptation for Simultaneous Multiple Desired Signals

In the presence of multiple simultaneous desired signals, the weight updating equation as in Ref. [28] given by Equation 24.50 changes to

$$W(n+1) = P[W(n) - \mu\, g(W(n))] + \frac{S_1}{S_1^H S_1} + \frac{S_2}{S_2^H S_2} + \cdots + \frac{S_m}{S_m^H S_m} \qquad (24.78)$$

Similarly, P, the projection operator, takes the following form:

$$P = I - \frac{S_1 S_1^H}{S_1^H S_1} - \frac{S_2 S_2^H}{S_2^H S_2} - \cdots - \frac{S_m S_m^H}{S_m^H S_m} \tag{24.79}$$

where $g(W(n))$ is an unbiased estimate of the gradient of $W^H(n)RW(n)$ w.r.t. $W(n)$, R being the auto-correlation matrix. I is the identity matrix and $S_1, S_2, \ldots, S_m$ represent the steering vectors corresponding to m desired signals impinging the array at different angles. This is the modified version of the improved LMS algorithm.

The ULA employs these weights for processing the received data $x(t)$ to obtain the array output

$$y(t) = w^H x(t) \tag{24.80}$$

The received signal is given by

$$x(t) = As(t) + n(t) \tag{24.81}$$

where $A = \left[a(\theta_1) a(\theta_2) \ldots a(\theta_k) \right]$ is the response matrix, $a(\theta_k) = e^{-jkd(i-1)\sin\theta_k}$ is the complex response of the sensor to the kth signal, d is the spacing between the sensors, c is the propagation speed of the signal, $s(t) = \left[s_1(t) s_2(t) \ldots s_k(t) \right]^T$ is the signal source vector and $n(t) = \left[n_1(t) n_2(t) \ldots n_k(t) \right]^T$ is the noise vector. The superscript T denotes the transpose operation.

24.8.2 Simulation Results

Simulations have been carried out to analyse the response of a *uniform linear array* (ULA) towards different signal environments. It is assumed that ULA has 10 isotropic elements with a uniform spacing of half wavelength between the array elements. The array employs the modified improved LMS algorithm for weight optimization, discussed in Section 24.8.1. For simulations, the power of the desired signal is assumed to be lower than the power of interfering sources. Results for both linear and planar antenna array demonstrate the efficacy of the modified improved LMS algorithm in active cancellation of the probing effect even when simultaneous multiple desired signals are required. This can have application towards *active RCS reduction*.

24.8.2.1 Narrowband Probing Sources

A simple scenario with two desired signals (60°, 120°; 0 dB each) and two interfering signals (80°, 150°; 30 dB each) has been considered. The resultant adapted beam pattern obtained is shown in Figure 24.44. The quiescent pattern, that is, pattern without probing sources, is also included in the plot. The adapted pattern using the standard LMS and recursive LMS algorithm is also shown. It can be observed that the adapted pattern of the standard LMS and recursive LMS algorithm do not show suppression of the probing sources, although the main lobes correspond to each of the desired directions accurately. However, this is not the case of the proposed *modified improved LMS* algorithm. The antenna array along with the modified improved LMS algorithm is found to be efficient enough to maintain sufficient gain towards the desired directions and simultaneously placing deep nulls in the directions of probing sources.

The variation of noise power level with a number of snapshots is shown in Figure 24.45. The results are compared for the algorithms, viz., standard LMS, recursive LMS and the proposed modified improved LMS algorithm. It can be seen that with time the output noise power of the array decreases drastically with time. This signifies that the complex weights adapt to their optimum values within a few iterations. The lowest output noise power is achieved by the modified improved

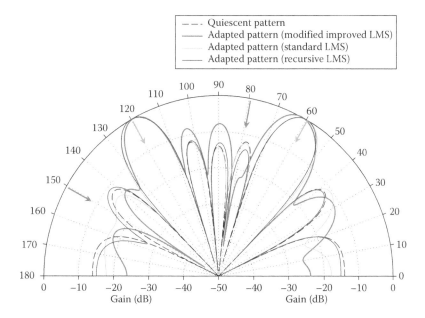

FIGURE 24.44 Adapted beam pattern using *modified improved LMS*, *standard LMS* and *recursive LMS* *algorithms*. Two desired signals (60°, 120°, 0 dB each) with two narrowband probing sources (80°, 150°, 30 dB each) are considered.

LMS algorithm for a given signal scenario. Moreover, the convergence rate of the proposed algorithm is also excellent.

Next, the array radiation pattern in the presence of two desired signals and three interfering sources is presented in Figure 24.46. The desired signals are assumed to arrive at the same directions with the same input powers as in the previous case. Three sources are considered to be probing from the directions of 90°, 140° and 145°, each with 30 dB power levels. Good suppression capability of the array employing the modified version of the improved LMS algorithm is demonstrated by the nulls obtained in the directions of the probing sources. Figure 24.47 illustrates the variation of output noise power with probing direction. Keeping the angle of arrival, power of the desired signals and step size the same as the previous case, the number of interfering sources is reduced to 1. The probing source having a power ratio of 1000 is taken with variable angle of arrival. It can be seen

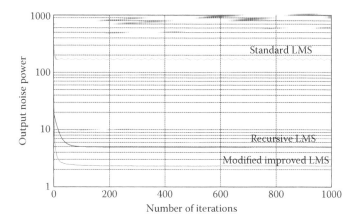

FIGURE 24.45 Output noise power of 10-element uniform linear array in the presence of two desired signals (60°, 120°, 0 dB each) and two probing sources (80°, 150°, 30 dB each).

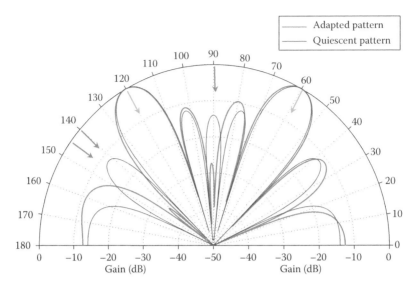

FIGURE 24.46 Adapted beam pattern for signal environment consisting of two desired signals (60°, 120°, power ratio of 1 each) and three probing sources (90°, 140°, 145°, power ratio of 1000 each).

from the graph that the output noise power level increases suddenly when the probing source falls in the main lobes of the array. This signifies that the array performance with the use of the improved LMS algorithm degrades drastically if the probing angle coincides with the signal direction. Figure 24.48 presents a 3-D plot where probing direction and input power of the probing source are considered as two variable parameters. It can be observed that as the power of the interfering source increases the noise power level also increases and a peak is observed at 60° and 120° for maximum probing power.

For further analysis of the performance of adaptive planar antenna array, the number of probing sources is increased to eight. The DOA of probing source and its power level are given in Table 24.5. It is apparent from the adapted pattern (Figure 24.49) that even for eight sources, 16 × 10 antenna array is capable of maintaining the main lobes towards each of the desired directions (−30° and 50°) and placing sufficient deep nulls towards the probing sources. Figure 24.50 shows the adapted pattern of 16 × 10 antenna array for two widely spaced desired signals (−60° and 60°) and 10 closely spaced unequal-powered probing sources. The detailed description of the entire signal scenario is

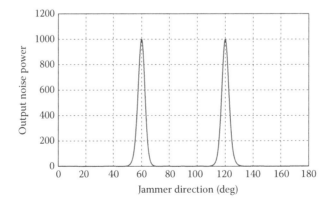

FIGURE 24.47 Output noise power of the antenna array versus the jammer direction. Two desired signals (60°, 120°, 0 dB each) and one probing source (30 dB) with variable angle of arrival are considered.

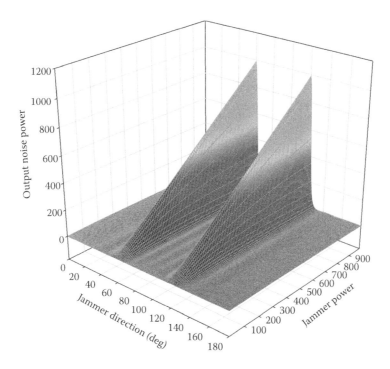

FIGURE 24.48   Output noise power of the antenna array versus jammer direction and jammer power. Two desired signals (60°, 120°, 0 dB each) and one probing source with variable angle of arrival and variable power are considered.

tabulated in Table 24.6. Once again the adapted pattern is along expected lines. As the next case, the number of desired signals was increased to four. The desired directions are assumed to be at −10°, −20°, 10°, 20°, whereas the three hostile (probing) sources are arbitrarily kept at −40°, 42°, 55° with power ratio of 1000, 100, 100, respectively. The power level of each desired signal is kept the same at 0 dB. The resultant adapted pattern (Figure 24.51) shows that the array is capable of maintaining sufficient gain in the desired signal directions and placing accurate deeper nulls towards each of probing direction.

Figure 24.52 presents the adapted pattern for closely spaced three probing sources (−20°, −5°, 3°; 100 each) in the presence of four widely spaced desired signals (20°, 60°, −40°, −60°). It is apparent that even in such a complex signal environment, the antenna array is capable of generating the desired adapted pattern. Keeping the number of the desired signals four (−10°, −20°, 10°, 20°), the number of probing sources is increased to four, that is, (−40°, 42°, 55°, 75°; 1000, 100, 100, 100).

TABLE 24.5

Signal Scenario of Two Desired Signals (−30°, 50°) and Eight Probing Sources Impinging 16×10 Planar Array

DOA of Probing Source	Power Ratio of Probing Source	DOA of Probing Source	Power Ratio of Probing Source
−5°	10	20°	50
−45°	180	27°	10
−12°	50	35°	180
3°	50	12°	50

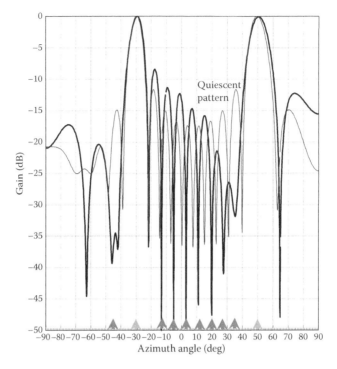

FIGURE 24.49 Adapted beam pattern of 16×10 antenna array. Two desired signals at $(50°, -30°$; power ratio of 1 each) and 8 probing sources at $(-5°, -45°, -12°, 3°, 20°, 27°, 35°, 12°; 10, 180, 50, 50, 50, 10, 180, 50)$.

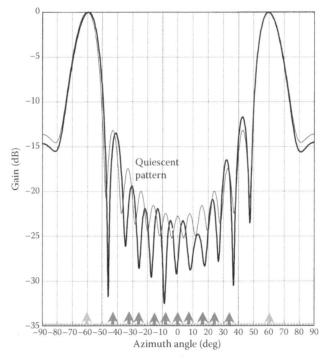

FIGURE 24.50 Adapted beam pattern of 16×10 antenna array. Two desired signals at $(60°, -60°$; power ratio of 1 each) and 10 arbitrary located probing sources at $(-43°, -33°, -26°, -15°, -8°, 0°, 7°, 16°, 24°, 34°; 100, 50, 40, 30, 60, 40, 50, 50, 40, 60)$.

TABLE 24.6

Signal Scenario of Two Widely Spaced Desired Signals (−60°, 60°) and 10 Probing Sources Impinging 16×10 Planar Antenna Array

DOA of Probing Source	Power Ratio of Probing Source	DOA of Probing Source	Power Ratio of Probing Source
−43°	100	0°	40
−33°	50	7°	50
−26°	40	16°	50
−15°	30	24°	40
−8°	60	34°	60

The resultant adapted pattern (Figure 24.53) demonstrates the accurate nullification of the probing sources without any effect on the main lobes. This demonstrates the capability of a narrowband antenna array (linear/planar) along with an efficient modified improved LMS algorithm to cater to the narrowband/wideband signal environment, having either single or multiple probing sources.

24.8.2.2 Overlap of Desired and Probing Directions

Let us consider a signal scenario where the DOA of the desired signal matches with the probing direction. It is assumed that the direction of one probing source coincides with that of the desired signal direction (−20°). It can be observed from the resultant adapted pattern (Figure 24.54) that there is a slight shift of about 0.5° to 1° in the main lobe towards the desired signal direction. However, the deeper nulls (−50 dB approximately) are placed accurately towards the probing directions. Yet, another

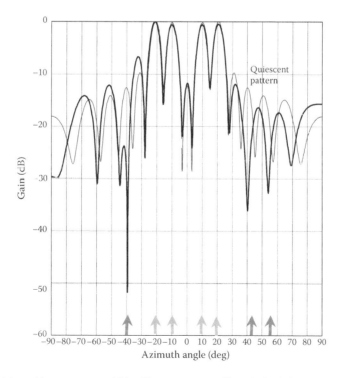

FIGURE 24.51   Adapted beam pattern of 16 × 10 antenna array. Four desired signals at (−10°, −20°, 10°, 20°; power ratio of 1 each) and 3 probing sources at (−40°, 42°, 55°; 1000, 100, 100).

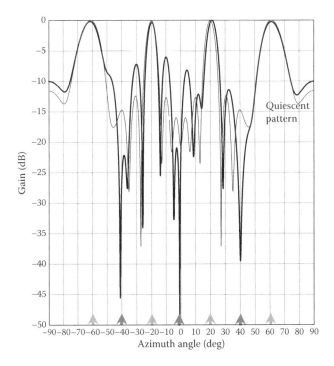

FIGURE 24.52 Adapted beam pattern of 16 × 10 antenna array. Four desired signals at (20°, −20°, 60°, −60°; power ratio of 1 each) and 3 probing sources at (40°, −40°, 2°; 1000, 1000, 1000).

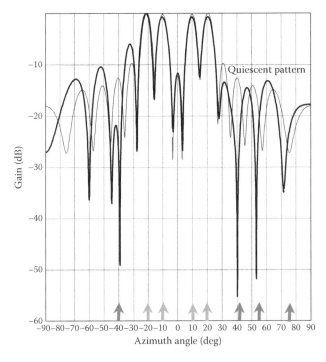

FIGURE 24.53 Adapted beam pattern of 16 × 10 antenna array. Four desired signals at (−10°, −20°, 10°, 20°, power ratio of 1 each) and 4 probing sources at (−40°, 42°, 55°, 75°; 1000, 100, 100, 100).

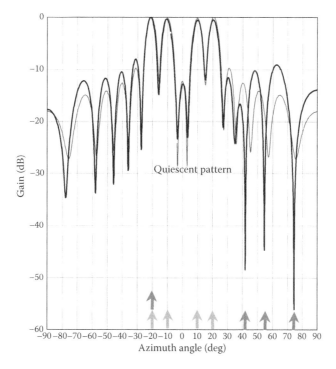

FIGURE 24.54 Adapted beam pattern of 16 × 10 antenna array. Four desired signals at (–10°, –20, 10, 20°; power ratio of 1 each) and 4 jammers at (–20°, 42°, 55°, 75°; 1000, 100, 100, 100).

case (Figure 24.55) is considered for confirmation where the probing source is assumed to coincide with the desired signal at 20°. A similar effect is observed in the resultant adapted pattern. Figures 24.56 and 24.57 demonstrate relatively complex cases where two and three probing sources coincide with the desired signal directions. It is apparent from the patterns that there is a negligible angular shift in the main lobes of the adapted pattern along with the accurate nulls towards the probing directions. This shows the additional capability of the modified improved LMS algorithm in maintaining the pattern even when the probing source coincides with the desired direction. This demonstrates the efficiency of the antenna array along with the adaptive algorithm in nullifying the jamming effect.

24.8.2.3 Wideband Probing Sources

In this section, the performance of the narrowband antenna array employing the modified improved LMS algorithm is analysed in the presence of wideband probing sources. A wideband source is assumed to be equivalent to many narrowband sources depending upon the number of spectral lines [22]. The phase factor and the covariance matrix corresponding to the wideband probing source are calculated considering equal spread in frequency around a centre frequency. Simulations have been carried out considering all the desired signals to be narrowband.

In Figure 24.58a, the adapted pattern is shown when two desired signals (60°, 1; 120°, 1) and single probing source (80°, 1000, 2% BW, 3 lines) are assumed to be incident on the antenna array. The step size is taken to be 1.0×10^{-6}. It can be seen that null up to –48 dB can be obtained. In Figure 24.58b, the bandwidth of the probing source is increased to 15% and the rest of all other parameters are kept the same as in Figure 24.58a. It can be seen that although the placement of nulls and gains is appropriate in both the cases, a broader null is obtained for the larger bandwidth case. Further in Figure 24.59, the number of probing sources is increased to two. Two desired signals (60°, 1; 120°, 1) and two wideband probing sources (80°, 1000; 150°, 100; 2% BW, 3 lines each) are considered to impinge on the array with unequal power ratios. It had been studied earlier in the case

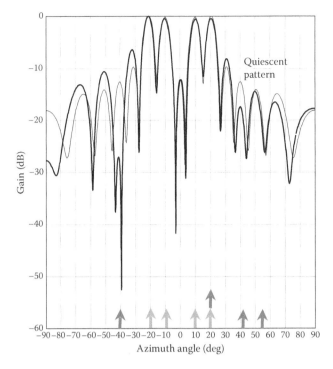

FIGURE 24.55 Adapted beam pattern of 16×10 antenna array. Four desired signals at $(-10°, -20°, 10°, 20°$; power ratio of 1 each) and 4 probing sources at $(-40°, 42°, 20°, 75°$; 1000, 100, 100, 100).

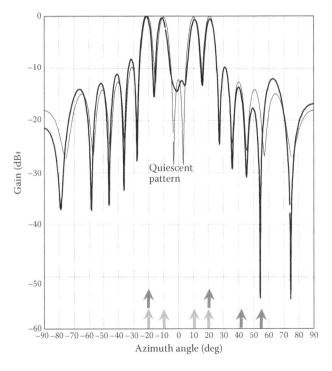

FIGURE 24.56 Adapted beam pattern of 16×10 antenna array. Four desired signals at $(-10°, -20° \ 10°, 20°$; power ratio of 1 each) and 4 probing sources at $(-20°, 20°, 55°, 75°$; 1000, 100, 100, 100).

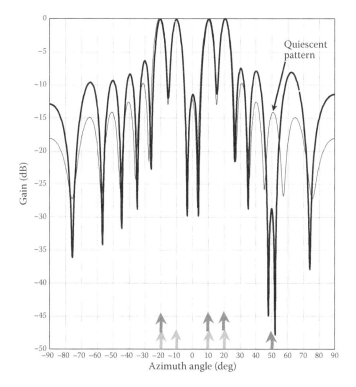

FIGURE 24.57 Adapted beam pattern of 16×10 antenna array. Four desired signals at $(-10°, -20°, 10°, 20°;$ power ratio of 1 each) and 4 probing sources at $(-20°, 10°, 55°, 20°; 1000, 100, 100, 100)$.

of one desired signal that when the array nulls an interference signal, the null depth is not infinitely deep but in fact depends on interference power. The result obtained for the present scenario with multiple simultaneous desired signals is also according to expectation lines. It can be well observed from the pattern that the source with higher power is suppressed more. Stronger source gets suppressed to -62 dB whereas the weaker one gets suppressed only to -44 dB.

A similar case but with equal power ratio (1000 each) of the two wideband probing sources is considered. All the parameters except the power of the sources are kept the same as in Figure 24.59. In the earlier studies with the single desired signal, the suppression of the equal-powered probing sources was found out to be lesser. In the signal scenario with two desired signals, the same effect is observed. The corresponding adapted beam pattern is presented in Figure 24.60. It can be seen that the interference suppression capability of the array degrades if the strengths of the sources are considered to be equal. The reason for this is the inability of the array to resolve the probing sources owing to the same eigenvalues obtained. In Figure 24.61, the adapted beam pattern for a signal scenario, having three closely spaced wideband probing sources, is shown. These probing sources are assumed to arrive arbitrarily from $88°$, $90°$ and $92°$ having 2% BW, 3 lines; 5% BW, 6 lines; 2% BW, 3 lines, respectively. The power ratio of the sources are unequal, that is, 600, 1200, 400, respectively. It can be seen that the sidelobe within the probing directions is suppressed extensively while the main lobes in the pattern remain undistorted.

Now, a 16×10 planar antenna array is considered. Figure 24.62 represents the case of three desired signals $(-50°, -10°, 30°)$ and two probing sources $(5°, -30°)$. One of the probing source $(5°)$ is narrowband while the one probing at $-30°$ is wideband with 5% bandwidth and 6 spectral lines. The resultant pattern consists of multiple main lobes each pointing towards desired directions along with accurate and deep nulls towards the probing direction. The difference in pattern for the case of two narrowband and wideband sources probing at $5°$ and $-30°$ shows the same trend as in the

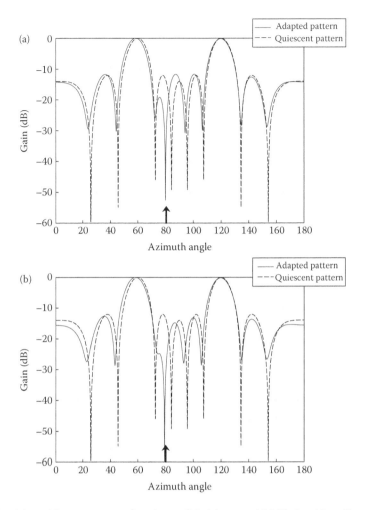

FIGURE 24.58   Adapted beam pattern using the modified improved LMS algorithm. Two desired signals (60°, 120°, power ratio of 1 each) with single wideband probing source (80°, power ratio of 1000) are considered. (a) 2% BW, 3 lines; (b) 15% BW, 16 lines.

previous cases. Let us consider a slightly complicated case of three desired signals (−60°, 10°, 30°) and three closely spaced wideband probing sources with different bandwidths (−25°, 5%, 6 spectral lines; −35°, 2%, 3 spectral lines; −20°, 10%, 5 spectral lines). It can be seen from Figure 24.63 that each probing source is suppressed efficiently, resulting in a deep and wide null in the pattern.

24.9   EFFECTIVE PROBE CANCELLATION IN COHERENT SIGNAL ENVIRONMENT

Interference cancelling capabilities of the antenna array along with adaptive algorithms are discussed so far, assuming that the desired signals arriving from different directions and the probing sources from other directions are not correlated. In practice, however, significant difficulties arise when the signals are correlated.

Two signals are said to be coherent when they are completely correlated, that is, if one is a scaled and delayed replica of the other [4]. Coherent interference can arise in the case of multipath propagation or when smart jammers deliberately induce coherent interference by retrodirecting the signal energy to the receiver. Coherence can completely destroy the performance of adaptive array

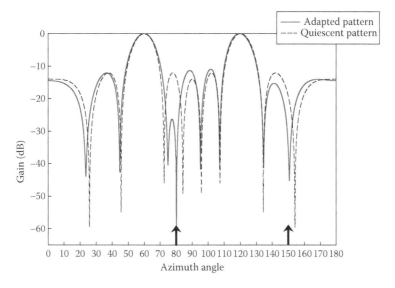

FIGURE 24.59 Adapted beam pattern using the modified improved LMS algorithm. Two desired signals (60°, 120°, power ratio of 1 each) and two wideband probing sources (80°, 2% BW, 3 lines, 1000; 150°, 2% BW, 3 lines, 100) with *unequal power ratios* are considered.

systems. Conventional beamformers like Frost beamformer and Howells–Applebaum arrays completely fail to operate as a receiver unit in such scenarios. Techniques like *subaperture sampling* or *spatial smoothing* have been proposed to overcome the effects of coherence. However, these techniques assume certain special circumstances and do not provide a clear general procedure [29]. Conventional adaptive beamforming approaches, which assume uncorrelated signal sources, suffer from signal cancellation in the presence of coherent signals. Such array response is attributed to the non-Toeplitz structure of the estimated covariance matrix in the presence of coherent signals. This may lead to fault detection in biomedical signal processing, aerospace applications and various

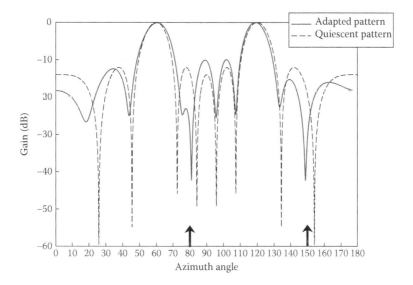

FIGURE 24.60 Adapted beam pattern using the modified improved LMS algorithm. Two desired signals (60°, 120°, power ratio of 1 each) and two wideband probing sources (80°, 2% 3 lines, 1000; 150°, 2% 3 lines, 1000) with *equal power ratios* are considered.

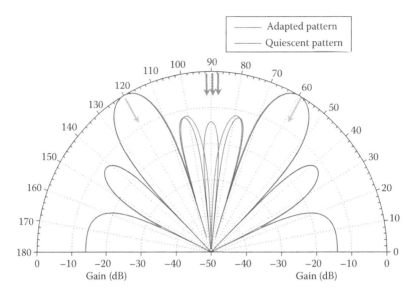

FIGURE 24.61   Adapted beam pattern for two desired signals (60°, 120°, 0 dB each) and three closely spaced wideband probing sources (88°, 600, 2% BW, 3 lines; 90°, 1200, 5% BW, 6 lines; 92°, 400, 2% BW, 3 lines) are considered.

other industrial applications. Certain techniques of redundancy averaging and enhanced redundancy are proposed to put nulls in the DOAs of coherent interferences [30]. However, steering the nulls in the directions of the coherent signals is not desirable. A beamformer should be able to constructively combine these signals instead of cancelling all but one of them, so as to avoid any loss of information. Gonen and Mendel [31] developed a cumulant-based blind beamformer to overcome this problem in the presence of coherent multipath propagation. Another approach consisting of matrix reconstruction scheme along with an iterative algorithm is proposed by Lee and Hsu [32] to maintain the array efficiency even in the presence of coherent signals.

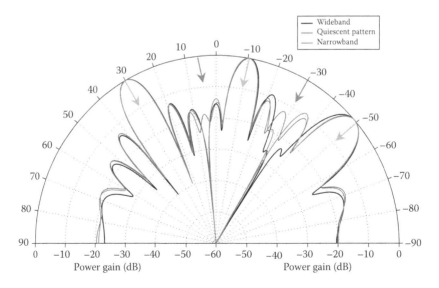

FIGURE 24.62   Adapted beam pattern of 16×10 planar antenna array for narrowband and wideband probing sources. Three desired signals at (−50°, −10°, 30°) and 2 probing sources. 1 narrowband probing source at (5°) and 1 wideband probing source at (−30°; 5%, 6 spectral lines).

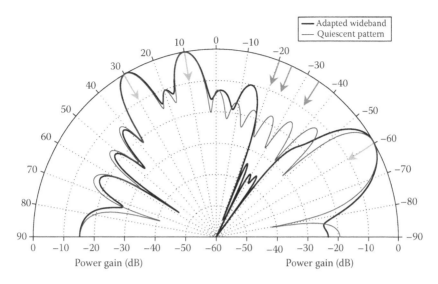

FIGURE 24.63   Adapted beam pattern of 16 × 10 planar antenna array. Three desired signals at (−60°, 10°, 30°) and 3 wideband probing sources at (−25°, 5%, 6 spectral lines; −35°, 2%, 3 spectral lines; −20°, 10%, 5 spectral lines).

24.9.1   Adaptive Array Processing with Multiple Beam Constraints

In conventional beamformers, the weights are estimated by minimizing the mean output power subjected to the various constraints. However, in the presence of correlation between the signals, the conventional adaptive schemes cannot be employed for weight estimation. Let us consider a case where an interfering signal is correlated with the desired signal. Then, in the presence of a directional constraint, the processor adjusts the phase of the correlated interference induced on each antenna element for optimal weight estimation. It aims at minimizing the total output power, which is the sum of the powers of both the desired signal and correlated interfering signal. This leads to the desired signal cancellation problem [4]. The correlation matrix loses its structure in the presence of correlated signals (Figure 24.64), thereby leading to signal cancellation. It should be noted that the signals $s_2(t)$ and $s_3(t)$ are correlated to each other owing to the multipath effect. In order to restore the desired structure of the correlation matrix, a matrix reconstruction scheme along with an iterative algorithm is employed for the simulations. The resultant matrix obtained after finite number of iterations is then used to calculate the weight vector. Then, the optimal weight vector is combined with the received data vector to form the array output signal.

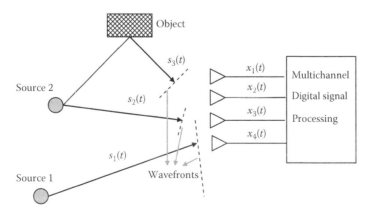

FIGURE 24.64   Schematic for coherent signal environment. Signals $s_2(t)$ and $s_3(t)$ are correlated to each other.

24.9.1.1 Steering Vector and Signal Representation

Consider a *uniform linear array* (ULA) composed of N identical sensors and inter-element spacing equal to $\lambda/2$, where λ is the smallest signal wavelength. Let us assume that p ($p < N$) far-field narrowband signals, centred at frequency ω_o impinge on the array from directions $\{\theta_1, \theta_2, \ldots, \theta_p\}$. Using complex signal representation, the received signal at the ith sensor can be expressed as

$$x_i(t) = \sum_{k=1}^{p} a_k s_k(t) e^{-j\omega_o(i-1)\sin\theta_k(d/c)} + n_i(t), \ i = 1,2,\ldots,N \tag{24.82}$$

where $s_k(t)$ is the complex waveform of the kth signal, a_k is the complex response of the sensor to the kth signal, d is the spacing between the sensors, c is the propagation speed of the signal and $n_i(t)$ is the spatially white noise with mean zero and variance σ^2 received at the ith sensor.

Assuming that the sensors are omnidirectional, that is, $a_k = 1$, Equation 24.82 may be expressed in vector notation as

$$x(t) = \sum_{k=1}^{p} a(\theta_k)s_k(t) + n(t) \tag{24.83}$$

where $a(\theta_k) = e^{-j\omega_o(i-1)\sin\theta_k(d/c)}$.

In order to simplify the notation further, Equation 24.83 may be re-written as

$$x(t) = As(t) + n(t) \tag{24.84}$$

where $A = \left[a(\theta_1)a(\theta_2)\ldots a(\theta_k) \right]$, $s(t) = \left[s_1(t)s_2(t)\ldots s_k(t) \right]^T$ is the signal source vector and $n(t) = \left[n_1(t)n_2(t)\ldots n_k(t) \right]^T$ is the noise vector. The superscript T denotes the transpose operation.

The Toeplitz–Hermitian correlation matrix of $x(t)$ of size $N \times N$ can be expressed as

$$R_x = E\left\{ x(t)x^H(t) \right\} = ASA^H + \sigma_n^2 I, \quad S = E\left\{ s(t)s(t)^H \right\} \tag{24.85}$$

H denotes the transpose of complex conjugate of the matrix and $E\{\ \}$ is the expectation value. The source correlation matrix S is diagonal when the signals are uncorrelated. However, the presence of correlation between the signals makes the matrix S different. It becomes non-conditional and non-singular when the signals are partially correlated, and non-diagonal but singular for coherent, that is, fully correlated signals [33]. Assume for simplicity that two out of k incident signals are coherent, that is, $s_2(t) = \alpha s_1(t)$, with α as the correlation coefficient denoting the gain phase relationship between the two coherent signals. In this case, $s(t)$ becomes a $(k-1) \times 1$ vector, that is,

$$s(t) = \left[(1+\alpha)s_1(t)s_3(t)\ldots s_k(t) \right]^T \tag{24.86}$$

and A becomes a $\{(k-1) \times N\}$ matrix, that is,

$$A = \left[a(\theta_1) + \alpha a(\theta_2)a(\theta_3)\ldots a(\theta_k) \right] \tag{24.87}$$

From Equation 24.85 it follows that R_x becomes a $(k-1) \times (k-1)$ nonsingular matrix. Thus, in coherent signal environment, the eigenstructure of the correlation matrix will be destroyed, leading to the inconsistent performance of the conventional eigenstructure technique.

24.9.1.2 Iterative Algorithm for Matrix Reconstruction

An iterative matrix reconstruction method [32] is used to obtain the Toeplitz–Hermitian matrix with the desired eigenstructure of R_x. The iterative matrix reconstruction method involves the following steps:

Step I. Estimate R_x from the received signals using Equation 24.85.

Step II. Reconstruct R_x to obtain a Toeplitz–Hermitian structure

1. Consider only the N diagonals of the upper triangular matrix. For each diagonal, the diagonal elements are averaged. The new value for each diagonal element is the mean value thus calculated.
2. Then, consider the $N-1$ diagonals of the lower triangular matrix. In each diagonal, every element is taken as the complex conjugate of the value in the corresponding diagonal of the upper triangular matrix so that the new matrix (ACM) is Hermitian.
3. Let $\hat{R}_x$ be the reconstructed correlation matrix.

Step III. Use this $\hat{R}_x$ to compute the matrix $\tilde{R}_{xs}$ given by

$$\tilde{R}_{xs} = \sum_{j=1}^{p} \lambda_j e_j e_j^H + \lambda_{av} \sum_{j=p+1}^{N} e_j e_j^H \tag{24.88}$$

where $\lambda_1 \geq \lambda_2 \geq \cdots \geq \lambda_N$ and $e_j, j = 1,2, \ldots, N$ are the eigenvalues and eigenvectors of matrix $\hat{R}_x$, respectively. λ_{av} is the average of eigenvalues.

Step IV. If the matrix norm $\left| \tilde{R}_{xs} - R_x \right| > \varepsilon$, where ε is a preset positive real number, then let $R_x = \tilde{R}_{xs}$ and repeat process from Step II. Otherwise, stop with the final form of $\tilde{R}_{xs}$ as computed in the above step.

Thus, the resulting correlation matrix $\tilde{R}_{xs}$ possesses both the Toeplitz–Hermitian and the desired eigenstructure properties and hence it can be used for the calculation of the optimal weight vector.

24.9.1.3 Weight Estimation

Assuming the gain/null requirements in p directions, that is, p denotes the number of signals with gain/null constraint ($p \leq K$), the optimal weight [32] can be expressed as

$$w_o = \tilde{R}_{xs}^{-1} M (M^H R_{xs}^{-1} M)^{-1} c \tag{24.89}$$

where M is the constraint matrix given by $M = \left[a(\theta_1) a(\theta_2) \ldots a(\theta_p) \right]$ and c is the corresponding gain vector written as $c = \left[c_1 c_2 \ldots c_p \right]$. The dimension of the constraint matrix M depends on the number of elements in the gain vector c. The ULA uses this weight for processing the received data $x(t)$ to obtain the array output $y(t) = w^H x(t)$.

In the present work, it will be shown that with the use of this scheme for weight estimation, multiple signals can be received with the desired gains and all the correlated as well as uncorrelated probing sources can be suppressed simultaneously.

24.9.2 PERFORMANCE ANALYSIS OF ARRAY IN COHERENT SIGNAL ENVIRONMENT

Here, the simulations are carried out using the basic steps of Lee and Hsu's algorithm [32], but for more generic signal environment. A ULA with 10 isotropic antenna elements is taken. The spacing between the elements is fixed to half a wavelength. The adapted radiation pattern is obtained using the weight vector computed as discussed in the previous section in order to analyse the array performance in the presence of correlated interferences and coherent desired signals. It is assumed that DOA of all the desired signals and the interfering sources are known *a priori.*

24.9.2.1 Correlated Probing Sources

In this section, the correlation between probing sources and any one of the desired signals is considered and the array capability to resolve the signals is analysed in different scenarios. The array is expected to decorrelate the signals and place nulls towards the probing directions and at the same time maintain peaks in the desired directions.

As a first case, two desired signals are considered to arrive from $-30°$ and $0°$. A single interfering source is assumed to probe the antenna array at $30°$. The power level of all the three sources is kept at 0 dB. The interfering source is correlated to the signal arriving at $-30°$. The value of the complex correlation coefficient α is taken to be $(-0.15, 0.05)$. The gain vector c considered for two desired signals is [1 1]. It can be readily seen from the adapted radiation pattern, shown in Figure 24.65a, that the two main lobes are maintained in the desired signal directions and a null up to -37 dB (approximately) is placed towards the probing direction. This signifies that irrespective of the correlation between an interfering source and a desired signal, the array is capable enough to maintain gain in the desired directions and place a null towards the probing direction.

Figure 24.65b shows the adapted beam pattern corresponding to a situation where all parameters are kept the same as in Figure 24.65a except for the gain vector. The gain vector c is taken as [1 1 0] corresponding to two desired signals and one unwanted signal. It can be observed that the array successfully preserves the desired signals as the main lobes are maintained as such and the null towards the probing source becomes deeper, that is, -70 dB (cf. Figure 24.65a). This further shows the capability of the present scheme. Figure 24.66 presents the adapted beam pattern and the corresponding quiescent pattern of an antenna array when two uncorrelated desired signals are considered to impinge on the array at $-30°$ and $0°$ simultaneously. Two interfering sources try to probe the array at $30°$ and $50°$. The probing signal arriving at $30°$ is correlated to the desired signal at $-30°$ $\{\alpha = (-0.15, 0.05)\}$. The power levels of all the signals are taken as 0 dB. It can be seen that the correlation between the desired signal and the unwanted interference is tackled efficiently by the adaptive beamformer as it successfully forms nulls in the directions of the coherent interferences without cancelling the desired signals in the output.

In order to analyse the probe suppression capability of the array for different values of correlation coefficient, a similar situation as in Figure 24.66 is considered to obtain the adapted beam pattern with the additional condition that the interfering signal arriving at $50°$ is correlated to signal at $0°$. Keeping the phase delay as 0, different values of gain in correlation coefficient are considered $(-0.08, -0.1, -0.2)$. Figure 24.67 shows that as the absolute value of the correlation coefficient increases, suppression of the probing source decreases. Thus, it can be inferred that with the increase in the correlation between interferences and the desired signal, the probe cancellation capability of the antenna array deteriorates. Now, two desired signals $(-30°, 0°)$ and two correlated interfering sources $(30°, 50°)$ are considered to impinge on the antenna array. Both the interfering sources are correlated to the signal arriving at $-30°$ with correlation coefficient $(-0.1, 0.09)$ and $(-0.1, -0.05)$. The gain vector, c, is taken as [1 1] corresponding to the two desired signals. The resultant adapted beam pattern and the quiescent pattern (black curve) in shown in Figure 24.68a. Interfering sources at $30°$ and $50°$ are suppressed to -38 dB and -29 dB, respectively. Thus, the antenna array resolves the desired signals efficiently from the unwanted ones, even in the presence of correlation.

Further, taking the correlation coefficients and all the other parameters the same as in Figure 24.68a, the gain matrix c is changed to [1 1 0 0] so as to obtain the desired gain towards $-30°$ and $0°$ while nulls towards $30°$ and $50°$. The output radiation pattern obtained along with the quiescent pattern is shown in Figure 24.68b. It can be seen that on imposing constraint on the correlated

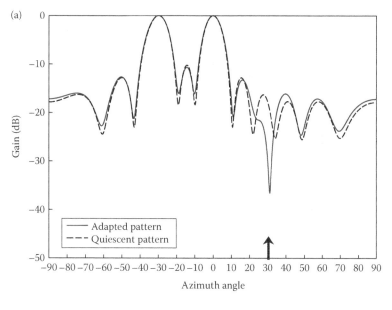

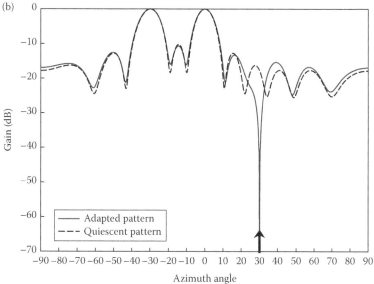

FIGURE 24.65 Adapted beam pattern of an antenna array with two desired signals ($-30°$, $0°$) and one interfering source ($30°$). Interfering signal is correlated to signal at $-30°$. $\alpha = (-0.15, 0.05)$, (a) $c = [1\ 1]$; (b) $c = [1\ 1\ 0]$ cf. [32].

signals, deeper nulls can be obtained towards probing directions. Figure 24.69 illustrates the radiation pattern of the array when two desired signals ($-20°$, $20°$) and four interfering sources ($-60°$, $-40°$, $5°$, $40°$) are considered. Signal at $-40°$ and signal at $5°$ are both correlated to the desired signal at $-20°$ with correlation coefficients (0.1, -0.23) and (-0.15, -0.08), respectively. The gain vector is taken as [1 1 0 0]. It is apparent that the uncorrelated interferences get suppressed more as compared to correlated ones; however, nulls are placed efficiently towards all the probing directions. This shows the capability of antenna array in decorrelating the signals and maintaining the gain/null corresponding to the desired/unwanted signals.

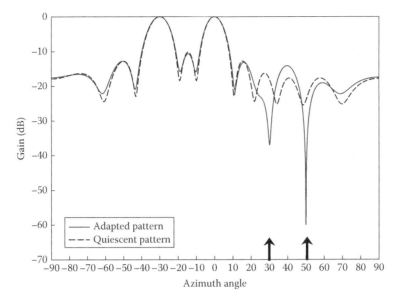

FIGURE 24.66 Adapted beam pattern of an antenna array with two desired signals ($-30°$, $0°$) and two probing sources ($30°$, $50°$). Probing signal at $30°$ is correlated to signal at $-30°$. $\alpha = (-0.1, 0.09)$, $c = [1\ 1\ 0]$.

24.9.2.2   Coherent Desired Signals

So far, the correlation between interfering sources and desired signals was being considered. In this section, a scenario with coherent desired signals and uncorrelated interfering sources is assumed for the simulations [34]. Cases with both the single and multiple interfering sources are considered.

Two coherent sources and one uncorrelated interfering source are assumed to be incident on the array at $-20°$, $20°$ and $60°$, respectively. The power level of all the three signals is kept constant

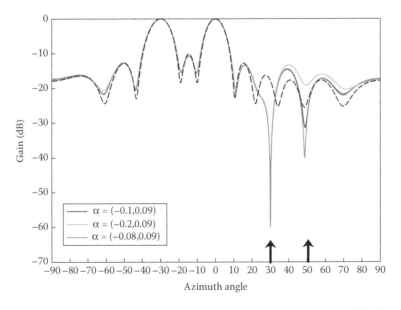

FIGURE 24.67 Adapted beam pattern of an antenna array with two desired signals ($-30°$, $0°$) and two probing sources ($30°$, $50°$). Probing signal at $50°$ is correlated to signal at $-30°$. $\alpha = (-0.08, 0)$, $(-0.1, 0)$, $(-0.2, 0)$, $c = [1\ 1\ 0]$.

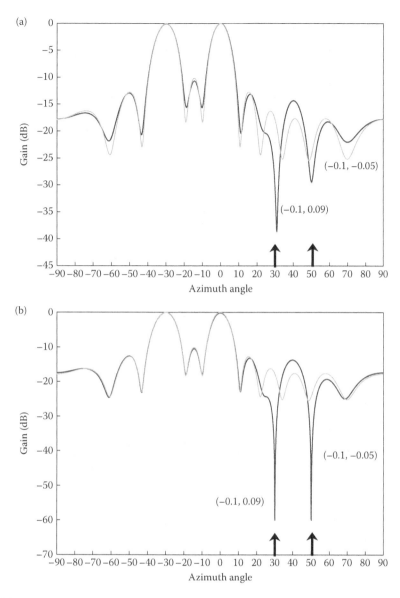

FIGURE 24.68 Adapted beam pattern of an antenna array when two desired signals (−30°, 0°) and two interfering sources (30°, 50°) are considered. Both the interfering signals at 30° and 50° are correlated to signal at −30° with α = (−0.1, 0.09) and (−0.1, 0.05), respectively. (a) c = [1 1]; (b) c = [1 1 0 0].

at 0 dB. The value of correlation coefficient is fixed at (1.0, −0.2). The gain vector is taken to be [1 0]. The values 1 and 0 correspond to the two coherent desired signals and one jammer, respectively. It can be observed from the adapted radiation pattern (Figure 24.70) that the array fosters simultaneous reception of the desired signals with an accurate and deep null placed towards the probing direction. In order to further analyse the array performance in the presence of multiple probing sources, the number of interfering sources is increased to two. The two coherent desired signals are kept the same while two uncorrelated interfering sources are assumed to probe the array at −55° and 60°, the gain vector given by c = [1 0 0]. It can be observed (Figure 24.71) that while peaks are maintained towards the desired directions, both the interferences are suppressed up to −60 dB.

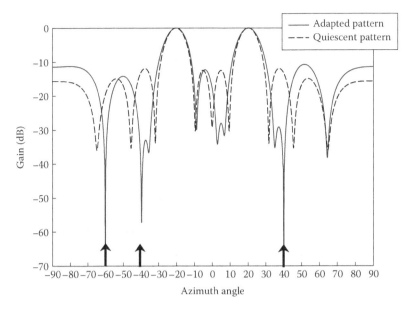

FIGURE 24.69   Adapted beam pattern of an antenna array with two desired signals ($-20°$, $20°$) and four probing sources ($-60°$, $-40°$, $5°$, $40°$). Both the probing signals at $-40°$ and $5°$ are correlated to signal at $-20°$ with $\alpha = (0.1, -0.23)$ and $(-0.15, 0.08)$, respectively. $c = [1\ 1\ 0\ 0]$.

In Figure 24.72, adapted beam pattern and the quiescent pattern for an array are shown when four desired signals and two interfering signals (S_1: $-40°$, S_2: $-20°$, S_3: $20°$, S_4: $40°$; I_1: $-60°$, I_2: $50°$) are considered to impinge on the antenna array. Two desired signals arriving at $-20°$ and $20°$, that is, S_2 and S_3, are coherent to each other $\{\alpha = (1.0, 0.0)\}$ and the other two arriving at $-40°$ and $40°$, that is, S_1 and S_4, are also coherent to each other $\{\alpha = (1.0, 0.0)\}$. The power ratios are kept 1 for all

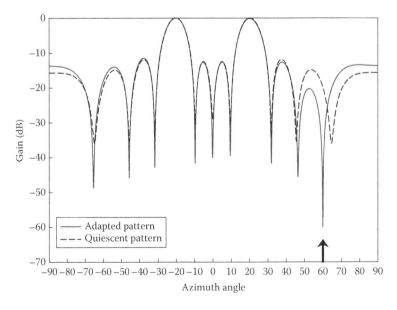

FIGURE 24.70   Adapted beam pattern of an antenna array when two desired signals ($-20°$, $20°$) and an interfering source ($60°$) are considered. The two desired signals are coherent. $\alpha = (1, -0.2)$, $c = [1\ 0]$.

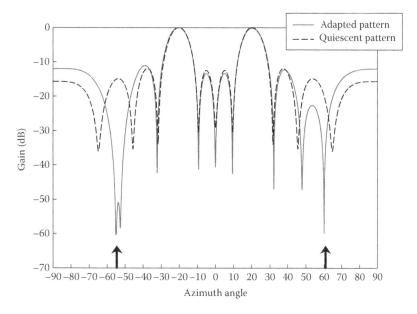

FIGURE 24.71 Adapted beam pattern of an antenna array with two desired signals (−20°, 20°) and two uncorrelated interfering sources (−55°, 60°). The two desired signals are coherent. $\alpha = (1.0, -0.2)$, $c = [1\ 0\ 0]$.

the signals and the gain vector is taken as [1 1 0 0], where the first two 1s correspond to the four coherent signals and the 0s are for the probing sources. It can be seen that the deep nulls are placed towards the uncorrelated probing directions. However, sufficient gain is maintained towards all the four desired signals irrespective of their coherence. Thus, one can conclude that with the use of the proposed signal model and the weight estimation technique, multiple signals can be received and all

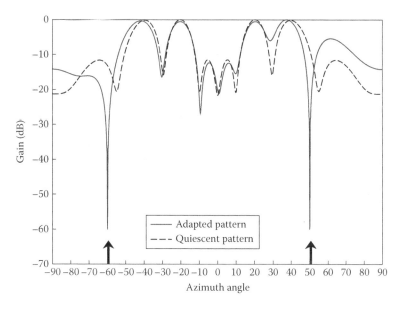

FIGURE 24.72 Adapted beam pattern of an antenna array when four desired signals (−20°, 20°, −40°, 40°) and two uncorrelated probing sources (−60°, 50°) are considered. The desired signals at −20° and 20°, $\alpha_1 = (1.0, 0.0)$; −40° and 40°, $\alpha_2 = (1.0, 0.0)$ are coherent. $c = [1\ 1\ 0\ 0]$.

the correlated as well as uncorrelated probing sources can be suppressed simultaneously. The approach used finds a direct application in aerospace engineering.

24.9.3 SUPPRESSION OF WIDEBAND CORRELATED SOURCES

As a first case, a 10-element linear phased antenna array is considered with two desired signals (−30°, 0°) and two interfering sources (30°, 50°). The spacing between the antenna elements is taken as half-wavelength. Interfering signal at 30° is correlated to the desired signal at −30° {$\alpha = (-0.1, 0.09)$}. The constraint vector is taken as [1 1 0]. The corresponding adapted pattern along with the quiescent pattern is shown in Figure 24.73. If the interfering signal at 50° is considered wideband with 5% bandwidth and 6 spectral lines, the corresponding adapted pattern is also according to the expectations. In simple words, each spectral component of the probing source is suppressed efficiently without any distortion in the main lobes. In both the cases, the antenna array is able to decorrelate the two signals and maintain the main lobes towards each of the desired directions and places accurate and deep nulls towards the probing directions. However, the depth of the wideband signal is found to be less than the narrowband signal.

Figure 24.74 presents adapted beam pattern of an antenna array with two desired signals (−20°, 20°) and one interfering source (60°). In this case, the two desired signals are coherent {$\alpha = (1.0, -0.2)$}. The constraint vector is taken as [1 1 0]. Interfering signal at 60° is considered wideband with 2% bandwidth and 3 spectral lines. The adapted pattern has undistorted main lobes with accurate and deep nulls towards the probing source, even when it has finite spectral distribution. In order to analyse the effect of bandwidth and spectral lines of the probing source, bandwidth of the probing source (60°) is increased to 5%, 6 lines and 15%, 9 lines. The performance of the antenna array is compared in Figure 24.75. It can be seen that in both the cases, the null placement is accurate but the depth of the null degrades with increase in the bandwidth and spectral lines.

Next, the number of desired signals is increased to four (−20°, 20°, −40°, 40°). The signal environment consisting of two probing sources (−60°, 50°) is considered. The desired signals at (−20°, 20°)

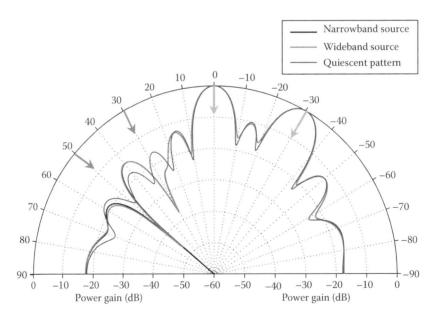

FIGURE 24.73 Adapted beam pattern of a 10-element antenna array with two desired signals (−30°, 0°) and two probing sources (30°, 50°). Probing signal at 30° is correlated to signal at −30°. Interfering signal at 50° is wideband (5%, 6 spectral lines). $c = [1\ 1\ 0]$.

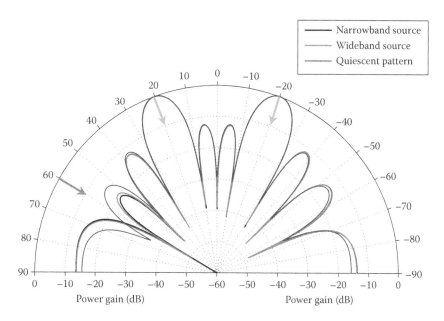

FIGURE 24.74 Adapted beam pattern of an antenna array with two desired signals (−20°, 20°) and one probing source (60°). Two desired signals are coherent. Probing signal at 60° is wideband (2%, 3 spectral lines). $c = [1\ 1\ 0]$.

and (−40°, 40°) are taken as coherent with correlation coefficient $\alpha = (1, 0)$. Two uncorrelated probing sources at (−60° and 50°) are wideband (2%, 3 spectral lines each). The constraint vector is taken as $c = [1\ 1\ 0\ 0]$. Figure 24.76 shows the adapted pattern for such a complex signal scenario. It is apparent that the null placement is accurate even when the desired signals are coherent. There is a small shift (less than 0.1°) in the main lobes at 40° and −40°. This may be due to the fact that the interfering

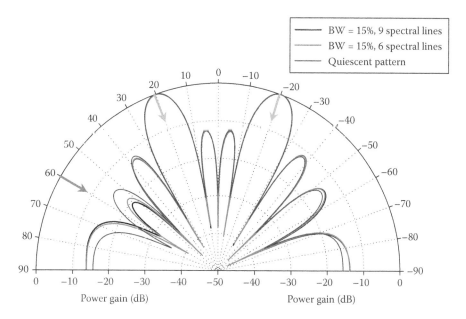

FIGURE 24.75 Adapted beam pattern of a 10-element antenna array with two desired signals (−20°, 20°) and one interfering source (60°). Two desired signals are coherent. Probing signal at 60° is wideband (15%, 9 lines; 5% and 6 lines). $c = [1\ 1\ 0]$.

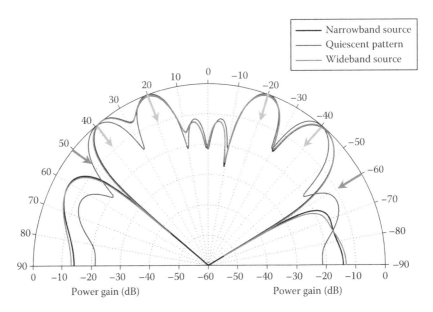

FIGURE 24.76 Adapted beam pattern of a 10-element antenna array with four desired signals (−20°, 20°, −40°, 40°) and two interfering sources (−60°, 50°). The desired signals at (−20°, 20°) and (−40°, 40°) are coherent with $\alpha = (1, 0)$. Two uncorrelated interfering sources (−60° and 50°) are wideband (2%, 3 spectral lines each). $c = [1\ 1\ 0\ 0]$.

sources probing at −60° and 50° are wideband with multiple spectral lines and also spatially close to the desired signals. However, the antenna array is capable of decorrelating the coherent signals and generating the desired adapted pattern. Figure 24.77 represents the case when the two probing sources are wideband with different bandwidths (−60°, 2%, 3 lines; 50°, 11%, 6 lines). Adapted beam pattern of an antenna array with four desired signals (−20°, 20°, −40°, 40°) and two interfering sources (60°, 50°) is shown along with the quiescent pattern. Four desired signals are coherent, $\alpha = (1, 0)$. The

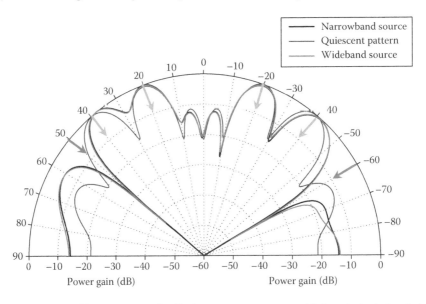

FIGURE 24.77 Adapted beam pattern of a 10-element antenna array with four desired signals (−20°, 20°, −40°, 40°) and two interfering sources (60°, 50°). Four desired signals are coherent. Wideband probing signals (−60°, 2%, 3 spectral lines; 50°, 11%, 6 spectral lines). $c = [1\ 1\ 0\ 0]$.

constraint vector is taken as $c = [1\ 1\ 0\ 0]$. In this case also the antenna array is able to maintain the main lobes of the pattern towards each of the desired signals even when they are coherent and the probing sources are wideband. The nulls placed are also accurate and deep (−60 dB).

24.10 CONCLUSIONS

The interference suppression capabilities of the *generalized sidelobe canceller* (GSC) and the DF-GSC for a uniformly spaced array are studied in different signal scenarios. The GSC implementation is of importance as it has less computational complexity. Addition of a blind equalizer with decision-feedback in conventional GSC provides better robustness, faster convergence and high-output SINR. The resultant design is referred to as the decision-feedback sidelobe canceller. It can be inferred from the results obtained that the DF-GSC can achieve higher SINR value for the same convergence rate. Moreover, on increasing the desired signal power, the SINR output is enhanced. In case of DF-GSC, deeper nulls are obtained which shows that DF-GSC has better suppression capabilities. It is shown that as compared to the standard LMS algorithm, the convergence rate and the output SINR are enhanced in the case of the structured LMS and the improved LMS algorithms. Further, the effect of step size and power level of the desired signal on GSC and DF-GSC are analysed.

In order to enhance its robustness furthermore, some derivative constraints are added to DF-GSC. However, these additional constraints require some initial training which is difficult to acquire in actual practice. Thus, to overcome this problem, a new approach known as the blind DF-GSC is implemented where three stages of simulations are performed. In this case, initial training is not required and higher output SINR is achieved with better convergence rate. Further, a new approach that uses only two stages is proposed which is called the semi-blind DF-GSC. The performance of DF-GSC is further enhanced by this approach and much deeper nulls are obtained. The performance analysis of blind DF-GSC and semi-blind DF-GSC scheme was carried out with different forms of LMS algorithm. The jammer suppression capabilities of sidelobe cancellers can be exploited in stealth techniques. Mounting these cancellers in suitable configurations on aircraft and missiles will allow the reception of only the desired signal, simultaneously suppressing the unwanted components. This will make them invisible to the radars at the enemy base or in their aircraft, leading to an *active RCS reduction*.

When the antenna array tracks multiple sources simultaneously, its pattern should have main lobes towards each of the sources. Thus, the steering vector, the projection vectors and hence the weight coefficients should be calculated accordingly. The modified improved LMS algorithm provides the optimal weight coefficients from which the desired adapted pattern is obtained. If the radar sources are wideband, then their spectral distribution can be exploited to estimate the array correlation matrix. It is based on the concept of considering each spectral line as a single narrowband source. This modifies the weight coefficients accordingly, leading to the suppression of each source over an entire spectral distribution. In simple words, a wider null is placed towards the probing direction. The reason behind this is that each spectral line is considered as separate source and hence the eigenvalues and eigenvectors of the resultant array correlation matrix will be distinct and widespread. This leads to the placement of wider null towards the probing direction.

If the correlation between interferences and the desired signal increases, the probe cancellation capability of the antenna array deteriorates. Further, it is inferred that on imposing constraint on the correlated signals in the resultant, adapted pattern, deeper nulls are obtained towards probing directions. It is not only the correlation but also the gain vector that plays a significant role in the performance of the array. It is shown that the antenna array along with matrix reconstruction algorithm is efficient in decorrelating the signals and maintaining the gain/null corresponding to the desired/unwanted signals. In other words, the antenna array resolves the desired signals efficiently from the unwanted ones, even in the presence of correlation. This establishes the capability of the linear/planar phased antenna array along with efficient adaptive algorithm in catering varied signal environments, consisting of narrowband, wideband, single or multiple sources, uncorrelated, correlated or coherent signals. However, for correlated signals, matrix reconstruction algorithm is required for restoring the eigenstructure of the array correlation matrix. Such adaptive/smart antenna array when mounted on any platform of robotics, satellite or aerospace category, can be exploited for the desired performance in any practical signal scenario.

REFERENCES

1. K. Nonami, Prospect and recent research & development for civil use autonomous unmanned aircraft as UAV and MAV, *Journal of System Design and Dynamics*, 1, 120–128, 2007.
2. S. Chandran, *Adaptive Antenna Arrays: Trends and Applications.* New York: Springer, 2004.
3. Y. Lee and W.-R. Wu, A robust adaptive generalized sidelobe canceller with decision feedback, *IEEE Transactions on Antennas and Propagation*, 53, 3822–3832, 2005.
4. L. C. Godara, *Smart Antennas.* Boca Raton, Florida: CRC Press, 2004.
5. H. L. V. Trees, *Detection, Estimation, and Modulation Theory: Optimum Array Processing.* New York: Wiley, 2002.
6. H. F. Harmuth, *Transmission of Information by Orthogonal Functions.* New York: Springer-Verlag, 1969.
7. J. B. Anderson and A. Svensson, *Coded Modulation Systems.* New York: Kluwer Academic/Plenum Publishers, 2003.
8. N. K. Jablon, Steady state analysis of the generalized sidelobe canceller by adaptive noise canceling techniques, *IEEE Transactions on Antennas and Propagation*, AP-34, 330–337, 1986.
9. L. J. Griffiths and K. M. Buckley, Quiescent pattern control in linearly constrained adaptive arrays, *IEEE Transactions on Acoustics, Speech, and Signal Processing*, ASSP-35, 917–926, 1987.
10. A. M. Vural, Effects of perturbation on the performance of optimum/adaptive array, *IEEE Transactions on Aerospace and Electronics Systems*, AES-24, 585–599, 1976.
11. S. P. Applebaum and D. J. Chapman, Adaptive arrays with main beam constraints, *IEEE Transactions on Antennas and Propagation*, AP-34, 650–662, 1986.
12. O. L. Frost III, An algorithm for linearly constrained adaptive array processing, *Proceedings of IEEE*, 60, 926–935, 1972.
13. K. M. Buckley and L. J. Griffiths, Adaptive generalized sidelobe canceller with derivative constraints, *IEEE Transactions on Antennas and Propagation*, AP-34, 311–319, 1986.
14. K.-C. Huarng and C.-C. Yeh, Performance analysis of derivative constraint adaptive arrays with pointing errors, *IEEE Transactions on Antennas and Propagation*, AP-40, 975–981, 1992.
15. A. K. Steele, Comparison of directional constraints for beamformers subject to multiple linear constraints, *IEE Proceedings*, pt. H, 130, 41–45, 1983.
16. K. Takao, M. Fujita and T. Nishi, An adaptive antenna array under directional constraint, *IEEE Transactions on Antennas and Propagation*, AP-24, 662–669, 1976.
17. B. Widrow, K. M. Duvall, P. R. Gooch and W. C. Newman, Signal cancellation phenomena in adaptive arrays: Causes and cures, *IEEE Transactions on Antennas and Propagation*, 30, 469–478, 1982.
18. L. C. Godara, Performance analysis of structured gradient algorithm, *IEEE Transactions on Antennas and Propagation*, AP-38, 1078–1083, 1990.
19. L. C. Godara and D. A. Gray, A structured gradient algorithm for adaptive beam forming, *Journal of the Acoustical Society of America*, 86, 1040–1046, 1989.
20. L. C. Godara, Improved LMS algorithm for adaptive beam forming, *IEEE Transactions on Antennas and Propagation*, AP-38, 1631–1635, 1990.
21. W. F. Gabriel, Adaptive arrays: An introduction, *Proceedings of IEEE*, 64, 239–273, 1976.
22. H. Singh, S. Sharma and R. M. Jha, Parametric study of suppression capabilities of adaptive arrays against multiple wideband radars, *International Symposium on Electromagnetics Theory, EMTS-2007*, Ottawa, ON, Canada, 3 p., July 26–28, 2007.
23. H. J. Visser, *Array and Phased Array Antenna Basics.* New York: John Wiley & Sons, 2005.
24. C. A. Balanis, *Antenna Theory: Analysis and Design.* New York: John Wiley & Sons, 1982.
25. B. Allen and M. Ghavami, *Adaptive Array Systems: Fundamentals and Applications.* New York: John Wiley & Sons, 2005.
26. Z. N. Chen and M-Y. W. Chia, *Broadband Planar Antennas: Design and Applications.* New York: John Wiley & Sons, 2006.
27. R. S. Elliott, *Antenna Theory and Design.* Englewood Cliffs, NJ: Prentice-Hall, 1981.
28. H. Singh and R. M. Jha, Efficacy of active phased arrays in maintaining desired multi-beam radar signals, *International Symposium on Aerospace Science and Technology, INCAST-2008*, Bangalore, p. 4, June 26–28, 2008.

29. T. J. Shan and T. Kailath, Adaptive beamforming for coherent signals and interference, *IEEE Transactions on Acoustics, Speech, and Signal Processing*, ASSP-33, 527–536, 1985.

30. M. H. E.-Ayadi, E. K. A.-Hussaini and E. A. E.-Hakeim, A combined redundancy averaging signal enhancement algorithm for adaptive beamforming in the presence of coherent signals and interferences, *Signal Processing*, SP-55, 285–293, 1996.

31. E. Gonen and J. M. Mendel, Applications of cumulants to array processing—Part III: Blind beamforming for coherent signals, *IEEE Transactions on Signal Processing*, SP-45, 2252–2264, 1997.

32. J. H. Lee and T. F. Hsu, Adaptive beamforming with multiple-beam constraints in the presence of coherent jammers, *Signal Processing*, SP-80, 2475–2480, 2000.

33. T. J. Shan, M. Wax and T. Kailath, On spatial smoothing for direction-of-arrival estimation of coherent signals, *IEEE Transactions on Acoustics, Speech, and Signal Processing*, ASSP-33, 806–811, 1985.

34. N. Purswani, H. Singh and R. M. Jha, Active cancellation of probing in the presence of multiple coherent desired radar sources, *IEEE Applied Electromagnetics Conference AEMC 2009*, Kolkata, India, Paper No.: ATT-36-7160, 4 p., ISBN: 978-1-4244-4819-7/09, December 14–16, 2009.

25 Integrated Modelling, Simulation and Controller Design for an Autonomous Quadrotor Micro Air Vehicle

Nitin K. Gupta and N. Ananthkrishnan

CONTENTS

25.1 INTRODUCTION

In recent decades, there have been several programmes for the development of micro and mini aerial vehicles, and many of these have been deployed in various roles. Yet, there is much scope for innovation, and this is now a full-fledged area of research and development [1,2]. While the early push came more from military applications, recent interest has been fuelled by the prospect of their use in various civilian forums [3].

Modelling and simulation technologies for aerospace systems have today advanced to the level where much of the uncertainty in the development process can be ironed out during ground-based simulations, thus obviating the need for repeated trials and modifications [4]. Integrated modelling and simulation of all systems and sub-systems is a key milestone in any aerospace vehicle design and development programme [5]. Multidisciplinary design optimization methods may also be employed [6], provided appropriate fidelity tools for each disciplinary analysis suitable for micro air vehicles are available. Modelling and simulation, both software-in-loop and hardware-in-loop, is of course a prerequisite for control system design and evaluation [7]. Not surprisingly, there is much work in developing modelling and simulation tools that may be used in the micro air vehicle design and development cycle.

In this chapter, we describe the modelling, simulation and controller design for a quadrotor micro aerial vehicle that was designed and developed following a systematic approach under severe time and cost constraints. Beginning with a baseline configuration, based on existing components and components readily available in the market, all sub-systems are modelled and integrated into a single system model. This is used to test the performance of the vehicle, its stability and response to external disturbances and control inputs. The model is validated against an existing quadrotor system for which data are available. In parallel with the modelling and simulation exercise, components are procured and the quadrotor assembled. The model is then used to design a controller which is implemented in software and integrated with the model. Simulations of the closed-loop system are then carried out to verify its stability and response characteristics. Where necessary, design modifications are made, such as adjusting the vertical location of the centre of mass, and the exercise is repeated. The digital simulation is then integrated with a joystick and an open-source flight simulator to evaluate the vehicle flying qualities before being cleared. Multidisciplinary design optimization was not employed due to lack of time and non-availability of the requisite tools. The control law is embedded into an autopilot board and integrated with the vehicle airframe. The quadrotor is then mounted on a 3-DOF roll–pitch–yaw test rig specially built for this purpose. The test rig is used to assess the stabilizability and controllability of the vehicle in roll, pitch and yaw, and tune the controller PID gains, and also to correct for thrust asymmetry between the four rotors. Once the rig tests are satisfactory, the vehicle is test flown indoors, first manually, and then in autonomous mode. Some adjustment of the PID controller gains will be necessary as the pivot point on the test rig does not coincide with the vehicle centre of mass. Finally, flight tests are conducted outdoors in an open ground and in a built-up area.

25.2 QUADROTOR MICRO AIR VEHICLE

A quadrotor is a vertical take-off and landing (VTOL) vehicle which uses four rotors, typically attached at four ends of a cross frame. As shown in Figure 25.1, one pair of opposite rotors of the

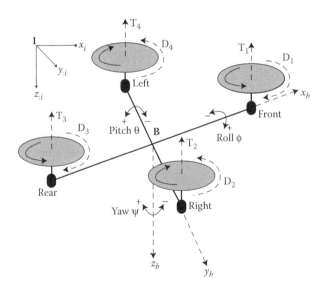

FIGURE 25.1 Configuration of quadrotor. (From Gupta, N.K., Goel, R. and Ananthkrishnan, N. In *Sp. Issue on Aerospace Avionics and Related Technologies* (Eds.: Raol, J.R. and Ajith Gopal), *Def. Sc. Jl.*, 61, 4, 337–345, 2011. With permission.)

quadrotor rotates clockwise, whereas the other pair rotates anticlockwise. Quadrotors enjoy a significant advantage over conventional helicopters as their rotors can use fixed-pitch blades, and no tail rotor is needed as it is able to avoid the yaw drift due to reactive torques. This makes their mechanical arrangement very simple, reduces gyroscopic effects, making them easy to build and test. They are also safer than other helicopters and more resistant to damage during flight testing. Quadrotor motion can be controlled by varying the relative RPM of each rotor, which changes its thrust and torque, as discussed below. On the flip side, the quadrotor configuration is inherently quite unstable and cannot be flown without automatic stabilization of angular motion about all three axes. However, for many unmanned flight vehicle tasks, quadrotors are ideal because of their many advantages over other helicopters and over fixed-wing aircraft. Consequently, there have been several references in the recent literature to quadrotor micro and mini aerial vehicle development [8,9]. Control of the quadrotor is achieved by commanding different speeds to different rotors, which in turn produces differential aerodynamic forces and moments. For hovering, all four rotors rotate at the same speed; for vertical motion, the speed of all four rotors is increased or decreased by the same amount, simultaneously. In order to pitch and move laterally in that direction, speed of rotors 1 and 3 is changed conversely. Similarly, for producing roll and corresponding lateral motion, speed of rotors 2 and 4 is changed conversely. To produce yaw, the speed of one pair of two oppositely placed rotors is increased while the speed of the other pair is decreased by the same amount. In this way, overall thrust produced is the same but differential drag moment creates yawing motion.

The quadrotor is designed using X-UFO (commercially available model) as the baseline. The brushed motors are replaced with brushless motors, and three-cell lithium polymer battery is used in order to improve the lifting capability and endurance of the vehicle. Further, to reduce the weight, the rotor shrouds and other non-critical structural components are removed. The electronics is replaced with custom autopilot hardware (including brushless motor controllers). A snapshot of the completed vehicle is shown in Figure 25.2. The autopilot used is a product developed in-house by Coral Digital Technologies (P) Ltd., Bangalore, India. The autopilot uses a single 16-bit 24 HJ series PIC microcontroller for all the computations, communication, and switching between manual and automodes. The navigation algorithm uses three gyros and two accelerometers to estimate orientation. Altitude and velocity are estimated using pressure sensors, and position is obtained using the GPS. A Zigbee modem is used for communication with the ground control station (GCS) during flight. The GCS receives data from the autopilot in order to monitor vehicle trajectory and other key parameters during flight, and is capable of updating way-points and PID gains during flight. An onboard SD card records several parameters at 50 Hz during flight for post-flight analysis of flight data. The autopilot hardware along with key interfaces and communication protocols used is illustrated in Figure 25.3. The final vehicle, capable of lifting autopilot and suitable battery pack, weighs 320 g (including battery, autopilot) and has an additional 40 g payload capability. The horizontal dimensions of this prototype are 65 cm each way, and the vertical dimension is about 15 cm.

FIGURE 25.2 Quadrotor platform: Modified X-UFO. (From Gupta, N.K., Goel, R. and Ananthkrishnan, N. In *Sp. Issue on Aerospace Avionics and Related Technologies* (Eds.: Raol, J.R. and Ajith Gopal), *Def. Sc. Jl.*, 61, 4, 337–345, 2011.)

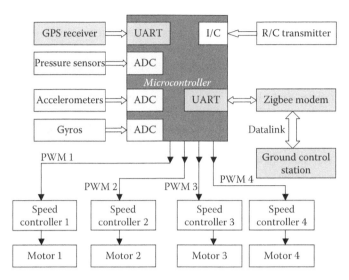

FIGURE 25.3 Hardware architecture and communication of the autopilot. (From Gupta, N.K., Goel, R. and Ananthkrishnan, N. In *Sp. Issue on Aerospace Avionics and Related Technologies* (Eds.: Raol, J.R. and Ajith Gopal), *Def. Sc. Jl.*, 61, 4, 337–345, 2011. With permission.)

25.3 DYNAMIC MODEL OF THE VEHICLE

In Figure 25.1, **I** is the inertial frame (subscript 'i') and **B** is the body fixed frame (subscript 'b'). The dynamic model is derived under the following assumptions [7,10], and is briefly presented below: (i) structure is rigid and has roll–pitch symmetry, (ii) centre of mass of the vehicle and the origin of **B** axis system coincide and (iii) the rotors are rigid in plane.

25.3.1 KINEMATICS MODEL

Using Euler angle parameterization, the orientation of the vehicle in space is given by the rotation of matrix A from frame **B** to **I**:

$$A = \begin{pmatrix} C\psi C\theta & C\psi S\theta S\varphi - S\psi C\varphi & C\psi S\theta C\varphi + S\psi S\varphi \\ S\psi C\theta & S\psi S\theta S\varphi + C\psi C\varphi & S\psi S\theta C\varphi - C\psi S\varphi \\ -S\theta & C\theta S\varphi & C\theta C\varphi \end{pmatrix}$$

where $C\varphi$, $S\varphi$, and so on are cos φ, sin φ, and so on.

Euler time derivatives are related to body angular rate as

$$[\dot{\varphi}\,\dot{\theta}\,\dot{\psi}]^T = M^{-1}[\omega_{xi} \quad \omega_{yi} \quad \omega_{zi}]^T$$
$$= M^{-1}A[\omega_{xb} \quad \omega_{yb} \quad \omega_{zb}]^T$$

(25.1)

where

$$M = \begin{pmatrix} \dfrac{C\psi}{C\theta} & \dfrac{S\psi}{C\theta} & 0 \\ -S\psi & C\psi & 0 \\ 0 & 0 & 1 \end{pmatrix}$$

Since we are only concerned about the velocity of the centre of mass located at origin of B, we can directly get body frame velocities from inertial frame velocities, using the transformation matrix as

$$[\dot{x}_b \; \dot{y}_b \; \dot{z}_b]^T = A^{-1}[\dot{x}_i \; \dot{y}_i \; \dot{z}_i]^T \tag{25.2}$$

25.3.2 FORCE EQUATIONS

Aerodynamic force (thrust) of a rotor can be shown to be proportional to the square of its rotational speed, and square of its radius, using momentum theory [11].

It is modelled as

$$T_i = C_1 \left(\frac{1 - 2\pi LCS}{P\alpha_i} + 2\pi \frac{\dot{z}_b - \omega_{zb}}{p\alpha_i} \right) \tag{25.3}$$

where $C_1 \; k_t \rho A_p \; \alpha_i^2 R_p^2$.

In Equation 25.3, $C = 1$ if $i = 1$ or 4, or $C = -1$ if $i = 2$ or 3, and $S = \omega_y b$ if $i = 1$ or 3, or $S = \omega_x b$ if $i = 2$ or 4. Forces due to translational velocity of quadrotor and wind disturbances are modelled as

$$F_{WI} = A[k_s(w_{xb} - \dot{x}_b)k_s(w_{yb} - \dot{y}_b)k_u(w_{zb} - \dot{z}_b)] \tag{25.4}$$

Hence, linear momentum balance in inertial frame gives

$$\begin{bmatrix} \ddot{x}_i \\ \ddot{y}_i \\ \ddot{z}_i \end{bmatrix} = - \begin{bmatrix} \omega_{xb} \\ \omega_{yb} \\ \omega_{zb} \end{bmatrix} \begin{bmatrix} \dot{x}_i \\ \dot{y}_i \\ \dot{z}_i \end{bmatrix} + g \begin{bmatrix} 0 \\ 0 \\ 1 \end{bmatrix} + \frac{F_{\omega i l}}{M} - \frac{T_1 + T_2 + T_3 + T_4}{m} A \begin{bmatrix} 0 \\ 0 \\ 1 \end{bmatrix} \tag{25.5}$$

25.3.3 MOMENT EQUATIONS

Aerodynamic drag moment of a rotor can be shown to be proportional to square of its rotational speed, and cube of its radius, using momentum theory [11].

It is modelled as

$$D_i = C_2 \left(1 - \frac{2\pi LCS}{P\alpha_i} + 2\pi \frac{\dot{z}_b - \omega_{zb}}{P\alpha_i} \right) \tag{25.6}$$

where $C_2 = k_d \rho A_p \alpha_i^2 R_p^3$.

Inertial counter torque, which is the reaction torque produced by a change in rotational speed of rotors is modelled as

$$I_{ct} = J_p(-\dot{\alpha}_1 + \dot{\alpha}_2 - \dot{\alpha}_3 + \dot{\alpha}_4) \tag{25.7}$$

Friction torque due to rotational motion is modelled as [12]

$$M_f - k_r[\dot{\phi} \; \dot{\theta} \; \dot{\psi}]^T \tag{25.8}$$

Disturbance torque due to uncontrollable factors (wind, etc.) is modelled as

$$\tau_d = [\tau_{xb}\,\tau_{yb}\,\tau_{zb}]^T \tag{25.9}$$

Gyroscopic moments, caused by a combination of rotations of four rotors and vehicle frame, are modelled as

$$M_g = J_p[\dot{\theta}_\alpha \quad \dot{\varphi}_\alpha \quad 0] \tag{25.10}$$

where $\alpha = -\alpha_1 + \alpha_2 - \alpha_3 + \alpha_4$.

Hence, the angular momentum balance in the body frame gives

$$\begin{bmatrix} \dot{\omega}_{xb} \\ \dot{\omega}_{yb} \\ \dot{\omega}_{zb} \end{bmatrix} = -J^{-1}\omega X J \begin{bmatrix} \omega_{xb} \\ \omega_{yb} \\ \omega_{zb} \end{bmatrix} - J^{-1}(M_f + \tau_d + M_g) + J^{-1} \begin{bmatrix} L(T_4 - T_2) \\ L(T_1 - T_3) \\ D_1 - D_2 + D_3 - D_4 + I_{ct} \end{bmatrix} \tag{25.11}$$

where

$$\omega X = \begin{bmatrix} 0 & -\omega_{zb} & \omega_{yb} \\ \omega_{zb} & 0 & -\omega_{xb} \\ -\omega_{yb} & \omega_{xb} & 0 \end{bmatrix}$$

25.3.4 MOTOR DYNAMICS

A standard DC motor with negligible inductance is modelled as

$$\begin{aligned} \tau_{mi=k_i}&\left(V_i - \frac{k_\upsilon \alpha_i}{G}\right)/R \\ \alpha_t &= \frac{G\tau_{mi} - D_i}{J_p} \end{aligned} \tag{25.12}$$

The dynamic model was coded in MATLAB® and a verification exercise was carried out using data from Ref. [13], reproduced in Table 25.1.

25.4 CONTROLLER DESIGN ASPECTS

In spite of having four independent actuators, the quadrotor is still an underactuated system. Hence, the controller design uses a two-loop structure: inner and outer. This is depicted in Figure 25.4. In the inner loop, four parameters: θ, φ, ψ and h, are independently controlled by suitably adjusting the speed of the four rotors. As described previously, speed of rotors 1 and 3 needs to be adjusted for controlling θ, 2 and 4 for controlling φ, and all four rotors for controlling ψ and h. In the outer loop, forward and sideward velocities ($\dot{X}_i'$ and $\dot{Y}_i'$) are controlled. These are velocities in a frame which is obtained by rotating the inertial frame by ψ, as shown below:

$$\begin{aligned} \dot{X}_i' &= \dot{X}_i \cos\psi + \dot{Y}_i \sin\psi \\ \dot{Y}_i' &= \dot{X}_i \sin\psi + \dot{Y}_i \cos\psi \end{aligned} \tag{25.13}$$

TABLE 25.1

Quadrotor Data for the Verification Case

Design Variable	Value	Units
m	4.493	kg
L	0.38	m
G	80/12	–
J_p	1.46E-3	kg/m²
I_{xx}, I_{yy}	0.177	kg/m²
I_{zz}	0.334	kg/m²
R_p	0.228	M
P	0.152	m
V	5	Volt
R	0.3	Ohm
k_l	0.008	–
k_d	0.0013	–
k_i	3.87E-3	Nm/ohm
k_v	0.0004	Volt/RPM
k_r	0.35	Nms/rad
k_s, k_u	1	Ns/m

Source: From Gupta, N.K., Goel, R. and Ananthkrishnan, N. In *Sp. Issue on Aerospace Avionics and Related Technologies* (Eds.: Raol, J.R. and Ajith Gopal), *Def. Sc. Jl.*, 61, 4, 337–345, 2011. With permission.

This choice is helpful when using a joystick to navigate the quadrotor on a simulator screen. In addition, actuator saturation and rate limits are also modelled. One of the objectives of the controller design is to avoid hitting these limits. Many different control design methodologies have been described in the literature for controlling the motion of the quadrotor [14–17]. However, from a practical viewpoint, PID controllers are the simplest and can be designed quickly. Also, standard procedures for tuning the PID gains are available and are well known [18]. Hence, four separate PID blocks are built corresponding to each variable to be controlled in the inner loop. Based on the error signals between the commanded and measured value a variable, the inner loop PIDs command differential voltages in order to reach the set point.

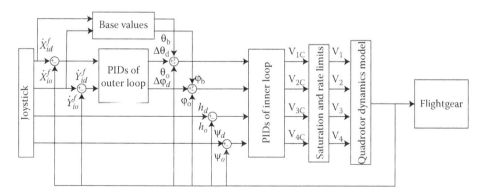

FIGURE 25.4 The structure of the two-loop controller. (From Gupta, N.K., Goel, R. and Ananthkrishnan, N. In *Sp. Issue on Aerospace Avionics and Related Technologies* (Eds.: Raol, J.R. and Ajith Gopal), *Def. Sc. Jl.*, 61, 4, 337–345, 2011. With permission.)

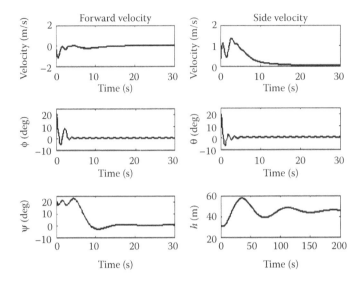

FIGURE 25.5 Tuning of inner loop parameters using simulation. (From Gupta, N.K., Goel, R. and Ananthkrishnan, N. In *Sp. Issue on Aerospace Avionics and Related Technologies* (Eds.: Raol, J.R. and Ajith Gopal), *Def. Sc. Jl.*, 61, 4, 337–345, 2011. With permission.)

For the outer loop, based on trim calculations or simulation results, look-up tables are formulated, to find base θ and φ required to fly at a particular velocity. These base angles (θ_b and φ_b) act as set points for the inner loop θ and φ controllers. The outer loop PIDs command additional pitch ($\Delta\theta$) and roll ($\Delta\varphi$) angles based on error velocities, so that the desired set point is maintained. This is shown in the block diagram of Figure 25.4. The PID gains are first tuned using the Ziegler–Nichols method and then tuned manually based on the desired simulation response. Some typical outputs from the simulation exercise are presented next. Figure 25.5 shows the simulation response used to tune the inner loop parameters. The task of the controller was to stabilize the orientation angles at zero and attain a height of 45 m starting from an initial condition of $h = 30$ m, $\varphi = \theta = \Psi = 18°$, $w = \tau = \omega = 0$. After the performance of inner loop controller is found satisfactory, simulations are carried out to tune the parameters of the outer loop controller. In the simulation response shown in Figure 25.6, the task of the controller is to obtain a forward velocity $\dot{X}_i' = 10$ m/s and $h = 50$ m. Initial altitude and rotor parameters are the same as above, but initial Euler angles are kept at zero. Due to symmetry, identical control parameters may be used for sideward velocity ($\dot{Y}_i'$). The PID gains finally selected are presented in Table 25.2. Interestingly, despite best efforts, altitude control remain sluggish, as seen from the h versus *time* sub-plots (please note the different time scale on this sub-plot) of Figures 25.5 and 25.6.

25.5 FLIGHT SIMULATION

The simulation model was implemented in real time on Simulink. For 3D visualization, Flightgear [19], an open-source flight simulator under GNU license is used. Interfacing Flightgear with MATLAB essentially requires sending the output vector from MATLAB/Simulink to Flightgear, as seen in the block diagram of Figure 25.4. A preconfigured interface block with Aerospace Blockset of MATLAB is used for this purpose. For visualization, the inertial frame coordinates are converted into latitude, longitude and height at that location, and orientation is specified by passing on the three Euler angles. We use the inbuilt model of a helicopter, Eurocopter Bo105, for 3D visualization due to lack of a quadrotor model in Flightgear. The leftmost block in Figure 25.4 represents joystick interface. A standard force feedback joystick is used for setting the desired values of $\dot{X}_i'$, $\dot{Y}_i'$, ψ and h. In this manner, a real-time simulator is set up which can be flown with a joystick and the quadrotor flight

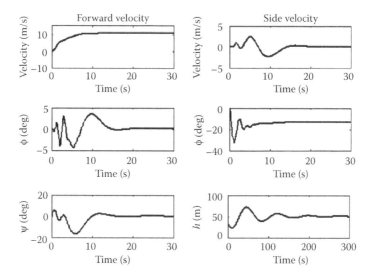

FIGURE 25.6 Tuning of outer loop parameters using simulation. (From Gupta, N.K., Goel, R. and Ananthkrishnan, N. In *Sp. Issue on Aerospace Avionics and Related Technologies* (Eds.: Raol, J.R. and Ajith Gopal), *Def. Sc. Jl.*, 61, 4, 337–345, 2011. With permission.)

TABLE 25.2

PID Gains Table

Control	Output	K_p	K_i	K_d
Θ	+V1, −V3	0.05	0	0.02
Φ	+V4, −V2	0.05	0	0.02
ψ	+V1, −V2, +V3, −V4	0.005	0	0.004
H	+V1, +V2, +V3, +V4	0.01	0.0007	0.01
$\dot{X}_i'$	$-\Delta\theta$	−1.5	−0.06	0
$\dot{Y}_i'$	$\Delta\varphi$	1.5	0.06	0

Source: From Gupta, N.K., Goel, R. and Ananthkrishnan, N. In *Sp. Issue on Aerospace Avionics and Related Technologies* (Eds.: Raol, J.R. and Ajith Gopal), *Def. Sc. Jl.*, 61, 4, 337–345, 2011. With permission.

can be observed on the Flightgear screen. A snapshot of the simulator screen is shown in Figure 25.7 with the model at $h = 1000$ m. For better speed and performance, MATLAB and Flightgear are run on separate PCs.

25.6 SIMULATION RIG AND FLIGHT TESTING

Following the simulator studies, the control law is loaded on to the autopilot, which is then integrated with the quadrotor airframe as described earlier. The autopilot design incorporates a switching logic between auto and manual model for safety during testing. The flow diagram of the autopilot in manual and auto modes is shown in Figure 25.8. Note that in either case, the rate feedback loop is part of the vehicle dynamics, that is, even in the manual mode the rate feedback controller continues to work. Due to its inherent instability, it would be impossible to fly any quadrotor manually if such a rate feedback is not provided to assist the pilot. The architecture also allows one to selectively

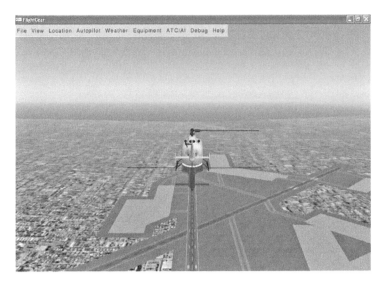

FIGURE 25.7 Simulation in Flightgear: A snapshot. (From Gupta, N.K., Goel, R. and Ananthkrishnan, N. In *Sp. Issue on Aerospace Avionics and Related Technologies* (Eds.: Raol, J.R. and Ajith Gopal), *Def. Sc. Jl.*, 61, 4, 337–345, 2011. With permission.)

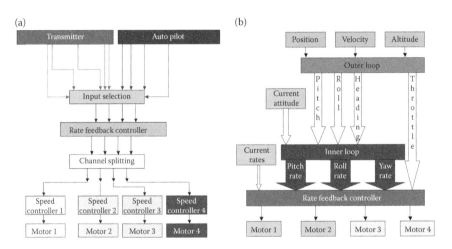

FIGURE 25.8 Schematic diagram of autopilot in (a) manual mode and (b) autonomous mode. (From Gupta, N.K., Goel, R. and Ananthkrishnan, N. In *Sp. Issue on Aerospace Avionics and Related Technologies* (Eds.: Raol, J.R. and Ajith Gopal), *Def. Sc. Jl.*, 61, 4, 337–345, 2011. With permission.)

assign any of the inputs to manual or auto modes. The outputs of the autopilot correspond to pitch, roll, yaw and thrust commands, which are converted into motor RPM commands by the *channel splitting* block.

25.6.1 Tests Using Rigs

The first set of experiments is conducted on a 3-DOF test rig, allowing only rotations, to test the attitude stabilization and orientation control of the vehicle. The test rig also allows the PID gains to be further fine tuned. The vehicle with autopilot is mounted on the 3-DOF test rig as shown in Figure 25.9. The autopilot (mounted on the vehicle) is connected to the PC to monitor attitudes and control action in real time. Sample results from the test rig for checking system stability in pitch and

FIGURE 25.9 Quadrotor vehicle on a 3-DOF test rig. (From Gupta, N.K., Goel, R. and Ananthkrishnan, N. In *Sp. Issue on Aerospace Avionics and Related Technologies* (Eds.: Raol, J.R. and Ajith Gopal), *Def. Sc. Jl.*, 61, 4, 337–345, 2011. With permission.)

roll are shown in Figure 25.10. In Figure 25.10, starting with autopilot in manual mode, the switch to auto mode is made at $t = 19$ s. During the entire experiment, the desired roll and pitch angles are kept at zero, and a constant thrust is maintained. The spikes in Figure 25.10 are the manual disturbance imparted to the system to test the controller robustness. The controller is observed to perform well and reject disturbances suitably in pitch and roll.

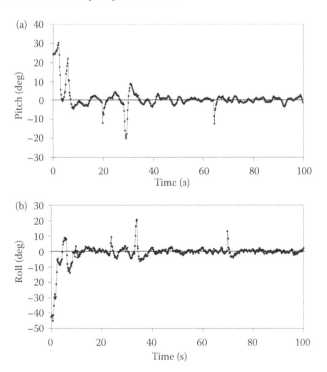

FIGURE 25.10 Test rig signals: (a) pitch angle and (b) roll angle. (From Gupta, N.K., Goel, R. and Ananthkrishnan, N. In *Sp. Issue on Aerospace Avionics and Related Technologies* (Eds.: Raol, J.R. and Ajith Gopal), *Def. Sc. Jl.*, 61, 4, 337–345, 2011. With permission.)

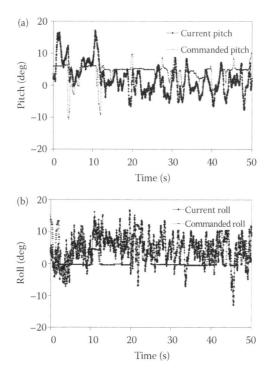

FIGURE 25.11   Signals measured in flight: (a) pitch angle and (b) roll angle signals. Dotted line: commanded value; solid line: measured value. (From Gupta, N.K., Goel, R. and Ananthkrishnan, N. In *Sp. Issue on Aerospace Avionics and Related Technologies* (Eds.: Raol, J.R. and Ajith Gopal), *Def. Sc. Jl.*, 61, 4, 337–345, 2011. With permission.)

25.6.2   FLIGHT TEST EXPERIMENTS

Following the rig tests, the controller is qualified to be flown in free flight. The free flight experiments are conducted first with the aim of achieving autonomous hover (attitude stabilization and control), and later for up and away flight. Only the thrust is controlled in manual mode by the pilot. Sample results for attitude stabilization and control are shown in Figure 25.11. The dotted lines represent the commanded value and the solid lines represent the measured attitude of the vehicle. Satisfactory attitude stabilization is achieved in the presence of disturbances. It is observed that the vehicle is able to keep itself afloat in hover with no active pilot inputs. However, a slow drift in position is observed because an outermost navigation loop is not yet implemented.

25.7   CONCLUDING REMARKS

This chapter has described an elaborate modelling and simulation exercise for a micro quadrotor which was then built and successfully flown. In addition to its use for control law design, the model formed an integral part of a ground-based simulation system. For this, the model was integrated with a joystick and a flight simulator and run in real time to make the experience quite realistic. It is believed that the modelling and simulation exercise helped cut down development time and cost by avoiding intermediate design changes and also by reducing the time taken for flight tests.

APPENDIX: NOMENCLATURE

A rotation matrix
A_p area of rotor

$\mathbf{B}$	body fixed frame
D	drag moment
DOF	degree of freedom
$F_w\mathbf{I}$	forces due to translational velocity and wind
G	gear ratio
h	altitude of vehicle
$\mathbf{I}$	inertial frame
I_{xx}, I_{yy}, I_{zz}	quadrotor moments of inertia about body x, y, z axes, respectively
I_{ct}	inertial counter torque
J	moment of inertia matrix
J_p	moment of inertia of a single rotor
k_d	aerodynamic drag moment coefficient
k_i	current constant of motor
k_r	friction coefficients due to rotational velocities
k_s, k_u	friction coefficients due to translational velocities
k_t	aerodynamic thrust coefficient
k_v	speed constant of motor
L	distance of the centre of the rotor from the origin
M	matrix relating Euler time derivatives with body angular rates
M_f	friction torque
M_g	gyroscopic moments
m	mass of the quadrotor vehicle assembly
P	pitch of the rotor blade
R	resistance of the motor
R_p	radius of the rotor
T	thrust force
V	voltage applied to the motor
w	wind velocity
$\dot{X}'_i\,\dot{Y}'_i$	forward and sideward velocity in the horizontal plane
$\dot{X}_\upsilon, \dot{Y}_i$	velocities in inertial frame
α	angular speed of rotor
φ, θ, ψ	Euler angles
φ_b, θ_b	commanded base Euler angles
τ	torque
τ_t	disturbance torque
τ_m	motor torque
ω	angular rates of quadrotor in body frame
1, 2, 3, 4	rotor number
b	coordinates in the body frame
c	commanded value
d	desired value
i	coordinates in the inertial frame
o	obtained value

ACKNOWLEDGEMENTS

The vehicle integration, rig and flight testing were carried out at Coral Digital Technologies (P) Ltd, Bangalore, India. The authors would like to acknowledge the contributions by Cdr. V. S. Renganathan and his colleagues for the development of autopilot hardware, for the modelling and simulation exercises, for the controller development, and for the flight tests. The autopilot used was an in-house product developed by Coral Digital Technologies (P) Ltd, Bangalore, India.

REFERENCES

1. Mueller, T.J., Kellogg, J.C., Ifju, P.G. and Shkarayev, S.V. (Eds.), *Introduction to the Design of Fixed-Wing Micro Air Vehicles: Including Three Case Studies*, AIAA Education Series, AIAA, Reston, Virginia, January 2007.
2. Mueller, T.J., On the birth of micro air vehicles, *International Journal of Micro Air Vehicles*, 2009, 1(1), 1–12.
3. Nonami, K., Prospect and recent research and development for civil use autonomous unmanned aircraft as UAV and MAV, *Journal of System Design and Dynamics*, 2007, 1(2), 120–128.
4. Reed, J.A., Follen, G.J. and Afjeh, A.A., Improving the aircraft design process using web-based modeling and simulation, *ACM Transactions on Modeling and Computer Simulation*, 2000, 10(1), 58–83.
5. Kumar, P.B.C., Gupta, N.K., Ananthkrishnan, N., Renganathan, V.S., Park, I.S. and Yoon, H.G., Modeling, Dynamic Simulation, and Controller Design for an Air-Breathing Combustion System, AIAA 2009-708, 47th AIAA Aerospace Sciences Meeting, 5–8 January 2009, Orlando, Florida.
6. Rohani, M.R. and Hicks, G.R., Multidisciplinary design and prototype development of a micro air vehicle, *Journal of Aircraft*, 1999, 36(1), 227–234.
7. Castillo, P., Lozano, R. and Dzul, A.E., *Modeling and Control of Mini-Flying Machines*, Springer-Verlag, New York, 2005, pp. 39–60.
8. Bouabdallah, S., Murrieri, P. and Siegwart, R., Design and Control of an Indoor Micro Quadrotor, International Conference on Robotics and Automation, New Orleans, USA, 2004.
9. Roberts, J.F., Stirling, T.S., Zufferey J.-C. and Floreano, D., Quadrotor Using Minimal Sensing for Autonomous Indoor Flight. 3rd US–European Competition and Workshop on Micro Air Vehicle Systems (MAV07) and European Micro Air Vehicle Conference and Flight Competition (EMAV2007), 17–21 September 2007, Toulouse, France.
10. Hamel, T., Mahony, R., Lozano, R. and Ostrowski, J., Dynamic Modeling and Configuration Stabilization for an X4-Flyer, 15th Triennial World Congress of International Federation of Automatic Control, Barcelona, Spain, 2002.
11. Bramwell, A.R.S., Done, G. and Balmford, D., *Bramwell's Helicopter Dynamics*, 2nd ed., Butterworth Heinemann, Oxford, UK, 2001.
12. Mahony, R., Altug, E. and Ostrowski, J.P., Control of a Quadrotor Helicopter Using Visual Feedback, Proceedings of 2002 IEEE Conference on Robotics and Automation, Washington DC, 2002, pp. 72–77.
13. Nice, E.B., Design of a Four Rotor Hovering Vehicle, MS thesis, Cornell University, 2004.
14. Tomlin, C.J., Jang, J.S., Waslander, S.L. and Hoffmann, G.M., Multi-Agent Quadrotor Testbed Control Design: Integral Sliding Mode vs. Reinforcement Learning, IEEE International Conference on Intelligent Robots and Systems, Alberta, Canada, 2005, pp. 468–473.
15. Tayebi, A. and McGilvray, S., Attitude stabilization of a VTOL quadrotor aircraft, *IEEE Transactions on Control Systems Technology*, 2006, 14, 562–571.
16. Kendoul, F., Lara, D., Coichot, I.F. and Lozano, R., Real-time nonlinear embedded control for an autonomous quadrotor helicopter, *Journal of Guidance, Control, and Dynamics*, 2007, 30(4), 1049–1061.
17. Das, A., Lewis, F. and Subbarao, K., Backstepping approach for controlling a quadrotor using Lagrange form dynamics, *Journal of Intelligent and Robotic Systems*, 2009, 56(1–2), 127–151.
18. Wang, Q.G., Lee, T.H., Fung, H.W., Bi, Q. and Zhang, Y., PID tuning for improved performance, *IEEE Transactions on Control Systems Technology*, 1999, 7(4), 457–465.
19. Anon. *Introduction to FlightGear: Open-Source Flight Simulator*, Under the GNU General Public License; FlightGear, http://www.flightgear.org, accessed December 3, 2009.
20. Gupta, N.K., Goel, R. and Ananthkrishnan, N. Design/development of mini/micro air vehicles through modeling and simulation: Case of an autonomous quadrotor. In *Sp. Issue on Aerospace Avionics and Related Technologies* (Eds.: Raol, J.R. and Ajith Gopal), *Def. Sc. Jl.*, 61(4), 337–345, 2011.

26 Determination of Impact and Launch Points of a Mobile Vehicle Using Kalman Filter and Smoother

Jitendra R. Raol and V. P. S. Naidu

CONTENTS

26.1 INTRODUCTION

Many defense situations require weapon delivery systems that should have an accurate prediction of launch and impact points of targets detected during a part of their trajectory. It is important to know from where the target had been launched and whether the target is from an enemy territory or friendly territory. Also it is important to know where the target would impact, for example, in the ocean or in a major city. If we are able to predict this with a very good accuracy, then any appropriate action to destroy the target could be initiated, if required. Traditionally, Kalman filter-based filters have been used to predict such trajectories. If the entire flight trajectory is observed then the estimation of the launch and impact points can be made straightforward using the Kalman filter. However, there are situations where the flight trajectory is available only for a part of the flight. In this case, the estimation of impact and launch points require an extrapolation of the estimated trajectory backward in time to predict the launch point and forward in time to predict the impact point. This can be accomplished using smoother and a filter. The process of determination of the launch and impact points can be carried out in two phases: (a) in phase I for the moving object that is being observed by sensors, the trajectory is estimated using point mass model and Kalman filter [1]. Since a smoothing technique can significantly improve the initial condition of the state estimates [2], this could be used to predict the Launch point very accurately and hence smoothed state estimates are obtained using Rauch–Tung–Striebel (R–T–S) smoother during phase I; and (b) in phase II, where the target is unobservable by sensors, the trajectory is estimated by using point mass model and covariance propagation calculation by forward propagation to get information about the impact point of the target [3]. The smoothed state estimate (at the launch point) is used to predict the launch point of the trajectory using backward integration. In this chapter, we use Kalman filter and forward prediction to predict the impact point of the flight trajectory. Similarly, Kalman filter and fixed interval R–T–S smoother [4] and backward integration are used to predict the launch point of the flight trajectory. PC MATLAB® is used to implement the algorithms and these algorithms are validated using simulated data of a target moving with constant acceleration. Simulated test results are presented in terms of the accuracy of prediction of the launch and impact points, time history comparisons, autocorrelation of the residuals with their theoretical bounds, innovation sequence with the theoretical bounds [5,6] and state error with bounds for the simulated data.

26.2 LAUNCH AND IMPACT POINT ESTIMATION

The two phases involved in the estimation of the launch and impact points of a flight trajectory are illustrated in Figure 26.1. A typical trajectory of a target from A to D is shown with A as the launch point and D as the impact point. It is assumed that measurements are available only between points B and C where the target is observed by the sensors, and it is proposed to estimate the impact point D and the launch point A. In the 1st phase, data between B and C are used to generate estimated target states and covariance matrices using Kalman filter. An R–T–S smoother is applied between points B and C utilizing the outputs of the Kalman filter to generate smoothed target states in a backward pass (i.e., going from C to B) operation. In the 2nd phase, point A is estimated by backward integration starting with the smoother output at point B and estimates at point D are obtained using the forward prediction of the Kalman filter output at point C.

26.2.1 KALMAN FILTER

Since the Kalman filter was originally developed by R. E. Kalman in 1960 [7], it has become a standard estimation algorithm that is extensively used in the development of tracking algorithms. Kalman filter provides minimum mean square error when the measurements are in Cartesian coordinates, measurements are independent and Gaussian distribution and target behaviour (i.e., target mathematical model) is known. The goal of the target tracking system is to form and maintain track on target of interest from the measurements provided by the sensors. Figure 26.2 shows an information flow

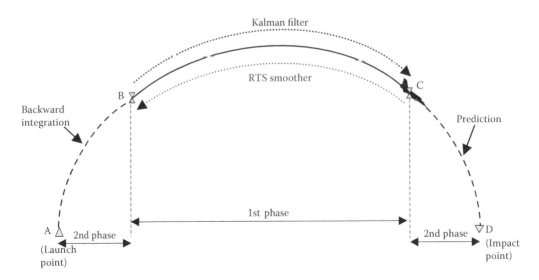

FIGURE 26.1 Typical flight trajectory.

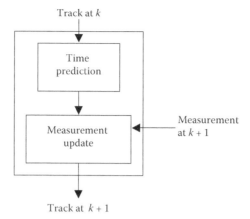

FIGURE 26.2   Information flow diagram of recursive target state estimation.

diagram of a typical recursive target tracking system. Its basic elements are time prediction and measurement update. First, the time prediction is carried out. Tracks from the frame k are predicted to frame $k + 1$ using the target process model. In measurement update, the measurements from frame $k + 1$ are incorporated into the predicted state estimate to obtain an improved estimate using measurement model.

26.2.1.1 Process Model

The dynamic model of the stochastic system must be constructed in the form of state-space representation in order to utilize it into the Kalman filter. Here, the Kalman filter is used as a target state estimator and does not perform any control functions. Hence, the control input to the system is not considered in the formation of system model. The stochastic process model provides dynamic relations among the states of a stochastic process. The model consists of a set of the first-order differential equations driven by random input noise. It can be represented in a generalized formulation in the discrete time form as [6,7]

$$X(k) = FX(k-1) + Gw(k), \quad X(0) = X_0 \tag{26.1}$$

where F: state transition matrix
     $X(k)$: state vector at time kth sample
     G: process noise gain matrix
     $w(k)$: process noise
     X_0: initial state vector

Kalman filter assumes that the process noise $w(k)$ is a white Gaussian with zero mean and known covariance. Also it is assumed that the process noise is independent from the system state $X(k)$. The initial system state $X(0)$ has known mean and covariance $P(0)$. We have the additional information as follows:

$$X(k) \approx N(\hat{X}(k), \hat{P}(k)) \tag{26.2}$$

$$E\left\{ \left[\hat{X}(k) - X(k) \right]\left[\hat{X}(k) - X(k) \right]^{\mathrm{T}} \right\} = \hat{P}(k)$$

$$w(k) \approx N(0, Q)$$

$$E\left\{w(k)w^{\mathrm{T}}(k)\right\} = Q\delta$$

$$E\left\{w(k)X^{\mathrm{T}}(k)\right\} = 0$$

where $E\{\}$: expectation operator
    $\hat{X}(k)$: estimated or mean value of $X(k)$
    $\hat{P}(k)$: symmetric, semidefinite covariance matrix of the state errors
    Q: process noise covariance matrix
    δ: Kronecker delta function

26.2.1.2   Measurement Model

The measurement model provides the relations between the system states and physical quantities measured by the sensor. It can be represented in a generalized formulation in the discrete time as

$$z_{\mathrm{m}}(k) = HX(k) + v(k) \tag{26.3}$$

where $z_{\mathrm{m}}(k)$: measurements obtained from the sensor
    H: observation matrix/sensor mathematical model
    $v(k)$: measurement noise

The measurement noise accounts for effects of the measurement system errors on the measured physical quantities. It is assumed that $z_{\mathrm{m}}(k)$ is a Gaussian-distributed random variable at each ample time and the measurement noise sequence $v(k)$ is a white Gaussian process with zero mean and known covariance and is independent of process noise sequence $w(k)$ as well as state vector $X(k)$. The following assumptions are pertinent:

$$v(k) \approx N(0, R) \tag{26.4}$$

$$E\left\{v(k)v^{\mathrm{T}}(k)\right\} = R\delta$$

$$E\left\{v(k)X^{\mathrm{T}}(k)\right\} = 0$$

$$E\left\{v(k)w^{\mathrm{T}}(k)\right\} = 0$$

where R: positive-definite covariance matrix of the measurement noise.
    Kalman filtering algorithm is given and carried out in two steps recursively:
    Step 1: Time update (effects of tracking process dynamics)

$$\tilde{X}(k\,|\,k-1) = F\hat{X}(k-1\,|\,k-1)\text{: state prediction} \tag{26.5}$$

$$\tilde{P}(k\,|\,k-1) = F\hat{P}(k-1\,|\,k-1)F^{\mathrm{T}} + GQG^{\mathrm{T}}\text{: prediction error covariance} \tag{26.6}$$

Step 2: Measurement update (effect of measurement)

$$\vartheta = z_{\mathrm{m}}(k) - H\tilde{X}(k\,|\,k-1)\text{: innovations} \tag{26.7}$$

$$S = H\tilde{P}(k\,|\,k-1)H^{\mathrm{T}} + R\text{: innovation covariance} \tag{26.8}$$

$$K = \tilde{P}(k \mid k-1)H^{\mathrm{T}}S^{-1}: \text{Kalman gain} \tag{26.9}$$

$$\hat{X}(k \mid k) = \tilde{X}(k \mid k-1) + K\vartheta: \text{state estimation/filtering} \tag{26.10}$$

$$\hat{P}(k \mid k) = (I - KH)\tilde{P}(k \mid k-1): \text{estimation error covariance} \tag{26.11}$$

The term ϑ is known as innovation because it provides new information when new measurement is available. It is observed that, Kalman gain K depends only on the statistics of process noise $w(k)$ and measurement noise $v(k)$ but not on the measurement $z_{\mathrm{m}}(k)$.

26.2.2 RTS: RAUCH, TUNG AND STREIBAL SMOOTHER

Smoothing is a non-real-time data processing scheme that uses all measurements starting from T (final time index) to 0 (to some initial time point) to estimate smoothed state of a target at a certain time t, where $0 \le t \le T$. The smoothing technique used in this work is a backward-pass sequel to a forward pass Kalman filter and utilizes the KF outputs: filter covariance, and Kalman gains. The RTS smoother formulation is used because of the simplicity [2,4,6]. The RTS recursions that generate smoothed target state estimates and corresponding error covariance are

$$\hat{X}(k \mid N) = F^{-1}\hat{X}(k+1 \mid N) \tag{26.12}$$

$$\hat{P}(k \mid N) = \hat{P}(k \mid k) + \hat{G}(k)(\hat{P}(k+1 \mid N) - \tilde{P}(k+1 \mid k))\hat{G}^{\mathrm{T}}(k) \tag{26.13}$$

where

$$\hat{G}(k) = \hat{P}(k \mid k)F^{\mathrm{T}}(k)\tilde{P}^{-1}(k+1 \mid k) \tag{26.14}$$

The recursion is a backward sweep from $k = (N_C - 1), \ldots, N_B$. The state estimates $\tilde{X}(k+1 \mid k)$ and the corresponding covariance matrix $\tilde{P}(k+1 \mid k)$ are found using the Kalman filter in the forward pass.

26.2.3 IMPACT POINT PREDICTION

Forward prediction is used (during 2nd phase) to predict the impact point D (whose index is N_D) of the flight trajectory [6]:

$$\tilde{X}(k+1 \mid k) = F\hat{X}(k \mid k) \tag{26.15}$$

$$\tilde{P}(k+1 \mid k) = F\hat{P}(k \mid k)F^{\mathrm{T}}, \quad k = N_C, \ldots, N_D \tag{26.16}$$

The initial values of $\hat{X}(k \mid k)$ and $\hat{P}(k \mid k)$ are the output of the Kalman filter at point C.

26.2.4 LAUNCH POINT PREDICTION

Backward integration is carried out (during 2nd phase) to predict the launch point of the flight trajectory [6]:

$$\hat{X}(k \mid N) = F\hat{X}(k+1 \mid N) \tag{26.17}$$

$$\hat{P}(k\,|\,N) = F^{-1}\hat{P}(K+1\,|\,N)(F^{\mathrm{T}})^{-1}, \quad k = (N_B - 1),\dots,1 \tag{26.18}$$

The initial values of $\hat{X}(k+1\,|\,N)$ and $\hat{P}(k+1\,|\,N)$ are the smoother or Kalman filter outputs at point B.

26.3   FILTER PERFORMANCE EVALUATION

The target tracking filter performance is ascertained by checking: (i) the estimated states and the bounds for convergence, (ii) residuals and their bounds for convergence and (iii) the autocorrelation of the residuals for whiteness. The tracking performance of Kalman filter has been evaluated by checking whether [5,6]: (i) the state error ($\Delta X = X - \hat{X}$) falls within the theoretical bounds of $\pm 2\sqrt{\hat{P}}$ (for simulated data only), (ii) the innovation sequence $\vartheta = z_\mathrm{m} - H\tilde{X}$ falls within the theoretical bounds of $\pm 2\sqrt{S}$, where $S = H\tilde{P}H^{\mathrm{T}} + R$ and (iii) the autocorrelation (cor) of residuals falls within the theoretical bounds of $\pm \dfrac{1.96}{\sqrt{N}}$, where '$N$' is the number of samples.

26.4   RESULTS AND DISCUSSION

The Kalman filter and RTS smoother techniques are implemented and demonstrated using PC MATLAB. The performance of the algorithm is evaluated using simulated test data of a target moving with constant acceleration. For generating the simulated test data, the state variables of the target considered are x-position, x-velocity, x-acceleration; y-position, y-velocity, y-acceleration and z-position, z-velocity, z-acceleration, that is, the target state vector is represented by $X = [x, \dot{x}, \ddot{x}, y, \dot{y}, \ddot{y}, z, \dot{z}, \ddot{z}]$. Equations 26.1 and 26.3 give the model of the system considered. Sampling time $T = 0.1$ s is used to generate simulated data for the duration of 50 s. The transition matrix and the other related matrices are given by

$$
F = \begin{pmatrix}
1 & T & \frac{T^2}{2} & 0 & 0 & 0 & 0 & 0 & 0 \\
0 & 1 & T & 0 & 0 & 0 & 0 & 0 & 0 \\
0 & 0 & 1 & 0 & 0 & 0 & 0 & 0 & 0 \\
0 & 0 & 0 & 1 & T & \frac{T^2}{2} & 0 & 0 & 0 \\
0 & 0 & 0 & 0 & 1 & T & 0 & 0 & 0 \\
0 & 0 & 0 & 0 & 0 & 1 & 0 & 0 & 0 \\
0 & 0 & 0 & 0 & 0 & 0 & 1 & T & \frac{T^2}{2} \\
0 & 0 & 0 & 0 & 0 & 0 & 0 & 1 & T \\
0 & 0 & 0 & 0 & 0 & 0 & 0 & 0 & 1
\end{pmatrix}
\qquad
G = \begin{pmatrix}
\frac{T^3}{6} & 0 & 0 \\
0 & \frac{T^2}{2} & 0 \\
0 & 0 & T \\
\frac{T^3}{6} & 0 & 0 \\
0 & T & 0 \\
0 & 0 & T \\
\frac{T^3}{6} & 0 & 0 \\
0 & T & 0 \\
0 & 0 & T
\end{pmatrix}
$$

$$
H = \begin{pmatrix}
1 & 0 & 0 & 0 & 0 & 0 & 0 & 0 & 0 \\
0 & 0 & 0 & 1 & 0 & 0 & 0 & 0 & 0 \\
0 & 0 & 0 & 0 & 0 & 0 & 1 & 0 & 0
\end{pmatrix}
$$

$$
R = \begin{pmatrix}
r_x^2 & 0 & 0 \\
0 & r_y^2 & 0 \\
0 & 0 & r_z^2
\end{pmatrix}
\qquad
Q = \begin{pmatrix}
q_x^2 & 0 & 0 \\
0 & q_y^2 & 0 \\
0 & 0 & q_z^2
\end{pmatrix}
$$

The initial state vector is $X(0) = [1200.0, 20, 0, 3000.0, 0, 0, 0, 20, -0.4]$. Measurements of x-position, y-position and z-position are generated using Equation 26.3 by adding random noise with variance of $r_x^2 = r_y^2 = r_z^2 = 1$. The process noise variance used in the simulation is $q_x^2 = q_y^2 = q_z^2 = 0.0001$. Figure 26.3 shows the x position state estimates using the data between the points B and C assuming that measured data is available between 10 and 40 s. The state estimates with prediction of launch point (of point A) using the RTS smoother output and impact point (point D) using the filter output are compared with the true states. It is clear from Figure 26.3 that the launch and impact point predictions are very close to the true values. Similar inferences are made for y and z positions from Figures 26.4 and 26.5, respectively. The innovations, state errors and autocorrelation of residuals with bounds are shown in Figures 26.6 through 26.8. The innovations, state errors and auto correlation of all axes are within the theoretical bounds, which indicate satisfactory filter performance. Although, it is observed from Figure 26.6 that some residuals are out of bounds, the bounds are 1-sigma. However, these residuals are fluctuating around zero as shown in Table 26.1 (mean of residuals). The state errors in x-, y- and z-positions using Kalman filter and RTS smoother are shown in Figure 26.9. It is clear that smoother provides very smooth states compare to KF states. The percentage fit error and mean state error in x, y and z positions are shown in Table 26.1. These errors are very small indicating good filter performance. A number of cases are presented where it is assumed that measured data are available for different time segments (indicated by points B and C in Tables 26.1 and 26.2). The results of impact point prediction using the Kalman filter output at point C and forward prediction (FP) is shown in Table 26.2. Results are presented in terms of the predicted impact point and the standard deviation of the impact point prediction. It is clear that even with only 10 s of data (case 6) the impact point prediction is very accurate thereby proving that when the model of the target and the noise statistics are known accurately, the prediction of the impact point using Kalman filtered output would be accurate. The results of launch point prediction using the RTS smoother output at point B and backward integration (BI) are presented in Table 26.2. The true impact point, since the simulation data are known, (point D at 50th Sec.) and the launch point (point A at 0th Sec.) are shown within

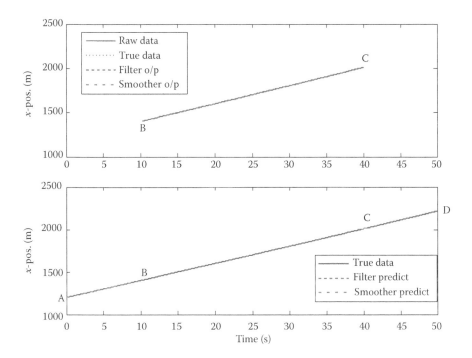

FIGURE 26.3 Estimated x-position using Kalman filter and RTS smoother.

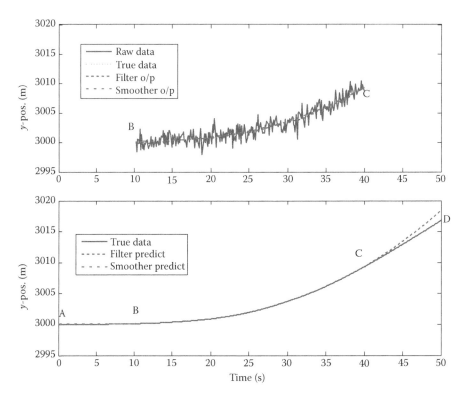

FIGURE 26.4 Estimated y-position using Kalman filter and RTS smoother.

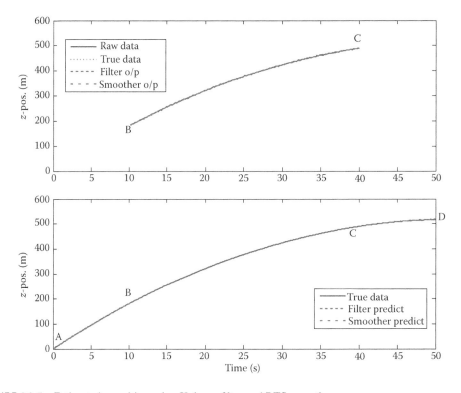

FIGURE 26.5 Estimated z-position using Kalman filter and RTS smoother.

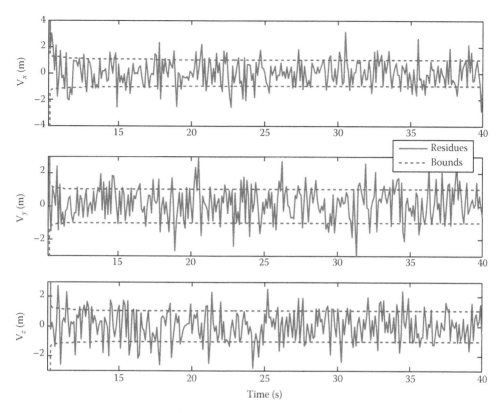

FIGURE 26.6 Innovation sequence with theoretical bounds.

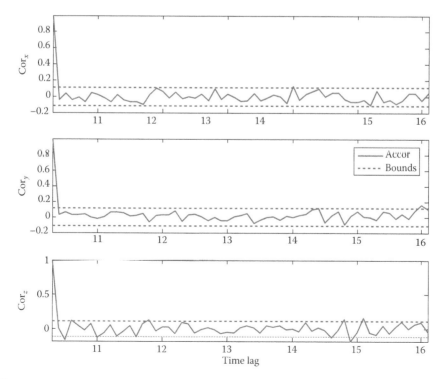

FIGURE 26.7 Autocorrelation with theoretical bounds.

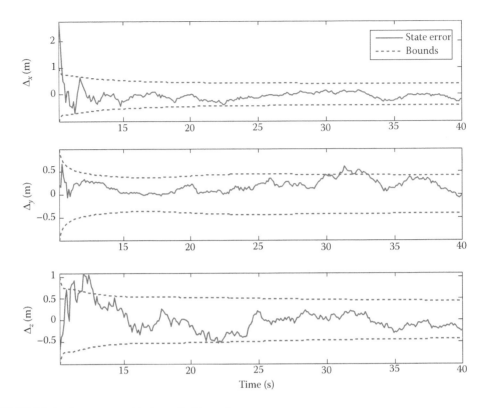

FIGURE 26.8 State errors with theoretical bounds.

parentheses in Table 26.2. The predicted launch or impact point's average uncertainties with respect to the duration of measurements available are presented in Table 26.2. It is observed from Table 26.2 that the prediction accuracy degrades when the availability duration of the measurements is less or/and the duration of prediction of launch/impact point (i.e., the length of the time for which the prediction is required to be made) is more (see Case 6). It is clear that in all the cases, the use of the RTS smoother results in prediction with better accuracy. The state errors along with their theoretical bounds are shown in Figures 26.10 and 26.11. The errors are within the bound during filtering and smoothing (1st phase) since measurements are available and the errors are crossing the bounds during 2nd phase due to lack of measurements hence no updation. The simu-

TABLE 26.1
Percentage Fit Error and Mean State Error

Case	Time at (s) B	C	PFE in x-pos.	y-pos.	z-pos.	Mean of Residuals x-pos.	y-pos.	z-pos.
1	0	50	0.056	0.033	0.261	0.039	0.080	−0.038
2	0.5	49.5	0.056	0.033	0.261	−0.018	0.072	−0.042
3	2.5	47.5	0.055	0.033	0.258	−0.010	0.056	−0.027
4	5	45	0.055	0.033	0.257	0.042	0.074	−0.088
5	10	40	0.054	0.033	0.253	−0.006	0.171	0.022
6	20	30	0.051	0.033	0.242	0.0182	−0.175	0.074

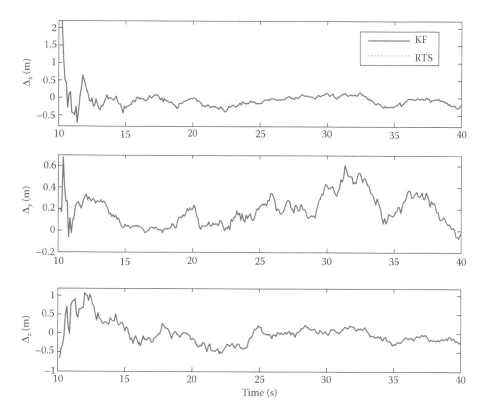

FIGURE 26.9 State errors with Kalman filter and RTS smoother.

TABLE 26.2
Estimated Launch and Impact Points

	Time at (s)		Impact point Point D (2216.0, 3016.0, 516.8), at 50 s			Launch Point—Point A (1201, 3000, 1.998) at 0 s		
Case	B	C	x-pre.	y-pre.	z-pre.	x-pre.	y-pre.	z-pre.
1	0	50	2216.796 ± 0.153	3016.758 ± 0.154	516.876 ± 0.168	1202.000 ± 0.105	3000.033 ± 0.095	1.912 ± 0.115
2	0.5	49.5	2216.743 ± 0.171	3016.705 ± 0.172	516.883 ± 0.187	1201.949 ± 0.109	3000.051 ± 0.095	1.943 ± 0.148
3	2.5	47.5	2216.792 ± 0.268	3016.958 ± 0.267	517.06 ± 0.28	1201.841 ± 0.148	2999.986 ± 0.1	1.796 ± 0.19
4	5	45	2217.164 ± 0.432	3017.278 ± 0.432	517.138 ± 0.450	1201.973 ± 0.176	2999.956 ± 0.099	1.832 ± 0.263
5	10	40	2218.598 ± 0.941	3018.415 ± 0.941	518.529 ± 0.982	1201.936 ± 0.251	3000.087 ± 0.119	1.311 ± 0.571
6	20	30	2214.194 ± 2.705	3013.802 ± 2.819	529.428 ± 7.044	1204.748 ± 1.212	3002.515 ± 0.589	18.508 ± 6.923

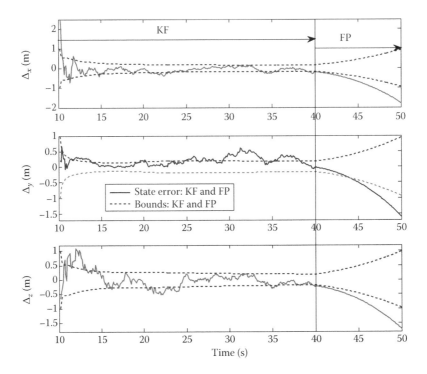

FIGURE 26.10 State errors with theoretical bounds during KF and forward prediction.

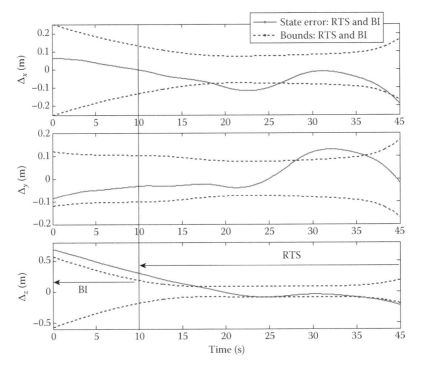

FIGURE 26.11 State errors with theoretical bounds during RTS and backward integration.

lated data of a target moving with constant velocity were chosen for validation of the technique/ algorithm. The application of this technique to real data could involve nonlinear state equations and manoeuvring targets. This problem would require an extensive modelling effort.

26.5 CONCLUSIONS

Kalman filter and RTS smoother were implemented and demonstrated in PC MATLAB and their performance was studied using simulated test data. The RTS smoother was found to generate more accurate target state estimates, which led to better launch point prediction accuracies. This requires further study. It will be worthwhile to explore the application of the presented approach to the cases where the state equations are highly non-linear and the data spans are shorter. The MATLAB code developed as a part of this work is available to users at http://www.crcpress.com/product/ isbn/9781439863008.

REFERENCES

1. Eli, B., *Tracking and Kalman Filtering Made Easy*, John Wiley & Sons, New York, 1998.
2. Arthur, G., *Applied Optimal Estimation*, 7th Edition, The MIT Press, MA, 1982.
3. Candillo, G.P., Mrstik, A.V., Plambeck, Y., A track filter for reentry objects with uncertain drag, *IEEE Transactions on Aerospace and Electronic Systems*, 35(2), 394–409, 1999.
4. Bierman, G.J., A new computationally efficient fixed—Interval, discrete-time smoother, *Automatica* 19(5), 505–511, 1983.
5. James, V.C., *Signal Processing, The Model Based Approach*, Second printing. McGraw-Hill International Edition, Singapore, 1987.
6. Naidu, V.P.S., Girija, G. and Raol, J.R., Estimation of launch and impact point of a flight trajectory using U-D Kalman filter/smoother, *Defense Science Journal*, 5(4), 451–463, 2006.
7. Kalman, R.E., A new approach to linear filtering and prediction problems, *Trans. of the ASME, Journal of Basic Engineering*, 82 (Series D), 35–45, 1960.

27 Novel Stability Augmentation System for Micro Air Vehicle
Towards Autonomous Flight

M. Meenakshi and Seetharama M. Bhat

CONTENTS

27.1 INTRODUCTION

In the last decade, the desire for low-cost, portable, low-altitude aerial surveillance has driven the development and testing of smaller and smaller aircrafts, known as micro air vehicles (MAVs) [1,2]. MAVs have a wide range of applications both in commercial and in military applications. Payloads ranging from cameras to acoustics to weapons, MAVs allow for a various applications within the three-dimensional freedom of the sky. MAVs are remotely controlled by radio control system and are difficult to fly due to its unconventional design and sometimes due to unpredictable flight characteristics. Another limitation of radio-controlled aircraft is the range of the pilot's sight. Hence, to alleviate the necessity of an expert pilot and to operate the MAV in a wide variety of missions, a robust optimal flight controller that provides an acceptable stability and performance over the entire flight envelope, is a must. Hence, design and realization of the robust performance controller to achieve the complete autonomy of an MAV is an active field of the research [3].

Over the last three decades, the quest for robust controllers resulted in a widespread search for controllers, which robustly stabilize the system [4,5]. Abundant literature is available on the design of robust controller using H_∞ [6,7] methods and μ synthesis [8,9]. However, H_∞ controller is the result of the worst-case design technique; hence, such methods lean too heavily on stability robustness and sacrifice an adequate view of performance. Very often, stable and smooth flight of an MAV is essential for surveillance and aerial survey missions. H_2 performance level is an indicator of real-life performance of a robust controller [10,11]. However, a significant shortcoming in Refs. [10,11] is that these techniques generally lead to higher-order controllers. However, Stability of the reduced-order controller and that of the closed-loop system cannot be guaranteed, and optimality may also be lost [12]. Very little effort is directed towards the design and development of a fixed low-order robust H_2 controller, particularly for the MAVs. Robust performance is a major concern for the proper functioning of the MAVs. This chapter extends [13] and documents the design and evaluation of a robust fixed-order H_2 controller for a discretized longitudinal dynamics of Sarika-1. In Refs. [14,15], an optimal steady-state fixed-order dynamic compensator is obtained from the solution of four matrix equations coupled through a projection operator, whose rank is precisely equal to the order of the compensator. In Ref. [15], four matrix equations known as strengthened discrete optimal projection equations (SDOPEs) are used to design a reduced-order LQG compensator in time domain. The weighting matrices for state and control variables are constant. In contrast, selection of frequency-dependent weights enables the designer to shape the responses tighter in pre-specified frequency ranges by giving them larger weights, clearly at the expense of larger errors at other frequency ranges that are of lesser importance. Therefore, in this chapter, the SDOPEs [15] are used to design robust fixed-order H_2 controller with the frequency-dependent weights. The design allows the required trade-off between sensitivity and control sensitivity of the closed-loop system. The main contribution of this chapter is design and real-time validation of a single controller, which meets the flying qualities, disturbance rejection and noise attenuation specifications for the entire flight envelope (flight speeds ranging from 15 to 25 m/s) of Sarika-1.

To appraise the performance and robustness of the closed-loop system in the face of plant uncertainty, *a posteriori* robustness analysis is done. To obtain the quantitative assessment of stability and performance of the closed-loop system, Lyapunov-based quadratic stability/performance tests involving parameter-dependent Lyapunov functions are performed using linear matrix inequalities (LMIs). Corresponding affine parameter-dependent state matrices [16] are developed by assuming uncertainties on the aerodynamic parameters. Finally, the controller is validated by means of offline simulation and real-time hardware-in-loop simulation (HILS) experimental setup. This chapter presents a generic design methodology of robust fixed-order H_2 controller and onboard computer for MAVs. The efficacy of the proposed method is demonstrated by designing a fixed-order robust H_2 stability augmentation system for an MAV, named Sarika-1. Strengthened discrete optimal projection equations, which approximate the first-order necessary optimality condition, are used for the controller design. Effect of low-frequency gust disturbance and high-frequency sensor noise is alleviated through the output sensitivity and control sensitivity minimization. Digital signal processor (DSP)-based onboard computer named Flight Instrumentation Controller (FIC) is designed to operate under automatic or manual mode. The controller is ported on to the flight computer, and subsequently, it is validated through the real-time hardware-in-loop-simulation. The responses obtained from the hardware-in-loop-simulation compares well with those obtained from the off-line simulation.

27.2 STANDARD H_2 OPTIMAL CONTROL FORMULATION

Figures 27.1 and 27.2 show the block diagram of the standard H_2 optimal control formulation. The signals w and z are the exogenous inputs and performance variables, respectively, y is the measured

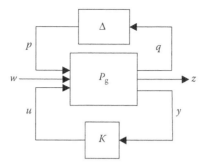

FIGURE 27.1   Standard H_2 control formulation.

variable, and u is the control input. G represents the plant and P_g is the generalized plant representing the concatenated vehicle dynamics with all the weighting functions. K_1 represents the sensor plus amplifier gain and K is the controller to be designed. The set of all possible uncertainties is Δ, grouped into a single block-diagonal finite-dimensional linear time-invariant system. W_1 and W_2, respectively, are the weighting matrices for the output sensitivity matrix function S_o and control sensitivity transfer function $S_i K$ (S_i = input sensitivity matrix). Let T_{zw} denote the resulting closed-loop transfer function from w to z. The objective of H_2 design is to find an optimal controller amongst all admissible controllers that yield internally stable closed-loop system so as to minimize the norm-2 of T_{zw}, that is,

$$\min \left\| T_{zw} \right\|_2 = \min \left\| \begin{bmatrix} W_1 S_0 \\ W_2 S_i K \end{bmatrix} \right\|_2 \tag{27.1}$$

where $S_o = (I + GKK_1)^{-1}$ and $S_i = (I + KK_1G)^{-1}$.

At low frequencies (where a satisfactory knowledge of the plant is assumed), performance requirements are most important. The weighting function W_1 determines the performance properties of the closed-loop system. In other words, W_1^{-1} gives an upper bound on the sensitivity.

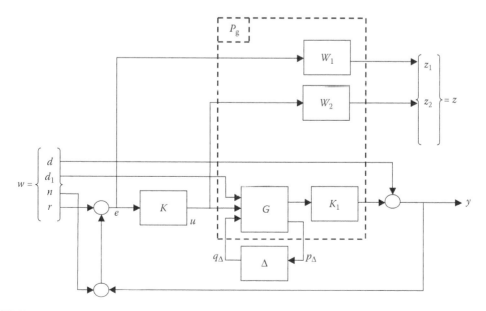

FIGURE 27.2   Interconnections of plant, controller and weighting matrices.

The weighting function W_2 determines robustness characteristics of the closed-loop system to unstructured additive uncertainties; that is, W_2^{-1} acts as an upper bound on the controller gain (S_1K). The larger the S_1K at any given complex frequency $s = j\omega$, the smaller the additive model error that will be required to destabilize the system [17]. Therefore, knowledge of the likely size of the additive model error dictates the safe upper bound on S_1K.

Let the state-space representation of the weights W_1 and W_2 are given by

$$x_{w1}(k+1) = A_{w1}x_{w1} + B_{w1}e; \quad z_1(k) = C_{w1}x_{w1} + D_{w1}e \tag{27.2}$$

$$x_{w2}(k+1) = A_{w2}x_{w2} + B_{w2}u; \quad z_2(k) = C_{w2}x_{w2} + D_{w2}u \tag{27.3}$$

where e is the input to the controller K. Using linear fractional transformation theory [17], the generalized/augmented plant with weighting cost functions are reformulated as

$$\begin{bmatrix} z \\ \overline{y} \end{bmatrix} = \begin{bmatrix} P_{g11} & P_{g12} \\ P_{g21} & P_{g22} \end{bmatrix}\begin{bmatrix} w \\ u \end{bmatrix} = \begin{bmatrix} W_1 & -W_1K_1G \\ 0 & W_2 \\ \hline I & -K_1G \end{bmatrix}\begin{bmatrix} w \\ u \end{bmatrix} \tag{27.4a}$$

or

$$\begin{bmatrix} x(k+1) \\ z(k) \\ y(k) \end{bmatrix} = \begin{bmatrix} \Phi & B_1 & \Gamma \\ \hline E_1 & E_3 & E_2 \\ C & D_1 & D \end{bmatrix}\begin{bmatrix} x(k) \\ w(k) \\ u(k) \end{bmatrix} \tag{27.4b}$$

where Φ, Γ, C, D, B_1 and D_1 are $n \times n$, $n \times m_1$, $p_1 \times n$, $p_1 \times m_1$, $n \times m_2$ and $p_1 \times m_2$ generalized plant matrices respectively, given by

$$\Phi = \begin{bmatrix} A_p & 0 & 0 \\ -B_{w1}K_1C_p & A_{w1} & 0 \\ 0 & 0 & A_{w2} \end{bmatrix}; \quad B_1 = \begin{bmatrix} 0 & 0 & 0 & B_{p2} \\ B_{w1} & -B_{w1}K_1 & -B_{w1} & -B_{w1}K_1D_{p2} \\ 0 & 0 & 0 & 0 \end{bmatrix} \tag{27.5a}$$

$$\Gamma = \begin{bmatrix} B_{p1} \\ -B_{w1}K_1D_{p1} \\ B_{w2} \end{bmatrix}; \quad E_1 = \begin{bmatrix} -D_{w1}K_1C_p & C_{w1} & 0 \\ 0 & 0 & C_{w2} \end{bmatrix}; \quad E_2 = \begin{bmatrix} -D_{w1}K_1D_{p1} \\ D_{w2} \end{bmatrix} \tag{27.5b}$$

$$E_3 = \begin{bmatrix} D_{w1} & -D_{w1}K_1 & -D_{w1} & -D_{w1}K_1D_{p2} \\ 0 & 0 & 0 & 0 \end{bmatrix}; \quad C = [-K_1C_p \quad 0 \quad 0] \tag{27.5c}$$

$$D_1 = [I \quad -K_1 \quad I \quad -K_1D_{p2}]; \quad D = [-K_1D_{p1}] \quad \text{and} \quad w = [r \quad d \quad n \quad w_g]^{\mathrm{T}} \tag{27.5d}$$

For a given matrix P_g, the requirements for the existence of H_2 controller is [14,17]

1. (Φ, Γ) is stabilizable, and (Φ, C) is detectable.
2. The feed-forward matrix E_3 must be zero.
3. E_2 is full column rank and D_1 is full row rank matrix.

Since fixed-low-order controller designs do not address the optimal order of a compensator, for the initial design exercise, normally, controllers of varying orders are designed and the least-order controller is selected so that the stability and performances are not compromised. For proving the adequacy of optimality of controller, performance of low-order compensator is compared with the full/high-order compensator. Hence, the goal of the optimal reduced-order stable dynamic compensation problem is to design an n_cth-order stable dynamic compensator for unknown A_c, B_c, C_c and D_c:

$$x_c(k+1) = A_c x_c(k) + B_c y(k) \tag{27.6a}$$

$$u(k) = C_c x_c(k) + D_c y(k) \tag{27.6b}$$

so as to satisfy the following two design criteria:

1. The closed-loop system corresponding to Equations 27.4 and 27.6 given by

$$x_{CL}(k+1) = A_{CL} x_{CL}(k) + D_{CL} w(k); \quad z(k) = E_{CL} x_{CL} \tag{27.7}$$

   where

$$x_{CL} = \begin{bmatrix} x \\ x_c \end{bmatrix}; \quad A_{CL} = \begin{bmatrix} \Phi + \Gamma D_c C & \Gamma C_c \\ B_c C & A_c \end{bmatrix}; \quad D_{CL} = \begin{bmatrix} \Gamma D_c D_1 \\ B_c D_1 \end{bmatrix} \quad \text{and} \quad E_{CL} = \begin{bmatrix} E_1 & E_2 C_c \end{bmatrix} \tag{27.8}$$

   is asymptotically stable.
2. The norm-2 of the closed-loop transfer function matrix, T_{zw}, where

$$T_{zw} = E_{CL}(zI - A_{CL})^{-1} D_{CL} \tag{27.9}$$

   is minimized. Objective 2 can be rewritten as minimization of

$$J(A_c, B_c, C_c, D_c) = \lim_{N \to \infty} \frac{1}{N} \left\{ \sum_{k=1}^{N} T_{zw}^*(e^{j\omega k}) T_{zw}(e^{j\omega k}) \right\},$$

   or

$$J(A_c, B_c, C_c, D_c) = \lim_{N \to \infty} \frac{1}{N} \left\{ \sum_{k=1}^{N} (x^*(e^{j\omega k}) Q(e^{j\omega k}) x(e^{j\omega k}) + u^*(e^{j\omega k}) R(e^{j\omega k}) u(e^{j\omega k})) \right\} \tag{27.10}$$

Q and R are obtained from the frequency-dependent weighting matrices W_1 and W_2 [18]. The controller (A_c, B_c, C_c, D_c), which minimizes the objective function (Equation 27.10), is obtained by solving the SDOPEs as defined in Theorem 1.1.

Theorem 1.1 [15]
A stabilizing compensator (A_c, B_c, C_c, D_c) satisfies the first-order necessary optimality conditions for optimal reduced-order LQG compensation and is minimal if and only if there exist non-negative symmetric $n \times n$ matrices $P, S, \hat{P}, \hat{S}$ such that for some projective factorization (G, M, H) of ($\hat{P}\hat{S}$),

$$A_c = H[\Phi - K_p C - \Gamma L_s]G^T; \quad B_c = HK_p; \quad C_c = L_s G^T; \quad D_c = 0 \tag{27.11}$$

(Note: For better computational efficiency, D_c is assumed to be zero.) and such that $P, S, \hat{P}, \hat{S}$ and τ satisfy the SDOPEs,

$$P = \Phi P \Phi^T - \Sigma_P^1 + V + \tau_\perp \Psi^1_{S,\hat{P},P} \tau_\perp^T \tag{27.12a}$$

$$S = \Phi^T S \Phi - \Sigma_S^2 + Q + \tau_\perp^T \Psi^2_{P,\hat{S},S} \tau_\perp \tag{27.12b}$$

$$\hat{P} = \frac{1}{2}\left[\tau \Psi^1_{S,\hat{P},P} + \Psi^1_{S,\hat{P},P} \tau^T \right] \tag{27.12c}$$

$$\hat{S} = \frac{1}{2}\left[\tau^T \Psi^2_{P,\hat{S},S} + \Psi^2_{P,\hat{S},S} \tau \right] \tag{27.12d}$$

provided, rank $(\hat{P}) = $ rank $(\hat{S}) = $ rank $(\hat{P}\hat{S}) = n_c$, and $\tau^2 = \tau = (\hat{P}\hat{S})(\hat{P}\hat{S})^{\#}$ (idempotent matrix) with $V \geq 0$, $W > 0 = $ Covariance of $B_1 w(k)$ and $D_1 w(k)$,

$$\Sigma_P^1 = K_p W_p K_p^T; \ \Sigma_S^2 = L_s^T R_s L_s; \ \Phi_P^1 = \Phi - K_p C; \ \Phi_S^2 = \Phi - \Gamma L_s$$

$$\Psi^1_{S,\hat{P},P} = \Phi_S^2 \hat{P}(\Phi_S^2)^T + \Sigma_P^1; \ \Psi^2_{P,\hat{S},S} = (\Phi_P^1)^T \hat{S}\Phi_P^1 + \Sigma_S^2 \ \tau_\perp = I_n - \tau$$

$$W_p = W + CPC^T; \ R_s = R + \Gamma^T S \Gamma; \ K_p = \Phi P C^T W_p^{-1} \text{ and } L_s = R_s^{-1} \Gamma^T S \Phi$$

The optimality exists if the steady-state cost, $J(A_c, B_c, C_c, D_c)_\infty = J(A_c, B_c, C_c, D_c)_{Q,R} = J(A_c, B_c, C_c, D_c)_{V,W}$, where

$$J(A_c, B_c, C_c, D_c)_{Q,R} = \text{tr}\left[QP + (Q + L_S^T R_S L_S)\hat{P} \right] \tag{27.13a}$$

$$J(A_c, B_c, C_c, D_c)_{V,W} = \text{tr}\left[VS + (V + K_P W_P K_P^T)\hat{S} \right] \tag{27.13b}$$

Equation 27.12 represents a coupled set of two modified Riccati and Lyaponov equations. The coupling by means of τ, whose rank is precisely equal to the order of the compensator, represents a graphic portrayal of the demise of the classical separation principle for the reduced-order controller. Though the coupled non-linear projection equations are rather difficult to solve, a number of efficient algorithms are available in literature [14,15]. In this chapter, SDOPEs with frequency-dependent weighting matrices are solved by using the iterative algorithm [15].

27.3 MAV DESCRIPTION AND SPECIFICATIONS

27.3.1 MAV Description

An MAV named Sarika-1 is shown in Figure 27.3, which is a remotely piloted small flying vehicle of about 1.28 m span and 0.8 m length and weights around 1.75 kg at takeoff. It has a rectangular wing of plan form area of 0.2688 m^2 and a constant area square section fuselage of width 0.06 m. The control surfaces are elevators, ailerons and rudder. The power plant is a 4 cc propeller engine (OSMAX—25LA), which uses methanol plus castor oil as fuel, with 10–15% nitro-methane to boost the engine power. Sarika-1 has a provision to carry video camera and sensor payloads.

FIGURE 27.3 Micro air vehicle: Sarika-1 (vehicle developed by the authors' team at Indian Institute of Science [IISc], Bangalore).

27.3.2 ON-BOARD FLIGHT SYSTEM

The selection of the onboard computer is based on many constraints. Size, speed of computation, available space, power and allowable weight are few among them. Keeping these constraints in mind, the FIC in Figure 27.4 is designed around two TI-based DSPs, namely, TMS320LF2407 and TMS320VC33. TMS320LF2407 DSP is used for data acquisition and motor control application while TMS320VC33 DSP is used for mathematical computations and data memory management. Communication between the two DSPs is through inbuilt synchronous serial channel operating at 5 MBPS. Owing to the complexity involved, simple errors in the software development can often lead to critical failures that result in crashes and loss of MAV airframe. To mitigate this problem, hardware digital switching logic is provided so that the pilot can switch-over from automatic to manual mode at his/her own will and can safely land the aircraft. This will assure that the command signals received from the ground bypass the controller block, and directly get linked to the actuator servo. Manual mode of flight is also useful for the estimation of the aircraft flight parameters since some of these parameters get camouflaged on closing the feedback loop.

To perform augmented control, information must be known about the current state of the aircraft, as well as the pilot inputs from the transmitter. The state data are obtained from two on-board sensors, a rate gyro and accelerometer. These sensors, shown in Figure 27.5, are directly interfaced to the on-board flight system, the FIC.

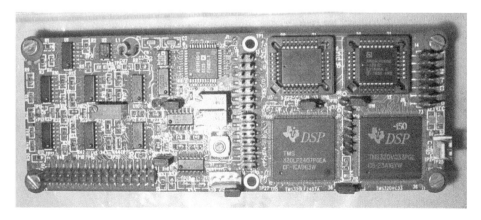

FIGURE 27.4 Flight instrumentation computer (the FIC card integrated/assembled in the authors' lab by the team at Indian Institute of Science [IISc], Bangalore).

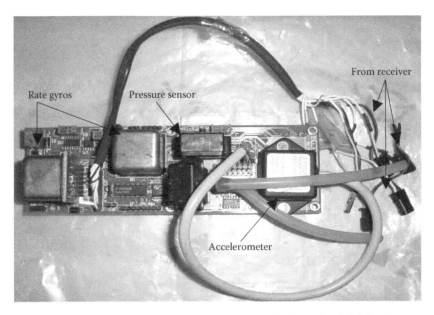

FIGURE 27.5 Sensor card (the card was integrated/assembled in the authors' lab by the team at Indian Institute of Science [IISc], Bangalore).

27.3.3 Longitudinal Dynamics of Sarika 1

Linearized state-space model representing small perturbation longitudinal dynamics are developed for a straight and level flight at an assumed constant altitude of 100 m above ground level at Bangalore (or 1000 m above the sea level) trimmed at five operating points in the speed range of 15–25 m/s. The longitudinal state variables are $x = [\Delta u \; \alpha \; q \; \theta]^T$, where, Δu, α, q and θ indicate forward speed, angle-of-attack, pitch rate and pitch attitude angle, respectively. The linearized longitudinal state equations [19] are

$$\Delta \dot{u} = \left(X_u + X_{Tu} \right) \Delta u + X_\alpha \left(\alpha + \frac{w_g}{U_1} \right) + \frac{Z_u}{U_1} q - g \left(\cos \theta_1 \right) \theta + X_{\delta e} \delta_e \qquad (27.14a)$$

$$\dot{\alpha} = \frac{Z_u}{\left(U_1 - Z_{\dot\alpha} \right)} \Delta u + \frac{Z_\alpha}{\left(U_1 - Z_{\dot\alpha} \right)} \left(\alpha + \frac{w_g}{U_1} \right) + \left(\frac{\left(Z_q + U_1 \right)}{\left(U_1 - Z_{\dot\alpha} \right)} \right) q + \frac{Z_{\delta e}}{\left(U_1 - Z_{\dot\alpha} \right)} \delta_e \qquad (27.14b)$$

$$\dot{q} = \left(\frac{M_{\dot\alpha} \times Z_u}{\left(U_1 - Z_{\dot\alpha} \right)} \right) \Delta u + \left(M_\alpha + M_{\dot\alpha} \frac{Z_\alpha}{\left(U_1 - Z_{\dot\alpha} \right)} \right) \left(\alpha + \frac{w_g}{U_1} \right) + \left(M_q + M_{\dot\alpha} \frac{Z_q}{\left(U_1 - Z_{\dot\alpha} \right)} \right) q$$

$$+ \left(M_{\delta e} + M_{\dot\alpha} \frac{Z_{\delta_e}}{\left(U_1 - Z_{\dot\alpha} \right)} \right) \delta_e \qquad (27.14c)$$

$$\dot{\theta} = q \qquad (27.14d)$$

where $w_g/U_1 = \alpha_g$ is the angle of attack due to vertical wind gust w_g. U_1 is the steady-state velocity and δ_e is the elevator deflection in radians. The elevator is actuated by an electro-mechanical servo system. The measured output variables are normal acceleration, a_z (using TAA-3804-100) and pitch rate (using Micro Gyro 100) of the vehicle. The normal acceleration at the centre of gravity of the vehicle [20] is given by

$$a_z = U_1(\dot{a} - q)$$

The dynamic derivatives are calculated using analytical approach [21], while static and control derivatives are calculated based on the wind tunnel generated data [22]. One of the controller design requirement is to design a single controller at the central operating point of the vehicle and using for all flight conditions. Hence, for the controller design purpose, a flight speed of 18 m/s is selected. At 18 m/s flight speed, the state-space representation of the longitudinal dynamics of Sarika-1 is given by

$$\begin{bmatrix} \Delta\dot{u} \\ \dot{\alpha} \\ \dot{q} \\ \dot{\theta} \end{bmatrix} = \begin{bmatrix} -0.2585 & 6.8252 & 0 & -9.81 \\ -0.0578 & -6.7463 & 0.9487 & 0 \\ 0.2675 & -89.6339 & -0.4115 & 0 \\ 0 & 0 & 1.00 & 0 \end{bmatrix} \begin{bmatrix} u \\ \alpha \\ q \\ \theta \end{bmatrix} + \begin{bmatrix} -0.0748 \\ -0.5111 \\ -72.482 \\ 0 \end{bmatrix} \delta_e \qquad (27.15)$$

Measured variables, normal acceleration a_z (with reference to the c.g.) and pitch rate q, are given by

$$\begin{bmatrix} a_z \\ q \end{bmatrix} = \begin{bmatrix} -0.3253 & 82.8876 & 1.36 & 0 \\ 0 & 0 & 1.00 & 0 \end{bmatrix} \begin{bmatrix} u \\ \alpha \\ q \\ \theta \end{bmatrix} + \begin{bmatrix} 71.97 \\ 0 \end{bmatrix} \delta_e \qquad (27.16)$$

Similarly, the mathematical models are developed at other operating points, that is, at 15, 20, 22 and 25 m/s for further analysis.

27.3.4 OPEN-LOOP ANALYSIS

Table 27.1 gives the frequency and damping ratios of longitudinal dynamics at five different operating points in the flyable speed range of 15–25 m/s [23]. As the speed increases, un-damped natural frequency of the short-period mode of the plant increases from 4.71 to 7.65 rad/s and the damping ratio

TABLE 27.1
Plant Poles, Frequency and Damping Ratios

Speed (m/s)	Eigenvalue	ξ	ω (rad/s)	Comment
15	$-3.00 \pm 6.75i$	0.4	7.39	Short period
	$-0.0847 \pm 1.11i$	0.08	1.12	Phugoid
18	$-3.61 \pm 8.65i$	0.38	9.37	Short period
	$-0.103 \pm 0.877i$	0.11	0.883	Phugoid
20	$-4.01 \pm 9.82i$	0.38	10.6	Short period
	$-0.12 \pm 0.772i$	0.15	0.781	Phugoid
22	$-4.42 \pm 11.0i$	0.37	11.8	Short period
	$-0.136 \pm 0.689i$	0.19	0.702	Phugoid
25	$-5.03 \pm 12.6i$	0.37	13.6	Short period
	$-0.159 \pm 0.59i$	0.26	0.612	Phugoid

remains constant around 0.27. On the other hand, un-damped natural frequency of the phugoid mode decreases from 0.821 to 0.503 rad/s and the damping ratio increases from 0.138 to 0.204. Though the longitudinal dynamics of Sarika-1 is designed to be stable, there might be large variations of the stability derivatives during the flight. The fast short-period response with poor damping makes the flying difficult with a radio control (at times, getting close to stall or other flight failure modes) due to poor handling qualities. Hence, pilot correction of wind gust is very difficult, if not impossible, during its open-loop flight. These difficulties demand tight requirements on flight stabilization at all operating conditions. Since, Sarika-1 does not have sensors to measure true or indicated airspeed and it uses non-inertial quality sensors; gain/controller scheduling is not feasible.

27.3.5   Design Specifications

The closed-loop design specifications for the longitudinal dynamics are determined from its expected responses to pilot/command inputs sent from the elevator joystick. Ground pilot adjusts the engine throttle based on visual cues; hence, throttle control is not a part of the feedback control. Thus, the main requirement of the stability augmentation system (SAS) towards the improvement of handling qualities is summarized in Section 27.3.5.1.

27.3.5.1   Specifications for Level-1-Flying Qualities

Short-period damping ratio: $0.35 \leq \zeta_{sp} \leq 1.3$ and phugoid damping ratio: $\zeta_{ph} \geq 0.5$.

27.3.5.2   Disturbance Rejection Specification

The standard turbulence parameters, L_v and σ_v [24] of the side wind gust are modified to $L_v = 3$ m and $\sigma_v = 5.5$ m/s^2 to take into account the worst-case scenario, when all the dynamic modes of Sarika-1 are excited. The gust spectral density is evaluated as a function of cruise speed from 15 to 25 m/s and it is found that [23], the spectral bandwidth of gust increases with increase in speed (to 13.6 rad/s at 25 m/s). Hence, the disturbance rejection specification is: Minimize the sensitivity function below 0 dB for $\omega < 14$ rad/s.

27.3.5.3   Sensor Noise Attenuation Specification

Experimentally, it is found that low-cost sensors like rate gyros (Micro Gyro 100 from Gyration, USA) and accelerometers (TAA-3804-100 from Neuw Ghent Technology, USA) have high-frequency noise content concentrated above 15 rad/s. Therefore, to achieve high-frequency noise attenuation the specifications are: Obtain—40 dB/decade roll-off above frequency of 15 rad/s.

27.3.5.4   Robustness Specification

The controller should be robust to structured and unstructured uncertainty in plant models at all flight conditions. Structured parametric uncertainty in the plant model arises due to varying flight conditions and fluctuations in aerodynamic stability derivatives. Larger uncertainty levels are placed on the dynamic derivatives, which are obtained analytically using DATCOM [21], since the wind tunnel generated data is more reliable than those calculated using the analytic approach. The uncertainty levels assumed on different dimensional derivatives are given in Table 27.2. These uncertainty levels on non-dimensional derivatives are assumed constant at all flight conditions. Thus, robustness specifications are: The controller designed at the central operating point should be robust at all flight conditions (by providing at least 6 dB gain-margin and 60° phase margin) in order to avoid gain/controller scheduling.

27.3.5.5   Robustness against the Computational Delay

The closed-loop system should also be robust to maximum expected time delays that may arise due to computational complexity. The controller for Sarika-1 is implemented on FIC built around TMS320LF2407 and TMS320VC33. The real-time computation of control signals at the current

TABLE 27.2
Uncertainty Levels on Stability Derivatives

Parameter	Uncertainty Level (%)	Parameter	Uncertainty Level (%)
X_α	20	X_{Tu}	40
Z_α	20	$X_{\delta e}$	20
Z_q	40	$Z_{\delta e}$	20
M_α	20	M_q	40
X_u	40	$M_{\delta e}$	20

instant is based on the measured signals in the previous instant. This introduces a delay of one sampling period for the control signals to reach the servos. Therefore, apart from the above robustness requirements, the closed-loop system should also be robust to a nominal computational delay of 20 ms.

27.3.5.6 Requirements on Control Surface Deflections

The maximum deflection of the control surfaces is limited by the mechanical linkages that connect the actuators to the control surfaces. Experimentally it is found that the maximum deflection of the elevator is limited to ±16°. Thus in the closed loop, the control surface deflection should not exceed its full-scale deflection of ±16°.

Thus, the objective is to design and validate a robust discrete optimal H_2-controller at a single operating point (18 m/s flight condition) such that it achieves simultaneous stabilization, disturbance rejection and noise rejection requirements at all flight conditions.

27.4 RESULTS AND DISCUSSIONS

The robust fixed-order H_2 controller is designed at the central operating point of Sarika-1 (i.e., at 18 m/s) and is used for all flight conditions. The weighting matrices W_1 and W_2 need to be strictly proper in order to make the feed through term, E_3 zero. Following the guidelines [18] to select the weighting matrices, the final choice of the weighting matrices, W_1 and W_2 (in discrete domain) are

$$W_1(z) = \begin{bmatrix} \dfrac{1.1881}{(z-0.9802)} & 0 \\ 0 & \dfrac{0.198}{(z-0.9802)} \end{bmatrix}; \quad W_2(z) = \begin{bmatrix} \dfrac{8.75(z-0.9139)}{(z-0.2466)} \end{bmatrix} \quad (27.17)$$

with the choice of the weighting matrices (Equation 27.17), the order of the generalized plant is nine. Hence, the fixed-order controllers of varying orders ($n_c = 9$ to 1) are designed and found that a third-order controller is sufficient to meet the design specifications. Thus, the elements of the third- longitudinal controller are

$$K(z) = \frac{1}{\Delta}\left[-0.0089(z-0.2524)(z-0.9735) \qquad 0.013(z-1.445)(z-1.051)\right] \quad (27.18a)$$

where

$$\Delta = (z-0.1323)(z-0.8509)(z-0.9915) \quad (27.18b)$$

TABLE 27.3

Closed-Loop Frequency and Damping Ratios

Cruise Speed (m/s)	ξ	Frequency (rad/s)	Comment
15	0.5	1.33	Phugoid
	0.68	10.1	Short period
18	0.51	1.06	Phugoid
	0.52	11.8	Short period
22	0.6	0.95	Phugoid
	0.5	13.2	Short period
25	0.9	0.53	Phugoid
	0.41	15.9	Short period

Table 27.3 gives the short-period and phugoid frequency of the closed-loop system at different flight conditions (using the controller designed at the flight speed of 18 m/s). At all flight conditions, short-period and phugoid damping remains greater than 0.35 and 0.5, respectively; hence, the stringent level-one flying quality requirement is met.

27.4.1 ROBUSTNESS AND PERFORMANCE ANALYSIS

To assess the robustness of the controller, singular values of the objective function (Equation 27.1) is analysed. A plot of the singular values of S_o and S_iK of the closed-loop system is shown in Figures 27.6 and 27.7. Singular values of S_o for the normal acceleration and pitch rate show slight deviation from one another at low frequencies (below 10 rad/s) having a magnitude in the neighbourhood of 0 dB in the negative side. This trend of variation is seen at all flight conditions. Thus, at all flight conditions, the closed loop is able to reject all low-frequency gust inputs that can occur below $\omega < 10$ rad/s. Note from Figure 27.7 that the maximum singular value of S_iK remains lower than 0 dB over the entire frequency range at all flight conditions. This indicates that the controller is robust against the possible additive uncertainties.

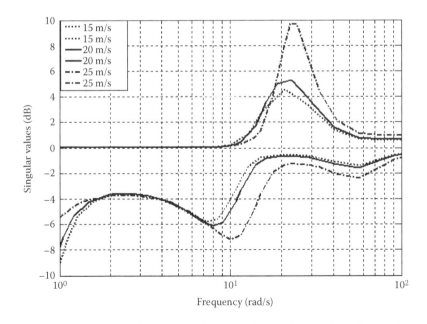

FIGURE 27.6 Frequency response of output sensitivity transfer function (S_o). (Plot from authors' printer.)

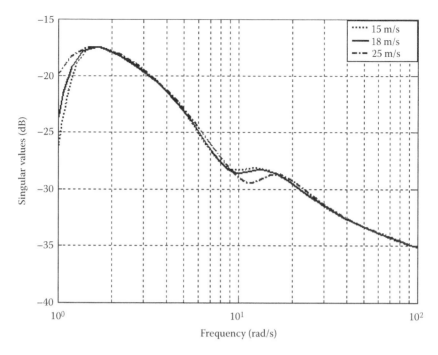

FIGURE 27.7 Frequency response of control sensitivity transfer function (S_iK). (Plot from authors' printer.)

27.4.2 PERTURBATION ANALYSIS

The robust stability and performance of the closed-loop system at all flight conditions (using a controller designed at the nominal flight speed of 18 m/s) are further demonstrated by using LMI-based tests [16,25]. For this, affine parameter-dependent plant models are developed by assuming structured real-valued parametric uncertainties (i.e., by perturbing the stability derivatives of the vehicle) at each of the flight conditions. At 18 m/s of cruise speed, the quadratic stability of the closed-loop system is established with 125.0133% of the prescribed uncertain parameter box. Likewise, at 15 and 25 m/s of cruise speeds, the quadratic stability of the closed-loop system is preserved with 103.185% and 101.861% of the uncertain parameter box, respectively. For the same uncertainty levels, μ upper bound is around 0.5549, 0.6454 and 0.6745 at cruise speeds of 18, 15 and 25 m/s, respectively. Thus, the largest amount of uncertainty Δ that can be tolerated without losing stability is 1.802, 1.549 and 1.483, which is greater than the required lower bound of 1.0.

27.4.3 TIME-RESPONSE STUDY

The most realistic pilot command in actual flight, a pulse elevator command input of 0.1 ms amplitude and the duration of 2 s is used to simulate the closed-loop time responses. Figures 27.8 through 27.11 show the responses of the plant and closed-loop systems to the pulse input at different cruise speeds. Again, the short period responses are seen to be fast and well damped with a settling time of 5 s at cruise speeds of 22–25 m/s. However, at the cruise speed of 15–18 m/s the settling time increases to 10 s. Conversely, the settling time of its plant responses are more than 50 s and are highly oscillatory. The settling time of the phugoid states is less than 9 s at cruise speeds of 22–25 m/s and increases to 11 s at 15–18 m/s. This increase in settling time can be endorsed to decreased phugoid damping. However, the settling time is still much faster than the open-loop responses which are oscillatory even after 50 s. The steady-state value of both q and a_z is zero in closed loop, as in the case of open-loop system. This is expected since an autopilot control system is not designed. However, the simulations in this study does not include the effect of sensor noise and gust inputs (the responses which are

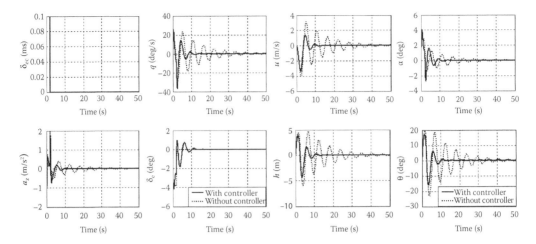

FIGURE 27.8 Pulse response at 15 m/s cruise speed.

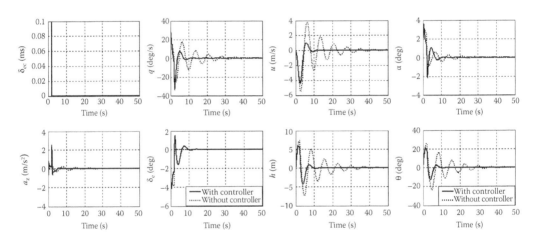

FIGURE 27.9 Pulse response at 18 m/s cruise speed.

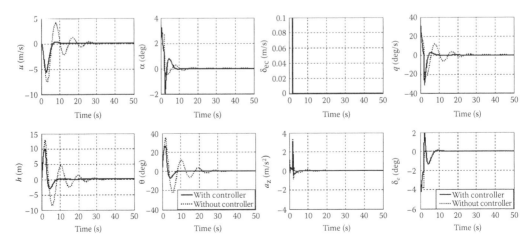

FIGURE 27.10 Pulse responses at 22 m/s cruise speed.

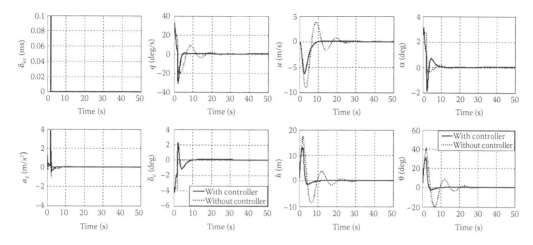

FIGURE 27.11 Pulse responses at 25 m/s cruise speed.

given in the next section) in order to ease the understanding of the normal aircraft behaviour. The trajectory followed by the aircraft is smooth with the controller while it is highly oscillatory without the controller. The introduction of time delay does not result in deterioration of responses which proves that the closed-loop system is also robust to the expected computational time delays.

27.4.4 TIME RESPONSE WITH GUST AND SENSOR NOISE INPUT

The simulation results of the closed-loop system and plant are considered now in the presence of sensor noise and gust in order to assess the gust and noise rejection capabilities of the closed-loop system. In order to simulate sensor noise with frequency contents for $\omega > 15$ rad/s, a high-pass filter of cut-off frequency 15 rad/s is used. Gust input is generated using the procedure given in Ref. [24]. Thus, injected sensor noise input and gust inputs are as shown in Figure 27.12. The closed-loop simulation is initiated using the pulse input mentioned in the previous section. Figures 27.13 and 27.14

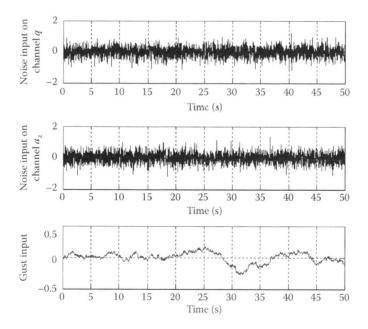

FIGURE 27.12 Injected sensor noise and gust inputs.

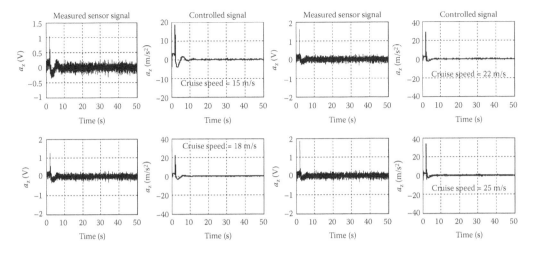

FIGURE 27.13 Time response on a_z channel with sensor noise and gust input.

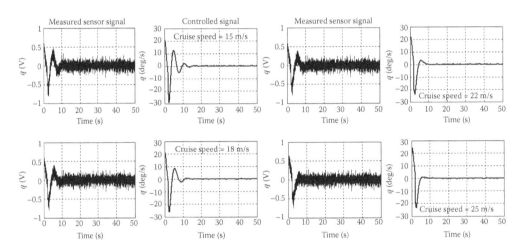

FIGURE 27.14 Time response on q channel with sensor noise and gust input.

show the actual responses of the airframe a_z and q at different flight conditions. As can be seen from Figures 27.13 and 27.14, the gust and sensor noise are completely rejected, by the closed-loop system at all flight conditions, as per the design specification.

27.5 REAL-TIME VALIDATION OF THE CONTROLLER: HILS

The process of HILS is illustrated in Figure 27.15. To perform augmented control, information must be known about the current state of the aircraft, as well as the pilot inputs from the transmitter. The state data are obtained from two on-board sensors, a rate gyro and accelerometer. These sensors are directly interfaced to the on-board flight system, the FIC. Human control signals from a radio control system are decoded by the on-board computer, calculations are performed based on sensor feedback, and actuation is performed. Thus, a HILS experimental setup includes a simulation computer and an on-board flight instrumentation computer. Onboard computer provides four primary functions: program and external data storing in a memory, high-speed data acquisition (with 10 Hz low-pass filter and necessary signal conditioning), internal data logging, generation and capture of PWM signals. The data can be logged to a maximum of 4 MBytes (FPROM). The simulation computer is Pentium4

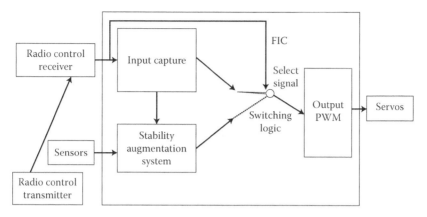

FIGURE 27.15   HILS Process.

PC (Windows 2000 OS) operating at 3.4 GHz with dSPACE DS1104RTI/RTW. The DS1104 control board is a complete real-time system based on a 603 Power PC floating-point processor running at 250 MHz. The PWM generation units of onboard computer are interfaced to the capture units of DS1104. The constitutive building blocks of HILS are given in Figure 27.15. The longitudinal flight dynamic model, running on the desk-top computer, calculates the vehicle's pitch rate and normal acceleration response and DS1104 system subsequently generates the equivalent analog sensor signal in volts. Sensitivities of MEMs-based sensors used for Sarika-1 are rate gyros: 1.11 mV/deg/s (with an offset of 2.25 V), accelerometers: 500 mV/g (with a bias of 2.5 V). Hence, the sensor outputs are generated according to the relation

$$V_{eqq} = 0.063603 \; q + V_{offsetq} \tag{27.19a}$$

$$V_{eqa \, z} = 0.05 \; a_z + V_{offsetaz} \tag{27.19b}$$

The analogue-to-digital converter (ADC) of the on-board computer unit receives the simulated sensor signal (Equations 27.19a,b) from the digital-to-analogue converter (DAC) of DS 1104. Ground pilot control signals from the radio-control system are decoded by the capture unit of the on-board computer. Alternatively, an input sequence (e.g., step, pulse, constant, etc.) to the controller under HILS can be produced by the utility's software-based function generator at the time instants specified for each scheduled variable. In the closed-loop test, on-board computer accepts controller inputs from the on-board receiver/scheduler and simulated sensor output from DAC. The actuator outputs are computed based on the embedded control logic and are linked back through the hardware RTI and the aircraft's response is simulated. The advantage of this strategy resides in the dynamic coupling of the vehicle dynamics and electronic controller to create an environment similar to the actual vehicle.

27.5.1 CONTROLLER IMPLEMENTATION AND VALIDATION

The control law is coded on to the on-board computer using floating-point programming. Since the inputs to the controller are in volts and the output needs to be in-terms of milliseconds of PWM signal a relation between the voltage and PWM signal width is found experimentally as

$$PWM = (V_{signal} - V_{offset}) \times 1.18 \; ms \tag{27.20}$$

A computer-supervised scheduler enables the controller inputs and driving scenarios, in a time-indexed manner. To record the controller's response to the given time indexed stimuli, data are logged from both the simulator and controller with the integrated data acquisition system.

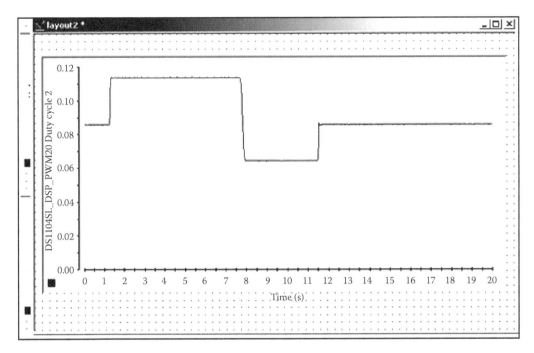

FIGURE 27.16   Elevator command input from the RF transmitter. (Plot from authors' printer.)

An extensive closed-loop test is conducted at different flight conditions in order to validate the controller in real time. Figures 27.16 and 27.17 show the traces of the elevator command signal in real time (given from the RF transmitter), and the corresponding closed-loop responses recorded for a short interval of time at 18 m/s. Note from Figure 27.16 that, at $t = 1$ s, a positive going pulse of 0.028 ms (with respect to 0.082 m) PWM signal is sent from the transmitter and is changed to -0.022 m (with respect to 0.082 m) at $t = 8$ s and is lasted at $t = 12$ s. Corresponding feedback signals which are simulated in real time (dSPACE 1104) is given in Figure 27.17. The responses are fast and well damped and are matched well with offline simulation responses. Note that, during the positive going pulse of 0.028 ms PWM command input, the vehicle shows an excursion of 1.5–1 m/s^2 on normal acceleration, a_z, response and settles down at its steady state within 3 s. Similarly at $t = 8$ s a_z shows a peak response of 2.25 m/s^2 with respect to its steady-state value. These values are comparable to that obtained during offline simulation. Figures 27.18 and 271.19 show the responses of the state variables, simulated in real time. Again responses are fast and well damped and match well with its offline simulation responses. The difference is due to the fact that, initial values of the variables, at the capturing instant in real-time simulation differs from its assumed initial values during its offline simulation. Similarly, the controller is tested in real time at other flight conditions (15, 22 and 25 m/s) and the results are found to be satisfactory.

27.6   CONCLUSIONS

A new approach on design of discrete robust frequency-shaped H_2 optimal flight stabilization system for MAV named Sarika-1 is developed. The design technique uses strengthened discrete optimal projection equations. Detailed time- and frequency-domain analysis shows that the H_2 optimal controller is able to meet all design specifications. A single controller robustly stabilizes Sarika-1 with margins better than a minimum requirement of 6 dB gain margin and 60° phase margin at all flight conditions within the speed range of 15–25 m/s. Off-line time responses demonstrates that flight stabilization performs well up to the designer's expectation. Subsequently, real-time validation of the controller, which is implemented on the digital signal processor-based onboard computer substantiates

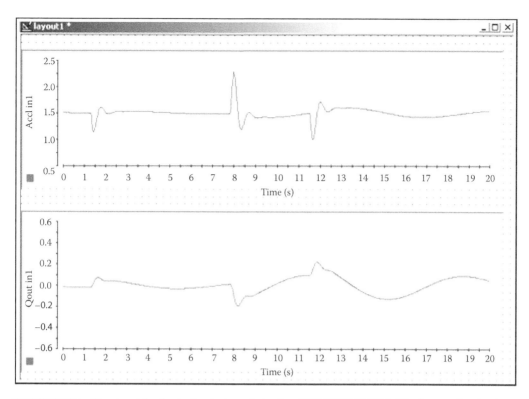

FIGURE 27.17   Simulated feedback signals (a_z and q) using dSPACE RTI/RTW. (Plot from authors' printer.)

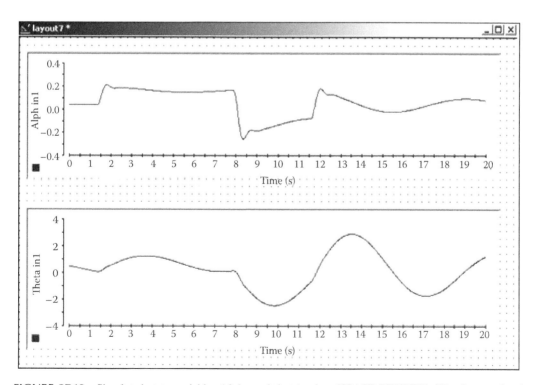

FIGURE 27.18   Simulated state variables (alpha and theta) using dSPACE RTI/RTW. (Plot from authors' printer.)

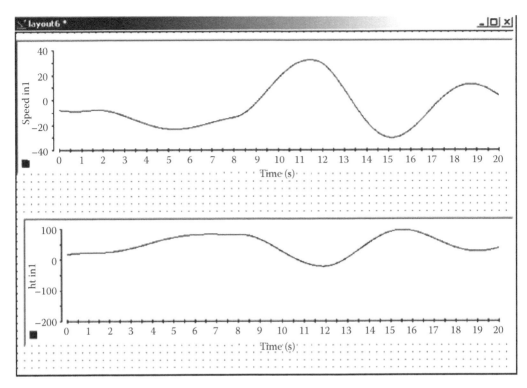

FIGURE 27.19 Simulated state variables (u and corresponding trajectory, (height, ht)) using dSPACE RTI/RTW. (Plot from authors' printer.)

the off-line simulation results. The real-time HILS responses match well with those of the desktop simulation responses. Thus, robust H_2 stability augmentation system can be successfully used towards equipping the MAVs with autonomous capabilities that could significantly enhance the utility of the MAVs for a wide range of missions.

REFERENCES

1. McMicheal JM, Francis MS, Micro-air vehicles—Toward a new dimension in flight world wide web, http://www.darpa.mil/tto/mav/mav_auvsi.html 3/21/2003.
2. Grasmeyer JM, Keennon MT. Development of the black widow micro air vehicle. AIAA Paper: 2001-0127.
3. Ruffier F, Franceschini N, Visually guided micro aerial vehicles: Automatic take off, terrain following, landing, and wind reaction. *Proceedings of the IEEE International Conference on Robotics and Automation*, New Orleans. 2004:2339–2346.
4. Stoorvogel AA. The singular H_∞ control problem with dynamic measurement feedback. *SIAM Journal of Control and Optimization* 1991; 29(1): 160–184.
5. Hyde RA, Glover K. The application of scheduled H_∞ controllers to VSTOL aircraft. *IEEE Transactions on Automatic Control* 1993; 38(7): 1021–1039.
6. Shiau JK, Tseng CE. A discrete H_∞ low-order controller design using coprime factors. *Tamkang Journal of Science and Engineering* 2004; **7**(4): 251–258.
7. Kannan N, Seetharama Bhat M. Longitudinal H_∞ stability augmentation system for a thrust vectored unmanned aircraft. *Journal of Guidance Control and Dynamics* 2005; 28(6): 1240–1250.
8. Aouf N, Boulet B, Botez R. Model and controller reduction for flexible aircraft preserving robust performance. *IEEE Transactions on Control Systems Technology* 2002; 10(2): 229–237.
9. Francesco A., Raffaele I. µ Synthesis for a small commercial aircraft: Design and simulator validation. *Journal of Guidance, Control, and Dynamics* 2004; **27**(3): 479–490.

10. Stoorvogel AA. The robust H_2 control problem: A worst-case design. *IEEE Transactions on Automatic Control* 1993; 38(9): 1358–1370.

11. Goh K-C, Wu F. Duality and basis functions for robust H_2—performance analysis. *Automatica* 1997; 33(11): 1949–1959.

12. Liu Y, Anderson BDO. Controller reduction: concepts and approaches. *IEEE Transactions on Automatic Control* 1989; 34(8): 802–812.

13. Meenakshi M. Seetharama Bhat M. Robust fixed order H_2 controller for micro air vehicle—design and validation. *International Journal of Optimal Control and Application Methods* 2006; 27: 183–210 (Published online on 15th February, 2006 in Wiley InterScience), DOI: 10.1002/oca.774.

14. Haddad WM, Huang HH, Bernstein DS. Robust stability and performance via fixed-order dynamic compensation: The discrete-time case. *IEEE Transactions on Automatic Control* 1993; 38(5): 776–782.

15. Van Willigenburg LG, De Koning WL. Numerical algorithms and issues concerning the discrete time optimal projection equations. *European Journal of Control* 2000; 6(1): 93–110.

16. Feron E, Apkarian P, Gahinet P. Analysis and synthesis of robust control systems via parameter-dependent Lyapunov functions. *IEEE Transactions on Automatic Control* 1996; 41(7): 1041–1046.

17. Green M, Limebeer DJN. *Linear Robust Control*. Prentice-Hall, Englewood Cliffs, NJ 1995: 131–178.

18. Hu J, Bohn C Wu HR, Systematic H_∞ weighting function selection and its application to the real-time control of a vertical take-off aircraft. *Control Engineering Practice* 2000; 8, 241–252.

19. Jan R. Flight dynamics part 4. Roskam Aviation and Engineering Corporation, Box 274, Ottawa, Kansas 66067, USA 1971:5.1–6.118.

20. Stevens BL, Lewis FL. *Aircraft Control and Simulation*. John Wiley and Sons, New York, Second Edition, 2003.

21. Jan R. *Airplane Design: Part VI—Preliminary Calculation of Aerodynamic, Thrust and Power Characteristics*. University of Kansas, Kansas, 1990.

22. Srinivasa Rao BR, Surendranath V, Prasanna HRS. *Wind Tunnel Test Results of SARIKA Airplane*, Rep. No: AE/WT/IRR/16, Department of Aerospace Engineering, Indian Institute of Science, Bangalore, 2001.

23. Meenakshi M. Design and real time validation of discrete fixed order robust H_2 controller for micro aerial vehicle. PhD thesis, Aerospace Engineering, Indian Institute of Science, Bangalore, July 2005.

24. *U.S. Military Handbook MIL-HDBK-1797*, 19 December 1997.

25. Gahinet P, Nemirovski A, Laub AJ, Chilali M. *LMI Control Toolbox for Use with MATLAB*. 1995:1.1–3.26.

28 Neuro-Fuzzy Fault-Tolerant Aircraft Autoland Controllers

Abhay A. Pashilkar, Rong Haijun and N. Sundararajan

CONTENTS

28.1 INTRODUCTION

It is a well-known fact that the landing and take-off phases of flight is the most important from the point of view of safety. Most modern airports have radio landing aids for severe weather conditions. Loss of control effectors can also create serious problems for aircraft landing. The conventional approach taken during the design of such systems is to make the probability of this occurrence very remote ($<10^{-9}$). The design for reliability and safety relies heavily on the use of multiple redundant actuators and control surfaces. While this is an effective way to achieve reliability, the development and maintenance of such systems cause the life cycle costs to go up enormously. Therefore, it is appropriate to study reconfigurable or intelligent systems which can detect failure and use the available aerodynamic redundancy effectively in completing the mission safely. In this chapter, we study the automatic landing problem from this point of view.

A popular approach taken to the problem of handling the failure of control surfaces and sensors is the failure detection and identification (FDI) approach. As the name suggests, in this approach, a predetermined set of failure scenarios are detected. Based on the failure observed, the control system is reconfigured in a deterministic manner. Therefore, this scheme has three distinct problems to be tackled, namely, detection, identification and reconfiguration. This scheme has the distinct advantage that predetermined failures can be handled well. On the other hand, there are no guarantees that the scheme will behave gracefully in the event of an unknown failure. In case of damage to the control surface which can range from minor loss of effectiveness (e.g., battle damage to one of the elevators) to a complete loss of the surface, the FDI algorithm will have to address this problem suitably. Therefore, in these situations, an online adaptive controller with guaranteed performance is a plausible solution. Some approaches based on feedback linearization and online mechanisms have been proposed [1–3]. Kim and Calise [1] and Pesonen et al. [3] proposed the feedback linearization approach to address this problem. While the former demonstrates application to the F-18 aircraft, the latter is applied to a general aviation aircraft. The primary difficulty associated with the use of feedback linearization for aircraft control is that a detailed knowledge of the non-linear plant dynamics is required. Napolitano et al. [2] have presented an approach to the online learning mechanism for dynamic control system reconfiguration for an aircraft that has sustained extensive damage to a vital control surface. A neural controller uses the extended back propagation algorithm (EBPA) to bring the aircraft back to equilibrium after failure. Kim and Lee [4] describe a non-linear adaptive flight control system using backstepping and neural network controllers and analyse the stability of the proposed control system using Lyapunov theory. Li et al. [5] present a fully tuned radial basis function network (RBFN) controller that uses only online learning to represent the local inverse dynamics of the aircraft system. Adapting tuning rules are derived based on the Lyapunov synthesis approach to guarantee closed-loop stability. Fully tuned RBFNs are used where the number of hidden units, centres and widths are updated along with the weights of the network. In the aircraft model as described in Ref. [6], a simple architecture originating from Kawato's learning feedback–error–learning scheme [7] has been utilized. This control architecture uses a conventional PID/H_∞ controller in the inner loop to stabilize the system dynamics, and the neuro-controller acts as an aid to the conventional controller. Krstic et al. proposed the integrator backstepping technique for the design of non-linear control systems [8]. An adaptive backstepping neural controller (ABNC) has been developed using this approach for the aircraft autolanding problem in Ref. [9]. This controller has a cascaded architecture. It assumes that all the aircraft states are available for feedback.

28.2 AIRCRAFT AUTOLANDING PROBLEM

The autolanding problem studied in this chapter consists of a high-performance fighter aircraft following a flight path consisting of flight segments such as a wing-level flight, a coordinated turn, descent on glide slope and finally the flare manoeuvre and touchdown on the runway. The trajectory segments corresponding to these phases have to be flown in the presence of severe winds. A detailed description of the wind model used can be found in Section 28.2.2. The winds cause deviation of the aircraft from the

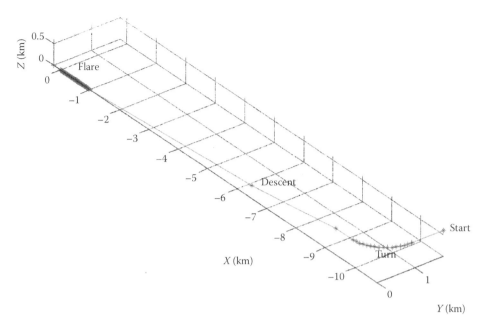

FIGURE 28.1 Segments of the autolanding. (From A. A. Pashilkar, N. Sundararajan and P. Saratchandran, *Journal of Guidance, Control and Dynamics*, 30(3), 835–847, 2007.)

specified trajectory. The touchdown conditions are given with tight specifications, named for convenience as the *touchdown pillbox*. The baseline controller is first designed to meet all these specifications for the touchdown phase under no failure conditions of the actuators. The neural controller helps to improve performance of the basic trajectory following controller (BTFC) controller when the aircraft undergoes actuator failures of different types. The path followed by the aircraft starts off with level flight and ends with the aircraft safely on the ground, within the boundaries of the runway. We analyse the problem, by dividing the entire flight plan into seven distinct segments defined below (see Figure 28.1).

1. Level flight: Altitude of 600 m, heading from east to west, that is, −90 degrees. Aircraft velocity maintained at 83 m/s.
2. Coordinated turn: The aircraft now makes a right turn, with a bank angle of 40 degrees at 600 m altitude, to align itself with the runway (which is in the north–south direction) vertically at 0 degrees. Velocity is still 83 m/s.
3. Level flight: After completion of the turn, the aircraft maintains its current heading and velocity at the same altitude for some time.
4. 1st Descent: The aircraft makes its first descent at a glide slope of −6 degrees and the altitude is reduced to 300 m.
5. 2nd Descent: A second descent at a lesser glide slope of −3 degrees causes the aircraft altitude to reduce further to 12 m.
6. Flare and touchdown: The aircraft velocity is reduced from 83 m/s to 79 m/s during the flare. The manoeuvre starts when the aircraft is at an altitude of 12 m. Airspeed reduces from 82.88 m/s to 76.19 m/s and the altitude decreases exponentially as

$$h = h_0 e^{-t/\tau} \tag{28.1}$$

where $h_0 = 12$ m, $\tau = h_0/\sin \gamma = 3$ s.
7. The touchdown pillbox conditions are evaluated at the point where the aircraft first touches down on the ground.

28.2.1 Safety and Performance Criteria

The controller is designed to meet several safety and performance criteria during landing. These include bounds on the flight parameters such as lateral deviation, altitude, heading angle, sink rate and airspeed. We define the *touchdown pillbox* region, that is, the boundary within which the aircraft must land under the specified conditions (Table 28.1). The controller is designed to operate even during the occurrence of failure and is said to have successfully taken care of failures in a given region only if the aircraft lands within the pillbox. The runway origin is situated at the end to which the aircraft is approaching. Thus, the x and y distance specifications in Table 28.1 restricts the required landing area of the aircraft to within a rectangle of length 400 m and width 10 m. The minimum velocity specification is to ensure that the aircraft does not go into a stall during its final approach. The sink rate limit is required so that landing gear loads are within acceptable limits. A limit on the bank angle ensures that the wing tips do not touch the ground. Under moderate turbulence conditions, the actuator rates for aileron, elevator, and rudder should not saturate at 60 degree/s for a long period of time. This puts an upper limit to the feedback gains.

28.2.2 Wind Profiles

There are three wind components which we consider during landing along the three axes of the aircraft. These simulate a microburst encounter.

28.2.2.1 Vertical Component

To produce a wind shear in the vertical direction (microburst) at a given height, the direction of the vertical wind is suddenly changed (from up to down), that is,

$$w_w = -w_0(1 + \ln(h/510)/\ln 51) \tag{28.2}$$

where w_0 is 12 m/s above the height h_{shear} and -12 m/s under the height h_{shear}. Figure 28.2 shows the wind profile as described, where h_{shear} is 91. The wind shear vanishes above 150 m.

28.2.2.2 Forward Component of Wind

Turbulence is modelled using a Dryden distribution. It is assumed that the turbulence exists only in the horizontal direction. Mathematically, the longitudinal wind is represented by

$$u_w = u_{g1} + u_{gc} \tag{28.3}$$

$$u_{g1} = 0.2 \, |u_{gc}| \, \sqrt{2a_u} N_1 - a_u u_{g1} \tag{28.4}$$

TABLE 28.1
Touchdown Pillbox Specifications

x-distance, m	$-100 \leq x \leq 300$
y-distance, m	$|y| \leq 5$
Total velocity, m/s	$V_T \geq 60$
Sink rate, m/s	$\dot{h} \leq 1.0$
Bank angle, degree	$|\Phi| \leq 10$

Source: A. A. Pashilkar, N. Sundararajan and P. Saratchandran, *Journal of Guidance, Control and Dynamics*, 30(3), 835–847, 2007.

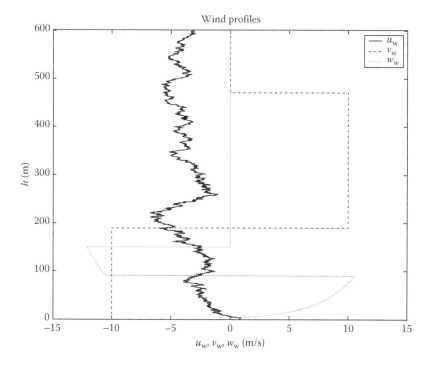

FIGURE 28.2 Wind profiles during the landing phase. (From A. A. Pashilkar, N. Sundararajan and P. Saratchandran, *Journal of Guidance, Control and Dynamics*, 30(3), 835–847, 2007.)

where u_{g1} is the turbulence component and u_{gc} is the mean wind given by

$$u_{gc} = \begin{cases} -u_0[1 + \ln(h/510)/\ln(51)], & h \geq 3\,\text{m} \\ 0, & h < 3\,\text{m} \end{cases} \tag{28.5}$$

where u_0 is 6 m/s and N_1 is the Gaussian random noises with mean zero and variance 100, and

$$a_n = U_0/(100\sqrt[3]{h}), \quad h > 70\,\text{m}$$
$$a_\eta = U_0/600, \qquad h \leq 70\,\text{m} \tag{28.6}$$

where U_0 is 72 m/s.

28.2.2.3 Side Component of Wind

The side wind is a step change of magnitude 7 m/s and -14 m/s at altitudes 470 m and 190 m, respectively.

$$v_w = \begin{cases} -7, & h < 190\,\text{m} \\ 7, & 190 \leq h < 470\,\text{m} \\ 0, & h \geq 470\,\text{m} \end{cases} \tag{28.7}$$

The winds are a combination of Dryden turbulence along the x-body axis and deterministic winds in the other two axes (Figure 28.2). The sharp step changes in the v_w (at 470 m and 190 m) and w_w

(150 m and 90 m) are particularly noticeable. These profiles represent a large horizontal and vertical wind shear, respectively.

28.2.3   AIRCRAFT MODEL AND FAILURE SCENARIOS

The aircraft model used in this study is derived from Ref. [6]. It is noted that this aircraft is open-loop-unstable in the longitudinal axis. Therefore, the control system must augment the stability as well as the performance of the aircraft to achieve the landing task.

The linearized F-16 model at the straight-level flight condition ($h = 600$ m, $V = 82.88$ m/s) can be described using the following equations:

$$\Delta \dot{x} = A\Delta x + B\Delta u$$
$$\Delta y = C\Delta x + D\Delta u$$

(28.8)

where

$$\Delta x = [\Delta u \quad \Delta w \quad q \quad \Delta \theta \quad \Delta h \quad v \quad p \quad r \quad \phi \quad \psi \quad y]^T$$
$$\Delta u = [\Delta \delta_e \quad \delta_a \quad \delta_r \quad \Delta \delta_{thr}]^T$$
$$\Delta y = [\gamma \quad q \quad \Delta \theta \quad \Delta h \quad \Delta V_T \quad p \quad r \quad \phi \quad a_y \quad \Delta \chi]^T$$

Note that the symbol Δ indicates the deviation of the particular quantity above its trim value. This is only applicable for those quantities which have a non-zero value at trim.

The control design is further simplified if one treats the longitudinal dynamics of the aircraft as decoupled from the lateral dynamics. Accordingly, the six-degrees-of-freedom equations are separated into the longitudinal- and lateral-directional set. This is also the traditional approach. The longitudinal system matrices are

$$A = \begin{bmatrix} -0.0183 & 0.1023 & -15.3342 & -9.6192 & -0.0002 \\ -0.1060 & -0.6485 & 73.6202 & -1.9024 & 0.0009 \\ -0.0025 & 0.0098 & -0.6491 & 0 & 0 \\ 0 & 0 & 1.0000 & 0 & 0 \\ 0.1940 & -0.9810 & 0 & 82.8421 & 0 \end{bmatrix}$$

$$B = \begin{bmatrix} 0.0049 & 7.1522 \\ -0.1359 & 0 \\ -0.0598 & 0 \\ 0 & 0 \\ 0 & 0 \end{bmatrix}$$

$$C = \begin{bmatrix} 0.0023 & -0.0118 & 0 & 1.0000 & 0 \\ 0 & 0 & 1.0000 & 0 & 0 \\ 0 & 0 & 0 & 1.0000 & 0 \\ 0 & 0 & 0 & 0 & 1.0000 \\ 0.9810 & -0.1938 & 0 & 0 & 0 \end{bmatrix}$$

$$D = \begin{bmatrix} 0 & 0 \\ 0 & 0 \\ 0 & 0 \\ 0 & 0 \\ 0 & 0 \end{bmatrix}$$

where the state and control vectors are given as

$$\Delta x = [\Delta u \quad \Delta w \quad q \quad \Delta\theta \quad \Delta h]^T$$
$$\Delta u = [\Delta\delta_e \quad \Delta\delta_{thr}]^T$$
$$\Delta y = [\gamma \quad q \quad \Delta\theta \quad \Delta h \quad \Delta V_T]^T$$

Similarly, the lateral-directional system matrices for the same flight condition are

$$A = \begin{bmatrix} -0.1487 & 16.2718 & -80.6519 & 9.6216 & 0 & 0 \\ -0.1708 & -1.7533 & 0.8792 & 0 & 0 & 0 \\ 0.0227 & -0.0482 & -0.2424 & 0 & 0 & 0 \\ 0 & 1.0000 & 0.1978 & 0 & 0 & 0 \\ 0 & 0 & 1.0194 & 0 & 0 & 0 \\ 1.0000 & 0 & 0 & -16.0768 & 82.8632 & 0 \end{bmatrix}$$

$$B = \begin{bmatrix} -0.0024 & 0.0336 \\ -0.1684 & 0.0340 \\ -0.0014 & -0.0169 \\ 0 & 0 \\ 0 & 0 \\ 0 & 0 \end{bmatrix}$$

$$C = \begin{bmatrix} 0 & 1.0000 & 0 & 0 & 0 & 0 \\ 0 & 0 & 1.0000 & 0 & 0 & 0 \\ 0 & 0 & 0 & 1.0000 & 0 & 0 \\ -0.0152 & 0.0195 & 0.0666 & 0 & 0 & 0 \\ 0.0121 & 0 & 0 & -1940 & 1.0000 & 0 \end{bmatrix}$$

$$D = \begin{bmatrix} 0 & 0 \\ 0 & 0 \\ 0 & 0 \\ -0.0002 & 0.0034 \\ 0 & 0 \end{bmatrix}$$

where the state vector, input vector and the outputs are

$$\Delta x = \begin{bmatrix} v & p & r & \phi & \psi & y \end{bmatrix}^T$$

$$\Delta u = \begin{bmatrix} \delta_a & \delta_r \end{bmatrix}^T$$

$$\Delta y = \begin{bmatrix} p & r & \phi & a_y & \Delta\chi \end{bmatrix}^T$$

It is seen that there is coupling between the longitudinal control inputs and the lateral-directional equations of motion. The elevators are to be operated in symmetric manner (i.e., both together in the same direction). Due to a failure of one of the elevator surfaces, they will no longer do so, thereby causing a contribution to the aerodynamic rolling moment. Thus, an elevator failure results in simultaneous disturbance in the pitch and roll channels. The contribution of differential elevators to the rolling motion is used in the two-aileron failure case to control the aircraft in roll. The actuator model for the aircraft consists of a first-order lag element given by

$$\frac{y(s)}{u(s)} = \frac{20}{s + 20} \tag{28.9}$$

The elevator deflections are limited to ±25 degrees, the aileron surface deflections to ±20 degrees and the rudder surface deflection does not exceed ±30 degrees. All primary surface actuators rate limit at 60 degree/s.

28.2.3.1 Scenarios

In this chapter, we have considered five types of failures including a single control surface failure as well as the failure of a combination of control surfaces. We have ignored the case of the failure scenario where both the elevators fail because this case is in general not recoverable. When the control surface(s) fail, their resultant hard over position could be one from among any value within the permissible range of deflections. Clearly, all possible hard over positions are not feasible because in some cases the resulting moments cannot be trimmed out for the landing manoeuvre. To determine the regions in which a single and two failures can be tolerated, the open-loop aircraft was trimmed for [10]: (i) region of level flight trim ($p = q = r = \gamma = 0$, 6 dof accelerations = 0), (ii) region of level descent trim ($p = q = r = 0$, $\gamma = -6$ degrees, 6 dof accelerations = 0) and (iii) region of level turning trim ($\phi = 40$ degrees, 6 dof accelerations = 0). The intersection of these trim regions was used to obtain the *feasible region* from where the faults can be recovered. It is to be noted that this computation is still conservative, since in a realistic situation, the aircraft has to have control authority over and above the trim to manoeuvre (e.g., for disturbance rejection). There is no margin for manoeuvering at the boundaries of these regions. In realistic conditions, the actual boundaries are expected to lie within the feasible region. Furthermore, the trim regions have been computed for no winds and may differ somewhat for the varying wind conditions imposed during the flight. The five different cases of actuator failures that are of interest to us are tabulated in Table 28.2. A pictorial representation of the feasible regions will be shown in the results section along with the performance of the individual controller.

28.2.3.2 Failure Injection Point

As stated in Section 28.2, on problem definition, there are seven segments in the flight path that the aircraft goes through before touchdown. The two critical stages of the flight path are (1) the level turn and (2) the descent phase. If the flight controller handles these phases robustly, successful landing can be assured. Failures can occur at any point when the aircraft is in flight, but it would be most

TABLE 28.2
Types of Actuator Failures

Type of Failure	Description of Specific Failure
Type I	Failure of any one elevator
Type II	Failure of any one aileron
Type III	Failure of left elevator and left aileron
Type IV	Failure of left elevator and right aileron
Type V	Failure of left and right ailerons

Source: A. A. Pashilkar, N. Sundararajan and P. Saratchandran, *Journal of Guidance, Control and Dynamics*, 30(3), 835–847, 2007.

detrimental if it occurs before these manoeuvres. Thus, in all the simulation results presented in this chapter, the failures occur at 10 s for the elevator and 8 s for either aileron, both of which occur before the turn is initiated.

28.3 NEURAL-AIDED TRAJECTORY FOLLOWING CONTROLLER

In this section, the improvement of fault tolerance of a baseline controller with the aid of a neural controller in a feedback error learning mode is studied with reference to actuator failures during the autonomous landing task. In Figure 28.3, the feedback error structure for a neural-aided controller scheme is presented. This controller processes the measured aircraft outputs and the reference signals from the classical controller to generate the error signal. The error signal is used to adapt the neural network parameters and obtain the neural network output u_{nn}. The complete control scheme is designed in two parts (Figure 28.3). The outer loop consists of a tracking command generator (to generate the command signals based on trajectory deviations), which is the navigation loop. The inner controller is the BTFC. The neural controller employs a dynamic radial basis function neural network called extended minimal resource allocating network (EMRAN), which uses

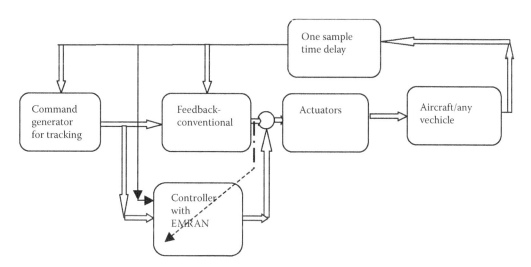

FIGURE 28.3 Neural-aided trajectory following controller. (From H. Gomi and M. Kawato, *Neural Networks*, 6(7), 933–946, 1993.)

only online learning and does not need prior training [5]. Both controllers use the reference signal 'r' (Figure 28.3) generated by the tracking controller to command the aircraft control surfaces. The neural controller uses the reference signals and the aircraft outputs to generate its command signal. It also uses the output of the BTFC to learn the inverse dynamics of the plant (in this case, the aircraft).

28.3.1 BASIC TRAJECTORY FOLLOWING CONTROLLER DESIGN

The BTFC is designed using the classical loop shaping single-input single-output (SISO) design techniques. It is assumed that the angle of attack and sideslip is not available for feedback. The classical cascade feedback controller processes the four reference signals generated by the tracking controller in the following way: (i) Altitude control is achieved by generating the desired attitude command. The flight path angle error is also used for feedback in order to provide damping to the altitude signal. The innermost loop is the pitch rate loop; (ii) Airspeed control is achieved by manipulating the throttle; (iii) Tracking of trajectories in the x–y plane is achieved using angular ground track deviation as the inner loop. This loop in turn commands the bank angle loop with the roll rate loop as the innermost; (iv) Sideslip is minimized by commanding the rudder using the estimated sideslip rate given by $(r - p*\tan\alpha)$ and lateral acceleration as feedback signals. All the loops are designed with a minimum gain margin of 6 dB and phase margin of at least 45 degrees. The final controller structure of the longitudinal controller is in the form of a cascade (see Figure 28.4) [11]. The 'from roll axis' is used for the two-aileron failure case where in it is used to induce differential elevator behaviour. When a Type V failure (i.e., both the left and right ailerons fail) occurs, the standard BTFC-based controller presented in Ref. [10] is unable to handle this situation. Failure of both ailerons implies that there is no control surface present to regulate the roll motion. The elevators of the aircraft used in differential mode are 60% as effective as the ailerons in inducing roll. This characteristic is used to control the aircraft for Type V failure. The final scheme for lateral-directional control laws is as shown in Figure 28.5 [11].

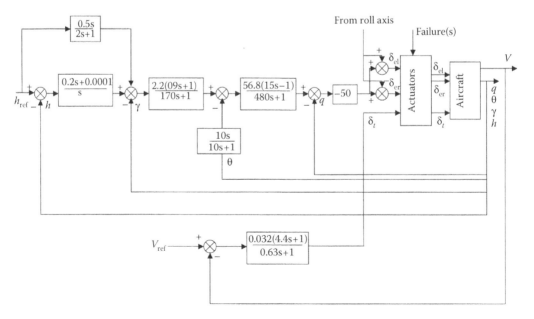

FIGURE 28.4 Longitudinal axis design for BTFC.

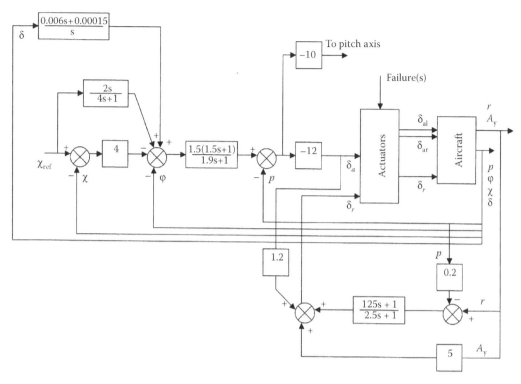

FIGURE 28.5 Lateral-directional axis design of BTFC.

28.3.2 EXTENDED MINIMUM RESOURCE ALLOCATING NETWORK

EMRAN is a fast implementation of the sequential learning radial basis feed forward (RBF) network of Lu et al. [12] referred to as MRAN. MRAN is a sequential learning algorithm and utilizes a compact RBF neural network. A brief description of MRAN and its extension EMRAN is given here. For more details, see Ref. [5]. The outputs of an RBF network with Gaussian function Φ are given by

$$f(\xi_n) = a_0 + \sum_{i=1}^{h} \Phi_i(\xi_n) \quad \xi \in R^m, \quad f \in R^p \tag{28.10}$$

where

$$\Phi_i(\xi_n) = \alpha_i \exp\left(-\frac{1}{\sigma_i^2}\|\xi_n - \mu_i\|^2\right)$$

where ξ is the input vector of the network, h indicates the total number of hidden neurons, μ_i, σ_i refers to the centre and width of the ith hidden neuron and n is the time index. $\|\ \|$ denotes the Euclidean norm. The function f is the output of the RBF network, which represents the network approximation to the desired output y_n. In the feedback error learning architecture, the desired output is that of the BTFC (u_c), as shown in Figure 28.3. The coefficient α_i denotes the connection weights of the ith hidden neuron to the output layer and a_0 is the bias term and both are vectors. The network starts with no hidden neurons. As input data are received sequentially, the network adds or prunes hidden neurons based on certain criteria. The sequential learning algorithm for MRAN is summarized as follows [12]: (i) Obtain an input and calculate the network outputs and the corresponding error; (ii) Create a new RBF hidden neuron if the following three conditions are satisfied: (a) the error exceeds a minimum threshold value, (b) the mean square error of the network for a series of past data has been above a certain threshold value and (c) the new input is sufficiently far from the existing hidden neurons and (iii) If condition (ii) is not met, adjust the weights and widths of the existing RBF network using the extended Kalman filter algorithm.

As discussed previously, EMRAN differs from the MRAN in this last step [5]. Instead of updating all the parameters (representing weights, centres and widths) of all the hidden neurons, it only updates the parameters of the nearest neuron. There is only slight difference between the performance of EMRAN and MRAN in terms of the approximation error, but in terms of speed, EMRAN outperforms MRAN significantly. In addition to the addition of neurons, a pruning strategy is adopted: (i) If a hidden neuron's normalized contribution to the output for a certain number of consecutive inputs is found to be below a threshold value, that hidden neuron is pruned; and (ii) If two hidden neurons are found to be close to each other, as defined by a threshold value, then they are combined into a single hidden neuron. The dimensions of the extended Kalman filter (EKF) are adjusted and the next input is processed. The pruning strategy ensures that the resulting radial basis network uses a minimum number of neurons. The lower number of neurons makes the algorithm computationally efficient. Consider the aircraft dynamics represented by the equation

$$\dot{x} = f(x, u) \tag{28.11}$$

With f assumed to be smooth and having bounded first derivatives in the neighbourhood of the trajectory. Bugajski and Enns [13] have shown that it is possible to invert the non-linear dynamic equations for an aircraft and generate a robust non-linear controller. The inversion process begins with the calculation of the control surface deflections based on the desired angular accelerations by inverting the three angular rate equations simultaneously. The desired angular accelerations are computed using the inversion of the force equations. This process eventually gives a controller, which is able to receive as reference signals, the desired velocity, altitude and heading and generate the elevator, aileron, rudder and throttle control signals to track the reference signals. One may represent the inversion by the equations:

$$u = f^{-1}(\dot{x}, x) \tag{28.12}$$

where f^{-1} represents the inversion of the equations of motion. In practice, if $\dot{x}_d$ is the desired value of the derivative of the state vector, it is beneficial to substitute $\dot{x}_d - G \cdot (x - x_d)$ in Equation 28.12 and obtain the control input. An explanation for this is provided in Ref. [11]. Thus, we must synthesise this set of multivariable functions using the states and their derivatives to obtain the control inputs required to make the aircraft follow the desired trajectory. Further, if this function representing the inverse aircraft dynamics is changing over a period of time, we can exploit the learning ability of the neural network to generate immediate corrective action when such changes take place. In this chapter, we make the assumption that the aerodynamic control inputs (elevator, aileron and rudder) do not have significant coupling. This means that the elevator primarily generates pitching moments, the aileron contributes the rolling moment and the rudder mainly results in yawing moments. This assumption allows us to postulate two inversion controllers for the aircraft, one each for the longitudinal- and lateral-directional axes. With this as the motivation, Gomi and Kowato [7] have proposed a feedback error learning strategy. This postulates the classical feedback controller output as the signal to be learned ($u_c = u_t = u_{nn}$). Over a period of time the EMRAN learns the total control signal ($u_{nn} \to u_t$) which results in driving the BTFC control output to zero (Figure 28.3). This signifies that the EMRAN has generated the inverse of the plant by learning the inverse functions represented by Equation 28.12. There are two such blocks used in the trajectory simulation, one each for the

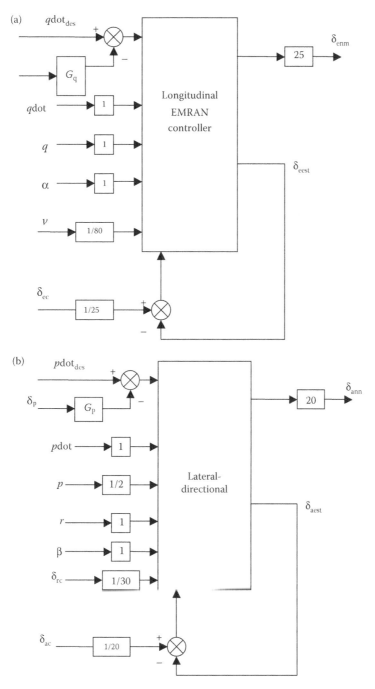

FIGURE 28.6 (a) EMRAN implementation for the longitudinal axis. (b) EMRAN implementation for the lateral-directional axis.

longitudinal- and lateral-directional axes. If EMRAN learns from the classical controller, there is the possibility that it will not achieve anything better than this controller. We have modified the error signal used by the EMRAN by adding scaled trajectory error signal to the classical controller output to remedy this drawback. Figure 28.6a shows the scaling factors used for the longitudinal- and lateral-directional axis EMRAN controllers [11]. Similar scheme is also used for the lateral EMRAN block (see Figure 28.6b) [11]. Taken together, the longitudinal- and lateral-directional EMRAN blocks represent the neuro-controller in Figure 28.3. It is noted that the inputs, errors and outputs of the longitudinal EMRAN block are

$$\xi = \left[\dot{q}_d - G_q \cdot (q - q_d) \quad \dot{q} \quad q \quad \hat{\alpha} \quad V_T \right]$$
$$e = \left[\delta_{ec}/25 - \delta_{eest} \right]$$
$$y = \left[\delta_{enm}/25 \quad \delta_{eest} \right]$$

In this case, $G_q = -0.48967$. A similar scheme is also used for the lateral EMRAN block. The inputs, error and the outputs for this are given as follows:

$$\xi = \left[\dot{p}_d - G_p \cdot (p - p_d) \quad \dot{p} \quad p/2 \quad r \quad \hat{\beta} \quad \delta_{rc}/30 \right]$$
$$e = \left[\delta_{ac}/20 - \delta_{aest} \right]$$
$$y = \left[\delta_{ann}/20 \quad \delta_{aest} \right]$$

In this case, $G_p = 0.14971$.

28.3.3 Performance Evaluation of BTFC and Neural-Aided BTFC under Failures

To generate the failure tolerance region for these controllers, simulations have been conducted at one–two degree intervals for both single (Type I and Type II) and double failures (Types III, IV and V) of control surface actuators. Success is judged by touchdown pillbox specifications discussed earlier. Figures 28.7 through 28.13 present the results for these cases that are analysed in detail below. Figure 28.7 shows the lateral position (Y), sideslip angle (beta), altitude and the velocity during the landing phase for both the EMRAN controller against the reference for the failure case of left elevator stuck at 16 degrees at 10 s. This is a significantly large failure position. Figure 28.8 shows the elevator and aileron deflections along with the control signals from the BTFC and EMRAN and also the neuron history for EMRAN. Figure 28.8a shows the deflection of the left elevator ($\delta_{e\text{-left-t}}$) and right elevator ($\delta_{e\text{-right-t}}$) along with the BTFC ($\delta_{e\text{-right-c}}$) and NN components ($\delta_{e\text{-right-nn}}$) of the control signals. The neuron history ($nn_{\delta e}$) based on EMRAN for right elevator is also shown. Figure 28.8b shows similar results for the ailerons. The neural-aided controller is not only able to land the aircraft but also achieves the touchdown performance requirements. The asymmetry due to the left elevator being stuck results in a persistent negative sideslip deviation subsequent to the occurrence of the failure. For learning, EMRAN requires 1 neuron each for the longitudinal axis and 15 neurons for the lateral axis. In Figures 28.9 through 28.13, the range of tolerance of the EMRAN controller to the different types of failures is shown. It is observed that there is a significant increase in the range of failures that can be successfully handled by the EMRAN controller as compared to the BTFC. Figure 28.9 shows that a single elevator failure in the range of −8 to 17 degrees (excluding failures at −2 degrees) can be tolerated by the EMRAN controller, whereas the BTFC has a much more restricted range of −2 to 6 degrees. Similarly, we observe a large improvement in performance for other failure cases as well. Furthermore, the case of two-aileron failure (Type V) can be handled by the EMRAN controller for a substantially large part of the feasible region. In the case of combination failures, there are some gaps in the overall fault-tolerant region. Some of these gaps can be attributed to the y-deviation from the pillbox conditions not being met. The neural-aided (EMRAN) controller shows a larger area of performance across all types of failures. An online neural network learning scheme called EMRAN was applied to the controller in a feedback error learning architecture. The non-linear dynamic inversion (NDI) design

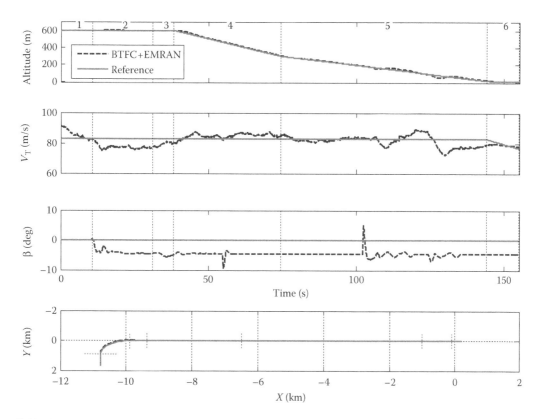

FIGURE 28.7 Time response of neural aided (EMRAN) + BTFC controller (single elevator failure at 16 degrees at 10 s). (From A. A. Pashilkar, N. Sundararajan and P. Saratchandran, *Journal of Aerospace Science & Technology*, 10(1), 49–61, 2006. With permission.)

technique is the basis for such neural augmentation. Results show that although the fault tolerance of the BTFC is increased, the region of failure tolerance exhibits several gaps (i.e., it is not singly connected). This poses problems in providing performance guarantees for the EMRAN controller. In the next section, we discuss an adaptive backstepping design which works on the principle of non-linear inversion which does not suffer from these problems.

28.4 ADAPTIVE BACKSTEPPING NEURAL CONTROLLER

The ABNC design for aircraft is developed in Ref. [9]. The control scheme uses radial basis function neural networks in an adaptive backstepping architecture with full state measurement for trajectory following. The requirement for stability is separated from the network learning part. This allows us to use any function approximation scheme (including neural networks) for learning. For the RBF neural networks, a learning scheme in which the network starts with no neurons and adds new neurons based on the trajectory error is developed. Stable tuning rules are derived for the update of the centres, widths and weights of the RBF neural networks. Using Lyapunov theory, a proof of stability in the ultimate bounded sense is presented for the resulting controller. The fault-tolerant controller design is illustrated for an unstable high-performance aircraft in the terminal-landing phase subjected to multiple control surface hard over failures and severe winds. The design uses the full six-degrees-of-freedom non-linear aircraft model and the simulation studies show that the above controller is able to successfully stabilize and land the aircraft within tight touchdown dispersions. Bugajski and Enns [13] have used the backstepping approach for NDI control. A feature of this

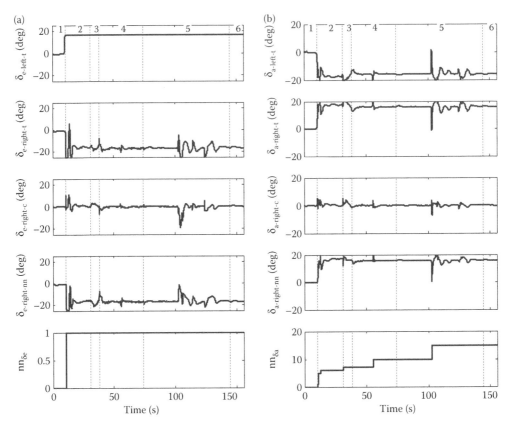

FIGURE 28.8 Elevator and aileron deflections and neuron history for EMRAN + BTFC (single elevator failure at 10 s). (From A. A. Pashilkar, N. Sundararajan and P. Saratchandran, *Journal of Aerospace Science & Technology*, 10(1), pp. 49–61, 2006. With permission.)

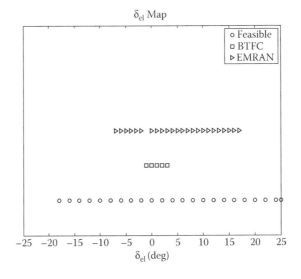

FIGURE 28.9 Left elevator failure tolerance of the EMRAN controller. (From A. A. Pashilkar, N. Sundararajan and P. Saratchandran, *Journal of Aerospace Science & Technology*, 10(1), 49–61, 2006. With permission.)

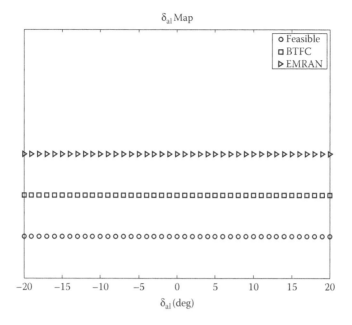

FIGURE 28.10 Left aileron failure tolerance of the EMRAN controller. (From A. A. Pashilkar, N. Sundararajan and P. Saratchandran, *Journal of Aerospace Science & Technology*, 10(1), 49–61, 2006. With permission.)

approach is to do away with the demanded derivative of the state at each step of the dynamic inversion process and replace it with a signal proportional to the error in the state. The NDI is made robust by these proportional feedback gains. This removes the problem of high-frequency signals arising out of numerical differentiation propagating down the neural network controller and causing undesirable actuator saturation. The approach requires that the control designer separate the plant dynamics into

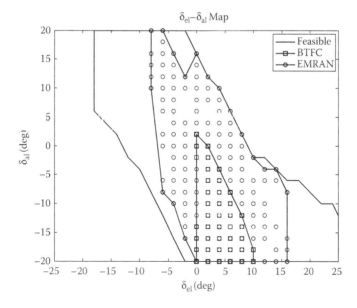

FIGURE 28.11 Left elevator and left aileron failure tolerance of the EMRAN controller. (From A. A. Pashilkar, N. Sundararajan and P. Saratchandran, *Journal of Aerospace Science & Technology*, 10(1), 49–61, 2006. With permission.)

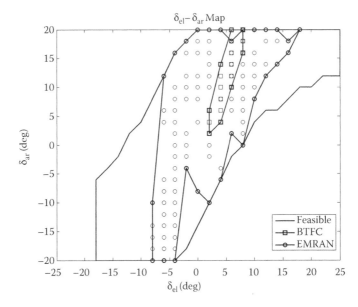

FIGURE 28.12 Left elevator and right aileron failure tolerance of the EMRAN controller. (From A. A. Pashilkar, N. Sundararajan and P. Saratchandran, *Journal of Aerospace Science & Technology*, 10(1), 49–61, 2006. With permission.)

'fast' and 'slow' time scales. The stability of the system depends on achieving adequate separation in the time scales. However, under large changes in the aircraft dynamics, the feedback gains of the system may not be sufficient to stabilize the system because the inverse dynamic model of the aircraft is not adaptive. Higher gains could be used to offset this, but may excite high-frequency unmodelled dynamics in the plant. In Ref. [9], we provide results applicable to the full six-degrees-of-freedom non-linear dynamics (including both longitudinal and lateral dynamics) and show that an online

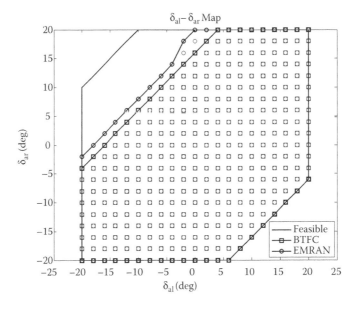

FIGURE 28.13 Left and right aileron failure tolerance of EMRAN controller. (From A. A. Pashilkar, N. Sundararajan and P. Saratchandran, *Journal of Aerospace Science & Technology*, 10(1), 49–61, 2006. With permission.)

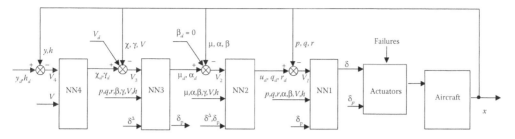

FIGURE 28.14 Adaptive backstepping neuro-controller architecture. (From A. A. Pashilkar, N. Sundararajan and P. Saratchandran, *Journal of Guidance, Control and Dynamics*, 30(3), 835–847, 2007.)

backstepping neural controller can be designed with stable adaptive tuning rules for trajectory tracking control of aircraft executing a landing manoeuvre under control surface actuator failures (elevator and ailerons). Both single and double actuator stuck-up failures have been considered.

The architecture of the backstepping neural controller is shown in Figure 28.14 and is called ABNC. It can be seen from Figure 28.14 that the scheme has four neural networks in a cascade. It is noted that Johnson and Kannan [14] proposed a cascade structure for adaptive control of helicopters. The number of neural networks required depend on the relative degree of the output (i.e., the number of times the output is differentiated to make the control input appear in the equation). All the networks receive the aircraft state vector as input apart from their individual inputs. Network NN1 is innermost and produces the elevator, aileron and rudder control signals. The cascade terminates in NN4 whose input is the desired aircraft output we wish to control during the landing phase (here, altitude and cross track deviation). This is the integrator backstepping process and is described in detail in Ref. [9]. The individual network learns the inverse non-linear dynamics of each stage. The parameters of each of the neural networks are adapted by using suitable error signals derived from Lyapunov theory. The stability proof presented here guarantees the ultimate boundedness of all signals in the closed-loop system.

28.4.1 NON-LINEAR DYNAMIC INVERSION

In this section, we present the design methodology for the adaptive neural network aircraft controller using the backstepping approach. The integrator backstepping design technique was proposed by Kanellakopoulos et al. [15]. We follow the formulation of the equations of motion as given in Ref. [13]. They are

$$\dot{x} = f(x, u)$$
$$Y = l(x) \tag{28.13}$$

where the state vector is

$$x = [x_1, \ldots, x_{11}]^T = \begin{bmatrix} (p, q, r)^T \\ (\mu, \alpha, \beta)^T \\ (\chi, \gamma, V)^T \\ (y, h)^T \end{bmatrix} = \begin{bmatrix} X_1 \\ X_2 \\ X_3 \\ X_4 \end{bmatrix}$$

the control vector is

$$u = \begin{bmatrix} (\delta_e, \delta_a, \delta_r)^T \\ \delta_p \end{bmatrix} = \begin{bmatrix} \delta \\ \delta_p \end{bmatrix}$$

and the output is $Y = [\beta\ y\ h\ V]^T$. The components of the state vector retain their usual meaning [13]. The control vector u consists of the elevator, aileron, rudder and throttle in that order. The objective is to design a controller to track a given smooth reference trajectory Y_d with bounded errors using state feedback. In shorthand vector notation, Equation 28.13 can be rewritten as

$$\dot{X}_1 = F_1(p,q,r,\alpha,\beta,V,h,\delta,\delta_p)$$
$$\dot{X}_2 = F_2(p,q,r,\mu,\alpha,\beta,\gamma,V,h,\delta,\delta_p)$$
$$\dot{X}_3 = F_3(p,q,r,\mu,\alpha,\beta,\gamma,V,h,\delta,\delta_p)$$
$$\dot{X}_4 = F_4(\chi,\gamma,V)$$

(28.14)

It should be noted that the uncertain terms in Equation 28.14 are due in the aerodynamic force and moment coefficients. Further, the throttle (δ_p) to thrust (T) mapping appearing in these equations is also assumed to be uncertain. The dependence of the aerodynamic and thrust force and moment coefficients with respect to the elements of the control vector is assumed to be non-affine.

28.4.2 Controller Design Methodology

Equation 28.14 can be written implicitly in terms of x and u as

$$h_1(x,u,\dot{X}_1) = \dot{X}_1 - F_1(x,u) = 0$$
$$h_2(x,u,\dot{X}_2) = \dot{X}_2 - F_2(x,u) = 0$$
$$h_3(x,u,\dot{X}_3) = \dot{X}_3 - F_3(x,u) = 0$$
$$h_4(x,\dot{X}_4) = \dot{X}_4 - F_4(x) = 0$$

(28.15)

The functions $h_1, \ldots, h_4$ in Equation 28.15 are the implicit form of the equations of motion. Using the implicit function theorem on these equations, there exist functions $K_1, \ldots, K_4$ such that

$$[\chi,\gamma]^T = K_4(\dot{X}_4,V)$$
$$[\mu,\alpha,\delta_p]^T = K_3(\dot{X}_3,p,q,r,\beta,\gamma,V,h,\delta)$$
$$[p,q,r]^T = K_2(\dot{X}_2,\mu,\alpha,\beta,\gamma,V,h,\delta,\delta_p)$$
$$[\delta_e,\delta_a,\delta_r]^T = K_1(\dot{X}_1,p,q,r,\alpha,\beta,V,h,\delta_p)$$

(28.16)

The above equations represent the exact inverse functions of Equation 28.14. The controller consists of evaluating the above multivariable vector functions starting from K_4 through K_1 in that order to obtain the various control signals (u). This amounts to stepping back through the equations of motion as they are normally written; hence, the name backstepping. The functions have to be evaluated with the desired state derivatives $\dot{X}_1, \ldots, \dot{X}_4$. The desired state vector is partly obtained from the output vector Y and from Equation 28.16. As discussed in Refs. [13,16], the NDI controller consists of evaluating the inverse functions K_i by setting the desired state derivatives equal to signals proportional to the state trajectory error. Towards this end, we define the pseudo-control variables v_i, $i = 1, \ldots, 4$.

$$V_1 = \begin{bmatrix} v_1 \\ v_2 \\ v_3 \end{bmatrix} = \begin{bmatrix} \Gamma_1 e_1 \\ \Gamma_2 e_2 \\ \Gamma_3 e_3 \end{bmatrix}, \quad V_2 = \begin{bmatrix} v_4 \\ v_5 \\ v_6 \end{bmatrix} = \begin{bmatrix} \Gamma_3 e_3 \\ \Gamma_4 e_4 \\ \Gamma_5 e_5 \end{bmatrix}, \quad V_3 = \begin{bmatrix} v_7 \\ v_8 \\ v_9 \end{bmatrix} = \begin{bmatrix} \Gamma_7 e_7 \\ \Gamma_8 e_8 \\ \Gamma_9 e_9 \end{bmatrix}, \quad V_4 = \begin{bmatrix} v_{10} \\ v_{11} \end{bmatrix} = \begin{bmatrix} \Gamma_{10} e_{10} \\ \Gamma_{11} e_{11} \end{bmatrix} \quad (28.17)$$

where Γ_i are the gains to be designed and e_i are the components of the error vector

$$e_i = x_i - x_i^d, \quad i = 1, \ldots, 11 \tag{28.18}$$

This ensures that the states converge to the desired values and the controller is robust to external disturbances. The inverse functions in Equation 28.16 are replaced by neural network stages to arrive at ABNC architecture for uncertain systems (Figure 28.14). The stages are implemented as

$$
\begin{aligned}
\left[\chi^d, \gamma^d\right]^T &= NN_4(V_4, V) = \hat{W}_{04} + \hat{W}_4^T \hat{\Phi}_4(V_4, V) \\
\left[\mu^d, \alpha^d, \delta_p^d\right]^T &= NN_3(V_3, p, q, r, \beta, \gamma, V, h, \delta) = \hat{W}_{03} + \hat{W}_3^T \hat{\Phi}_3(V_3, p, q, r, \beta, \gamma, V, h, \delta^\Delta) \\
\left[p^d, q^d, r^d\right]^T &= NN_2(V_2, \mu, \alpha, \beta, \gamma, V, h, \delta, \delta_p) = \hat{W}_{02} + \hat{W}_2^T \hat{\Phi}_2(V_2, \mu, \alpha, \beta, \gamma, V, h, \delta^\Delta, \delta_p) \\
\left[\delta_e^d, \delta_a^d, \delta_r^d\right]^T &= NN_1(V_1, p, q, r, \alpha, \beta, V, h, \delta_p) = \hat{W}_{01} + \hat{W}_1^T \hat{\Phi}_1(V_1, p, q, r, \alpha, \beta, V, h, \delta_p)
\end{aligned}
\tag{28.19}
$$

where $\hat{W}_{0i}$ are estimates of the network bias terms, $\hat{W}_i$ are the estimates of optimal RBF weight vectors and $\hat{\Phi}_i$ are estimates of the optimal radial basis function vectors. The quantities $[\sigma_1, \ldots, \sigma_{11}]^T = \chi^d, \gamma^d, \mu^d, \alpha^d, \delta_p^d, p^d, q^d, r^d, \delta_e^d, \delta_a^d, \delta$ are time-varying neural network outputs representing the desired intermediate control signals in the approximating neural network cascade shown in Figure 28.14. It is noted that the neural network biases and weights in Equation 28.19 represent multivariable vector functions. Thus, each bias and weight in Equation 28.14 is a vector and matrix, respectively. The update rules for these networks must be designed in such a way that the system is driven to follow the desired trajectory. The functions NN_i represent the neural networks arranged in a cascade. The sequence of evaluation starts on the left-hand side from NN_4 and ends in NN_1. The latter produces the aerodynamic control vector (δ) that drives the aircraft actuators. It is seen that the networks NN_3 and NN_2 require the aerodynamic control deflections as input. This gives rise to an algebraic loop as these controls are computed downstream in network NN_1. One sample delayed aerodynamic controls as inputs to these neural networks (denoted by δ^Δ) are used for solving this problem. It is assumed that if the sampling rate is sufficiently high, the delay will not affect the network approximation process. For the purpose of illustration, we have used the radial basis function neural networks (RBFNN) for approximation. Any other neural network can also be used as the universal approximator. It is proved in Ref. [9] that stability is guaranteed if we ensure that the neural network tuning rule is designed to give the following rate of change:

$$
\begin{aligned}
\begin{bmatrix} \dot{\sigma}_1 \\ \dot{\sigma}_2 \\ \dot{\sigma}_3 \end{bmatrix} &= - \begin{bmatrix} \Lambda_{11} & \Lambda_{12} & \Lambda_{13} \\ \Lambda_{21} & \Lambda_{22} & \Lambda_{23} \\ \Lambda_{31} & \Lambda_{32} & \Lambda_{33} \end{bmatrix} \cdot \begin{bmatrix} e_1 \\ e_2 \\ e_3 \end{bmatrix} - \begin{bmatrix} \Omega_1 \cdot \sigma_1 \\ \Omega_2 \cdot \sigma_2 \\ \Omega_3 \cdot \sigma_3 \end{bmatrix} \\
\begin{bmatrix} \dot{\sigma}_4 \\ \dot{\sigma}_5 \\ \dot{\sigma}_6 \end{bmatrix} &= - \begin{bmatrix} \Lambda_{44} & \Lambda_{45} & \Lambda_{46} \\ \Lambda_{54} & \Lambda_{55} & \Lambda_{56} \\ \Lambda_{64} & \Lambda_{65} & \Lambda_{66} \end{bmatrix} \cdot \begin{bmatrix} e_4 \\ e_5 \\ e_6 \end{bmatrix} - \begin{bmatrix} \Omega_4 \cdot \sigma_4 \\ \Omega_5 \cdot \sigma_5 \\ \Omega_6 \cdot \sigma_6 \end{bmatrix} \\
\begin{bmatrix} \dot{\sigma}_7 \\ \dot{\sigma}_8 \\ \dot{\sigma}_9 \end{bmatrix} &= - \begin{bmatrix} \Lambda_{77} & \Lambda_{78} & \Lambda_{79} \\ \Lambda_{87} & \Lambda_{88} & \Lambda_{89} \\ \Lambda_{97} & \Lambda_{98} & \Lambda_{99} \end{bmatrix} \cdot \begin{bmatrix} e_7 \\ e_8 \\ e_9 \end{bmatrix} - \begin{bmatrix} \Omega_7 \cdot \sigma_7 \\ \Omega_8 \cdot \sigma_8 \\ \Omega_9 \cdot \sigma_9 \end{bmatrix} \\
\begin{bmatrix} \dot{\sigma}_{10} \\ \dot{\sigma}_{11} \end{bmatrix} &= - \begin{bmatrix} \Lambda_{10,10} & \Lambda_{10,11} \\ \Lambda_{11,10} & \Lambda_{11,11} \end{bmatrix} \cdot \begin{bmatrix} e_{10} \\ e_{11} \end{bmatrix} - \begin{bmatrix} \Omega_{10} \cdot \sigma_{10} \\ \Omega_{11} \cdot \sigma_{11} \end{bmatrix}
\end{aligned}
\tag{28.20}
$$

The above translates to the rate of change of RBFNN parameters via the following equation:

$$\dot{w}_i = \left[\nabla \hat{g}_i\right]^+ \cdot \dot{\sigma}_i, \quad i = 1, \dots, 11 \tag{28.21}$$

where $\left[\nabla \hat{g}_i\right]^+$ is the pseudo-inverse of the gradient vector of the RBFNN given by

$$\left[\nabla \hat{g}_i\right] = \left[I, \hat{\phi}_{i1} \cdot I, 2\hat{\phi}_{i1}/\hat{\sigma}_{i1}^2 \cdot \hat{W}_{i1} \cdot (x - \hat{\mu}_{i1})^T, 2\hat{\phi}_{i1}/\hat{\sigma}_{i1}^3 \cdot \hat{W}_{i1} \cdot \left\| x - \hat{\mu}_{i1} \right\|^2, \right.$$

$$\dots,$$

$$\left. \hat{\phi}_{ik} \cdot I, 2\hat{\phi}_{ik}/\hat{\sigma}_{ik}^2 \cdot \hat{W}_{ik} \cdot (x - \hat{\mu}_{ik})^T, 2\hat{\phi}_{ik}/\hat{\sigma}_{ik}^3 \cdot \hat{W}_{ik} \cdot \left\| x - \hat{\mu}_{ik} \right\|^2 \right]^T$$

The index I represents the ith network and the second subscript runs over the number of neurons $(1, \dots, m)$ in the network. The parameter vector $w_i = \left[W_{0i1}, \hat{W}_{i1}, \hat{\mu}_{i1}, \hat{\sigma}_{i1}, \dots, \hat{W}_{ik}, \hat{\mu}_{ik}, \hat{\sigma}_{ik} \right]^T$ consists of the RBFN biases, weights and centre widths. The MRAN scheme [12] described previously in which the neurons are allowed to adaptively grow based on certain criteria is used for the ABNC. Pruning of neurons is not used, and the maximum number of neurons is limited by an upper bound. When the number of neurons reaches this limit, no more neurons are added but the updates of the parameters are carried out. All signals in the controller are ultimately bounded and approach a compact set which depends on the network approximation error. To prevent loss of stability when the trajectory enters the compact set, a standard solution is to apply a deadzone [17]. As the network learns, the approximation error reduces. Hence, the deadzone is larger initially (ε_{max}) and subsequently reduced to a lower value (ε_{min}). The instantaneous value of the deadzone is given by $\varepsilon = \max(\varepsilon_{max} \cdot r^n, \varepsilon_{min})$, where n is the iteration count and the rate of change r is chosen to lie between $(0,1]$. The control law is implemented in discrete form. The computations to be performed by the controller are given below:

```
Begin
Compute the RBF network output
Compute the deadzone threshold ε
if ‖e‖ > ε then
        Update the network parameters using Equations 28.20 and 28.21.
    if m < N_max then
            Add a neuron at the current location with fixed overlap (κ_p)
            to the nearest neuron in input space (see Ref. [11] for
            details).
    End
End
End
```

The gains for the ABNC controller are chosen by exploiting the cascade structure of the controller. The innermost gains are designed first to achieve the desired speed of response and stability. Diagonal entries of the gain matrix multiplying the trajectory error (Λ_{ii}) primarily affect the speed of response, while the gain multiplying the output term (Ω_i) affects the stability of the closed loop. The gain Γ_i is the backstepping correction as discussed in the above section. Examination of the networks indicates that NN4 can be dispensed with in favour of the exact analytical inverse $\gamma^d = \sin^{-1}(\dot{h}^d/V), \chi^d = \sin^{-1}[\dot{y}^d/(V \cos \gamma)]$. Further, to exploit the full power of the approach, it is possible to design the neural network NN_1 to output the five control surface deflections separately, namely left elevator, right elevator, left aileron, right aileron and rudder. This is defined as multiple surface redundant control (MSRC) in Ref. [9]. The advantage of this approach is that by proper

TABLE 28.3
ABNC Design Gains for the MSRC Case

Network	Λ	Ω	Γ	ε_{min}	ε_{max}	r	κ_p
NNW	$\begin{bmatrix} 60 & 0 & 0 \\ 60 & 0 & 0 \\ 7 & -70 & 0 \\ -7 & -70 & 0 \\ 0 & 0 & -5 \end{bmatrix}$	$\begin{bmatrix} 60 \\ 60 \\ 37.5 \\ 37.5 \\ 5 \end{bmatrix}$	$\begin{bmatrix} 5 \\ 5 \\ 5 \end{bmatrix}$	0.001	0.002	0.88	2.5
NNW2	$\begin{bmatrix} 18 & 0 & 0 \\ 0 & 25 & 0 \\ 0 & 0 & -1 \end{bmatrix}$	$\begin{bmatrix} 13.5 \\ 7 \\ 1 \end{bmatrix}$	$\begin{bmatrix} 5 \\ 4 \\ 5 \end{bmatrix}$	0.001	0.002	0.88	2.5
NNW3	$\begin{bmatrix} 12 & 0 & 0 \\ 0 & 15 & 0 \\ 0 & 0 & 1.5 \end{bmatrix}$	$\begin{bmatrix} 7.5 \\ 16 \\ 5 \end{bmatrix}$	$\begin{bmatrix} 5 \\ 0.5 \\ 5 \end{bmatrix}$	0.002	0.003	0.88	2.5

Source: A. A. Pashilkar, N. Sundararajan and P. Saratchandran, *Journal of Guidance, Control and Dynamics*, 30(3), 835–847, 2007.

choice of the controller gains, it is possible to exploit the redundancy when multiple control effectors are available. For example, in this aircraft, the elevators can be used in differential mode (i.e., like the ailerons) to achieve control in the lateral axis. This means that it is possible to handle two aileron failures using the healthy elevators. The gains selected for evaluation of the ABNC controller are presented in Table 28.3 for the MSRC case.

28.4.3 PERFORMANCE EVALUATION OF ABNC UNDER FAILURES

Figure 28.15a shows the time history for the complete landing trajectory for both normal and failure cases for the MSRC implementation of the ABNC. Figure 28.15b shows the control surface deflections due to the MSRC. The failure case corresponds to the left elevator stuck at −8 degrees at 10 s and the left aileron stuck at 8 degrees at 8 s. From the altitude and velocity time histories, it can be seen that the performance of ABNC under elevator failure is close to the normal case, except for a residual sideslip in the failure condition. The result is a successful touchdown under failure meeting all the specifications. The MSRC implementation has the ability to handle both left and right aileron failure. This is demonstrated for the MSRC in the simulation plots shown in Figure 28.16a and b, where the left aileron is failed to 8 degrees and simultaneously the right aileron is also failed to −10 degrees at 8 s. It is seen that the MSRC is able to handle the two-aileron failure case. The ability of the neural network to apply differential elevator control to achieve fault tolerance in this case arises from the off-diagonal entries in the gain matrix (Table 28.3).

28.5 NEURO-FUZZY-AIDED TRAJECTORY FOLLOWING CONTROLLER

Fuzzy systems hold the capability to approximate non-linear systems with uncertainty over a wide range of operating conditions by means of a series of if–then fuzzy inference rules that represent the experience of human experts. Fuzzy logic-based controller design is a powerful tool to solve the aircraft landing problem under severe winds [18,19]. However, these methods create the fuzzy inference rules for fuzzy controller according to the designer experience and have no online learning

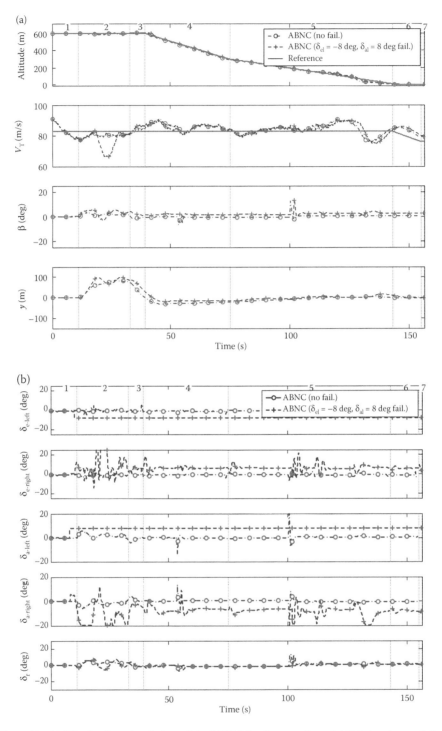

FIGURE 28.15 (a) The landing simulation of the MSRC implementation of ABNC under no failures and with left elevator and left aileron failed to −8 degrees at 10 s and 8 degrees at 8 s, respectively. (b) The control deflections of the MSRC implementation of ABNC with no failures and with left elevator and left aileron failed to −8 degrees at 10 s and 8 degrees at 8 s, respectively. (From A. A. Pashilkar, N. Sundararajan and P. Saratchandran, *Journal of Guidance, Control and Dynamics*, 30(3), 835–847, 2007.)

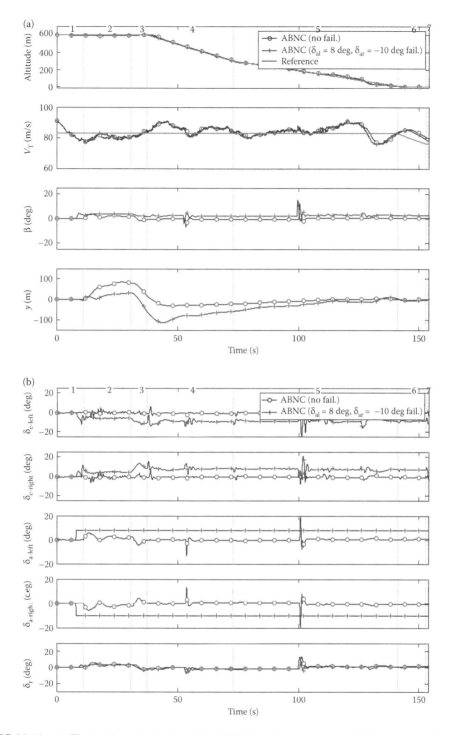

FIGURE 28.16 (a) The landing simulation of the MSRC implementation of ABNC under no failures and with left aileron and right aileron failed to 8 degrees at 8 s and −10 degrees at 8 s, respectively. (b) The control deflections of the MSRC implementation of ABNC with no failures and with left aileron and right aileron failed to 8 degrees at 8 s and −10 degrees at 8 s, respectively. (From A. A. Pashilkar, N. Sundararajan and P. Saratchandran, *Journal of Guidance, Control and Dynamics*, 30(3), 835–847, 2007.)

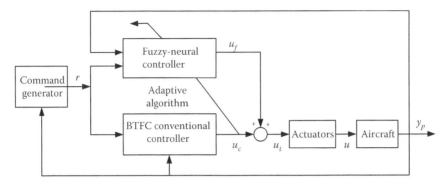

FIGURE 28.17   Fuzzy neural controller-aided BTFC control strategy for the autolanding problem.

ability. When these fuzzy rules are difficult to be preconstructed, an efficient control cannot be realized. Fuzzy neural networks (FNNs) can overcome the shortcomings of the conventional fuzzy logic systems in that they utilize the learning ability of neural networks to tune the shape of the fuzzy membership functions and their output weights. Fuzzy control schemes based on FNNs have been developed by many researchers for an aircraft landing problem [20,21]. These fuzzy neural control schemes achieve the desired performance under different wind patterns as shown in their simulation results. However, in all these works on landing, the failure of the actuators is not considered, although severe winds have been taken into account. Actuator failure during landing is a severe condition resulting in instability. Also, since one cannot predict the type of failures beforehand, we propose to use only online learning schemes compared to off-line training schemes that have been used in earlier works. When an FNN is utilized to solve practical problems, one has to consider two issues, viz., structure identification and parameter adjustment. Structure identification is related with the determination of number of fuzzy rules while parameter adjustment is to adjust the parameters existing in the antecedent and consequent parts of fuzzy rules. Recently, we have developed two efficient fuzzy neural algorithms named sequential adaptive fuzzy inference system (SAFIS) [22], where the number of fuzzy rules is determined online according to the input data, and an online sequential fuzzy extreme machine learning (OS-Fuzzy-ELM) [23], where the parameters for the fuzzy rules are updated at an extremely high speed. Two fault-tolerant fuzzy neural control strategies based on the two algorithms for the landing problem described above have been developed [24,25]. The control structure is based on the feedback–error–learning scheme which is illustrated in Figure 28.17. In the scheme, the proposed SAFIS and OS-Fuzzy-ELM algorithms are utilized as the fuzzy neural controller together with the BTFC controller for the aircraft autolanding problem. In this chapter, their performance is compared with the EMRAN-aided BTFC controller and the BTFC controller. The simulation results show that the proposed SAFIS-aided BTFC and OS-Fuzzy-ELM-aided BTFC control schemes have a clear improvement for the fault-tolerant envelope for both single and double faults compared to the conventional control methods used in the BTFC controller. Also, a direct performance comparison between the two fuzzy neural controllers for the autolanding fault-tolerant problem is presented in this chapter.

28.5.1   Neuro-Fuzzy Network

FNNs are built by incorporating the fuzzy inference process in the structure of neural networks and then the learning ability of neural networks are used to adjust the fuzzy rules. Except for some special fuzzy neural systems which made use of fuzzy neurons and fuzzy weights [26,27], most of the recent FNNs [22,23,28] have been built based on the standard feedforward network with local fields to approximate the fuzzy inference systems with local properties. Neural networks can handle the fuzzy inference systems, including the most commonly used Mamdani type of fuzzy models where the antecedent (if) part and the consequent (then) part are both described by the fuzzy sets, and Takagi–Sugeno–Kang (TSK) type of fuzzy models where only the antecedent part is described by fuzzy sets and the consequent part is described by real numbers. TSK fuzzy model can achieve better performance with fewer rules and has been widely applied in the non-linear system control problem [29,30]. The SAFIS and OS-Fuzzy-ELM algorithms are developed based on FNN with TSK fuzzy model.

The TSK fuzzy model is given by the following rules [31]:

$$Rule\, i : if\, (x_1\, is\, A_{1i})AND(x_2\, is\, A_{2i})AND \cdots AND(x_n\, is\, A_{ni}), then(y_1 is \omega_{i1}) \cdots (y_m is \omega_{im})$$

where A_{ji} ($j = 1, 2, \ldots, n$; $i = 1, 2, \ldots, \tilde{N}$) are the fuzzy sets of the jth input variable x_j in rule i, n is the dimension of the input vector $x(x = [x_1, \ldots, x_1]^T)$, m is the dimension of the output vector $y(y = [y_1, \ldots, y_m]^T)$ and $\tilde{N}$ is the number of fuzzy rules. ω_{ik} ($k = 1, 2, \ldots, m$; $i = 1, 2, \ldots, \tilde{N}$) is the crisp value that may be any function of the input variables but the commonly used is a constant value or a linear combination of the input variables. In the case of a linear function, it is given by $\omega_{ik} = q_{ik,0} + q_{ik,1}x_1 + \cdots + q_{ik,n}x_n$.

Based on the FNN structure illustrated in Ref. [32], the system output is calculated by

$$\hat{y} = \frac{\sum_{i=1}^{\tilde{N}} \omega_i R_i(x; c_i, a_i)}{\sum_{i=1}^{\tilde{N}} R_i(x; c_i, a_i)} = \sum_{i=1}^{\tilde{N}} \omega_i G(x; c_i, a_i) \tag{28.22}$$

where $\omega_i = (\omega_{i1}, \omega_{i2}, \ldots, \omega_{im})$ can be a constant value and a linear combination of input variables. R_i is the firing strength (if part) of the ith rule. If the Gaussian membership function is employed it will simply be

$$R_i(x; c_i, a_i) = \exp\left(-\frac{1}{a_i^2} \|x_t - c_i\|^2\right) \tag{28.23}$$

Or, if the Cauchy membership function is applied, it is given by

$$R_i(x; c_i, a_i) = \prod_{j=1}^{n} \frac{1}{1 + \left((x_i - c_{ij}/a_i)\right)^2} \tag{28.24}$$

where c_i and a_i are the parameters of the fuzzy membership function.

28.5.2 Sequential Adaptive Fuzzy Inference System

A SAFIS is developed to realise a compact fuzzy system with lesser number of rules by using the concepts from growing and pruning-RBF (GAP-RBF) neural network [33]. The SAFIS algorithm has two aspects: determination of the fuzzy rules and the adjustment of the premise and consequent parameters in the fuzzy rules. SAFIS uses the concept of influence of a fuzzy rule to add and remove rules during learning. Similar to the significance of a neuron in GAP-RBF, the influence of a fuzzy rule is also a mark of its contribution to the system output in a statistical sense when the input data are uniformly distributed. The learning procedure of the SAFIS algorithm is similar to that of the GAP-RBF algorithm, that is, addition and removal of a fuzzy rule is equivalent to the addition and removal of a neuron. The difference between them is that the growing and pruning criteria in SAFIS are based on the influence of a fuzzy rule while the GAP-RBF makes use of the significance of a neuron to add and remove the hidden neurons. However, conceptually they are utilized as the mark to add and delete fuzzy rules and neurons. In SAFIS, the TSK fuzzy model with the constant consequence is applied and its firing strength represented by the antecedent part (if part) of fuzzy rules is given by the Gaussian functions of the RBF network. Thus, the system output and the membership function are given by Equations 28.22 and 28.23, respectively. SAFIS uses the concept of influence of a fuzzy rule to add and remove rules during learning. It is given by

$$E_{\mathrm{inf}}(i) = |\omega_t| \frac{(1.8a_i)^n}{\sum_{i=1}^{\tilde{N}}(1.8a_i)^n} \tag{28.25}$$

Influence of a rule is utilized for the addition and deletion of a fuzzy rule in the SAFIS algorithm as indicated below. The sequential learning process of SAFIS is briefly summarized below. For details, see Ref. [10]. SAFIS begins with no fuzzy rules. When the first input x_1, y_1 is received, it is translated into the first rule whose parameters are given as $c_1 = x_1$, $\omega_1 = y_1$, $a_1 = \kappa \, \|x_1\|$. Then, as new inputs, x_t,y_t ($(t > 1)$ is the time index) are received sequentially during learning, growing of fuzzy rules is based on the following two criteria which are the distance criterion and the influence of the newly added fuzzy rule $\tilde{N} + 1$:

$$\begin{cases} \|x_t - c_{nr}\| > \varepsilon_t \\ E_{\mathrm{inf}}(\tilde{N}+1) = |e_t| \dfrac{(1.8k \, \|x_t - c_{nr}\|)^n}{\sum_{i=1}^{\tilde{N}+1}(1.8a_i)^n} > e_g \end{cases} \tag{28.26}$$

where ε_t and e_g are thresholds to be selected appropriately, x_t is the latest input data, c_{nr} is the centre of the fuzzy rule nearest to x_t, e_g is the growing threshold and is chosen according to the desired accuracy of SAFIS. $e_g = y_t - \hat{y}_t$, where y_t is the true value, $\hat{y}_t$ is the approximated value. κ is an overlap factor that determines the overlap of fuzzy rules in the input space, ε_t is the distance threshold which decays exponentially and is given by

$$\varepsilon_t = \max\{\varepsilon_{\max} \times \gamma^t, \varepsilon_{\min}\} \tag{28.27}$$

where $\varepsilon_{\max}$ and $\varepsilon_{\min}$ are the largest and smallest length of interest and γ is the decay constant. This equation shows that initially it is the largest length of interest in the input space which allows fewer fuzzy rules to coarsely learn the system and then it decreases exponentially to the smallest length of interest in the input space which allows more fuzzy rules to finely learn the system. When the new fuzzy rule $\tilde{N} + 1$ is added, its corresponding antecedent and consequent parameters are allocated as follows:

$$\begin{cases} \omega_{\tilde{N}+1} = e_t \\ c_{\tilde{N}+1} = x_t \\ a_{\tilde{N}+1} = \kappa \, \|x_t - c_{nr}\| \end{cases} \tag{28.28}$$

When no new fuzzy rule is added, SAFIS modifies the parameters of the nearest fuzzy rule which is the closest to the current input data. The parameter vector for all the fuzzy rules is given by

$$\theta_t = [\theta_1 \cdots \theta_{nr} \cdots \theta_{\tilde{N}}]^T = [\omega_1, c_1, a_1, \ldots, \omega_{nr}, c_{nr}, a_{nr}, \ldots, \omega_{\tilde{N}}, c_{\tilde{N}} a_{\tilde{N}}] \tag{28.29}$$

where $\theta_{nr} = [\omega_{nr}, c_{nr}, \alpha_{nr}]$ is the parameter vector of the nearest fuzzy rule and its gradient is derived as follows:

$$\dot{\omega}_{nr} = \frac{\partial \hat{y}_t}{\partial \omega_{nr}} = \frac{\partial \hat{y}_t}{\partial R_{nr}} \frac{\partial R_{nr}}{\partial \omega_{nr}} = \frac{R_{nr}}{\sum_{i=1}^{\bar{N}} R_i}$$

$$\dot{c}_{nr} = \frac{\partial \hat{y}_t}{\partial c_{nr}} = \frac{\partial \hat{y}_t}{\partial R_{nr}} \frac{\partial R_{nr}}{\partial c_{nr}} = \frac{\omega_{nr} - \hat{y}_t}{\sum_{i=1}^{\bar{N}} R_i} \frac{R_{nr}}{\sum_{i=1}^{\bar{N}} c_{nr}}$$

$$\dot{a}_{nr} = \frac{\partial \hat{y}_t}{\partial a_{nr}} = \frac{\partial \hat{y}_t}{\partial R_{nr}} \frac{\partial R_{nr}}{\partial a_{nr}} = \frac{\omega_{nr} - \hat{y}_t}{\sum_{i=1}^{\bar{N}} R_i} \frac{R_{nr}}{\sum_{i=1}^{\bar{N}} a_{nr}} \qquad (28.30)$$

$$\frac{\partial R_{nr}}{\partial c_{nr}} = 2R_{nr} \frac{x_t - c_{nr}}{a_{nr}^2}$$

$$\frac{\partial R_{nr}}{\partial \alpha_{nr}} = 2R_{nr} \frac{\|x_t - c_{nr}\|^2}{a_{nr}^3}$$

After obtaining the gradient vector of the nearest fuzzy rule, that is, $B_{nr} = [\omega_{nr}, C_{nr}, \alpha_{nr}]^T$, EKF is used to update its parameters as follows:

$$K_t = P_{t-1} B_t [R_t + B_t^T P_{t-1} B_t]^{-1}$$
$$\theta_t = \theta_{t-1} + K_t e_t \qquad (28.31)$$
$$P_t = [I - K_t B_t^T] P_{t-1} + Q_0 I$$

When a new rule is added, the dimension of P_t increases to

$$\begin{pmatrix} P_{t-1} & 0 \\ 0 & p_0 I_{z_1 \times z_1} \end{pmatrix} \qquad (28.32)$$

where Z_1 is the dimension of the parameters introduced by the newly added rule and P_0 is an initial value of the uncertainty assigned to the newly allocated rule. After parameter adjustment, the influence of the nearest fuzzy rule is computed. The deletion of fuzzy rules is based on the following criterion where the influence of the nearest fuzzy rule is compared with a pruning threshold e_p, If

$$E_{\text{inf}}(i) = |\omega_i| \frac{(1.8a_i)^n}{\sum_{i=1}^{\bar{N}} (1.8a_i)^n} < e_p \qquad (28.33)$$

the influence of the nearest rule to the output is small and can be removed.

28.5.3 Performance Evaluation of SAFIS Controller for Autolanding Problem

28.5.3.1 Single Surface Failure

In this section, a single failure of elevator or aileron is studied along with severe winds. First, we present the results for elevator failure. Here, the left elevator is stuck at −10 degrees at the beginning of the turn. Figures 28.18 and 28.19 show the altitude (h), velocity (V_T), sideslip angle (β) and lateral position (Y) during the landing phase for the SAFIS-aided BTFC, EMRAN-aided BTFC and BTFC.

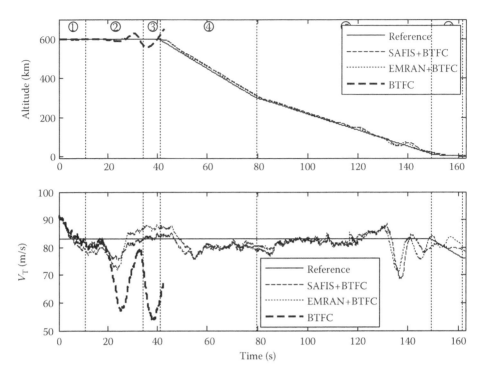

FIGURE 28.18 Comparison of SAFIS-aided BTFC, EMRAN-aided BTFC and BTFC for altitude (h) and velocity (V_T) under the left elevator stuck at −10 degrees.

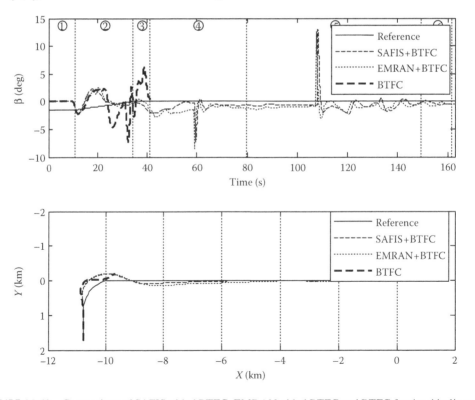

FIGURE 28.19 Comparison of SAFIS-aided BTFC, EMRAN-aided BTFC and BTFC for the sideslip angle (β) and lateral position (Y) under the left elevator stuck at −10 degrees.

TABLE 28.4

Performance Comparison between SAFIS Aided BTFC and EMRAN Aided BTFC Control Schemes under the Left Elevator Stuck at –10 Degrees

Methods	SAFIS Aided BTFC	EMRAN Aided BTFC
Trajectory error (RMSE)	8.8573	8.9206
Velocity error (RMSE)	3.4243	3.8146
Number of rules/neurons for δ_e	6	28
Number of rules/neurons for δ_a	14	15

The numbers at the top of Figures 28.18 and 28.19 represent the different segments of the trajectory. BTFC alone is unable to cope with this failure and achieve a safe landing as the altitude drops around 50 s. However, from Figures 28.18 and 28.19 it can be seen that SAFIS-aided BTFC and EMRAN-aided BTFC controllers are not only able to land the aircraft but also satisfy the touchdown performance requirements. Furthermore, it can be seen from Figures 28.18 and 28.19 that SAFIS-aided BTFC and EMRAN-aided BTFC control schemes are able to follow the reference trajectory closely. A large sideslip seen in Figure 28.19 around 110 s is due to the abrupt step inputs in the side gust profile but these excursions are quickly damped out by the controllers. Table 28.4 shows the RMSE trajectory errors along with the number of rules/neurons for SAFIS-aided BTFC and EMRAN-aided BTFC schemes. From the table, it can be seen that the trajectory error and the number of rules for SAFIS are smaller than those of EMRAN. To analyse the control scheme performance in more detail, Figure 28.20 shows the deflection of the left elevator $\delta_{e\text{-left-t}}$ and the right elevator $\delta_{e\text{-right-t}}$ along with the BTFC $\delta_{e\text{-right-c}}$ and SAFIS/EMRAN components $\delta_{e\text{-right-f/n}}$ of the control signals. The rule and neuron history based on SAFIS and EMRAN for the elevator δ_e is also shown. In Figure 28.20, the solid line represents the results for SAFIS-aided BTFC and the dotted

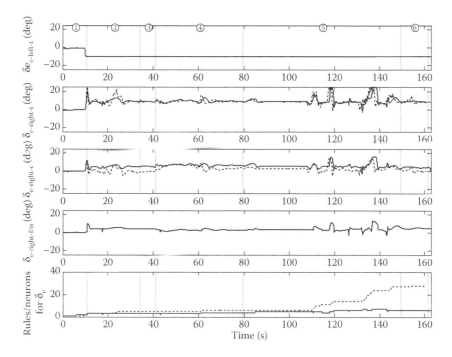

FIGURE 28.20 Left and right elevator control signals for SAFIS-aided BTFC and EMRAN-aided BTFC under the left elevator stuck at –10 degrees (solid line: SAFIS-aided BTFC; dot line: EMRAN-aided BTFC).

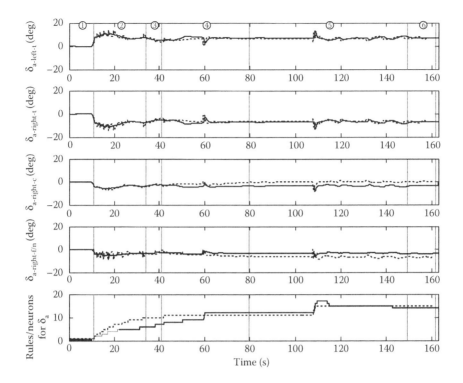

FIGURE 28.21 Left and right aileron control signals for SAFIS-aided BTFC and EMRAN-aided BTFC under the left elevator stuck at −10 degrees (solid line: SAFIS-aided BTFC; dot line: EMRAN-aided BTFC).

line represents the results from EMRAN-aided BTFC. From Figure 28.20, it is to be noted that the control signals for SAFIS-aided BTFC and EMRAN-aided BTFC are similar. However, the control signal for SAFIS-aided BTFC is less oscillatory than that of EMRAN-aided BTFC. Also, SAFIS only needs six rules for learning while EMRAN needs 28 neurons for learning.

The left and right aileron signals are given in Figure 28.21 and they are similar. The fault-tolerant capabilities of SAFIS-aided BTFC, EMRAN-aided BTFC and BTFC for a single elevator surface failure are given in Figure 28.22. Each point in Figure 28.22 indicates a successful landing meeting the touchdown pillbox requirements. From Figure 28.22, it can be seen that SAFIS-aided BTFC is able to meet the pillbox requirements for a wider range of deflections (−12 to 18 degrees) compared to EMRAN-aided BTFC (−12 to 12 degrees). No points corresponding to BTFC are shown in Figure 28.22 because it cannot meet the pillbox requirements for the entire range of failures from −18 to +25 degrees. Furthermore, SAFIS-aided BTFC is able to meet the pillbox requirements during continuous elevator deflections without 'gaps' but EMRAN-aided BTFC is unable to meet the pillbox requirements at −8, 0 and 8 degrees of elevator deflections. In the previous chapter, a detailed analysis has been given at the −8, 0 and 8 degree cases and we can observe that the three cases violate only one condition of the pillbox, namely the y-deviation at touchdown. Further, the amount of deviation from the pillbox condition is not large, being less than 1.5 m. The pillbox in the x–y plane is shown in Figure 28.23 along with the touchdown points for SAFIS. It can be seen from Figure 28.23 that all these touchdown points lie inside the pillbox. Similar results for the case of single aileron failure are shown in Figure 28.24. From Figure 28.24, one can note that the SAFIS-aided BTFC has a wider aileron failure tolerance range (−14 to 14 degrees) compared to BTFC (−7 to 4 degrees) and a slightly wider failure tolerance range than EMRAN-aided BTFC (−7 to 20 degrees) with the exception of failure at 2 degree (for EMRAN). As to the range from 15 to 20 degrees where EMRAN can tolerate the failures, SAFIS still is able to land the aircraft successfully

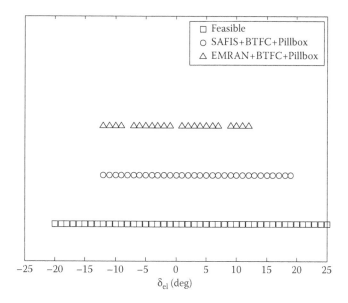

FIGURE 28.22 Failure tolerance for SAFIS-aided BTFC, EMRAN-aided BTFC and BTFC under left elevator stuck conditions.

but is unable to satisfy the touchdown pillbox requirements by only violating the y-distance criterion. Figure 28.25 gives the touchdown points for SAFIS together with the pillbox in the x–y plane and we can observe that these touchdown points are inside the pillbox.

28.5.3.2 Two Surface Failures

In this section, two failure cases are considered. Specifically, for the first case, the left elevator and left aileron are stuck at different deflections. For the second case, the left elevator and right aileron are considered.

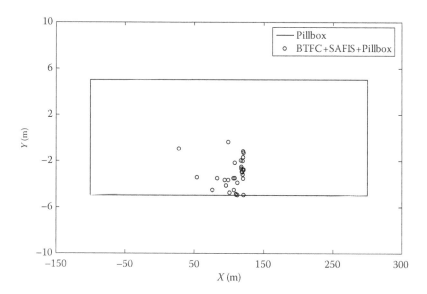

FIGURE 28.23 Touchdown points for SAFIS-aided BTFC under left elevator stuck conditions.

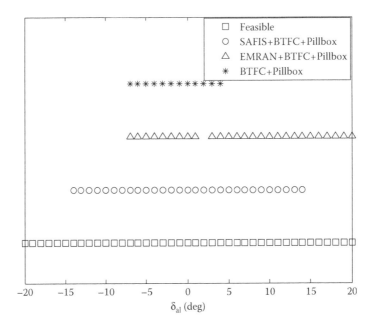

FIGURE 28.24 Failure tolerance for SAFIS-aided BTFC, EMRAN-aided BTFC and BTFC under left aileron stuck conditions.

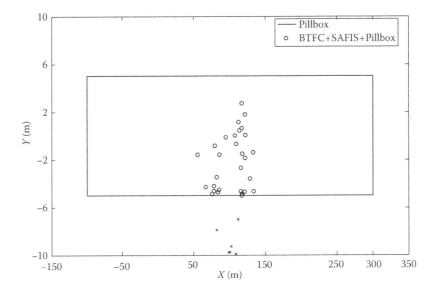

FIGURE 28.25 Touchdown points for SAFIS-aided BTFC under left aileron stuck conditions.

28.5.3.3 Left Elevator and Left Aileron Stuck at Different Deflections

We first present a typical trajectory result when the left elevator is stuck at −10 degrees and the left aileron is stuck at +10 degrees. Figures 28.26 and 28.27 show the altitude (h), the velocity (V_T), sideslip angle (β) and lateral position (Y) during the landing phase for the SAFIS-aided BTFC, EMRAN-aided BTFC and BTFC along with the reference trajectory. From Figures 28.26 and 28.27, it can be seen that BTFC alone is unable to cope with this failure and achieve a safe landing as the altitude drops around 30 s. However, SAFIS-aided BTFC and EMRAN-aided BTFC controllers are not only able to land the aircraft but also satisfy the touchdown performance requirements and also

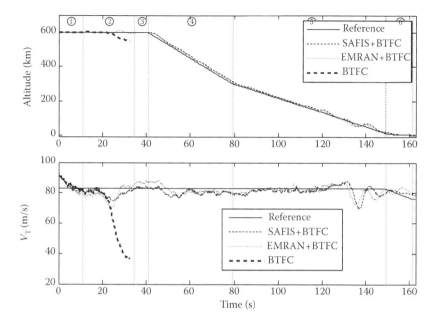

FIGURE 28.26 Comparison of SAFIS-aided BTFC, EMRAN-aided BTFC and BTFC for the altitude (*h*) and velocity (*V*_T) under the left elevator stuck at −10 degrees and the left aileron stuck at 10 degrees.

follow the reference trajectory closely. Table 28.5 shows the RMS trajectory errors along with the number of rules/neurons for SAFIS-aided BTFC and EMRAN-aided BTFC schemes. It can be seen from the table that the trajectory error and the number of rules for SAFIS are smaller than those of EMRAN. Figure 28.28 shows the deflection of the left elevator ($\delta_{\text{e-left-t}}$) and right elevator ($\delta_{\text{e-right-t}}$). In Figure 28.28, the solid line represents the results from SAFIS-aided BTFC and the dotted line

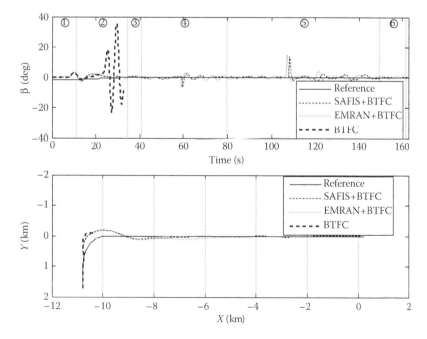

FIGURE 28.27 Comparison of SAFIS-aided BTFC, EMRAN-aided BTFC and BTFC for the sideslip angle (β) and lateral position (*Y*) under the left elevator stuck at −10 degrees and the left aileron stuck at 10 degrees.

TABLE 28.5

Performance Comparison between SAFIS Aided BTFC and EMRAN Aided BTFC Control Schemes under the Left Elevator Stuck at −10 Degrees and the Left Aileron Stuck at 10 Degrees

Methods	SAFIS Aided BTFC	EMRAN Aided BTFC
Trajectory error (RMSE)	8.8978	8.9127
Velocity error (RMSE)	3.2761	3.8097
Number of rules/neurons for δ_e	8	32
Number of rules/neurons for δ_a	22	25

represents the results from EMRAN-aided BTFC. Figure 28.28 also shows the control signals for the BTFC component ($\delta_{e\text{-right-t}}$) along with the SAFIS/EMRAN components ($\delta_{e\text{-right-t}}$). The rule and neuron update process based on SAFIS and EMRAN for the elevator δ_e is also shown. It is seen that EMRAN scheme needs around 30 neurons for learning whereas SAFIS requires only around 10 rules indicating that SAFIS can do the job with a compact network.

Similar results for the left and right ailerons are given in Figure 28.29. The fault-tolerant capabilities of SAFIS-aided BTFC, EMRAN-aided BTFC and BTFC for the whole range of deflections are given in Figure 28.30. The edge of the fault tolerance envelopes achieved by SAFIS-aided BTFC, EMRAN-aided BTFC and BTFC is given by the solid line, dashed line and dash–dot line. In their respective fault-tolerant regions, each point which indicates a successful landing meeting the touchdown pillbox

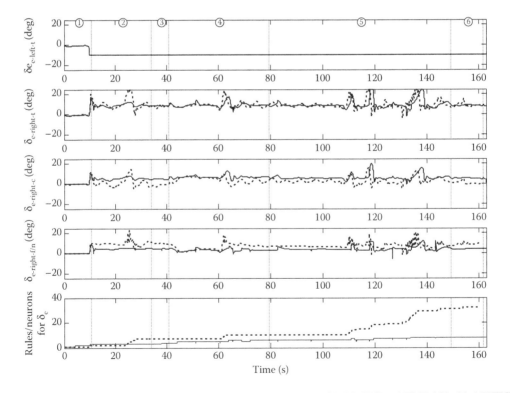

FIGURE 28.28 Left and right elevator control signals for SAFIS-aided BTFC and EMRAN-aided BTFC under the left elevator stuck at −10 degrees and the left aileron stuck at 10 degrees (solid line: SAFIS-aided BTFC; dot line: EMRAN-aided BTFC).

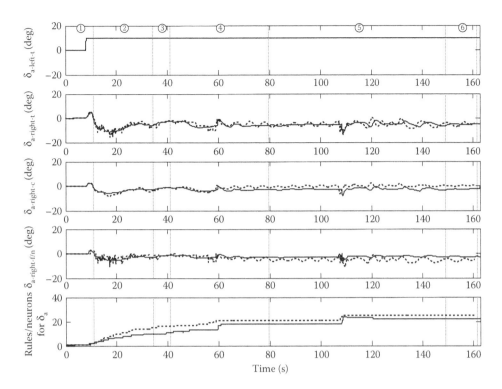

FIGURE 28.29 Left and right aileron control signals for SAFIS-aided BTFC and EMRAN-aided BTFC under the left elevator stuck at −10 degrees and the left aileron stuck at 10 degrees (solid line: SAFIS-aided BTFC; dot line: EMRAN-aided BTFC).

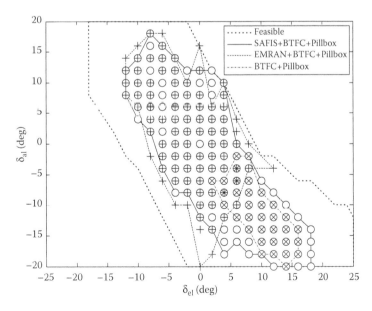

FIGURE 28.30 Failure tolerance for SAFIS-aided BTFC, EMRAN-aided BTFC and BTFC under left elevator and left aileron stuck conditions.

requirements is represented by o, + and ×. From Figure 28.30, it can be seen that SAFIS-aided BTFC is able to meet the pillbox requirements for a wider range of deflections compared to EMRAN-aided BTFC as well as BTFC. It can be seen from Figure 28.30 that the fault tolerance region achieved by SAFIS-aided BTFC covers those of EMRAN-aided BTFC and BTFC. But EMRAN-aided BTFC controller does not cover the entire BTFC controller region due to its inability to meet the tight pillbox requirements. It can also be observed from Figure 28.30 that SAFIS-aided BTFC has no 'gaps' in the whole region whereas the regions covered by EMRAN-aided BTFC and BTFC have 'gaps' indicating their inability to meet the tight pillbox requirements at those stuck deflections.

28.5.3.4 Left Elevator and Right Aileron Stuck at Different Deflections

The fault-tolerant envelope achieved by SAFIS-aided BTFC, EMRAN-aided BTFC and BTFC under both left elevator and right aileron failures is given in Figure 28.31. Their fault tolerance envelope edge is given by the solid line, dashed line and dash–dot line, respectively. In their respective fault-tolerant regions, each point which indicates a successful landing meeting the touchdown pillbox requirements is represented by the symbols: o, + and ×. From Figure 28.31, it can also be seen that SAFIS-aided BTFC is able to meet the pillbox requirements for a wider range of deflections compared to EMRAN-aided BTFC as well as BTFC. It can be seen from Figure 28.31 that the fault tolerance region achieved by SAFIS-aided BTFC covers those of EMRAN-aided BTFC and BTFC. But EMRAN-aided BTFC controller does not cover the entire BTFC controller region due to its inability to meet the tight pillbox requirements. It can be further observed from Figure 28.31 that SAFIS-aided BTFC has no 'gaps' in the whole region whereas the regions covered by EMRAN-aided BTFC and BTFC have 'gaps', indicating their inability to meet the tight pillbox requirements at those stuck deflections. One disadvantage of the SAFIS algorithm is that it requires many control parameters to be determined by trial and error before learning. In addition, SAFIS only can be applied for a specified fuzzy membership function that is a Gaussian membership function. In the next section, a new fast training algorithm which includes a smaller number of control parameters and which can be applied for any fuzzy membership function satisfying bounded non-constant piecewise continuous functions is introduced to train the FNN and is also utilized to solve the high-performance fighter autolanding problem.

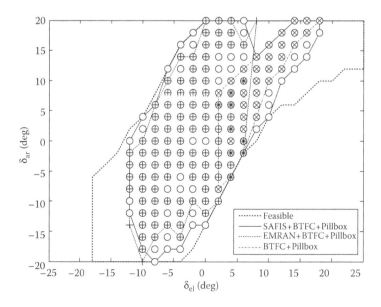

FIGURE 28.31 Failure tolerance for SAFIS-aided BTFC, EMRAN-aided BTFC and BTFC under left elevator and right aileron stuck conditions.

28.6 ONLINE SEQUENTIAL FUZZY EXTREME LEARNING MACHINE

A fast fuzzy neural algorithm called OS-Fuzzy-ELM is developed based on ELM [34–39] which combines the advantages of the neural network (learning) and the fuzzy inference system (approximate reasoning). OS-Fuzzy-ELM handles the fuzzy inference systems (including both TSK and Mamdani type of fuzzy models) with any bounded non-constant piecewise continuous membership function. Further, the learning in OS-Fuzzy-ELM can be done in a one-by-one or chunk-by-chunk (a block of data) mode with fixed or varying chunk sizes. In OS-Fuzzy-ELM, the parameters of the fuzzy membership functions need not be adjusted during training and one can randomly assign the values to them and then analytically determine the consequent parameters. The OS-Fuzzy-ELM consists of two main phases. The initialization phase is to train the FNNs using the Fuzzy-ELM* method with some batch of training data in the initialization stage and these initialization training data will be discarded as soon as the initialization phase is completed. For this, the required number of training data is very small, which can be equal to the number of rules. For example, if there are 10 rules, we may need 10 training samples to initialize the learning. After the initialization phase, the OS-Fuzzy-ELM will learn the train data one by one or chunk by chunk (with fixed or varying size) and then all the training data will be discarded once the learning procedure on these data is completed. The learning process of OS-Fuzzy-ELM for the TSK fuzzy model with the linear consequence is briefly summarized below. For details, see Ref. [23].

Proposed OS-Fuzzy-ELM algorithm: Given certain membership function g and rule number $\tilde{N}$ for a specific application, the data $\aleph = \{(x_i, t_i) \mid x_i \in R^n, t_i \in R^m, i = 1, \ldots\}$ arrives sequentially.

Step 1. Initialization phase: Initialize the learning using a small chunk of initial training data $\aleph_0 = \{(x_i, t_i)\}_{i=1}^{N_0}$ from the given training set $\aleph = \{(x_i, t_i) \mid x_i \in R^n, t_i \in R^m, i = 1, \ldots\}, N_0 \geq \tilde{N}$

1. Assign random membership function parameters (c_i, a_i), $i = 1, \ldots, \tilde{N}$.
2. Calculate the initial matrix H_0 for the TSK model:

$$H_0 = H(c_1, \ldots, c_{\tilde{N}}, a_1, \ldots, a_{\tilde{N}}; x_1, \ldots, x_{N_0})$$

where the details of H are referred to in [23].
3. Estimate the initial parameter matrix $Q^{(0)} = P_0 H_0^T T_0$, where $P_0 = (H_0^T H_0)^{-1}$ and

$$T_0 = [t_1, \ldots, t_{N_0}]^T$$

4. Set $k = 0$.

Step 2. Sequential learning phase: Present the $(k+1)$th chunk of new observations: $\aleph_{k+1} = \{(x_i, t_i)\}_{i=\left(\sum_{j=0}^{k} N_j\right)+1}^{\sum_{j=0}^{k+1} N_j}$, where N_{k+1} denotes the number of observations in the $(k+1)$th chunk.

1. Calculate the partial matrix H_{k+1} for the $(k+1)$th chunk of data $\aleph_{k+1}$ for the TSK model:

$$H_{k+1} = H\left(c_1, \ldots, c_{\tilde{N}}, a_1, \ldots, a_{\tilde{N}}; x_{\left(\sum_{j=0}^{k} N_j\right)+1}, \ldots, x_{\sum_{j=0}^{k+1} N_j}\right)$$

where the details of H are referred to in [23].

---

* Fuzzy-ELM is a specific version of OS-Fuzzy-ELM when all the training data are learned in a batch mode.

Set $T_{k+1} = \left[t_{\left(\sum_{j=0}^{k} N_j \right)+1}, \ldots, t_{\sum_{j=0}^{k+1} N_j} \right]^T$

2. Calculate the parameter matrix $Q(k+1)$:

$$P_{k+1} = P_k - P_k H_{k+1}^T (I + H_{k+1} P_k H_{k+1}^T)^{-1} H_{k+1} P_k$$

$$Q^{(k+1)} = Q^{(k)} + P_{k+1} H_{k+1}^T (T_{k+1} - H_{k+1} Q^{(k)})$$

3. Set $= k + 1$. Go to step 2.

The proposed OS-Fuzzy-ELM is also utilized as the fuzzy neural controller to aid the basic trajectory following controller (BTFC) for the aircraft autolanding problem based on the feedback–error–learning control strategy. In the study, both Gaussian form and Cauchy form membership functions are studied for the problem based on TSK fuzzy model under the one-by-one implementation mode. Similarly, the performance of OS-Fuzzy-ELM-aided BTFC control strategy is evaluated on the fault scenarios described in the previous chapter and is compared with those of EMRAN-aided BTFC and BTFC controllers. The results show that OS-Fuzzy-ELM-aided BTFC improves the fault-tolerant capabilities compared to the other two schemes. The details can be referred to in Ref. [32]. In the following section, the proposed OS-Fuzzy-ELM and SAFIS algorithms are further compared based on the aircraft autolanding fault-tolerant control problem and the same fault scenarios. As described in Ref. [32], OS-Fuzzy-ELM-aided BTFC has been studied based on the two types of membership functions, viz., the Gaussian membership function and Cauchy membership function, and similar results from the two kinds of membership functions have been obtained. Thus, for convenience, in the following section, for comparing OS-Fuzzy-ELM and SAFIS, only the results obtained from the Cauchy form membership function are presented.

28.6.1 COMPARISON OF OS-FUZZY-ELM AND SAFIS FOR AIRCRAFT AUTOLANDING PROBLEM

28.6.1.1 Single Surface Failure

In this section, a single failure of elevator or aileron is considered. First, we present the results for elevator failure. Here, the left elevator is stuck at −10 degrees at the beginning of the turn. Figures 28.32 and 28.33 give the altitude (h), the velocity (V_T), sideslip angle (β) and lateral position (Y) during the landing phase for the OS-Fuzzy-ELM-aided BTFC and SAFIS-aided BTFC.

The numbers at the top of Figures 28.32 and 28.33 represent the different segments of the trajectory. As observed from Figures 28.32 and 28.33, OS-Fuzzy-ELM-aided BTFC and SAFIS-aided BTFC are able to handle the failure and land the aircraft successfully for they utilize online learning to learn the desired signals quickly and generate a larger control signal to drive the aircraft to follow the desired outputs. Figures 28.32 and 28.33 further illustrate that OS-Fuzzy-ELM-aided BTFC and SAFIS-aided BTFC are able to follow the reference trajectory closely. OS-Fuzzy-ELM and SAFIS produce the large sideslip seen in Figure 28.33 around 110 s due to the abrupt step inputs in the side gust profile but these excursions are quickly damped out by the controllers. Table 28.6 shows the RMS trajectory error together with the number of rules for OS-Fuzzy-ELM-aided BTFC and SAFIS-aided BTFC and we can observe that the trajectory error of OS-Fuzzy-ELM is smaller than that of SAFIS and the number of rules for OS-Fuzzy-ELM is lesser than that of SAFIS for the lateral direction motion. In the longitudinal direction motion, they require the same number of fuzzy rules. Figure 28.34 shows the deflection of the left elevator ($\delta_{e\text{-left-t}}$) and the right elevator ($\delta_{e\text{-right-t}}$) along with the BTFC ($\delta_{e\text{-right-c}}$) and OS-Fuzzy-ELM/SAFIS components ($\delta_{e\text{-right-f}}$) of the control signals achieved by OS-Fuzzy-ELM-aided BTFC and SAFIS-aided BTFC control schemes. In Figure 28.34, the solid line represents the results from OS-Fuzzy-ELM-aided BTFC and the dot line represents the results from SAFIS-

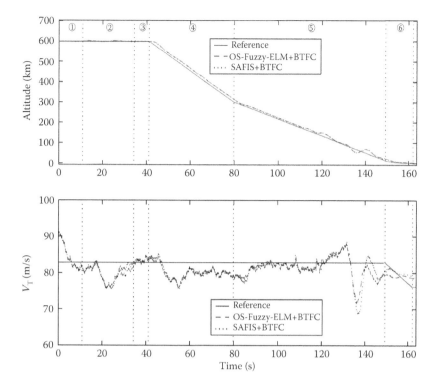

FIGURE 28.32 Comparison of OS-Fuzzy-ELM-aided BTFC and SAFIS-aided BTFC for the altitude (h) and velocity (V_T) under the left elevator stuck at −10 degrees.

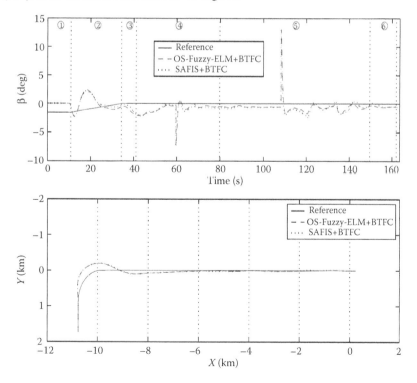

FIGURE 28.33 Comparison of OS-Fuzzy-ELM-aided BTFC and SAFIS-aided BTFC for the altitude (β) and velocity (Y) under the left elevator stuck at −10 degrees.

TABLE 28.6

Performance Comparison between OS-Fuzzy-ELM Aided BTF and SAFIS Aided BTFC Control Schemes under the Left Elevator Stuck at −10 Degrees

Methods	OS-Fuzzy-ELM Aided BTFC	SAFIS Aided BTFC
Trajectory error (RMSE)	8.8364	8.8573
Velocity error (RMSE)	3.1929	3.4243
Number of rules/neurons for δ_e	6	6
Number of rules/neurons for δ_a	6	14

aided BTFC. As observed from Figure 28.34, control signals achieved from OS-Fuzzy-ELM-aided BTFC and SAFIS-aided BTFC are very similar. The left and right aileron signals are given in Figure 28.35 and they are very similar. The fault-tolerant capabilities of OS-Fuzzy-ELM-aided BTFC and SAFIS-aided BTFC for a single elevator surface failure is given in Figure 28.36. Each point in Figure 28.36 indicates a successful landing meeting the touchdown pillbox requirements. From Figure 28.36, it can be seen that OS-Fuzzy-ELM-aided BTFC is able to tolerate the same continuous range of elevator failures (−12 to 18 degrees) compared to SAFIS-aided BTFC (−12 to 18 degrees). Figure 28.37 gives the case of single aileron failure, and one can note from this figure that the OS-Fuzzy-ELM-aided BTFC has a slightly wider continuous aileron failure, tolerance range (−20 to 9 degrees) than SAFIS-aided BTFC (−14 to 14 degrees).

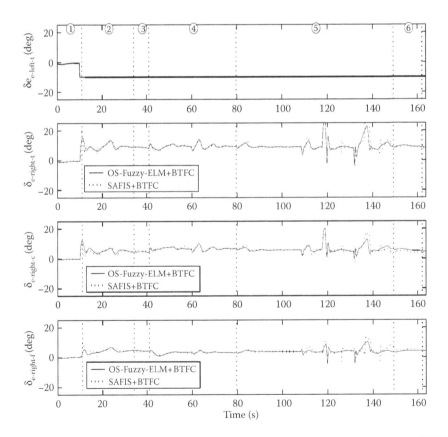

FIGURE 28.34   Left and right elevator control signals for OS-Fuzzy-ELM-aided BTFC and SAFIS-aided BTFC under the left elevator stuck at −10 degrees.

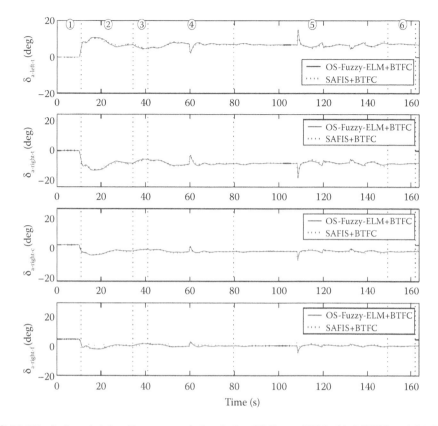

FIGURE 28.35 Left and right aileron control signals for OS-Fuzzy-ELM-aided BTFC and SAFIS-aided BTFC under the left elevator stuck at −10 degrees.

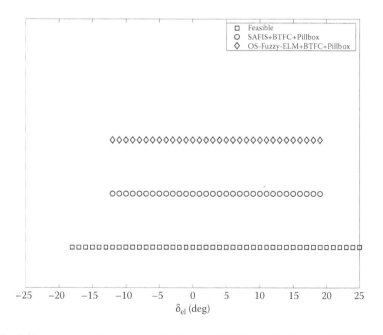

FIGURE 28.36 Failure tolerance for OS-Fuzzy-ELM-aided BTFC and SAFIS-aided BTFC under left elevator stuck conditions.

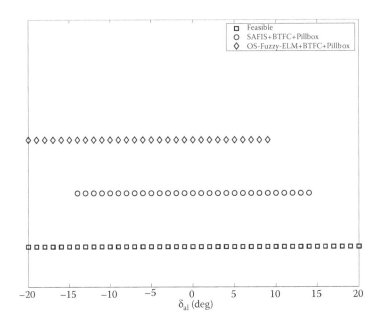

FIGURE 28.37   Failure tolerance for OS-Fuzzy-ELM-aided BTFC and SAFIS-aided BTFC under left aileron stuck conditions.

28.6.1.2   Two Surface Failures

In this section, two failures are considered. Specifically, for the first case, the left elevator and left aileron are stuck at different deflections. For the second case the left elevator and right aileron are considered.

28.6.1.3   Left Elevator and Left Aileron Stuck at Different Deflections

In this section, we first present a typical trajectory result when the left elevator is stuck at −10 degrees and the left aileron is stuck at +10 degrees. Figures 28.38 and 28.39 give the altitude (h), the velocity (V_T), sideslip angle (β) and lateral position (Y) during the landing phase for the OS-Fuzzy-ELM-aided BTFC and SAFIS-aided BTFC. The numbers at the top of Figures 28.38 and 28.39 represent the different segments of the trajectory. As observed from Figures 28.38 and 28.39, OS-Fuzzy-ELM-aided BTFC and SAFIS-aided BTFC both are able to handle the failure and land the aircraft successfully by utilizing their online learning ability. Figures 28.38 and 28.39 further illustrate that OS-Fuzzy-ELM-aided BTFC and SAFIS-aided BTFC are able to follow the reference trajectory closely. Table 28.7 shows the RMS trajectory error together with the number of rules for OS-Fuzzy-ELM-aided BTFC and SAFIS-aided BTFC and we can observe that the trajectory error and the number of rules for OS-Fuzzy-ELM are smaller than those of SAFIS.

Figure 28.40 shows the deflection of the left elevator ($\delta_{e\text{-left-}t}$) and right elevator ($\delta_{e\text{-right-}t}$) along with the BTFC ($\delta_{e\text{-right-}c}$) and OS-Fuzzy-ELM/SAFIS components ($\delta_{e\text{-right-}f}$) of the control signals achieved by OS-Fuzzy-ELM-aided BTFC and SAFIS-aided BTFC control schemes. In Figure 28.40, the solid line represents the results from OS-Fuzzy-ELM-aided BTFC and the dot line represents the results from SAFIS-aided BTFC. As observed from Figure 28.40, control signals achieved from OS-Fuzzy-ELM-aided BTFC and SAFIS-aided BTFC are very similar. The left and right aileron signals are given in Figure 28.41 and they are very similar. The fault-tolerant capabilities of OS-Fuzzy-ELM-aided BTFC and SAFIS-aided BTFC for the whole range of deflections are given in Figure 28.42. The edge of the fault tolerance envelopes achieved by OS-Fuzzy-ELM-aided BTFC and SAFIS-aided BTFC is given by the solid line and dash–dot line. In their respective fault-tolerant regions, each point which indicates a successful landing meeting the touchdown pillbox

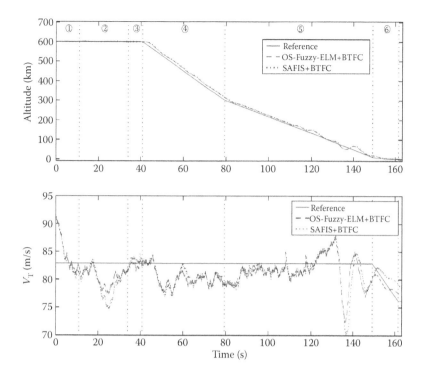

FIGURE 28.38 Comparison of OS-Fuzzy-ELM-aided BTFC and SAFIS-aided BTFC for the altitude (*h*) and velocity (*V*_T) under the left elevator stuck at −10 degrees and the left aileron stuck at 10 degrees.

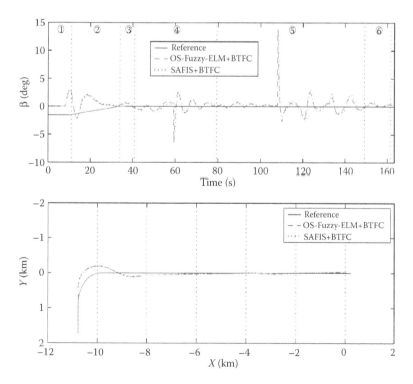

FIGURE 28.39 Comparison of OS-Fuzzy-ELM-aided BTFC and SAFIS-aided BTFC for the sideslip angle (β) and lateral position (*Y*) under the left elevator stuck at −10 degrees and the left aileron stuck at 10 degrees.

TABLE 28.7

Performance Comparison between OS-Fuzzy-ELM Aided BTFC and SAFIS Aided BTFC Control Schemes under the Left Elevator Stuck at −10 Degrees and the Left Aileron Stuck at 10 Degrees

Methods	OS-Fuzzy-ELM Aided BTFC	SAFIS Aided BTFC
Trajectory error (RMSE)	8.8926	8.8978
Velocity error (RMSE)	2.8663	3.2761
Number of rules/neurons for δ_e	6	8
Number of rules/neurons for δ_a	6	22

requirements is represented by an open circle (o). From Figure 28.42, it can be seen that OS-Fuzzy-ELM-aided BTFC is able to meet the pillbox requirements in a slightly wider range of deflections than SAFIS-aided BTFC and their fault-tolerant envelopes largely overlap.

28.6.1.4 Left Elevator and Right Aileron Stuck at Different Deflections

The fault-tolerant capabilities of OS-Fuzzy-ELM-aided BTFC and SAFIS-aided BTFC for the whole range of deflections is given in Figure 28.43. The edge of the fault tolerance envelopes

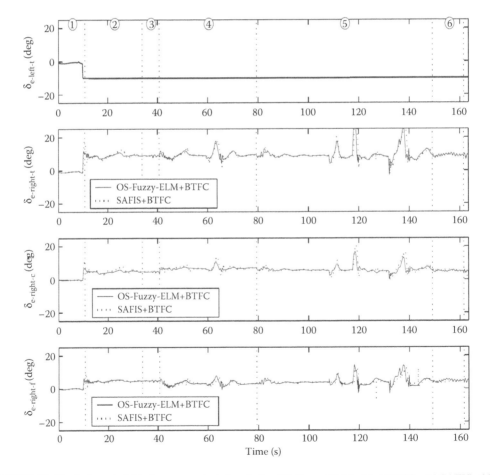

FIGURE 28.40   Left and right elevator control signals for OS-Fuzzy-ELM-aided BTFC and SAFIS-aided BTFC under the left elevator stuck at −10 degrees and the left aileron stuck at 10 degrees.

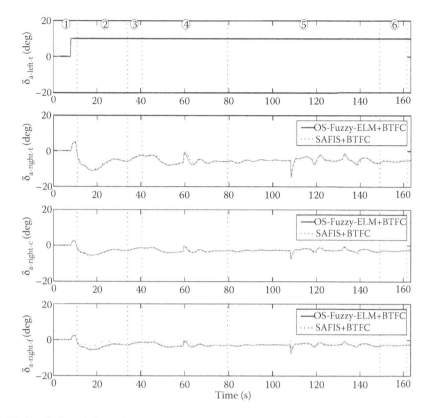

FIGURE 28.41 Left and right aileron control signals for OS-Fuzzy-ELM-aided BTFC and SAFIS-aided BTFC under the left elevator stuck at −10 degrees and the left aileron stuck at 10 degrees.

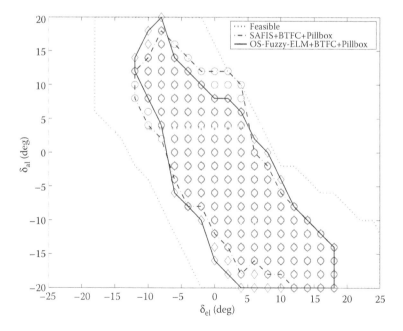

FIGURE 28.42 Failure tolerance for OS-Fuzzy-ELM-aided BTFC and SAFIS-aided BTFC under left elevator and left aileron stuck conditions.

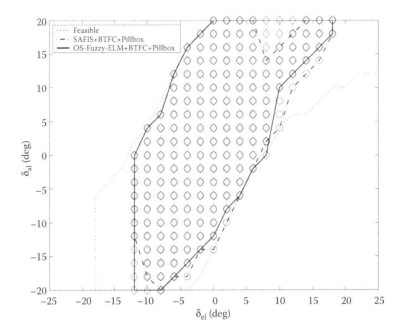

FIGURE 28.43 Failure tolerance for OS-Fuzzy-ELM-aided BTFC and SAFIS-aided BTFC under left elevator and right aileron stuck conditions.

achieved by OS-Fuzzy-ELM-aided BTFC and SAFIS-aided BTFC is given by the solid line and dash–dot line. In their respective fault-tolerant regions, each point which indicates a successful landing meeting the touchdown pillbox requirements is represented by an open circle (o). From Figure 28.43, it can be seen that OS-Fuzzy-ELM-aided BTFC is able to meet the pillbox requirements in a slightly wider range of deflections than SAFIS-aided BTFC and its fault-tolerant envelope has much overlap with that of SAFIS-aided BTFC.

28.7 CONCLUDING REMARKS

The autolanding problem has been set up in its entirety without any approximations. The aim of posing this problem is to demonstrate the capability of various neural and fuzzy controllers to handle single and multiple control surface failures during the landing phase. Severe winds are also injected close to the ground to test the efficacy of the control schemes. A landing is considered successful only if the aircraft meets stringent conditions of a touchdown in terms of the touchdown point, sink rate, bank angle and touchdown velocity. The performance of various adaptive/ reconfigurable controls schemes based on neural and fuzzy ideas have been evaluated. The first control scheme is the neural-aided controller based on augmenting the control signal computed by a classical control system (BTFC) with an additional signal computed by the EMRAN in parallel. The neural controller attempts to exploit its learning ability to overcome the control surface failure scenarios. The EMRAN is able to increase the failure tolerance of the BTFC. However, due to the failure tolerance regions not being singly connected, it is difficult to give performance guarantees. The second control scheme studied is the ABNC. This is completely a neural network-based controller. ABNC has a cascaded structure mimicking the equations of motion. This scheme has the ability to handle multiple redundant controls to overcome failures.

The remaining two control schemes studied in this chapter are based on neural and fuzzy concepts. The SAFIS and OS-Fuzzy-ELM have been developed. Both the schemes are used to augment the basic BTFC controller. In general, the OS-Fuzzy-ELM has shown a smaller trajectory error compared to the SAFIS. In the lateral-directional axis, the OS-Fuzzy-ELM also requires lesser number of rules. The OS-Fuzzy-ELM demonstrates a marginally better performance in terms of increased failure envelope. Based on the above results, it may be concluded that there is a significant scope for enhancing the failure envelope for actuator failures by augmenting the basic controllers with controllers inspired by neural and fuzzy concepts. The authors feel that further research in this direction is needed to provide performance guarantees in the failure envelope.

REFERENCES

1. B. S. Kim and A. J. Calise, Nonlinear flight control using neural networks, *Journal of Guidance, Control and Dynamics*, 20(1), 26–33, 1997.
2. M. R. Napolitano, S. Naylor, C. Neppach and V. Casdorph, On-line learning nonlinear direct neurocontrollers for restructurable control systems, *Journal of Guidance, Control and Dynamics*, 18(1), 170–176, 1995.
3. U. J. Pesonen, J. E. Steck, K. Rokhsaz, H. S. Bruner and N. Duerksen, Adaptive neural network inverse controller for general aviation safety, *Journal of Guidance, Control and Dynamics*, 27(3), 434–443, 2004.
4. T. Kim and Y. Lee, Nonlinear adaptive flight control using backstepping and neural networks controller, *Journal of Guidance, Control and Dynamics*, 24(4), 675–682, 2001.
5. Y. Li, N. Sundararajan and P. Saratchandran, Analysis of minimal radial basis function network algorithm for real-time identification of nonlinear dynamic systems, *IEE Proceedings on Control Theory Applications*, 147(4), 476–484, 2000.
6. L. T. Nguyen, M. E. Ogburn, W. P. Gilbert, K. S. Kibler, P. W. Brown and P. L. Deal, Simulator study of stall/post-stall characteristics of a fighter airplane with relaxed longitudinal static stability, NASA Technical Paper 1538, December 1979.
7. H. Gomi and M. Kawato, Neural network control for a closed-loop system using feedback-error-learning, *Neural Networks*, 6(7), 933–946, 1993.
8. M. Krstic, I. Kanellakopoulos and P. V. Kokotovic, *Nonlinear Adaptive Control Design*, Wiley, New York, 1995.
9. A. A. Pashilkar, N. Sundararajan and P. Saratchandran, Adaptive nonlinear neural controller for aircraft under actuator failures, *Journal of Guidance, Control and Dynamics*, 30(3), 835–847, 2007.
10. A. A. Pashilkar, N. Sundararajan and P. Saratchandran, A fault-tolerant neural aided controller for aircraft auto-landing, *Journal of Aerospace Science & Technology*, 10(1), 49–61, 2006.
11. A. A. Pashilkar and N. Sundararajan, Enhanced-tolerant neural controller for aircraft autolanding, Final Report DSOCLO1144, School of Electrical & Electronic Engineering, Nanyang Technological University, Singapore, February 2005.
12. L. Yingwei, N. Sundararajan and P. Saratchandran, A sequential learning scheme for function approximation using minimal radial basis function neural networks, *Neural Computation*, 9, 461–478, 1997.
13. D. J. Bugajski and D. F. Enns, Nonlinear control law with application to high angle-of-attack flight, *Journal of Guidance, Control and Dynamics*, 15(3), 761–767, 1992.
14. E. N. Johnson and S. K. Kannan, Adaptive trajectory control for autonomous helicopters, *Journal of Guidance, Control and Dynamics*, 28(3), 524–538, 2005.
15. I. Kanellakopoulos, P. V. Kokotovic and A. S. Morse, Systematic design of adaptive controllers for feedback linearizable systems, *IEEE Transactions on Automatic Control*, 36(11) 1241–1253, 1991.
16. J. Park and I. W. Sandberg, Universal approximation using radial basis function networks, *Neural Computation*, 3, 246–257, 1991.
17. J.-J. E. Slotine and W. Li, *Applied Nonlinear Control*, Prentice-Hall, Englewood Cliffs, NJ, 1991.
18. K. Nho and R. K. Agarwal, Automatic landing system design using fuzzy logic, *Journal of Guidance, Control and Dynamics*, 23(2), 298–304, 2000.
19. K. Nho and R. K. Agarwal, Glideslope capture in wind gust via fuzzy logic controller, in *37th AIAA Aerospace Sciences Meeting and Exhibit*, Reno, NV, pp. 1–11, 1999.
20. J.-G. Juang, K.-C. Chin and J.-Z. Chio, Intelligent automatic landing system using fuzzy neural networks and genetic algorithm, in *Proceedings of the 2004 American Control Conference*, Vol. 6, Boston, MA, pp. 5790–5795, 2004.
21. S. M. B. Malaek, N. Sadati, H. Izadi and M. Pakmehr, Intelligent autolanding controller design using neural networks and fuzzy logic, in *Proceedings of the 5th Asian Control Conference*, Vol. 1, Melbourne, Australia, pp. 365–373, 2004.

22. H.-J. Rong, N. Sundararajan, G.-B. Huang and P. Saratchandran, Sequential adaptive fuzzy inference system (SAFIS) for nonlinear system identification and prediction, *Fuzzy Sets and Systems*, 157(9), 1260–1275, 2006.

23. H.-J. Rong, G.-B. Huang, P. Saratchandran and N. Sundararajan, On-line sequential fuzzy extreme learning machine for function approximation and classification problems, *IEEE Transactions on Systems, Man, and Cybernetics: Part B*, 39(4), 1067–1072, 2009.

24. H.-J. Rong, N. Sundararajan, P. Saratchandran and G.-B. Huang, Adaptive fuzzy fault-tolerant controller for aircraft autolanding under failures, *IEEE Transactions on Aerospace and Electronic Systems*, 43(4), 1586–1603, 2007.

25. H.-J. Rong, G.-B. Huang, N. Sundararajan and P. Saratchandran, Fuzzy fault tolerant controller for actuator failures during aircraft autolanding, in *2006 IEEE International Conference on Fuzzy Systems*, Vancouver, BC, Canada, pp. 1200–1204, July 16–21, 2006.

26. M. M. Gupta and D. H. Rao, On the principles of fuzzy neural networks, *Fuzzy Sets and Systems*, 61(1), 1–18, 1994.

27. W. Pedrycz and A. F. Rocha, Fuzzy-set based models of neurons and knowledge-based networks, *IEEE Transactions on Fuzzy Systems*, 1(4), 254–266, 1993.

28. J-S. R. Jang, ANFIS: Adaptive-network-based fuzzy inference system, *IEEE Transactions on Systems, Man, and Cybernetics*, 23(3), 65–685, 1993.

29. R.-J. Wai and Z.-W. Yang, Adaptive fuzzy neural network control design via a T-S fuzzy model for a robot manipulator including actuator dynamics, *IEEE Transactions on Systems, Man, and Cybernetics, Part B: Cybernetics*, 38(5), 1326–1346, 2008.

30. K. M. Passino, *Biomimicry for Optimization, Control, and Automation*, Springer-Verlag, London, UK, 2005.

31. T. Takagi and M. Sugeno, Fuzzy identification of systems and its applications for modeling and control, *IEEE Transactions on Systems, Man, and Cybernetics*, 15(1), 116–132, 1985.

32. R. Haijun, Efficient sequential fuzzy neural algorithms for aircraft fault-tolerant control, Nanyang Technological University, PhD thesis, 2007.

33. G.-B. Huang, P. Saratchandran and N. Sundararajan, An efficient sequential learning algorithm for growing and pruning RBF (GAP-RBF) networks, *IEEE Transactions on Systems, Man, Cybernetics, Part B: Cybernetics*, 34(6), 2284–2292, 2004.

34. G.-B. Huang, L. Chen and C.-K. Siew, Universal approximation using incremental constructive feedforward networks with random hidden nodes, *IEEE Transactions on Neural Networks*, 17(4), 879–892, 2006.

35. G.-B. Huang and C.-K. Siew, Extreme learning machine: RBF network case, in *Proceedings of the Eighth International Conference on Control, Automation, Robotics and Vision (ICARCV 2004)*, Kunming, China, pp. 6–9, 2004.

36. G.-B. Huang, Q.-Y. Zhu, K. Z. Mao, C.-K. Siew, P. Saratchandran and N. Sundararajan, Can threshold networks be trained directly? *IEEE Transactions on Circuits and Systems II*, 53(3), 187–191, 2006.

37. G.-B. Huang, Q.-Y. Zhu and C.-K. Siew, Extreme learning machine: Theory and applications, *Neurocomputing*, 70, 489–501, 2006.

38. G.-B. Huang, Q.-Y. Zhu and C.-K. Siew, Extreme learning machine: A new learning scheme of feedforward neural networks, In *Proceedings of International Joint Conference on Neural Networks (IJCNN2004)*, Budapest, Hungary, pp. 985–990, July 25–29, 2004.

39. N.-Y. Liang, G.-B. Huang, P. Saratchandran and N. Sundararajan, A fast and accurate on-line sequential learning algorithm for feedforward networks, *IEEE Transactions on Neural Networks*, 17(6), 1411–1423, 2006.

29 Reconfiguration in Flight Critical Aerospace Applications

C. M. Ananda

CONTENTS

29.1   INTRODUCTION

Advances in computer technology have encouraged the avionics industry to utilize the increased processing power, communication bandwidth and hosting of multiple federated applications into a single integrated platform. This has been realized as integrated modular avionics (IMA). The IMA has emerged as a platform for integrating multiple avionics applications on a shared computing environment. A computing environment with common hardware and system resource platform will be powerful enough to meet the computing demands of multiple applications. IMA has advantages such as reduced weight, efficient shared computing resource, lesser looming volume, reduced electrical interface complexity and physical maintenance. However, the federated system lacks the above advantages of IMA. In civil aircraft applications, the effective use of resource management of hardware and software is the key to the success. Avionics architecture plays a very important role in design, development, maintenance, service and certification in aircraft programme. Therefore, the use of integrated architectures has grown quite rapidly in recent times.

The aerospace systems demand very high system availability under varied aircraft conditions. The continued availability of the avionics functions over a period of time under limited failure conditions is very essential. This forms a measure of the system's robustness. The reconfiguration provides enhancement in continued availability of the applications under limited failure conditions. This is an improvement on the existing system that has become a topic of current research. This chapter aims to improve the availability of the avionics application in a time and memory partitioned platform using reconfiguration technique. Reconfiguration in real-time avionics flight critical system is a process of task or process management having provision to reconfigure the functionality in near real time to the accuracy of one major frame. In the present proposal, the faulty task is removed from the schedule in the next major frame and continued with the reconfigured schedule. The algorithm uses critical control metrics for reconfiguration of a task or a process without impacting the safety of the rest of the schedule. These control metrics proposed are used in reconfiguration and help the algorithm in proper decision making.

Reconfiguration involves four phases: (i) control parameter identification, (ii) error detection and correction, (iii) control parameter validation and (iv) reconfiguration of the identified task. The algorithm is based on reconfiguration to improve the availability of avionics flight critical application. The avionics hardware and software architecture of federated systems are good in fault containment, fault tolerance and has been a foolproof architecture. Each hardware functional box called line replaceable unit (LRU) has its own resources in terms of processing system, memory and peripheral systems. It has excellent fault-tolerant capabilities against the disadvantage of duplication of resources. However, the federated system has disadvantages such as increased weight, redundant computer resources in each LRU, higher looming volume, electrical interface complexity and physical maintenance when compared to the current-day integrated systems. The growth of advanced technology in computational platform enabled the avionics industry to utilize the increased processing power, communication bandwidth and hosting of multiple federated applications into a single integrated platform [1]. IMA emerged as a platform for integrating multiple avionics applications on a shared integrated computing environment [2]. This common computing environment will be powerful enough to service all the applications with required real-time requirements.

Figure 29.1 shows typical system architecture of IMA [2] having three components: core system, core hardware and core processor. Also, the crucial component of such architecture is the application executive (APEX) [3]. Application partitions are ARINC 653-based partitions with spatial or time and memory protection mechanism. Because of this, the data integrity of the application software is quite high which is very essential for flight critical applications.

29.2   IMPORTANCE AND RELEVANCE

In the recent past, aerospace avionics applications are more open architecture computing systems (OACS) and have already migrated towards an integrated approach to take advantage of the plat-

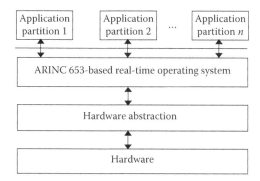

FIGURE 29.1   IMA system architecture.

form architecture. Therefore, it is more relevant to address the issues related to enhancement of safety, availability and reliability.

Avionics applications are realized using avionics cabinets which host multiple avionics applications using the ARINC 653 platform. This means that multiple LRUs of federated architecture (FA) are combined into one single system hosting multiple software applications. Hence, the failure of the avionics cabinet may lead to a catastrophic condition of safety because of single point failure. One such failure, if not handled properly, will cause serious impact on the rest of the functionalities. The availability of avionics functionalities reduces to a great extent defeating the integrated architecture if such failures are not handled properly. The reconfiguration mechanism in civil aerospace applications is the state-of-the-art technology in the industry.

29.3   CIVIL AIRCRAFT AVIONICS

Avionics is the term used in aerospace for aircraft electronics or the aviation electronics. Avionics in an aircraft is the key element of man–machine interface (MMI) or human–machine interface (HMI) for aircraft navigation. Traditionally, avionics systems are more functionality centric with dedicated algorithmic and architectural methods for performance enhancement [4]. Avionics architectures are classified into two major sectors: (i) federated architectures and (ii) integrated architectures.

29.3.1   Federated Architecture

FA is a 1980s technology. This architecture consists of number of individual systems, each intended for specific functions. These individual systems called LRU are integrated on an avionics platform using the applicable interfaces. Functionalities in the federated architectures are realized by dedicated hardware LRUs. Federated architecture does not share any resource across the LRUs except the external communication data bus. A typical federated avionics [5] has independent hardware units or LRUs for each of the functionality. Avionics system can be broadly classified based on functionality [6,7] into (i) communication system, (ii) navigation system, (iii) display system, (iv) radar system, (v) engine instrumentation system and (vi) data acquisition and recording system. Civil aircraft avionics comprising LRUs for the above functional systems are interconnected by conventional digital bus such as ARINC 429 as shown in Figure 29.2. Each LRU is responsible for a single functionality and is self-contained with reference to that function. Integration testing of avionics flight critical systems is quite critical and important for the success of the avionics suite [8,9]. Fault containment and fault-handling capability with sufficient resources for each functionality is the key advantage of a federated system.

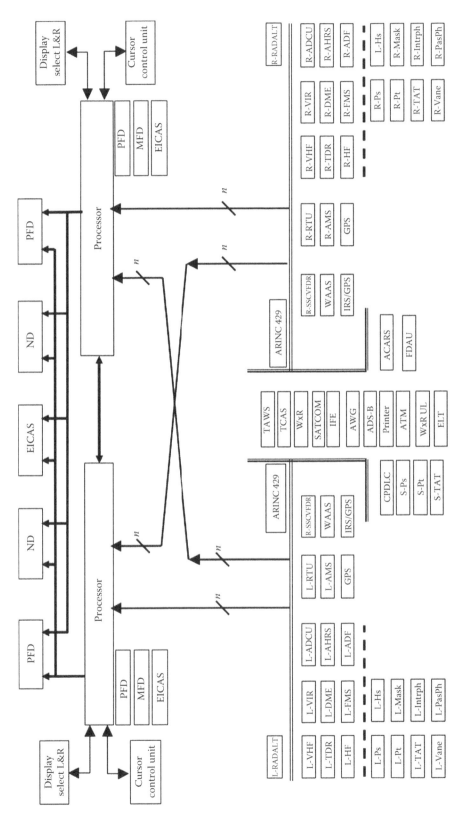

FIGURE 29.2 Federated avionics architecture for civil aircraft application.

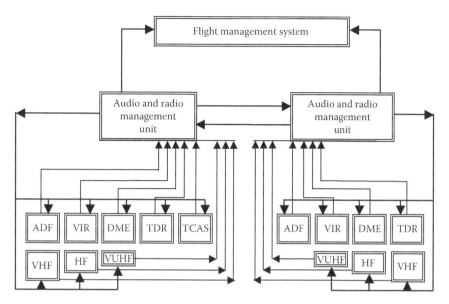

FIGURE 29.3 Typical COMNAV interface in federated system.

The major disadvantage of federated architecture is duplicated resource in each LRU irrespective of the functionality. This increases power, weight and volume of the avionics suite. Power, weight and volume in civil aircraft industry is more critical.

Avionics is built on the digital communication mode with current requirements of traffic collision avoidance system (TCAS), digital autopilot and AMLCD multi-purpose glass displays. To illustrate the complexity of interconnections in a federated system, the communication and navigation system (COMNAV) is taken as an example from the federated avionics architecture and shown in Figure 29.3. Hence, current-day architectures are already moving towards integrated architecture. If the same is implemented in integrated architecture, the interfaces will be simplified and is as shown in Figure 29.4. The systems are interconnected by a networked kind of global bus with drop connects and the data transfer is address-based delivery with well-defined source and destinations.

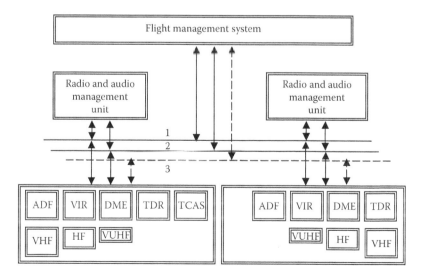

FIGURE 29.4 Typical COMNAV interface in integrated system with maintenance bus (dotted).

29.3.2 Integrated Architecture

Electronics systems on aircraft or the avionics plays a major role in aircraft functionalities such as communication, navigation and flight safety. Dedicated computing resources for each one of the disjoint functionalities were assigned independently. Such architectures lack in efficient utilization of resources at run time. With the advancement in aerospace and in particular avionics technologies across the globe, the federated architecture is found to be less advantageous as compared to the present integrated architecture, in terms of resource management.

In the recent past, avionics system architectures have evolved substantially, migrating from federated architecture to the integrated architecture [10,11]. Unlike a lot of dials and gauges, the pilot interacts with multi-function displays (MFD). The systems are coupled with MFD, communication and navigation radios with control units, multi-mode interactive instruments for control and navigation, recording and fault management systems, airframe and health-monitoring diagnostic capability. Pilot vehicle interface (PVI) is an important measure of good avionics and cockpit layout, which implies the optimization of MMI, enhancement of economy and safety of flight operations. Integrated architectures are implemented using avionics cabinets or processing cabinets. These house the hardware modules and software applications in partitions and are called integrated modular avionics (IMA).

29.4 INTEGRATED MODULAR AVIONICS

Traditionally, automated aircraft control has been realized using well-defined functions that are implemented as independent systems. Each function has its own resources with fault containment as a federated architecture, where the faults are strongly contained within each functional system and will not propagate across other functional systems. The integrated architecture has common hardware and underlying system software to accommodate multiple avionics functionalities. Partitioning uses appropriate hardware and software mechanism to restore strong fault containment to the maximum extent in such integrated architecture [12]. Time and memory are the major resource management in integrated architecture. One such mechanism is the IMA [2,13,14] applications supported by distributed multi-processor architecture as shown in Figure 29.5. These are with shared memory and network communication called avionics application software standard interface or application executive (APEX)—ARINC 653 [1]. APEX standard consists of a set of operating system application programming interfaces (API) [3], which can be used by the application software. Time and memory are shared across the same platform with good protection mechanisms provided by ARINC 653 among multiple avionics functionalities. This has been enabled by ARINC 653, as part of real-time operating systems (RTOS) supporting the partitioning protections using specific defined application executive (APEX) application programming interfaces (API). APEX supports structuring of applications as per IMA principles: (a) physical memory is divided into sections called as *partitions* and (b) avionics application consists of a set of processes and they do communicate inside and outside the partition. This is allowed with the APEX and IMA guidelines for safety. The basic definition of APEX is relatively associated with ARINC 659 backplane mechanism [15] and the global communication protocols such as ARINC 629 data bus [16], avionics full duplex ethernet (AFDX) [17]/ARINC 664 [18–22] and time triggered ethernet [23]. IMA has the advantage of better dispatch reliability as compared to the federated architecture. To achieve required dispatch reliability, flight critical functions are run on replicated hardware with the penalty of higher cost of complexity. In IMA systems, the replicated processors are not dedicated to specific functions and hence the dependency is very less. With this approach, functions can be allocated as required and hence the maintenance could be deferred as long as minimum required functionality is realized within the available processors. Full realization of IMA requires adoption of good real-time concepts and methods [24]. Therefore, for the IMA partitioning, it is better to be implemented within and across the distributed processor, so that the fault tolerance is efficient and the interaction across processors is more robust.

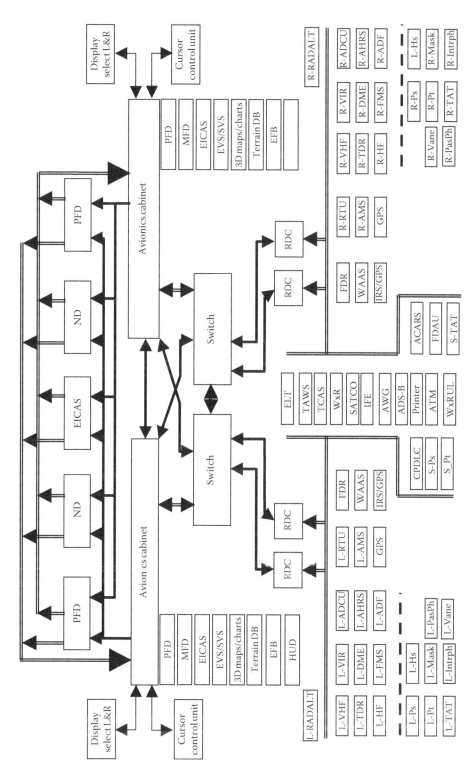

FIGURE 29.5 Typical integrated modular avionics architecture system.

29.4.1 PARTITIONS

Partitions in IMA have better fault containment and are very effective if architected well. The very intent of partitioning is to control the hazard of functions sharing a common processor as against the separate processor in federated systems. One argument persists in the industry: if two critical functions which are interdependent are shared by a single processor and if the processor fails, then the entire functionality could fail which is not so in the case of federated systems. Therefore, as detailed above, the allocation of functions to the relevant processors and partitions is quite critical and needs careful attention and analysis based on failure modes. IMA architecture has vital properties which play an important role in the definition of the IMA and flight critical system. Partitions are major constituents of the IMA architecture and hence have vital attributes such as

- *Criticality level*: Airborne systems have varied criticality levels based on the functionality, and hence, such functionalities can be constructed on the same hardware using IMA partitions. Also, the criticality is spread over Level A to Level E of DO 178B [25] for airborne application software development.
- *Duration*: It is the amount of time assigned to the partition in each period. During this time, partition has exclusive use of the processing resources.
- *Lock level*: Lock level indicates whether pre-emption is currently permitted among the processes with a partition. In flight critical systems, pre-emption is not allowed for deterministic behavioural requirements. The simplest way of defining the IMA architecture with two partitions supporting two flight applications is as shown in Figure 29.6.

Partitions across APEX-IMA platform are scheduled using cyclic, static table-driven pre-defined fixed priority and non-pre-emptive schedule. This provides resource access to all the partitions based on preset duration. Schedules are created offline constraining all partitions at least once. Some may appear more than once, depending upon the relationship between a partition period and length of the schedule. Partition can contain one or more application processes and each process having additional attributes such as

- *Period*: Processes can be periodic or aperiodic. Periodic process have fixed *period* defined between successive releases of the process. Aperiodic processes are set with a unique value to indicate they are not periodic and hence they do not have fixed *period*.
- *Time capacity*: Each process has fixed time for execution and has a deadline by which time the process should complete execution, which is a constant value.
- *Priority*: Each process has fixed priorities. This is the default and current priority of the process, based on the selection. The priority of the processes is pre-fixed as part of static scheduling.
- *State*: Process state can be dormant, ready, waiting or running based on the actual execution condition.

With the above discussions, it is mandatory that the partitioning in IMA is critical and hence there is a guideline for partitioning if not a contract rule. Jon Rushby [26] details two types of gold standards for partitioning as spatial partitioning (memory partition and protection) and temporal partitioning (time partition and protection).

Partition A	Partition B
ARINC 653 compliant real-time operating system	
Hardware	

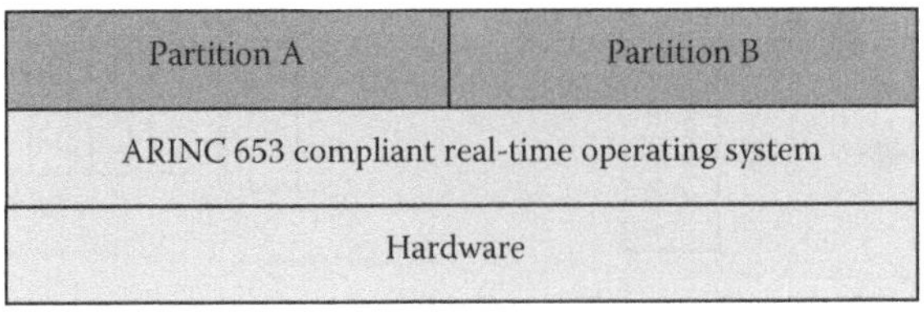

FIGURE 29.6 Simple IMA partitioning architecture.

29.4.2 GOLD STANDARD

A partitioned system should provide fault containment equivalent to an idealized system in which each partition is allocated an independent processor and associated peripherals, and all inter-partition communications are carried on dedicated lines.

Alternate gold standard is

the behaviour and performance of software in one partition must be unaffected by the software in other partitions.

However, these gold standards are not followed as rules and contract with the implementation by the users. However, this is a good reference. Partitioning of the system has two classifications:

1. Spatial partition (memory partition and protection)
2. Temporal partition (time partition and protection)

Jon Rushby [26] defines and details spatial and temporal partition as follows.

Spatial partition: 'Spatial partitioning must ensure that software in one partition cannot change the software or private data of another partition (either in memory or in transit), nor command the private devices or actuators of other partitions'.

Temporal partition: 'Temporal partitioning must ensure that the service received from shared resources by the software in one partition cannot be affected by the software in another partition. This includes the performance of the resource concerned, as well as the rate, latency, jitter and duration of scheduled access to it'. Spatial and temporal partitioning must block the fault propagation across partitions and functions so that fault containment is strictly adhered to.

If there is no functional partitioning supported by time and memory protection, then the level of software design development must be compliant to the highest level (Level A) for all functionality in an application. Figure 29.7 shows IMA with ARINC 653 compliant platform to host different application software having varying software levels from A to E on the same hardware and system software resources. Consider an example of typical aircraft display system flight critical application software to be hosted on federated and IMA architecture. The implementation is for level A based on the failure hazard analysis (FHA) and failure mode effect analysis (FMEA) requirement. The display system has the following major functionalities: (i) primary critical functionality, (ii) secondary critical functionality, (iii) caution and warning functionality, (iv) ground test

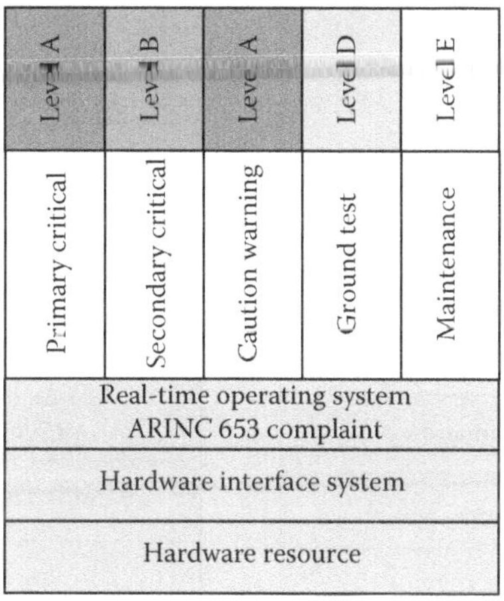

Level A	Level B	Level A	Level D	Level E
Primary critical	Secondary critical	Caution warning	Ground test	Maintenance
Real-time operating system ARINC 653 complaint				
Hardware interface system				
Hardware resource				

FIGURE 29.7 IMA with ARINC 653 supporting multiple levels of software hosted on the same hardware platform.

TABLE 29.1

Comparison of Severity Requirements of DO 178B Levels

Sl. No.	Functionality (Level Required)	SLOC	Required DO 178 B Levels		
			Based on Standards in Federated	Based on Standards in IMA	Effort Saving in IMA
1	Primary critical (A)	10K	A	A	
2	Secondary critical (B)	15K	A	B	15K
3	Caution warning (A)	05K	A	A	
4	Ground test (D)	22K	A	D	22K
5	Maintenance (E)	14K	A	E	14K

functionality and (v) maintenance functionality. Out of these major functionalities, critical functions such as primary, secondary and caution warning requires software to be Level A compliant. The rest of the functionalities do not require Level A and can be Level B, Level D or Level E compliance. Software lines of code (SLOC) corresponding to each of the functionality along with the comparison of display system functionality implementation using federated and integrated implementation is as shown in Figure 29.8a and b. Figure 29.8a describes a flight critical application being hosted on the federated architecture-based system. This is a monolithic executable-based application with highest demanded software criticality level for all the functionalities. Figure 29.8b details a flight critical application hosted on an IMA-based system. This has separate partitions for each function having varied software criticality levels from level A to level E. The large effort of level A is limited to only for level A demanded functionality, and hence saving large life cycle costs. The effort saving doing so in IMA is shown in Table 29.1 for a typical case study of display system.

If even a small part of the application demands Level A criticality, then the complete application in federated system is to be designed to Level A as shown in Figure 29.8a. However, in IMA, only required functionalities are implemented in Level A and the others are B, D and E as shown Figure 29.8b. This provides a mechanism to save enormous amount of time, money and effort for flight critical software design, development testing, and independent verification and validation

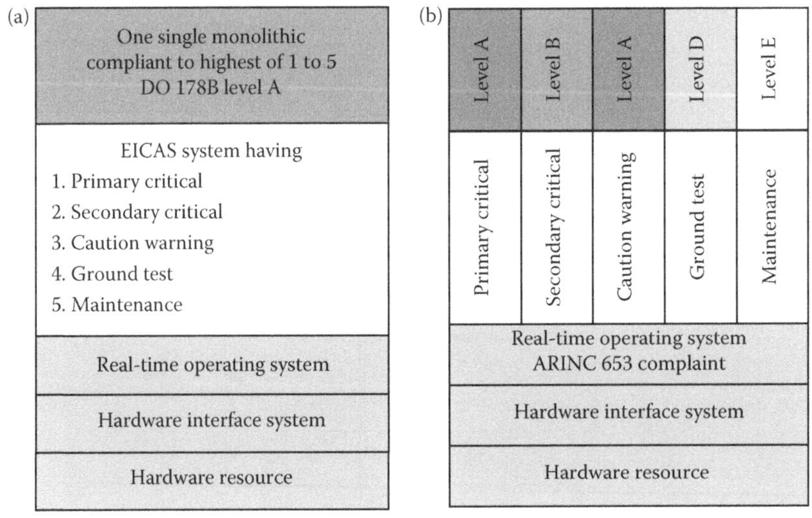

FIGURE 29.8 Comparison of federated (a) and IMA (b) implementation of typical EICAS application.

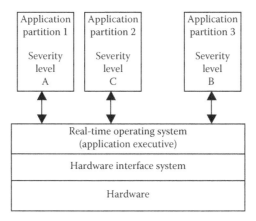

FIGURE 29.9 Application partitions in integrated avionics architecture.

(IV&V). Even in integrated architecture, the fault management to some extent depends on the type of application, mode of interface and methodology for fault management in each partition. The architecture is efficient if built using a powerful processor with an ARINC 653-compliant real-time operating system along with necessary device drivers to support the hardware interface. Aerospace systems demand very high system availability under various limited aircraft operating conditions. The continued availability of avionics functions is an important measure of the system's robustness. Even with the integrated architectures, the reconfiguration capability is quite limited and most of the applications developed on IMA currently do not have reconfiguration. Therefore, the integrated architecture has scope for improvement. The reconfiguration is an improvement on the existing system which provides enhancement in continued availability of the applications under limited failure conditions. The feature of time and memory partitioning protection and control mechanism is utilized to enhance the availability by means of reconfiguration methods. Existing mechanism of system behaviour in the event of a task failure is to declare system failure resulting in non-availability of either part or full functionality in a partition. Here, the failure recovery is not exercised; rather, the application aborts. Figure 29.9 shows the typical application partition of avionics integrated architecture [12,2] with varying severity levels.

In order to avoid an application being aborted due to a fault, the application has to be reconfigured in the context of an identified fault. The reconfiguration of an application depends on the then running task, its status, dependency and functional requirement. If by some means, the application is reconfigured and the application continues successfully, then the system architecture is to a limited extent fault-tolerant. This aids in the enhanced availability requirements of the flight critical systems. The reconfiguration algorithm addresses this issue using control metrics parameter validation checks in real time to reschedule or reconfigure the task or a process to achieve the required functionality.

29.4.3 Architecture and Its Capability for Reconfiguration

Reconfiguration of a task in architecture is strongly supported by its platform features for the safety requirements. Therefore, it is quite critical that the platform features and capabilities aid in successful reconfiguration of a task. The partitioning mechanism of ARINC 653 is quite significant and effective in adopting the reconfiguration algorithm. Applications in IMA platform typically stress more on intended functionality rather than system-related checks and monitors. Applications use the platform information and fuse with the application health information for decision making. Typically, system health monitoring in a flight critical system is derived using built-in-test (BIT). BIT in aerospace applications is mainly categorized into (i) power up BIT (PUBIT) or simply PBIT, (ii) continuous BIT (CBIT), (iii) initiated BIT (IBIT) and (iv) maintenance BIT (MBIT). The results of application BIT and platform health management (HM) form the basic data for system-level fault management.

TABLE 29.2
Failure Condition Categorisation

Categorisation	Definition	Probability	Software Level
Catastrophic	Failure condition that causes loss of safe operation of the system. For example, preventing continued safe flight	<1E-9	A
	State: aircraft destroyed and significant number of fatalities		
Hazardous/severe major	Failure condition, which significantly reduces the safety margins or functional capabilities of the system, could cause fatal injuries	<1E-7	B
	State: damage to aircraft, occupants hurt, some deaths. More work load		
Major	Failure condition which reduces the safety margins of the system	<1E-5	C
	State: pilots busy, interference with ability to cope with emergencies		
Minor	Failure condition which could significantly reduce system (or aircraft) safety	1E-3	D
	State: little or no effect on occupants or airplane capabilities		
No effect	Failure condition which does not affect the operation capability of the system aircraft	N/A	E
	State: no effect on occupants or airplane capabilities		

29.4.4 TECHNOLOGY DEMAND FOR AVAILABILITY AND RELIABILITY

Airborne systems in the past encompassed the complete requirements of safety, reliability and availability inside the same LRU as part of federated architecture. However, this has many demerits of redundant, inefficient resource utilization in run time, duplication of hardware and software capability in each hardware box. The avionics or flight controls application demands the maximum availability with reliability numbers of the order of 1 failure in 10^9 flight hours (1E-09). As per civil aircraft requirements, the severity failure condition categorization [27] is based on the type of failure and its functionality as detailed in Table 29.2.

As the technology demand for efficient systems increases, the use of common resources for multiple functionalities becomes more efficient and affordable. This is being implemented using integrated modular avionics architecture with base platform support from ARINC 653 APEX based real-time operating system (RTOS).

29.4.5 NEED FOR RELIABLE PROCESSING PLATFORM

Implementation of IMA partitioning system of having time and memory protection needs to have an enhancement to the platform. This is not a drawback of the IMA, rather an additional enhancement to the applications running on the IMA platform. Therefore, the additional enhancement in terms of reconfigurable features provides enhancement in terms of the availability of applications in case of failure of single or multiple partitions or functionalities. This feature of safety monitoring is used for the implementation of reconfigurable algorithm. Since the monitoring software need not be at the same level as the application or monitored software, it is much simpler and easier to implement using partitioning in the IMA platform. Monitoring feature of the integrated architecture does much more than just monitoring along with system safety requirements.

29.5 REVIEW OF RECONFIGURABLE MECHANISM

Sturdy et al. [28] describe the application of IMA platform for tankers and transport platforms. The IMA architecture was implemented for the first time in the 777 aircraft information management system (AIMS) where the time and memory partitioning was implemented. This has been proved to be the first of its kind called versatile integrated avionics (VIA) and created a boost to the aviation industry. The VIA provides a commercially available computational processing platform, which can be adapted to integrate military and other unique functions. VIA design is based on ARINC 651, external interfaces meet civil and military standards such as ARINC 429, Mil 1553, backplane meets ARINC 659, operating system meet ARINC 653 and test and maintenance interfaces meet IEEE 1149 standards. The tanker/transport implementation approach has also been addressed using VIA architecture. VIA is implemented in several advanced commercial flight deck applications, including military platforms [28]. Basic integrated modular avionics does provide strict partition for memory and time with crucial platform integrity checks. However, research is still on to reconfigure tasks dynamically to achieve a high level of availability even beyond the ARINC 653 IMA platform. This has potential danger of safety in flight critical systems. Architectures or platforms even without ARINC 653 are also being reconfigured with well-defined protection boundaries.

29.5.1 Reconfiguration Issues

Distributed system and control reconfiguration are the main issues in reconfiguration [29]. System reconfiguration is more of structure driven compared to the algorithm driven in control reconfiguration. Reconfiguration can be classified into two methods, namely, *static* and *dynamic*. Static reconfiguration is more concerned with a set of pre-defined offline approaches and has been in use in many aircraft applications. This is based on the pre-defined, prioritized, static table-driven schedules. Dynamic reconfiguration is more relevant and important for real-time execution of configurations. However, in such cases, the maximum time allowable is to the tune of one major frame. Various approaches of reconfiguration are addressed such as dynamic reconfiguration [30], control reconfiguration [31] and strategy-based fault-tolerant control [32]. Also, other reconfiguration schemes are based on functionality [33], resources [34], static configurations [35] and BIT [36]. Of course, the basic methodology in reconfiguration is the hardware redundancy which is being adopted in various applications [37]. However, this kind of redundancy is very efficient but at the same time it is very expensive and complex. In many cases, the redundancy is achieved with the computational processing nodes [38] and also using the network management [39]. However, each algorithm has its own merits and demerits based on the application and context features. The major issues of reconfiguration are (i) memory, (ii) timing, (iii) functionality, (iv) schedule modification and (v) availability. Reconfiguration algorithms address the above issues to ensure that the real-time and safety requirements of the system are carefully treated. The reconfiguration algorithm is more of functional redundancy than the processor or network and efforts are made to address all the above issues of reconfiguration. Also, it is very critical and important that the fault tolerance and redundancy management are two key issues relevant to the reconfiguration of a task in a flight critical application. These two issues are addressed in brief in the next sections.

29.5.2 Fault Tolerance

Advanced technology and rapid development of computer processing technology have placed extreme safety requirements on flight critical systems such as automatic flight control system (AFCS) of a commercial aircraft [40]. Commercial aircraft failure probability requirements are several fold higher than the military aircraft requirements [41]. The major problems related to extremely reliable flight controls can be addressed using the software mechanism such as software implemented fault tolerance (SIFT) computer systems [42,43]. This type of fault tolerance mechanisms have overhead of heavy throughput utilization and is not desirable in flight critical systems as the resource management is quite critical and expensive. To overcome this limitation, the alternate approach is to have multiple processors for fault tolerance. This architecture of multiple processor is called fault-tolerant multiple processor (FTMP) detailed in Ref. [44]. FTMP provides hardware-centric fault tolerance and added advantage of system functionality for voting and synchronization. In this method too, throughput consumption is appreciably high and is not desirable for the critical system. However, there is one more approach using single processor called the fault-tolerant processor (FTP) of advanced information processing system (AIPS) programme [45]. Since the FTP is based on uniprocessor, the processing bandwidth is limited to a single processor and fault tolerance is provided by the redundant processing channels. So far, the fault tolerance was using a single computer with a combination of software, hardware processor and its combinations. The architecture is called the multicomputer architecture for fault tolerance (MAFT) [46–48]. MAFT caters to the extremely reliable and higher performance in a distributed computer system. The main characteristic of a real-time system is the determinism [49] in terms of time utilization. There are two main sources of information that should be referred for introduction to real-time systems: one is by Kopetz [50] and other is by Krishna et al. [51]. Both have reviewed several basic concepts that are integrated to give a coherent overview of real-time systems, such as fault tolerance strategies, the most common protocols, the most common clock synchronization algorithms as well as performance measures. So far, various architectures such as the use of multiple processes, multiple processor and multiple channels voting were discussed. The advanced platform-driven mechanism is supported by the systems architecture and is realized with the current-day integrated modular avionics architectures [3,12] by which the rapid obsolescence can be managed in a cost-effective way. The basic concept is the use of commercially off-the-shelf (COTS) flexible hardware and software resources to manage the least dependency to the rapid changing hardware systems. IMA provides the advantage of reduction in size, weight and cost of the systems with capability to add or modify the hardware and software resources at all time with a caution of appreciable re-certification efforts for aerospace applications [52]. Fault tolerance of IMA is very effective [12] against the common faults of the federated systems such as fault propagation leading to memory corruption, denying of services to other systems or inappropriate command to critical systems and so on. It is difficult for a function to protect itself against fault scenarios unless it is implemented in a protected environment for both time and resources. So, IMA realization must provide the capability of partitioning to ensure that fault propagation from one to another is well controlled and protected, which is inherent in the federated architecture. Typical IMA with ARINC 653 compliance provides two major partition mechanisms, *time partitioning* and *memory partitioning*, as detailed earlier. With this partitioning mechanism, the fault containment capability of IMA is extremely effective and the applications are enormous.

29.5.3 REDUNDANCY MANAGEMENT

Fault tolerance is addressed in real-time systems using the configuration strategy to accommodate failures in many cases. Most current strategies are related to the redundancy approach based on (a) hardware, (b) software and (c) time redundancy. Hardware redundancy is basically duplication of hardware or part of the resources using voting algorithm named N-modular redundancy (NMR) [53] as described in Figure 29.10.

The reliability of the NMR system [53] is expressed as

$$R_{\text{NMRV}} = R_{\text{V}} \left[\sum_{i=0}^{m} \left(\frac{N!}{(N-i)!\,i!} \right) R_{\text{m}}^{N-i}(t)[1 - R_{\text{m}}]^i \right] \tag{29.1}$$

where R_{NMRV} is the reliability of the NMR system with the voter, R_{V} is the voter reliability and R_{m} is the module reliability.

However, NMR is not the optimized method of achieving redundancy as the number of module elements increases. Redundant system implementations typically use a voting method to determine which outputs are correct. This is easily implemented using *r-out-of n* network model [32] of reliability and mean time to failure (MTTF) computation.

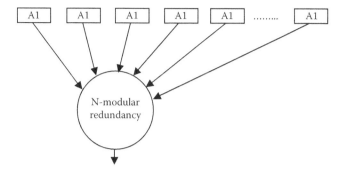

FIGURE 29.10 N-modular redundancy model.

29.5.3.1 Voting Mechanism

Some of the most common voting mechanisms are majority voter, weight average voter and median voter. Majority voter is more popular in safety critical systems. Consider x_n inputs with a limit threshold ε to evaluate the difference between two inputs $d(x_i, x_j)$. A group G_r is formed whose inputs fall below ε. This mechanism can be explained in Figure 29.11: (i) difference ε between two inputs as $d(x_i, x_j) = |x_i, x_j|$, (ii) any two inputs belong to group G_r if $d(x_i, x_j) < \varepsilon$ and (iii) majority of the group having $< \varepsilon$ is the voted output. The validation voting mechanism is addressed using state machine logic having multiple states. State machine variables are tuned for each parameter specific to the application. A typical state machine for signal validation is shown in Figure 29.11. This has five states and each state has definite transitions based on the signal FAIL_COUNT and PASS_COUNT.

Each parameter produced by input has a data field and an associated status field. The data field contains the value of a parameter and the status field contains the validity status of a parameter. There may not always be a one-to-one association between data and status fields of parameters. In some cases (i.e., packed discrete signals), several parameters share a common status, but each parameter has a unique data field. These parameter data/status fields are updated by input after validation is performed.

29.5.3.2 Architecture Platform-Level Redundancy

Typical methods of increasing the availability of systems in federated architecture are by adding more number of hardware for critical functional requirements. However, in the integrated approach, the duplication of hardware is not by a physical box but by increasing the number of

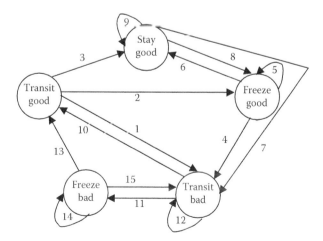

FIGURE 29.11 Five state signal voting scheme.

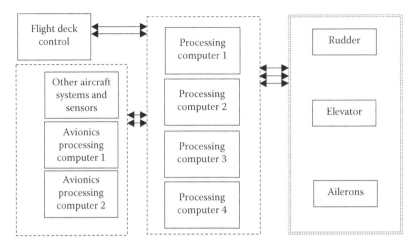

FIGURE 29.12 Typical architecture with four computers for redundancy.

processing modules or processors and its related hardware and software. An example is the flight control system of an aircraft which duplicates the processing system and also the drive electronics for the actuators along with the actuator channels themselves. Thus, hardware duplication depends on the function being realized whereas levels of redundancy will vary from system to system. With the development of safety protection architectures such as ARINC 653 supported by real-time operating systems, the physical hardware redundancy is being re-looked into to reduce the level of hardware and to increase software redundancy. Figures 29.12 and 29.13 show the typical redundancy mechanism [54] being followed for regional category civil commercial aircraft. Figure 29.12 shows the architecture with hardware redundancy in flight control computer (FCC) interfacing with electronics processing and drive system in the same hardware. FCC uses four modules each having command and monitor architecture and hence there are eight lanes. Therefore, the sensor interfaces also connect to all the lanes for synchronous data to all processing systems.

Figure 29.13 shows an alternate architecture where the drive electronics is separate from the main processing system. This has one dual-channel, one single-channel flight control module (FCM) and three dual-channel actuator control electronics (ACE). Therefore, there are six lanes of FCM and six lanes of ACE.

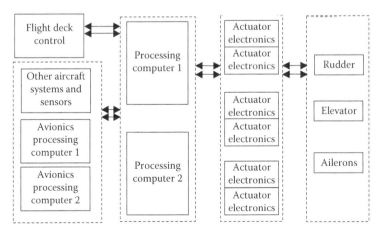

FIGURE 29.13 Typical architecture with split computers for redundancy for processor and drive electronics.

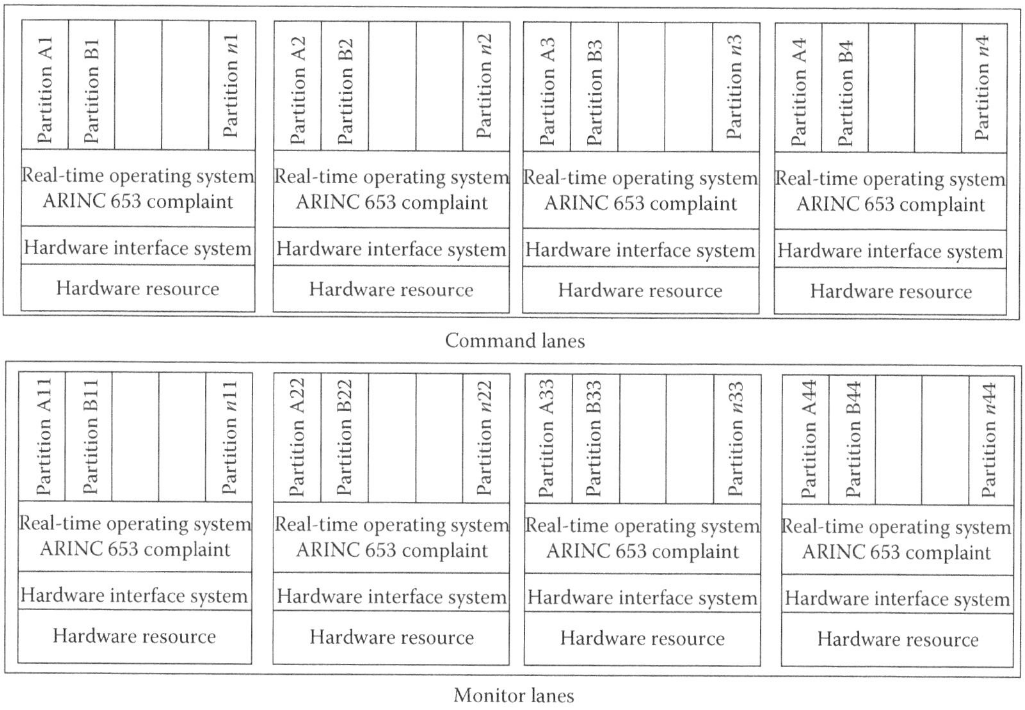

FIGURE 29.14 Processor architecture with partitioning.

In integrated architectures, the processor or lane redundancy is addressed rather than the complete channel. Consider the processing power of the first architecture having four computing modules, the platform supporting ARINC 653 time and memory partitioning. It has eight independent lanes and each lane having time and memory partitioning capability. Figure 29.14 shows the ARINC 653 platform with partitioning for computing module and the actuator electronics in each of the processor lanes both for command and monitor.

The redundancy is built on processors, lanes, partitioning and interfaces. However, in this case, the fault management is not configurable with restricted scenarios of system. Reconfiguration of an application depends on the then running task, its status, its functionality, criticality and other application requirements. If by some means the application is reconfigured and successful, the continuation of application happens, and then the architecture is to a limited extent fault-tolerant and aids in the enhanced availability requirements of the flight critical systems. This can be done by the reconfiguration algorithm using control parameter validation checks in real time to re-schedule or reconfigure the task or a process to achieve the required functionality. Using the ARINC 653 architecture and its platform, the IMA architecture can be implemented in two basic modes: (i) using independent processing nodes and (ii) using common processing nodes. Figure 29.14 shows the architecture with independent processing nodes having ARINC 653 platform. The functionality is split into separate processing nodes. The number of processing nodes, the bandwidth and the throughput are well addressed. However, it is quite expensive in terms of implementation. Figure 29.15 shows the architecture with common processing nodes having the ARINC 653 platform. This can be implemented in one hardware processor for command and monitor, respectively (limitation based on processor loading and bandwidth traffic). It is seen that the applications for different criticality are hosted in different partitions for multiple systems. In summary, looking at the aerospace systems, the functional availability is very crucial. Failure management in integrated platform is quite essential for enhanced functional availability. Also, integrated modular avionics systems do have scope for improvement in the area of task reconfiguration.

Partition A1	Partition B1	Partition A2	Partition B2	SPARE	SPARE	SPARE	Partition A3	Partition B3	Partition A4	Partition B4

Real-time operating system ARINC 653 complaint

Hardware interface system

Hardware resource

Command lanes

Partition A11	Partition B11	Partition A22	Partition B22	SPARE	SPARE	SPARE	Partition A33	Partition B33	Partition A44	Partition B44

Real-time operating system ARINC 653 complaint

Hardware interface system

Hardware resource

Monitor lanes

FIGURE 29.15   Implementation of Figure 29.16 using one single-processor hardware for command and monitor.

29.6   RECONFIGURATION ALGORITHM FOR ENHANCED AVAILABILITY OF A PROCESS OR A TASK

Reconfiguration in real-time aerospace flight critical system is a method of task or process management having provision to reconfigure the functionality in an event of failure in near real time. This is executed to the accuracy of one major frame so that the faulty task is reconfigured as part of the schedule in the next major frame. The system uses the concept of major frames, multiple partitions and each partition having multiple processes to schedule the tasks. Typical avionics systems are scheduled with *major frames*. Each major frame has a set of partition. Each partition or defined functionality is grouped into a set of *processes* as depicted in Figure 29.16, each having modular sub-functionalities in each process. As software components, each process consists of a set of low-level functions called *tasks*.

Consider a major frame M having a set of partitions $P_{ti} \dots P_{tn}$ based on functionalities as shown in Figure 29.16. Each partition P_{ti} consists of a set of process $P_{si} \dots P_{sn}$ based on the application's sub-functionalities. Each P_{si} consists of a set of tasks ($\tau_1 \dots \tau_2$). The number of partitions and the number of processes in each partition are a trade-off to get the real-time response based on capabilities of the hardware and software together.

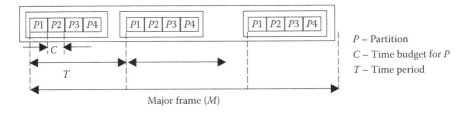

P – Partition
C – Time budget for P
T – Time period

FIGURE 29.16   Typical IMA partitioned schedule table with frames, process and tasks.

$$\text{Each major frame}(M) = \left\{ P_{t_1}, P_{t_2}, P_{t_3}, P_{t_4}, \ldots, P_{t_n} \right\}$$

$$\text{Each partition}(P_t) = \left\{ P_{s_1}, P_{s_2}, P_{s_3}, P_{s_4}, \ldots, P_{s_n} \right\} \tag{29.2}$$

$$\text{Each task}(P_s) = \left\{ \tau_1, \tau_2, \tau_3, \tau_4, \ldots, \tau_n \right\} \quad (\tau \text{ is single task})$$

Figure 29.16 shows the set of partitions P, which are scheduled across a major frame M consisting of a set of partitions P_{t_n} and each partition having a set of task/process τ_n / P_{s_n} [3]. Typical integrated avionics ARINC 653-based applications [1,2] have a major frame, partitions and a minor frame. The reconfiguration algorithm [55,56] uses the concept of major frame, minor frame and task schedule to improve the availability. The major frames, partition, processes and each process with the number of tasks [56–58] are represented as

$$\text{Major frame} = \begin{bmatrix} Pt_1 \\ Pt_2 \\ Pt_3 \\ \cdot \\ Pt_n \end{bmatrix} = \begin{bmatrix} Ps_{11} & Ps_{12} & Ps_{13} & \cdot & Ps_{1n} \\ Ps_{21} & Ps_{22} & Ps_{23} & \cdot & Ps_{2n} \\ Ps_{31} & Ps_{32} & Ps_{33} & \cdot & Ps_{3n} \\ \cdot & \cdot & \cdot & \cdot & \cdot \\ Ps_{m1} & Ps_{m2} & Ps_{m3} & \cdot & Ps_{mn} \end{bmatrix} \tag{29.3}$$

Each process P_s consists of a set of tasks $\tau_1, \ldots, \tau_n$ and the sequence of tasks are pre-defined and priorities are fixed as per the static table scheduling represented in Equation 29.4 for Ps_{11}.

$$\begin{bmatrix} Ps_{11} \end{bmatrix} = \begin{bmatrix} \tau_1, \tau_2, \ldots, \tau_n & \tau_1, \tau_2, \ldots, \tau_n & \tau_1, \tau_2, \ldots, \tau_n \\ \tau_1, \tau_2, \ldots, \tau_n & \tau_1, \tau_2, \ldots, \tau_n & \tau_1, \tau_2, \ldots, \tau_n \\ \tau_1, \tau_2, \ldots, \tau_n & \tau_1, \tau_2, \ldots, \tau_n & \tau_1, \tau_2, \ldots, \tau_n \\ \tau_1, \tau_2, \ldots, \tau_n & \tau_1, \tau_2, \ldots, \tau_n & \tau_1, \tau_2, \ldots, \tau_n \end{bmatrix} \tag{29.4}$$

Each task τ_i has definite timing characteristics [3]:

$$C_i \leq D_i \leq T_i$$

where C_i is the task worst-case execution time (WCET), D_i is the task deadline and T_i is the period.

Each task τ_i has critical timing characteristics which are examined for real-time requirements such as worst-case blocking, worst-case partition delay, worst-case process jitter and operating systems (OS) overheads. During the execution of a process, WCET and worst-case process jitter (J_i) are the two important timing characteristics to be considered for realistic estimation of execution time. A brief description of these timing characteristics is provided here for completeness.

A. Worst-Case Blocking (Bi)   Worst-case blocking relates to the time that τ_i is prevented from executing by processes of lower priority [3].

Worst-case blocking is given by

$$(B_i) = (TO_i * b_i) + b_i \tag{29.5}$$

where b_i is the maximum single blocking delay due to a lower process and TO_i is the maximum number of timeouts that τ_i may issue during execution.

B. Worst-Case Partition Delay (L_i)   In multi-partition environment, it is possible for one partition to delay the next scheduled partition. Partition delay is very crucial for flight critical application in the aerospace industry. APEX implementation ensures that this delay is not cumulative and can be as long as the operating system critical section of the partitions that could immediately precede Pt_i.

C. Worst-Case Process Jitter (WCPJ) (J_i)   WCPJ quantifies the maximum difference of the response time with the execution times for each period [59]. Jitter depends on the kernel overheads and partition jitter. Worst-case process jitter (J_i) depends on the release jitter on release time P_s due to kernel overhead and absolute start time of the execution P_t that will recognize the release time of P_s at time t. Jitter measurements for typical medium-range embedded target were experimented [60] and this timing measurements help to characterize the delays and execution non-linearity in the algorithm. However, the response time of a task or a process encompasses the various delays and execution times. They are

L_i	C_s	S_i	A_i
ns	μs	μs	ms

where L_i is the interrupt latency time, C_s is the context save time, S_i is the schedule time and A_i is the process time.

Therefore, the response time is expressed as

$$R_i = L_i + C_s + S_i + A_i \tag{29.6}$$

All the above timing issues are addressed as part of criticality analysis in the reconfiguration algorithm for aerospace applications.

29.6.1   CONTROL PARAMETERS FOR RECONFIGURATION ALGORITHM

The reconfiguration algorithm used in an integrated platform is based on the past and present results on the system being updated continuously by an observer or any other mechanism called reconfiguration monitoring and control unit (RCMU). The criteria of control parameter selection are important in the definition of the role of the control parameter in reconfigurable algorithms. Detailed study, analysis and simulations were carried out to formulate the identification of control parameters to make decisions on the real-time reconfiguration. Based on the above study, the following control parameters are identified and used for reconfigurable algorithms. With the use of these control parameters, the application software process or task gets reconfigured and continues to function under identified failure conditions. In some cases, the extent of reconfiguration depends on the type of fault and depends on the context variables at that time. However, there is definitely an appreciable non-reconfigurable system. The following control metrics are defined for use in reconfigurable algorithms in flight critical applications on integrated platforms: (i) reconfigurability information factor (RI), (ii) schedulability test/TL/UF (TL), (iii) context adaptability and suitability (CAS) and (iv) context flight safety factor (CFS). A novel metric-based reconfiguration algorithm is proposed using four control metrics. The algorithm uses these critical control metrics for graceful reconfiguration or graceful degradation of a task or a process. These control metrics help the algorithm in efficient decision making during the reconfiguration process. Based on the control metrics, the reconfiguration Go/NoGo is declared. The above-listed control metrics are detailed in subsequent sections.

29.6.1.1 Reconfigurability Information Factor

RI is defined as the ratio of re-scheduled task functional credit point (FCP) to the original scheduled task FCP. FCP is the measure of functionality and is assigned to each function of every partition in the range of 0.0–1.0. FCP values are assigned based on the following measures: (i) execution time, (ii) resource consumption and (iii) complexity of the functionality. For every selected critical task (τ_s) in a frame consisting of a number of scheduled lists, there can be at least one configurable task (τ_r). The selection of replaceable task is based on the RI, that is, a process P_s or task τ_s can be reconfigured by a process P_r or a task τ_r if and only if the RI of the new process P_r or task τ_r is at least equal to or greater than the RI of the faulty process P_s or task τ_s and is expressed as

$$(\tau_s = \tau_r) \leftrightarrow (RI(\tau_r) \geq RI(\tau_s)) \quad \text{or} \quad (P_s = P_r) \leftrightarrow (RI(P_r) \geq RI(P_s)) \tag{29.7}$$

For every task (τ_s) or (P_s), there is exactly one task (τ_r) (denoted by E!) or process (P_r) such that

$$[RI(\tau_r) \geq RI(\tau_s)] \quad \text{or} \quad [RI(P_r) \geq RI(P_s)]$$

$$((\forall \tau_s)(E! \tau_r)(RI(\tau_r) \geq RI(\tau_s))) \quad \text{or} \quad ((\forall P_s)(E! P_r)(RI(P_r) \geq RI(P_s)))$$

FCP is derived based on the type of task, criticality of the task and phase of application envelope with identified evaluation parameters such as time, complexity and resource. For all critical tasks, task τ in a process P_s scheduled in a partition P_t has a defined FCP. Every element of Equation 29.4 has a corresponding functional credit point matrix as denoted in matrix (29.8). FCP elements in matrix (29.8) are derived from the system requirements, design limits and failure mode effect analysis and testing guidelines.

$$\begin{bmatrix} f_1, f_2, \ldots, f_n & f_1, f_2, \ldots, f_n & \cdot & \cdot & f_1, f_2, \ldots, f_n \\ f_1, f_2, \ldots, f_n & f_1, f_2, \ldots, f_n & \cdot & \cdot & f_1, f_2, \ldots, f_n \\ f_1, f_2, \ldots, f_n & f_1, f_2, \ldots, f_n & \cdot & \cdot & f_1, f_2, \ldots, f_n \\ & \cdot & \cdot & \cdot & \cdot & \cdot \\ f_1, f_2, \ldots, f_n & f_1, f_2, \ldots, f_n & \cdot & \cdot & f_1, f_2, \ldots, f_n \end{bmatrix} \tag{29.8}$$

29.6.1.2 Schedulability Test (Time Loading or Utilization Factor)

Schedulability Test is the standard method of time loading or utilization for a task to be scheduled

$$(\tau_s = \tau_r) \leftrightarrow \left(\text{WCET}(\tau_s) \leq \text{WCET}(\tau_r) \right)$$

$$(\tau_s = \tau_r) \leftrightarrow \left[\left(\sum_{i=1}^{s_n} \left(\frac{C_{s_i}}{T_{s_i}} \right) \leq \sum_{i=1}^{r_n} \left(\frac{C_{r_i}}{T_{r_i}} \right) \right) \right] \qquad (29.9)$$

Similarly for a process, the faulty process shall be replaceable if and only if schedulability test as per expression (29.9) passes for a process. For all cases of task phasing, a set of n tasks will always meet their deadlines [3] if

$$\sum_{i=1}^{n} \left(\frac{C_i}{T_i} \right) \leq U(n) = n(2^n - 1) \leq 0.69 \qquad (29.10)$$

where C_i is the worst-case execution time of task i and T_i is the time capacity. The selection of replaceable task is based on the TL that is, a process P_s or task τ_s can be reconfigured by a process P_r or a task τ_r if and only if the TL of new process P_r or task τ_r is at least equal to or less than the TL of the faulty process P_s or task τ_s and is expressed as

$$(\tau_s = \tau_r) \leftrightarrow (\text{TL}(\tau_s) \geq \text{TL}(\tau_r)) \quad \text{or} \quad (P_s = P_r) \leftrightarrow (\text{TL}(P_s) \geq \text{TL}(P_r))$$

Most of the aerospace flight critical systems do follow this guideline of loading process not more than 70% and typically the loading is around 50% to take care of growth potential in the project life cycle. Expressions 29.9 and 29.10 are strictly enforced in algorithm for static computation of time loading in each schedule table for every partition. Execution time or utilization is an important dataset for efficient selection of a task or process to reconfigure. Each task is benchmarked for execution time using industry practice of source stubbing and the same is used by the algorithm. The corresponding matrix is

$$\begin{bmatrix} t_1, t_2, \ldots, t_n & t_1, t_2, \ldots, t_n & \cdot & \cdot & t_1, t_2, \ldots, t_n \\ t_1, t_2, \ldots, t_n & t_1, t_2, \ldots, t_n & \cdot & \cdot & t_1, t_2, \ldots, t_n \\ t_1, t_2, \ldots, t_n & t_1, t_2, \ldots, t_n & \cdot & \cdot & t_1, t_2, \ldots, t_n \\ & \cdot & & \cdot & \cdot & \cdot & \cdot \\ t_1, t_2, \ldots, t_n & t_1, t_2, \ldots, t_n & \cdot & \cdot & t_1, t_2, \ldots, t_n \end{bmatrix} \qquad (29.11)$$

For selected critical tasks, the reference execution time dataset is compiled and generated in accordance with expression 29.4. Reconfigurable algorithm, which uses matrix in (29.11) as one control parameter input, is tested using the data captured from a flight critical programme. The algorithm checks this reference data for task selection criteria.

29.6.1.3 Context Adaptability and Suitability

CAS metric decides acceptability of the faulty task replacement in real time. This involves checking the CAS table and context sensitivity (CS) table to decide whether the reconfiguration is permissible. Hence, the context of the scenario is verified and validated for the functionality and suitability of the task. CAS are expressed as

$$(\text{CAS} = \text{TRUE}) \leftrightarrow \begin{pmatrix} \text{Re-scheduled task or process context flag is equal to or} \\ \text{greater than the original task or process context flag} \end{pmatrix} \quad (29.12)$$

And it is expressed as: For every task (τ_s) or process (P_s), there can be a replaceable task (τ_r) or process (P_r) such that

$$(\text{CAS}(\tau_s, \tau_r) \text{ is TRUE}) \quad \text{or} \quad (\text{CAS}(P_s, P_r) \text{ is TRUE})$$

$$((\forall \tau_s \, (\text{CAS}(\tau_s, \tau_r) \text{ is TRUE})) \quad \text{or} \quad ((\forall P_s \, (\text{CAS}(P_s, P_r) \text{ is TRUE}))$$

CAS uses the CAS factor and CS for final CAS Go/NoGo decision. CAS factor is derived using analytical methods. Also, the CAS factor used in the algorithm is derived based on the system functionality and inter-system reconfiguration dependencies using failure mode effect analysis (FMEA), failure hazard analysis (FHA) and system safety assessment (SSA). The selection of the replaceable task is based on the CAS, that is, a process P_s or task τ_s can be reconfigured by a process P_r or a task τ_r if and only if the CAS of a new process P_r or task τ_r is at least equal to or greater than the CAS of the faulty process P_s or task τ_s and is expressed as

$$(\tau_s = \tau_r) \leftrightarrow (\text{CAS}(\tau_r) \geq \text{CAS}(\tau_s)) \quad \text{or} \quad (P_s = P_r) \leftrightarrow (\text{CAS}(P_r) \geq \text{CAS}(P_s))$$

Also, the CAS directly depends on the CS with reference to the identified critical failure of the system. Hence, the final CAS is derived as

$$\text{CAS}_{\text{Ptn}} = \text{CAS}_{\text{Ptn}} \quad \text{and} \quad \text{CS}_{\text{Ptn}} \text{ (failure), that is,}$$

$$(\tau_s = \tau_r) \leftrightarrow [(\text{CAS}(\tau_r) \geq \text{CAS}(\tau_s)) \quad \text{and} \quad (\text{CS}(\tau_r)_{n \text{ failure}} == 1)]$$

29.6.1.4 Context Flight Safety Factor

It is very vital in aerospace flight critical applications to check the safety of the system before and after reconfiguration. After validating the above three control parameters, the system is checked for safe state to initiate reconfiguration. For aircraft systems in closed loop control, a wrong function being reconfigured can lead to catastrophic failure. Hence, any action carried out in real time is verified and validated thoroughly by all the control parameter artifacts along with the system information.

Context flight safety factor (CFS) is defined as

$$(\text{CFS} = \text{TRUE}) \rightarrow ((\text{Re-scheduled task or process safety factor/original}$$
$$\text{scheduled task or process safety factor}) \geq 1.0)$$

Also process or task is replaceable only if

$$(P_s = P_r) \leftrightarrow (\text{CFS}(P_r) \geq \text{CFS}(P_s))$$

$$(\tau_s = \tau_r) \leftrightarrow (\text{CFS}(\tau_r) \geq \text{CFS}(\tau_s))$$

and is described as: For every critical task (τ_s) or process (P_s) there can be a replaceable task (τ_r) or process (P_r) such that

$$(\text{CFS}\,(\tau_r, \tau_s) \ge 1.0) \quad \text{or} \quad (\text{CFS}\,(P_r, P_s) \ge 1.0)$$

$$((\forall \tau_s\,(\text{CFS}\,(\tau_r, \tau_s) \ge 1.0)) \quad \text{or} \quad ((\forall P_s\,(\text{CFS}\,(P_r, P_s) \ge 1.0))$$

The selection of replaceable task is based on the CFS, that is, a process P_s or task τ_s can be reconfigured by a process P_r or a task τ_r if and only if the CFS of a new process P_r or task τ_r should be at least equal to or greater than the CFS of the faulty process P_s or task τ_s and is expressed as

$$(\tau_s = \tau_r) \leftrightarrow (\text{CFS}\,(\tau_r) \ge \text{CFS}\,(\tau_s)) \quad \text{or} \quad (P_s = P_r) \leftrightarrow (\text{CFS}\,(P_r) \ge \text{CFS}\,(P_s))$$

Also, the CFS directly depends on the safety unit (Su) with reference to the flight phases of the aircraft. Su is a Boolean value assigned to each task of every partition and is assigned based on the flight phase. Task reconfiguration is safety critical and hence cannot be reconfigured in all phases of the flight. Hence, the final CFS is derived as

$$\text{CFS}_{\text{Ptn}} = \text{CFS}_{\text{Ptn}} \quad \text{and} \quad \text{SU}_{\text{Ptn}}(\text{Phase}), \text{ that is,} \qquad (29.13)$$

$$(\tau_s = \tau_r) \leftrightarrow [(\text{CFS}\,(\tau_r) \ge \text{CFS}\,(\tau_s)) \quad \text{and} \quad (\text{SU}\,(\tau_r)_{n\ \text{Phase}} == \text{TRUE})]$$

CFS is derived from both the RI and the SU based on the failure hazard analysis (FHA), failure mode effect analysis (FMEA) and system safety analysis (SSA) [61]. Every element of Equation 29.4 has a corresponding SU matrix, which will be used by CFS. SU is a measure of margin of system safety to reconfigure a task with the prevailing dynamic context of the flight. Finally, the task reconfiguration is decided by the algorithm with consistent decision of all four control metrics. Reconfiguration is successful if (RI & TL & CAS & CFS) == TRUE. In case of failure of the task after reconfiguration, the system can have a degraded mode for limited functionality. The extent of degraded performance allowed in such safety critical systems is decided based on CAS and the degradation level derived specific to the application objectives and functionalities for both process and task. The degradation level (DL) is the measure of allowed degraded performance or functionality in the selected envelope of the system being reconfigured. If degraded functionality is not allowed, then the degradation factor is 1. The reference dataset of DL is captured from functionality requirements under dynamic pre-defined scenarios. Each scenario is analysed statically and the functionality is simulated for varying degraded functions. Finally, the data is compiled for each functional requirement with the measure of allowed degradation. The degradation factor is not critically used in the reconfiguration algorithm.

29.6.2 RECONFIGURATION ALGORITHM

A non-reconfigurable system either shuts down or performs a degraded functionality in the event of a task failure. In some cases, this may lead to infinite loops or crash of the application leading to serious failure. Here, the fault is not resolved; rather, the system enters failed/degraded state. The reconfiguration algorithm overcomes the above fault scenario by reconfiguration of faulty task resulting in the recovery of fault either completely or partially. The identification of a task failure is by the monitors, which are called system monitors or RCMU. System monitor continuously monitors the state and status of critical tasks with reference to the context of control parameters. The algorithm replaces a faulty process or task by a replaceable, suitable and safe substitute after extensive checks and validation. The reconfigured task or process performs the required operation without any safety impact to the system and aircraft. After critical validation of the state data using control metrics, the reconfiguration algorithm decides to reconfigure a task or process if and only if ((RI & TL & CAS & CFS) == TRUE). The algorithm has the fail off path in case the algorithm enters the fault loop with multiple reconfigurations without effective output. This is handled by a reconfiguration counter, which avoids the repetitive reconfiguration for the same failure. The steps of the proposed algorithm are listed below:

- *If a task fails*
- Capture the task (τ_s) status, functionality, priority, criticality to identify the faulty task
- Identify the most suitable substitution task (τ_r) after validating the following metrics for feasibility
- Reconfigurability I-Factor (RI)
- Schedulability test/TL/UF (TL)
- Context adaptability/suitability (CAS)
- Context flight safety unit (CFS)
- Reconfigure the task table or process (Ps) before the next major frame after system assessment of functional state of the partition
- *If reconfigured task fails*
- *If the system can run in degraded mode*
- Revert all tasks to its original state
- Identify a set of tasks which need to be removed from the schedule
- Reschedule the task set with degraded performance using dead task removal techniques (all the failed tasks are removed from the task set)
- The rest of the task set continues to run provided no safety impact after reschedule
- *In case degraded mode is not feasible*
- Declare failure
- Shut down the system

Figure 29.17 shows the abstracted flowchart of the reconfiguration algorithm. The algorithm is well suited for an open architecture multiple process and multiple schedule static table mechanism. The algorithm plays the role of a high-level real-time monitor software continuously monitoring the status of the running tasks of a schedule table. The algorithm is explained in the five phases as status capture, control parameter validation, reconfiguration algorithm, degraded performance and fail off procedure.

29.6.3 Phases of Algorithm

29.6.3.1 Phase I: Status Capture

Critical avionics applications are typically embedded with lot of monitoring functionalities. The data so monitored and captured are processed to extract the abstract information required for health management process. Here, the algorithm starts with continuously monitoring the status and health of a task during execution. The data capture is part of the application software and the algorithm receives the information on demand call or through global shared resources. When the algorithm detects a task failure or malfunction, the algorithm initiates the Phase II execution. Validation of the control data or the captured data for metrics parameter enhances the performance of the reconfigurable algorithm. Validation is achieved by the following methods: (i) data threshold averaging for variable input data, (ii) efficient assignment of functional credit point for RI, (iii) flight phase identification and definition and (iv) context sensitivity analysis. Aircraft sensor inputs are of different types, levels, sample rates and protocols. The data so received from multiple sources having different threshold and persistence values need to be processed with special care to ensure the proper sequencing and

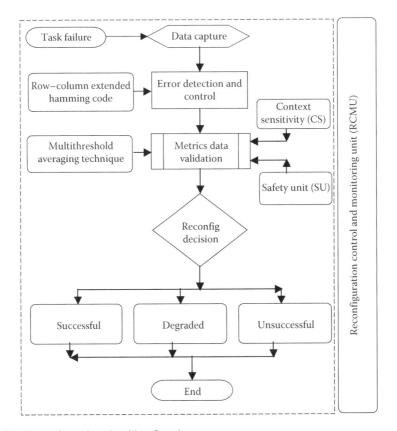

FIGURE 29.17   Reconfiguration algorithm flowchart.

synchronization of data across samples. Averaging-based algorithm is proposed with multi-level threshold scheme. Therefore, the averaging of the collected data is very much essential before it is used in algorithm. Typical validation involves the range, resolution, level, polarity and threshold. Apart from these inputs, the multi-signal averaging [57] is used in flight critical application. Multi-threshold data-averaging technique is used to process the input data of the system before the same is processed by the control metrics parameters. Functional credit point (FCP) is assigned to each function of original and re-schedulable tasks in the range of 0.0 to 1.0. The reconfiguration of task or process with control metrics cannot be applied to all phases of flight even though the control data is satisfied in terms of the validation. Flight phase validation is an additional dimension to validate the control metrics. Typically, flight data analysis is carried out with identification of phases and sectors. A tool called NAL's Flight Operations Quality Assurance software (NALFOQA) [62,63] is designed and developed for identification and dynamic definition of control metrics validation. NALFOQA is configured with aircraft-specific phase parameters and their trigger limits of various phases.

29.6.3.2   Phase II: Control Parameter Validation

Having collected the required data to be used in the algorithm, the data is then applied to the control metrics to aid the reconfiguration algorithm for decision making. The main objective of Phase II is to identify the most suitable task using the results of control parameter validations (RI, TL, CAS and CFS).

29.6.3.2.1   Reconfigurability Index

RI uses the FCP of each task to identify the suitable task for reconfiguration. FCP is derived based on the execution time, resource consumption and complexity of the functionality.

The FCP for each process is derived as a function of FCP (Process) = $0.[T_f][S_f][C_f]$, where T_f = time factor, S_f = space factor and C_f = complexity factor.

29.6.3.2.2 Time Loading

TL provides the actual time consumed by each task having maximum allowable task utilization limit of TL ≤ 0.69 as per industry standards for flight critical system. A typical process involves the estimation of worst-case execution time of each task in every partition and schedules the tasks appropriately. This is to ensure that under the worst-case scenario, the schedule does not hit the time budget of the defined period.

29.6.3.2.3 Context Adaptability and Suitability

CAS is validated against analytical methods and the measure of CS metrics and is expressed as CS = [failures][partitions][tasks]. In a given implementation, the set of context-sensitive probable failures are identified, and based on these failures, the sensitivity of each task is derived. This matrix is used to ensure that for a given defined failure, the CS is critical or non-critical. This is a three-dimensional matrix with defined failures, number of partitions and number of tasks in each partition. CS is weighed for a limit value and is graded between 0 and 1 for Go/NoGo analysis during algorithm execution. Under a failure scenario, in case if the CS is GO, then the reconfiguration status is set TRUE, else FALSE.

29.6.3.2.4 Context Flight Safety

CFS is a critical metric and is based on the safety of the system. CFS uses FCP along with the Su to identify the potential non-safety scenario. Su is derived with reference to the phases of the flight for every task in a partition. The reconfiguration of a task cannot be executed in all critical phases of the flight such as take-off, climb, descent and landing because of the safety and criticality of the flight conditions. Therefore, the detection of flight phases based on the aircraft dynamic performance is quite important and challenging. Input data are sampled and processed for control parameters in the selected phases only. Hence, the phase of the flight is used in defining the safety consideration of reconfiguration. Such critical phases of flight do not allow any operation other than the minimum required to complete the mission. In these cases, the system does not allow very critical functions to reconfigure as this operation will worsen the situation compared to continuing with the failure. CFS defines the metrics called Su as a three-dimensional matrix with Su_{task} = [partitions][process/task][phases]. Under a task failure scenario, if Su_{task} is TRUE then the reconfiguration Flag is set to TRUE, else FALSE. In any case, if one of the control parameter fails to comply with safe values, the algorithm returns to the system without any action resulting in restoration to the original state or transits to phase IV. On successful completion of control parameter checking and validation, the algorithm initiates the Phase III execution.

29.6.3.3 Phase III: Reconfiguration

On successful validation of the control metrics, the reconfiguration is performed. However, the reconfiguration algorithm uses all the control metrics state before a task is reconfigured as per condition (29.14).

$$\text{Reconfiguration is TRUE if } ((RI \& TL \& CAS \& CFS) == TRUE) \qquad (29.14)$$

During the reconfiguration process, the global state of the system, process or partition is not altered. Only the selected task or process gets altered for their respective state variables. On successful reconfiguration, the reconfigured task starts execution in the next major frame as task reconfiguration is activated only in the next major frame.

29.6.3.4   Phase IV: Degraded Performance

Algorithm has safe exits along with degraded performance. This means that if the reconfigured task fails again in the next major frame, then the algorithm reverts to its earlier state by reverting the reconfigured task. If the degraded performance or functionality is allowed for that particular function, then the algorithm, the faulty task or process from the schedule table and allows it to continue. In this case, the system continues to execute without the functionality of the removed task. This is still an improved mechanism instead of totally shutting down the application with many others tasks in good state. Also, this is not valid for all failures subjected to context of phase and failures.

29.6.3.5   Phase V: Fail off Procedure

During the execution of the algorithm, if the degraded functionality is not allowed then the algorithm behaves similar to the normal process of entering failed state. The reconfiguration algorithm is applied only for the critical task or process, which improves the availability as an effect of reconfiguration. Configuration of a task or a process in aerospace flight critical system is a crucial event concerned with the safety of the aircraft and availability of systems. Identification of task for reconfiguration requires serious judgement, methodical analysis and extensive cross-comparison across various relative parameters in real time. The control parameters are checked for their states, status and validation before re-scheduling the sequence of tasks. Failure data collection for various scenarios is based on standards, equipment life cycle and quality control data management [60].

29.7   ERROR DETECTION AND CONTROL

29.7.1   Extension to the Hamming Code-Extended Row–Column Hamming Code

There have been extensions to the basic hamming code [61] in the past. A few of the modifications to the hamming code were based on the distribution of check bits in the stream and hence reduction of decoding efforts. However, to overcome the limitation of the basic hamming code, extended hamming code called as row–column (RC) extended hamming code is being introduced and adopted for multiple bit error detection and correction. The RC extended hamming code [64] is used for the error detection and correction of control data transmission across the partitions and also partition to RCMU. Each of the control metrics data is of 4×4 matrixes and hence the data are used as 16 element stream with 4 rows and 4 columns. This extension to the hamming code is implemented by rearranging the bits in the form of a 4×4 matrix that is able to correct the burst of errors compared to conventional hamming code. The extended hamming code takes 16 bits of data in the form of a 4×4 matrix and encodes each row first followed by each column, respectively. The parity bit for each row is embedded between the actual data as per the regular hamming code technique and the parity bits generated for the column is padded at the end. At decoding end, each of the horizontal rows is decoded and corrected for single bit error. After decoding all the four rows in the matrix, the vertical columns are decoded and corrected for single-bit errors. This takes more time when compared to regular hamming code but this will correct more than 1 bit of error in a single row. RC extended hamming code works as follows: (i) First the 16 bits of data stream is arranged in a 4×4 matrix; (ii) Apply the RC hamming algorithm for one row at a time. Basic hamming code corrects one-bit error in case there are two errors in a row and (iii) Again apply the RC hamming algorithm for each column of the matrix. Hamming code column algorithm detects and corrects the left-out errors of row algorithm as part of RC extended hamming code algorithm. Considering a situation where all the 4 bits in the row are toggled (introduced error), when decoded by regular technique, this is difficult to correct. By using the extended hamming technique, at one go the errors are not detected but first each horizontal rows are detected for errors and corrected, respectively, and then while detecting the errors for the vertical columns all the errors are corrected with the regular hamming code technique. This technique will take a lot of time but its result is appreciable. However, this system does have few limitations. Consider the following example:

$$\text{Original data} = \begin{bmatrix} a & b & c & d \\ e & f & g & h \\ i & j & k & l \\ m & n & o & p \end{bmatrix} \quad (29.15)$$

After transmission, the bits a, b, c, d, e, i, m are toggled.

$$\text{Received data} = \begin{bmatrix} \bar{a} & \bar{b} & \bar{c} & \bar{d} \\ \bar{e} & f & g & h \\ \bar{i} & j & k & l \\ \bar{m} & n & o & p \end{bmatrix} \quad (29.16)$$

Original data and received data are shown in matrices (Equations 29.15 and 29.16), respectively. When horizontal decoding is done, bits a, b, c, d are checked for errors which will not be corrected. Then, in the next row bit, e, f, g, h are checked for error and bit e is corrected. Similarly, bit i and m will also be corrected. In second run bits a, e, i, m is decoded and bit **a** is corrected. Similarly bits **b**, **c** and **d** will be corrected. Using this technique, a maximum of 7 bits can be corrected for every 16 bits of data.

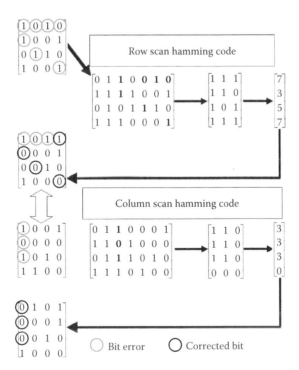

FIGURE 29.18 Demonstration of row–column extended hamming code algorithm.

Figure 29.18 shows the partition 1 task matrix and demonstration of the RC extended hamming code algorithm with a typical bit error scenarios and correction of the same. The RC extended hamming code is implemented in two phases called row operation and column operation. To demonstrate the CFS with Su, for taxi out phase is used here. Extended hamming code is demonstrated for 4×4 matrix dataset. Let us take 4×4 data as

$$\begin{bmatrix} 0 & 1 & 0 & 1 \\ 0 & 0 & 0 & 1 \\ 0 & 0 & 1 & 0 \\ 1 & 0 & 0 & 0 \end{bmatrix}$$

This matrix data will be coded as per hamming code protocol and the final data sent are

$$\begin{bmatrix} 0 & 1 & 0 & 0 & 1 & 0 & 1 \\ 1 & 1 & 0 & 1 & 0 & 0 & 1 \\ 0 & 1 & 0 & 1 & 0 & 1 & 0 \\ 1 & 1 & 1 & 0 & 0 & 0 & 0 \end{bmatrix}$$

When the receiver receives the data, the *dataset* if received without error is

$$\begin{bmatrix} 0 & 1 & 0 & 0 & 1 & 0 & 1 \\ 1 & 1 & 0 & 1 & 0 & 0 & 1 \\ 0 & 1 & 0 & 1 & 0 & 1 & 0 \\ 1 & 1 & 1 & 0 & 0 & 0 & 0 \end{bmatrix}$$

However, consider a case where a couple of bits get corrupted as shown in bold numbers, where 0 is corrupted as 1. It is seen that in each row (bold bits) of data, only one bit is corrupted and this can be detected and corrected by the hamming code easily as shown with identification of position of the error bit

$$\begin{bmatrix} 0 & 1 & \mathbf{1} & 1 \\ \mathbf{1} & 0 & 0 & 1 \\ 0 & \mathbf{1} & 1 & 0 \\ 1 & 0 & 0 & \mathbf{1} \end{bmatrix}$$

The corresponding hamming code matrix is

$$\begin{bmatrix} 0 & 1 & 0 & 0 & 1 & \mathbf{1} & 1 \\ 1 & 1 & \mathbf{1} & 1 & 0 & 0 & 1 \\ 0 & 1 & 0 & 1 & \mathbf{1} & 1 & 0 \\ 1 & 1 & 1 & 0 & 0 & 0 & \mathbf{1} \end{bmatrix}$$

Applying the hamming code process for each row, we get Position of error bit $= 2^2 b_2 + 2^1 b_1 + 2^0 b_0$, where

$$b_0 = a_1 + a_3 + a_5 + a_7 \;(\text{mod } 2)$$
$$b_1 = a_2 + a_3 + a_6 + a_7 \;(\text{mod } 2)$$
$$b_2 = a_4 + a_5 + a_6 + a_7 \;(\text{mod } 2)$$

we get 4 * 3 matrix for b_2, b_1 and b_0.

$$\begin{bmatrix} 0 & 1 & 1 \\ 1 & 1 & 0 \\ 1 & 0 & 1 \\ 1 & 1 & 1 \end{bmatrix}$$

Applying hamming code process for each row, we get

$$\begin{bmatrix} 6 \\ 3 \\ 5 \\ 7 \end{bmatrix}$$

This means that the first row error is at bit 6, the second row error is at bit 3, the third row error is at bit 5 and the fourth row error is at bit 7. Having known the error bit, it is easy to correct it. However, assume a case when the error is more than one bit per row, then the normal hamming code cannot correct the bits. Therefore, RC extended hamming code is used.

Consider the case again with multiple errors. In the first row, all four bits have error

$$\begin{bmatrix} \mathbf{1} & \mathbf{0} & \mathbf{1} & \mathbf{0} \\ \mathbf{1} & 0 & 0 & \mathbf{1} \\ 0 & \mathbf{1} & 1 & 0 \\ 1 & 0 & 0 & \mathbf{1} \end{bmatrix}$$

The corresponding hamming code matrix is

$$\begin{bmatrix} 0 & 1 & \mathbf{1} & 0 & \mathbf{0} & \mathbf{1} & \mathbf{0} \\ 1 & 1 & \mathbf{1} & 1 & 0 & 0 & \mathbf{1} \\ 0 & 1 & 0 & 1 & \mathbf{1} & 1 & 0 \\ 1 & 1 & 1 & 0 & 0 & 0 & \mathbf{1} \end{bmatrix}$$

We will apply the extended hamming code. The extended hamming code is a combination of hamming code applied to rows first, then to the columns of the first processed result matrix. By computing b_2, b_1 and b_0 from the above error matrix, we get b matrix along with the error position

$$\begin{bmatrix} 1 & 1 & 1 \\ 1 & 1 & 0 \\ 1 & 0 & 1 \\ 1 & 1 & 1 \end{bmatrix}$$

Applying the hamming code process, we get

$$\begin{bmatrix} 7 \\ 3 \\ 5 \\ 7 \end{bmatrix}$$

Using this error positions, the matrix is corrected as shown below

$$\begin{bmatrix} \mathbf{1} & \mathbf{0} & \mathbf{1} & 1 \\ 0 & 0 & 0 & 1 \\ 0 & 0 & 1 & 0 \\ 1 & 0 & 0 & 0 \end{bmatrix}$$

The corresponding hamming code matrix is

$$\begin{bmatrix} 0 & 1 & \mathbf{1} & 0 & \mathbf{0} & \mathbf{1} & 1 \\ 1 & 1 & 0 & 1 & 0 & 0 & 1 \\ 0 & 1 & 0 & 1 & 0 & 1 & 0 \\ 1 & 1 & 1 & 0 & 0 & 0 & 0 \end{bmatrix}$$

RC extended hamming code row operation has corrected only one out of four error bits in row 1. Now apply the same hamming code to the column matrix. For this we will transpose the matrix and generate the hamming code matrix as shown below

$$\begin{bmatrix} \mathbf{1} & 0 & 0 & 1 \\ \mathbf{0} & 0 & 0 & 0 \\ \mathbf{1} & 0 & 1 & 0 \\ 1 & 1 & 0 & 0 \end{bmatrix}$$

The corresponding hamming code matrix is

$$\begin{bmatrix} 0 & 1 & 1 & 0 & 0 & 0 & 1 \\ 1 & 1 & 0 & 1 & 0 & 0 & 0 \\ 0 & 1 & 1 & 1 & 0 & 1 & 0 \\ 1 & 1 & 1 & 0 & 1 & 0 & 0 \end{bmatrix}$$

By computing b_2, b_1 and b_0 from the above error matrix, we get b matrix along with the error position.

$$\begin{bmatrix} 1 & 1 & 0 \\ 1 & 1 & 0 \\ 1 & 1 & 0 \\ 0 & 0 & 0 \end{bmatrix}$$

Applying the hamming code process, we get

$$\begin{bmatrix} 3 \\ 3 \\ 3 \\ 0 \end{bmatrix}$$

Therefore, we can now correct the remaining error bits based on the second-level detection of error positions. Hence, all the error bits are corrected with the RC extended hamming code and transpose the matrix back to obtain the original data as shown below.

$$\begin{bmatrix} 0 & 1 & 0 & 1 \\ 0 & 0 & 0 & 1 \\ 0 & 0 & 1 & 0 \\ 1 & 0 & 0 & 0 \end{bmatrix}$$

Similarly, the datasets for each parameter are processed for errors and corrected before it is used in the algorithm. Hence, the integrity of the data used by RCMU is quite good and hence the reliability of the reconfiguration is enhanced with this error detection and correction method using RC extended hamming code. Control metrics is methodically validated as shown in Figure 29.19.

Now, all four metrics are derived, validated and evaluated for failure scenarios, and each metric has a Go/NoGo decision for a given task failure. An example is shown for task12 failure and the individual metric's decision of Go/NoGo is RI(TRUE), TL(TRUE), CAS(TRUE) and CFS(TRUE). With clearance from all the four metrics for reconfiguration, the algorithm executes the reconfiguration process and the result is based as follows:

Reconfiguration is TRUE if ((RI & TL & CAS & CFS) == TRUE)

Hence, task12 is cleared for reconfiguration and the same is shown in Figure 29.20.

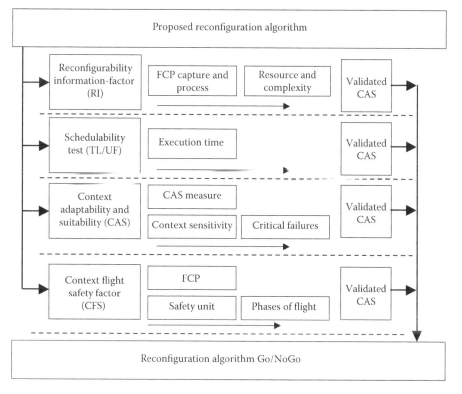

FIGURE 29.19 Schematic of reconfiguration control metrics analysis.

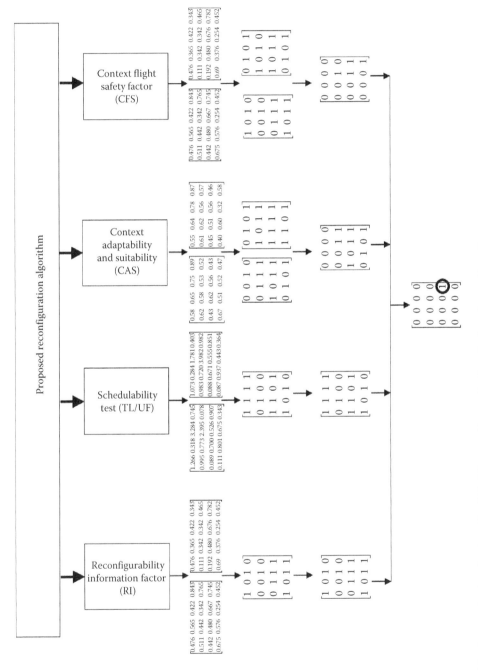

FIGURE 29.20 Summarized evaluation of task12 failure and hence the functioning of reconfiguration.

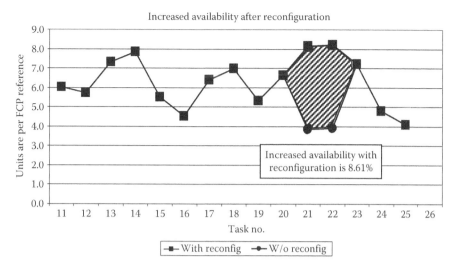

FIGURE 29.21   Increased availability of partition 1 with task12 reconfiguration.

Figure 29.20 shows the summarized evaluation of the reconfiguration algorithm for task12 failure scenario. With successful reconfiguration, the availability of the intended function has increased as shown in Figure 29.21.

The algorithm was evaluated with actual flight data captured during flight with 25 datasets. It is seen that the reconfiguration algorithm cleared task12 for reconfiguration. Figure 29.22 shows the results of the reconfiguration algorithm as seen by pilots in the cockpit for a typical failure scenario of task12. The figure shows with and without failure for the symbology availability, which is very

$$
\begin{array}{l}
\text{Reconfiguration} = \quad\quad\text{RI} \quad\quad \& \quad\quad \text{TL} \quad\quad \& \quad\quad \text{CAS} \quad\quad \& \quad\quad \text{CFS} \\
\text{algorithm}
\end{array}
$$

$$
\begin{array}{l}
\text{Reconfiguration} = \\
\text{algorithm}
\end{array}
\begin{bmatrix} 1 & 0 & 1 & 0 \\ 0 & 0 & 1 & 0 \\ 0 & 1 & 1 & 1 \\ 1 & 0 & 1 & 1 \end{bmatrix}
\&
\begin{bmatrix} 1 & 1 & 1 & 1 \\ 0 & 1 & 0 & 0 \\ 1 & 1 & 0 & 1 \\ 1 & 0 & 1 & 0 \end{bmatrix}
\&
\begin{bmatrix} 0 & 0 & 0 & 0 \\ 0 & 0 & 1 & 1 \\ 1 & 0 & 1 & 1 \\ 0 & 1 & 0 & 1 \end{bmatrix}
\&
\begin{bmatrix} 0 & 0 & 0 & 0 \\ 0 & 0 & 1 & 0 \\ 0 & 0 & 1 & 1 \\ 0 & 0 & 0 & 1 \end{bmatrix}
$$

$$
\text{Output} =
\begin{bmatrix} 0 & 0 & 0 & 0 \\ 0 & 0 & 0 & 0 \\ 0 & 0 & 0 & 1 \\ 0 & 0 & 0 & 0 \end{bmatrix}
$$

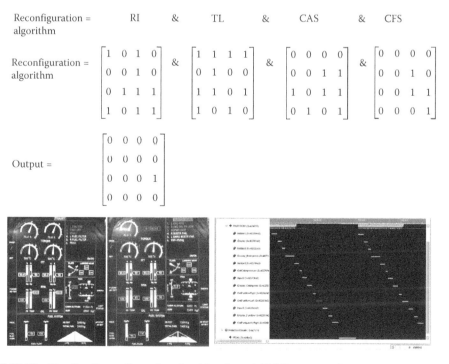

FIGURE 29.22   Results of reconfiguration algorithm for task12 failure scenario.

critical to the pilot for aircraft navigation. The implementation is purely on ground systems and this requires a lot more safety considerations to be adopted on aircraft systems. Hardware, software, algorithm and the complete system need to be looked into in terms of safety, reliability and certification to implement for airborne systems. The screen capture of the actual aircraft display system with successful reconfiguration of task is shown in Figure 29.22 (left). Also, it is found that the display non-availability without reconfiguration, where the symbology is failed. With the reconfiguration algorithm, the availability of the system enhanced by 8.61% which is quite substantial for flight critical systems. It is possible to enhance the availability of the avionics flight critical system for predefined fault scenario along with defined phases of flight. This can be extended to other systems also. However, care has to be taken in terms of safety, reliability, certification and airworthiness requirements as per civil aircraft applications where the reconfiguration algorithm is implemented.

APPENDIX: NOMENCLATURE

P_s	Original process
P_r	Reconfigurable process
M	Major frame
P_t	Partition
τ	Task
J	Jitter
B_i	Worst-case blocking
L_i	Worst-case partition delay
P_h	Phase of flight
θ_{cd}	Cumulative defect ratio of design
α	Total number of reviews
N_i	Total number of unique defects in the ith design review
L	Source lines of design reviewed in thousands
γ_{cd}	Cumulative defect ratio of code
M_i	Total number of unique defects in the ith code review
SL	Source lines of code reviewed in thousands
N_l	Maximum likelihood of the unseeded faults
N_{sf}	Number of seeds faults
m_{fu}	Number of unseeded faults uncovered
m_{sf}	Number of seeded faults discovered
ACE	Actuator control electronics
ANDAS	Aircraft nodes discovery and analysis system
APEX	Application executive
ARINC	Aeronautical Radio Incorporation
BIT	Built-in-test
CAS	Context adaptability and suitability factor
CBIT	Continuous built-in-test
COMNAV	Communication navigation
COTS	Commercially off the shelf
CS	Context sensitivity
DF	Degradation factor
DL	Degradation level
DME	Distance measuring equipment
EFB	Electronic flight bag
EVS	Enhanced vision system
FA	Federated architecture

FTMP	Fault-tolerant multiple processor
FTP	Fault-tolerant processor
GUI	Graphical user interface
HM	Health management/monitoring
HMI	Human–machine interface
HUD	Head up display
IBIT	Initiated built-in-test
IDE	Integrated development environment
IVHM	Integrated vehicle health management system
LRU	Line replaceable unit
MAFT	Multicomputer architecture for fault tolerance
MBIT	Maintenance built-in-test
MFD	Multi-function display
NALFOQA	NAL flight operations quality assurance
ND	Navigation display
NMR	N-modular redundancy
OACS	Open architecture computing system
OS	Operating system
PBIT	Power up built-in-test
PVI	Pilot vehicle interface
RCMU	Reconfiguration control and monitoring unit
RTOS	Real-time operating system
SIFT	Software implemented fault tolerance
SLOC	Software lines of code
SVS	Synthetic vision system
TCAS	Traffic and collision avoidance system
TL	Time loading
UF	Utilization factor
VIA	Versatile integrated avionics
VL	Virtual link
WAAS	Wide area augmentation system
WCET	Worst-case execution time

REFERENCES

1. ARINC Specification 653–1, *Avionics Application Standard Interface*, Aeronautical Radio Inc. Software, October 2003.
2. ARINC Report 651, *Design Guide for Integrated Modular Avionics*, Aeronautical Radio Inc., November 1991, Annapolis, MD.
3. N. Audsley and A. Wellings, Analyzing APEX Applications, *IEEE Real Time Systems Symposium RTSS*, Washington D.C., USA, 1996.
4. R.W. Duren, Waco, Algorithmic and architectural methods for performance enhancement of avionics systems, *28th Digital Avionics Systems Conference (DASC)*, Orlando, FL, 25–29 October 2009, pp. 1D4–1–1D1–6.
5. C.M. Ananda, General aviation aircraft aviation aircraft avionics: Integration and system system tests, *IEEE Aerospace and Electronic Systems Magazine*, 2009, ISSN 0885–8985, 25, 19–25.
6. C.M. Ananda, Civil aircraft advanced avionics architectures—An insight into SARAS avionics, present and future, *Conference on Civil Aerospace Technologies*, 2003, National Aerospace Laboratories, Bangalore, India.
7. C.M. Ananda, Avionics for general aviation light transport aircraft: An insight into the avionics architecture and integration, *AIAA Southern California Aerospace Systems and Technology Conference*, Santa Anna, CA, USA.

8. C.M. Ananda, General aviation transport aircraft avionics: Integration and systems tests, *26th Digital Avionics Systems Conference (DASC) on 4D Trajectory Based Operations—Impact on Future Avionics and Systems*, Dallas, TX, 21–25 October 2007, pp. 2.A.3–1–2.A.3–7.

9. C.M. Ananda, Avionics testing and integration for general aviation Light Transport Aircraft: Practical functional and operational integration activities for LTA, *AIAA Southern California Aerospace Systems and Technology Conference*, Santa Anna, CA, USA.

10. C.B. Watkins, and R. Walter, GE Aviation, Grand Rapids, Michigan, Transitioning from federated avionics architectures to integrated modular avionics, *26th Digital Avionics Systems Conference (DASC)*, Dallas, TX, USA, 21–25 October 2007, pp. 2.A.1-1–2.A.1-10.

11. J. López, P. Royo, C. Barrado and E. Pastor, Modular avionics for seamless reconfigurable UAS missions, *27th Digital Avionics Systems Conference (DASC)*, St. Paul, Minnesota, 26–30 October 2008, pp. 1A3–1–1A3–10.

12. J. Rushby, *Partitioning in Avionics Architectures: Requirements, Mechanisms and Assurance*, Computer Science Laboratory, SRI International, NASA Contractor Report CR-1999-209347, USA, March 1999.

13. R. Nadesakumar, M. Crowder and C.J. Harris. Advanced system concepts or future civil aircraft-an overview of avionic architectures. *Proceedings of the Institution of Mechanical Engineers, Part G: Journal of Aerospace Engineering*, 1995, 209: 265–272.

14. R. Garside and F. Pighetti, Jr., GE Aviation, Grand Rapids, Michigan, integrating modular avionics: A new role emerges, *26th Digital Avionics Systems Conference (DASC)*, Dallas, TX, USA, 21–25 October 2007, pp. 2.A.2–1–2.A.2.-5.

15. *ARINC 659: Backplane Data Bus*, Airlines Electronic Engineering Committee (AEEC), December 1993.

16. *ARINC 629: IMA Multi-Transmitter Data Bus Parts 1–4*, Airlines Electronic Engineering Committee (AEEC), October 1990.

17. GE Fanuc, *AFDX/ARINC 664 Protocol, Tutorial.*

18. AFDX®/®/ARINC 664, *Tutorial, (700008_TUT-AFDX-EN_1000 _29/08/2008)*, techSAT.

19. J.L. Mauff and J. Elliott, *Architecting ARINC 664, Part 7 (AFDX) Solutions*, Xilinx.

20. B. Pickles, *Avionics Full Duplex Switched Ethernet (AFDX)*, SBS Technologies.

21. R. Collins, Inc, *Users Manual for the Avionics Full Duplex Ethernet (AFDX) End-System.*

22. ARINC 664, *Aircraft Data Network, Part 7– Avionics Full Duplex Switched Ethernet (AFDX) Network.*

23. Time Triggered Ethernet: Tutorial.

24. DO-178B: *Software Considerations in Airborne Systems and Equipment Certification*, RTCA, http://www.rtca.org.

25. H. Kopetz, *Real Time Systems—Design Principles for Distributed Embedded Applications*, Kluwer and Academic Publishers, Norwell, MA, USA, 1998.

26. J. Rushby, *Partitioning in Avionics Architectures: Requirements, Mechanisms and Assurance*, SRI International, Menlo Park, California, NASA/CR-1999–209347, June 1999.

27. J. Ganssle, *The Firmware Handbook—The Definitive Guide to Embedded Firmware Definitive Guide to Embedded Firmware Design and Applications*, Oxford Newness, Oxford, UK, 2004.

28. J. Sturdy, T. Redling and P. Cox, *An Innovative Commercial Avionics Architecture Military Tanker/ Transport Platforms, Digital Avionics Systems Conference*, St. Louis, Missouri, 1999.

29. H. Benitez-Perez and F. Garcia-Nocetti, *Re-Configurable Distributed Control*, Springer-Verlag, London, 2005.

30. R. Alves and M.A. Garcia, Communication in distributed control environment with dynamic configuration, *IFAC 15th Trennial World Congress*, Spain, 2002.

31. S. Kanev and M. Verhaegen, Reconfigurable robust fault tolerant control and state estimation, *IFAC 15th Trennial World Congress*, Spain, 2002.

32. M. Balnke, M. Kinnaert, J. Lunze and M. Staroswiecki, *Diagnosis and Fault Tolerant Control*, Springer-Verlag, Berlin Heidelberg, 2003.

33. E.A. Strunk and J.C. Knight, Assured reconfiguration of embedded real-time software, *Proceedings of the 2004 International Conference on Dependable Systems and Networks (DSN)*, Florence, Italy, July 2004,.

34. J. López, P. Royo, C. Barrado and E. Pastor, Modular avionics for seamless reconfigurable UAS missions, *27th Digital Avionics Systems Conference*, 26–30 October 2008, pp. 1.A.3.1–1.A.3.10.

35. E. A. Strunk, C. Knight and M. A. Aiello, Assured reconfiguration of fail-stop systems, *Proceedings of the 2005 International Conference on Dependable Systems and Networks (DSN'05)*, Yokohama, Japan © 2005 IEEE.

36. K.A. Seeling, *Reconfiguration in an Integrated Avionics Design Lockheed Martin Aeronautical Systems Company*, 0–7803–3385–3/96 © IEEE 1996.

37. E. Sharif, A. Richardson and T. Dorey, *A Diagnostic Reconfiguration Methodology for Integrated Systems*, IEE, Savoy Place, London WCPR OBL, UK, 1997 The Institution of Electrical Engineers.
38. M.D. Derk and L.S. DeBrunner, *Dynamic Reconfiguration For Fault Tolerance For Critical, Real-Time Processor Arrays*, University of Oklahoma, Normaq Oklahoma 730 19, pp. 1058–1062.
39. P.M. Sonwane, D.P. Kadam and B.E. Kushare, Distribution system reliability through reconfiguration, fault location, isolation and restoration, *International conference on control, automation, communication and energy conservation-2009*, Kongu Engineering College, Erode, India, 4–6 June 2009, pp. 1–6.
40. D.P. Gluch and M.J. Paul, Fault-tolerance in distributed digital flyby-wire flight control systems, *in Proceedings AIAA/IEEE Seventh Digital Avionics Syst. Conf.*, Fort Worth, Texas, 13–16 October 1986.
41. M.W. Johnston et al., *AIPS System Requirements (System Requirements (Revision 1)*, CSDL-C-5738, Charles Stark Draper Lab., Inc., Cambridge, MA, August 1983.
42. J.H. Wensley et al., SIFT: Design and analysis of a fault-tolerant computer for aircraft control, *Proc. IEEE*, 66, 1240–1255, 1978.
43. J. Goldberg et al., *Development and Analysis of the Software Implemented Fault-Tolerance (SIFT) Computer*, Final Rep. NASA Contract NASA-CR-172146, February 1984.
44. A.L. Hopkins et al., FTMP-A highly reliable fault-tolerant multiprocessor for aircraft, *Proc. IEEE*, 66, 1221–1239, 1978.
45. J.C. Knight et al., *A Large Scale Advanced Information Processing System (AIPS) System Requirements (Revision 1)*, Rep. *A Large Scale Advanced Information Processing System (AIPS) System Requirements (Revision 1)*, Rep. CSDL-C-5709, Charles Stark Draper Lab., Inc., Cambridge, MA, October 1984.
46. C.J. Walter et al., MAFT: A multicomputer architecture for fault tolerance in real-time control systems, *in Proc. IEEE Real-Time Syst. Symp.*, San Diego, California, USA, 3–6 December 1985.
47. R.M Kieckhafer, C.J. Walter, A.M. Finn and P.M Thambidurai, The MAFT Architecture for distributed fault distributed fault tolerance, *IEEE Transactions on Computers*, 37(4), 398–405, 1988.
48. C.J. Walter, *Evaluation and Design of an Ultra Reliable Distributed Architecture for Fault Tolerance*, Allied signal Aerospace Company, Columbia, 1990.
49. A.M.K. Cheng, *Real-Time Systems: Scheduling, Analysis, and Verification*, Wiley-Interscience, Canada, 2001.
50. H. Kopetz, *Real Time Systems—Design Principles for Distributed Embedded Applications,* Kluwer and Academic Publishers, the Netherlands, 1998.
51. C.M. Krishna and K.G. Shin. *Real-Time Systems*. McGraw-Hill, New York, 1997.
52. C. Wilkinson, *Prognostics and Health Management for Improved Dispatchability of Integrated Modular Avionics Equipped Aircraft*, CALCE Electronic Products and Systems Center, Department of Mechanical Engineering, University of Maryland, College Park, MD 20742.
53. B.S. Dillon, *Design Reliability—Fundamentals and Applications*, CRC Press, Boca Raton, FL, 1999.
54. NAL and HTSL team, *Regional Transport Aircraft, Integrated Flight Control System (IFCS)*, September 23, 2008.
55. C.M. Ananda, Re-configurable avionics architecture algorithm for embedded applications. *IADIS International Conference on Intelligent System and Agents 2007*, Lisbon Portugal. pp 154–160,
56. C.M. Ananda, *Improved Availability and Reliability Using Reliability Using Re-configuration Algorithm for Task or Process in a Flight Critical Flight Critical Software*, (F. Saglietti and N. OsterEds): Spriner LNCS publication, © Springer-Verlag Berlin Heidelberg 2007, pp. 532–545, 2007.
57. C.M. Ananda, Re-configuration of task in flight critical system—Error detection and control, *28th Digital Avionics Systems Conference (DASC) on Modernization of Avionics and ATM—Perspectives from the Air and Ground*, The Florida Conference Center, Orlando, FL, 25–29 October 2009, USA, IEEE, pp. 1.A.4-1–1.A.4.-10.
58. C.M Ananda, Improved availability using re-configuration algorithm in a flight critical system, *26th Digital Avionics Systems Conference (DASC) on 4-D Trajectory-Based Operations: Impact on Future Avionics and Systems*, Dallas, TX, 21–25 October 2007, USA, IEEE, pp. 2.C.3–1–2.C.3–11.
59. L.P. Briand and D.M. Roy, *Meeting Deadlines in Hard Real-Time Systems the Rate Monotonic Approach*, 1999, IEEE Computer Society Press, USA.
60. IEC 60812, 1985, *Analysis Techniques for System Reliability—Procedure for Failure Mode and Effects Analysis (FMEA)*, IEC 60812 Ed. 1.0 b: 1985.
61. U.K. Kumar and B.S. Umashankar, *Improved Hamming Code for Error Detection and Correction*, 1-4244-0523-8/07, 2007, IEEE.

62. C.M. Ananda, R. Kumar and M. Ghoshhajra, Configurable flight data analysis for trends and statistics analysis—An embedded perspective of an efficient flight safety system, *International Conference on Aerospace Electronics, Communications and Instrumentation (ASECI-2010)*, V.R. Siddhartha Engineering College and P.V.P.S.I.T, Vijayawada.
63. C.M. Ananda, *Configurable Flight Safety System for Incident and Accident Investigation*, REQUEST 2008, Bangalore, India.
64. C.M. Ananda, Sheshank, *Implementation of Improved Hamming Code in Xilinx Platform*, NAL-ALD/2009/010, CSIR-National Aerospace Laboratories, November 2009.

30 Study of Three Different Philosophies to Automatic Target Recognition

Classical, Bayesian and Neural Networks

Vishal C. Ravindra, Venkatesh K. Madyastha and Girija Gopalratnam

CONTENTS

30.1   INTRODUCTION

This chapter is aimed at providing the reader with some of the recent scientific advances in automatic target recognition (ATR) from the perspectives of dynamic modelling, algorithmic advances and sensor modelling. We start by asking some simple questions such as 'Where and why is ATR required today?', 'What are the robustness metrics that a properly designed ATR systems should satisfy?', 'What is image processing and why is it important?', 'What is an ATR system and what are the related issues?', 'What is meant by artificial intelligence based ATR?' and 'What are the various approaches/methodologies to solving an ATR problem?'. In the process of answering the questions just listed we might come up with more questions, and hopefully this chapter will do justice in terms of either answering those questions or pointing the reader to the relevant reference. In addition to this, we attempt to provide a clear and concise understanding of the classical as well as artificial intelligence-based approaches for solving the ATR problem and the so-called *unified approach* for solving the ATR problem which is based on the assumption of the existence of a Bayesian prior model (predicted/modelled behaviour of the real world) and accordingly propagated/weighted depending on the error between the real measured world and its estimated value. Furthermore, this chapter also makes a point about the futile exercise of relying on one single approach to solve the ATR problem, that is, either purely classical or entirely Bayesian, given the current days stringent requirements from an automatic target identifier and tracker and vast amount of data which needs to be made available in user-defined time lines. Hence, this calls for an efficient ATR algorithmic approach that combines the efficacy of the above-mentioned approaches into one single approach which can be termed as a *hybrid* ATR approach. This section addresses certain issues of an ATR system and the robustness that an ATR system must have in order to be efficient and successful for a variety of images (static) and scenes (dynamic) in extracting the desired target(s) information from the observed image or scene and passing it to a decision maker, preferably in real time.

Section 30.1 discusses different sources of sensor imagery. Section 30.2 presents an overview of classical-based approaches to ATR, focusing on the various stages and the algorithms proposed in the literature for these stages. In Section 30.3, the ATR problem is discussed from the perspective of Bayesian statistics, discussing the advantages that a Bayesian-based approach can bring to an ATR problem. Finally, Section 30.4 discusses artificial intelligence for an ATR system from the perspective of designing and implementing neural networks. This section also presents a brief overview of neural networks, discussing the neural network structure and some associated mathematical models for some of the basic element of a neural network. Before proceeding further, for the benefit of the reader we will devote a few words to image processing as this leads into ATR, the main topic of this chapter.

30.2   WHAT IS IMAGE PROCESSING?

Image processing, succinctly put, is a form of signal processing where the input is an image that is typically noisy or distorted and the output is in the form of an image or a set of parameters extracted from the input image. Here, signal processing is the analysis of a signal in discrete or continuous time in order to derive useful conclusions about the signal [1]. The typical operations involved in image processing are image denoising, deblurring, reconstruction, restoration, registration and restoration, among others. In one of the most common image processing applications, called image denoising, the input image is either a noisy/cluttered image of a desired object(s) like a photograph (in the static case) or a video frame (in the dynamic case) and the output is a reconstruction of the input image but with the noise/clutter sufficiently removed such that a cost function is minimized. Sometimes, the output could also refer to a set of indicators, parameters or features (e.g., orientation, quantity, location) which are related to the objects being imaged and they contain information that usually explains something about either about the coordinates or the attitude of the object(s). The input images from the camera or the video frames are treated as measurements from which the ideal/true behaviour of the object is to be determined. The technique of obtaining such input images which serve as measurements for an image processing algorithm is called as imaging. Throughout this chapter we refer to an object as a target and the original image/signal as the raw image/signal, while the measurement data are referred to as the sensor data. Typically, sensor data are either a stream of images obtained with a specific type of sensor mechanism such as a regular camera or an infrared camera, or a continuity of information such as range or distance to target as obtained from a radar. Electromagnetic wave-based (in particular, radio wave-based) object detection system is called RAdio Detection And Ranging and sound propagation-based navigation (and communication) for detecting other vessels is called sonar, SOund Navigation And Ranging.

30.3   WHAT IS ATR AND WHY IS IT REQUIRED?

An ATR algorithm is a mathematical framework for detecting, recognizing and tracking objects such as animate or inanimate beings (constrained to land, sea or air) by processing sufficient amounts of relevant information/data obtained from one or multiple sensors via an onboard computer or a ground station central monitoring unit [2,3]. The objective of an ATR system is to serve to reduce the workload of human operators such as that of pilots (especially pilots of fighter aircrafts), submarine pilots/co-pilots, tank commanders and so on, by acting as the eye and the brain for the human. This implies that an ATR system should continuously collect and process sensor images/data and finally represent these processed images in a manner that makes the decision or aids the end user (e.g., an image analyst) to make a decision. Such a dedicated ATR system can aim to alleviate or reduce the excessive burden placed on these human operators who are continuously tasked with other very critical mission-specific activities such as monitoring military activities over large spans of a battlefield, sea space or airspace for enemy movement. Specifically, a good ATR system can provide a detailed and thorough understanding of a battlefield and a reliable classification of the objects captured in the scene. Furthermore, large volumes of data produced by surveillance, possibly at high update rates, makes it infeasible and uneconomical for human interpretation of this data. Thus, a reliable ATR system that makes/aids in making decisions such as reacquiring targets onto the field of view (FOV) of the onboard tracking instrument is an absolute must, especially while surveying hostile enemy territory and making mission-specific critical decisions.

30.4 HOW ROBUST SHOULD AN ATR SYSTEM BE?

A fully functional robust ATR system can detect, recognize, track and deliver enormous volumes of processed data to the human analyst for further analysis and yet adhere to extremely short timelines dictated by strict target acquisition scenarios. That a highly efficient and effective ATR system is an absolute requirement for military purposes is quite apparent [4]. In the very least, an ATR system that is tolerably accurate, when it comes to classifying objects of a battlefield survey captured via imaging or video streaming, is required for appropriate countermeasures. Thus, for the successful operation and deployment of an ATR system, particularly in military operations, it must be robust to a suite of conditions that are most likely to be encountered, mostly in real time. Thus, a robust ATR system should:

1. Be able to classify target/targets irrespective of the distance to the target, that is, insensitive to distance between the onboard sensor and the target.
2. Be orientation sensitive, that is, not only detect the presence of a tank, for instance, but also the direction in which the turret of the tank is pointed.
3. Also be scale insensitive so that whether the target appears large or small on the image plane of the measurement recording device depending on the distance to the target, the algorithm should correctly classify the target. This is especially required of a classical ATR approach (see Section 30.2) where no extra intelligence or learning is employed.
4. Allow for variations in the angle in which the scene image is captured by an onboard camera.
5. Be capable of segregating the desired target from its background and from possibly other similar looking benign targets/objects. This situation wherein a target of interest is camouflaged or partially hidden from the ATR system is known as *partial occlusion*.
6. Maintain low false alarm rates due to varying backgrounds or due to partial occlusion of the desired target, while operating in real time.
7. Be unaffected by environmental conditions such as heavy rain, fog, snow, smoke, smog and sand storms, to name a few.
8. Suitably handle uncertainty in the type of target make/model to be detected. This is an important feature that any robust ATR must have since a potential enemy's new models (planes, ships and tanks) might not be a part of the model database of the tracker. Hence, with the practical limitation of existing or known data sets that simply cannot represent the real changing world accurately, the ATR system should be robust enough to detect and track any variant of these existing models, while operating on imagery from the real world.
9. Be able to detect a target model despite changes/modification/inclusions to its basic structure such as targets with or without gunships/military armory.
10. Have a sufficiently exhaustive database containing heat signatures of different targets under different operating conditions, particularly when using infrared sensors to generate images. This is a feature that is *necessary for classical ATR systems* which do not rely on any intelligence as an aiding mechanism to the conventional ATR system.

30.5 WHAT ARE THE TYPICAL SENSORS OF AN ATR SYSTEM?

Some of the types of sensors that are commonly used for an ATR application are discussed next.

30.5.1 INFRARED

The light is electromagnetic radiation (EM is a form of energy exhibiting wave–like behaviour as it travels through space and which has both electric and magnetic field components), with a wavelength between 0.7 and 300 µm, implying a frequency range approximately between 1 and 430 THz, as

TABLE 30.1

Electromagnetic Spectrum: Wavelength and Frequency Range

Electromagnetic Spectrum Waves	Wavelength Range in Nanometer (nm) (1 nm = 10^{-9} m)	Frequency Range in Terahertz (THz) (1 THz = 10^{12} Hz)
Long	$>10^{12}$	$<10^{-6}$
Radio and television	$\approx 10^8$ to 10^{10}	10^{-6} to 10^{-3}
Ultra high frequency (UHF)	$\approx 10^8$ to 10^9	5×10^{-4} to 10^{-3}
Very high frequency (VHF)	$\approx 10^9$ to 10^{10}	3×10^{-5} to 3×10^{-4}
Super high frequency (SHF)	$\approx 10^7$ to 10^8	3×10^{-3} to 3×10^{-2}
Extremely high frequency (EHF)	$\approx 10^6$ to 10^7	3×10^{-2} to 3×10^{-1}
Microwaves (RADAR)	10^7 to 10^8	3×10^{-4} to 3×10^{-1}
Infrared	700 to 3.5×10^5	$1-430$
Near infrared	700 to 5×10^3	$100-430$
Mid infrared	5×10^3 to 4×10^4	$1-100$
Far infrared	4×10^4 to 3.5×10^5	$0.1-1$
Visible light	390–750	400–790
Ultraviolet light	10–400	10^3 to 5×10^4
X-rays	0.01–10	3×10^4 to 3×10^7
Gamma-rays	$<10^{-2}$	$>10^7$

Source: Adapted from D. Halliday, R. Resnick and J. Walker, *Fundamentals of Physics*, 7th ed., John Wiley & Sons, Hoboken, NJ, 2005.

shown in Table 30.1 (1 THz, denoted as THz, is equal to 10^{12} Hz) [5,6]. IR imaging is used extensively for military operations such as target detection, identification, surveillance, homing and tracking, as well as for civilian purposes such as remote temperature sensing, short-range wireless communication, spectroscopy and weather forecasting. Furthermore, IR sensor-based telescopes are used to penetrate regions of space (molecular clouds) and detect other celestial objects such as planets and stars.

30.5.2 FORWARD-LOOKING INFRARED

Forward-looking infrared (FLIR) senses infrared (IR) radiation and relies on its ability to detect the thermal/heat energy from a target in its field of view and construct an image of the target which can be output as a video image [7]. FLIRs can aid pilots, tank commanders, drivers and so on, especially at night or during hazy/foggy conditions, to steer their vehicles without colliding with obstacles along the way. Furthermore, since FLIRs rely on sensing heat emitted from a target they can be used to detect warm targets against cold backgrounds—a situation quite common during the night. Typically, FLIRs are used for population surveillance, low visibility flying, detection of insulation loss in buildings, detection of leaks of natural gas and/or other gasses, search and rescue operations especially in thickly wooded areas, marshy swamps or water, to name a few.

30.5.3 SYNTHETIC APERTURE RADAR

A form of radar, relies on the relative motion between a target and the tracker antenna at different antenna positions (by mounting the antenna on a moving platform such as an aircraft) to reconstruct the image of a target region [8,9]. The movable antenna radiates a beam whose wave-propagation direction has a substantial component perpendicular to the flight path direction, so as to illuminate the terrain from beneath the aircraft out towards the horizon with pulses of radio waves at wavelengths anywhere from a meter down to millimetres. SAR-based images have widespread applications in remote sensing and mapping of the surfaces of Earth and other planets.

30.5.4 Inverse Synthetic Aperture Radar

The inverse synthetic aperture radar (ISAR) is a technique to generate a high-resolution image of a target by observing a moving target over a substantial length of time with a stationary antenna, thus creating the synthetic aperture [10–14]. ISAR images have sufficient resolution and enough detail to distinguish between a civilian and military aircraft and identify missile systems. Furthermore, ISAR is used in maritime surveillance to classify ships and other objects. While both SAR and ISAR have the same underlying theory, SAR imaging is accomplished with the radar moving and the target stationary, whereas ISAR imaging is performed with the target moving and the radar stationary. Thus, ISAR images are accompanied with errors such as defocusing and geometric errors, antenna aberrations, signal leakage due to range and azimuth compression errors (side-lobe generation), to name a few.

30.5.5 Light RADAR

Light RADAR (LIDAR), referred to in civilian applications, is an optical remote sensing technology that measures the properties of scattered light to estimate, with sufficient accuracy, the range to a distant target [15]. Range detection via LIDAR differs from that of a radar system in that LIDAR systems use wavelengths in the electromagnetic spectrum that correspond to ultraviolet or visible light. These are shorter wavelengths compared to the wavelength spectrum employed by RADAR which correspond to radio waves, having the highest wavelength of any spectra. Owing to the fact that only those features of target(s) whose size is greater than or equal to the wavelength of the emitting source can be imaged, LIDAR is very sensitive to cloud particles and hence has potentially many applications in meteorology.

30.5.6 Laser RADAR

Laser RADAR (LADAR), referred to in military operations, is an optical remote sensing technology that uses laser pulses and measures the properties of scattered light to estimate, with sufficient accuracy, the range to a distant target [16,17]. A LADAR emits beams of wavelengths in the region of ultraviolet which are many times smaller than those of radio waves thereby ensuring that these waves are then reflected by very small objects or particles in the air such as aerosols, molecules, which are invisible to radar frequencies. This type of reflection of waves is called as backscattering, of which some of the most common are Rayleigh scattering, Mie scattering and Raman scattering.

30.5.7 Millimeter Wave Radar

Millimeter wave radar (MMWR) is a sensing technique based on the range detection principles of a radar system. The millimeter waves, which are defined from 30 GHz to about 300 GHz, have smaller components and greater bandwidths than those of microwave radars. Furthermore, an MMWR sensing device is usually marked with high speed, high resolution and less signal attenuation than what is typically observed with microwave radars [18]. The basic types of MMWR devices are continuous-wave radar (CWR), frequency-modulated continuous-wave radar (FMCWR), and pulsed-wave radar (PWR) [19]. Ground-based MMWRs can be used to conduct a detailed study of the behaviour and movement of clouds in the atmosphere [20], especially cloud's three-dimensional structures. Meteorological radars are generally pulsed radars and transmit electromagnetic pulses in the frequency range of 3–10 GHz for detecting, mapping and measuring precipitation intensity.

30.5.8 Multi-Spectral Imaging

Multi-spectral imaging (MSI) originally developed for space-based imaging, is an image processing technique that captures image data at specific frequencies across the electromagnetic spectrum [21].

The wavelengths may be separated by filters or by the use of instruments that are sensitive to particular wavelengths, including light from frequencies beyond the visible light range, such as infrared. Due to its inherent property of operating at specific wavelengths across the electromagnetic spectrum, multi-spectral imaging aids in the extraction of additional information that the naked eye fails to capture with its receptors for red, green and blue. Remote sensing devices such as radiometers operate by acquiring multi-spectral images. Typically, each radiometer onboard a satellite acquires a digital image, also called a scene, within a small band of visible spectra ranging from 400 nm to 700 nm (visible light wavelength range) which is called the red–green–blue region and progressing towards higher wavelengths into the domain of infrared.

30.5.9 HYPER-SPECTRAL IMAGING

Hyper-spectral imaging (HSI) also called *ultra-spectral* imaging, is an imaging technique which can be thought of multi-spectral imaging with a wider spectral coverage or finer spectral resolution [22]. Contrary to the human eye (which can only see the visible light spectrum), hyper-spectral imaging, in addition to being able to identify visible light spectrum, is also sensitive to the region of the electromagnetic spectrum from ultraviolet wavelengths to as high as infrared wavelengths. Since hyper-spectral imaging sensors look at targets using a wider spectral bandwidth across the electromagnetic spectrum, it is a very popular technique in the identification of different materials that make up a scanned target. This is because certain targets/objects, especially composites, leave unique fingerprints, also called spectral signatures, across the electromagnetic spectrum, which in turn can be identified via hyper-spectral imaging. A typical application is in the area of oil field discovery.

30.5.10 LOW-LIGHT TELEVISION

The low-light television (LLTV) is a type of electronic sensing device, with a frequency detection range extending above the visible wavelength and into the short-wave IR wavelength. This allows viewing of objects in extremely low light levels, which, otherwise, cannot be seen by the naked eye [23]. A typical device for sensing used here is the charge-coupled device (CCD) [24,25]. A CCD, invented at the AT&T Bell Laboratories, is a device for the movement of electrical charge, usually from within the device to an area where the charge can be manipulated. CCDs have widespread scientific applications and medical usage where high-quality imaging techniques are a priority.

30.5.11 VIDEO

Video is the science of electronically capturing, processing, reconstructing and transmitting a sequence of still images to represent scenes in motion. The frequency of the video imaging (number of pictures per unit time), also called frame rate, ranges from about 6 to 8 (frames/s) for old mechanical cameras to about >120 (frames/s) for the state-of-the-art cameras. Video images can be interlaced (improve the picture quality of a video signal without consuming extra bandwidth) or progressive, where the moving images are displayed, stored and transmitted such that all lines of each frame are drawn in sequence. In interlacing (traditional television systems), the odd lines are drawn first followed by the even lines of each frame.

The wavelength and frequency ranges of the various waves in the electromagnetic spectrum, such as radio, micro wave, infrared, visible, ultraviolet, x-rays and gamma-rays, are presented in Table 30.1. The wavelength and frequency ranges of the visible light spectrum (violet, blue, cyan, green, yellow, orange and red) are presented in Table 30.2. (*The names of the colours in Table 30.2 that appear as* **bold** *font in the first column denote the primary colours, i.e., red, green and blue [RGB].*)

TABLE 30.2

Visible Light Spectrum: Wavelength and Frequency Range

Visible Light Spectrum	Wavelength Range in Nanometer (nm) ($1 \ nm = 10^{-9} \ m$)	Frequency Range in Terahertz (THz) ($1 \ THz = 10^{12} \ Hz$)
Violet	390–450	668–790
Blue	450–475	631–668
Cyan	476–495	606–630
Green	495–570	526–606
Yellow	570–590	508–526
Orange	590–620	484–508
Red	620–750	400–484

Source: Adapted from D. Halliday, R. Resnick and J. Walker, *Fundamentals of Physics*, 7th ed., John Wiley & Sons, Hoboken, NJ, 2005.

Remark 1

Some sensors such as IR, FLIR and a few SONARs are passive sensors since these sensors only per-form the role of receiving information that is generated from a different source. Hence, these sensors are not themselves easily detectable by enemy countermeasures as they do not send any signal to the target from or about which information is desired. On the contrary, sensors such as SAR, ISAR, LADAR, LIDAR and MMWR are active sensors, which means that these sensors play the part of sending and receiving signals in order to process information about the desired target. Therefore, these sensors are easily detectable and run the risk of being susceptible to enemy countermeasures.

30.6   CLASSICAL APPROACH TO ATR

The ATR problem has been approached either from the perspective of a classical pattern recognition methodology or from the perspective of an artificial intelligence-based approach such as a neural network-based learning methodology. It should be noted that while classical approaches attempt to solve the ATR problem possibly in a global sense albeit for a restrictive class of system descriptions, approaches based on artificial intelligence can be seen to provide solutions, intuitively only local, for a class of problems which are less restrictive in its assumptions compared to a classical ATR algorithm. Hence, the proper choice of the ATR algorithm depends not only on the type of problem to be solved but also on the degree of uncertainty/discrepancy between the models in the database and the actual scene.

30.6.1   WHAT ARE THE DIFFERENT STAGES OF A CLASSICAL ATR SYSTEM?

A typical pattern recognition and target classification-based [26] classical ATR system consists of the various stages shown in Figure 30.1. These stages, which start with preprocessing the raw captured image and culminating in identification and tracking of the target can be described as follows.

30.6.1.1   Noise/Clutter Removal

This is called the *preprocessing* stage and is the first stage of any ATR system. It is an essential first step to eliminating unwanted, noisy information that can otherwise leak into the various other stages of the target recognition process, thus making the subsequent stages more cumbersome in terms of noise and/or clutter removal and highly cost ineffective. The goal of this stage is to improve the contrast of the captured target image and to remove, to a sufficient extent, any unwanted noise and clutter. Clutter refers to real objects that are captured in the image such as buildings, cars,

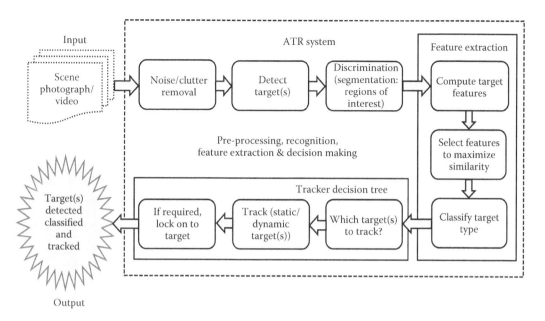

FIGURE 30.1   Different stages of an ATR algorithm.

trucks, grass, trees and other objects but are not the targets of interest. Clutter can be naturally occurring such as grass, trees and so forth, or man made such as buildings, vehicles and so on, [3]. Whatever be the type of clutter it is clear that clutter tends to dominate the imagery simply because targets are generally sparse compared to the environment in which they operate. Noise, on the other hand, refers to electronic noise in the sensor as well as inaccuracies introduced in the computations by a signal processor. Depending on the ATR application, the problem may be one of extracting a signal from noise or separating a target from its background clutter. Typically, a median filter, which is a nonlinear digital filtering technique, is used in this stage because of its edge detection and preservation property [27]. The main idea of the median filter is to scan the image pixel by pixel, replacing each pixel with the median of neighbouring pixels. This has an effect of reducing the influence of outlier pixels such as dark pixels that could otherwise render the subsequent stages of ATR problematic, hence, having a smoothening type of effect on the noisy image.

30.6.1.2   Detect Target

In this stage, the focus is on localizing areas in the captured image where a target is potentially present. Some of the techniques such as those reported in refs. [28–30] present a detailed evaluation of target detection techniques where the detection of either light or dark targets is discussed. Generally, targets that are hotter than its background tend to appear as bright objects on an FLIR-based image. In Ref. [31] a technique is developed for estimating the target range via scene dynamics by employing a sequence of passive sensor images such as FLIR- or SAR-based imagery. This is an important aspect since range can be very useful in detecting and classifying targets. The risk of using an active sensor such as a LADAR or LIDAR to obtain target range is that there is a chance of being detected by the enemy.

30.6.1.3   Discrimination/Segmentation

Segmentation is the procedure of extracting, as accurately as possible, a localized target from its background by partitioning an image into multiple segments or sets of pixels. The ultimate goal of segmentation is to represent the captured image in a form that is more meaningful and easier to analyse. Thus, segmentation is the procedure by which targets and their respective boundaries such as lines, curves and so forth, are located. A popular algorithm here is the *supersplice* algorithm which assumes that the target edges distinguish the target from its surrounding [32]. Due to this, detecting different targets in a given scene depends solely on how accurately the target's edge or edges are recovered during the segmentation procedure. Edge detection is a well-developed field of image processing which aims at identifying points in an image at which there is a sharp change in the image brightness called as image discontinuities. These discontinuities in the image brightness more often than not correspond to either depth or surface orientation discontinuities, possible changes in material properties (in the case of a particular target) and/or variations in scene illumination (a hot target such as a tank against the background of a cold tree or a less hot tank). Some of the practical applications of segmentation are in the field of medical imaging to locate tumours, measure tissue volume, diagnosis and anatomical structure study, to name a few. Furthermore, segmentation also finds its uses in location of targets in satellite imagery (roads, forests, rivers, etc.), face and fingerprint recognition, among many others.

30.6.1.4 Target Feature Extraction

This step consists of computing the features of the target, selecting the desired features and classifying the target type based on the chosen set of features.

30.6.1.5 Compute Target Features

The segmentation procedure in the classical ATR system identifies and extracts a set of features for the desired target. The reliability of these extracted features, such as spectral, geometric or topological, is essential for the success of any classical ATR system. Furthermore, these features should be invariant under translation, changes in scale and rotation. In this regard, image moments (weighted average or moment of the image pixels' intensities) are useful metrics to describe objects after segmentation. A few properties of the image which are found via moments are area (total intensity), centroid and orientation information. The method reported in Ref. [33] introduces *seven* moment-based image invariants called Hu moments that exhibit invariance due to rotation, scaling and translation. The first Hu moment is analogous to the moment of inertia around the image's centroid, where the pixels' intensities are analogous to physical density. The last Hu moment is skew invariant, which enables it to distinguish mirror images of otherwise identical images. However, computing higher order Hu moments is quite complex, thus, making this approach less preferable in today's ATR schemes. To overcome the complexities of Hu's moments, the approach proposed in Ref. [34] introduces the concept of Zernike moments based on the theory of orthogonal polynomials. Zernike moments, like Hu moments, are invariant under translation, scaling and rotation. Zernike moments have been shown to be extremely insensitive to image noise and information content and can provide an accurate representation of the image. Zernike moments have been used in face, gait and moving recognition, biometrics and signature authentication, to name a few applications.

30.6.1.6 Select Appropriate Target Features

In the phase of target feature selection, the criterion is to obtain such target features that not only maximize the similarity of same class objects but also maximize the dissimilarity of different class objects. In the context of ATR problems, feature selection via histogram examination is reported in Ref. [32], via Bhattacharya measure (distance metric measuring the similarity of two discrete or continuous probability distributions) is reported in Ref. [35], and via F-statistic (fixation indices which describe the level of gene similarity for a trait in a population) is reported in [36].

30.6.1.7 Classify Target Type

Target-type classification techniques are widely researched with some of the most popular ones such as linear, quadratic, structural and tree-based classifiers reported in Refs. [32,36–38]. The approach in Ref. [38] is based on tree comparisons and requires a preliminary automatic determination of the form class to select the associated model and then process the form. Target classification has also been done via—nearest-neighbour (K-NN) algorithm and is reported in Refs. [35,36]. The K-NN algorithm (a type of instance-based learning) is a method for classifying objects based on closest training examples in the feature space—an abstract space in which each pattern sample is represented as a point in n-dimensional space. Similar samples are grouped together, which allows the use of density estimation for finding patterns. The rules for the nearest-neighbours' algorithm assist in implicitly computing the decision boundary. In the K-NN algorithm, the constant K is a user-defined constant and an unclassified test feature is classified by assigning the label which is most frequent among the K training samples nearest to that test point.

30.6.1.8 Tracker Decision Tree

In this step the target is prioritized and tracked or locked on to.

30.6.1.9 Target Priority

Assigning a probability or probabilities to a target or targets in the field of view is based on the type of target and the probability of its correct classification. For example, in Ref. [39] an application such as digging/mining for metallic sources is addressed where the target is the metallic source and the clutter is any non-metallic source, a target prioritization routine is developed based on the likelihood of a metallic source. Once a target is prioritized, that is, desired targets are ranked in a descending order based on its priority to indicate the importance level, the information is communicated to a target tracker.

30.6.1.10 Tracking the Target (Static/Dynamic) and Lock On

An efficient target tracker is necessary for the success of an ATR system. A target tracker is used in many military systems such as for guiding anti-aircraft missiles, smart bombs and so on, and in all these systems the tracker obtains target location information at timed intervals and 'locks on' to the target. While this happens on the tracker side, the potential targets attempt to break this lock by the tracker. This procedure employed by potential targets is called *breaklock*. Thus, a target tracker, in addition to being able to track more than one target when required, should also not be influenced by frequent target *breaklocks* due to severe clutter and low signal-to-noise ratio situations. Trackers based on a combination of correlation, feature, intensity and contrast are always desirable since they complement one another and the tracker can switch from one approach, say feature, to another, say contrast, when the confidence level with the tracked feature is lower than that of the contrast. This can help to increase the confidence interval of successful target tracking and prevent frequent breaklock situations since obtaining the target once a breaklock has occurred is not an easy task. Finally, target aimpoint selection requires determination of a critical point(s) of a tracked target so that once an onboard missile system locks on to the target, it can carry out its mission by either pointing towards an interior point of the target or on the boundary.

30.7 A UNIFIED BAYESIAN-BASED APPROACH TO ATR

The main motivation of Bayesian-based approaches is the unique ability of Bayesian statistics to handle limited as well as conflicting pieces of information in a fully consistent manner [4]. Bayesian statistics is based on manipulating probabilities assigned to observed data. A Bayesian approach is also friendly to the integration of various approaches and fusion of data from multiple sensors. Another important capability of a Bayesian technique is to incorporate prior information on a target class based on the target dynamics, terrain, situation, context and so forth. It also helps in constructing bounds, confidence intervals and other statistics for estimated parameters. Bayes' statistics provides the probability of the data belonging to a certain class given the observed data, from which there are various ways to make an inference, in contrast to traditional classifiers which directly output the decision instead. Bayesian statistics is a well-evolved and elegant way to tackle the ATR problem mathematically instead of depending on training data and heuristics that are inherently not robust to variations in the scene. Another advantage of developing an ATR algorithm from the perspective of a Bayesian approach is that it is possible to incorporate good physical sensor models into the classification procedure. Model-based approaches such as the ones discussed in the subsequent section construct likelihoods for data based on the physics of the sensors, while neural networks can be looked at as model-free approaches where a mathematical model is not constructed for received data. There is also an option of combining the two approaches whenever reliable models are not known for data by combining feature-based classifiers with model-based classifiers. A big advantage of using model-based Bayesian techniques is that it is possible to obtain various information-theoretic bounds on performance. The following section motivates Bayesian-based ATR through a few papers that have appeared since the mid–late 1990s through early 2000s. The book chapter [40] provides an excellent introduction and overview of Bayesian object recognition. The biggest advantage of a Bayesian-based approach to ATR is that it is flexible. What this means is that the user has plenty of degrees of freedom while defining the problem to be solved. For example, it is invariant to the number of targets of interest, classes of targets can be added and deleted during each scan, pose of targets can be estimated, it could be made invariant to scale, it is perfectly suited for multisensor fusion, sensor models can be changed in the middle of the algorithm depending on various scenario circumstances and so on. This chapter aims to describe and provide an outline for Bayesian object recognition.

Generally, ATR scenarios unfold as follows: A number of sensors (cameras, RADARs, LADARs, etc.) observe and report information (in their field of view (FOV)) from an unfolding scene on the ground, for example, a battlefield scene consisting of a number of tanks, trucks and jeeps on desert terrain, or an airport runway scene consisting of several kinds of planes and cars and vans. The goal is to, at each scan, detect and classify targets and dynamically track them across the sensor observation frames. Bayesian-based ATR can be divided into three main areas of focus, which can be listed accordingly as: (i) problem formulation from a Bayesian perspective, (ii) sensors and sensor modelling and (iii) inference.

30.7.1 Bayesian Problem Formulation

Classical ATR approaches detect possible targets using various image processing procedures and then extract features to be fed into a pattern classifier for target recognition. On a high level, the classifier '*compares*' the extracted features with the features of various established classes of targets in its database in order to take decisions regarding the class of targets in the observed image. This could be thought of as a top-down approach. The Bayesian approach, however, adopts more of a bottom-up philosophy. Scenes are hypothesized and generated using known backgrounds of the terrain and 3-D templates of targets. These are then mapped nonlinearly to the observed image. Subsequently, the (hypothetical) scene that best matches the observed image is the solution to the ATR problem. This non-linear mapping is the key to solving the problem. The non-linear map between the hypothesized scene and observed image, also known as the sensor model, has to account for: (i) the number of targets, (ii) the classes of targets, (iii) the positions and scales of the targets, (iv) the orientation or pose of the targets, (v) atmospheric effects such as illumination, temperature, (vi) noise and other effects caused by the sensor medium and (vii) clutter and occlusions. In the current subsection we focus on the first four factors, while in the subsequent subsections the remaining three factors are considered. For the purpose of this chapter, we assume that the background for the scene is known to the user. Hence, the focus will be on targets only. A three-dimensional (3D) CAD model of standard size and scale, such as those shown in Figure 30.2, is created to form a template for each target of interest to the user. A mathematical tool called *deformable template theory* [40] is used to model the basic variations of the templates that are needed for scene building, that is, translation, rotation and/or scaling. A paper by Lanterman et al. [7] also discusses incorporating thermal variability for ATR using FLIR image data. Building a scene using multiple templates requires the ability to model the variability. Each scene consisting of a finite

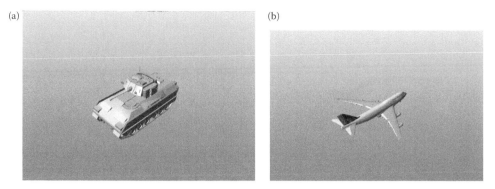

FIGURE 30.2 CAD-model templates of a tank and an aircraft. (a) Bradle tank and (b) Boeing 747.

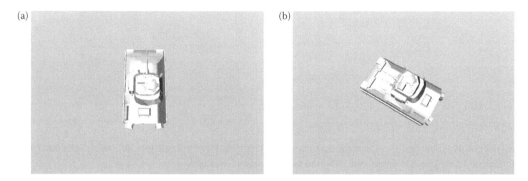

FIGURE 30.3 Template of a Bradle tank used for hypothetical scene generation along with a rotation by $-60°$ about its centroid to generate another pose. (a) Template of a tank and (b) template rotated by $-60°$.

number of target templates is assumed sufficiently (in the sense of a sufficient statistic) represented by a parameter vector. Miller et al. in Ref. [41] provide a framework for the mathematical representation of scenes. The goal is to mathematically hypothesize a scene that could best 'explain' the observed image from a sensor such as FLIR camera. To do so, one has to construct several hypotheses and then allow the ATR algorithm to 'pick' (in a statistical sense) the best hypothetical scene based on the observation. As a result, the construction of mathematical structures around which one can build hypothetical scenes is important and one has to account for two classes of variability: (i) Variability associated with the templates, that is, they can undergo translation, rotation and scaling [42]. Scenes will also contain variable number of targets and varying identities (classes) of the targets. Lie group theory [41,43] provides a good framework for the representation of such variability. Figure 30.3 shows a CAD-model of a tank that is rotated about its centroid; and (ii) Variability associated with the sensor, that is, various kinds of imaging sensors, as well as ranging sensors, are used in ATR and the key is to come up with a good model for the sensor being used.

30.7.2 SCENE GENERATION AND TARGET REPRESENTATION

Given a known background such as a desert battlefield or an airport runway, one can construct scenes using templates of targets. All the possible configurations of the targets in a scene can be accounted for by translating and rotating the templates. The goal is to come up with the templates and the transformations to match the occurrence of targets in the observed scene. There are two key aspects to target representation: the first one is the target template itself and the second one is the set of transformations it undergoes during scene generation. Two articles [40,44] provide a rigorous background in target representation.

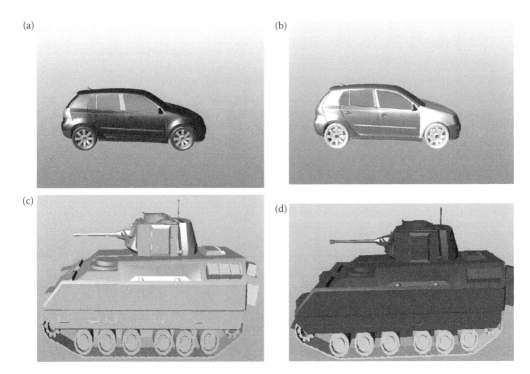

FIGURE 30.4   CAD models of a car and a tank along with their IR templates. (a) CAD model of a car, (b) IR template of a car, (c) CAD model of a tank and (d) IR template of a tank.

30.7.2.1   Template Creation for Targets

As mentioned earlier, a (rigid) target template is created using a realistic CAD-model representation. This rigid-body template can have various attributes depending on the sensor and environment of interest, for example, a thermal profile in the case of FLIR, texture information in the case of video sensors, electromagnetic reflectivity in the case of magnetic radars and so on. Figure 30.4 shows CAD-models of a car and a tank augmented by their respective thermal *profiles,** respectively, to form IR templates for ATR using FLIR images. A set of target labels, denoted as $\mathcal{A}$ can be defined as

$$\mathcal{A} = \big\{\text{car, truck, jeep, tank, airplane,}\ldots\big\} \tag{30.1}$$

The elements of the set $\mathcal{A}$ are known as classes. This set can be extended, curtailed or refined based on the user's requirements. Mathematically, a target template may be defined as I^{α} where $\alpha \in \mathcal{A}$, in other words, the template corresponding to class 'car' is denoted by I^{car}, the template corresponding to class 'airplane' is denoted by $I^{airplane}$ and so on.

30.7.2.2   Transformations

Rigid transformations such as translation and rotation are applied on the template I^{α} to match the occurrence of the target of class α in the observed scene. The transformations applied on multiple templates in the case of multi-target scenarios are generally modelled as independent. However, in cases where there is dependence between the motion of two or more targets, as in the case of military formations, future research might be on including dependence in the modelling of transformations. There are two categories of transformations: rigid and non-rigid. Translation and

---

* IR templates from the Prism database [45] have been used in Refs. [40,41]. In this chapter, thermal profiles have manually been generated by whitening and darkening various parts of the CAD modelling using Ref. [42].

(a)  (b)

FIGURE 30.5 Illustration of the template of a tank with its turret rotated independently to the body of the tank. (a) CAD model of a tank and (b) CAD model of the tank with its turret rotated by $180°$.

rotation are rigid transformations, while, thermal characteristics, change in light conditions and so forth, are examples of non-rigid transformations. We focus on rigid transformations here; rigid translation is denoted by a vector $p \in \Re^n$, where $n = 2$ (ground vehicles) and $n = 3$ (air vehicles in air). Rotation is represented by a matrix O that could be of dimension 2×2 or 3×3 for ground or air scenarios, respectively. There are special cases, as illustrated in Figure 30.5, where a template undergoes two/multiple independent rotations. Both the transformation operations, that is, translation and rotation, belong to groups. The translation group is $\Re^n$ and the rotation group is special orthogonal $SO(n)$, where n denotes dimensionality.[*]

Definition 1

A group is a set that is always associated with a group operation (called product and denoted by 'o') in such a way that the product of any two elements from a group always lies in the same group.

For instance, the group operation in $\Re^n$ is a vector addition while in $SO(n)$ is a matrix multiplication. An additional important property of groups is that they have an identity element—zero vector in case of $\Re^n$ and identity matrix in case of $SO(n)$. A rotation matrix $O \in SO(n)$ has the property that

$$OO^H = I$$
$$\det(O) = +1$$

(30.2)

where $\det(\cdot)$ denotes the determinant operator[†] and H denotes Hermitian or complex conjugate transpose. The 2nd equation in 30.2 can be easily seen by taking the determinant on both sides of the 1st equation of 30.2, that is, $\det(OO^H) = \det(O)\det(O^H) = (\det(O))^2 = \det(I) = 1$. The type of rotation matrices, that form a subgroup called special orthogonal group, for which $\det(O) = +1$ are also called as *proper* rotation matrices. Another interesting property of groups in the scene building context is that one transformation s_1 applied before another transformation s_2 has the combined effect of a third transformation $s_3 = s_1 \cdot s_2$ applied alone. Translation and rotation can be combined to form a single transformation that is a member of the special Euclidean group $SE(n)$. Consider a point $x \in \Re^2$ on a template in an image. If a translation operation $p \in \Re^2$ is applied, it results in new coordinates $x + p$. In addition, if a rotation $O \in SO(2)$ is applied, it results in the point x being rotated to result in new coordinates Ox. Together, the operation could be combined and represented by $Ox + p$. Generally, the joint translation–rotation can be represented in matrix form as follows: for any $n \times n$ matrix $U \in \Re^{n \times n}$ such that

$$U = \begin{bmatrix} O & p \\ 0 & 1 \end{bmatrix}$$

(30.3)

---

[*] Dimensions $n = 2, 3$ are typically of interest in ATR.
[†] If O were just an orthogonal matrix, its determinant would be $\det(O) = \pm 1$.

(a)    (b)

Frame 1   Frame 2

FIGURE 30.6 Illustration of scene generation using a known desert background in conjunction with CAD models of two different tanks, where (a) represents an earlier time frame and (b) represents a time frame where both translation and rotation have been applied on both templates.

and a vector $x_1 \in \Re^{n+1}$

$$x_1 = \begin{bmatrix} x \\ 1 \end{bmatrix} \tag{30.4}$$

the operation

$$s = Ux_1 = \begin{bmatrix} Ox + p \\ 1 \end{bmatrix} \tag{30.5}$$

results in an $n \times 1$ dimensional vector whose first n elements constitute the coordinates of the transformed point x [40]. The set of all matrices U is denoted by $SE(n)$, the special Euclidean group. Hence, the set of all transformations s of a template I^α is given by

$$O^\alpha = sI^\alpha \tag{30.6}$$

for all $s \in \Re^n$, $s \in SO(n)$ or $s \in SE(n)$. Figure 30.6 shows an example of scene creation using a known background (desert battlefield) where Figure 30.6a and b represent two distinct frames, separated in time, in such a way that both translation and rotation have been applied independently on both templates, in the time interval in between the two frames.

30.7.3 SENSOR MODELLING

As mentioned in Section 30.1, there are two kinds of variability in a hypothetical scene generation. The first one, that is, template transformation was discussed in Section 30.3.2, and the other variability is in sensor modelling and is discussed in this subsection. An indepth treatment of sensor modelling for various sensors in ATR applications is provided in Refs. [17,40,41,46]. Hypothetical scenes using known backgrounds (see Figure 30.6 for an illustration of a desert battlefield scene) and multiple target templates are formed with the intention of arriving at the best match to the observed image. The map between the hypothesized scene and the observed image is the sensor model. The sensor model takes into account the physics of the sensor as different sensors might produce different observations of the same scene. As discussed earlier, the templates used to construct scenes are three-dimensional (3D) in nature, however, image sensors produce 2D observations and other sensors such as the high-resolution range (HRR) radars produce observations that are in the form of 1D arrays. Imaging sensors use a projection mechanism to map real 3D scenes to 2D camera detector spaces. In essence, they accumulate responses from scene elements and project them to a pixel in the image [40]. Denoting the observation space by $I^D \in \Re^d$ where d represents the dimensionality of the sensor observation space (2 for images and 1 for HRRs), we can represent a sensor model as a nonlinear mapping T between the transformed template sI^α and the detector space I^D as follows:

$$I^D = TsI^\alpha \tag{30.7}$$

In addition to the non-linear mapping, the sensor may also give rise to random noise w which can be modelled to be additive

$$I^D = TsI^\alpha + w \tag{30.8}$$

Remark 2

The goal of an ATR algorithm is, given observations I^D, an appropriate sensor model T and templates I^α, estimate the transformation parameters given by s.

The sensors typically used for ATR can be very broadly divided into two categories (with some room given for overlap): (a) high-resolution sensors such as imaging devices like FLIR, synthetic aperture radar (SAR), electro-optical cameras (EO) and so on, and (b) low-resolution sensors suck as millimetre-wave radar (MMWR), other ranging radars, acoustic sensors (for direction of arrival observations) and so forth. Typically, the high-resolution sensors are used for identification and are not traditionally used to obtain ranging or location information of targets, while the lower resolution sensors are used for ranging and target localization applications. One can see that the two categories of sensors are complementary in nature. Hence, a good ATR algorithm is one that fuses information from both categories of sensors in order to exploit their complementary nature. Some of the commonly used imaging sensors in ATR and the models used to simulate their data are described below, while some of the likelihood functions for sensors in the literature that govern their statistical models are provided in Section 30.3.4.3.

1. *Video Imager*: A video or an optical imager [41] provides high-resolution 2D real-valued images of rigid targets. The detector space forms a lattice $I^O = \{I^O (l):l \in \mathcal{L}^O\}$ where $l \in \Re^2$ and $\mathcal{L}^O$ represents the lattice that forms the detector space. As the templates used for scene creation are 3D, imaging sensors such as the video imager produce output images that contain projections of the real-world input scene. The projections could be either orthographic or perspective. In an orthographic projection, a point (x, y, z) in the 3D input space is projected onto the 2D detector space by a simple mapping $\{(x, y, z) \mapsto (x, y)\}$. In perspective projection, the input-to-output mapping is in the form $\{(x, y, z) \mapsto (x/z, y/z)\}$ [41]. This creates the so-called vanishing point effect where objects that are further away appear closer. The video imager output can be modelled as a Gaussian random field with mean field being the projection of the input scene onto the camera [41]. The input scene is constructed as described in Section 30.3.2.

2. *FLIR*: The FLIR camera is a sensor employed by most ATR systems with military and aircraft landing applications, as it can operate during the night and also under foggy or cloudy conditions as it detects the heat radiated by objects. To simulate FLIR images, true scenes are constructed using 3D CAD model templates of targets which also contain their thermal profile. A database of such templates along with their thermal profiles can be

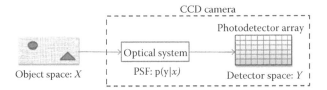

FIGURE 30.7 Object viewed with a CCD camera.

obtained from the PRISM database [45]. One can also assume that the target surfaces radiate known intensities and superimpose a thermal map on the surfaces of the templates by rendering as shown in Figure 30.4. The practical pattern-theoretical ATR algorithms for infrared imagery are discussed in Ref. [46]. The charge-coupled device (CCD) camera model given by Snyder et al. [47] has frequently been used to model and generate (synthetic) FLIR images. According to Srivastava et al., and Miller et al. [40,41] (among others), an FLIR camera captures the thermodynamic profile of a target body via CCD detectors. CCD cameras are commonly used for acquiring images of incoherently radiating objects. Their wide spectral response make them ideal for acquiring image data in the visible (ultraviolet) UV and IR ranges [47]. From Figure 30.7 one can see that the detector array is preceded by the optical system of the camera. The optical system consisting of the field stops and the lenses could limit resolution and introduce aberrations. The point spread function (PSF) describes the response of an imaging system to a point source or an object. It can be seen as the impulse response of an optical system.

A Poisson random field model has been used in Refs. [40,41] among others, to simulate FLIR images while, a Gaussian random field model has been used in Ref. [17]. Both the Poisson as well as the Gaussian models assume a common mean field. A blurred version of the input image is used to model the mean field for both models. Figure 30.8 shows the images simulated using a known background corresponding to the real FLIR scene shown in Figure 30.9. It has to be noted that the background in the scene is assumed known *a priori* and is the grey-scale version of the background of the image obtained by an electro-optical camera, that has been placed adjacent to the FLIR camera (used to capture the frame in Figure 30.9) and time synchronized with it. Figure 30.8a shows a synthetic FLIR image obtained via a Poisson model while Figure 30.8b shows the same generated using a Gaussian model. A CAD-model of a car rendered in order to include its thermal profile is superimposed on the known background for the purpose of input scene

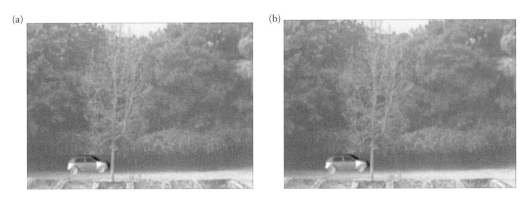

FIGURE 30.8 Simulated IR images via (a) Poisson and (b) Gaussian models, with mean field for both being the blurred input image.

FIGURE 30.9   Real FLIR camera image.

construction. Figure 30.10 shows FLIR simulation results using the Poisson model and the Gaussian model with a low standard deviation and a high standard deviation. It was observed during various simulation experiments that while the differences in the results achieved by both the Poisson and the Gaussian model tend to look small when the blurring effect in the mean field is significant, the differences tend to be apparent when the blur is not pronounced as can be inferred from Figure 30.10. A blurring effect can be achieved by the convolution between the true object and the PSF which is also known as a blurring function. The degree of spreading (blurring) of a point object can be seen as a measure of the quality of the imaging system. A 2D Gaussian kernel is commonly used as PSF [47,48].

3. *LADAR*: Coherent laser radar (LADAR) systems can collect 2D intensity, range or Doppler images by scanning a field of view. Ideal LADAR imagery of a scene can be generated in the following two steps:
   i. Building templates of targets via CAD models as described in Section 30.3.2
   ii. Simulating range data using the XPATCH range radar simulation software. XPATCH is a computer software package developed jointly by US Air Force Wright Laboratory and DEMACO, Inc., that is designed to simulate the radar return from a given target [49].

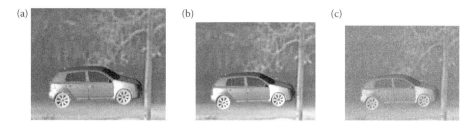

FIGURE 30.10   Simulated IR image of cropped scene from Figure 30.8 showing a Poisson model (a) and two Gaussian models, one with a low standard deviation (b) and the other with a high standard deviation (c).

Furthermore, to project the true scene from the 3D space (as generated via CAD models) to the 2D detector space, an approach such as *perspective projection* has been used as reported in Ref. [17]. The above-mentioned steps do not account for model inaccuracies due to noise and clutter. However, in the real world, LADAR imagery is degraded by uncertainty due to target speckle, atmospheric turbulence and radar-beam jitter, as well as finite carrier-to-noise ratio (CNR). The most significant out of these are finite CNR and target speckle for CO_2 LADARs with modest sized optics, that is, 5–20 cm diameter optics [17]. Their modelling at the pixel level has been discussed in Ref. [50]. Thus, the simulated LADAR model needs to account for these noise sources in order to depict a realistic scenario. The LADAR output at each pixel l is modelled as a Gaussian with mean given by the true range at pixel l and standard deviation given by the local range accuracy [17,51].

4. *MMWR*: An outline for the model of an active MMWR is provided in Ref. [52]. The sensor output data is in the form of a 2D lattice where each data point represents an azimuth/range cell (for ground targets). Corresponding to each cell, the radar takes N frequency samples.[*] The square root of the sum of the squared magnitudes of these frequency samples constitutes the data.

5. *High-Resolution Range*: The problem of joint tracking and recognition of a target using a sequence of high-resolution range HRR range profiles is addressed in Ref. [49]. Data produced by an HRR is significantly different to the output of the other sensors in this list. An HRR provides the range profile of targets in its FOV in the form of a 1D array. Data are collected from targets by illuminating them with coherent radar waves, and then sensing the reflected wave with an antenna [49]. The reflected waves are received at times proportional to the round-trip distance travelled. The received signal at each range bin represents the superposition of the echoes from all the reflectors corresponding to that particular range [40]. Range profiles can be computed from the XPATCH tool given the target pose [49]. However, effects such as the flexing or vibrations in ground targets, physical effects due to moisture or dirt on the vehicle and control surface movement need to be incorporated into the model [49]. Hence, two models have been considered in Ref. [49] within the family of Gaussian random process models. In the first model, the range profiles are modelled as being deterministic and assumed to be completely known given the target type, position and orientation. Here, the mean of the Gaussian random process is in the form of the modelled range profiles from software such as XPATCH, with the covariance being zero. In the other model, the range profiles are assumed independent from one orientation to another and modelled as independent complex Gaussian random variables at each orientation. The University Research Initiative Synthetic Dataset (URISD) is a collection of simulated range profiles produced by XPATCH [49].

6. *SAR*: Statistical modelling of SAR images has become an active research field [53]. The book [54] by C. J. Oliver, provides a good treatment of SAR image modelling. Combining SAR statistical models with an ISAR target database can simulate SAR images. Variability in aspect, terrain content, region and signal-to-clutter ratio (SCR) can be added to the simulated SAR images. An extensive review of the various statistical models used for SAR images has been carried out in Ref. [53]. Modelling of real SAR data has been broadly classified into two groups in Refs. [53,55,56]: parametric and non-parametric models. When dealing with parametric models, several known distributions can be used to model the SAR image. However, if the parameters of these distributions are unknown they can be estimated using the known real image. The best/optimal distribution that fits the real data can be arrived at using certain metrics such as a goodness-of-fit test. Non-parametric modelling implies that no distributions are assumed and

---

* $N = 64$, in the radar used in Ref. [52].

Ref. [53] provides data-driven techniques to estimate statistical nonparametric models that best fit the data. Since, nonparametric modelling is computationally intensive and time consuming the parametric modelling approach is commonly used for SAR statistical modelling. Parametric models can be classified into four categories: (a) empirical distributions, (b) models developed from the product model, (c) models developed from the generalized central limit theorem and (d) other models. Of these, the product model is the most widely used parametric model in SAR image modelling [53]. The product model is in turn developed from the speckle model. The speckle model presented in Ref. [57] assumes that under the ideal circumstance the imaged scene has a constant radar cross section. The details of this speckle model are given in Ref. [53]. Based on the speckle model the product model for SAR images is given in Ref. [58]. The product model combines an underlying RCS component σ with an uncorrelated multiplicative speckle component n, so that the observed intensity I in an SAR image can be expressed as the product

$$I = \sigma \cdot n \tag{30.9}$$

30.7.4 Bayesian Paradigm

Bayesian inference is a statistical inference methodology in which some kind of evidence or observation is used to calculate the probability that a hypothesis, denoted as $\mathcal{H}$, may be true. The observation can also be used to correct a previously predicted probability. The afore-mentioned calculation makes use of the Bayes' theorem which quantifies the posterior probability of $\mathcal{H}$ in terms of the prior probability of $\mathcal{H}$ and a likelihood correction which can be treated as the prior probability of the available observation given information about the hypothesis $\mathcal{H}$. To this end, given a set of available measurements, or in the context of ATR a set of m images denoted as $I = \{I_1, I_2, \ldots, I_m\}$, the ATR problem can be cast as estimating an unknown quantity such as the target pose denoted by θ in the presence of unwanted noise and/or clutter. Before proceeding further, we state Bayes' theorem and give a simple proof in the setting of scalar variables and extend the result to vector variables. According to Bayes [59], Bayes' theorem states that 'the probability of any event is the ratio between the value at which an expectation depending on the happening of the event ought to be computed, and the value of the thing expected upon its happening'. Simply put, given two random variables I and θ, Bayes' theorem states that *the probability of θ given I is equal to the ratio of the product the probability of I given θ and the marginal probability of θ to the marginal probability of I*. The marginal probability of a collection of random variables, also called the unconditional probability, is the probability distribution of the variables under consideration. We begin with the case of scalar variables. Mathematically, Bayes' theorem can be stated as

$$p(\theta \mid I) = \frac{p(I \mid \theta) p(\theta)}{p(I)} \tag{30.10}$$

where $p(\theta \mid I)$ is called the conditional probability of θ given I, also called the *posterior probability*, $p(I \mid \theta)$ is the conditional probability of I given θ, also called the *likelihood*, $p(\theta)$ is the *prior probability* of θ, also called the marginal probability of θ and $p(I)$ is the *prior probability* of I, which also acts as a normalizing constant. To see the relation in Equation 30.10 note that the joint probability of θ and I, denoted as $p(\theta, I)$, is given by the relation

$$p(\theta, I) = p(\theta \mid I) p(I) \tag{30.11}$$

Furthermore, the joint probability of θ and I can also be denoted as $p(I, \theta)$, and is given by

$$p(\theta, I) = p(I \mid \theta) p(\theta) \tag{30.12}$$

Given that $p(\theta, I) = p(I, \theta)$ and comparing Equations 30.11 and 30.12, immediately results in Equation 30.10. Thus, the posterior probability is proportional to the product of the prior probability and the likelihood. To generalize, Equation 30.10 can be extended to the case of vector of random variables. Consider the following sequence of random variables as $\theta = \{\theta_1, \theta_2, \ldots, \theta_n\}$ and $\mathbf{I} = \{I_1, I_2, \ldots, I_m\}$, where, $\{\theta_i\}_{i=1 \text{ to } n}$ denotes the unknown quantity to be estimated and $\{I_j\}_{j=1 \text{ to } m}$ denotes the observed quantity. Furthermore, the bold font for θ and I implies that these are vector quantities. Assume that the unknown parameters, $\{\theta_1, \theta_2, \ldots, \theta_n\}$, follow a Markov process, such that

$$p(\theta_k \mid \theta_{k-1}, \theta_{k-2}, \ldots, \theta_2, \theta_1) = p(\theta_k \mid \theta_{k-1}), \quad \forall k \tag{30.13}$$

where k denotes the time instant. From Equation 30.10, the marginal of the posterior density of θ can be expressed as

$$
\begin{aligned}
p(\theta_k \mid \mathbf{I}_{1:k}) &= \frac{p(\mathbf{I}_1, \mathbf{I}_2, \ldots, \mathbf{I}_{k-1}, \mathbf{I}_k \mid \theta_k)\, p(\theta_k)}{p(\mathbf{I}_1, \mathbf{I}_2, \ldots, \mathbf{I}_{k-1}, \mathbf{I}_k)} \\
&= \frac{p(\mathbf{I}_k, \mathbf{I}_{1:k-1} \mid \theta_k)\, p(\theta_k)}{p(\mathbf{I}_k, \mathbf{I}_{1:k-1})}
\end{aligned}
\tag{30.14}
$$

Using Equation 30.11, an expression for the joint probability of $p(\mathbf{I}_k, \mathbf{I}_{1:k-1} \mid \theta_k)$ in Equation 30.14 can be written as

$$p(\mathbf{I}_k, \mathbf{I}_{1:k-1} \mid \theta_k) = p(\mathbf{I}_k \mid \mathbf{I}_{1:k-1}, \theta_k)\, p(\mathbf{I}_{1:k-1} \mid \theta_k) \tag{30.15}$$

and an expression for the joint probability of $p(\mathbf{I}_k, \mathbf{I}_{1:k-1})$ in Equation 30.14 can be written as

$$p(\mathbf{I}_k, \mathbf{I}_{1:k-1}) = p(\mathbf{I}_k \mid \mathbf{I}_{1:k-1})\, p(\mathbf{I}_{1:k-1}) \tag{3.16}$$

Substituting Equations 30.15 and 30.16 into Equation 30.14, we obtain

$$p(\theta_k \mid \mathbf{I}_{1:k}) = \frac{p(\mathbf{I}_k \mid \mathbf{I}_{1:k-1}, \theta_k)\, p(\mathbf{I}_{1:k-1} \mid \theta_k)\, p(\theta_k)}{p(\mathbf{I}_k \mid \mathbf{I}_{1:k-1})\, p(\mathbf{I}_{1:k-1})} \tag{30.17}$$

Applying Bayes' theorem to the term $p(\mathbf{I}_{1:k-1} \mid \theta_k)$ in Equation 30.17, we obtain

$$p(\mathbf{I}_{1:k-1} \mid \theta_k) = \frac{p(\theta_k \mid \mathbf{I}_{1:k-1})\, p(\mathbf{I}_{1:k-1})}{p(\theta_k)} \tag{30.18}$$

Substituting Equation 30.18 into Equation 30.17 yields

$$p(\theta_k \mid \mathbf{I}_{1:k}) = \frac{p(\mathbf{I}_k \mid \mathbf{I}_{1:k-1}, \theta_k)\, p(\theta_k \mid \mathbf{I}_{1:k-1})\, p(\mathbf{I}_{1:k-1})\, p(\theta_k)}{p(\mathbf{I}_k \mid \mathbf{I}_{1:k-1})\, p(\mathbf{I}_{1:k-1})\, p(\theta_k)} \tag{30.19}$$

Cancelling the common term $p(\mathbf{I}_{1:k-1})\, p(\theta_k)$ in the numerator and the denominator of Equation 30.19, we finally obtain

$$p(\theta_k \mid \mathbf{I}_{1:k}) = \frac{p(\mathbf{I}_k \mid \mathbf{I}_{1:k-1}, \theta_k)\, p(\theta_k \mid \mathbf{I}_{1:k-1})}{p(\mathbf{I}_k \mid \mathbf{I}_{1:k-1})} \tag{30.20}$$

where $p(\theta_k \mid \mathbf{I}_{1:k})$ is the posterior density. Thus, the posterior density is proportional to the product of the prior density and the likelihood.

30.7.4.1 Bayesian ATR Structure

The most fundamental problem that any ATR algorithm must address can be defined as follows:

Definition 2

Given observations of a time-evolving scene, consisting of a single non-cooperative target, by a passive imaging sensor and an active tracking (ranging) sensor, is it possible to recognize the target and then track its position and pose (orientation)?

The scope of the problem can then be extended by throwing into the mix multiple sensors with differing sensor models and outputs, multiple targets, scenarios where new targets appear and existing targets die out (leave the FOV of interest), scenarios that include clutter and occlusions, and any other parameters of interest that provide new and required information (based on the application) about the targets. Grenander's *Pattern Theory* [60–62] provides the necessary framework for the formulation of the ATR problem in the format given above. In this theory, data are modelled as transformations of underlying templates to the observation space. The transformations may be entirely deterministic or could contain in-built randomness (in addition to additive noise in some cases). The ATR problem when formulated in this manner can be solved using Bayes' theorem given in Section 30.3.4 by converting it into an estimation problem. This section provides the outline for the formulation and the Bayesian solution to ATR. There are four key steps in building the underlying structure for the Bayesian solution to the ATR problem given in Definition 2: (a) parameterization of the time-evolving scene, (b) physics-based sensor modelling using the likelihood function, (c) using information from target kinematics in the prior function and (d) inference using the maximum *a posteriori* (MAP) estimator or the maximum likelihood (ML) estimator.

30.7.4.2 Scene Parameterization

If observations are made at times $k = 1, 2, \ldots, K$ within a particular time interval of the motion of a single ground target, its position can be parameterized at time k by $\{\mathbf{p}_k \in \Re^2; k = 1, 2, \ldots, K\}$ and its pose at time k can be parameterized by the rotation matrix $\{\theta_k \in SO(2); k = 1, 2, \ldots, K\}$ as described in Section 30.3.2. Hence, the motion of a rigid-body ground target can be represented by a sequence of translations and rotations (where the rotation is around the z-axis in a classical (x, y, z) Cartesian system)

$$\begin{pmatrix} x \\ y \end{pmatrix} \mapsto \begin{pmatrix} \cos\theta & \sin\theta \\ -\sin\theta & \cos\theta \end{pmatrix} \begin{pmatrix} x \\ y \end{pmatrix} + \begin{pmatrix} p_1 \\ p_2 \end{pmatrix} \tag{30.21}$$

where $[p_1\ p_2]^T$ represents the translation parameter vector and the angle θ represents the rotation angle.* The target type is denoted by $\alpha \in \mathcal{A}$ (see Section 30.3.2 for further details). Hence, for a given set of observations $\mathbf{Z}_k = \{\mathbf{z}_1\ \mathbf{z}_2 \ldots \mathbf{z}_k\}$ up to time k, the parameter set to be estimated consists of the target position, pose and class, and can be represented as an element of the space

$$(\Re^2 \times SO(2))^K \times \mathcal{A} \tag{30.22}$$

One can observe that two of the elements of the parameter set in Equation 30.22 are continuous-valued variables and the third is discrete valued. At the kth observation time let the sequence of translations be

$$T_{l_k} = \left\{ t_{l_1}\quad t_{l_2}\quad \cdots\quad t_{l_k} \right\} \tag{30.23}$$

---

* Time indices have been omitted for notational convenience.

and let the sequence of target orientations up to the time k be

$$\Theta_k = \left\{ \theta_1 \quad \theta_2 \quad \ldots \quad \theta_k \right\} \tag{30.24}$$

then the joint posterior density of the parameters at time k is given by Bayes' theorem

$$\underbrace{p\left(\mathbf{t}_{l_k}, \theta_k, \alpha \mid Z_k\right)}_{\text{posterior}} \propto \underbrace{p\left(\mathbf{z}_k \mid \mathbf{t}_{l_k}, \theta_k, \alpha\right)}_{\text{likelihood}} \underbrace{p\left(\mathbf{t}_{l_k}, \theta_k, \alpha \mid T_{l_{k-1}}, \Theta_{k-1}\right)}_{\text{prior}} \tag{30.25}$$

Furthermore, the proportionality sign in Equation 30.25 can be replaced with the equality sign by dividing the right-hand side of Equation 30.25 by the term $p(\mathbf{z}_k \mid \mathbf{z}_{k-1})$ as shown in Equation 30.20.

30.7.4.3 Likelihood Functions

Generally, the likelihood function is defined as the conditional density of measurement $\mathbf{z}$ given target state $\mathbf{x}$, that is, $\Lambda(\mathbf{x}) = p(\mathbf{z} \mid \mathbf{x})$. In the context of ATR the likelihood function represents the mapping from the underlying scene, parameterized by the state vector $\mathbf{x}$, to the sensor observation $\mathbf{z}$. It can be interpreted as the statistical model of the sensor based on its physics. Clutter models can also be incorporated into the likelihood function as described in Ref. [40]. The likelihood function $p(\mathbf{z}_k \mid \mathbf{t}_{l_k}, \theta_k, \alpha)$ given in Equation 30.25, characterizing the problem stated in Definition 2, can be interpreted as the probability that a target of class α translated by $\mathbf{p}$ and rotated by θ at time k gives rise to the observed image contained in measurement $\mathbf{z}$. The likelihood function is derived from the physical characteristics of the sensor map T (see Equations 30.7 and 30.8). The likelihood functions used in the Bayesian ATR literature for some commonly used imaging sensors listed in Section 30.3.3 are as follows:

1. *Video Imager*: Let the output image produced by the video or optical imager be denoted by $\mathbf{I}^O = \{\mathbf{I}^O(l) : l \in \mathcal{L}^O\}$, where l represents a pixel such that $l \in \mathfrak{R}^2$, $\mathcal{L}^O$ is the lattice that makes up the detector space, and the superscript O denotes an image from an optical imager. Assuming that the output pixel intensities are independent, the likelihood function of the optical imager can be modelled as Gaussian [40]

$$p(\mathbf{I}^O \mid \mathbf{I}^x) = \prod_{l \in \mathcal{L}^O} \frac{1}{\sqrt{2\pi\sigma_O^2}} \exp\left(-\frac{\left[\mathbf{I}^O(l) - \mathbf{I}^x(l)\right]^2}{2\sigma_O^2} \right) \tag{30.26}$$

   where $\mathbf{I}^x$ represents the input scene parameterized by $x \in \mathcal{X}$. Here, $\mathcal{X}$ represents the parameter space and σ_O denotes the standard deviation of the distribution.

2. *FLIR*: There have been two different sensor models used, in recent literature, to build likelihood functions for FLIR camera. A Poisson model has been used in Refs. [40,41] while a Gaussian likelihood model has been used in Ref. [17]. Suppose the output image produced by the FLIR imager is denoted by $\mathbf{I}^F = \{\mathbf{I}^F(l) : l \in \mathcal{L}^F\}$, where l represents pixel such that $l \in \mathfrak{R}^2$, $\mathcal{L}^F$ is the lattice that makes up the detector space, and the superscript F implies that the image was generated from an FLIR camera. Assuming that the output pixel intensities are independent, the likelihood function of an FLIR camera can be modelled as Poisson [40,41,48] where the blurred input image

$$I_{\text{blur}}^x(l) = \sum_{i \in I, \, l \in \mathcal{L}^F} p(l \mid i) I^x(i) \tag{30.27}$$

formed by the convolution between a point spread function (PSF) and the input image parameterized by $x \in X$ forms the mean field of the Poisson process. In Equation 30.27, $p(l \mid i)$ denotes the point spread function and $\mathbf{I}$ denotes the lattice that the input image $\mathbf{I}^x$ forms. In the Poisson model presented above, the sensor is assumed to be calibrated to give specific photon counts. According to Dixon and Lanterman [46], this assumption might be valid when it comes to applications such as astronomic imaging, however, it is typically not the case for FLIR sensors. Hence, a more appropriate model might be the Gaussian model presented in [17,63]

$$p(\mathbf{I}^{\mathrm{F}} \mid \mathbf{I}^x) = \prod_{l \in \mathcal{L}} \frac{1}{\sqrt{2\pi(NE\Delta T)^2}} \exp\left(\frac{\left[\mathbf{I}^{\mathrm{F}}(l) - \mathbf{I}^x(l)\right]}{2(NE\Delta T)^2}\right) \qquad (30.28)$$

where $\mathbf{I}^{\mathrm{F}} \equiv \mathbf{I}^{\mathrm{F}}(l) \in \mathrm{L}^{\mathrm{F}}$ is the output image, $\mathbf{I}^x \equiv \mathbf{I}^x(l) \in I$ is either the ideal noiseless input image or the blurred input image, and $(NE\Delta T)$ is the FLIR's noise-equivalent temperature difference that forms standard deviation of the distribution [46].

3. *LADAR*: Let the output range image produced by the LADAR be denoted by $\mathbf{I}^{\mathrm{L}} = \{\mathbf{I}^{\mathrm{L}}(l) : l \in \mathcal{L}^{\mathrm{L}}\}$ where l represents a pixel such that $l \in \mathfrak{R}^2$, $\mathcal{L}^{\mathrm{L}}$ is the lattice that makes up the detector space, and the superscript L implies that the image was generated from a LADAR. Assuming that the output pixel intensities are independent, the likelihood function of a LADAR is modelled as a Gaussian density in [17]

$$p\left(\mathbf{I}^{\mathrm{L}} \mid \mathbf{I}^x\right) = \prod_{l \in \mathcal{L}^{\mathrm{L}}} \left([1 - \mathrm{Pr}(A)] \frac{1}{\sqrt{2\pi(\delta D)^2}} \exp\left(-\frac{\left[\mathbf{I}^{\mathrm{L}}(l) - \mathbf{I}^x(l)\right]^2}{2(\delta D)^2}\right) + \frac{\mathrm{Pr}(A)}{\Delta D} \right) \qquad (30.29)$$

where $\mathbf{I}^x \equiv \{\mathbf{I}^x(l) \in I : l \in \mathcal{L}^{\mathrm{L}}\}$ represents the input range image parameterized by $x \in X$. Here, $\mathrm{Pr}(A)$ is the single pixel anomaly probability—the probability that speckle and shot-noise effects combine to yield a range measurement that is more than one range resolution cell from the true range, and is assumed to be same for all pixels [17]. The first term in the product in Equation 30.29 represents the case where pixel l is not anomalous and the second term represents the case where pixel l is anomalous. The range when pixel l is not anomalous is modelled as a Gaussian distribution with standard deviation δD, the local range accuracy. The range when pixel l is anomalous is modelled as a uniform distribution over the entire range of the uncertainty interval, $\Delta D = D_{R_{\max}} - D_{R_{\min}}$ that is much larger than the local range accuracy δD. The terms $D_{R_{\max}}$ and $D_{R_{\min}}$ denote the upper and lower limits of the uncertainty interval ΔD, respectively.

4. *HRR*: In the HRR model presented in Ref. [17], an additive white complex Gaussian noise model is used for the complex envelope of the observed signal. Hence, the magnitude of the observed signal is modelled as Rice distributed. Suppose $\mathbf{I}^{\mathrm{H}} \equiv \{\mathbf{I}^{\mathrm{H}}(r) \in \mathcal{R}\}$, where $\mathbf{r} \in \mathfrak{R}$ and $\mathcal{R}$ the range profile lattice is the random range profile produced by the HRR, the superscript H implies that the image was generated from an HRR source, and $\mathbf{I}^x \equiv \{\mathbf{I}^x(r) \in \mathcal{R}\}$ be the input range profile parameterized by $x \in X$. Then the likelihood function is of the form

$$p(\mathbf{I}^{\mathrm{H}} \mid \mathbf{I}^x) = \prod_{r \in \mathcal{R}} \left\{ 2I_0\left(\frac{2(\mathbf{I}^{\mathrm{H}}(r))^2 \left|\mathbf{I}^x(r)\right|}{\left(\sigma_r^2 + 1\right)^2}\right) \exp\left(-\frac{(\mathbf{I}^{\mathrm{H}}(r))^2 + \left|\mathbf{I}^x(r)\right|^2}{\sigma_r^2 + 1}\right) \right\} \qquad (30.30)$$

where I_0 is the zeroth-order-modified Bessel function of the first kind and σ_r is the standard deviation of the distribution.

30.7.4.4 Prior Distributions

Generally, the prior function is defined as the unconditional density of the target state $\mathbf{x}$, that is, $p(\mathbf{x})$. In the context of ATR, the prior represents the *a priori* information about the parameters that constitute the vector $\mathbf{x}$. The prior density in the problem defined in Definition 2 is $p(\mathbf{t}_{l_k}, \theta_k, \alpha \mid T_{l_{k-1}}, \Theta_{k-1})$. It implies the prior knowledge had by the user on the translation p and rotation Θ that the target of class α undergoes at time instant k. The prior density can be thought of as a weight attached to the likelihood to form the posterior density, which implies, for instance, that a moving target may be present in a certain region with a higher probability as compared to some other regions. The prior is derived using dynamic models for target motion. Hence, recognition and tracking become inseparable. Miller et al. [41] provide an extensive treatment of dynamics-based priors for aircraft. As can be seen from Equation 30.25, the prior density is nothing but the joint density of the predicted position vector and rotation matrix. Hence, the prior is induced from the nonlinear equations governing the equations of motion of the vehicle, so that velocity and angular velocity can be modelled using the Gauss–Markov probabilistic model. In case there are multiple targets, Miller et al. [41] introduce a penalty term in the prior against targets occupying the same physical space at any one time. Assuming there are two targets denoted by m_1 and m_2, respectively, and the targets are parameterized by x^m where $x \in \chi$ and $m \in \{m_1, m_2\}$, the intersection penalty takes the Gibbs form

$$p\left(x\left(m_1, m_2\right)\right) = \frac{1}{Z} e^{\left(-\alpha\left(\sum_{m,m':m\neq m'} \left|V\left(x^{(m)} \cap V\left(x^{(m')}\right)\right)\right|\right)\right)}, \quad \alpha > 0 \tag{30.31}$$

where $V(x^{(m)}) \subset \Re^2$ denotes the volume occupied by a target parameterized by x^m, α denotes a design parameter in the distribution of $\mathbf{x}$, the notation $\cap$ denotes intersection and Z is the normalizing term.

30.7.4.5 Inference

Sections 30.3.4.1 through 30.3.4.4 provided the formalism for the Bayesian approach to ATR. This approach is based on pattern theory and aims to build scenes that explain the sensor output data via sensor modelling. Inference is the aspect of Bayesian ATR algorithms where the scene that best explains the observed sensor output is recreated by statistically estimating the parameters that have been designed to fully parameterize it. It can be seen from Equation 30.25 that the parameter space of typical scenes encountered in ATR is continuous as well as discrete. Furthermore, the parameter representations are high-dimensional rotation group products. Hence, the estimation of the joint continuous-discrete parameters requires an iterative optimization technique of the continuous and discrete type [41]. Continuous search accommodates the various poses, scales and positions, while the discrete search is able to search over the various classes and the number of targets. Hence, the inference is organized over *jump-diffusion* random processes containing both discrete and continuous trajectories. The jump-diffusion framework was presented in Refs. [41,64]. This algorithm is designed around a reversible jump Markov process that accounts for the continuous and discrete aspects of the ATR problem; a flowchart for this algorithm has been provided in Ref. [46]. The way the algorithm works, in a nutshell, is: the discrete components in the parameter space are handled by *jumps*, that is, by adding or deleting targets or by changing the class of a target, and the continuous space is searched using the *diffusion* process. Bayesian-based ATR can be divided into two paradigms: (a) when the parameters that characterize a scene assumed to be nonrandom, that is, do not have a prior model, and (b) when the parameters are random, that is, can be modelled using a prior distribution [65]. The inference boils down to maximum likelihood (ML) estimation in case (a), and maximum *a priori* (MAP) estimation in case (b). Brief outlines of the jump-diffusion algorithms used in both cases are given below:

a. *ML-based jump-diffusion*
    i. Initialize the algorithm with an empty scene.
    ii. Initiate *jump* by adding (birth), deleting (death) and/or metamorphing (class change) targets. Generate a number of alternate hypotheses to form candidate scenes consisting of varying number and classes of targets.
    iii. Choose the configuration or the candidate scene that maximizes the log-likelihood function such as the one given in Section 30.3.4.3.
    iv. Initiate *diffusion* on the chosen configuration *after* jump to match the continuous parameters such as pose, position and so on, given the configuration of targets selected at the jump step.
    v. Diffusion is carried out by small perturbation of the continuous parameters so as to compute the numerical derivatives of the log-likelihood function and further maximizing it using a steepest ascent approach.
b. *MAP-based jump diffusion*
    i. Initialize the algorithm with an empty scene.
    ii. Initiate *jump* by adding (birth), deleting (death) and/or metamorphing (class change) targets. Generate a number of alternate hypotheses to form candidate scenes.
    iii. The posterior probabilities are computed for each of the candidate hypotheses generated and one candidate is randomly chosen with a probability proportional to its posterior probability. This acceptance probability has its roots in the Metropolis–Hastings algorithm [66].
    iv. Initiate *diffusion* on the chosen configuration *after* jump to refine and match the continuous parameters such as pose, position, and so on, to the observation.
    v. Diffusions are accomplished using the Langevin stochastic differential equation:

$$dX_N(\tau) = \nabla_{X_N} H\left(X_N(\tau) \mid D\right)d\tau + \sqrt{2}\ dW_N(\tau) \tag{30.32}$$

where $X_N \in \chi_N$ is the composite parameter vector for an N-target scenario with fixed target classes (after the jump)

$$X_N = [x(1), \quad x(2), \quad \ldots, \quad x(N)] \tag{30.33}$$

where χ_N is the parameter space for N targets. In Equation 30.32, $W_N(\tau)$ is a Wiener process in χ_N, and $H(X_N(\tau) \mid D)$ is the log posterior where D represents the sensor data.*

30.7.4.6 Multi-Sensor Data Fusion

An often understated but powerful feature of the Bayesian-based approach to ATR is the ease with which it lends itself to multi-sensor data fusion. Sensor fusion occurs automatically in this framework [41], as no matter how many sensors are considered only one inference problem is solved. In order to demonstrate this, let a generic likelihood function be given by $L(D \mid X)$, where $D \equiv \{D_1, D_2, \ldots, D_n\}$ represents the data obtained from n different sensors (such as FLIR, LADAR, HRR, etc.), and X is the parameter that represents the input image, for instance. Assuming that the sensors are independent, the likelihood function can be written as

---

* One should note that τ is the algorithmic time and not the scenario time.

$$L(\mathcal{D} \mid X) = L\left(\{D_1, D_2, \cdots, D_n\} \mid X\right) = \prod_{i=1}^{n} L\left(D_i \mid X\right) \qquad (30.34)$$

Hence, the joint log-likelihood function reduces to the sum of n joint log-likelihood functions and can be used in the inference problem discussed in Equation 30.25

$$\log\left(L(\mathcal{D} \mid X)\right) = \sum_{i=1}^{n} \log\left(L\left(D_i \mid X\right)\right) \qquad (30.35)$$

Hence, while the multi-sensor observations provide additional information in the posterior distribution, yet only one inference problem needs to be solved.

30.7.4.7 Error Bounds

In addition to multi-sensor data fusion, another thing that the Bayesian-based ATR technique naturally lends itself to is the derivation of statistical bounds on estimation errors. Optimal estimates can be derived in classical Bayesian theory using the minimum mean-squared error (MMSE) estimator that minimizes the mean-squared error (MSE), which is a generally accepted measure of quality of an estimator. However, due to the nonflat geometry of $SO(n)$, the standard MMSE estimator requires modifications, that is, an MMSE estimator is needed that is defined on $SO(n)$. The Hilbert–Schmidt estimator (HSE) is such an estimator that is an MMSE estimator restricted to $SO(n)$. The paper by Grenander et al. [44] derives the Hilbert–Schmidt lower bound (HSLB) that is the error associated with the optimal HSE which forms a lower bound on the error associated with any estimator. The derivations of HSLBs are beyond the scope of this chapter; however, the derivations can be found in Ref. [44] for pose estimation using various commonly used sensors such as FLIR, HRR and video. Hence, just like in the case of Euclidean parameters the Cramer–Rao lower bound [65] is often used to establish the performance of estimators, the HSLB provides a similar platform to compare the performance of various algorithms in the context of ATR [40].

30.8 ARTIFICIAL INTELLIGENCE FOR ATR

This section describes an artificial intelligence-based approach to the ATR problem. In the context of this chapter, we discuss artificial intelligence-based ATR systems from the perspective of neural network (NN)-based intelligent adaptation by either augmenting an NN to an existing classical ATR system or developing a fully functional supervised learning-based NN that acts as a standalone detector, recognizer and classifier.

30.8.1 WHAT IS A NEURAL NETWORK?

To this end, we begin with the most basic question: 'What is a neural network?' To quote from Haykin [67], 'An NN is a massively parallel distributed processor made up of simple processing units, which have a natural propensity for storing experiential knowledge and making it available for use as and when required. It resembles the brain in two respects: (i) knowledge is acquired by the network from its environment through a learning process, and (ii) interneuron connection strengths, known as synaptic weights, are used to store the acquired knowledge'. The procedure used to perform the learning process is called as the *learning algorithm*, the function of which is to modify the synaptic weights of the network in an orderly fashion so as to attain the desired design objective such as functional reconstruction (to enhance conventional Bayesian prior modelling), pattern classification, target identification, to name a few. Some of the benefits that NNs offer are functional approximation (highly useful for non-linear functions), input–output mapping, adaptivity, neurobiological analogy,

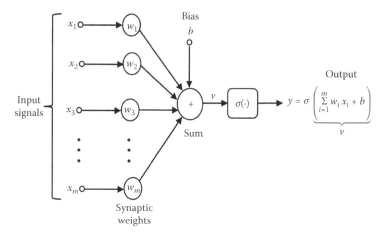

FIGURE 30.11   A typical nonlinear model of a nueron.

pattern classification, contextual information and fault tolerance (graceful degradation of NN performance in the presence of adverse operating conditions), to name a few. A typical non-linear model of a neuron is shown in Figure 30.11, where the m inputs to the NN are denoted as $(x_1, x_2, ..., x_m)$, the m-dimensional input layer NN weights are denoted as $(w_1, w_2, ..., w_m)$, the neuron activation function, which is essentially an output amplitude limiter, is denoted as $\sigma(\cdot)$ and the NN output is denoted as y and is given by

$$y = \sigma\left(\sum_{i=1}^{m} w_i x_i + b\right)$$
(30.36)

where b is the external neuron model bias and has the effect of either increasing or decreasing the input argument to the activation function σ depending on whether the bias is positive or negative. Figure 30.12 shows different models of activation functions such as thresholding function (Figure 30.12a), piecewise-linear function (Figure 30.12b) and sigmoidal function (Figure 30.12c) and hyperbolic tangent function (Figure 30.12d). Notice from Figure 30.12c, as the value of a in the sigmoid function increases, the function approaches a step function. Furthermore, note that while the hyperbolic tangent function can take on negative values as seen in Figure 30.12d, the sigmoid function never goes negative irrespective of the value of the argument and approaches 0 for very large negative values of the argument. Thus, by controlling the value of the parameter a the sigmoid function can approximate the behaviour of functions ranging from step to flat functions. Hence sigmoids are a preferred choice for the activation function in a neural network.

30.8.2   What Is an Artificial Intelligence-Based ATR System?

In the setting of automatic target detection, identification and tracking, artificial intelligence based on the adaptive capability of NNs have been proposed as a standalone, powerful tool to solving the real-world image processing problems. Over the years of understanding and implementing NNs, ATR problems have been benefited from NNs in such aspects as target recognition, class identification, feature extraction/computation, target orientation (also known as target pose) and target tracking scenarios, to name a few. To this end, NNs can be viewed as a robust solution to highly uncertain problems, which cannot be otherwise solved via classical or Bayesian methods alone. Furthermore, NNs can be viewed from the perspective of designing and implementing an ATR system that has sufficient amount of intelligence built within so as to quickly and efficiently adapt

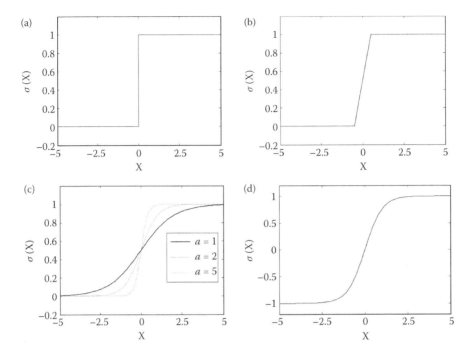

FIGURE 30.12 Different models of activation functions. (a) Treshold function, (b) piecewise-linear function, (c) Sugmoid function: $\sigma(X) = 1/(1 + e^{-ax})$ and (d) hyperbolic tangent function: $\tanh(X)$.

to either rapid changes in the environment which is being imaged or to the severe mismatch between the target models in the database of an onboard air vehicle and the real world which the air vehicle is imaging. Thus, NNs provide a number of tools which have the potential to form the basis for a computationally efficient and robust approach to the ATR problem. Neural networks offer potentially powerful techniques (computationally) for designing special-purpose hardware which can implement fast optimization for a number of computational vision and multisensor fusion problems.

In Ref. [68], NNs have been proposed from the perspective of target classification. The architecture here is called as the modular neural network classifier, which consists of several independent NNs trained on local features extracted from portions of the target image. The classification decisions from each of these independent NNs is collected together by a method called as *stacked generalization* [69,70] and the final classification is based on the outcome of this process. Stacked generalization is a general method of using a high-level model to combine lower level models to achieve greater predictive accuracy. It is a way of combining multiple models that have been designed to learn a specific classification task. The approach proposed in Ref. [68] has been tested on exhaustive data obtained from real FLIR imagery. In Ref. [71], NNs are used from the perspective of target detection and classification based on a multiresolution foveal image. Fovea is a part of the eye, located in the centre of the macula region of the retina. Here, a Hopfield-based NN is used to design an energy function (cost function), the minimization of which leads to decisions about the target identification process. In Ref. [72], the NN-based target detection problem is solved by employing linearly parameterized and non-linearly parameterized NNs. A quadratic gamma detector (QGD), which is a non-parametrically trained classifier based on local image intensity is used. A QGD is a target discrimination module for synthetic aperture radar imagery based on an extension of gamma functions to two-dimension, thereby estimating the intensity of image pixels under test and real conditions. These intensity estimates is used to create a quadratic discriminant called as the quadratic gamma discriminator, which is known to be optimal for a classification of Gaussian distributed classes. In Ref. [73], an automatic fingerprint classification system is designed to create an efficient database for fingerprint matching. A four-layered NN that performs feature extraction given two-dimensional data is designed. Each category of the fingerprint identification scheme is powered by an individual NN. The NNs are trained by supervised learning via a back propagation algorithm with a two-step training method.

In the realm of target segmentation, NNs have been vastly implemented as reported in Refs. [74–76]. NN-based segmentation relies on processing small areas of an image by either using a single NN or a swarm of NNs. Once processing is complete, markings are made corresponding to the areas of the image that are identified by the NN or NNs. A type of network designed especially for this is the Kohonen map. In Ref. [76], a pulse-coupled neural network (PCNN) is proposed and developed for high-performance biometric image processing. A PCNN, a two-dimensional NN, is very useful for image processing technique. Each neuron in the network corresponds to a pixel in an image, receiving its corresponding pixel's colour information (e.g., intensity) as an external stimulus. Furthermore, each neuron connects with its neighbouring neurons, receiving local inputs from them, as is standard with any NN-based detection, segmentation scheme. The external and local inputs are combined in an internal activation system, resulting in a pulse output. Through iterative computation, PCNN neurons produce temporal series of pulse outputs, which contain information of input images and can be utilized for various image processing applications, such as image segmentation and feature generation. Compared with conventional image processing means, PCNNs have several significant advantages such as robustness against noise, independence of geometric variations in input patterns (similar to Zernike or Hu moments), the capability of bridging minor intensity variations in input patterns, to name a few. In 1989, Eckhorn introduced a neural model to emulate the mechanism of the visual cortex of a cat. This model proved to be an effective tool in analysing such visual cortex images and was soon found to have a significant application potential in image processing. In 1994, the Eckhorn model was adapted to be an image processing algorithm by Johnson, who termed this algorithm PCNN. The approach in Ref. [77] proposes an NN-based automatic segmentation and classification method for outdoor images, by segmenting the captured images using self-organizing feature maps (SOFM) based on texture and colour information of the targets. Features such as average colour, position, size, rotation, texture (Gabor filter—named after Dennis Gabor, is a linear filter used for edge detection. Frequency and orientation representations of Gabor filter are similar to those of human visual system, and it has been found to be particularly appropriate for texture representation and discrimination) and shape are extracted from each segmented region. Classification is then performed using a Multi Layer Perceptron. In feature detection, hidden neurons play a pivotal role as they assume the role of the feature detectors themselves [67]. As the learning process progresses and as training data becomes available, these hidden neurons begin to adapt to the hitherto unknown features, thereby piecing together information, constructively, that characterize the training data. This is accomplished by a non-linear transformation applied on the input space or the data space that takes the input space into a new space called the hidden space. The output of the NN is so calculated as to minimize, with respect to the synaptic weights of the NN, the difference between the value of this output and the desired target output pattern or response. This minimization is similar, in principle, to maximizing a discriminant function, which is a function of the trace of the product of two matrices, both related to the inputs to the NN and the desired output response after these are de-biased by removing the respective means. A discriminant-based approach estimates the 'decision boundaries' directly bypassing the intermediate stage of conditional density estimation. The decision boundaries in the measurement or feature space are based on discriminants which are either linear or non-linear functions of measurements or features. In Ref. [78], an approach that augments an ATR system with an NN for the purpose of visual pattern recognition in a hybrid model is proposed. This NN forms the feature extraction front end of the ATR system and is derived from the Neocognitron network first proposed in Refs. [79,80]. For complex target recognition, modifications to the basic Neocognitron network paradigm were required to enhance robustness against image distortions due to under sampling (aliasing) and poor feature selection during training. This was addressed in Ref. [78], where the focus is on the enhancements and also on the use of the network as a self-organizing feature extraction element of an ATR system. The Neocognitron (inspired from the model proposed by Hubel and Wiesel in 1959) is a hierarchical multi-layered NN proposed by Fukushima [79]. It has been used for handwritten character recognition and other pattern recognition tasks.

In summary, the augmentation of NNs to ATR-based problems has the advantage of bringing in massively parallel computation hardware and improved programmability by using powerful models and learning algorithms, such as the back propagation technique [67]. Given that the basic processing elements of an NN structure are relatively simple, less cumbersome hardware can be used to design the building blocks of the NN structure [81]. Furthermore, since NNs are robust universal functional approximators, particularly useful for highly nonlinear functions, and added to that if an appropriate database exists of input and output examples, NN learning can be used to calculate the connection synaptic weights so that the network can approximate the function. Such learning can be used for automatic ATR knowledge acquisition and system refinement [81]. Ultimately, NNs could be used to address the ATR needs for efficient and intelligent adaptation to target and environment changes, selection of good target features and integration of *a priori* knowledge about target signatures, backgrounds, orientation and position into the ATR approach.

30.9   CONCLUDING REMARKS

To summarize, in this chapter we have presented an overview of three different philosophies to solving the ATR problem. These philosophies are based on classical techniques, Bayesian-based methodologies and neural network-based approaches. Classical-based ATR approaches perform target recognition and detection via the five stages (strictly uni directional) which are pre-processing, detection, segmentation, feature extraction and tracking. Hence, any error(s) in the initial stages has an adverse cascading effect on the subsequent classical ATR stages. Furthermore, classical-based ATR approaches provide accurate results so long as the features from the image of the target lie in easily classifiable regions. Hence, the results of such approaches are more likely to be influenced by the presence of occlusions and clutter that are not successfully removed via the pre-processing stage. Bayesian-based approaches unify the various stages in a classical-based ATR approach by recasting the classical ATR problem as a predictor–corrector or a parameter estimation problem. The prediction and correction steps in a Bayesian approach allow the predicted estimate of an unknown parameter to be corrected via a known set of scene measurements, thereby increasing the probability of accurately describing the desired target(s). Furthermore, Bayesian approaches allow for model and/or problem-specific parameterization of the real scene in terms of efficient sensor models (including the effects of sensor noise) or robust prior models by accounting for the target dynamics. For instance, for ATR of tanks, the recognition of the direction in which the turret is pointed might be valuable, as in the case of a battlefield scenario illustrated in Figure 30.6. In such cases, a Bayesian-based ATR problem can formulate the parameter vector in such a way that an additional parameter that belongs to the $SO(2)$ group (see Section 30.3.2) can be added alongside the $SO(2)$ parameter that already exists for the pose estimation of the main body of the tank itself. Hence, the rotation of the turret can be modelled as independent of the rotation of the main body of the tank, as illustrated in Figure 30.5. Being able to detect the direction in which the tank's turret is pointed is essential the functioning of a reliable ATR system [82] in the context of battlefield scenarios. Neural network-based ATR approaches unlike classical ATR, rely on the availability of an efficient set of training data. These data are required to learn the ever-changing environment in order to adapt to it. However, this training set needs to be exhaustive in order to account for the vagaries of the environment in which it operates. This requirement can serve to hinder the usability of neural network approaches when an extensive training data set is unavailable. The above-mentioned approaches work well in the scenarios and requirements that they were built to address. For example, if the scenario under consideration is simple with less clutter and the requirement is only from the perspective of classification, one would choose a classical-based ATR approach. However, if one of the key requirements is estimating the orientation (pose) [82] of a target(s) and tracking it and further in cases where sensor models are well understood and available, a Bayesian-based approach is a more prudent choice for the designer. Finally, if the target(s) and its surrounding environment undergo random and unpredictable changes which are tough to model purely via mathematical equations, neural network-based approaches can be used for efficient and intelligent adaptation of these changes. However, realistic scenarios are more often than not a combination of the above-mentioned scenarios. Thus, no single approach is likely to solve this combination, rather a mix and match of these approaches to form a hybrid approach is the key to successful target detection, recognition and tracking. One such hybrid approach could be augmenting a neural network to a Bayesian-based approach in order to account for the prior modelling errors or learning the statistics of a given sensor model. Furthermore, enhancing the prior Bayesian models to include contextual information, and enhancing likelihood models to incorporate clutter models (see Ref. [40]) could render Bayesian approaches more intelligent/adaptable to varying terrains.

REFERENCES

1. A. C. Bovik, *The Handbook of Image and Video Processing*, Second Edition, New York, New York: Academic Press, pp. 1341–1353, June 2005.
2. B. Bhanu, Automatic target recognition: State of the art survey, *IEEE Transactions on Aerospace and Electronic Systems*, 22(4), 364–378, July 1986.
3. D. E. Dudgeon and R. T. Lacoss, An overview of automatic target recognition, *The Lincoln Laboratory Journal*, 6(1), 3–10, 1993.
4. K. Copsey, Bayesian approaches for robust automatic target recognition, PhD Thesis, University of London, 2004.
5. J. Miller, *Principles of Infrared Technology*, Van Nostrand, Reinhold, 1992.
6. E. Friedman and J. L. Miller, *Photonics Rules of Thumb: Optics, Electro-Optics, Fiber Optics, and Lasers*, 2nd ed., McGraw-Hill, New York, 2004.
7. A. D. Lanterman, M. I. Miller and D. L. Snyder, Representations of thermodynamic variability in the automated understanding of FLIR scenes, in *Automatic Object Recognition VI, Proceedings of the SPIE*, Vol. 2756, Ed. Firooz A. Sadjadi, pp. 26–37, April 1996.
8. L. J. Cutrona, Synthetic aperture radar, *Radar Handbook*, 2nd ed., M. Skolnik, ed., McGraw-Hill, New York, 1990.
9. E. N. Leith, A short history of the Optics Group of the Willow Run Laboratories, in *Trends in Optics: Research, Development, and Applications*, Anna Consortini, Academic Press, San Diego, CA, 1996.
10. E. C. Botha, Classification of aerospace targets using super resolution ISAR images, in *the Proceedings of IEEE COMSIG*, pp. 138–145, 1994.
11. J. Li, Inverse synthetic aperture radar imaging, *Technical Report*, Electrical and Computer Engineering, University of Texas at Austin, 1998.
12. B. Borden, Some issues in inverse synthetic aperture radar image reconstruction, *Inverse Problems*, 13, 571–584, June 1997.
13. H. Wu, D. Grenier, G. Y. Delisle and D. G. Fang, Translational motion compensation in ISAR image processing, *IEEE Transactions on Image Processing*, 4, 1561–1570, November 1995.
14. M. Soumekh, A system model and inversion for synthetic aperture radar imaging, *IEEE Transactions on Image Processing*, 1, 64–76, January 1992.
15. T. D. Wilkerson, G. K. Schwemmer and B. M. Gentry, LIDAR profiling of aerosols, clouds, and winds by Doppler and non-Doppler methods, *NASA International H_2O Project*, 2002.
16. T. J. Green, Jr. and J. H. Shapiro, Detecting objects in 3D laser radar range image, *Optical Engineering*, 33, 865–873, March 1994.
17. J. K. Kostakis, M. Cooper, T. J. Green Jr., M. I. Miller, J. A. O'Sullivan, J.H. Shapiro and D. L. Snyder, Multispectral sensor fusion for ground-based target orientation, in *Automatic Target Recognition IX, Proceedings of the SPIE*, Vol. 3718, pp. 14–24, Orlando, FL, April 1999.
18. N. C. Currie and C. E. Brown, *Principles and Applications of Millimeter Wave Radar*, Artech House, Boston, MA, 1988.

19. O. Yildirim, Millimeter wave RADAR design considerations, *Istanbul University—Journal of Electrical & Electronics Engineering*, 3(2), 983–986, 2003.

20. E.E. Clothiaux, T.P. Ackerman and D.M. Babb. 1996. Ground-based remote sensing of cloud properties using millimeter-wave radar. In: *Remote Sensing of Processes Governing Energy and Water Cycles in the Climate System. NATO International Scientific Exchange Programmes*, Advanced Study Institute, Plön, Germany.

21. H. Hough, Satellite surveillance, *Loompanics Unlimited*, ISBN 1-55950-077-8, 1991.

22. J. Ellis, Searching for oil seeps and oil-impacted soil with hyperspectral imagery, *Earth Observation Magazine*, January 2001.

23. J.L. Grossman, Thermal infrared vs. active infrared: A new technology begins to be commercialized, http://www.irinfo.org/articles/03_01_2007_grossman.html

24. W. S. Boyle and G. E. Smith, Charge coupled semiconductor devices, *Bell Systems Tech. Journal*, 49(4), 587–593, April 1970.

25. J. R. Janesick, T. Elliott, S. Collins, M. M. Blouke and J. Freeman, Scientific charge-coupled devices, *Optical Engineering*, 26, 692–714, 1987.

26. A. R. Webb, *Statistical Pattern Recognition*, John Wiley & Sons, Chichester, 2nd edition, August 2002.

27. P. M. Narendra, A separable median filter for image noise smoothing, *IEEE Transactions on Pattern Analysis and Machine Intelligence*, *PAMI* 3(1), 20–29, January 1981.

28. M. Burton and C. Benning, Comparison of imaging infrared detection algorithms: Infrared technology for target detection and classification, *Proceedings of the SPIE*, 302, 26–32, 1981.

29. B. J. Schachter, A survey and evaluation of FLIR target detection/segmentation algorithms, in *Proceedings of DARPA Image Understanding Workshop*, pp. 49–57, Palo-Alto, CA, September 1982.

30. A. S. Politopoulos, An algorithm for the extraction of target-like objects in cluttered FLIR imagery, *IEEE Transactions on Aerospace and Electronic Systems Society Newsletter*, 23–37, November 1980.

31. W. B. Lacina and W. Q. Nicholson, Passive determination of three dimensional form from dynamic imagery, *Proceedings of the SPIE*, 186, 178–189, May 1979.

32. D. L. Milgram and A. Rosenfeld, Algorithms and hardware technology for image recognition. Final report to U. S. Army Night Vision and Electro-Optics Lab., Fort Belvoir, VA, March 1978.

33. M. K. Hu, Visual pattern recognition by moment invariants, *IRE Transactions on Information Theory*, IT-8, 179–187, 1962.

34. A. Khotanzad and Y. Hong, Invariant image recognition by Zernike moments, *IEEE Transactions on Pattern Analysis and Machine Intelligence*, 12(5), 489–497, 1990.

35. B. Bhanu, A. S. Politopoulos and B. A. Parvin, Intelligent autocueing of tactical targets in FLIR images, in *Proceedings of the IEEE Conference on CVPR*, pp. 502–503, Arlington, VA, USA, 1983.

36. D.E. Soland and P.M. Narendra, Prototype automatic target screener: Smart sensors, in *Proceedings of the SPIE*, 178, 175–184, 1979.

37. B. A. Parvin, A structural classifier for ship targets, in *Proceedings of the 7th Conference on Pattern Recognition*, pp. 550–552, Montreal, Canada,1984.

38. P. Héroux, S. Diana, E. Trupin and Y. Lecourtier, A structural classifier to automatically identify form classes, *Advances in Pattern Recognition Lecture Notes in Computer Science*, 1451/1998, 429–436, 1998.

39. D. B. Hall, S. C. MacInnis, J. Dickerson and J. Hare, Target prioritization in TEM surveys for sub-surface Uxo investigations using response amplitude, decay curve slope, signal to noise ratio, and spatial match filtering, http://www. zonge.com/PDF_Papers/ UXO_ TEM_Target.pdf, accessed July 2011.

40. A. Srivastava, M. I. Miller and U. Grenander, Statistical models of targets and clutter for use in Bayesian object recognition, in *The Handbook of Image and Video Processing*, 2nd edition, A. C. Bovik, Eds., New York, New York: Academic Press, pp. 1341–1353, June 2005.

41. M. I. Miller, U. Grenander, J. A. O'Sullivan and D. L. Snyder, Automatic target recognition organized via jump-diffusion algorithms, *IEEE Transactions on Image Processing*, 6(1), 157–174, January 1997.

42. Dassault Systèmes (3DVIA), http://www.3dvia.com

43. A.W. Knapp, *Lie Groups Beyond an Introduction, Progress in Mathematics*, 2nd edn., Birkhäuser, Boston, vol., 140, 2002, http://en.wikipedia.org/wiki/Lie group, accessed July 2011.

44. U. Grenander, M. I. Miller and A. Srivastava, Hilbert-Schmidt lower bounds for estimators on matrix Lie groups for ATR, *IEEE Transactions on Pattern Analysis and Machine Intelligence*, 20(8), 790–802, August 1998.

45. *Prism 3.1 User's Manual*. Keweenaw Research Center, Michigan Technological University, 1987.

46. J. H. Dixon and A. D. Lanterman, Toward practical pattern-theoretical ATR algorithms for infrared imagery, in *Proceedings of the SPIE, Automatic Target Recognition XVI*, ed. F.A. Sadjadi, Vol. 6234, pp. 62340R-1-62340R-9, 2006.

47. D. L. Snyder, A. M. Hammoud and R. L. White, Image recovery from data acquired with a charge-coupled-device camera, *Journal of the Optical Society of America*, 10(5), 1014–1023, May 1993.

48. M. J. Smith, Bayesian sensor fusion: A framework for using multi-modal sensors to estimate target locations and identities in a battlefield scene, *Doctoral dissertation, Florida State University*, Tallahassee, FL, 2003.

49. S. P. Jacobs and J. A. O'Sullivan, Automatic target recognition using sequences of high resolution radar range-profiles, *IEEE Transactions on Aerospace and Electronic Systems*, 36(2), 364–382, April 2000.

50. S. M. Hannon and J. H. Shapiro, Active-passive detection of multipixel targets, in *Laser Radar V*, in *Proceedings of the SPIE*, Ed. R.J. Becherer, Vol. 1222, pp. 2–23, May 1990.

51. T. J. Green, Jr. and J. H. Shapiro, Detecting objects in 3d laser radar range image, *Optical Engineering*, 33, 865–873, March 1994.

52. A. Lanterman, M. Miller, D. Snyder and W. Miceli, The unification of detection, tracking and recognition for millimeter wave and infrared sensors, in *RADAR/LADAR Processing, Proceedings of the SPIE*, Ed. W.J. Miceli, Vol. 2562, pp. 150–161, San Diego, CA, August 1995.

53. G. Gao, Statistical modelling of SAR images: A survey, *Sensors*, 10, 775–795, 2010.

54. C. J. Oliver, *Understanding Synthetic Aperture Radar Images*, Boston: Artech House, 1998.

55. G. Moser, J. Zerubia and S. B. Serpico, SAR amplitude probability density function estimation based on a generalized Gaussian model, *IEEE Transactions on Image Processing*, 15, 1429–1442, 2006.

56. G. Moser, J. Zerubia and S. B. Serpico, Dictionary-based stochastic expectation–maximization for SAR amplitude probability density function estimation, *IEEE Transactions on Geoscience and Remote Sensing*, 44(1), 188–200, 2006.

57. H. Arsenault and G. April, Properties of speckle integrated with a finite aperture and logarithmically transformed, *Journal of the Optical Society of America*, 66, 1160–1163, 1976.

58. K. D. Ward, Compound representation of high resolution sea clutter, *Electronics Letters*, 7, 561–565, 1981.

59. T. R. Bayes, Essay towards solving a problem in the doctrine of chances, *Philos. Trans. Roy. Soc. Lond.*, 53, 370–418, 1763. Reprinted in Biometrika, 45, 1958.

60. U. Grenander, *General Pattern Theory*, New York: Oxford University Press, 1993.

61. U. Grenander and M. I. Miller, Representations of knowledge in complex scenes, *Journal of the Royal Statistical Society B*, 56(3), 549–603, 1994.

62. D. Mumford, Pattern theory: A unifying perspective, in *Proceedings of the 1st European Congress of Mathematics*, Birkhauser, Germany, 1994.

63. A.E. Koskal, J.H. Shapiro and M.I. Miller, Performance analysis for ground-based target orientation estimation: FLIR/LADAR sensor fusion, in *Conference Record of the Thirty-Third Asilomar Conference on Signals, Systems, and Computers*, Vol. 2, pp. 1240–1244, Pacific Grove, CA, 1999.

64. A. Srivastava, U. Grenander, G.R. Jensen and M.I. Miller, Jump-diffusion Markov processes on orthogonal groups for object pose estimation, *Journal of Statistical Planning and Inference*, 103(1–2), 15–27, 2002.

65. Y. Bar-Shalom, X. Li and T. Kirubarajan, *Estimation with Applications to Tracking and Navigation*, Wiley-Interscience, New York, 2001.

66. A. Lanterman, M. Miller and D. L. Snyder, General Metropolis-Hastings jump-diffusion for automatic target recognition in infrared scenes, *Optical Engineering*, 36(4), 1123–1137, 1997.

67. S. Haykin, Neural *Networks: A Comprehensive Foundation*, 2nd edn. Pearson Education, Delhi, India, 2001.

68. L. C. Wang, S. Z. Der and N. M. Nasrabadi, Automatic target recognition using a feature decomposition and data decomposition modular neural network, *Image Processing: Special Issue on Applications of Artificial Neural Networks to Image Processing*, 7(8), 1113–1121, 1998.

69. D. H. Wolpert, Stacked generalization, *Neural Networks*, 5, 241–259, Pergamon Press, 1992.

70. L. Breiman, Stacked regressions, *Machine Learning*, 24, 49–64, 1996.

71. S. S. Young, P. D. Scott and C. Bandera, Foveal automatic target recognition using a multiresolution neural network, *Image Processing: Special Issue on Applications of Artificial Neural Networks to Image Processing*, 7(8), 1122–1135, 1998.

72. J. C. Principe, M. Kim and J. W. Fisher, III, Target discrimination in synthetic aperture radar using artificial neural networks, *Image Processing: Special Issue on Applications of Artificial Neural Networks to Image Processing*, 7(8), 1122–1135, 1998.

73. M. Kamijo, Classifying fingerprint images using neural network: Deriving the classification state, in *Proceedings of the International Conference on Neural Network*, Vol. 3, pp. 1932–1937, San Francisco, CA, USA, April 1993.

74. M. Pathegama and Ö Göl, Edge-end pixel extraction for edge-based image segmentation, *Transactions on Engineering, Computing and Technology*, 2, 213–216, 2004.

75. J.M. Kinser, K. Waldemark, T. Lindblad, and S.P. Jacobsson. Multidimensional pulse image processing of chemical structure data, *Chemometrics and Intelligent Laboratory Systems*, 51(1), 115–124, May 2000.

76. T. Lindblad and J. M. Kinser, *Image Processing Using Pulse-Coupled Neural Networks*, Second revised edition, Springer, Berlin, Heidelberg, New York, 2005.

77. N. W. Campbell, B. T. Thomas and T. Troscianko, Automatic segmentation and classification of outdoor images using neural networks, *International Journal of Neural Systems*, 8(1), 137–144, 1997.

78. J. G. Landowski and B. Gil, Application of a vision neural network in an automatic target recognition system, in *the Proceedings of the SPIE*, Vol. 1709, pp. 34–43, *Applications of Artificial Neural Networks III*, S. K. Rogers, Ed., Washington, 1992.

79. K. Fukushima, Neocognitron: A self-organizing neural network model for a mechanism of pattern recognition unaffected by shift in position, *Biological Cybernetics*, 36(4), 93–202, 1980.

80. K. Fukushima, S. Miyake and T. Ito, Neocognitron: A neural network model for a mechanism of visual pattern recognition, *IEEE Transactions on Systems, Man, and Cybernetics*, SMC-13(Nb. 3), 826–834, 1983.

81. M. W. Roth, Survey of neural network technology for automatic target recognition, *IEEE Transactions on Neural Networks*, 1(1), 28–43, 1990.

82. A. Srivastava, Bayesian filtering for tracking pose and location of rigid targets, in *the Proceedings of the SPIE Signal Processing, Sensor Fusion, and Target Recognition, IX*, I. Kadar Ed., Vol. 4052, pp. 160–171, Orlando, FL, April 2000.

83. D. Halliday, R. Resnick and J. Walker, *Fundamentals of Physics*, 7th ed., John Wiley & Sons, Hoboken, NJ, 2005.

31 Real-Time Implementation of a Novel Fault Detection and Accommodation Algorithm for an Air-Breathing Combustion System

Rahee Walambe, Nitin K. Gupta, Niteen Bhange,
N. Ananthkrishnan, Ik Soo Park, Jong Ho Choi and
Hyun Gull Yoon

CONTENTS

31.1 INTRODUCTION

Failure of sensors used to provide a feedback signal in a control system can cause serious deterioration in performance of the system, and even instability may be observed. Based on knowledge of aircraft engine systems, fault in such Air-breathing Combustion Systems (ACS) with no rotating parts is primarily due to the pressure sensors. Fast online detection and accommodation of faults before the error becomes very large is critical to the success of the mission. However, at the same time, it is necessary to avoid false alarms. Hence, early detection of small-magnitude faults with acceptable reliability is very challenging, especially in the presence of sensor noise, unknown engine-to-engine variation and deterioration and modelling uncertainty. This chapter discusses the novel fault detection and accommodation (FDA) algorithm based on the analytical redundancy-based technique for ACS. This chapter is an extension of Ref. [1].

A controller for an ACS was designed initially [2]. The controller's main objective is to regulate the thrust so that the desired acceleration is obtained at all flight conditions while maintaining supercritical intake operation. The performance of the controller is tested by simulating a nominal flight trajectory involving an accelerated climb from 2.1 Mach at 1.4 km altitude to 3.0 Mach at 14.5 km altitude, followed by cruise at that condition. A unique feature of this controller is that it only requires measurement of a single variable internal to the engine, that is, the intake backpressure. Accurate, reliable and fail-safe measurement of backpressure is a key to this controller design. An FDA algorithm is required for this purpose. The FDA algorithm reported here employs an 'intelligent' analytical redundancy-based FDA algorithm working over triplex redundant backpressure sensor hardware.

31.2 REVIEW AND BACKGROUND

For the purpose of simulation, the combustion system is modelled as three sub-systems: intake, combustor and nozzle (a separate fuel supply system is also modelled). These sub-systems are linked to produce a global model that correctly represents the physics of the combustion system. A detailed description of the model development and implementation is provided in Refs. [3,4].

31.2.1 DESCRIPTION OF MODEL AND CONTROLLER

A schematic of the ACS with station numbers marked is shown in Figure 31.1. Only intake backpressure, P_4, is assumed to be measurable. The controller design of the ACS is shown in Figure 31.2 (the upper section consisting of static map, PID, plant etc.). Separate PID controllers for fuel flow rate and throat area, with P_{4_margin} as the commanded variable, are designed at several operating points on the acceleration and cruise segments of the flight. A separate PID controller is designed for the fuel supply system. Various components of the controller are individually tested extensively. Finally, a composite closed-loop simulation is successfully carried out taking the system through the acceleration phase to the desired cruise condition with a smooth switching between the acceleration and cruise segments.

The next stage of the research consisted of implementation of an appropriate FDA algorithm.

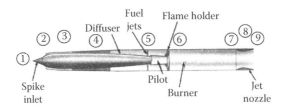

FIGURE 31.1 Air-breathing combustion system. 1: Spike. 2: start of cowl. 3: intake throat. 4: intake exit. 5: fuel injection. 6: ignition point. 7: combustor exit. 8: exhaust nozzle throat. 9: exhaust exit. (From Walambe, R.A. et al., *Defence Science Journal Special Issue—Mobile Intelligent Autonomous Systems*, 60(1), 61–75, 2010. With permission.)

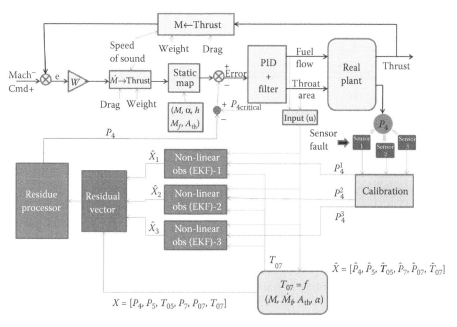

FIGURE 31.2 P_4 sensor—Model-based FDA algorithm scheme. (From Walambe, R.A. et al., *Defence Sciences Journal Special Issue—Mobile Intelligent Autonomous Systems*, 60(1), 61–75, 2010. With permission.)

31.2.2 SURVEY OF ANALYTICAL REDUNDANCY

Numerous approaches to FDA in dynamical systems have been reported in the literature. Generally, FDA techniques are classified into three categories—those based on, (i) hardware redundancy, (ii) analytical redundancy and (iii) knowledge-based redundancy. Although for the current work, we concentrate on analytical redundancy management, it is important to note that analytical redundancy is not used as a substitute for hardware redundancy; instead, we use it to provide a measure of 'intelligence' to an FDA algorithm that uses triplex hardware redundancy for the P_4 sensors.

The field of model-based FDA for linear systems is well studied [5–11]. Traditionally, analytical redundancy-based methods [12] have been used to provide an indirect measurement of the variable of interest. However, when variables measured by different sensors are related by physical equations, the principle of analytical redundancy can also be used as a diagnostic tool to test whether the sensor outputs satisfy these known relationships. A set of sensors is healthy if an equation which relates them is verified, and (at least) one sensor of the set fails if the relationship is violated [13–15].

Analytical redundancy is an attractive option for high-performance aerospace vehicles that need to operate safely, reliably and with a high-level of mission success, even in the presence of faults. Weight and volume considerations often make it difficult to justify redundant sensors and actuators in aerospace applications. Analytical redundancy management holds the promise of reduction in price, weight and power consumption when implemented on board. For air-breathing engines, multiple redundancies are harder to achieve due to lack of operating space, cost, engineering complexity and added maintenance requirements. This motivates us to propose the use of an analytical redundancy algorithm as a higher level functionality, on top of the hardware redundancy, to provide 'intelligence' to the FDA algorithm and also improve its reliability.

31.2.3   FDA FOR BACKPRESSURE SENSOR

First step in development of an observer-based FDA algorithm is a realistic representation of the physical system, which includes system dynamics, faults and all kinds of possible unknown inputs. Residual generation using observer-based method is presented here.

Figure 31.2 (lower part) shows model-based (analytical-redundancy-based) FDA for P_4 sensors. Triplex redundant static pressure sensor by using Kulite pressure sensor (located at axial stations between 4 and 5 in intake duct/combustor entry, well before fuel injection station 5) can be seen. Calibration is necessary as pressure falls in duct due to boundary layer growth. Modelling of various possible sensor faults and noise is included, and calibration for the sensor location is done based on CFD data.

The intake backpressure, P_4 value is measured by a triplex system. These three estimates are termed as $\hat{P}_4^1, \hat{P}_4^2, \hat{P}_4^3$. These measured P_4 values may not be accurate owing to various sensor faults. To detect, isolate and accommodate the faults (if any), analytical redundancy-based FDA algorithm is implemented. The P_4 measurement from each sensor is fed to separate extended Kalman filter (EKF) (forming an EKF bank). The EKF removes the disturbances from the measurements and output of EKF consists of the 'clean' P_4 estimates along with P_5, T_{05}, P_7, T_{07}, P_{07}, $\dot{m}_5$ and $\dot{m}_7$. The EKF bank then generates the residual vector which is used by the residual processor for determination of accurate value of P_4 to be fed back to the system. Review of the EKF theory and design as implemented for ACS is presented in next section. To accommodate the faults, adaptive weightage-based fault accommodation algorithm is developed which is discussed in the later sections. The underlining assumption to this development is that not more than one sensor fails at any given time.

31.3   EKF: DESIGN AND IMPLEMENTATION

For estimating correct system states and filtering the noise and disturbances present in the P_4 measurements, the EKF [16] is applied. To estimate the states, EKF needs two independent measurements. In this case, one measurement is obtained by the actual P_4 sensor and the second independent measurement is obtained from the modelled T_{07} (discussed in Section 31.4). These two measurements are fed to the KF. Thus, there are three KF in all, one for each of the three backpressure sensors. The continuous time version of KF is used. Since the system is non-linear, the EKF which uses the (linearized) Jacobian matrices at specific operating conditions along the trajectory is implemented. The Jacobian matrices are obtained by the small perturbation theory.

31.3.1   KF IN DISCRETE TIME

The discrete time Kalman filter (DTKF) model assumes the true state at time k is evolved from the state at $(k-1)$ according to [17]

$$X_k = F_k X_{k-1} + B_k U_k + w_k \tag{31.1}$$

where
   F_k is the state transition model which is applied to the previous state x_{k-1};
   B_k is the control-input model which is applied to the control vector u_k;
   w_k is the unknown process noise which is assumed to be drawn from a zero mean multivariate normal distribution with process noise covariance Q_k.

$$w_k \approx N(0, Q_k) \tag{31.2}$$

w_k acts as a random disturbance to the plant. It represents effects of unmodelled high-frequency plant dynamics that is modelled as zero-mean, white Gaussian noise. At time k, an observation (or measurement) z_k of the true state x_k is made according to

$$Z_k = H_k X_k + V_k \tag{31.3}$$

where H_k is the non-linear observation model which maps the true state space into the observed space and v_k is the observation noise which is assumed to be zero mean Gaussian white noise with covariance R_k.

$$v_k \approx N(0, R_k) \tag{31.4}$$

The initial state and the noise vectors at each step $\{x_0, w_1, \ldots, w_k, v_1 \ldots v_k\}$ are all assumed to be mutually independent.

In KF, only the estimated state from the previous time step and the current measurement are needed for computing the estimate for the current state. No history of observations and/or estimates is required. The state of the filter is represented by two variables:

$\hat{X}_{k|k}$, the estimate of the state at time k given observations up to and including time k;
$P_{k|k}$, the error covariance matrix (a measure of the estimated accuracy of the state estimate).

The discrete KF has two distinct phases: *Predict* and *Update*. The predict phase uses the state estimate from the previous time step to produce an estimate of the state at the current time step. In the update phase, measurement information at the current time step is used to refine this prediction to arrive at a new, (hopefully) more accurate state estimate, again for the current time step.

31.3.2   DTKF State Estimation

Optimal state estimate $\hat{x}$ and the state covariance matrix P are propagated from measurement time $(k-1)$ to measurement time (k), based on the previous value, the system dynamics and the previous control input and error of actual system. This is done by numerical integration of the following equations:

Predicted state and estimate covariance:

$$\hat{X}_{k|k-1} = F_k \hat{X}_{k-1|k-1} + B_{k-1} u_{k-1}$$

$$P_{k|k-1} = F_k P_{k-1|k-1} F_k^{\mathrm{T}} + Q_{k-1} \tag{31.5}$$

where F_k is the state transition model which is applied to the previous state X_{k-1}.

Data update step:

Innovation or measurement residual

$$\tilde{Y}_k = Z_k - H_k(\hat{X}_{k|k-1})$$

Innovation (or residual) covariance

$$S_k = H_k P_{k|k-1} H_k^{\mathrm{T}} + R_k$$

Optimal Kalman gain

$$K_k = P_{k|k-1} H_k^{\mathrm{T}} S_k^{-1} \tag{31.6}$$

Updated state estimate

$$\hat{X}_{k|k} = \hat{X}_{k|k-1} + K_k \tilde{Y}_k$$

Updated estimate covariance

$$P_{k|k} = (I - K_k H_k) P_{k|k-1}$$

Values of F, H, Q and R matrices are application dependent.

The next section discusses the KF applied to the continuous time systems. The equations are obtained from the original KF equations for discrete KF.

31.3.3 KALMAN–BUCY CONTINUOUS TIME FILTER

The CTKBF [18,19] is a continuous time version of the KF. It is based on the state-space model:

$$\frac{d}{dt} x(t) = F(t)x(t) + w(t)$$
$$Z(t) = H(t)x(t) + v(t)$$
(31.7)

where the covariances of the noise terms $w(t)$ and $v(t)$ are given by $Q(t)$ and $R(t)$ (in fact these are called spectral densities), respectively. The filter consists of two differential equations, one for the state estimate and the other for the state-covariance matrix:

$$\frac{d}{dt} \hat{x}(t) = F(t)\hat{x}(t) + K(t)(z(t) - H(t)(\hat{x}(t)$$

$$\frac{d}{dt} P(t) = F(t)P(t) + P(t)F^{\mathrm{T}}(t) + Q(t) - K(t)R(t)K^{\mathrm{T}}(t)$$
(31.8)

where the Kalman gain is given by

$$K(t) = P(t)H(t)R^{-1}(t)$$
(31.9)

Note that in this expression for $K(t)$ the covariance of the observation noise $R(t)$ (again it is a spectral density matrix) represents at the same time the covariance of the prediction error (or *innovation*) $\tilde{y}(t) = z(t) - H(t)(\hat{x}(t))$; these covariances are equal only in the case of continuous time. The distinction between the prediction and update steps of discrete-time Kalman filtering does not exist in continuous time. The second differential equation, for the state-covariance matrix is an example of a Riccati equation.

31.4 EKF FOR NON-LINEAR SYSTEMS

The basic KF is limited to a linear assumption. However, most non-trivial systems are non-linear. The non-linearity can be associated either with the process model or with the observation model or with both. A KF that linearizes about the current mean and covariance is referred to as an *extended Kalman filter* or EKF. EKF linearizes all non-linear models so that the traditional linear KF can be applied. An EKF design has to be carried out to estimate states, while it should filter sensor noise and model measurement parameters.

31.4.1 FORMULATION OF EKF

In the EKF, the state transition and observation models need not be linear functions of the state but may instead be (differentiable) functions.

For discrete systems:

$$X_k = f(X_{k-1}, u_k) + w_k$$
$$Z_k = h(X_k) + v_k$$
(31.10)

For continuous systems:

$$\frac{d}{dt} x(t) = f(x(t), u(t)) + w(t)$$
$$z(t) = h(x(t)) + v(t)$$

(31.11)

The function, f, can be used to compute the predicted state from the previous estimate. Similarly the function, h, can be used to compute the predicted measurement from the predicted state. However, f and h cannot be applied to the covariance directly. Instead a matrix of partial derivatives (the Jacobian) is computed. At each time step the Jacobian is evaluated with current predicted states. These matrices can be used in the KF equations. This process essentially linearizes the non-linear function around the current estimate.

The ACS system under consideration here is a non-linear system as can be seen from the dynamics equations presented in the following section. Hence, the EKF is implemented to estimate the states of the system.

31.4.2   ACS DYNAMICS

The dynamics of the system is represented by the relationships shown next:

$$\dot{\hat{P}}_4 = \frac{1}{\tau_{54}}(\hat{P}_{4_{ss}} - \hat{P}_4)$$

$$\dot{\hat{m}}_5 = \frac{1}{\tau_{45}}(\hat{m}_{5_{ss}} - \hat{m}_5)$$

$$\dot{\hat{P}}_5 = \frac{1}{\tau_{75}}(\hat{P}_{5_{ss}} - \hat{P}_5)$$

$$\dot{\hat{T}}_{05} = \frac{1}{\tau_{75}}(\hat{T}_{5_{ss}} - \hat{T}_5)$$

$$\dot{\hat{m}}_7 = \frac{1}{\tau_{57}}(\hat{m}_{7_{ss}} - \hat{m}_7)$$

$$\dot{\hat{P}}_7 = \frac{1}{B}\left(\hat{m}_7 - A_{th}\frac{\hat{P}_{07}\beta}{\sqrt{\hat{T}_{07}}}\right)$$

$$\dot{\hat{P}}_{07} = \frac{1}{\tau_{75}}(\hat{P}_{07_{ss}} - \hat{P}_{07})$$

$$\dot{\hat{T}}_{07} = \frac{1}{\tau_{57}}(\hat{T}_{07_{ss}} - \hat{T}_{07})$$

(31.12)

where

$$\beta = \frac{\sqrt{\gamma}}{\sqrt{R}}\left(\frac{2}{\gamma+1}\right)^{(\gamma+1)/2(\gamma-1)}$$

(31.13)

The above equations for the ACS dynamics are non-linear; hence, the EKF is employed. The EKF design as applied to ACS is discussed in detail in the next section.

31.4.3 EKF as Applied to ACS

The ACS system is modelled as a continuous time system and hence the continuous version of KF (Kalman–Bucy Filter) discussed above is applied to the ACS model. The corresponding Equation 31.11 are used with the following:

$X(t)$ = state vector = $[P_4\ P_5\ T_{05}\ P_7\ P_{07}\ T_{07}\ \dot{m}_5\ \dot{m}_7]$
$U(t)$ = control input = $[\dot{m}_f,\ A_{th}]$
$Z(t)$ = measured output = $[P_4]$
Modelled output = $[P_{07},\ T_{07}]$
$v(t)$ = observation noise

Similar to the KF theory explained earlier, signal $w(t)$ is an unknown process noise that acts as a random disturbance to the plant. It represents effects of unmodelled high-frequency plant dynamics that is modelled as zero-mean, white Gaussian noise. Process noise covariance matrix $Q(t)$ (referred to as Q) describing the random process is given as

$$w(t) \sim N(0,Q) \tag{31.14}$$

$v(t)$ is the observation noise which is assumed to be zero mean Gaussian white noise with covariance $R(t)$. $v(t)$ should be selected such that all failure modes are accounted in it. Measurement noise covariance matrix $R(t)$ (referred to as R) is given by

$$v(t) \sim N(0,R) \tag{31.15}$$

By applying the small disturbance theory, the dynamic relationships for ACS given in Equation 31.12 can be linearized. Here each variable is assumed to be composed of two parts; a constant component associated with the linear part and the perturbation associated with non-linear model. The state transition and observation matrices are defined to be the following Jacobians:

$$F = \left.\frac{\partial f}{\partial x}\right|_{\hat{X}_{t-1|t-1},u_t} = \frac{\partial f(x(t),u(t))}{\partial x} \quad H = \left.\frac{\partial h}{\partial x}\right|_{\hat{X}_{t,t-1}} \tag{31.16}$$

Note that for simplicity in the notation we do not use the time step subscript with the Jacobians F, H even though they are in fact different in every cycle.

$X_c(t)$ of $\Delta X(t)$ can be linearized about the central estimate as

$$X(t+1) \cong f_0(\hat{X}_c(t),u_0(t)) + \left.\frac{\partial f_0(x,u_0)}{\partial x}\right|_{x=\hat{x}(t)} (x(t)-\hat{x}_c(t)) \tag{31.17}$$

where $X_c(t)$ is central estimated value about which perturbation $\Delta X(t)$ is added. As in EKF, replacing and rewriting the above equation we have

$$X(t+1) \cong f_0(\hat{X}_c(t),u_0(t)) + F(\hat{x}_c(t))(x(t)-\hat{x}_c(t)) \tag{31.18}$$

31.4.4 Linerization of ACS Dynamics for EKF

The non-linear plant dynamics is complex and it consumes significant resources during simulations. Since, the FDA algorithm is to be implemented in real time it should be based on a simpler linear model that can be processed much faster. Hence, we develop a family of linearized model of the non-linear model at various operating conditions, which is suitably fast and provides close results to the non-linear plant when properly scheduled with a suitable parameter (Mach number in climb, and angle of attack in cruise). This linear model of the plant estimates the states using the state-space model:

$$\left. \begin{aligned} dx/dt &= Ax + Bu \\ y &= Cx + Du \end{aligned} \right\} \tag{31.19}$$

Since, we make use of the first-order partial derivatives (Jacobians) instead of the first-order differential equations, the above model can be rewritten as

$$\left. \begin{aligned} dx/dt &= \text{Amat}*x(t) + \text{Bmat}*u(t) \\ y &= \text{Cmat}*x(t) + \text{Dmat}*u(t) \end{aligned} \right\} \tag{31.20}$$

where Amat, Bmat, Cmat, Dmat are the Jacobian matrices corresponding to the given operating condition.

We obtain the exact Jacobians at a number of specific operating conditions and for intermediate operating conditions we obtain the corresponding Jacobians by linear interpolation. Let us assume that our process has a state vector $x = [P_4, P_5, T_{05}, P_7, P_{07}, T_{07}, \dot{m}_5, \dot{m}_7]$, but that the process is governed by the non-linear stochastic equation:

$$\left. \begin{aligned} \frac{d}{dt} x(t) &= f(x(t), u(t)) + w(t) \\ y(t) &= h(x(t)) + v(t) \end{aligned} \right\} \tag{31.21}$$

where the random variables $w(t)$ and $v(t)$ again represent the process and measurement noise. Note that $w(t)$ and $v(t)$ are inherently present in the system. By using the linear model of plant for state estimation the above equations can be modified as

$$\frac{d}{dt} x(t) = f(x(t), u(t)) + w(t) = [\text{Amat}.x(t) + \text{Bmat}.u(t)] + w(t) \tag{31.22}$$
$$y(t) = h(x(t)) + v(t) = \text{Cmat}.x(t) + \text{Dmat}.u(t) + v(t)$$

To estimate the process with non-linear difference and measurement relationships, we begin by writing new governing equations that linearize the estimate. The state transition matrix F is given by

$$f = \left. \frac{\partial F}{\partial x} \right|_{\hat{X}_{t-1|t-1}, u_t} = \frac{\partial F(x(t), u(t))}{\partial x} \tag{31.23}$$

$$f = \left. \frac{\partial F}{\partial x} \right|_{\hat{X}_{t-1|t-1}, u_t} = \frac{\partial F(x(t), u(t))}{\partial x} = \frac{\partial(\text{Amat}.x(t) + \text{Bmat}.u(t) + w(t))}{\partial x} \tag{31.24}$$

The observation matrix H is given by

$$H = \text{Hmat} = \left. \frac{\partial H}{\partial x} \right|_{\hat{X}_{t|t-1}} = \begin{bmatrix} 1 & 0 & 0 & 0 & 0 & 0 & 0 & 0 \\ 0 & 0 & 0 & 0 & 0 & 1 & 0 & 0 \end{bmatrix} \tag{31.25}$$

The determination or selection of the Q and R matrices is dependent on the application requirements, operating conditions and how trustworthy the measurements are.

31.5 FAULT ACCOMMODATION AND RESIDUAL GENERATION

The non-linear model of ACS is simulated for different flight conditions with varying fuel-to-air ratio (FAR), and the T_{07} and FAR values for each operating condition are tabulated. Figure 31.3 shows the 2D table in graphical form.

31.5.1 T_{07} MODELLING

The T_{07} (total temperature at the end of combustor exit) is used as a second independent measurement for the EKF implementation. While P_4 is obtained from actual sensor measurement, T_{07} is read from a look-up table using the FAR as a parameter for different flight conditions. Two different sets of T_{07} look-up tables are generated. The first is looked up during the acceleration phase and makes use of the Mach number and FAR as the two look-up parameters. During cruise phase, the second set of tables is used which makes use of Angle of Attack and FAR as the two look-up parameters. During the switching from acceleration to cruise, a switching function is used which blends the acceleration and cruise values linearly.

31.5.2 RESIDUAL PROCESSING

Using the estimated values of the modelled T_{07} obtained from the FAR-to-T_{07} static map and P_{07} from each EKF, along with the known bias setting of the nozzle throat area, it is possible to get an estimate of the choking mass flow rate at the nozzle throat. Thus, three estimated values of nozzle mass flow rate are available from the bank of three EKFs. It is also possible to independently estimate the nozzle mass flow rate from measurements of the free stream quantities, such as static pressure, total pressure and static temperature, by the air data system (ADS). Finding the difference

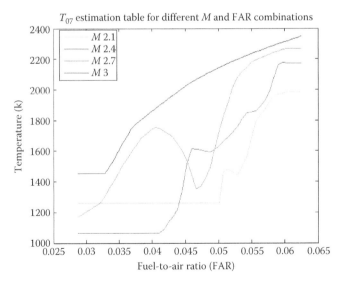

FIGURE 31.3 T_{07} plots for different M values and FAR. (From Walambe, R.A. et al., *Defence Sciences Journal Special Issue—Mobile Intelligent Autonomous Systems*, 60(1), 61–75, 2010. With permission.)

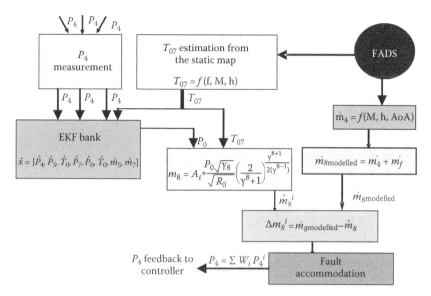

FIGURE 31.4 Scheme of residual processor for ACS. (From Walambe, R.A. et al., *Defence Sciences Journal Special Issue—Mobile Intelligent Autonomous Systems*, 60(1), 61–75, 2010. With permission.)

between each EKF-estimated nozzle mass flow rate and the ADS-estimated mass flow rate gives three values of error, also called residue or residual. In the absence of fault in a sensor, the residue generated by it is expected to be very small, ideally zero, whereas in the presence of a fault, the residue is expected to deviate significantly from zero. Thus, each residue will provide a good indication whether the sensor in that particular EKF channel is faulty or not, thus providing fault detection capability. Figure 31.4 shows the residual processor as part of the FDA algorithm. The present choice of residue is effective because changes in backpressure (such as due to a fault) correlate well with changes in total pressure at combustor exit, and hence with the estimated nozzle mass flow rate.

The inverse of the residue is also a measure of confidence in that particular measurement; hence, can be used for fault accommodation. The strategy explored here is to use the inverse of each residual to compute a normalized weighting factor that is dynamic, that is, changes with time. This can be thought of as an adaptive weighting scheme. The final P_4 value is then computed by applying the weighting factor to each of the three values, and is used as the feedback signal to the plant.

31.5.3 Accommodation of Faults

Present FDA algorithm does not require the faulty sensor to be identified. It also does not demand the faulty sensor to be eliminated. The fault accommodation part of the FDA employs an adaptive weightage assignment scheme without declaring and eliminating a faulty sensor. In each time step, an adaptive weightage is assigned to each P_4 sensor and the weighted output is obtained. An important advantage of using this approach is that weighted value of P_4 (to be fed back), computed using the three P_4 sensors, is as accurate as possible for each iteration. For any iteration, if the sensor has not entirely failed but measures a wrong P_4 value, the sensor is not discarded; instead, the faulty measurement for that iteration is suppressed by allocating less weightage to it. A variable weighting factor is assigned to each measurement. In case all three sensors are healthy, the mean value of the P_4 measurements is fed back to the plant.

The weighting factor W_i is computed as follows:

1. The deviation of each estimated m8dot (from the EKF channel) from the $\dot{m}_{8\text{modelled}}$ is computed.
   for $i = 1$ to 3 (for 3 EKF channels):
   $$\Delta\dot{m}_8^i = \text{abs}(\dot{m}_{8\text{modelled}} - \hat{\dot{m}}_8^i)$$

2. The ratio of each $\Delta\dot{m}_8^i$ to the $\dot{m}_{8\text{modelled}}$ is computed. This ratio decides how far or close estimated $\dot{m}_8$ is from the $\dot{m}_{8\text{modelled}}$.
   for $i = 1$ to 3 (for 3 EKF channels):
   $$\text{ratio}(i) = \Delta\dot{m}_8^i/\dot{m}_{8\text{modelled}}$$

3. Minimum of the three ratios (*ratio_min*) is determined.

4. The minimum ratio value is then subtracted from the individual ratio for each channel.
   for $i = 1$ to 3 (for 3 EKF channels):
   ratio_new(i) = ratio(i)-ratio_min;

5. Weightage for each channel is computed as:
   for $i = 1$ to 3 (for 3 EKF channels):
   $$W(i) = 10^{[-10(\text{ratio\_new}(i))]}$$

This makes sure that the channel with lowest ratio (zero) is given the highest weightage (one), and the others are assigned exponentially reducing weights based on their ratios. The exponential curve demonstrating the above relationship is shown in Figure 31.5. The exponential weightage formula $W(i) = 10^{[-10(\text{ratio\_new}(i))]}$ is obtained by the trial-and-error method by carrying out a number of simulations.

These weights are then normalized to find the final weights for each $\dot{m}_8^i$ (corresponding to each EKF channel):

$$W_{i\_\text{norm}} = W_i \Big/ \sum_{i=1}^{3} W_i \qquad (31.26)$$

This weighting reflects the faults in any sensor measurement and also accommodates it. Once the weighting factor is calculated, the P_4 mean value is calculated as

$$P_4 = \sum W_{i\_\text{norm}} P_4^i \qquad (31.27)$$

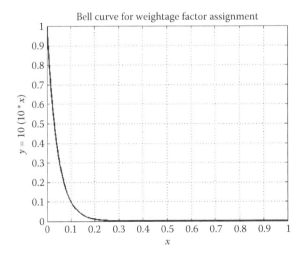

FIGURE 31.5 Weightage assignment. (From Walambe, R.A. et al., *Defence Sciences Journal Special Issue—Mobile Intelligent Autonomous Systems*, 60(1), 61–75, 2010. With permission.)

TABLE 31.1

Fault Accommodation Scheme Parameters

Mass Flow Rate at Station 8 (kg/s)	Case 1	W	Case 2	W	Case 3	W
$\dot{m}_{8_{modelled}}$	16.5	—	10.15	—	10.15	—
$\dot{m}_8^1$	6.2	4.77e–6	16.8	2.087e–7	18.5	6.647e–9
$\dot{m}_8^2$	14.98	~1	9.80	0.3361	5	9.449e–6
$\dot{m}_8^3$	7.2	1.927e–5	10.1	0.6639	10.1	~1

Source: From Walambe, R.A. et al., *Defence Sciences Journal Special Issue—Mobile Intelligent Autonomous Systems*, 60(1), 61–75, 2010. With permission.

This P_4 value is then fed back to the plant. The weightage factor computation for three hypothetical cases is shown in Table 31.1.

The W_i is the percentage of weightage of each measurement towards the P_4 value to be fed back. Above three cases demonstrate that the conceptual development for computation of W_i is suitable for range of $\dot{m}_8$ values and for extreme cases of deviations.

We needed to develop the FDA in the case of only one sensor failure, but it can be observed that the designed FDA algorithm allows for failure and accommodation of two sensors as shown in Table 31.1. If all three sensors fail and measure incorrect value of P_4, then the closest (although wrong) measurement (and corresponding EKF channel) will be assigned highest weightage and that will contribute the highest to the feedback P_4 value. But it is important to note that this value will not be a required P_4 value and the system may not work satisfactorily.

This concludes the discussion about the FDA algorithm. The next section presents the results from the closed-loop simulations for different Mach profiles after implementing the complete plant, controller and FDA algorithm. It demonstrates the effective working of the fault accommodation scheme along with the FDA as a whole.

31.6 EVALUATION OF FDA AND RESULTS

To evaluate and validate the FDA algorithm, a number of simulations are carried out. This section presents the results from the FDA algorithm in the presence of different faults. It is observed that the system operates inefficiently with fault. In fact, some faults may lead to complete failure of the system. For testing and simulation purposes, the fault is introduced once the system reaches steady state (at around 0.1 s). The fault introduction details are as explained below. Figures 31.6 through 31.9 present the results obtained by running the closed-loop simulations in the presence of some of the fault conditions given subsequently.

1. *Fault-free case*
2. *Ramp bias*: In this simulation, the output is made to slowly deviate from its nominal value. The simulation is fault-free for the first 0.1 s. At $t = 0.1$ s, bias value is added to the sensed value of voltage by a ramp change. This erroneous sensed value is fed back to the controller, which slowly decreases fuel flow rate until the lower limit of P_{4_margin} is attained.
3. *Pressure (P_4) stuck at a non-zero constant value*: The simulation is fault-free for the first 0.1 s. At $t = 0.1$ s, the sensed value at the sensor port is held constant at 1.5 bar by a step change. This erroneous sensed value is fed back to the controller creating a P_{4_margin} error. The controller, therefore, injects less fuel in trying to raise P_{4_margin} to the commanded value but ends up raising the actual P_{4_margin} to a high value. But, since sensor output is stuck, the P_{4_margin} error continues to persist.

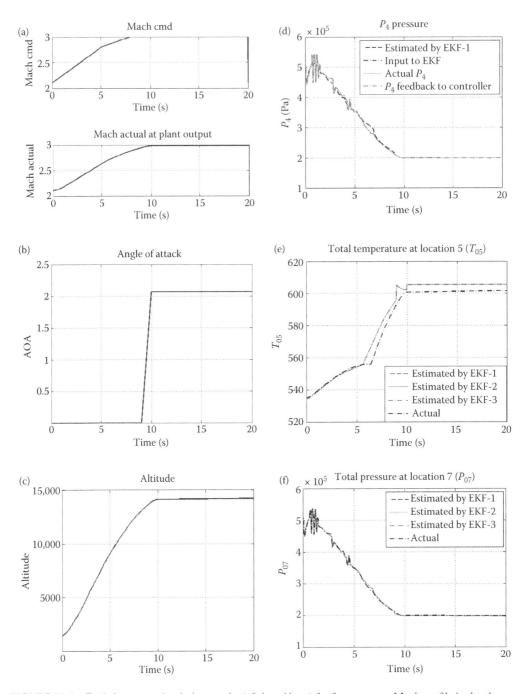

FIGURE 31.6 Fault free case, simulation results (of closed loop) for fast response Mach profile in the absence of fault (all EKF have exactly same P_4 plots in no fault condition). (a) Mach cmd and Mach actual vs. time, (b) AoA vs. time, (c) Altitude vs. time, (d) P_4 vs. time, (e) T_{05} vs. time, (f) P_{07} vs. time, (g) T_{07} vs. time, (h) mass flow rate at station 8 vs. time and (i) weightage vs. time. (From Walambe, R.A. et al., *Defence Sciences Journal Special Issue—Mobile Intelligent Autonomous Systems*, 60(1), 61–75, 2010. With permission.)

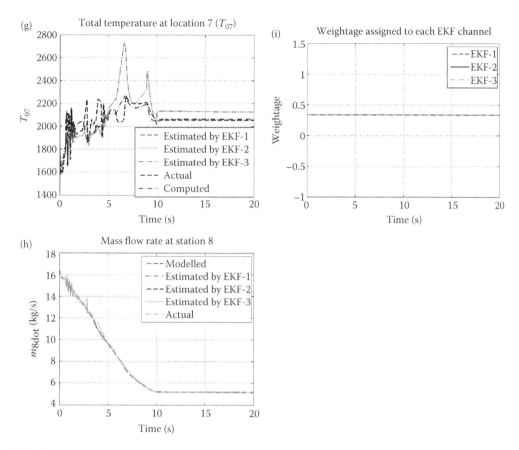

FIGURE 31.6 Continued.

4. *Mis alignment*: Due to misalignment of P_4 sensor, static pressure measured by pressure sensor will be larger and hence during simulation at time $= 0.1$ s, the sensed value at the sensor port is deliberately increased by 5% higher than the true pressure value by impulse change. This erroneous higher pressure is further fed back into the controller. The controller tries to maintain that erroneous P_4 value as its commanded value by decreasing fuel flow rate, and this in turn lowers the Thrust and Mach number.

5. *Temperature compensating system failure*: In this fault, response to the error introduced due to temperature bias shift and temperature sensitivity shift in case of failing of the temperature compensating system is considered. Error of about twice/thrice the magnitude of a function of $T4$ is introduced in P_4 measurement. Fault is introduced at 0.1 s. As T_4 temperature is higher than the temperature at which sensors are calibrated, output voltage is sensed at a higher value than the true value. This implies that the pressure is also sensed to be of a higher value. This erroneous higher pressure is further fed back into the controller. The controller tries to maintain that erroneous P_4 value as its commanded value by decreasing fuel flow rate, and this in turn lowers the Thrust and Mach number. Because of the decrease in Thrust, P_{4_margin} command gets decreased up to its lowest limit, and commanded thrust goes on increasing.

6. *Noise*: The simulation is fault-free for the first 0.1 s. At $t = 0.1$ s, white noise of power spectral density (PSD) 1×10^{-6} value is added in output voltage and the latter is seen to fluctuate about its true value. It is observed that sometimes P_{4_margin} hits its lower limit.

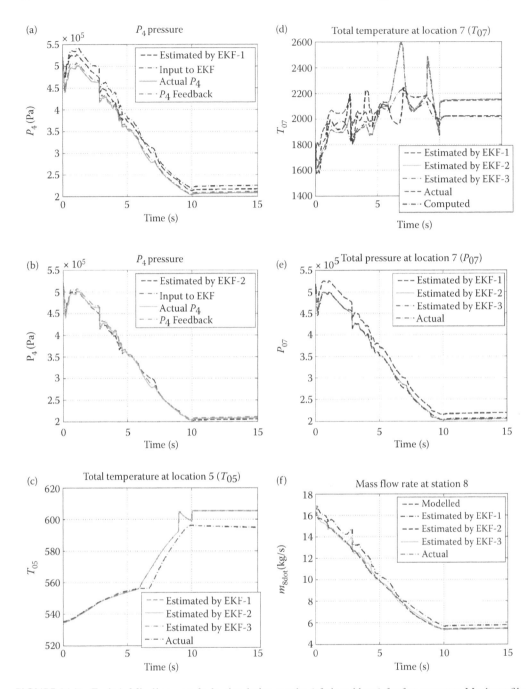

FIGURE 31.7 Fault 1: Misalignment fault, simulation results (of closed loop) for fast response Mach profile in the presence of misalignment fault. (a) P_4 vs. time—EKF 1, (b) P_4 vs. time—EKF 2, (c) T_{05} vs. time, (d) T_{07} vs. time, (e) P_{07} vs. time, (f) mass flow rate at station 8 vs. time and (g) weightage vs. time. (From Walambe, R.A. et al., *Defence Sciences Journal Special Issue—Mobile Intelligent Autonomous Systems*, 60(1), 61–75, 2010. With permission.)

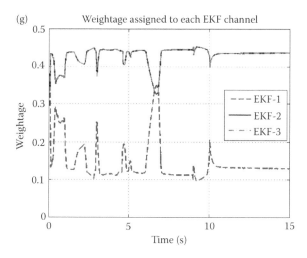

FIGURE 31.7   Continued.

31.6.1 EVALUATION OF PERFORMANCE OF THE FDA ALGORITHM FROM RESULTS

A novel concept of using the mass flow rate at station 8 has been developed, as two independent estimates of a variable are not directly available for use in analytical redundancy-based FDA algorithm. Use of $\dot{m}_8$ proves to be a promising approach as shown in the closed-loop response plots for different conditions (Figures 31.6 through 31.9). Following conclusions can be drawn from the plots for different operating conditions:

1. Bank of EKF is implemented as part of the FDA algorithm. EKF design for all operating conditions works well. It facilitates the separation of disturbances and also generates the estimated state vector. Capability of EKF to estimate the faulty values is observed through number of simulations for known sensor faults. Results show that the estimation of states by EKF represents the faults faithfully. From plots for state parameters P_4, P_{07}, T_{05}, T_{07} and $\dot{m}_8$ it can be seen that the estimated parameters from EKF deviate from the actual values after introduction of fault.

2. In the plots for P_4, it can be seen that for fault-free condition, all three P_4 values are same and they are also very close to the actual P_4 value as well as the P_4 value fed back to the controller after the application of FDA.

3. From plots for $\dot{m}_8$ at all operating conditions, it can be seen that the estimated $\dot{m}_8$ responds well to the fault. Thus a clear signature of the fault can be noted for each fault type and operating condition. The $\dot{m}_{8_{modelled}}$ $(\dot{m}_4 + \dot{m}_f)$ follows $\dot{m}_{8_{actual}}$. After the fault is introduced at 0.1 s, the estimated $\dot{m}_8$ $(\hat{\dot{m}}_8)$ slowly starts deviating from the $\dot{m}_{8_{modelled}}$ / $\dot{m}_{8_{actual}}$. Hence choice of $\dot{m}_8$ as a key parameter for fault detection proves to be a promising concept.

4. In plots for T_{07}, three values of T_{07}, namely; actual T_{07}, T_{07} fed to EKF (modelled T_{07} value), T_{07} estimated are plotted. As can be seen the modelled T_{07} is very close to the actual T_{07} value. Thus the method used for determining T_{07} (Section 31.4.1) is fairly accurate and acceptable.

5. It can be seen that the P_{07} follows the P_4 closely.

6. Result of the adaptive weightage assignment scheme is also shown. The plots demonstrate that the scheme works well. Very small weightage is assigned to the faulty P_4 value and hence the P_4 fed back is very close to the actual P_4 value.

It is important to note that the present algorithm does not require the type of fault to be detected in order to accommodate it. It is also not necessary to set threshold and declare a sensor faulty. These are significant advantages over previous purely hardware-redundant FDA schemes.

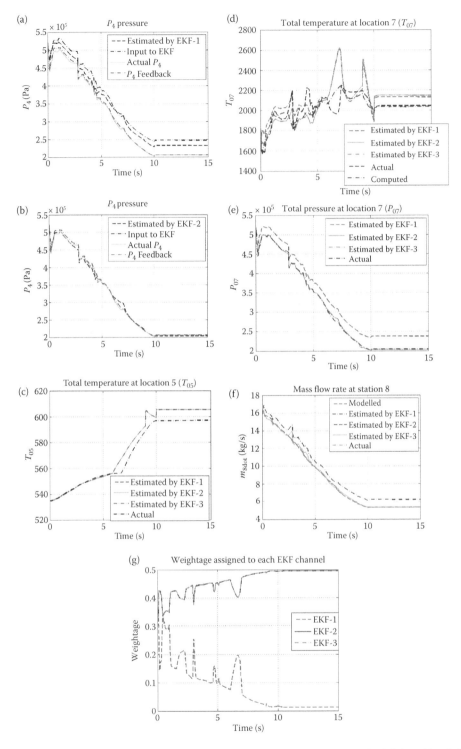

FIGURE 31.8 Fault 2: Temperature compensating system failure simulation results (of closed loop) for fast response Mach profile in the presence of temperature compensating system failure. (a) P_4 vs. time—EKF 1, (b) P_4 vs. time—EKF 2, (c) T_{05} vs. time, (d) T_{07} vs. time, (e) P_{07} vs. time, (f) mass flow rate at station 8 vs. time and (g) weightage vs. time. (From Walambe, R.A. et al., *Defence Sciences Journal Special Issue—Mobile Intelligent Autonomous Systems*, 60(1), 61–75, 2010. With permission.)

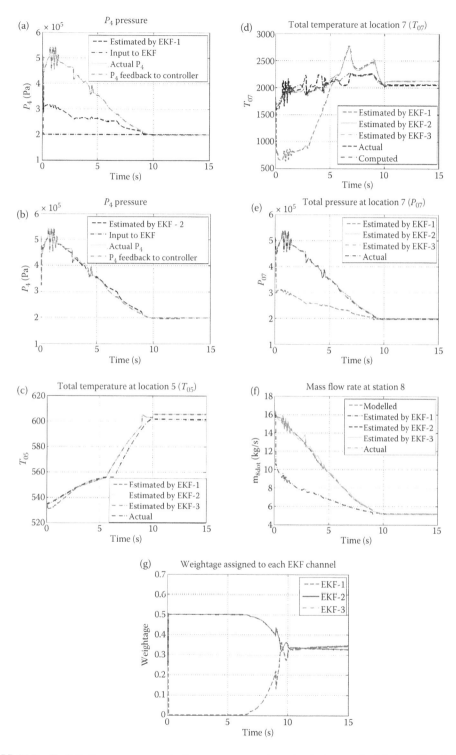

FIGURE 31.9 Fault 3: Pressure sensor stuck at a constant value; simulation results (of closed loop) for fast response Mach profile in the presence of pressure sensor stuck at a constant value. (a) P_4 vs. time—EKF 1, (b) P_4 vs. time—EKF 2, (c) T_{05} vs. time, (d) T_{07} vs. time, (e) P_{07} vs. time, (f) mass flow rate at station 8 vs. time and (g) weightage vs. time. (From Walambe, R.A. et al., *Defence Sciences Journal Special Issue—Mobile Intelligent Autonomous Systems*, 60(1), 61–75, 2010. With permission.)

Our simulations show that the FDA algorithm is able to successfully provide a good value of backpressure to the controller under a variety of fault cases in the sensors. The complete closed-loop simulation, with the air-breathing combustion system and the controller, using the backpressure from this FDA algorithm has demonstrated good results for both fault-free and different fault cases (Figures 31.6 through 31.9).

31.6.2 FDA's Real-Time Implementation

The FDA algorithm developed as part of this is required to run in real-time once it is deployed on the actual system, that is, it should complete the calculations for a time-step in time smaller than that time step, and wait for the inputs (sensor data) to be made available to begin processing for the next time step. We need to demonstrate that the algorithm is sufficiently simple and low in complexity so that it can be processed in real time on processors available today.

To maintain numerical stability, the time step should be smaller than the time constant used to model the system. From combustor look-up tables, we find that the lowest time constant used is around 0.002 s, putting a lower limit on the update frequency at 500 Hz. For real-time execution at 500 Hz, the processor is required to complete one time-step of the entire FDA algorithm within 0.002 s.

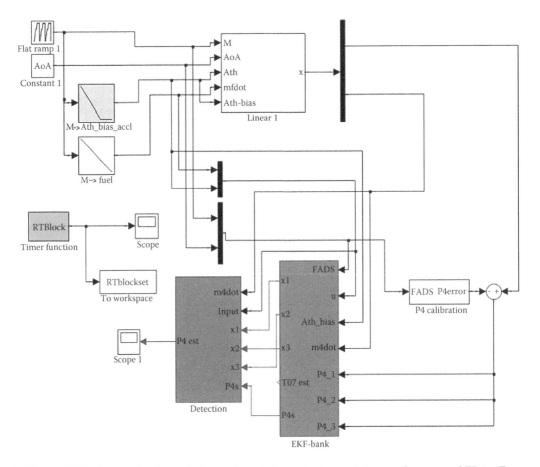

FIGURE 31.10 Schematic of Simulink model used for evaluating real-time performance of FDA. (From Walambe, R.A. et al., *Defence Sciences Journal Special Issue—Mobile Intelligent Autonomous Systems*, 60(1), 61–75, 2010. With permission.)

To test real-time performance, a Simulink® model with a linear plant, EKF bank and fault detection/accommodation block is prepared as shown in Figure 31.10. The linear and non-linear models are run for different operating conditions as explained below:

- Simulations are run for each operating condition, and a step input is given to fuel-flow-rate and the throat area. The results from the linear and non-linear plants for different operation show a good match indicating a good linear approximation of the model at each condition.
- After checking the performance of the linear plant at specific operation points, the performance is evaluated at other intermediate values of Mach number and angles of attack, using two different trajectory profiles:
  1. Fast response profile: for this case, the Mach number is varied as, $M = [2.1\ 2.1\ 2.4\ 2.4\ 2.7\ 2.7\ 3\ 3]$ corresponding to time, $t = [0\ 2\ 4\ 6\ 8\ 10\ 12\ 14\ 20]$. Results for this case are shown in Figure 31.11.
  2. Flat-ramp profile: for this case, the Mach number is varied as, $M = [2.1\ 2.8\ 3\ 3]$ corresponding to time, $t = [0\ 5\ 8\ 20]$. Results for this case are shown in Figure 31.12.

We need to check the real-time performance of the FDA algorithm and the calibration block. So the linearized plant model is used in order to provide the P_4 input to the FDA. We configure Simulink to run using fixed time stepping of 0.002 s and ODE1 (first-order Euler method) solver, on a computer with 1.7 GHz processor, 512 MB RAM, Windows XP, Simulink version 6.4. We observe that the simulation of 20 s of flight takes around 1–2 s. This clearly indicates that the processor power available on this system is adequate to execute this algorithm in real time. To further demonstrate this, we use Real-Time Blockset [20] to slow down the simulation to real time and to evaluate the amount of time CPU waited before it could begin the next time step. Figure 31.13 shows that the processing for each time step was completed well within 2 μs and the CPU has to wait for significant time before it could start processing for the next time step.

The key step in this process has been the use of a linear plant in place of the non-linear plant for the EKF which significantly reduced the computation complexity. The system used for this demonstration uses a non-real-time operating system (Windows) and has several tasks running in parallel with the simulation. The performance of the system would significantly improve when a real-time operating system is deployed and the processor is used for execution of the FDA algorithm alone. Since the model could run in real time on a non-optimized system, this is adequate to prove that the code is easily implementable in real time on an optimized system.

31.7 CONCLUDING REMARKS

The work of this chapter has focused on the development and testing of a novel FDA algorithm for an air-breathing combustion system which uses an innovative analytical redundancy based algorithm to provide 'intelligence' to a triplex redundant P_4 sensor measurement hardware. The following issues were addressed, and were successfully resolved and evaluated:

1. EKF bank was designed for state estimation and disturbance removal from the P_4 sensor measurement. The EKF bank also generated the residual vector, which is fed to the residual processor for FDA.
2. The residual processing algorithm consisting of the following has been thoroughly tested:
   - T_{07} computation from the known parameters using a look-up table
   - P_{07} estimation from the measured P_4
   - Estimation of $\dot{m}_8^i$ from three $\hat{P}_{07}$ and T_{07} values
   - Comparison of $\dot{m}_8^i$ with $\dot{m}_{8_{modelled}}^i$ for detection of fault

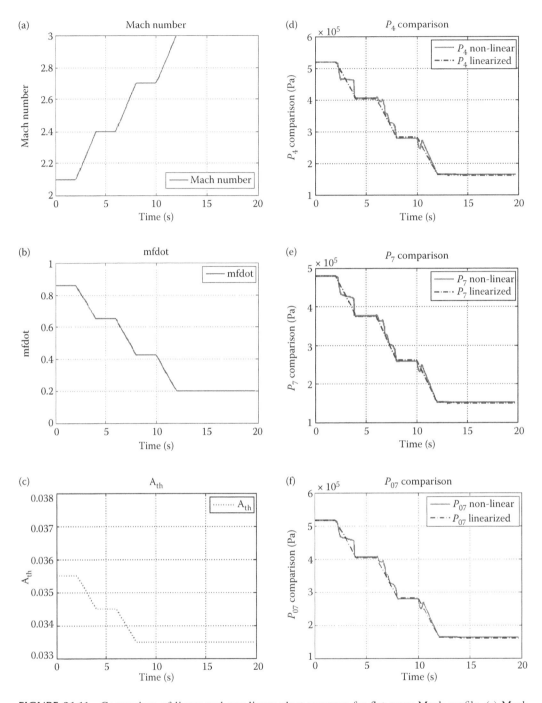

FIGURE 31.11 Comparison of linear and non-linear plant response for flat ramp Mach profile. (a) Mach number vs. time, (b) mfdot vs. time, (c) A_{th} vs. time, (d) P_4 (linear and non-linear) comparison, (e) P_7 (linear and non-linear) comparison, (f) P_{07} (linear and non-linear) comparison, (g) P_5 (linear and non-linear) comparison, (h) T_{07} (linear and non-linear) comparison and (i) m7dot (linear and non-linear) comparison. (From Walambe, R.A. et al., *Defence Sciences Journal Special Issue—Mobile Intelligent Autonomous Systems*, 60(1), 61–75, 2010. With permission.)

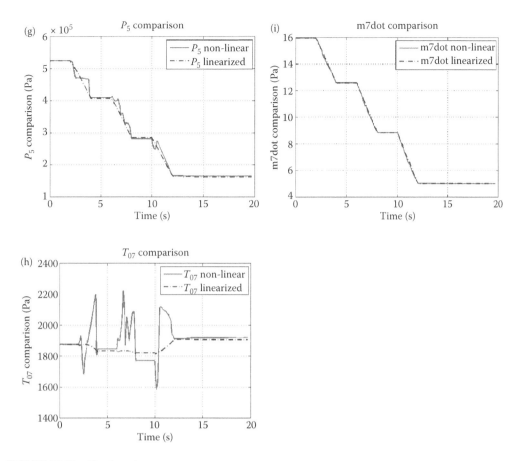

FIGURE 31.11   Continued.

- Weightage factor calculation for individual $\dot{m}_8^i$ value based on the deviations from the $\dot{m}_8^i{}_{\text{modelled}}$
- Demonstration of error plots for fault detection and sensor removal

The main contribution of this work is the use of analytical redundancy-based FDA algorithm for P_4 sensor. A novel concept of using the mass flow rate at station 8 is developed, as two independent estimates of a variable are not directly available for use in analytical redundancy-based FDA algorithm. Use of $\dot{m}_8^i$ proves to be a promising approach as shown in the closed-loop response plots for different conditions in Figures 31.6 through 31.9. The real-time implementation results (Figures 31.10 through 31.13) show that the present design of the FDA algorithm is suitable to run on an embedded hardware.

APPENDIX: NOMENCLATURE

A_{th}	Nozzle throat area, m^2
α	Angle of attack, deg
B	Backpressure factor
R	Universal gas constant, J/kmol K
τ_{ij}	Time constant between station i and j
$\dot{m}_{\text{f}}$	Fuel flow rate
P_1	Free stream pressure, Pa

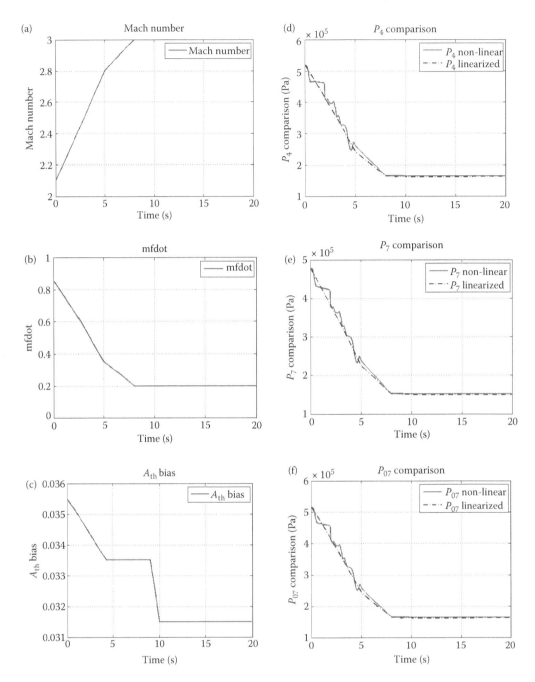

FIGURE 31.12 Comparison of linear and non-linear plant response for fast response Mach profile. (a) Mach number vs. time, (b) mfdot vs. time, (c) A_{th} vs. time, (d) P_4 (linear and non-linear) comparison, (e) P_7 (linear and non-linear) comparison, (f) P_{07} (linear and non-linear) comparison, (g) P_5 (linear and non-linear) comparison, (h) T_{07} (linear and non-linear) comparison and (i) m7dot (linear and non-linear) comparison. (From Walambe, R.A. et al., *Defence Sciences Journal Special Issue—Mobile Intelligent Autonomous Systems*, 60(1), 61–75, 2010. With permission.)

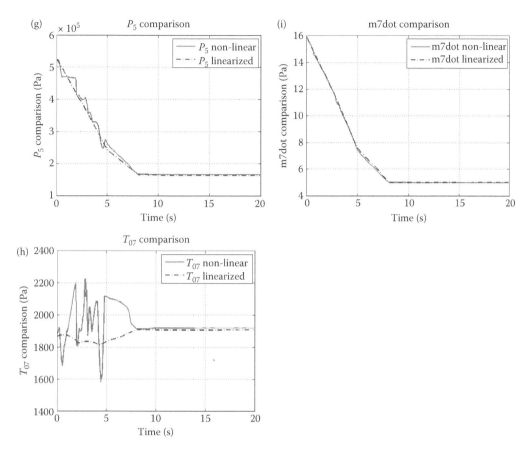

FIGURE 31.12   Continued.

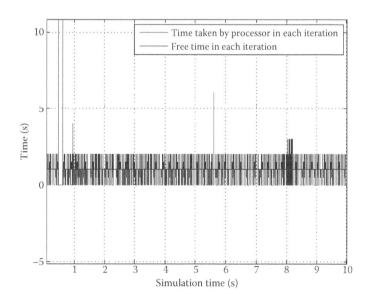

FIGURE 31.13   Time taken for computation of each time step of FDA. (From Walambe, R.A. et al., *Defence Sciences Journal Special Issue—Mobile Intelligent Autonomous Systems*, 60(1), 61–75, 2010. With permission.)

P_4	Static pressure at station 4, Pa
P_{4_margin}	Backpressure margin
P_5	Static pressure at station 5, Pa
T_{05}	Total temperature at station 5, K
P_7	Static pressure at station 7, Pa
T_{07}	Total temperature at station 7, K
P_{07}	Total pressure at station 7, Pa
$\dot{m}_4$	Fuel flow rate at station 4, kg/s
$\dot{m}_5$	Fuel flow rate at station 5, kg/s
$\dot{m}_7$	Fuel flow rate at station 7, kg/s
$\dot{m}_8$	Fuel flow rate at station 8, kg/s
$\hat{P}_4^1, \hat{P}_4^2, \hat{P}_4^3$	P_4 measured by three pressure sensors
P_{port}	Pressure at FADS orifice port
$Q_{k}, _k$	Covariance matrices
FAR	Fuel-to-air ratio
W_i	Weightage factor for ith EKF channel
$\dot{m}_{8modelled}$	Modelled value of $\dot{m}_8$, kg/s
q	Dynamic pressure
v_k	Unknown measurement noise
w_k	Unknown process noise
x_k	State at time k

The subscript 'ss' refers to the steady-state value of the variable. '$\hat{x}$' represents the estimated value of the variable x.

REFERENCES

1. Walambe, R.A., Gupta, N.K., Bhange, N., Ananthkrishnan, N., Park, I.S., Choi, J.H. and Yoon, H.G. Novel redundant sensor fault detection and accommodation algorithm for air-breathing combustion system and its real-time implementation, Raol, J.R. and Gopal, A. (Eds.), *Defence Sciences Journal Special Issue—Mobile Intelligent Autonomous Systems*, 2010.

2. Bharani Chandra, P., Gupta, N.K., Ananthkrishnan, N., Renganathan, V.S., Park, I.S. and Yoon, H.G. Modeling, dynamic simulation, and controller design for an air-breathing combustion system, *AIAA Paper 2009-708, 47th AIAA Aerospace Sciences Meeting*, Orlando, FL, Jan 2009.

3. O'Brian, T.F., Starkey R.P. and Lewis, M.J. Quasi-one-dimensional high-speed engine model with finite-rate chemistry, *Journal of Propulsion and Power*, 2001, 17(6), 1366–1374.

4. Gupta, N.K., Gupta, B.K., Ananthkrishnan, N., Shevare, G.R., Park, I.S. and Yoon, H.G. Integrated modeling and simulation of an air-breathing combustion system dynamics, AIAA Paper 2007-6374. *AIAA Modeling and Simulation Technologies, Conference and Exhibit*, Hilton Head, SC, August 2007.

5. Chen, J. and Patton, R.J. *Robust Model-based Fault Diagnosis for Dynamic Systems*, Kluwer Academic Publishers, Boston, USA, 1999.

6. Gertler, J.J. *Fault Detection and Diagnosis in Engineering Systems*, ed.1, Dekker, New York, 1998.

7. Isermann, R. and Ballé, P. Trends in the application of model-based fault detection and diagnosis of technical processes, *Control Engineering Practice*, 1997, 5, 709–719.

8. Patton, R.J. Fault tolerant control: The 1997 situation, *In Proc.: IFAC Safeprocess*. Hull, UK, 1997, pp. 1033–1055.

9. Frank, P.M. Analytical and qualitative model-based fault diagnosis—A survey and some new results, *European Journal of Control*, 1996, 2(1), 6–28.

10. Massoumnia, M.A., Verghese, G.C. and Willsky, A.S. Failure detection and identification, *IEEE Trans. Automat. Control*, 1989, 34(3), 316–321.

11. Willsky, A.S. A survey of design methods for failure detection in dynamic systems, *Automatica*, 1976, 12(6), 601–611.

12. Frank P. Fault diagnosis in dynamic system using analytical and knowledge based redundancy—A survey and some new results, *Automatica*, 1990, 26(3), 459–474.

13. Patton, R. Fault detection and diagnosis in aerospace system using analytical redundance, *IEE Computing and Control Engineering Journal*, 1990, 2, 127–136.
14. Patton, R., Frank, P., Clark, R. (eds.) *Fault Diagnosis in Dynamic Systems, Theory and Application*, Prentice Hall (Control Engineering Series), UK, 1989.
15. Welch, G. and Bishop, G. *An Introduction to the Kalman Filter*, TR 95-041, University of North Carolina, USA, 2004.
16. Kalman, R.E. A new approach to linear filtering and prediction problems, *Transactions of the ASME - Journal of Basic Engineering*, 1960, 82(1), 35–45.
17. Brown, R.G. *Introduction to Random Signal Analysis and Kalman Filtering*, John Wiley and Sons, New York, 1983.
18. Bucy, R.S. and Joseph, P.D. Filtering *for Stochastic Processes with Applications to Guidance*, John Wiley & Sons, 1968; 2nd Edition (2005), AMS Chelsea Publ.
19. Jazwinski, A.H. *Stochastic Processes and Filtering Theory*, Academic Press, New York, 1970.
20. Daga, L. http://leonardodaga.insyde.it/Simulink/RTBlockset.htm, 2008.

32 Fuzzy Logic–Based Sensor and Control Surface Fault Detection and Reconfiguration

Ambalal V. Patel and Shobha R. Savanur

CONTENTS

32.1 INTRODUCTION

Aircraft crash investigation agencies point out that one of the major factors for the aircraft accidents is the failure or damage of components such as actuators, sensors and other structural parts of aircraft. The Fault Detection Identification and Accommodation (FDIA) is important for the successful completion of a given flight mission. Modern flight control systems require fault tolerance that involves both passive and active techniques. Passive techniques are based on the design of robust controllers which guarantee the performance of the system even under faults, but do not provide sufficient fault tolerance when the effect of the faults is considerably large. In active approaches, the control systems are reconfigured according to the faults in the systems by using a pre-determined control law or combining a new control scheme online to maintain the stability and acceptable performance of the overall system. As a result, the active approaches have higher fault-tolerant capability compared to the passive approaches. An aircraft system in certain phases of its flight envelope can be considered a time-varying, non-linear system with process and measurement noises; therefore, the control of such a system can be attempted with an adaptive scheme. A Fault-Tolerant Control System (FTCS) is a control system which is capable of accommodating the component faults in dynamic system automatically. Such a control system has the capabilities of maintaining overall closed-loop stability for increasing the overall operational safety of the aircraft in the event of faults.

Typically, a reconfigurable FTCS consists of a fault detection, identification scheme and reconfiguration scheme. The fault detection and identification techniques generally involve the comparison of the expected values under normal conditions as determined by the state estimators and the actual measured outputs of the sensors. In general, a fault-tolerant flight control system needs to perform: Sensor Fault Detection, Identification and Accommodation or Reconfiguration (SFDIA/SFDIR) and Actuator Fault Detection, Identification and Accommodation or Reconfiguration (AFDIA/AFDIR). Sensor Failure Detection and Identification (SFDI) module monitor the fault in a sensor and identifies or isolates the faulty sensor. Sensor failure accommodation (SFA)/reconfiguration replaces the faulty sensor with an appropriate estimation in analytical redundancy or by another healthy sensor in hardware redundancy. For SFA purposes, most of today's high-performance military aircraft as well as commercial jetliners implement a triple physical redundancy in their sensor capabilities. However, when reduced complexity, lower costs and weight optimization are important factors in aircraft design, an analytical sensor redundancy approach is more appealing. In terms of the AFDIA problem, an actuator fault may imply a locked surface, a missing part of the control surface, or a combination of both. Actuator Fault Detection and Identification (AFDI) scheme detects significant abnormalities or faults and identifies the cause of the fault. Actuator fault accommodation/reconfiguration takes actions so as to recover the pre-fault system performance as much as possible. Generally, the size and weight of actuators due to their power delivering capability are large and hence use of multiple redundant actuators is limited in aircraft. Hence, actuator faults are considered as very important faults and must be attended immediately and effectively so as to minimize the damages. Here, a model-based scheme using Kalman filter (KF) is used for SFDIR [1,2]. Faults that change the system dynamics by abnormal measurements, sudden shifts and so on affect the characteristics of the normalized innovation sequence by changing its white-noise nature, displacing its zero mean and varying unit covariance matrix [1,2]. Hence, the objective is to detect any change of these parameters from their nominal values and provide the necessary remedies.

In studying control systems, one must be able to model dynamic systems and analyse dynamic characteristics. It is generally difficult to represent a complex process accurately by a mathematical model. Fuzzy control is a non-model-based technique which deals with knowledge of process behaviour and experience of people working with the process and it can handle non-crisp and incomplete information. Since the first successful application of the idea of fuzzy sets of Zadeh to the control of a dynamic plant by Mamdani and Assilian, *Fuzzy Control Systems Engineering* has gained worldwide interest [3,4]. It is possible to control many complex systems effectively by (experienced) human operators who have no knowledge of their underlying dynamics, while it is difficult to achieve the same with conventional controllers. Here, fuzzy logic is used for detection and reconfiguration of a sensor fault. It is also shown that fuzzy logic can be extended for multiple faults. One of the popular methods to detect and reconfigure the surface fault is a model-based approach, for example, extended Kalman filter (EKF) [5,6]. Here, the parameters of control distribution matrix are estimated as augmented states of the system using EKF which are subsequently used to compute feedback gain to reconfigure the impaired system using the pseudo inverse technique [2]. Detection and reconfiguration of surface fault in an elevator of an aircraft is demonstrated using non-model-based fuzzy logic in terms of determining the factor of effectiveness of control surface and in turn, new control gain for reconfiguration. A comparison study is carried out by using two T-norm operations namely intersection and algebraic product and different implication methods.

32.2 FAULT DETECTION, ISOLATION AND RECONFIGURATION OF SENSORS

The sensor fault detection and identification (SFDI) process is very important, particularly when the measurements from a faulty sensor are used in the feedback control loop of any dynamic system including aircraft. Since the aircraft control laws use sensor feedback to establish the current dynamic state of the airplane, even slight sensor inaccuracies and faults can lead to closed-loop instability, if this is not attended and may lead to unrecoverable flight conditions. Common sensor faults/failures are: (i) bias, (ii) drift, (iii) loss of accuracy of the sensors and (iv) freezing. The fault/failure terminology is used in this chapter synonymously, although strictly speaking it is not so. In particular, there is some fault (often hidden) in some hardware, and this fault (when activated by some means, under certain condition, inputs exceeding limits, etc.) causes system-state-error which in turn causes failure of the component, sub-system or the entire system.

32.2.1 Kalman Filter-Based SFDI and Reconfiguration

One of the many techniques of sensor fault detection available in the open literature involves the generation of residuals that carry information about the faults/failures. The most common approach using state estimators such as Kalman filter is based on the analysis of the innovation sequence. These approaches do not require *a priori* statistical characteristics of the faults and hence computational burden is not more. When the faults occur, the characteristics of the error signal which is the difference between the actual system output and the expected output will be changed. If the system operates normally, the normalized innovation sequence in Kalman filter is a Gaussian white noise with a zero mean and with a unit covariance matrix. When a fault occurs, the decision statistics change and its effect is more significant for the faulty sensor channel, hence the faulty sensor is identified. Subsequently, the KF is reconfigured by ignoring the measurement from the faulty sensor [1,2].

32.2.1.1 Sensor Fault Detection

Here, SFDIR (sensor fault detection, identification and reconfiguration) study is carried out using the longitudinal dynamics of an aircraft in the simulation. The state-space equations of the longitudinal motion of an aircraft in continuous domain are given by [5]

$$
\begin{aligned}
\dot{u} &= X_u\, u + X_w\, w - g\cos\gamma_0\,\theta + X_{\delta_E}\, u_c \\
\dot{w} &= Z_u\, u + Z_w\, w + U_0 q - g\sin\gamma_0\,\theta + Z_{\delta_E}\, u_c \\
\dot{q} &= M_u\, u + M_w\, w + M_q\, q + M_{\delta_E}\, u_c \\
\dot{\theta} &- q
\end{aligned}
\tag{32.1}
$$

where $X_u, X_w, Z_u, Z_w, M_u, M_w, M_q, X_{\delta_E}, Z_{\delta_E}$ and M_{δ_E} are stability derivatives of the aircraft [5]. The parameter U_0 is the flight speed of the vehicle, γ_0 is flight path angle and the parameter 'g' is the gravity acceleration. In matrix form, the state equation is given by [5]

$$
\dot{x} = Ax + Bu_c + \Gamma w_n
\tag{32.2}
$$

where

$$
A = \begin{bmatrix}
X_u & X_w & 0 & -g\cos\gamma_0 \\
Z_u & Z_w & U_0 & -g\sin\gamma_0 \\
M_u & M_w & M_q & 0 \\
0 & 0 & 1 & 0
\end{bmatrix}, \quad
B = \begin{bmatrix}
X_{\delta_E} \\
Z_{\delta_E} \\
M_{\delta_E} \\
0
\end{bmatrix}
\quad \text{and} \quad
x = \begin{bmatrix}
u \\
w \\
q \\
\theta
\end{bmatrix}
$$

and $u_c = \delta_E$ = perturbation elevator deflection = control input, w_n is a white Gaussian process noise (random) with zero mean and covariance Q. Γ is a perturbation noise transition matrix. Now, the B matrix elements for longitudinal motion represent: $B(1) = X_{\delta_E}$ = X-axis force derivative with respect to elevator control surface, $B(2) = Z_{\delta_E}$ = Z-axis force derivative with respect to elevator control surface, $B(3) = M_{\delta_E}$ = pitching moment derivative with respect to elevator control surface. The state vector 'x' is obtained by integrating Equation 32.2 using Runge–Kutta fourth-order integration method. The measurement equation (mathematical model of sensor) is obtained as

$$z = Hx + v \tag{32.3}$$

where z is measurement vector and v is a white Gaussian measurement noise with zero mean and covariance R and is uncorrelated with process noise w_n. It is assumed that all the states are measurable. Therefore, observation matrix H of a sensor is kept as an identity matrix of size 4×4. The measured states are fed to the Kalman filter so as to estimate the actual states. The Kalman filter equations are given as follows:

State and Covariance Time-Propagation

$$\dot{\hat{x}} = A\hat{x} + Bu_c \tag{32.4}$$

$$P(k/k-1) = F(k/k-1)P(k-1/k-1)F^T(k/k-1) \\ + \Gamma(k/k-1)Q(k-1)\Gamma^T(k/k-1) \tag{32.5}$$

where $F = e^{AT}$ is the state transition matrix and T is sampling time interval.

Measurement Update/Data update

$$\hat{x}(k/k) = \tilde{x}(k/k-1) + K(k)\gamma(k) \tag{32.6}$$

where $\gamma(k)$ is innovation sequence and is expressed as

$$\gamma(k) = z(k) - H(k)\tilde{x}(k/k-1) \tag{32.7}$$

$$K(k) = P(k/k-1)H^T(k)S(k)^{-1} \tag{32.8}$$

Here, S is the covariance matrix of innovations and is given by

$$S(k) = H(k)P(k/k-1)H^T(k) + R(k) \tag{32.9}$$

$$P(k/k) = [I - K(k)H(k)]P(k/k-1) \tag{32.10}$$

Here, $P(k-1/k-1)$ is a covariance matrix of estimate errors at the preceding step, $K(k)$ is the gain matrix of the Kalman filter and I is an identity matrix.

To detect the faults that would change the mean of the innovation sequence, the following statistical function is used [1]:

$$\beta(k) = \sum_{j=k-M+1}^{k} \tilde{\gamma}^T(j)\tilde{\gamma}(j) \tag{32.11}$$

Here, M is the number of the samples (window length). The two hypothesis tests used to detect the faults are [1]

$$\left.\begin{array}{ll} \beta(k) \le \chi^2_{\alpha, Ms} & \text{then } H_0 \text{ (no fault)} \\ \beta(k) > \chi^2_{\alpha, Ms} & \text{then } H_1 \text{ (fault)} \end{array}\right\} \tag{32.12}$$

$\chi^2_{\alpha, Ms}$ is a threshold taken from chi-square table, α is probability of confidence level and 'Ms' is degree of freedom (DOF) which is equal to M, multiplied by 's' (no. of sensors). If the mean of the innovation sequence exceeds this statistical function value then fault is detected.

32.2.1.2 Sensor Fault Isolation Algorithm

For the isolation of sensor fault the approach of Ref. [1] is used. The s-dimensional innovation sequence is transformed into s–one-dimensional sequences. The statistics of the faulty sensor, that is, the variables related to this faulty sensor, are assumed to be affected much more than those of the other healthy sensors. The statistics which is a ratio of sample and theoretical variances $\hat{\sigma}^2_i / \sigma^2_i$ is used to verify the variances of one-dimensional innovation sequences $\tilde{\gamma}_i(k)$, $i = 1, 2, ..., s$ [1]:

$$\hat{\sigma}^2_i(k) = \frac{1}{M-1} \sum_{j=k-M+1}^{k} \left[\tilde{\gamma}_i(j) - \overline{\tilde{\gamma}}(k) \right]^2$$

$$\overline{\tilde{\gamma}}_i(k) = \frac{1}{M} \sum_{j=k-M+1}^{k} \tilde{\gamma}_i(j) \tag{32.13}$$

when $\tilde{\gamma}_i = N(0, \sigma_i)$ it is known that $(v_i / \sigma^2_i) \sim \chi^2_{\alpha, M-1}$; $\forall i$, $i = 1, 2, ..., s$. Here, $v_i = (M-1) \hat{\sigma}^2_i$; $\forall i$, $i = 1, 2, ..., s$. As $\sigma^2_i = 1$ for a normalized innovation sequence, it follows that,

$$v_i \sim \chi^2_{\alpha, M-1}, \quad \forall_i, i = 1, 2, ..., s \tag{32.14}$$

Using Equation 32.14, any change in the mean of the normalized innovation sequence can be detected and hence the ith sensor, in which γ_i exceeds the threshold value can be identified as the faulty sensor.

32.2.1.3 Sensor Fault Reconfiguration

On occurrence of the sensor fault the effect of the sensor fault to its channel is more significant which needs reconfiguration. Once the fault is detected in a particular channel, Kalman filter is reconfigured by ignoring the feedback from the faulty sensor and using measurements from healthy sensors only. Therefore, there will be no more faulty measurements, and KF estimates the states with reduced healthy measurements, thus providing the necessary reconfiguration.

32.2.2 FUZZY LOGIC-BASED SENSOR FAULT DETECTION AND RECONFIGURATION

In this case, perturbation elevator deflection is the input to the fuzzy module of the plant and true states are estimated as outputs. Then, measured states are compared with true estimated states and if their difference exceeds the threshold value, then the fault is detected. The fault detection automatically triggers the sensor isolation algorithm to locate the faulty channel and once the fault is located then the accommodation algorithm bypasses the faulty measurements, that is, the particular sensor measurement is ignored and replaced by the true estimated state. Thus, the reconfiguration is provided for sensor fault. Figure 32.1 shows the scheme for sensor fault detection and

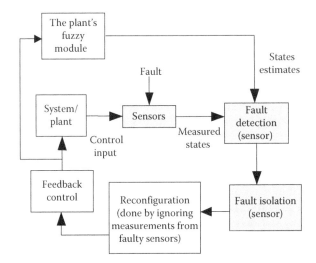

FIGURE 32.1 Schematic for fault detection, isolation and reconfiguration using fuzzy logic. (Reprinted from S. R. Savanur and A. V. Patel, *Defense Science Journal*, 60, 1, 76–86, January 2010. With permission.)

reconfiguration based on fuzzy logic. Input to the plant is perturbation elevator input deflection, that is, $u_c = \delta_E$ and universe of discourse (UOD) for the input is defined as UOD $u_c = [u_c^1 \; u_c^2 \; u_c^3 \; u_c^4 \; u_c^5]$. The triangular membership functions are constructed for input u_c on its UOD, as shown in Figure 32.2. The outputs are the four states of the system u, w, q and θ. Figure 32.3 shows the general shape of the membership functions for the states of the system (e.g., perturbation velocity along x-axis, u) which are unsymmetrical and triangular. To partition the outputs, triangular membership functions for each output are constructed on their respective UODs. For an output state x (representing w, q and θ) it is chosen as UOD $x = [x^1 \; x^2 \; x^3 \; x^4 \; x^5]$. Table 32.1 gives If–Then fuzzy rules described for output state 'u', and Table 32.2 gives Fuzzy Associative Memory (FAM) table for output states 'w', 'q' and 'θ'. Using the inference rules of Tables 32.1 and 32.2, a simulation of the (developed) fuzzy system is carried out. Each simulation cycle results in membership functions for the four outputs. The membership functions for the outputs are defuzzified based on inferred rule and intersection T-norm to get their crisp values. The true states are estimated. For detection and reconfiguration of sensor fault, the estimated states are compared with the respective measured states of the system (aircraft dynamics) and if the difference (or the residual) exceeds the pre-computed threshold of the measurement noise (for respective channel), then fault is detected in that sensor. For this purpose, the thresholds are computed using Monte–Carlo simulation of random noises (for all four sensors with a specified standard deviation) for 1000 runs. The minimum and maximum values of these noises are computed for each run and then average of all run are considered as thresholds for detection purpose. If the fault is detected then output of that faulty sensor is

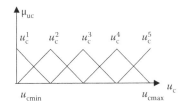

FIGURE 32.2 Fuzzy membership functions (input variable u_c). (Reprinted from S. R. Savanur and A. V. Patel, *Defense Science Journal*, 60, 1, 76–86, January 2010. With permission.)

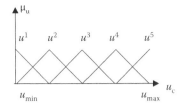

FIGURE 32.3 For u-fuzzy membership functions. (Reprinted from S. R. Savanur and A. V. Patel, *Defense Science Journal*, 60, 1, 76–86, January 2010. With permission.)

TABLE 32.1

If–Then Rules for Velocity u

If control input u_c is u_c^1 Then output u is u^1.

If control input u_c is u_c^2 Then output u is u^2.

If control input u_c is u_c^3 Then output u is u^3.

If control input u_c is u_c^4 Then output u is u^4

If control input u_c is u_c^5 Then output u is u^5.

Source: Reprinted from S. R. Savanur and A. V. Patel, *Defense Science Journal*, 60, 1, 76–86, January 2010. With permission.

TABLE 32.2

If–Then Rules for x (w, q and θ)

If control input u_c is u_c^1 Then output x is x^5.

If control input u_c is u_c^2 Then output x is x^5.

If control input u_c is u_c^3 Then output x is x^4.

If control input u_c is u_c^4 Then output x is x^2.

If control input u_c is u_c^5 Then output x is x^2.

Source: Reprinted from S. R. Savanur and A. V. Patel, *Defense Science Journal*, 60, 1, 76–86, January 2010. With permission.

---

ignored and it is replaced by the respective estimated state. Hence, the reconfiguration for sensor fault is carried out. This method is also capable of handling multiple sensor faults, that is, if two sensors become faulty, then fault is detected in the faulty sensors and reconfiguration is provided by replacing measurements of faulty sensors with estimated values. It must be emphasized here that we can consider this as a static-reconfiguration, since there is no dynamics involved, just that the faulty sensor is cut-out and the measurements from the healthy sensors are used for future computations.

32.3 CONTROL SURFACE FAULT DETECTION AND RECONFIGURATION

In many situations, the aircraft would become unstable due to a fault in an actuator or if there is a loss of control surface effectiveness due to damaged or blown off/out surfaces. Hence, these faults must be detected and accommodated immediately and effectively using control surface fault detection and reconfiguration (CSFDR).

32.3.1 EKF Implementation

An actuator surface fault detection algorithm based on EKF is studied and used for estimation of elements of the control-input matrix [6].

32.3.1.1 Control Distribution Matrix Identification

Figure 32.4 shows the CSFDR scheme. The $b_{i,j}$ ($i = 1, n; j = 1, m$) elements of the control distribution matrix B are identified using EKF to detect the actuator surface faults. For this purpose, the state vector x is augmented as follows:

$$x_a = [x_1, x_2, \ldots, x_n, b_{11}, b_{12}, \ldots, b_{ij}, \ldots, b_{nm}] \tag{32.15}$$

The augmented dynamic system is represented by

$$x_a(k+1) = \tilde{F}(k+1,k)x_a(k) + \tilde{\Gamma}(k+1,k)w(k) \tag{32.16}$$

The measurement equation is given as

$$\tilde{z}(k) = \tilde{H}(k)x_a(k) + v(k) \tag{32.17}$$

Here, x_a is an $(n + nm)$–dim. augmented system state vector, $\tilde{F}(K, K+1)$ is a $(n + nm)$ by $(n + nm)$ augmented system matrix, $\tilde{\Gamma}(K, K+1)$ is an $(n + nm)$ by $(n + nm)$ augmented perturbation noise transition matrix and $\tilde{z}(k)$ is an s by nm-dimensional system measurement matrix. The matrix $\tilde{F}$ is the discrete form of the system matrix $\tilde{A}$ where [6],

$$\tilde{A} = \begin{bmatrix} A_{n \times n} & \begin{bmatrix} u_1 \ldots u_m \; 0 \ldots\ldots\ldots\ldots 0 \\ 0 \ldots\ldots 0 \; u_1 \ldots u_m \; 0 \ldots\ldots 0 \\ \ldots\ldots\ldots\ldots\ldots\ldots\ldots\ldots\ldots\ldots \\ 0 \ldots\ldots 0 \; \ldots\ldots\ldots 0 \; u_1 \ldots\ldots u_m \end{bmatrix} \\ 0_{nm \times n} & I_{nm \times nm} \end{bmatrix}$$

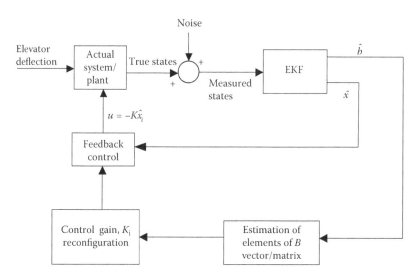

FIGURE 32.4 Control surface fault detection and reconfiguration scheme using EKF. (Reprinted from S. R. Savanur and A. V. Patel, *Defense Science Journal*, 60, 1, 76–86, January 2010. With permission.)

In discrete form of $\tilde{A}$ is given as

$$\tilde{F} = e^{\hat{A}T} \tag{32.18}$$

The EKF estimation algorithm is given as follows:
State and Covariance Time Propagation:

$$\tilde{x}_a(k,k-1) = \int \tilde{A}\hat{x}_a(k-1,k-1) \tag{32.19}$$

$$\begin{aligned} P(k,k-1) &= \tilde{F}(k,k-1)P(k-1,k-1)\tilde{F}^T(k,k-1) \\ &+ \tilde{\Gamma}(k,k-1)Q(k-1)\tilde{\Gamma}^T(k,k-1) \end{aligned} \tag{33.20}$$

Measurement Update/Data update:

$$\tilde{x}_a(k,k-1) = \int \tilde{A}\hat{x}_a(k-1,k-1)$$

$$\hat{x}_a(k,k) = \tilde{x}_a(k,k-1) + K(k)\gamma(k) \tag{32.21}$$

Here, $\gamma(k) = \tilde{z}(k) - \tilde{H}(k)\tilde{x}_a(k,k-1)$ is innovation sequence

$$K(k) = P(k,k-1)\tilde{H}^T(k)\left[\tilde{H}(k)P(k,k-1)\tilde{H}^T(k) + R(k)\right]^{-1} \tag{32.22}$$

$$P(k/k) = [I - K(k)\tilde{H}(k)]P(k/k-1) \tag{32.23}$$

32.3.1.2 Control Reconfiguration Algorithm

The state feedback method [7,8] can be used to improve the stability properties of the control system as follows:
Consider the basic state equation of the dynamic system:

$$\dot{x} = Ax + Bu \tag{32.24}$$

Then, the state feedback control law is expressed as

$$u = -Kx \tag{32.25}$$

Here, $K = [k_1 \, k_2 \, k_3 \, k_4 \, k_5]$ is a constant state feedback gain matrix. By substituting this state feedback control law (i.e., for u) into Equation 32.24, we obtain the closed-loop system described by the augmented state equation

$$\dot{x} = (A - BK)x \tag{32.26}$$

First, the feedback control law is designed for fault-free system. Then under surface fault condition, control reconfiguration is realized using the pseudo inverse technique described next. Let the dynamics of the closed system be expressed as

$$\dot{x}_0 = (A - B_0 K_0)x_0 \tag{32.27}$$

After an actuator surface fault occurs, the dynamics may be represented as

$$\dot{x}_i = (A - B_i K_i)x_i \tag{32.28}$$

To ensure that the closed-loop dynamics are the same, as before, the following condition must be satisfied:

$$B_0 K_0 = B_i K_i \qquad (32.29)$$

Here, B_0 is unimpaired control distribution matrix (of the healthy systems), K_0 is gain matrix for unimpaired system and B_i is estimated (by EKF) control distribution matrix, after the impairment (i.e., after the fault has occurred) and K_i is gain matrix for thus the impaired system. Then the gain matrix for the impaired system is obtained as

$$K_i = B_i^{\#} B_0 K_0 \qquad (32.30)$$

Here, the matrix $B_i^{\#}$ is the pseudo inverse of the matrix B_i. Thus, the new gain matrix for control surface fault reconfiguration is computed from Equation 32.30. The state feedback is then used for reconfiguration by computing the new control input according to Equation 32.25.

32.3.2 Fuzzy Logic-Based Control Surface Fault Detection and Reconfiguration

Figure 32.5 shows the scheme of actuator fault detection and reconfiguration. The perturbed elevator deflection is the input to the actual plant (with actuator fault). The differences in output states of nominal plant and those of actual faulty plant are used as the inputs to the fuzzy module. Correction factor for loss of effectiveness of control surface is used as the desired output. Seven triangular fuzzy membership functions, with two overlapping membership functions, are constructed for both the inputs. To partition the control output, that is, correction factor, seven triangular fuzzy membership functions, two membership functions overlapping, are constructed on its UOD in the range of 0–1. To determine the factor of effectiveness of control surface, in the case of loss of effectiveness of control surface fault, the errors are computed for different factors of effectiveness ranging from 0 to 1. The fuzzy membership functions for error inputs e_1 and e_2 are shown in Figures 32.6 and 32.7. The errors in the first and second channels (errors in u and w states denoted as e_1 and e_2, respectively) are used as inputs to fuzzy model in determining the factor of

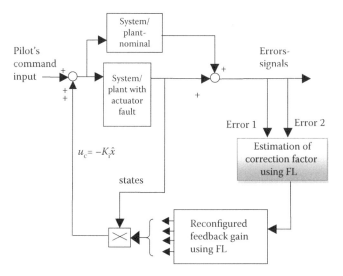

FIGURE 32.5 Parameter estimation and reconfiguration using FL. (Reprinted from S. R. Savanur and A. V. Patel, *Defense Science Journal*, 60, 1, 76–86, January 2010. With permission.)

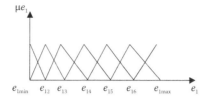

FIGURE 32.6 Error e_2-Fuzzy membership functions (error for w). (Reprinted from S. R. Savanur and A. V. Patel, *Defense Science Journal*, 60, 1, 76–86, January 2010. With permission.)

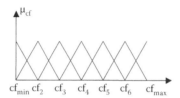

FIGURE 32.7 Output correction factor (cf)—Fuzzy membership functions. (Reprinted from S. R. Savanur and A. V. Patel, *Defense Science Journal*, 60, 1, 76–86, January 2010. With permission.)

TABLE 32.3

Fuzzy Associate Memory (FAM) for Correction Factor (cf)

		$e_2 \rightarrow$					
		e_2^1	e_2^2	e_2^3	e_2^4	e_2^5	e_2^6	e_2^7
	e_1^1	cf^4	cf^4	cf^4	cf^4	cf^4	cf^2	cf^1
	e_1^2	cf^5	cf^5	cf^4	cf^4	cf^4	cf^2	cf^2
	e_1^3	cf^6	cf^6	cf^4	cf^4	cf^3	cf^3	cf^2
$e_1 \downarrow$	e_1^4	cf^6	cf^4	cf^4	cf^4	cf^4	cf^3	cf^2
	e_1^5	cf^6	cf^5	cf^4	cf^4	cf^4	cf^3	cf^2
	e_1^6	cf^6	cf^4	cf^4	cf^4	cf^4	cf^3	cf^2
	e_1^7	cf^5	cf^4	cf^4	cf^4	cf^4	cf^1	cf^1

Source: Reprinted from S. R. Savanur and A. V. Patel, *Defense Science Journal*, 60, 1, 76–86, January 2010. With permission.

effectiveness. Since, it is found that the errors in u and w are comparatively much larger and sufficient to estimate the factor of effectiveness of the control surface, only these two errors are used in further computations. The Fuzzy-Inference rules are constructed in 7×7 Fuzzy Associative Memory (FAM) as shown in Table 32.3, where the entries are the control actions or reference points of membership functions of the output corresponding to particular rule. Using these rules as expressed in FAM the simulation of the problem is carried out for fuzzy-logic-related operations. Each simulation cycle will result in membership function for the two input variables. From the FAM the output contributed from each rule is inferred. The basic function of the fuzzy inference engine is to compute the overall value of the output variable based on the individual contributions of each rule from the rule base. Timothy Ross [9] and Driankov [10] have explained how each such individual contribution represents the value of the output variable as computed by a single rule. First, the degree of match between the crisp inputs and fuzzy sets describing the meaning of the rule antecedent is computed for each rule by using the triangular norm (T-norm intersection or algebraic product) as follows:

$$\mu_A(u) = \mu_{ant}(e_1, e_2)$$

$$\mu_{ant}(e_1, e_2) = \min(\mu(e_1), \mu(e_2)) \quad \text{intersection T-norm} \tag{32.31}$$

$$\mu_{ant}(e_1, e_2) = \mu(e_1) \cdot \mu(e_2) \quad \text{algebraic product T-norm} \tag{32.32}$$

Then, based on this degree of match, the clipped fuzzy set representing the value of the output variable is determined via one of the inference methods. As a sensitivity study, different implication methods are used in estimating the factor of effectiveness or correction factor and the results obtained are compared.

32.3.2.1 Effects of Different Implication/Inference Methods

The different implication methods [3,4,11] used for comparison are (i) Mamdani's minimum implication, (ii) Larsen's product implication, (iii) bounded difference and (iv) drastic product or intersection implication. The details on the fuzzy implication methods are omitted for brevity here. Table 32.4 gives the outcome of T-norm operator for rule antecedent, defined in Equations 32.31 and 32.32 for ith rule, control decision due to different implication methods and area of the clipped output fuzzy set [3,4]. Then, the fuzzy outputs suggested by each rule are aggregated. Defuzzification is then applied to get crisp values for the output or correction factor. The Centroid method is used in determining the defuzzified value of output. This is given by the expression

$$\text{crisp value of output} = \frac{\sum \text{value of output member} \times \text{inferred area of output member}}{\sum \text{inferred area of output member}} \tag{32.33}$$

Once the correction factor (output) is determined then the actual B-matrix elements under control surface fault conditions are determined and subsequently new control gain is determined by another fuzzy module. For this module, the input is the estimated correction factor and output is the feedback control gain matrix, K_i. The elements of matrix K_i, where $K_i = [K_1 \ K_2 \ K_3 \ K_4]$ are

TABLE 32.4
Fuzzy Implications/Inference Methods

Sl. No.	Fuzzy Reasoning Method/ Fuzzy Implication Method	Control Decision Due to ith Rule $\mu_{A_i}(u) = \mu_{ant_i}(e_1, e_2)$	Remarks
1	Mamdani's minimum (MM)	$\mu_{A_i}(u) \wedge \mu_{B_i}(v)$	$\mu_{A_i}(u)$ is the resulting membership of the
2	Larsen' product (LP)	$\mu_{A_i}(u) \ \mu_{B_i}(v)$	T-norm operation on the antecedent part of
3	Bounded product (BP)	$\max(0, \mu_{A_i}(u) + \mu_{B_i}(v) - 1)$	the ith rule and $\mu_{B_i}(v)$ is memberships of the fuzzy set B of the output variable for ith rule
4	Drastic product (DP)	$\mu_{A_i}(u) \quad \text{if } \mu_{B_i}(v) = 1$ $\mu_{B_i}(v) \quad \text{if } \mu_{A_i}(u) = 1$ $0 \qquad \text{otherwise}$	

Source: Reprinted from S. R. Savanur and A. V. Patel, *Defense Science Journal*, 60, 1, 76–86, January 2010. With permission.

TABLE 32.5

If–Then Rules for k_1 (K_1 and K_2)

If input cf is cf^1 Then output k_1 is k_i^7.

If input cf is cf^2 Then output k_1 is k_i^6.

If input cf is cf^3 Then output k_1 is k_i^5.

If input cf is cf^4 Then output k_1 is k_i^4.

If input cf is cf^5 Then output k_1 is k_i^3.

If input cf is cf^6 Then output k_1 is k_i^2.

If input cf is cf^7 Then output k_1 is k_i^1.

Source: Reprinted from S. R. Savanur and A. V. Patel, *Defense Science Journal*, 60, 1, 76–86, January 2010. With permission.

TABLE 32.6

If–Then Rules for k_2 (K_3 and K_4)

If input cf is cf^1 Then output k_2 is k_2^1

If input cf is cf^2 Then output k_2 is k_2^2

If input cf is cf^3 Then output k_2 is k_2^3

If input cf is cf^4 Then output k_2 is k_2^4

If input cf is cf^5 Then output k_2 is k_2^5

If input cf is cf^6 Then output k_2 is k_2^6

If input cf is cf^7 Then output k_2 is k_2^7

Source: Reprinted from S. R. Savanur and A. V. Patel, *Defense Science Journal*, 60, 1, 76–86, January 2010. With permission.

defined in their respective UOD and partitioned into seven fuzzy partitions, for example, UOD $K_1 = [K_1^1 \ K_1^2 \ K_1^3 \ K_1^4 \ K_1^5 \ K_1^6 \ K_1^7]$. The inference rules for determining new control gain matrix elements are as given in Tables 32.5 and 32.6. From the new computed control gain matrix, reconfiguration is provided using state feedback, that is, $u = -K_i \ x_i$, where x_i = states under fault conditions. Hence, the control law is reconfigured for actuator (surface) fault.

32.4 SIMULATION RESULTS AND DISCUSSIONS

For numerical simulation the longitudinal dynamics of a Delta-4 aircraft are considered [12]. The state-space matrices for the given model are

$$A = \begin{bmatrix} -0.033 & 0.0001 & 0.0 & -9.81 \\ 0.168 & -0.367 & 260 & 0.0 \\ 0.005 & -0.0064 & -0.55 & 0.0 \\ 0.0 & 0.0 & 1.0 & 0.0 \end{bmatrix}, \quad B = \begin{bmatrix} 0.45 \\ -5.18 \\ -0.91 \\ 0.00 \end{bmatrix},$$

$$C = I(4 \times 4), \quad \Gamma = I(4 \times 4)$$

The measurement noise vector is considered as: $v = [0.36*randn; 0.3*randn; 0.15*randn; 0.1*randn]$.

32.4.1 SIMULATION RESULTS FOR SENSOR FAULT DETECTION AND RECONFIGURATION

For the purpose of simulation, the sensor fault is modelled by adding to the appropriate sensor measurement a fixed bias shift. Thus, to introduce a sensor fault, a bias of 4 is added at the certain iteration/instance (say at 300th), changing the mean value of the innovation sequence in the first measurement channel as follows: $v = [1 + 0.36*randn; 0.3*randn; 0.15*randn; 0.1*randn]$. For multiple sensor faults, fixed bias is added in the first and third measurement channels as: $v = [1 + 0.36*randn; 0.3*randn; 1 + 0.15*randn; 0.1*randn]$. For the simulation the sampling interval is chosen as $T = 0.01$ s and the number of iterations is $N = 1000$. Figure 32.8 shows true (non-faulty), measured (with fault) and estimated reconfigured states. It can be seen that fault is detected in the first measurement channel at 301st iteration (just after 3 s) when the difference between measured and true values exceeds pre-computed threshold bound. The faulty measurement is ignored and the estimated values are used. Figure 32.9 shows the time histories of the filter's residuals in all channels and the respective threshold bounds when there is a fault in a single sensor, that is, in the 1st channel with and without reconfiguration using fuzzy logic. Figure 32.10 shows the time histories of true (non-faulty), measured (with fault) and estimated reconfigured states. Figure 32.11 shows the time histories of residuals in all channels and threshold bounds when the fault is in multi-sensors, that is, in 1st and 3rd channels with and without reconfiguration using fuzzy logic. It can be seen that the fault is detected in the first and third measurement channels at 301st iteration (fault is introduced at the same time in both channels) when the difference between measured and true values exceeds pre-computed threshold bounds for respective measurement channels. From these plots it is seen that detection and reconfiguration of sensor fault are effectively provided using fuzzy logic. It is seen from Figure 32.11 that detection and reconfiguration can be effectively provided for multi-sensor faults using fuzzy logic without additional computational burden. Figure 32.12 shows the errors between the true and measured states with and without reconfiguration using both fuzzy logic and KF schemes for sensor fault in the first measurement channel. It can be seen that the residual with reconfiguration lies within the threshold bounds, here the states with closed-loop state feedback are shown.

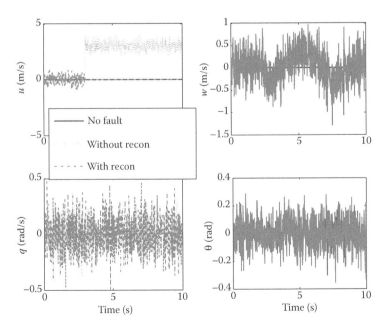

FIGURE 32.8 True, measured and reconfigured states using fuzzy logic for fault in the 1st channel. (Reprinted from S. R. Savanur and A. V. Patel, *Defense Science Journal*, 60, 1, 76–86, January 2010. With permission.)

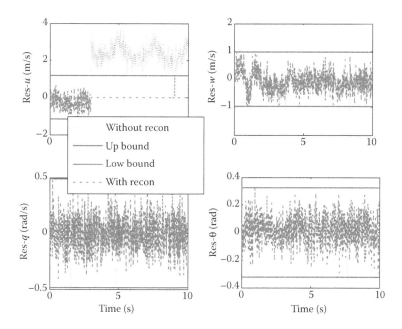

FIGURE 32.9 Time histories of residuals with/without reconfiguration using fuzzy logic (for fault in the 1st channel). (Reprinted from S. R. Savanur and A. V. Patel, *Defense Science Journal*, 60, 1, 76–86, January 2010. With permission.)

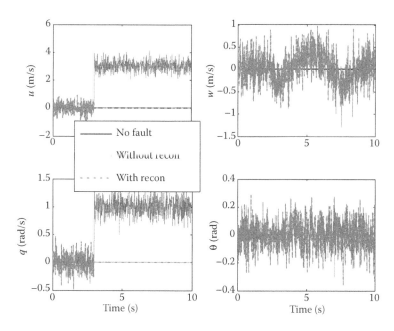

FIGURE 32.10 True, measured and reconfigured states using fuzzy logic (faults in the 1st and 3rd channels). (Reprinted from S. R. Savanur and A. V. Patel, *Defense Science Journal*, 60, 1, 76–86, January 2010. With permission.)

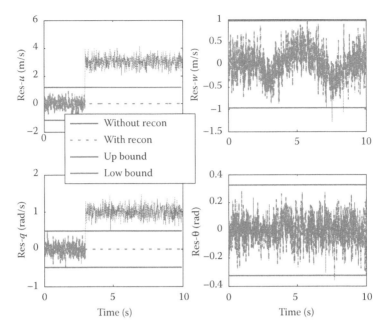

FIGURE 32.11 Time histories of residuals without/with reconfiguration using fuzzy logic (faults in the 1st and 3rd channels). (Reprinted from S. R. Savanur and A. V. Patel, *Defense Science Journal*, 60, 1, 76–86, January 2010. With permission.)

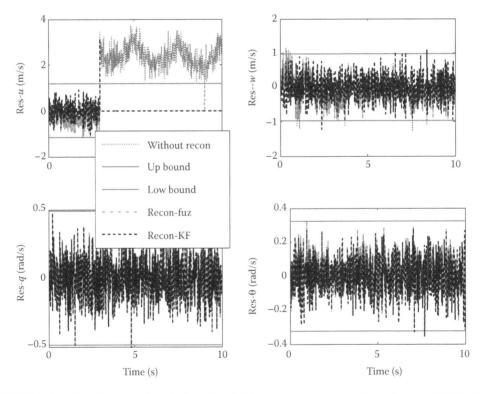

FIGURE 32.12 Time histories of residuals without/with reconfiguration using fuzzy logic and KF (faults in 1st channel). (Reprinted from S. R. Savanur and A. V. Patel, *Defense Science Journal*, 60, 1, 76–86, January 2010. With permission.)

32.4.2   Simulation Results for Control Surface Fault Detection and Reconfiguration

For the simulation, the longitudinal dynamics of a delta-4 aircraft are considered. To simulate the fault, factor of effectiveness is changed to 50% of the normal value (100%). Hence, B-matrix elements are multiplied by 0.5. In the EKF method, for the reconfiguration, feedback control gain K_0 for fault-free plant is determined using the LQR (linear–quadratic–Gaussian optimization) technique. The values for the LQR method are chosen as: QQ (Q in LQR) = zeros(4,4); QQ(1) = 0.01; Q(2,2) = 0.00001; QQ(3,3) = 0.00001; QQ(4,4) = 0.00001. With this, the control gain for fault-free aircraft 'K_0' value is obtained as: K_0 = [0.0887, 0.0046, −0.8978, −2.1077] which is used in reconfiguration. In fuzzy logic control, the correction factor is estimated under fault condition using the errors in states for a factor of effectiveness of 50% and then this is used in determining the control gain. Figures 32.13 through 32.16 show the comparison of the closed-loop responses of unimpaired, impaired without reconfiguration and reconfigured (after) impaired aircraft using fuzzy logic and KF schemes. It is observed from these plots that the reconfigured states converge (tend to do so) to those of the unimpaired one in both schemes. Figure 32.17 shows the estimated values of control distribution matrix using model-based EKF and non-model-based fuzzy logic schemes. It is seen that the estimated parameters are (nearly) close to the true values for both the schemes and the delay in estimation is noticed in fuzzy logic scheme, this delay indicates that some fine tuning of membership functions and inference rules is required. Figure 32.18 shows the error time histories of the estimated states for the cases with and without reconfiguration using both fuzzy logic and KF schemes. From the plots it can be interpreted that reconfigured aircraft states converge to the true states in both cases and hence the error between the true (unimpaired) and reconfigured states is reduced. For perturbation velocity along x-axis u, the error is less using KF; otherwise error is less in all other states using fuzzy logic.

32.4.3   Sensitivity Study Using Two T-Norms and Some Implication Methods

For this study the results using fuzzy intersection T-norm and different fuzzy implication methods for the estimation of factor of effectiveness for control surface fault are shown in Figure 32.19. It

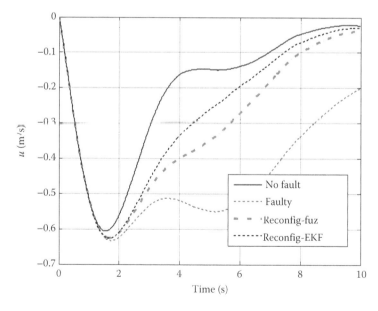

FIGURE 32.13   Velocity along X-axis for impaired, unimpaired and reconfigured (control gain for the aircraft) using FL and EKF. (Reprinted from S. R. Savanur and A. V. Patel, *Defense Science Journal*, 60, 1, 76–86, January 2010. With permission.)

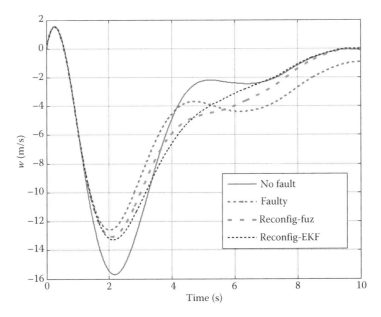

FIGURE 32.14 Velocity along Z-axis for impaired, unimpaired and reconfigured control gain (for the aircraft) using EKF and FL. (Reprinted from S. R. Savanur and A. V. Patel, *Defense Science Journal*, 60, 1, 76–86, January 2010. With permission.)

can be seen that the estimation is satisfactory using most of the methods. Using the estimated value of factor of effectiveness, the reconfiguration is carried out. Figure 32.20 shows the results using algebraic product T-norm and different implication methods for estimation of factor of effectiveness under control surface fault. It can be seen that the estimation is (mostly) satisfactory for different implication methods using the algebraic product T-norm also.

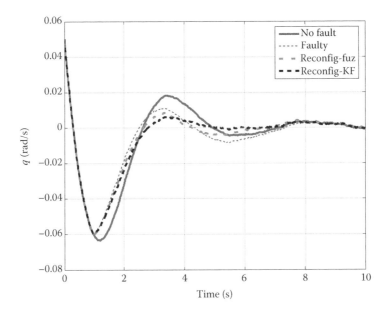

FIGURE 32.15 Pitch rate for impaired, unimpaired and reconfigured control gain (for the aircraft) using EKF and FL. (Reprinted from S. R. Savanur and A. V. Patel, *Defense Science Journal*, 60, 1, 76–86, January 2010. With permission.)

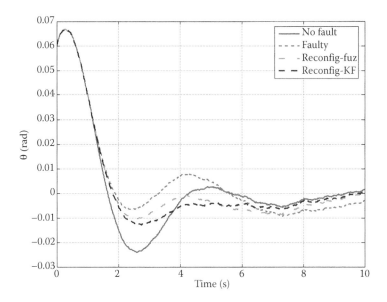

FIGURE 32.16 Pitch angle for impaired, unimpaired and reconfigured control gain (for the aircraft) using EKF and FL. (Reprinted from S. R. Savanur and A. V. Patel, *Defense Science Journal*, 60, 1, 76–86, January 2010. With permission.)

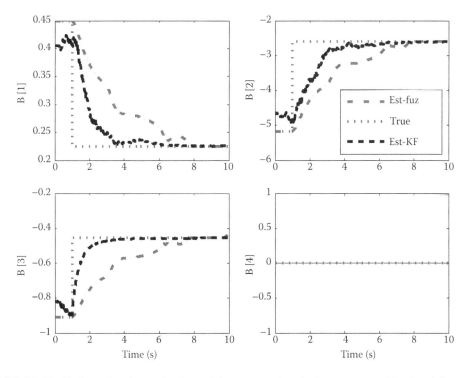

FIGURE 32.17 Estimated and actual values of input vector/matrix b-parameters. (Reprinted from S. R. Savanur and A. V. Patel, *Defense Science Journal*, 60, 1, 76–86, January 2010. With permission.)

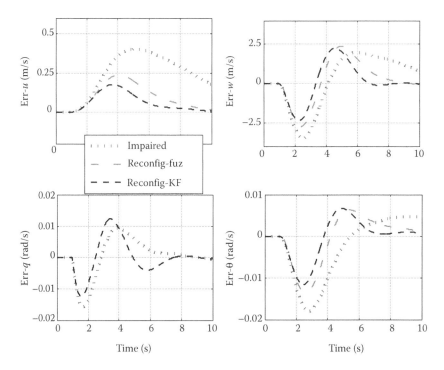

FIGURE 32.18 State–error time histories with state feedback control (without and with reconfiguration using FL and EKF). (Reprinted from S. R. Savanur and A. V. Patel, *Defense Science Journal*, 60, 1, 76–86, January 2010. With permission.)

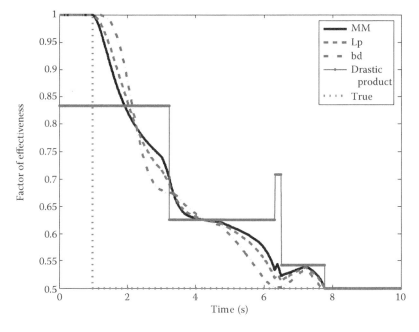

FIGURE 32.19 Estimated values of factor of effectiveness using different implication methods and intersection T-norm of fuzzy logic. (Reprinted from S. R. Savanur and A. V. Patel, *Defense Science Journal*, 60, 1, 76–86, January 2010. With permission.)

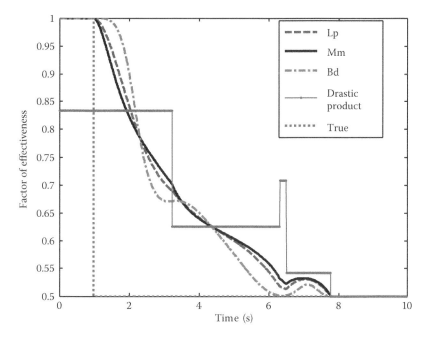

FIGURE 32.20 Estimated values of factor effectiveness using different implication methods and using of algebraic product T-norm of fuzzy logic. (Reprinted from S. R. Savanur and A. V. Patel, *Defense Science Journal*, 60, 1, 76–86, Jan. 2010, with permission.)

32.5 CONCLUDING REMARKS

In this chapter, we have studied and evaluated the performance of fuzzy logic-based methods for sensor fault detection and reconfiguration. The results are compared with those obtained using Kalman filter. It is also shown that using fuzzy logic, multi-sensor faults can also be detected and reconfiguration can be effectively provided. Model-based EKF is used in the parameter estimation for control surface fault. Fuzzy logic is used to estimate the factor of effectiveness in the case of a control surface fault and thereby new control gain is determined. In both the schemes, state feedback is adopted to reconfigure for control surface fault. For fuzzy logic-based methods, some fuzzy implication methods are used for the study and to determine the parameters of control distribution matrix, and a sensitivity study of performance has been carried out successfully.

REFERENCES

1. C. Hajiyev and F. Caliskan, Sensor/actuator fault diagnosis based on statistical analysis of innovation sequence and robust Kalman filter, *Aerospace Science and Technology*, 4, 2000, 415–422.
2. Ch. M. Hajiyev and F. Caliskan, Integrated sensor/actuator FDI and reconfigurable control for fault tolerant flight control system design, *The Aeronautical Journal*, 105, September 2001, 525–533.
3. A. V. Patel and B. M. Mohan, Analytical structures and analysis of the simplest fuzzy PI controllers, *Automatica*, 38, 2002, 981–993.
4. A. V. Patel, Analytical structures and analysis of the simplest fuzzy PD controllers with multi fuzzy sets having variable cross-point level, *Fuzzy Sets and Systems,* 129, 2002, 311–334.
5. J.R. Raol, G. Girija and J. Singh, Modeling and parameter estimation of dynamic systems, *IEE Control Series Book*, 65, IET/IEE, London, 2004.
6. C. Hajiyev and F. Caliskan, *Fault Diagnosis and Reconfiguration in Flight Control Systems*, Kluwer Academic Publishers, Boston, 2003.
7. G. F. Franklin and J. D. Powel, *Michael Workman, Digital Control of Dynamic Systems*, 3rd Edition, Pearson Education, Pte. Ltd, Indian Branch, Delhi, 2003.

8. K. Ogata, *Modern Control Engineering*, 4th Edition, Pearson Education, Pte. Ltd, Indian Branch, Delhi, 2005.

9. T. Ross, *Fuzzy Logic with Engineering Applications*, 3rd Edition, Wiley, USA, 1997.

10. D. Driankov, H. Hellendoorn and M. Reinfrank, *An Introduction to Fuzzy Control*, Narosa Publishing House, Delhi, India, 2001.

11. S. K. Kashyap and J. R. Raol, *Unification and Interpretation of Fuzzy Set Operations*, Project document, PDFCO502, Flight Mechanics and Control Division, National Aerospace Laboratories, Bangalore, March 2005.

12. D. McLean, *Automatic Flight Control Systems*, Prentice-Hall International, UK, 1990.

13. S. R. Savanur and A. V. Patel, Sensor/control surface fault detection and reconfiguration using fuzzy logic, In Sp. Issue, Mobile Intelligent Autonomous Systems, Eds. J. R. Raol and A. Gopal, *Defense Science Journal*, 60, 1, 76–86, January 2010.

33 Target Tracking Using a 2D Radar

Marelize Kriel and Herman le Roux

CONTENTS

33.1 INTRODUCTION

This chapter briefly outlines a few mathematical techniques to track targets in 3D using a 2D radar. 2D radars are relatively cheap and efficient sensors that often form the first line of defence in airspace control. In military applications they are often used as early warning devices because they can detect approaching enemy aircraft or missiles at great distances. In case of an attack, early detection of the enemy is vital for a successful defence against attack. Depending on the threat evaluation of tracked aircraft, the tracking process is passed along to 3D search radars or fire control tracking radars once it comes within range of those sensors. A key component in the above hierarchy is the threat evaluation component. It relies on many factors such as angle of incidence towards defended assets, time to approach to defended asset, speed of target and so forth. The normal 2D radar provides range and azimuth but the altitude of the target is omitted. This can be an important consideration as aircraft altitude limits the attack profiles a target can fly [1].

33.2 HEIGHT ESTIMATION

The current literature regarding height estimation restricts itself to computations involving two or more 2D radars where the height can be completely determined by simple geometric computations. In this section, we present some mathematical methods to infer aircraft altitude from two updates given by a single 2D radar [2]. A single 2D radar source cannot directly determine the altitude of aircraft, thus naturally, the method presented here is either coupled with a number of assumptions and limitations or is a mere approximation. The terms *height* and *altitude* are used interchangeably. Height often refers to the height of an aircraft above ground level, and altitude the height of the aircraft above mean sea level. The proposed techniques do not consider terrain,

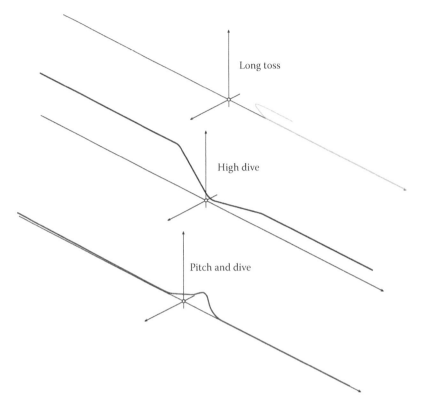

FIGURE 33.1   Flight profile examples.

terrain height, or height above mean sea level, but rather the difference in height of the sensor and observed aircraft.

It should also be noted that if the aircraft is flying perfectly tangential to the radar beam, then the radial speed component is zero and it is impossible to estimate its altitude. Conversely, however, it is more accurate to determine the altitude of an aircraft flying at great speeds at a 45° angle to the radar beam, than a slow flying aircraft that is flying towards the radar. The aircraft speed is instrumental in determining the aircraft altitude. The accuracy to which these speeds are known is directly proportional to the accuracy to which the altitude can be determined. Knowledge of aircraft speed can be obtained in a variety of ways. For example, due to the volatile nature of their payload, the speed at which bombers fly is usually controlled by doctrine, similarly cruise missiles fly at known speeds. We will make use of the three known flight profiles depicted in Figure 33.1 as examples. The defended asset as well as the sensor is located at the origin.

In the discussion that follows we will make use of two sequential sensor readings at time t_1 and t_2. The given data sets will consist of a slant range and azimuth reading, denoted (r_1, θ_1) and (r_2, θ_2). If the aircraft speed, denoted v_2, is known, then we can easily determine the distance travelled between time t_1 and time t_2 as $u_2 = v_2 (t_2 - t_1)$. In Figure 33.2, we will, without loss of generality, assume that $r_1 \geq r_2$.

33.2.1   Using Doppler Measurements

Modern 2D radar sensors allow Doppler measurements of one (radial) component of the velocity of the aircraft it is observing. In other words, Doppler measurements do not give us the velocity vector, $\bar{v}_2$, but only the magnitude of its radial component, denoted $\hat{v}_2$, at time t_2. From the measured radial

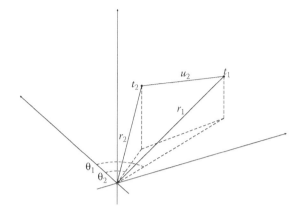

FIGURE 33.2 Radar tracking in 2D.

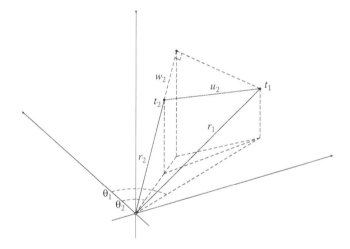

FIGURE 33.3 Radar tracking in 2D with Doppler data.

speed $\hat{v}_2$ we can easily determine the radial distance travelled $w_2 = \hat{v}_2(t_2 - t_1)$ and from that the total distance travelled is easily obtainable:

$$u_2 - \sqrt{r_1^2 - (t_2 + w_2)^2 + w_2^2} \tag{33.1}$$

It is clear from Figure 33.3 that the height can still not be directly calculated, even with Doppler measurements at hand, since the relations between these known values are valid at any height. In short, without any height-dependent data, there is no general way to directly compute the height.

33.2.2 SPECIAL CASE

If we make an assumption that the aircraft is flying radially towards/away from the radar at level height and known speed (Figure 33.4), then we can compute the height with simple trigonometry as

$$h_1 = h_2 = r_1 \sqrt{1 - \left(\frac{r_2^2 - r_1^2 - u_2^2}{-2r_1 u_2}\right)^2} \tag{33.2}$$

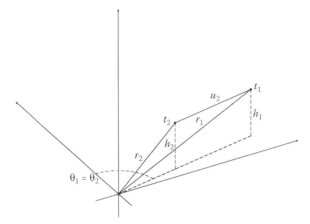

FIGURE 33.4 Simple case.

We will refer to this method as the AASC (Altitude Approximation via Simple Case) method to find an approximation to h_1 without knowing whether or not the assumptions of radial, straight flight hold true. The accuracy of this approximation is entirely dependent on the flight profile of the aircraft as illustrated in Figure 33.5. The model is usable without Doppler data as long as we have an accurate estimate of the speed of the aircraft, or more precisely, u_2, so it seems natural to proceed with a sensitivity analysis to see how sensitive this model is to changes in the given u_2.

Assume that the aircraft is flying level and radially towards the sensor. We have $0 < h = h_1 = h_2 < r_1$ and $\theta = \theta_1 - \theta_2 = 0$. Thus

$$u = u_2 = \sqrt{r_1^2 + r_2^2 - 2h^2 - 2\sqrt{r_1^2 - h^2}\sqrt{r_2^2 - h^2}} \tag{33.3}$$

Differentiation yields

$$\frac{\partial h}{\partial u} = \sqrt{r_1^2 - h^2}\sqrt{r_2^2 - h^2}\left(h\sqrt{r_1^2 + r_2^2 - 2h^2 - 2\sqrt{r_1^2 - h^2}\sqrt{r_2^2 - h^2}}\right)^{-1} \tag{33.4}$$

from which it is clear that

$$\frac{\partial h}{\partial u} \to \infty \ \text{as} \ h \to 0 \quad \text{and} \quad \frac{\partial h}{\partial u} \to 0 \ \text{as} \ h \to r_1$$

This implies that as the aircraft's height approaches zero, the estimated value of h will become infinitely sensitive to changes in u and as the aircraft's height approaches its range, the estimated height h will become infinitely insensitive to changes in u.

33.3   VERTICAL ACTIVITY ESTIMATION

Understanding the vertical activity of an aircraft enables airspace control to predict aircraft intent which means more accurate situation awareness. The problem of estimating the behaviour of an aircraft is well known with applications in airspace control and threat evaluation.

The existing literature typically makes use of two or more separate radars to compute aircraft altitude. The advantages are immediately obvious as an aircraft can be tracked in 3D and the flight path can be compared to known flight profiles to offer a strong basis for threat evaluation.

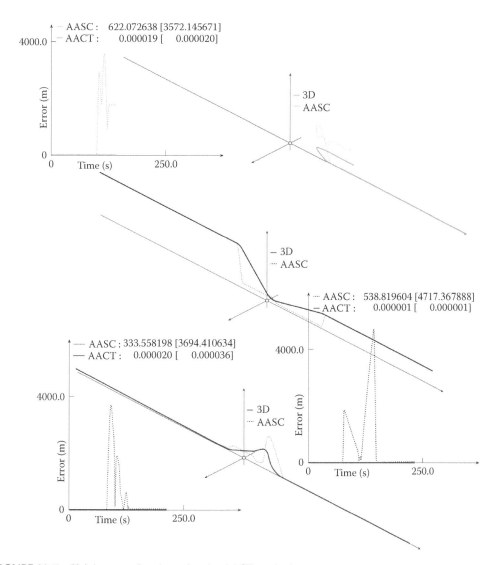

FIGURE 33.5   Height approximation using the AACT method.

33.3.1 Using Doppler Measurements

Vertical manoeuvres are easily and accurately recognizable when we have Doppler data available. Furthermore, it will give us a method to track the target in 3D.

If we have a relatively accurate estimation for h_1, then we can compute h_2 by solving the equation

$$-w_2 = r_2 - r_1\cos(\theta_1 - \theta_2)\cos(\epsilon_1)\cos(\epsilon_2) - r_1\sin(\epsilon_1)\sin(\epsilon_2) \tag{33.5}$$

for ϵ_2 once we know the value of ϵ_1 as illustrated in Figure 33.6. For simplicity, we define

$$c_1\sin(\epsilon_2) + c_2\cos(\epsilon_2) + c_3 = 0 \tag{33.6}$$

where $c_1 = r_1\sin(\epsilon_1) = h_1$

$c_2 = r_1\cos(\theta_1 - \theta_2)\cos(\epsilon_1)$

$c_3 = -w_2 - r_2$

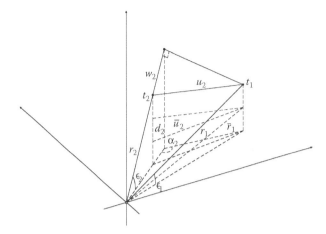

FIGURE 33.6   Calculation of change in altitude.

Then for $c_1^2 + c_2^2 - c_3^2 \geq 0$ we have

$$\epsilon_2 = \begin{cases} \epsilon_{2:1} = 2\arctan\left(\dfrac{c_1 - \sqrt{c_1^2 + c_2^2 - c_3^2}}{c_2 - c_3}\right) & \text{if } \alpha_2 > 90° \text{ and} \\[3mm] \epsilon_{2:2} = 2\arctan\left(\dfrac{c_1 - \sqrt{c_1^2 + c_2^2 - c_3^2}}{c_2 - c_3}\right) & \text{otherwise.} \end{cases} \tag{33.7}$$

We do not have enough information to calculate the projected angle α_2 so we define two functions covering all possible projection angles:

$$h_{2:1}(h) = r_2 \sin(\epsilon_{2:1}) \quad \text{and} \quad h_{2:2}(h) = r_2 \sin(\epsilon_{2:2}),$$

where $\epsilon_{2:1}$ and $\epsilon_{2:2}$ is computed as above with $h_1 = h$. We start by correcting the given initial height h if the error in range $\gamma(h) = |r_1 - \bar{r}_1(h)|$ is too big (e.g., >0.001). We do this by adjusting the fixed height h to

$$h = \begin{cases} h + \gamma(h) \\ h - \gamma(h) \end{cases} : \gamma(h \pm \gamma(h)) \text{ is a minimum.}$$

and continue to do so until $\gamma(h) \leq 0.001$.

Accuracy can further be increased by making this threshold smaller but at a computational cost. The absolute difference in height $d_2 = |h_2 - h_1|$ can be computed by projecting coordinates onto a plane of fixed height h and using the Euclidean distance $\bar{u}_2(h)$ between the projected coordinates to compute

$$d_2(h) = \sqrt{u_2 - \bar{u}_2(h)}. \tag{33.8}$$

Using this, we define the following two conditions which will assist us in deciding whether the aircraft gained or lost altitude.

$$\begin{aligned} \mu_1(h) &= \left\| h_{2:1}(h) - h \right| - d_2(h) \right| \\ \mu_2(h) &= \left\| h_{2:2}(h) - h \right| - d_2(h) \right| \end{aligned} \tag{33.9}$$

The vertical manoeuvre can be approximated if two acceptable conditions are found at a given fixed height h ($\sim h_1$). The method fails when there is no solution for ϵ_2 in the calculation of $h_{2:1}(h)$ and $h_{2:2}(h)$ or when the conditions are very close. This can be handled by using the previous change in altitude $h_1 - h_0$ and in such, assuming that the vertical manoeuvre graph is smooth. We have $h_2 \sim h + \delta_2(h)$, where

$$
\delta_2(h) = \begin{cases}
d_2(h)\,\text{sign}\,(h_1 - h_0) & \text{if } \nexists\,\text{a solution} \quad \text{or} \quad |\mu_2(h) - \mu_1(h)| < 0.001. \\
d_2(h)\,\text{sign}\,(h_{2:1}(h) - h) & \text{if } \exists\,\text{a solution} \quad \text{and} \quad \mu_1(h) < 0.001 \text{ and } \mu_2(h) > \mu_1(h) \\
d_2(h)\,\text{sign}\,(h_{2:2}(h) - h) & \qquad\qquad\qquad\qquad \mu_2(h) < 0.001 \text{ and } \mu_1(h) > \mu_1(h) \\
d_2(h)\,\text{sign}\,(h_{2:1}(h) - h) & \qquad\qquad\qquad\qquad \text{otherwise if } \mu_1(h) < \mu_2(h) \\
d_2(h)\,\text{sign}\,(h_{2:2}(h) - h) & \qquad\qquad\qquad\qquad\qquad \text{if } \mu_1(h) > \mu_2(h).
\end{cases}
\tag{33.10}
$$

The vertical manoeuvre graphs for our three examples using this method to calculate the change in altitude are illustrated in Figure 33.7. The numbers displayed in the legends of the error graphs are of the form Calculated: Average Error (m) [Maximum Error (m)]. The errors in the beginning

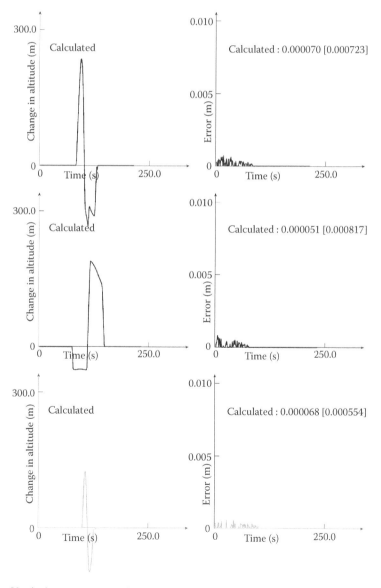

FIGURE 33.7 Vertical manoeuvre graphs.

TABLE 33.1
Numerical Values of the Various Error-Metrics (Change in Altitude Calculation)

Scenario	Sensor Offset Radius			Change in Altitude Errors		
	Range (m)	Dir (°)	Altitude (m)	Avg	Max	Errors
Pitch and dive	0–1000	0–360	0–1000	0.000003	0.001094	0.000000
Standard	0	0	0	0.000000	0.000034	0
Max average	330	260	110	0.000015	0.001047	0
Max single	600	110	280	0.000010	0.001094	0
High dive	0–1000	0–360	0–1000	0.000000	0.001126	0.000000
Standard	0	0	0	0.000000	0.000001	0
Max average	110	110	670	0.000005	0.001126	0
Max single	110	110	670	0.000005	0.001126	0
Long toss	0–1000	0–360	0–1000	0.000003	50.764000	0.000008
Standard	0	0	0	0.000000	0.000001	0
Max average	720	0	180	0.357493	50.764000	1
Max single	260	0	90	0.357493	50.764000	1

are due to rounding errors in the calculation of d_2 giving a false indication that there was a small change in altitude when the aircraft was actually in straight flight. The whole method was set up with a threshold of 1 mm which suggests that we should ignore changes in altitude smaller than this threshold. By doing this, we end up with average and maximum errors of less than 0.0001 mm. Moving the sensor away from the defended asset will affect the calculation of the vertical manoeuvre in the sense that the occasional error (not present in the standard setup) might arise. These errors refer to instances where the algorithm gave an increase in altitude when there was actually a decrease and vice versa. This is entirely dependent on the flight profile, though. Table 33.1 displays the average error, maximum error and error probability (or number of errors for specific scenarios) when moving the sensor within a 1 km radius away from the defended asset. Table 33.1 is further expanded to include the standard and extreme scenarios where maximums are found. Figure 33.8 illustrates this for a special case. All the errors were due to the system having no solution and were caused by using the previous update based on the assumption of continuous flight. Table 33.2 displays the average error, maximum error and error probability (indicating how likely the vertical manoeuvre

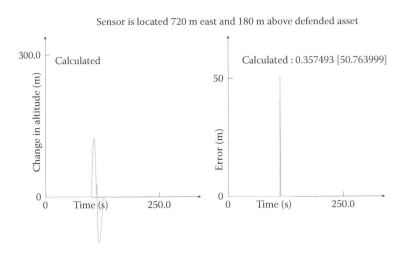

FIGURE 33.8 Vertical manoeuvre approximation.

calculation algorithm is to fail) for the three given profiles as the sensor moves further away from the defended asset. All vertical activity errors are still due to the system having no solution.

33.4 CONTINUOUS TRACKING

The previous section gives us a method to track a target in 3D from the data sets $(r_1, \theta_1, \hat{v}_1)$ and $(r_2, \theta_2, \hat{v}_2)$ accumulated from two consecutive sensor updates at time t_1 and t_2, respectively. We will refer to this method as the AACT (Altitude Approximation via Continuous Tracking) method. The method to approximate the change in altitude $h_2 - h_1$ at time t_2 described in the previous section takes as input an initial value h ($\sim h_1$) which is used as an initial approximation. The algorithm will adjust this height before trying to compute the change in altitude to yield better conditions $\mu_1(h)$ and $\mu_2(h)$. This is done by using the given range r_1 at time t_1 to find a more accurate approximation $h \sim h_1$ by mapping onto the range sphere.

The AACT method uses this idea to continuously track a target in 3D (at the nth sensor update) by providing: (i) an estimate δ_n for the change in altitude $h_n - h_{n-1}$ by using the approximated height h_{n-1} (computed at the previous sensor update) as input and (ii) an estimate for the height h_n. We will take h_0 as the AASC height whilst one may just as well take it as some other constant value (Figure 33.5). Moving the sensor away from the defended asset does affect the average AACT height errors for a specific flight profile but this is entirely dependent on the actual profile. Table 33.3 displays the average error and maximum error when moving the sensor within a 1 km radius away from the defended asset. The standard and extreme scenarios where maximums are found are also given. Table 33.4 displays the average error and maximum error for the AACT method for the three given profiles as the sensor moves further away from the defended asset. In the case of the above three scenarios we are better off when the sensor is 5 km away from the defended asset but this might not always be the case.

33.4.1 TURNING MANOEUVRES

More complex turning manoeuvres in the flight path will naturally increase the error probability when detecting vertical manoeuvres and a bad first AASC approximation might cause a scenario to

TABLE 33.2
Numerical Values of the Various Error-Metrics (for Given Profiles)

Scenario	Radius	Change in Altitude Errors		
		Avg	Max	Err Prob
Pitch and dive	0–1000	0.00000289	0.00109412	0.00000000
	1000–2000	0.00000288	0.00112229	0.00000000
	2000–3000	0.00000291	0.00126226	0.00000000
	3000–4000	0.00000307	6.53600000	0.00000278
	4000–5000	0.00000365	9.44600000	0.00000278
High dive	0–1000	0.00000005	0.00112551	0.00000000
	1000–2000	0.00000005	0.00113658	0.00000000
	2000–3000	0.00000056	20.73200000	0.00000556
	3000–4000	0.00000017	0.00154998	0.00000000
	4000–5000	0.00000039	0.00150087	0.00000000
Long toss	0–1000	0.00000302	50.76400000	0.00000833
	1000–2000	0.00000019	0.00151653	0.00000000
	2000–3000	0.00000049	0.00151914	0.00000000
	3000–4000	0.00000403	50.76400000	0.00001111
	4000–5000	0.00000439	50.76400000	0.00000833

TABLE 33.3

Numerical Values of the Various Error-Metrics (Altitude Approximation/for Three Scenarios)

Scenario	Sensor Offset Radius			AACT Errors	
	Range (m)	Dir (°)	Altitude (m)	Avg	Max
Pitch and dive	0–1000	0–360	0–1000	0.025447	1.014586
Standard	0	0	0	0.000020	0.000036
Max average	990	350	0	0.252607	1.014428
Max single	980	10	0	0.252499	1.014586
High dive	0–1000	0–360	0–1000	0.002799	0.014047
Standard	0	0	0	0.000001	0.000001
Max average	980	60	0	0.004149	0.013457
Max single	960	10	0	0.004039	0.014047
Long toss	0–1000	0–360	0–1000	0.038625	0.546741
Standard	0	0	0	0.000019	0.000020
Max average	990	350	0	0.242730	0.546741
Max single	990	10	0	0.242729	0.546741

TABLE 33.4

Numerical Values of the Various Error-Metrics or the AACT Method/Altitude Approximation

Scenario	Radius	AACT Errors	
		Avg	Max
Pitch and dive	0–1000	0.02544727	1.01458580
	1000–2000	0.00496901	0.03094935
	2000–3000	0.00304537	0.01616566
	3000–4000	0.00226141	0.01117682
	4000–5000	0.00183988	0.00870827
High dive	0–1000	0.00279944	0.01404736
	1000–2000	0.00210260	0.00998633
	2000–3000	0.00170983	0.00788764
	3000–4000	0.00146634	0.00655928
	4000–5000	0.00129752	0.00570002
Long toss	0–1000	0.03862496	0.54674133
	1000–2000	0.00890965	0.03202178
	2000–3000	0.00527871	0.01690597
	3000–4000	0.00377008	0.01177279
	4000–5000	0.00294581	0.00909983

have an average AACT height error that is above the average for the standard flight profile. The various results are shown in Figures 33.9 and 33.10, and Table 33.5.

33.4.2 Practical Aspects

It should be noted that the accuracy of the method to calculate a vertical manoeuvre is dependent on the accuracy to which we can measure (r_1, θ_1) and (r_2, θ_2) and w_2. It is well known that 2D radars have

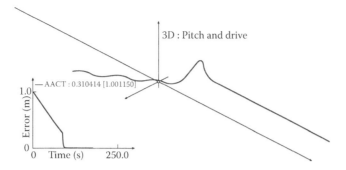

FIGURE 33.9   Height approximation using the AACT method: some results.

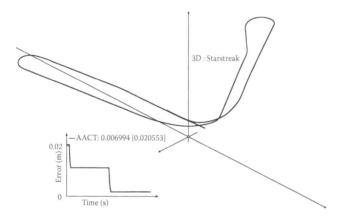

FIGURE 33.10   Height approximation using the AACT method: some more results.

TABLE 33.5
Additional Flight Profiles

	Change in Altitude Errors			AACT Errors	
Scenario	Avg	Max	Err Prob	Avg	Max
Pitch and dive	0.00248889	37.47600000	0.00248889	0.03543821	1.03319801
Starstreak	0.00000130	0.00082250	0.00000000	0.01112874	0.03089402

excellent slant range measurements but poor azimuth readings. The data displayed in Table 33.6 shows the effect of randomly introduced errors in azimuth readings. The errors we have encountered here with respect to the calculation of vertical manoeuvres are not only due to the system having no solution but due to the algorithm failing as a result of the errors in azimuth readings. As one would naturally expect, average errors and error probability increases as the azimuth error bias increases.

33.5   ALTERNATIVE METHOD

The solution suggested in [3] can be used when no Doppler data is available. It makes use of two separate motion models that independently estimate the behaviour of the tracked entity. The relative errors between predictions of these models and observations made by the 2D radar are then used to

TABLE 33.6
Adding Random Azimuth Errors

Scenario	Azimuth Error (mrad)	Change in Altitude Errors			AACT Errors	
		Avg	Max	Err Prob	Avg	Max
Pitch and dive	0.0000	0.000003	0.001094	0.000000	0.025447	1.014586
Radius: 0–1000	0.1000	0.016568	471.120000	0.002519	0.047542	1.741158
	0.2000	0.019945	471.120000	0.008628	0.046481	1.737893
	0.3000	0.022500	471.120000	0.016542	0.045727	1.767407
	0.4000	0.024488	471.120000	0.025508	0.045189	1.764944
	0.5000	0.026517	471.120000	0.034939	0.044771	1.761901
High dive	0.0000	0.000000	0.001126	0.000000	0.002799	0.014047
Radius: 0–1000	0.1000	0.000106	346.184000	0.000194	0.002799	0.014047
	0.2000	0.000376	346.184000	0.000411	0.002819	0.278816
	0.3000	0.001648	346.184000	0.000683	0.003003	0.499761
	0.4000	0.003521	346.184000	0.000753	0.003489	0.499910
	0.5000	0.005684	346.184000	0.000961	0.004194	0.499916
Long toss	0.0000	0.000003	50.764000	0.000008	0.038625	0.546741
Radius: 0–1000	0.1000	0.083851	194.536000	0.308378	0.040033	0.546741
	0.2000	0.132797	194.536000	0.647572	0.039996	0.546741
	0.3000	0.175531	194.536000	0.976175	0.039965	0.546741
	0.4000	0.214212	194.536000	1.255950	0.039937	0.546741
	0.5000	0.249601	194.536000	1.475436	0.039913	0.546741

make probabilistic statements on the current vertical behaviour of the aircraft. These models are based on inferring changes in the perceived velocity of a tracked target from a 2D radar track. The perceived changes may be due to changes in the target acceleration, changes in target altitude, or a combination of both. For this method to be useful we need to assume that the aircraft is at a fixed initial height h. We will proceed in vector notation to describe the 3D position vectors p_n (at time t_n) obtained by projecting the actual aircraft positions onto a 2D horizontal plane at height h.

Aircraft position: p_n
Aircraft velocity: $v_n = p_n - p_{n-1}$
Aircraft acceleration: $a_n = v_n - v_{n-1}$

The first model assumes that the aircraft is flying at a constant altitude. This implies that perceived deviations (on the projection plane) from the expected flight path must be due to changes in velocity of the target. We can associate a 2D position from a single 2D radar sensor update via the above mentioned projection. A second update can then be used to estimate the perceived velocity of the target and finally, a third update can be used to estimate the change in velocity of the target. We will not concern ourselves with the time term as it does not influence the result, provided updates occur at a constant rate. These values can then be used to predict the position of the target at the next update.

$$p_{n+1} = p_n + v_n + a_n. \tag{33.11}$$

The second model assumes that the estimated aircraft speed stays constant during updates, thus all perceived deviations from the expected flight path are due to changes in altitude. This model additionally assumes that the aircraft was in flying level during the prior update. Two predictions are made: one prediction for the case that the aircraft gained altitude and another for the case that the aircraft lost altitude. We iteratively find a new position p_n in each of the above cases. We start with a constant step size δ_0 which will be halved at every iteration and an initial value $q_0 = p_n$ for the aircraft position. We will continue to adjust the position q_i by increasing/decreasing its altitude by the current step size and mapping it onto the range sphere of the sensor until $|q_i - p_{n-1}| = |p_{n-1} - p_{n-2}|$ or the step size δ_i is small enough. At termination we update our position $p_n = q_i$ and make a prediction about where the target will be at the next update:

$$p_{n+1} = p_n + v_n. \tag{33.12}$$

We can include other predictions here as the aircraft might steepen its ascent or start to level out. The predictions of these models are compared with actual observations during the subsequent sensor update. This is done by projecting the prediction to two dimensions representing slant range and azimuth. The Euclidean distance between the projected 2D position and the observed position is taken to be our measure of error. It is assumed that the smaller the error, the more likely the manoeuvre. We will abide by the following set of rules to decide as to which model to follow:

1. If the sum of all three prediction errors (representing upwards, downwards and acceleration-based errors) is less than a threshold value, then the aircraft is considered to be in straight flight. This threshold should be in the range of 5–10% of typically observed errors.
2. If the sum of all three prediction errors exceeds an upper threshold, then the aircraft is assumed to be flying a complex manoeuvre which cannot be captured by the underlying models. This threshold should be around twice the typical observed errors during normal manoeuvres.
3. If the acceleration errors is at least 10% less than the average of the upward and downward errors, then the aircraft is considered to be accelerating.
4. After ruling out the above three situations, we will assume that the aircraft is flying a vertical manoeuvre. The probability of the manoeuvre being upward μ_U or downward μ_D is given by the ratio of the corresponding errors e_U and e_D, respectively:

$$\mu_U = \frac{e_D}{e_D + e_U} \tag{33.13}$$

$$\mu_D = \frac{e_U}{e_D + e_U}. \tag{33.14}$$

It is crucial to note that the distinction between upwards and downwards error is typically quite small, thus not a lot of weight should be given to probabilistic statements with respect to vertical manoeuvres being up or down. Unfortunately, this is exactly what is needed to allow accurate continuous tracking of a target.

33.6 CONCLUSIONS

We have given a method to estimate vertical manoeuvres of an aircraft using a single 2D radar when Doppler data is available and used it to track a target in 3D. The average error when tracking a target in 3D using this method depends on the actual flight profile, position of the sensor with respect to the defended asset, and the accuracy of the data we get from the sensor but on average never exceeded 0.05 ms^{-1} for the standard flight profile examples we considered even with an 0.03° error bias in the azimuth readings. The maximum error at any given time was never more than 1.8 m. An alternative method that can be used when we do not have Doppler data was briefly outlined but this method was statistically proven to be right only 50% of the time, making it equivalent to throwing a dice when we have to decide whether the aircraft gained or lost altitude.

REFERENCES

1. D. E. Manolakis, Aircraft vertical profile prediction based on surveillance data only. In *IEEE Proceedings on Radar, Sonar and Navigation*, 144(5), 301–307, 1997.
2. H. Hakl, E. Davies and W.H. Le Roux, Aircraft height estimation using 2-D radar. *Defence Science Journal*, 60(1), 100–105, 2010.
3. H. Hakl and W.H. Le Roux, Vertical activity estimation using 2-D radar. *Scientia Militaria: South African Journal of Military Studies*, 36(2), 60–76, 2008.

34 Investigating the Use of Bayesian Network and *k*-NN Models to Develop Behaviours for Autonomous Robots

Isaac O. Osunmakinde, Chika O. Yinka-Banjo
and Antoine Bagula

CONTENTS

34.1 INTRODUCTION

Bayesian network technology is very useful for encoding probabilistic knowledge as graphical structures. It is rapidly gaining popularity in modern artificial intelligence (AI) for solving real-life problems involving reasoning under uncertainty [1,2]. The most important benefit of using Bayesian networks in real-life applications is in carrying out probabilistic inference (or reasoning). Bayesian inference is a type of statistical inference in which probabilities are interpreted as degrees of belief and its fundamental computation is derived from Bayes' theorem [3]. The Network belief technology has been successfully used for reasoning in the areas of power transformer diagnosis [1], medical diagnoses [4], telecommunication networks [5] and so on. Knowledge is expensive to acquire and most of the time, there are no domain experts or knowledge engineers to interpret environments and model knowledge as Bayesian belief networks. Since data are cheap and contain useful information about the environments, Bayesian networks offer a great advantage that can capture and encode this hidden information as knowledge. *k*-Nearest neighbour (*k*-NN) is a nonparametric instance-based learning as it allows a hypothesis of model complexity to grow with data sizes. k-NN is based on minimum distance from a query instance to all training samples to determine the k-NN, which spans the entire input texture space. Prediction of the query instance is taken as majority votes of the k-NN. The k-NN model has been successfully used for prediction or reasoning in the areas of face recognition [2], traffic accident prediction [6] and fault detection [7].

The recent literature addresses robot slope-walking problem [8], using the machine learning technique of k-NN and robot localization problems [9,10] with wide applicability of the Bayesian theorem. There is not enough focus on the autonomous robot behaviour problem using the predictive power of k-NN and Bayesian network models as learning and reasoning techniques to manage autonomous robot navigation with respect to collision avoidance. The reasons for not using the models in this area could be as a result of their challenges, such as determination of appropriate kth value in k-NN and the computational intensity of Bayesian learning, in autonomous applications. This is a motivation for raising the research questions: (i) To what extent can a robot autonomously manage its behaviours when navigating without collision in a static environment where it was trained using teleoperation? and (ii) To what extent can a robot autonomously manage its behaviours when navigating without collision in a dynamic or different environment from where it was trained using teleoperation? An environment is said to be static when obstacles do not move from their positions while it is dynamic when obstacles are moving. To address these challenges and the questions, we propose an approach of training a robot to avoid obstacles through teleoperation and thereafter use the knowledge acquired to develop behaviours for autonomously navigating in various environmental sensing conditions using the learning and predictive capabilities of k-NN and Bayesian network models. The chosen behaviour or navigational direction of the robot determines the control command values of translational and rotational velocities the robot uses for navigation. This work integrates a kth set measure to k-NN for determining an appropriate kth value where robot behaviour to a new sensors reading is predicted based on majority voting. Using real-life publicly available ultrasound sensors minimum readings to obstacles on a number of comparative evaluations in static and dynamic environments, our experimental results show that both predictive learning paradigms developed as collision avoidance models (CAM) are capable of dealing with uncertainties during autonomous robot navigation. This excellent performance suggests a wider application of the behavioural models which learn tasks and command robot successfully without collisions in an unknown environment in industry.

The rest of this chapter is arranged as follows: Section 34.2 presents useful theories, which includes Bayesian network and k-NN models. Section 34.3 presents the proposed modelling for behavioural and collision avoidance for robots which includes perception of sensor data, learning and reasoning processes of the approaches. Section 34.4 critically presents experimental evaluations of the approaches on number of comparative evaluations in static and dynamic environments using publicly available minimum ultrasound sensors readings to obstacles. The average performance evaluation of the models is also compared based on a configuration of four sensors and related work is presented in Section 34.5. Section 34.6 concludes the chapter.

34.2   SOME USEFUL CONCEPTS AND MODELS

In this section some useful concepts and theories are discussed: (a) Bayesian networks modelling concepts and (b) nearest-neighbour model.

34.2.1   BAYESIAN NETWORK MODELS

A Bayesian belief network is formally defined as a directed acyclic graph (DAG) represented as $G = \{X(G), A(G)\}$, where $X(G) = \{X_1, \ldots, X_n\}$, vertices (variables) of the graph G and $A(G) \subseteq X(G) \times X(G)$, set of arcs of G. The network requires discrete random values such that if there exists random variables $X_1, \ldots, X_n$ with each having a set of some values $x_1, \ldots, x_n$ then, their joint probability density distribution is defined as

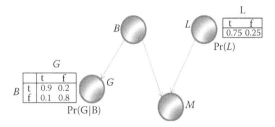

FIGURE 34.1 A simple Bayesian network model of a block lifting machine.

$$pr(X_1,...,X_n) = \prod_{i=0}^{n} pr(X_i \mid \pi(X_i)) \tag{34.1}$$

where $\pi(X_i)$ represents a set of probabilistic parent(s) of child X_i [3]. A parent variable otherwise refers to as *cause* has a dependency with a child variable known as *effect*. Every variable X with a combination of parent(s) values on the graph G captures probabilistic knowledge as conditional probability table (CPT). A variable without a parent encodes a marginal probability. For the purposes of illustrating BN, Figure 34.1 shows the DAG and the CPTs of a BN model as the core reasoning component of an intelligent system. In this case, it describes the operation of a block-lifting machine. The operation is monitored with the following attributes: battery (B), movement (M), liftable (L) and gauge (G) [3]. Each of the attributes contains states true (t) and false (f), with their associated probabilities captured as CPTs, such as L having $t = 0.75$ and $f = 0.25$. Figure 34.1 depicts conditional dependencies of the attributes which best describe the complexity of variables of the block lifting machine. For instance, in Figure 34.1, G is conditionally dependent on B and it is computed as $Pr(G|B)$. Also, M is conditionally independent of G which implies that it is computed as $Pr(M|B, L)$. The estimation of the probabilities, using the maximum likelihood estimate (MLE) algorithms, captured as CPTs results from the environment, for example, obstacle distances perceived by the robot sensors. A Bayesian network can be modelled by eliciting the probabilistic knowledge from domain experts, if the environment is small. For a more complex domain like robot environment, the most suitable BN is learned from the environment captured as samples using learning algorithms described in Refs. [11,12]. Having a BN model in place, a probabilistic inference is required for reasoning about any situation and the beliefs (or probabilities) of possible outcomes are propagated in a model based on the evidence of the situation. Understanding various obstacle distances is a possible situation that can be acted upon by the inference. The Bayesian inference accounts for the uncertainty capability of BNs through the Bayes' theorem shown in the following equation [13]:

$$Pr(X_i \mid X_j) = \frac{Pr(X_j \mid X_i) \times Pr(X_i)}{Pr(X_j)} \tag{34.2}$$

The constituents of Equation 34.2 are: (i) $Pr(X_i \mid X_j)$ is the posterior probability called the original degree of belief when the likelihood and prior are combined, (ii) $Pr(X_j \mid X_i)$ is the likelihood function which is referred to as the conditional probability of what we know (evidence) based on what we do not know (query) and (iii) $Pr(X_i)$ is the prior probability of X_i before making any observations; here, the marginal probability $Pr(X_j)$ is a measure of the impact that observations have on the degree of beliefs.

34.2.2 *k*-NN Models

k-NN [14] is a non-parametric instance-based learning as it allows a hypothesis of model complexity to grow with data sizes. k-NN is based on minimum distance from a query instance to all training samples to determine the k-NN, which span the entire input space. The Euclidean distance of lower dimensional space is commonly applied for computing the minimum distance in this step. The Euclidean distance for two-dimensional space, say points $x = (x_1, x_2)$ and $y = (y_1, y_2)$, is given as

$$d(x, y) = \sqrt{(x_1 - y_1)^2 + (x_2 - y_2)^2}$$ (34.3)

Prediction of the query instance is taken as majority votes of the k-NNs. The idea is that any point x is likely to be similar to those points in the neighbourhood of x. The choice of parameter value k is critical but k-NN is advantageously robust to uncertainty or noisy training samples. This is the simplest k-NN, but more sophisticated versions can be proposed.

34.3 BEHAVIOURAL AND CAM FOR ROBOTS

Figure 34.2 illustrates an unstructured indoor environment with scattered chairs and tables as obstacles where behaviour could be developed to support the navigation of a pioneer robot. We illustrate an experimental set-up with wall-following navigation task [15] of a robot, which uses 24 ultrasound sensors arranged circularly around its waist. The numbering of the ultrasound sensors starts at the front of the robot and increases in clockwise direction. Sensor readings are sampled at a rate of 9 samples/s and the data samples were collected at the same time step, as the robot navigates through a room following the wall in a clockwise direction, for 4 rounds. Ultrasound sensors send out an ultrasonic pulse and then wait for a response [16]. When the pulse leaves the device, it travels through the air until it collides with an object or obstacles, at which point an echo is reflected back. This echo is then sensed by the ultrasonic sensor. The sent pulse is anywhere from 40 kHz to 200 kHz, but is typically in the 40–50 kHz range.

34.3.1 PERCEPTION OF ULTRASOUND SENSOR DATA IN REAL LIFE

From the wall-following navigation task with the mobile robot [15], different data samples were captured from the environment based on three sensor configurations. The first configuration captures

FIGURE 34.2 Experimental in-door environment developing behaviours for a pioneer robot for avoiding obstacles of chairs, tables and wall. (Authors' Lab and Work environment.)

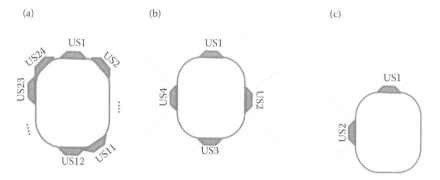

FIGURE 34.3   Three configurations of ultrasound sensors arranged circularly around a robot. (a) 24 Sensors. (b) 4 Sensors. (c) 2 Sensors.

the raw values of the measurements of all 24 ultrasound sensors, which are precisely tagged as US_1, $US_2,\ldots, US_{24}$, as shown in Figure 34.3a. This configuration consists of the minimum sensor readings among those within 15° arcs located around the robot. The second configuration captures four sensor readings named simplified distances. The simplified distances are referred to as the *front distance, left distance, right distance* and *back distance*. These distances consist, respectively, of the minimum sensor readings among those within 60° arcs located at the front, left, right and back parts of the robot, as shown in Figure 33.3b. The third configuration captures only the front and left simplified distances and consists of the minimum sensor readings among those within 60° arcs located at the front and left of the robot, as shown in Figure 34.3c. The robot is teleoperated for learning in the environment and behaviours were captured as it perceives obstacles. The ultrasound minimum sensor readings from the obstacles determine the behaviour or the navigational direction of the robot. The four directions defined for Figures 34.3a and b are Move-Forward, Slight-Right-Turn, Sharp-Right-Turn and Slight-Left-Turn, while two directions Move-Forward and Slight-Right-Turn are defined for the configuration in Figure 34.3c. The chosen direction of the robot determines the control command values, which are translational and rotational velocities the robot uses for navigation. Sensor readings and their associated robot actions captured from any of the three configurations are used for training the BN and k-NN models as they learn the environment.

34.3.2   Bayesian Learning and Reasoning Process to Robot Behaviour

Figure 34.4 illustrates the stages required to learn a BN model from an environment captured as sensor readings. As illustrated in Figure 34.4, learning such models from the environment can be decomposed as follows into sub-problems of: (a) data discretization as a pre-processing step, (b) learning a suitable network structure, (c) learning the associated conditional probability tables (CPTs) and (d) model visualization. Data discretization classifies numerical data into their corresponding interval values relative to the patterns in the data attributes. William and co-workers and Osunmakinde and Potgieter [11,12] have presented many algorithms including genetic and hill climbing algorithms to learn Bayesian networks from datasets. Its characteristics of capturing knowledge in dependency variables make it suitable for handling uncertainty problems, such as noisy sensor readings. Having a Bayesian network model in place after teleoperation of a robot, a probabilistic inference is required for reasoning about any positions of obstacles and the beliefs (or probabilities) of possible outcomes are propagated in a model based on the observations of the obstacles. In this research, we want to predict the probability Pr (or most likely) action, which is not known, based on the current knowledge of the ultrasound sensor readings d to obstacles that the robot understands as shown in Equation 34.4. For instance, for an obvious situation, a robot may navigate towards a freest direction, but preferential treatment is given to navigational targets.

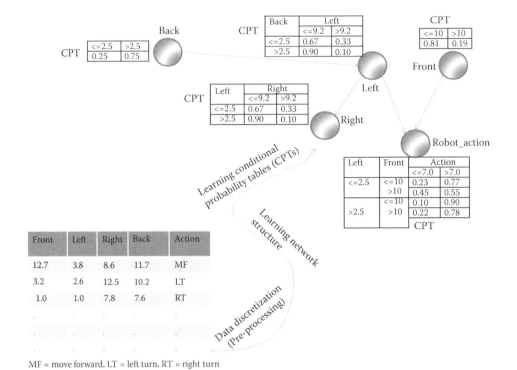

MF = move forward, LT = left turn, RT = right turn

FIGURE 34.4 Learning stages of a BN where front node = US_1, right = US_2, back = US_3 and left = US_4.

$$\mathrm{pr}\left(\mathbf{Action}\,?\mid US_1 = d_1, US_2 = d_2,\ US_3 = d_3, ..., US_n = d_n\right) \tag{34.4}$$

Using Bayes' theorem in Equation 34.2, Equation 34.3 implies

$$\Rightarrow \frac{\Pr(US_1 = d_1,\ US_2 = d_2,\ US_3 = d_3, ...,\ USn = d_n \mid \mathrm{Action}) \times \Pr(\mathrm{Action})}{\Pr(US_1 = d_1,\ US_2 = d_2,\ US_3 = d_3, ...,\ USn = d_n)}$$

This is a Bayesian inference problem with more information in Ref. [13]. If a robot is expected to keep moving towards forward directions, then the back sensors would not participate in the reasoning process even if the sensors read the freest. The robot then negotiates among the due *forward*, *slight-right-turn* and *slight-left-turn* as the next most likely behaviour from the BN model.

34.3.3 THE *k*-NN MODELLING AND REASONING PROCESS TO ROBOT BEHAVIOUR

The adaptation of the *k*-NN model to address the robot behaviour in collision avoidance relies heavily on (i) the minimum distance from the new query instance to all training obstacle instances captured by the robot's knowledge, (ii) the choice of *k*th value and (iii) biasness check in the training instances. With regard to the minimum distance, this research proposes *n*-dimensional Euclidean measure based on the number *n* of sensors' readings that would participate in the choice of action as shown

$$d(x, y) = \sqrt{(x_1 - y_1)^2 + (x_2 - y_2)^2 + (x_3 - y_3)^2 + \cdots + (x_n - y_n)^2} \tag{34.5}$$

where query instance $x = (x_1, ..., x_n)$ and training instance $y = (y_1, ..., y_n)$.

The minimum distance eventually indicates a specific direction at which, for many purposes, the robot's action behaves as if it were concentrated on that chosen sensor. Given these numerous distances computed from all training samples, it is reasonable to choose the kth value as a number of sets s of distances and dynamically increases s with a step whenever there is a tie during the voting scheme. With regard to the biasness check, this chapter introduces an idea of making the training instances having equal number of robot actions since its prediction of new obstacle distances is based on majority votes. Now the robot autonomously navigates and perceives a new instance of sensor readings from obstacles. Without another expensive teleoperation or training, can the robot accurately predict its behaviour towards this new instance? The adaptation of the k-NN model requires the implementation of the following algorithm (k-NN approach):

```
INPUT:   Training & query sets of obstacle distances
OUTPUT:  Predicted Robot Behaviours

Step 1: Specify training set
Step 2: Determine the neighbourhood size as kth set
Step 3: Compute the distance between a query instance and all training
        instances using Equation 34.5
Step 4: Determine nearest neighbours using the kth set minimum distance
Step 5: Assess the training actions/behaviours of the nearest Neighbours
Step 6: Predict robot behaviour for the query instance
Step 7: Repeat steps (3)-(6) for other query instances as perceived by the
        robot sensors
```

34.3.4 EVALUATION SCHEME

In this section, the performances of the approaches investigated are studied through an evaluation scheme commonly used in practice based on n-fold cross validation technique [13] as well as measuring the execution time. Cross validation sometimes called rotation estimation, is the most generally applicable strategy in model selection in machine learning since it does not rely on any probabilistic assumptions. Here the dataset is partitioned into a number n mutually disjoint folds and leave-one-out cross validation (LOO) for testing model performance while the remainder is used for training. This process is repeated n times to find the overall performance of the approaches. In this research, since learning from the environment and training of robot is carried out at the teleoperation phase, the execution speed of the model reasoning when a robot is autonomously reacting to obstacles is an important issue based on various sensor configurations.

34.4 EXPERIMENTAL EVALUATIONS OF THE CAM

One of the objectives of the investigation of the behavioural modelling approaches is to bring theory to practice with an emphasis on application to collision avoidance work. This section describes the experiments we conducted for evaluating the performances of the approaches for developing behaviours for robots using two machine learning models on three sets of real-life datasets based on sensor configurations. The models considered are BN and k-NN as described above. As described in the experimental setup of Section 34.3.2, the three datasets are captured from: (i) 24 ultrasound sensors, (ii) 4 ultrasound sensors and (iii) 2 ultrasound sensors. These datasets used are publicly available sensor readings from the University of California Irvine (UCI) machine learning repository, which most researchers use to validate their techniques. In practice, the major contributing factors that affect accuracy of robot behaviour to obstacle avoidance are the model learning process, number of sensors considered and speed of reasoning based on the sensor configurations. We conducted three main experiments to compare the performance achieved by the different models on the four ultrasound sensors configuration in terms of; (a) collision avoidance efficiency in static environments,

(b) collision avoidance efficiency in dynamic environments and (c) comparing average performance evaluations of the models. These experiments were carried out specifically on a machine processor by implementing the k-NN in MATLAB® and the BN in GeNile software [17]. The training samples extracted for robot to navigate towards the freest direction contains 200 samples, but going backwards is not an option since the robot is required to follow wall forward in a clockwise direction. Using three-fold cross validation, the full dataset was divided into three partitions as randomly. Some samples from one of the partitions were selected for testing and the others were used for training. Generally, while the training dataset is approximately 95% of all data, the rest of the 5% data are used as testing samples. Since the cross-validation technique was used, this process was repeated three times.

34.4.1 PERFORMANCE ACCURACY OF CAM IN A STATIC ENVIRONMENT USING BN AND k-NN

This section addresses our first research question presented at the introduction. Imagine a robot being allowed to autonomously avoid obstacles in an environment where it was trained-static. We therefore conducted experiments for finding the impact of the BN and the k-NN models on collision avoidance with the expectation of selecting a better model in terms of predicting accurate behaviour for robot when it perceives obstacles in similar positions already seen during training. It examines the consistency of the model in predicting robot behaviours. The BN model in Figure 34.5 is learned from the training sensor samples using the GeNle software and the results depicted by Table 34.1 are a summary of the average performance of the two models in terms of testing with five random samples repeated three times. The cross validation is modified differently here since the environment is static; the test samples form part of the training instances for evaluating the consistencies of the models. For each set of samples, the Ei indicates an instance of evidence of obstacle distances to the robot in the four sensor directions of front, left, right and back. Every instance is used to predict the robot behaviour or action using the reasoning processes of the BN in Equation 34.4 and the majority voting scheme of the k-NN described in Section 34.3.4. In all the sets, the expected robot behaviour (ERB) and the results of the predicted robot behaviour (PRB) revealed that autonomous collision avoidance using the BN and the k-NN are accurate in a static environment. One could observe that the readings of the back sensors formulates part of the learning proces, but do not participate in predicting the behaviour as the robot gives preference to its goal—follow a wall in a clockwise direction. Observe in Table 34.1 that the BN model tremendously indicates better beliefs or confidence for the robot's behaviour than the k-NN. For illustration, Figure 34.6 presents a pictorial representation of the belief results from the first partition. However, these results suggest that using the two models to develop behaviours for robot assist the controller in determining the translational and rotational velocities values for navigation.

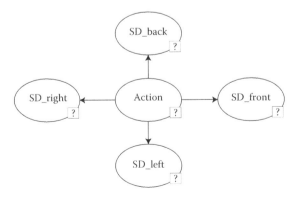

FIGURE 34.5 BN model for collision avoidance using real-life robot sensor readings.

TABLE 34.1

Robot Behaviour in a Static Environment with Real-Life Samples Using Three-Fold Cross Validation

ERB = Expected Robot Behaviour; PRB = Predicted Robot Behaviour

E_i	Percepts of Obstacle Distances				ERB	BN PRB (%)	k-NN PRB (%)
	SD_Front	SD_Left	SD_Right	SD_Back			
E_1	1.687	0.449	2.332	0.429	Slight-right	Slight-right (80.2)	Slight-right (78.69)
E_2	1.327	1.762	1.37	2.402	Slight-left	Slight-left (99.7)	Slight-left (90.0)
E_3	1.318	1.774	1.359	3.241	Slight-left	Slight-left (99.7)	Slight-left (90.0)
E_4	1.525	0.739	1.379	0.689	Move forward	Move forward (88)	Move forward (38.0)
E_5	0.786	0.661	2.748	0.689	Shift-right	Shift-right (89.2)	Shift-right (59.62)
Test Samples 1: Accuracy = 5/5 = 100%							*5/5 = 100%*
E_1	1.19	2.29	1.414	2.369	Slight-left	Slight-left (99.7)	Slight-left (90.0)
E_2	1.586	0.758	1.357	0.634	Front	Front (88.1)	Move forward (38.1)
E_3	0.762	0.482	1.697	0.473	Shift-right	Shift-right (85.4)	Shift-right (40.65)
E_4	1.637	0.474	1.715	0.465	Slight-right	Slight-right (94.2)	Slight-right (50.0)
E_5	0.79	0.779	1.345	0.67	Shift-right	Shift-right (96.2)	Shift-right (59.62)
Test Samples 2: Accuracy = 5/5 = 100%							*5/5 = 100%*
E_1	1.311	2.636	1.456	2.204	Slight-left	Slight-left (99.7)	Slight-left (90.0)
E_2	1.419	0.7	1.415	0.795	Move forward	Move forward (88.1)	Move forward (36.36)
E_3	0.799	0.664	2.478	1.239	Shift-right	Shift-right (99.7)	Shift-right (45.45)
E_4	1.636	0.475	1.707	0.47	Slight-right	Slight-right (94.2)	Slight-right (49.37)
E_5	2.644	0.69	2.084	0.955	Move forward	Move forward (98.7)	Move forward (63.29)
Test Samples 3: Accuracy = 5/5 = 100%							*5/5 = 100%*
Average Accuracy = 100%							

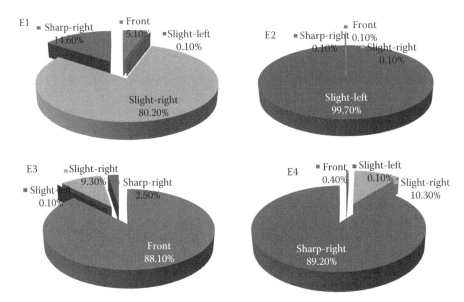

FIGURE 34.6 Presentation of the predicted behaviour using BN in a static environment from the first cross validation of Table 34.1.

34.4.2 PERFORMANCE ACCURACY OF CAM IN A DYNAMIC ENVIRONMENT USING BN AND *k*-NN

This section addresses our second research question presented at the introduction. Imagine a robot being allowed to autonomously avoid obstacles in a new environment different from where it was trained-dynamic. We also conducted experiments similar to the static above to evaluate the collision avoidance capability of the two models when navigating in a new environment. The results depicted in Table 34.2 are a summary of the average performance of the models in terms of three-fold cross validation, where the test set are separated from the training instances and appear as new obstacle readings to the models. In all the folds, the ERB and the results of the PRB revealed that autonomous collision avoidance using the BN and the *k*-NN are also promising in a dynamic environment. Observe in Table 34.2 that the BN model tremendously indicates average accuracy of 93.3% better beliefs for the robot's behaviour than the 73.3% of *k*-NN, probably due to the use of prior beliefs of the BN. The choice of *k*th value still needs more improvement. Figure 34.7 also presents a pictorial representation of the belief results from the first partition. However, the results suggest that using the two models to develop behaviours for robot assists the controller in determining the translational and rotational velocities values for navigation.

34.4.3 COMPARING AVERAGE PERFORMANCE EVALUATIONS OF THE MODELS

From the results of evaluation in Table 34.3, we specifically access the average performance accuracies of the BN and *k*-NN collision avoidance approaches with respect to static and dynamic environments ranging from one to six cross validations. In Figure 34.8, one can see that the trend of the error on BN is lowered in the dynamic case compared to the error trend on *k*-NN. This obviously implies that a higher trend of accuracy is better for predicting behaviours for robots.

34.5 RELATED WORK

The recent literature [18–20] addresses the behaviour-based systems some of which were originally inspired by biological systems, but more work on developing behaviours for robots was recommended in Ref. [21] to assist in the control architecture. In Ref. [18], modelling a biological behaviour is

TABLE 34.2

Robot Behaviour in a Dynamic Environment with Real-Life Samples Using Three-Fold Cross Validation

E_i	Percepts of Obstacle Distances				ERB	BN PRB (%)	k-NN PRB (%)
	SD_Front	SD_Left	SD_Right	SD_Back			
E_1	2.651	0.625	1.599	0.795	Move forward	Move forward (94.1)	Move forward (66.67)
E_2	2.885	0.623	1.606	0.814	Move forward	Move forward (99.2)	Move forward (72.58)
E_3	0.894	0.649	1.071	1.085	Shift-right	Shift-right (93)	Shift-right (43.59)
E_4	1.501	0.492	1.816	1.28	Slight-right	Slight-right (79.5)	Move forward (38.1)
E_5	1.523	0.485	1.8	1.069	Slight-right	Slight-right (51.6)	Slight-right (46.99)
Test Samples 1: Accuracy = 5/5 = 100%							*4/5 = 80%*
E_1	2.581	0.613	1.619	0.852	Move forward	Move forward (99.3)	Move forward (65.78)
E_2	2.828	0.607	1.626	0.871	Move forward	Move forward (99.2)	Move forward (74.24)
E_3	0.854	0.628	1.016	1.168	Shift-right	Move forward (49.6)	Move forward (36.51)
E_4	1.511	0.49	1.82	1.27	Slight-right	Slight-right (79.5)	Shift-right (37.59)
E_5	0.784	0.487	1.797	1.156	Shift-right	Shift-right (96.5)	Shift-right (40.65)
Test Samples 2: Accuracy = 4/5 = 80%							*3/5 = 60%*
E_1	2.549	0.599	1.633	0.889	Move forward	Move forward (99.3)	Move forward (64.1)
E_2	2.544	0.597	1.639	0.908	Move forward	Move forward (99.3)	Move forward (64.1)
E_3	0.873	0.642	1.053	1.105	Shift-right	Shift-right (93)	Move forward (35.38)
E_4	1.617	0.475	1.854	1.169	Slight-right	Slight-right (51.6)	Slight-right (54.55)
E_5	0.789	0.49	1.864	1.076	Shift-right	Shift-right (79.1)	Shift-right (70.42)
Test Samples 3: Accuracy = 5/5 = 100%							*4/5 = 80%*
Average Accuracy = 93.3%							*73.3%*

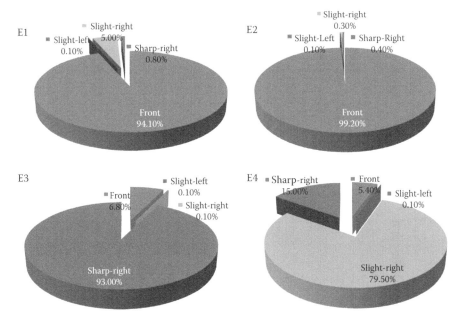

FIGURE 34.7 Presentation showing the predicted behaviour using BN in a dynamic environment from first cross validation of Table 34.2.

TABLE 34.3
Average Performance Evaluation Results of BN and *k*-NN Models

		Bayesian Networks (BN)		*k*-Nearest Neighbour (*k*-NN)	
Environment	CVn	Accuracy (%)	Error (%)	Accuracy (%)	Error (%)
Static	1	100	0	100	0
	2	100	0	100	0
	3	100	0	100	0
Dynamic	4	100	0	80	20
	5	80	20	60	40
	6	100	0	80	20

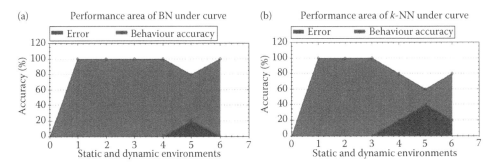

FIGURE 34.8 The BN accuracy in the lighter grey area in (a), at CV levels 4–6, is larger than that of the accuracy when compared to the k-NN model in (b).

studied by investigating problems in neuroethology by building physical robot models of biological sensorimotor systems. For instance, Robots are believed to mimic the behaviour of biological systems, but do they model complex behaviours very well, such as emotional expressions? It is argued that in building robot models biological relevance is more effective than loose biological inspiration. This reflects the view that biological behaviour needs to be studied and modelled in the context of real problems faced by real robots in real environments. A proposal of vision-based mobile robot, which can find the location of doors and can traverse doors in complex environments, is presented in Ref. [19]. A Principal Component Analysis (PCA) algorithm using a vision sensor and a fuzzy controller are used for obstacle avoidance and door traversal behaviours. Hongjun et al. [20], proposed a novel method for sensor planning using mobile robot localization based on Bayesian network inference. In their work, they proved that an autonomous robot cannot always determine its unique situation by local sensing information only. The reason is that, the sensor is prone to errors and a slight change of robotic behaviour deteriorates the sensing result. In [21], an attempt was made to obtain an adequate BN model for achieving a door-crossing behaviour using sonar sensors available on their robot platform. Concerning the performance of their obtained behaviour, they proposed that their results should be compared with other approaches, indicating that more work is required on developing behaviours for robots. As they pointed out that they are missing a common frame of reference with other approaches, this chapter experimented with a real-life publicly available navigation data. In our previous work [22], we mentioned that Bayesian network can be used for developing behaviours for autonomous robots for avoiding collisions in the environments. In the approach, an unstructured environment was simulated and information of the obstacles generated was used to build the BN model for avoiding collisions. A proposal to use a real-life robot data and test with more approaches were presented.

34.6 CONCLUDING REMARKS

In this chapter, BN and k-NN models have been successfully investigated and applied to developing behaviours for robots for avoiding collisions in static and dynamic environments. The performance of the learning mechanism of both models in the experiments using a real-life robot observation show that the predictive power of BN and k-NN models are valuable for their consistency in handling robot collision in a static environment. However, based on the scope and the data used, BN proved better with 93.3% compared to 73.3% of k-NN accuracy in the dynamic environment due to its probabilistic calculus in handling uncertainties. The choice of the kth minimum value in k-NN might need more improvement for an improved prediction to avoid collision. It is worth noting that the chosen direction of the robot assists in determining the control command values, which are translational and rotational velocities the robot uses for navigation. Hence, this investigation contributes to an attempt of using machine learning models for developing behaviours within the robotics domains. The results of this chapter extend our previous work in developing behaviours for a robot as reported in [22], which implemented only the BN model using simulated obstacle data samples. This research work can further be explored in different forms in future work: (i) developing behaviours for conditional collision avoidance by moving from a starting position to a goal position; (ii) develop behaviours based on other sensor configurations as the 24 sensors may be computationally intensive while the two sensors may be faster and (iii) develop cooperative behaviours for multi-robot systems to avoid conflicts among robot team members.

REFERENCES

1. J.G. Rolim, P.C. Maiola, H.R. Baggenstoss, A.R.G. da Paulo, Bayesian networks application to power transformer diagnosis, *IEEE Lausanne Power Tech*, pp. 999–1004, 2008.
2. A. Thamizharasi, Performance analysis of face recognition by combining multiscale techniques and homomorphic filter using fuzzy K nearest neighbor classifier, *IEEE International Conference on Communication Control and Computing Technologies (ICCCCT)*, pp. 394–401, 2010.

3. N. Nilsson, *Artificial Intelligence, a New Synthesis*, 1st edition. San Fransisco, USA: Morgan Kaufmann Publishers, 1998.

4. Y. Sun, S. Lv and Y. Tang, Construction and application of Bayesian network in early diagnosis of Alzheimer disease's system, *International Conference on Complex Medical Engineering, IEEE/ICME*, pp. 924–929, 2007.

5. I.O. Osunmakinde and A. Potgieter, Immediate detection of anomalies in call data—An adaptive intelligence approach. *Proceedings of the 10th Southern African Telecommunications Networks and Applications International Conference (SATNAC)*, Mauritius. 2007.

6. Y. Lv, S. Tang and H. Zhao, Real-time highway traffic accident prediction based on the k-nearest neighbor method, *International Conference on Measuring Technology and Mechatronics Automation, ICMTMA '09*, pp. 547–550, 2009.

7. G. Verdier and A. Ferreira, Fault detection with an adaptive distance for the k-nearest neighbors rule, *International Conference on Computers & Industrial Engineering, CIE*, pp. 1273–1278, 2009.

8. J. Nagasue, Y. Konishi, N. Araki, T. Sato and H. Ishigaki, Slope-Walking of a biped robot with k nearest neighbor method, *Fourth International Conference on Innovative Computing, Information and Control (ICICIC)*, pp. 173–176, 2010.

9. H. Zhou and S. Sakane, Sensor planning for mobile robot localization—A hierarchical approach using a Bayesian network and a particle filter, *IEEE Transactions on Robotics*, 24, 481–487, ISSN: 1552-3098, 2008.

10. S.I. Roumeliotis and G.A. Bekey, Distributed multirobot localization, robotics and automation, *IEEE Transactions on Robotics and Automation*, 18(5), 781–795, ISSN: 1042-296X, 2002.

11. H. William and G. Haipeng, P. Benjamin, & S. Julie, A Permutation genetic algorithm for variable ordering in learning Bayesian networks from data. *Proceedings of Genetic and Evolutionary Computation Conference*, Morgan Kaufmann Publishers Inc, San Francisco, CA, USA, pp. 383–390, 2002.

12. I.O. Osunmakinde and A. Potgieter, Emergence of optimal Bayesian networks from datasets without backtracking using an evolutionary algorithm. *Proceedings of the Third IASTED International Conference on Computational Intelligence*, Banff, Alberta, Canada, ACTA Press, pp. 46–51, 2007.

13. S. Russell and P. Norvig, *Artificial Intelligence, A Modern Approach*, 2nd edn., Prentice-Hall Series Inc. NJ, 2003.

14. J. Arroyo and C. Mate, Forecasting histogram time series with k-nearest neighbors methods, *International Journal of Forecasting*, 25, 192–207, 2009.

15. D. Newman, S. Hettich, C. Blake and C. Merz, UCI *Repository of Machine Learning Databases* (University of California, Department of Information and Computer Science, Irvine, CA). DOI = http://www.ics.uci.edu/$\sim$mlearn/MLRepository.html; (last accessed 2011).

16. N. Harper and P. McKerrow, Recognizing plants with ultrasonic sensing for mobile robot navigation, *Journal of Robotics and Autonomous Systems*, 34, 71–82, 2001.

17. GeNle 2.0, *Decision Systems Laboratory*, University of Pittsburgh, URL =http://genie.sis.pitt.edu, 2009.

18. B. Webb, Can robots make good models of biological behavior? *Journal of Behavioral and Brain Sciences*, 24, 1033–1050, 2001.

19. M.-W. Seo, Y.-J. Kim and M.-T. Lim, Door traversing for a vision based mobile robot using PCA, in: *Lecture Notes on Artificial Intelligence*, Springer-Verlag, pp. 525–531, 2005.

20. Z. Hongjun and S. Shigeyuki, Mobile robot localization using active sensing based on Bayesian network inference, *Journal of Robotics and Autonomous Systems*, 55, 292–305, 2007.

21. E. Lazkano, B. Sierra, A. Astigarraga and J.M. Martinez-Otzeta, On the use of Bayesian networks to develop behaviors for mobile robots, *Journal of Robotics and Autonomous Systems*, 55, 253–265, 2007.

22. C. Yinka-Banjo, I.O. Osunmakinde and A. Bagula, Collision avoidance in unstructured environments for autonomous robots: A behavioral modeling approach, In *Proceedings of the International Conference on Control, Robotics and Cybernetics (ICCRC)*, New Delhi, India, IEEE, pp. 297–303, 2011.

35 Modelling Out-of-Sequence Measurements

A Copulas-Based Approach

Bhekisipho Twala

CONTENTS

35.1 INTRODUCTION

Within the general framework of multi-sensor applications, a difficult step in target tracking and filtering (i.e., the process of maintaining state estimates of one or several instances or objects over a period of time) is the handling of out-of-sequence measurements (OOSM). Most of the work on tracking and filtering has been based on the assumption that measurements are immediately available to an agent. However, it is not difficult to conceive situations in which measurements are subject to non-negligible delays such that the lag between measurement and receipt is of sufficient magnitude to have an impact on estimation or prediction. These measurements can be classified as either constant delays or random delays with the resulting occurrence of the latter having the potential to cause OOSM.

Handling OOSM represents a challenge for engineers or researchers using multi-sensor target tracking data. The question is how to incorporate these OOSMs in a track that has already been updated with a later instance or observation in order to enhance the performance of the tracking system. A simple solution is to simply ignore (neglect) or discard the OOSM in the tracking process. The appropriateness of this approach is that it has a natural limitation in that critical information is lost due to the discarded OOSM and this can lead to degradation in tracking for time-critical targets. Other approaches for dealing with OOSM include data re-processing or rollback and data buffering. In the rollback approach, sensor reports are stored in the memory and the OOSM is used to re-order the sensor measurements in a track hypothesis. The data-buffering approach holds the incoming measurements in a buffer with the size of the buffer greater than the maximum expected delay of arriving measurements. Both approaches require significant memory and storage measurements. Also, since the tracker processing always lags behind the current time, both approaches pose potential problems for real-time target applications. The several aspects related to handling of time delays (in the context of OOSM) are considered in Refs. [1–8]. For time delays, one common approach for dealing with the OOSM problem is related to solving a partial differential equation and boundary condition equations that do not have an explicit solution in general [1,2,5,6]. For random delays, the problem has been investigated via a standard Kalman filtering and by augmenting the system accordingly [3,8]. Mallick et al. [9] address the OOSM problem by re-calculating the filter through the delayed period. In the same context, Ref. [6] proposes a measurement extrapolation approximation using past and present estimates of the Kalman filter (KF) and calculating an optimal gain for this extrapolated measurement (ME-KF). In Ref. [10], an iterative form of state augmentation for random delays with a random lag is considered.

The copulas modelling has found many useful applications in actuarial science, survival analysis, hydrology and finance [11]. This most important aspect in the copula framework is due to Ref. [12]. Ref. [13] examines the case of random delay under the name of fixed sampling and random delay filter (FSRD-KF) that is shown to be equivalent to constraining the lag to a value of 1. Later, in Ref. [6], the use of delayed measurements to calculate a correction term and adding this to the filter estimate was suggested. Ref. [14] relates OOSM to the incomplete (missing) data problem and uses statistical multiple imputation to deal with OOSM. In Ref. [15], the algorithms that try to minimize the information storage in an OOSM situation (MS-KF) were proposed. Ref. [4] formulated the OOSM problem in a Bayesian framework (BF-KF). Although the vast majority of the above methods understand the solution, most of them fail to recognize the theoretical basis of the conditional distribution between the delayed measurements and the measurements that are already available (which are sometimes referred to as history). The major contributions and uniqueness of the work presented in this chapter are as follows: (i) we show the robustness of one of the top five techniques for handling OOSM in terms of predictive accuracy for multi-target tracking and (ii) we further show how copulas could be used to deal with OOSM and how the use of copulas lead to a significant improvement in classification performance for multi-target tracking.

35.2　THE COPULAS STRATEGY

A motivation for copulas is that it exists as a multivariate distribution function and allows a consistent and flexible modelling of the dependence structure of dealing with OOSM. It offers a

---

1. Consider a sequence of measurements up to k instances $X_1, X_2, ..., X_k$ (where k is the delay point) with distribution function $H(x_1, x_2, ..., X_{k-1}) = P(X_1 \leq x_1, X_2 \leq x_2, ..., X_k \leq x_k)$ and univariate marginal distributions $F_1(x_1), F_2(x_2), ..., F_k(x_{k-1})$.
2. A copulas C represents the joint cumulative distribution function in terms of the margins such that $H(x_1, x_2, ..., X_{k-1}) = C(F_1(x_1), ..., F_k(x_k))$ for all values $x_1, x_2, ..., x_k$ (or $(X_1, X_2, ..., X_k \in \Re^k)$.
    - If $F_1, F_2, ..., F_k$ are continuous, C is unique for every fixed F and equals $C(u_1, ..., u_k) = F(F_1^{-1}(u_1), ..., F_k^{-1}(u_k))$, where $F_1^{-1}, ..., F_k^{-1}$ are the quantiles functions given marginals and are uniform $[0,1]$ variables.
    - If the sequence of measurements $(X_1, X_2, ..., X_k)$ are independent then the copula function that links their marginals is the product copula $C(F_1, F_2, ..., F_k) = F_1 * F_2 * \cdots * F_k$.
    - If C and $F_1, F_2, ..., F_k$ are differentiable, then the joint density $f(x_1, x_2, ..., x_k)$ corresponding to the joint distribution $F(x_1, x_2, ..., X_k)$ can be written as the product of the marginal densities and copula density $f(x_1, x_2, ..., x_k) = f_1(x_1) \times \cdots \times f_k(x_k) \times C(F_1, F_2, ..., F_k)$ where $f_i(x_i)$ is the density corresponding to F_i and the copula density is defined as $C = \partial^k C/(\partial F_1 ... \partial F_k)$.
3. Find the conditional distribution $F(x_k|H)$ for delayed measurements conditioned to the history of measurements available as a predictive distribution.
4. Predict the delayed measurement from the conditional distribution (copula could be used to find both the joint and conditional distributions even if the joint distribution is unknown).

FIGURE 35.1　The copulas algorithm for dealing with OOSM.

convenient representation of arbitrary joint distribution functions, with the key property being that the specification of the marginal distributions and the dependence structure is separated. This is the most important result in the copula framework and is due to Ref. [12]. In recent years, copulas modelling has found many successful applications in actuarial science, survival analysis, hydrology and with high intensity in finance [11]. The generalized copulas algorithm for handling OOSM is summarized in Figure 35.1. For a more detailed discussion on copulas, the reader is referred to Refs. [11,12].

35.3 EXPERIMENTS AND RESULTS

In order to empirically evaluate the performance of the proposed copula-based OOSM approach (which from now on we call COOSM) against existing approaches for dealing with OOSM (FSRD-KF, ME-KF, SARD-KF, MS-KF and BF-KF), experiments are used on simulated datasets in terms of root square mean error (RMSE). RMSE is a measure of the differences between values predicted by a model (or an estimator) and the values actually observed. The experiment is carried out to rank individual OOSM methods and also to assess the impact of delayed measurements (at various time and distance intervals) on a single delay against COOSM in terms of position error. Like in Ref. [4], we assume that the OOSM can only have a maximum of one lag delay and the data delay is uniformly distributed within the whole simulation period with probability P_r that the current measurement is delayed. All statistical tests were conducted using the MINITAB statistical software programme. Analyses of variance, using the general linear model procedure, were used to examine the main effects and their respective interactions. This was done using a three-way repeated measures design (where the effects were tested against its interaction with datasets). The main effects are OOSM methods, the probability of measurement and the manoeuvering index. All the main effects were found to be significant at 5% level of significance ($F = 18.9$, $df = 5$ for methods; $F = 29.4$, $df = 1$ for probability of measurement and $F = 31.2$, $df = 1$ for manoeuvering index; p-value < 0.05 for each effect). As shown in Figure 35.2, COOSM is the best method for handling OOSM with an error rate of 6.1%, closely followed by BF-KF, FSRD-KF and MEKF with excess error rates of 9.2%, 11.7% and 14.2%, respectively. The worst method is SARD-KF, which exhibits an error rate of 18.0%. Another poor performance (after SARD-KF) is by MR-KF with an error rate of 16.8%.

Tukey's multiple comparison tests showed significant differences in performances between all the methods at 5% level of significance. We now present results for the performances of the six methods (current and new) for single delay over 1000 runs. Due to space requirements,

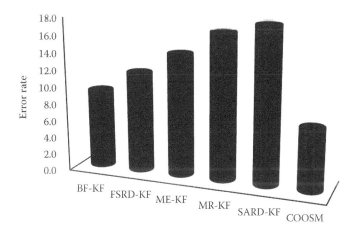

FIGURE 35.2 The performance of OOSM methods.

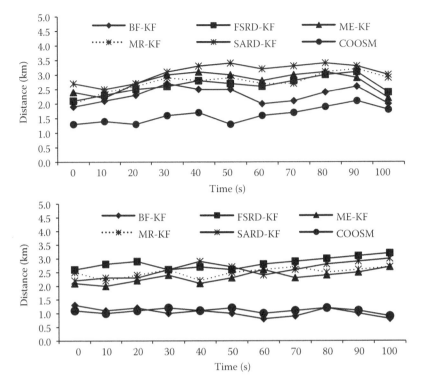

FIGURE 35.3 RMS performance in case of a highly manoeuvering target with single delay OSSM ($P_r = 0.5$ and 0.25; manoeuvering index = 0.1).

we only present results for one manoeuvering index. From Figure 35.3, the following results are observed:

1. For manoeuvering target tracking, COOSM outperforms all the other methods when the probability of measurement is 0.5. However, its performance with BF-KF is comparable when the probability of measurement is 0.25. The differences in performance among methods are mostly prominent at higher probabilities of measurement. Poor performances are observed for SARD-KF (for $P_r = 0.5$) and FSRD-KF (for $P_r = 0.25$).

2. Increases in probability measurement delay are associated with increases in performance differences between methods. In fact, the performance of all the methods degrades with increases in the probability of measurement.

3. From Figure 35.4, most of the methods have similar RMS performance regardless of OOMS. However, this is the case up until the 50 s time limit. In other words, OOSM does not seem to be critical for most of the methods for the first 50 s (with the exception of SARD-KF and FSRD-KF). Thereafter, the difference in RMS performances between the methods becomes quite prominent.

4. Overall, COOSM and BF-KF achieve higher accuracy rates with COOSM slightly outperforming BF-KF most of the time.

35.4 CONCLUDING REMARKS

Practical data fusion schemes are challenged by the inevitable appearance of delayed measurements. Accurate prediction of target tracking given delayed measurements can be very valuable to

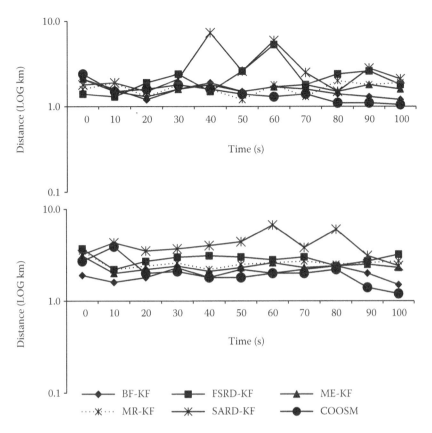

FIGURE 35.4 RMS performance in case of a highly manoeuvering target with multiple delays OOSM (P_r = 0.5 and 0.25).

engineers, especially those dealing with sensoring applications. This is important for minimizing cost and improving effectiveness of the multi-target tracking process. The major contribution of the chapter has been the application of copulas to predict multi-target tracking given that some measurements are out of sequence. Simulated datasets were utilized for this task. Individually, COOSM is the most effective method for handling OOSM with BF-KF not far behind. The worst performance is by SARD-KF.

Our results further show the probability of measurement delays as having an impact on the performance of methods with BF-KF more effective for the smaller probability. Bigger positional error rates were achieved by methods for high-probability delays with much bigger differences in performance among the methods. Also, given that the performance of each method varies by probability of measurement delay, it appears that the treatment of delayed measurements heavily depends not only on the probability of measurement delay but also on the range of manoeuvering target tracking. The use of copulas also deserves further investigation on a number of fronts, for example, in terms of the training parameters and the combination rules that can be employed. Also, empirical studies of the application of the copula to real-world datasets should be undertaken to assess its performance across a more general field. We leave the above issues to be investigated in the future.

ACKNOWLEDGEMENTS

This work was funded by the Department of Electrical Engineering and Electronic Engineering Science at the University of Johannesburg, South Africa. The comments and suggestions from my colleagues and the anonymous reviewers greatly improved this chapter.

REFERENCES

1. Anderson, B.D.O. and Moore, J.B., 1979. *Optimal Filtering*, Prentice-Hall, Englewood Cliffs, New Jersey, USA.

2. Bar-Shalom, Y., 2000. Update with out-of-sequence measurements in tracking: Exact solution, *Proceedings of the SPIE Conference on Signal and Data Processing of Small targets*, Orlando, FL, USA, pp. 51–556.

3. Bar-Shalom, Y. and Li, X.-R., 1993. *Estimation and Tracking: Principles, Techniques and Software*, Artech House, MA, USA.

4. Challa, S., Evans, R.H. and Wang, X., 2003. A Bayesian solution and its approximation to out-of-sequence measurement problems, *Information Fusion*, 4: 185–199.

5. Kalman, R.E., 1960. A new approach to linear filtering and prediction problems, *Transaction of the ASME—Journal of Basic Engineering*, 82(D): 33–45.

6. Larsen, T., Puolsen, N., Anderson, N. and Ravino, O., 1998. Incorporating of time delayed measurements in a discrete-time Kalman filter, *CDC '98*, Tampa, FL, USA.

7. Mallick, M., and Bar-Shalom, Y., 2002. Non-linear out-of-sequence measurement filtering with applications to GMTI tracking, *Proceedings of SPIE Conference Signal and Data Processing of Small Targets*, Orlando, FL, USA.

8. Mallick, M., Zhang, K. and Li, X.R., 2003. Comparative analysis of multiple-lag out-of-sequence measurement filtering algorithms, *Proceedings Signal and Data Processing of Small Targets*, San Diego, CA, USA.

9. Mallick, M., Krant, J. and Bar-Shalom, Y., 2002. Multi-sensor multi-target tracking using out-of-sequence measurements, *Proceedings of the 5th International Conference on Information Fusion*, August 8–11, Annapolis, MD, USA.

10. Matveev, A. and Savkin, A., 2003. The problem of state estimation via asynchronous communication channels with irregular transmission times, *IEEE on Automatic Control*, 48(4): 670–676.

11. Nelsen, R., 1999. *An Introduction to Copulas*, Springer-Verlag, New York.

12. Sklar, A., 1959. Fonctions de r'eparuition 'a n dimensionse leurs merges, *Publ. Inst. Statis. Univ. Paris*, 8: 229–231.

13. Thomopolous, S.C.A. and Zhang, L., 1994. Decentralize filtering with random sampling and delay, *Information Sciences*, 81: 117–131.

14. Twala, B., 2010. Handling out-of-sequence measurements: Kalman filter or statistical imputation? *Electronics Letters*, 46(4), 302–304.

15. Zhang, K., Li, X.R. and Zhu, Y., 2005. Optimal update with out-of-sequence measurements, *IEEE Transactions on Signal Processing*, 53(6), 1992–2004.

Appendix A: Statistical and Numerical Concepts

CONTENTS

Some concepts of statistical and numerical analysis are briefly presented here in order to facilitate better understanding of mathematical developments and performance analysis appearing at various places in several chapters in the present volume. Much of the material is adapted from Refs. [1–5] and in a few cases the mathematical formulae/expression/equations are avoided. In a few cases, these are given without derivations.

A.1   ALTERNATIVE HYPOTHESIS

This hypothesis is being tested in an experiment and a conclusion is reached when a null hypothesis is rejected.

A.2   ANALYSIS OF VARIANCE

This is a test for ascertaining significant differences between multiple means by comparing variances of the chosen time series. Analysis of variance (ANOVA) is a special case of multiple regression analysis. Two important assumptions in ANOVA are (i) the homogeneity of variances and (ii) normal probability distribution of the data within each test group. The variance estimated from the within-group random variability should be about the same as the variance estimated from between-groups variability, that is, if the null hypothesis is true, mean ESS/mean RSS (variance ratio) would be equal to 1. This is also known as the *F test* or variance ratio test. The ANOVA approach is based on the partitioning of sums of squares and degrees of freedom associated with the response variable.

A.3   ASYMPTOTICALLY UNBIASED

In the estimation cycle, the bias in estimated parameter or state approaches zero as the sample size (N) increases to infinity. Thus, estimation with this property (of the estimator) improves as N increases.

A.4   BIAS

If the expected value of the parameter is the true value of the parameter, then the estimator is called unbiased estimator: Bias $(\beta) = \beta - E(\hat{\beta})$. If the bias tends to decrease as N, the sample size gets larger; it is the property of asymptotic unbiasedness.

A.5   BINOMIAL DISTRIBUTION

It gives the probability of obtaining exactly r successes in n independent trials/experiments.

A.6   CENTRAL LIMIT THEOREM

The means of a relatively large (say, 30 or more) number of random samples from any population (not necessarily a normal distribution) will be approximately normally distributed with the attributes: (i) the population mean being their (the samples') mean, (ii) their variance (the samples') being the population variance/N) and (iii) the approximation will improve as the sample size (N) increases.

A.7   CHI-SQUARED DISTRIBUTION AND TEST

This distribution is derived from the normal distribution and the Chi-squared variable is distributed with n degrees of freedom with mean $= n$ and variance $= 2n$. This CS test is most frequently used for frequency data (analysis) and goodness-of-fit criterion. The Chi-square value is obtained by summing up the values (e.g., error-samples') (square {residual}/expected value). Here, the residual is the difference between the observed value and its expected value. Let x_i be the normally (Gaussian) distributed variables with zero mean and unit variance. If $\chi^2 = x_1^2 + x_2^2 + \cdots + x_n^2$, then the random variable χ^2 has the pdf (probability density function) with n degrees of freedom: $p(\chi^2) = 2^{-n/2} \, \Gamma(n/2)^{-1} (\chi^2)^{(n/2)-1} \exp(-\chi^2/2)$. Here, $\Gamma(n/2)$ is Euler's gamma function, and we also have $E(\chi^2) = n; \quad \sigma^2(\chi^2) = 2n$. In the limit, the χ^2 distribution approximates the Gaussian/normal distribution with mean n and variance $2n$. Let x_i be normally (Gaussian) distributed and mutually uncorrelated variables around mean m_i and with variance σ_i. Then the normalized sum of squares $s = \sum_{i=1}((x_i - m_i)^2/\sigma_i^2)$ is formed, then s follows the χ^2 distribution with n DoF. In estimation exercises, the χ^2 test is used for hypothesis testing to see if, for example, the order of the mathematical model determined by system identification is adequate or not.

A.8 CORRELATION

It is given by the following formula: $\rho_{ij} = (\text{cov}(x_i, x_j)/\sigma_{x_i}\sigma_{x_j})$ $-1 \le \rho_{ij} \le 1$, for certainly and fully correlated process, $\rho = 1$, this parameter defines the degree of correlation between the two random variables. In Kalman filter development, it is often assumed that the state errors, measurement errors and residuals are uncorrelated.

A.9 COVARIANCE

Covariance is a measure of the association/relationship between two variables, or two variables taken at a time if there are more than two variables, the respective means being removed from these variables. If the two variables are independent, then their covariance is zero. It is given as

$$\text{Cov}(x_i, x_j) = E\left\{ \left[x_i - E(x_i) \right] \left[x_j - E(x_j) \right] \right\}.$$

A.10 EIGENVALUES

This German word means latent values of a system or its dynamics. Let $Ax = \lambda x$; then this operation means a matrix operation on a vector x upgrades the vector x by scalar λ. We formulate the eigenvalues/eigenvector problem as $(\lambda x - Ax) = 0 \Rightarrow (\lambda I - A)x = 0$, then $|\lambda I - A| = 0$ and λ_i are the so-called eigenvalues of the matrix A.

A.11 EM ALGORITHM

EM is a method for computing maximum likelihood estimates with incomplete data. In the first step, that is, E (expectation)-step, the expected values for missing data are computed, and in the second step, that is, M (maximization)-step, the maximum likelihood estimates are computing as if the complete data were available.

A.12 F-DISTRIBUTION AND F-TEST

This distribution is a continuous probability distribution of the ratio of two independent random variables divided by their respective degrees of freedom. Each random variable has a Chi-squared distribution. In regression and least squares analysis and other estimation methods, the F-test is used to test the joint significance of all the variables of a model that is fitted to the test data. The F-test is used to check if the slope (say from the least squares fit to the data) is significantly different from 0, which is equivalent to testing whether the model-fit using non-zero slope is significantly better than the null model with 0 slope.

A.13 LEAST SQUARES METHOD

It is a principle of fitting a straight line or a curve based on the minimization of the sum of squares of differences (also called residuals) between the predicted and the observed points or samples. It is considered a deterministic approach to the parameter estimation problem. Choose an estimator of β that minimizes the sum of the squares of the error:

$$J \cong \frac{1}{2} \sum_{k=1}^{N} v_k^2 = \frac{1}{2} \ (z - H\beta)^T (z - H\beta)$$

where J is a cost function and v are the residual errors at time k, superscript T stands for the vector/matrix transposition. The minimization of J w.r.t. β yields $\partial J/\partial\beta = -(z - H\hat{\beta}_{LS})^T H = 0$, and simplification leads to $\hat{\beta}_{LS} = (H^T H)^{-1} H^T z$ giving the computational means to determine the least squares estimates.

A.14 LIKELIHOOD AND LIKELIHOOD TEST

The likelihood gives a probability of a set of measurements, given the value of some parameter. For example, the likelihood of a random sample x of N observations with probability distribution $f(x; b)$ is given by: $L = b\, f(x_i; b_0)$. It is the basis of maximum likelihood estimation. The likelihood test is a general test of hypothesis H_0 against an alternative hypothesis H_1 based on the ratio of two likelihood functions, one derived from each of H_0 and H_1. The statistics $b = -2\ln(L_{H_0}/L_{H_1})$ has approximately a b^2 distribution with DoF equal to the difference in the number of parameters in the two hypotheses.

A.15 LINEAR REGRESSION MODELS

Here, linear can also be taken to mean linear in the parameters (coefficients/LIP).

A.16 MAXIMUM LIKELIHOOD

The ML method is a method of finding estimated values of parameters based on the principles of maximum likelihood. The method yields values for the unknown parameters that maximize the probability of obtaining the measured values from the mathematical model that has these unknown parameters. A likelihood function is set up which expresses the probability of the measured data as a function of these unknown parameters. Then the maximum likelihood estimator of these parameters is chosen to be those values that maximize this function. The resulting estimator agrees most closely with the measured data.

A.17 MEAN, MEDIAN AND MODE

It is a measure of central location for a batch of data values and it is computed as a sum of all the data values divided by the number of elements/samples. The median is the value that divides the frequency distribution in half when all numerical data values are listed in order. Mode is the observed value that occurs with the greatest frequency. The mode is generally not affected by small numbers of extreme values.

A.18 MEAN SQUARES

In the mean squares estimation, the cost function is defined as $J = E\{\underline{x}^T(k)\underline{x}(k)\}$, where E is the mathematical expectation that takes into account the probability of occurrence of the favourable event.

A.19 MONTE CARLO METHOD

The idea is to study by means of computer simulations a complex relationship or a problem difficult to solve by mathematical analysis.

A.20 MULTIPLE REGRESSION

In multiple regression analysis, a quantification of the relationship between several independent variables is carried out. The coefficients of these relationships are estimated by the least squares method, which is a special case of ML estimation method discussed above. The multiple regression correlation coefficient (R^2) is a measure of the proportion of variability explained by or due to the regression (linear relationship) in a sample of data. It is a measure of the effect of X in reducing the uncertainty in predicting Y. When all measurements fall on the fitted regression line, $R^2 = 1$. When the fitted regression line is horizontal, $R^2 = 0$. The square root of R^2 is the correlation coefficient (r). The stepwise regression method seeks a model that balances a relatively small number of variables with a good fit to the data by seeking a model with high R^2. The method can be started from a null or a full model and can go forward or backward, respectively. At any step in the procedure, the statistically most important variable will be the one that produces the greatest change in the log-likelihood relative to a model lacking the variable.

A.21 NON-LINEAR REGRESSION

In this case, the fitted (or predicted) value of the response variable is a non-linear function of one or more X variables, that is, independent variables.

A.22 OUTLIER

An extreme measurement value/values that are very well separated from the remaining data samples. Most outliers values will have some influence on the fitted function.

A.23 POISSON DISTRIBUTION

Poisson distribution is the probability distribution of the number of (rare) occurrences of some random event in an interval of time or space. Poisson distribution is used to represent distribution of counts like number of defects in a piece of material, customer arrivals, insurance claims, incoming telephone calls or alpha particles emitted. A transformation that often changes Poisson data approximately normal is the square root.

A.24 PROBABILITY, PROBABILITY DISTRIBUTION FUNCTION AND PROBABILITY DENSITY FUNCTION

These are the ratio of the number of favourable outcomes to the number of total/possible outcomes of a random even. Probability distribution function is a function that gives for each number x the probability that the value of a continuous random variable X is less than or equal to x. For discrete random variables, the probability distribution function is given as the probability associated with each possible discrete value of the variable. When a curve is used to model the variation in a population and the total area between the curve and the x-axis is 1, then the function that defines the curve is a probability density function. In Gaussian pdf $p(x) = (1/\sqrt{2\pi}\ \sigma)\exp(-((x-m)^2/2\sigma^2))$, m is the mean and σ^2 is the variance of the distribution. When the state x (or parameters, β) is given the pdf for the measurements

$$p(z \mid x) = \frac{1}{(2\pi)^{n/2}|R|^{1/2}}\ \exp\left(-\frac{1}{2}(z - Hx)^T\ R^{-1}(z - Hx)\right),$$

where R is the covariance matrix of measurement noise.

A.25 STUDENT'S T-TEST

It is a statistical hypothesis test for the significance between means or between a mean and a hypothesized value. Here, the measurements should be normally distributed. The ratio of variances in two samples should not be more than three. The test for independent samples tests whether or not two

TABLE A.1

Type I and Type II Errors in Hypothesis Testing

	H_0—Null Hypothesis Is True	H_0—Null Hypothesis Is False
The null hypothesis is rejected	Type I error/false positive	Correct outcome/true positive
The null hypothesis is not rejected	Correct outcome/true negative	Type II error/false negative

means are significantly different from each other. It is defined as the difference of sample means divided by standard error of difference of sample means.

A.26 TYPE I AND TYPE II ERRORS

The Type I error is said to occur if the null hypothesis is true but it is rejected. This error is viewed as the error of excessive belief. If the null hypothesis is accepted when it is in fact wrong, Type II error is said to occur. Table A.1 further clarifies these errors.

A.27 GRADIENT DESCENT

It is a first-order optimization algorithm to find a (local) minimum of a function. The incremental steps for the variable/parameter are taken in some proportion to the negative of the gradient value evaluated at that point/variable/parameter. The method also known as a steepest descent method is relatively slow in convergence.

A.28 QUADRATIC FORM

It is a homogeneous polynomial of degree two in its number of variables/parameters. Often, this form is associated with a symmetric matrix. The definite quadratic form has either zero or positive numerical value, whereas indefinite quadratic form would have mixed values, including negative numerical values. The latter occur in H-infinity norm and H-infinity filter theory. The quadratic form can be used to derive a measure of a vector or matrix (of a signal) in order to have knowledge of their magnitudes and strengths. The distance measure or norm is defined as

$$L_p = \|x\|_p = \left(\sum_{i=1}^{n} |x_i|^p \right)^{1/p} ; \quad p \geq 1$$

If $p = 1$, the length of vector x is $\|x\|_1 = |x_1| + |x_2| + \cdots + |x_n|$, the centre of a probability distribution estimated using L_1 norm is the median of the distribution. If $p = 2$, we get an Euclidean norm which gives the length of the vector. For $p = 2$, the centre of a distribution estimated using L_2 norm is the mean of the distribution. The norm is used in many state/parameter estimation problems to define the cost functions in terms of state or measurement error. The optimization problems with this norm are mathematically tractable. This leads to the least squares or maximum likelihood estimator as the case may be.

REFERENCES

1. Tevfik Dorak M. http://www.dorak.info/mtd/glosstat.html, December 2010.
2. Stengel R. F., *Robotics and Intelligent Systems, A Virtual Textbook*, Princeton University, Princeton, NJ, January 25, 2010, http://www.princeton.edu/~stengel/RISVirText.html#Chapter, October 2011.

3. Raol J. R. *Multisensor Data Fusion with MATLAB*, CRC Press, Boca Raton, FL, USA, 2009.
4. Raol J. R., Girija G. and Singh, J. *Modelling and Parameter Estimation for Dynamic Systems*, IEE/IET Control Series, Vol. 65, IEE/IET, London, UK, 2004.
5. Raol J. R. and Singh, J. *Flight Mechanics Modeling and Analysis*, CRC Press, Boca Raton, FL, USA, 2008.

Appendix B: Notes on Software and Algorithms Related to Robotics

CONTENTS

This appendix offers a brief description of some software tools which would be useful for carrying out simulation and other analyses related to robotics and or aerospace vehicles is given.

B.1 USARSIM TUTORIAL

The USARSim is a game (game-theoretic)-based high-fidelity and interactive simulation of urban search and rescue (USAR) robots and environments [1]. It provides (i) models of the environments, (ii) models of the experimental and commercial robots, (iii) models of the sensors employed and (iv) auxiliary tools used for robot control. The USARSim is also considered as a research tool for the study of human–robot interaction (HRI) and multi-robot coordination.

B.1.1 USARSim System Architecture

The main components of the system architecture are (i) team cooperation, (ii) control, (iii) networks and (iv) robot's environment. In the control component, we have high-level and middle-level controls and control interface. The clients, control and the world interact with each other via the network. From the environment, the network inputs are maps, various models, unreal world and Gamebots (Game-robots). In addition, from the unreal clients there would be video feedback to the control block, which in turn, interact with the team cooperation block. The server connecting to the map, models, unreal world and the Gamebots (MMUG) maintains the states of objects and responds to clients. The client stage comprising team cooperation, controller, unreal client and the network visualizes objects, makes decisions and requests the server. Thus, the system architecture is divided into two parts: (a) the user's controller and (b) the USARSim. The system architecture has in effect (a) unreal engine, (b) Gamebots and (c) controller. The unreal engine is a multi-layered combat-oriented first-person game engine released by Epic Games consisting of 3D scene render, physic engine (Karma engine), script language and 3D authoring tool. The Gamebots is a modification to unreal tournament to bridge unreal engine with the outside applications consisting of TCP socket connection and message exchange. The controller is the application designed by users for research purposes and consists of robot control and data exchange.

B.1.2   Simulator Components

These are (i) environment simulation, (ii) sensor simulation, (iii) robot simulation and (iv) control simulation.

B.1.2.1   Environment Simulation

The environment simulation components are geometric models, obstacles, light, special effects and victim. Also, there are NIST arenas, the real arenas, which are coloured yellow, orange and red. The features of the yellow arena are simple to traverse with no agility requirements, planer 2D maze, isolates sensors with obstacle/targets, and reconfigurable in real time to test mapping. The orange arena features are more difficult to traverse, variable floorings, spatial 3D maze, stairs, ramp and holes, and physical obstacles including rubble, paper and pipes. The red arena is difficult to traverse, has unstructured environment, simulated rubble piles and shifting floors, and there would be problematic junk like plastic bags, pipes and so on.

B.1.2.2   Sensor Simulation

The sensor simulation aspects are the method, features, sensors and video feedback. This method includes calculation of the data from the ground truth database and noise/distortion aspects. The features could be hierarchical structure and configurability. The sensor aspects are robot camera, range sensing, sound sensing, human motion sensor and range scanner/sensor. The sensors used are state sensors for battery state, headlight state, location/rotation, velocity and perception sensors such as sonar, laser and pan-tilt-zoom (ptz) camera. Video feedback is via Web camera that captures scenes in unreal client. Then, the raw images and jpeg images are sent out through the network.

B.1.2.3   The Robot Simulation

The robot model is Karma engine-based configurable model. The features are encapsulate the programming details and building robot by assembling. The components are chassis, parts, joints and attached auxiliary items. The method is to connect the chassis and joints and attach the auxiliary items to chassis or parts.

B.1.2.4   Control Simulation

The control simulation components are (1) method, (2) communication data, (3) auxiliary tools and (4) player with USARSim drivers. In the method, we have Gamebots to communicate between controller and robot server (the virtual robot) that consistently sends out messages and responds to command (and may send out messages also). The communication data part consists of messages and commands. The messages are state message, sensor messages, geometry and configuration messages. The commands are robot spawning command, wheel/joint control command and query commands. In auxiliary tools, we have pyro with USARSim plug-in which has in turn a Python library, environment, GUI and low-level drivers used for explore AI and robotics, and USARSim robot to pyro robot. For player with USARSim drivers, we have player, 'a robot device server that gives users simple and complete control over the sensors and actuators on the robot'.

B.2   ROBOTICS TOOLBOX

This toolbox provides several functions useful in robotics analysis for [2] (i) kinematics, dynamics and trajectory generation, (ii) simulation and (iii) analysing results from experiments with real robots. It is based on a very general method of representing the kinematics and dynamics of serial-link manipulators. The mathematical models are provided for well-known robots such as the Puma 560 and the Stanford arm [2]. The merits of this toolbox include the following: (i) the programme code is quite mature, (ii) it provides a point of comparison for other implementations of the same algorithms and (iii) source code is available. It provides functions for manipulating data types (related to motion models and transformations discussed in Chapter 2). The toolbox facilitates graphical display of the pose of any robot. It has Simulink® and MATLAB® versions also.

B.3 ROBOT PATH PLANNING ALGORITHMS

Ref. [3] gives programme listings of several path planning algorithms: (1) breadth first search, (2) Dijkstra's algorithm, (3) dynamic bug algorithm, (4) minimal tangent bug algorithm, (5) rotational plane sweep algorithm, (6) algorithm: line intersects the positive x-axis, (7) algorithm: is a line an initial edge for the rotational plane sweep algorithm, (8) algorithm: is a line on the inside of an obstacle, (9) algorithm: is a point visible from another one, (10) algorithm: updates the rotational plane intersection list, (11) algorithm: rotational plane sweep algorithm, (12) A* algorithm, (13) D* algorithm and (14) D* lite algorithm. The Dijkstra's algorithm obtains an optimal path in a graph which has edges with possibly different positive costs. The listing of this algorithm is given below [3]:

Dijkstra's algorithm

```
/* Finds an optimal path from v to w and returns it or deter1 mines if
none exists and returns false */
2 function Dijkstra get path(V ertex v, V ertex w)
3 begin
4 /* Initialise the queue */
5 queue = an empty queue which orders its elements using their g values
6 g(v) = 0
7 t(v) = OPEN
8 queue.insert(v)
9 while not queue.isEmpty() do
10 u = queue.removeMinimum()
11 t(u) = CLOSED
12 if u == w then
13 return the back pointer path from v to w.
14 end
15 foreach neighbor n of u do
16 if t(n) 6= CLOSED then /* = → not equal to */
17 /* Either the vertex has not been visited before, or it's g value can be
improved upon */
18 if t(n) 6= OPEN or g(n) > g(u) + c(u, n) then /* = → not equal to */
19 /* If we are performing a relaxation */
20 if t(n) == OPEN then
21 queue.remove(n)
22 end
23 g(n) − g(u) + c(u, n)
24 b(n) = u
25 t(n) = OPEN
26 queue.insert(n)
27 end
28 end
29 end
30 end
31 return false
32 end
```

The above algorithm's listing is given in order to compare with the modified D* algorithm discussed in Chapter 14. The MATLAB code of the path planning A* algorithm for which some results were presented in Chapter 10 is given in Ref. [4].

B.4   SIMBAD ROBOT SIMULATOR

This is a Java 3D robot simulator for scientific and educational use. It provides a simple basis for studying [5] (a) AI, (b) machine learning and (c) more generally AI algorithms all in the context of autonomous robotics/agents. The programme enables the user to write his/her own robot controller, modify the environment and use the available sensors. The Simbad simulator package is free for use and modification. It has the following features: (i) 3D visualization and sensing, (ii) single/multi-robots simulation, (iii) vision sensors, colour monoscopic camera, (iv) range sensors: sonars and IR, (v) contact sensor: bumpers and (vi) swing user interface for control.

B.5   CADAC: COMPUTER-AIDED DESIGN OF AEROSPACE CONCEPTS IN C++

The CADAC programme provides an environment (in FORTRAN and C++) for the development of general-purpose, digital computer simulations of dynamic systems [6]. It (i) handles I/O, (ii) generates stochastic noise sources, (iii) controls state variable integration and (iv) provides post-processing data analysis (with display). The CADAC package can be adapted to missiles, aircraft and hypersonic vehicles. Its environment is quite suitable for 3DoF, 5DoF and 6DoF simulations of the dynamics of these vehicles. It supports deterministic and Monte Carlo simulation runs. It is IBM-compatible PC-based, OS Windows 2000 (>SP2) and Windows XP.

REFERENCES

1. Anon. USARSim Tutorial—Basic Session (simurobotenviron.ppt). University of Pittsburgh, School of Information Sciences, Pittsburgh, USA, http://www.usl.sis.pitt.edu/wjj/USAR/Release/Basic.ppt, January 2011.
2. Corke, P. I. A robotics tool box, *IEEE Robotics and Automation Magazine*, 3, 1, 24–32, March 1996, http://petercorke.com/Robotics_Toolbox.html, December 2008, Accessed January 2011.
3. Crous, C. B. *Autonomous Robot Path Planning*, M.S. thesis, University of Stellenbosch, Stellenbosch, South Africa, March 2009.
4. Paul, V. Path planning A* algorithm in MATLAB (code, April 2005), http://www.yasni.com/vivian+paul+premakumar/check+people, July 2011.
5. Louis, H. and Nicolas, B. *What is Simbad?*, http://simbad.sourceforge.net, October 2011.
6. Zipfel, P. H. Advanced 6DoF aerospace simulation and analysis in C++, AIAA Self-Study Series, www.aiaa.org, USA, 2005.

Index